367 CONSOLIDATION

255 AGGREGATE SIZE

603 CONC. INSP. HANDBOOK P.C.A.

451 OIL ON REBAR - REMOVE.

431 STEEL ID MARKS

629 SPEC. INSPECTOR'S TASKS

233 COLD JOINTS

301 BREAK TESTS 7 DAYS, 28 days

374 SLAB ON GROUND

160 94# BAGS; 1 cuft

CONCRETE MANUAL

2009 IBC® AND ACI 318-08

CONCRETE QUALITY AND FIELD PRACTICES

GERALD B. NEVILLE, P.E.
JOSEPH J. WADDELL, ORIGINAL AUTHOR

Concrete Manual

ISBN: 978-1-58001-871-5

Errata on various ICC publications may be available at www.iccsafe.org/errata.

First Printing: September 2010

Printed in the United States of America

Preface

This text on concrete inspection will provide the inspector with an understanding of the scientific principles that underlie sound practices and will assist the inspector in making rational rather than arbitrary decisions. The *Concrete Manual* provides the guidance and information that inspectors and related professionals need to become more proficient and professional in relating to concrete field practices and inspection. The information provided will increase the technical capabilities of jurisdictional inspectors in the performance of their inspection duties.

The *International Building Code*®, published by the International Code Council® (ICC®) presents a number of situations in which specially qualified inspectors are required to perform continuous inspection of construction. The special inspectors are individuals with highly developed and specialized skills who observe those critical building or structural features that they are qualified to inspect. The *Concrete Manual* serves as a valuable reference to assist the special inspector in performing the duties and responsibilities of special inspection for reinforced and prestressed concrete construction.

To perform inspection of any phase or part of construction, the inspector must be versed in the phase or part that he or she is inspecting. The inspector's knowledge of laws, codes and specifications will be of little value if he or she does not have an understanding of the construction to be inspected. This book provides the inspector with a source of that knowledge as applied to concrete construction.

The initial chapters of the book introduce the reader to concrete and explain what concrete is and why it behaves as it does. A brief history of portland cement and concrete is included. In the following chapters, materials are presented to the reader as a preparation for the final chapters, which cover construction. Conventional construction procedures as well as special techniques are covered in sufficient detail to enable the reader to understand and recognize them. Throughout all chapters, the reader will find information about unsatisfactory materials and methods, together with discussion of acceptable materials and methods. Actual control and inspection procedures are described and should be of immediate interest to the inspector.

The *Concrete Manual* provides basic information that can be related to the described inspection procedures. The descriptions are, by necessity, somewhat general, as the responsible building official will designate the details of inspection. Codes, specifications and other requirements differ from job to job. For example, what are the conditions under which the inspector is authorized to order the work stopped or refuse to accept certain material or construction? These administrative decisions must be made by the building official. The statistical quality control methods, although of limited value to most inspectors, are included for completeness as concrete mixture proportioning and strength test evaluation and acceptance are based on statistical methods of analysis and the mathematics of probability. The inspector should at least be aware of the basic concepts of statistical quality control and its applications to concrete construction. A "Resource References" list of the concrete industry and technical organizations is included at the back of this book. The reader is encouraged to contact a listed organization for additional information and/or a publication on a specific concrete subject.

Of special note, the International Code Council, in cooperation with the American Concrete Institute, offers an examination for reinforced concrete special inspectors with national certification opportunities. The International Code Council offers the examination dealing with the codes and standards involved with reinforced concrete inspection; ACI offers certification of field technicians. When combined, they provide a national certification for Reinforced Concrete Special Inspector. For more information on the "Reinforced Concrete Special Inspector Certification" the reader should contact ACI or the International Code Council.

In addition to the reinforced concrete special inspector certification, ICC, in cooperation with the Precast Prestressed Concrete Institute (PCI) and the Post-tensioning Institute (PTI), offers a "Prestressed Concrete Special Inspector Certification." The reinforced concrete certification is a prerequisite for obtaining a prestressed concrete certification.

A companion document, *Concrete Manual Workbook*, is also available. This self-study guide is intended to provide practical learning assignments for independent study of the *Concrete Manual*, chapter by chapter. Advancing through the guide, the learner can measure his or her level of knowledge by using the quizzes provided in each study session.

Acknowledgments

The initial author of the *Concrete Manual* was Joseph J. Waddell, noted concrete consultant. Special thanks are due to Mr. Waddell for his outstanding contribution to this unique publication addressing the special needs of the concrete field and laboratory inspector/technician.

For nearly three decades, the publication has been continuously updated to address new developments in concrete technology and construction practice. In addition, the text is revised on a regular basis to reflect ongoing changes in the building code and corresponding reference standards.

Primary responsibility for the text of this publication, since 1988, is with Gerald B. Neville, P.E., formerly of the ICBO/ICC technical staff. Special thanks go to Steven H. Kosmatka of the Portland Cement Association (PCA) for his continued help and reviews of the total text of the publication for conformance to current concrete technology and construction practice; to Connie Field of PCA for her help in securing the many new color photographs for the 7th edition; to Terry Collins of PCA for his review of Chapter 11 addressing formwork; to Anthony Felder of the Concrete Reinforcing Steel Institute (CRSI) for his critical review of Chapter 18 on steel reinforcement; to Roy Reiterman of the Wire Reinforcement Institute (WRI) for his special review of the welded wire reinforcement text in Chapter 18; to James Baty of the Tilt-Up Concrete Association (TCA) for updated information on tilt-up construction; and to Jason Krohn and George Nasser of the Precast/Prestressed Concrete Institute (PCI) for their critical review of Chapter 20 addressing precast and prestressed concrete.

About the International Code Council®

The International Code Council (ICC®), a membership association dedicated to building safety, fire prevention and energy efficiency, develops the codes and standards used to construct residential and commercial buildings, including homes and schools. The mission of ICC is to provide the highest quality codes, standards, products and services for all concerned with the safety and performance of the built environment. Most United States cities, counties and states choose the International Codes®, building safety codes developed by the International Code Council. The International Codes also serve as the basis for construction of federal properties around the world, and as a reference for many nations outside the United States. The Code Council is dedicated to innovation and sustainability, and Code Council subsidiary, ICC Evaluation Service, issues Evaluation Reports for innovative products and reports of Sustainable Attributes Verification and Evaluation (SAVE).

Headquarters: 500 New Jersey Avenue, NW, 6th Floor,
Washington, DC 20001-2070
District Offices: Birmingham, AL; Chicago, IL; Los Angeles, CA

1-888-422-7233
www.iccsafe.org

Table of Contents

CHAPTER

Fundamentals of Concrete

1.1. History of Cement and Concrete
- Early History
- Modern Usages

1.2. The Hydration Reaction
- Composition of Portland Cement
- Mechanism of Hydration
- Heat of Hydration

1.3. Characteristics of Concrete
- What Is Concrete?
- Fresh Concrete
- Hardened Concrete

1.4. The Water-Cement Ratio

1.5. The Role of Admixtures

1.6. Good Durable Concrete

1.7. Distress and Failure of Concrete

1.8. The Five Fundamentals

(1) Investigation of the Site
(2) Design of the Structure
(3) Selection of the Materials and Mix
(4) Workmanship in Handling the Materials and Concrete
(5) Maintenance of the Structure Throughout its Life
Application of the Five Fundamentals

1.1. History of Cement and Concrete

Early History. Shelter, from the very beginning of man's existence, has demanded the application of the best available technology of the contemporary era. In the earliest ages, structures consisted of rammed earth, or stone blocks laid one on another without benefit of any bonding or cementing medium. Stability of the stone structures depended on the regular setting of the heavy stones. The earliest masonry probably consisted of sun-dried clay bricks, set in regular courses in thin layers of moist mud. When the moist mud dried, a solid clay wall resulted. Construction of this kind was common in the dry desert areas of the world.

Burnt gypsum as a cementing material was developed early in the Egyptian period and apparently was used in construction of some of the pyramids. Later, the Greeks and the Romans discovered methods of burning limestone to produce quicklime, which was subsequently slaked for use in making mortar. Both the Greeks and the Romans learned that certain fine soil or earth, when mixed with lime and sand, produced a superior cementing material. The Greek material, a volcanic tuff from the island of Santorin, is still used in that part of the world. The best of the materials used by the Romans was a tuff or ash from the vicinity of Pozzuoli near Mt. Vesuvius; hence, the name pozzolan used to identify a certain type of mineral admixture used in concrete today.

The cement produced by the Romans was a hydraulic cement; that is, it had the ability to harden under water. Many Roman structures were constructed of a form of concrete, using these materials, and stone masonry was bonded with a mortar similarly composed. The Basilica of Constantine, an early example of the use of stone and broken brick or tile as an aggregate in concrete, and the Coliseum are two examples of Roman architecture of this period.

During the Middle Ages, the art of making good mortar was nearly lost, the low point having been reached in about the 11th century, when much inferior material was used. Quality of the lime started to improve from this time forward, and in the 14th century or later the use of pozzolans was again practiced.

One of the most famous projects in more recent times was the construction of the new Eddystone Lighthouse off the coast of England in 1757 – 1759. John Smeaton, the engineer and designer of the structure, investigated many materials and methods of bonding the stones for the building. According to Samuel Smiles,

> he bestowed great pains upon experiments, which he himself conducted, for the purpose of determining the best kind of cement to be used in laying the courses of the lighthouse, and eventually fixed upon equal quantities of the lime called *blue lias* and that called *terra puzzolano*, a sufficient supply of which he was fortunate enough to procure from a merchant at Plymouth, who had imported it on adventure, and was willing to sell it cheap.[1.1]

The *blue lias* referred to is an argillaceous, or clay, limestone, and the *terra puzzolano* was a pozzolan that had apparently been imported on a speculative basis from Italy.

Engineering and scientific development was beginning to move rapidly at this time, and researchers in several countries were investigating cementing agents made from gypsum, limestone and other natural materials. Lesage and Vicat in France, and Frost and Parker in England, were among these pioneer experimenters. One discovery was a method of making a cement by burning a naturally occurring mixture of lime and clay. Properties of the natural cement were very erratic because of variations in the proportions in the natural material, although use of this natural cement continued for many years.

In 1824 Joseph Aspdin, a brick mason of Leeds, England, took out a patent on a material he called portland cement, so called because concrete made with it was supposed to resemble the limestone quarried near Portland, England. Aspdin is generally credited with inventing a method of proportioning limestone and clay, burning the mixture at high temperature to produce clinkers, then grinding the clinkers to produce a hydraulic cement. His small kiln, producing about 16 tons of clinker at a time, required several days for each burn. Expansion and development of cement manufacturing was slow for a number of years. About 1850, however, the industry had become well established not only in England but also in Germany and Belgium.

Shipments to the United States begun in 1868 and reached a peak in about 1895, at which time production was well under way in the United States.

Meanwhile, the United States production of natural cement had been started early in the 19th century as a result of the demand for cement for construction of the Erie Canal and related works. The first portland cement made in the United States was produced by David Saylor at Coplay, Pennsylvania, in 1871. Subsequent development of the rotary kiln led to large-scale production of cement throughout the world.

The use of concrete was expanded by the construction of railroads, bridges, buildings and street pavements. Research in reinforcing concrete with steel rods had been started in France, and the year 1875 saw the first use of reinforced concrete in the United States. Much of the concrete at this time contained barely enough water to enable the concrete to be rammed into place by the application of much hand labor. There next ensued a period of wet concrete in which the concrete was flowed into place. Many users of concrete, however, realized the folly of wet mixes, and in 1918 Duff Abrams revealed the results of his research and observations. He stated that the quality of concrete was directly affected by the amount of water in relation to the amount of cement; within reasonable limits, the quality of the concrete decreases as the water-cement ratio goes up. This has become one of the basic laws of concrete technology.

The first third of the 20th century saw great expansion and improvement in the use of concrete besides the disclosure of the water-cement law. Test and control methods were being developed. Even before 1912 the pioneers in precasting were active, and by 1925 the ready-mix industry was well established. The introduction of high-frequency vibrators in 1928 permitted the use of relatively dry, harsh mixes. Investigation of materials for Hoover Dam in 1930 resulted in the development of low-heat cement for mass concrete. Further cement research gave the industry five standard portland cements. Research in admixtures was conducted by many researchers during the 1930s, revealing the advantages of air-entrainment, which came to be specified by many agencies during the 1940s. Pozzolans and other admixtures gained approval about this time. Problems with deterioration of concrete caused by reaction between certain aggregates and cements gave us low-alkali cement in 1941.

Modern Usages. Concrete has undergone a remarkable transformation in the last 60 years. In its early history and development, concrete was a gray and utilitarian construction material—used in dams, foundations, pavements, structural columns and beams. Rarely was advantage taken of its artistic potential. Today, however, it has reached new heights of service and beauty, thanks to the pioneering work of a few outstanding architects and engineers. Dramatic and striking structures offer exciting evidence of the freedom of aesthetic expression in textures, colors, shapes and sizes that enables the designer to impart elegance and artistry to concrete structures, utilizing bold and colorful techniques that were not even dreamed of a few years ago. High-rise building frames, hyperbolic paraboloids, barrels, precast and prestressed elements, tilt-up, slipforms, lift slabs, freeform shotcrete and plaster, all lend their unique characteristics to the construction scene. Transporting and placing concrete for buildings has been revolutionized by the concrete pump. Other developments have included:

- Procedures and equipment to adapt such techniques as slipforming and tilt-up to small as well as large buildings;
- Precasting of large and small, and plain and intricate, building components;
- Site precasting and prestressing;
- Availability of white portland cement;
- Development of expansive cement;
- New techniques for imparting color and texture to exposed concrete and plaster;
- New knowledge of lightweight aggregates and lightweight concrete;
- More realistic specifications on the part of architects and engineers;
- Improved methods of welding reinforcement;
- Availability of epoxy-coated, stainless steel and fiberglass reinforcement;
- Improved concrete ingredients and quality control enabling concrete of very high strength to be produced—in excess of 15,000 psi; and
- Evolving technology in the field of admixtures—superplasticizers, silica fume, chemical systems to control cement hydration, etc.

1.2. The Hydration Reaction

To lead to a better understanding of the chapters that follow, we should first examine and understand the reaction that takes place when portland cement comes in contact with water. The reaction is called hydration, which is defined as the process of reacting with water to form new compounds.

When cement and water first come together, they become a paste. It is this paste that binds the particles of aggregate together to form concrete. Every particle, from the smallest grain of sand to the largest stone, must be coated with this paste. After an initial period in a plastic condition, the paste starts to stiffen or set. After several hours it reaches a point in which the paste completely loses its plasticity. If the paste is disturbed after this point is reached, it will be seriously damaged.

The initial setting time is an arbitrary period of time measured from the time water and cement are combined until a small pat of cement paste will just support a certain size and weight of steel needle. Final set is determined with a heavier and larger needle.

The rate of setting is not necessarily the same as the rate of hardening. A high-early-strength cement can have nearly the same setting times as a common Type I cement, but it develops strength much more rapidly once it has set.

Composition of Portland Cement. Cement is a complex mixture of several compounds that react with water. All of the compounds are anhydrous; that is, they are completely devoid of water. When combined with water, they actively react with the water, forming new hydrated compounds. These hydration compounds are all of low solubility, which gives concrete its durability.

Because of its complexity, cement cannot be shown by a simple chemical formula. Instead, a cement analysis report will show the amounts of all constituents of the cement. See Chapter 7.

Mechanism of Hydration. At the instant cement comes in contact with water the hydration reaction begins. This reaction can continue for years. The rate of hydration is affected by the composition of the cement, fineness of the cement, temperature, the amount of water present and the presence of admixtures. By varying the relative proportions of the compounds composing the cement, the manufacturer can alter the rate at which the cement develops strength.

A fine-ground cement will hydrate more rapidly than a coarse cement, merely because the smaller particles permit the water to penetrate into the particles faster. As in most chemical reactions, hydration of cement is accelerated under higher temperatures. At a temperature close to freezing of water, the reaction practically stops. High-temperature curing of concrete is practiced in products plants and casting yards to speed up the development of strength.

Water must be available for the hydration to continue. The process of hardening and strength development of concrete is not a process of drying out. If the concrete dries out completely, the reaction stops; however, it will recommence at a reduced rate if the concrete is again wetted.

Heat of Hydration. The reaction of cement and water is exothermic; that is, heat is generated during the reaction. Depending on the kind of structure in which the concrete is being used, this heat can be an advantage or disadvantage. In ordinary construction, members are of such size that release of this heat is not a problem. In dams and other massive structures, means must be taken to reduce the rate of heat liberation by modifying the composition of the cement or by special design and construction provisions to remove it. By insulating the forms during cold weather, the heat can be used to advantage to protect the concrete from freezing until it develops sufficient strength to withstand freezing.

1.3. Characteristics of Concrete

What Is Concrete? One thing should be made clear at this point: Cement and concrete are not the same thing. It is erroneous to speak of cement driveways because there is no such thing. When sand, broken rock or gravel, cement and water are all mixed together, then placed in a mold, the mixture becomes hard and is known as concrete. Figure 1-1 is a section sawed through a specimen of concrete. When the aggregate consists of sand-size material only, all less than about $^1/_4$ inch in diameter, the mixture is called mortar. The cement-water portion of the concrete is called paste.

Grout is a mixture of cement and water, either with or without fine aggregate, containing sufficient water to produce a pouring consistency without segregation. Grout has a wetter consistency than mortar. All of these mixtures can be made with or without admixtures.

Ordinary or plain concrete is brittle and not suitable to resist tension, or pulling apart. It is used in structures that are subject primarily to compressive, or squeezing, loads. Reinforced concrete is concrete in which steel bars or wires are embedded when it is cast to give it strength in tension. The steel and concrete complement each other, acting as a unit, while the concrete also protects the steel from rusting and from fire.

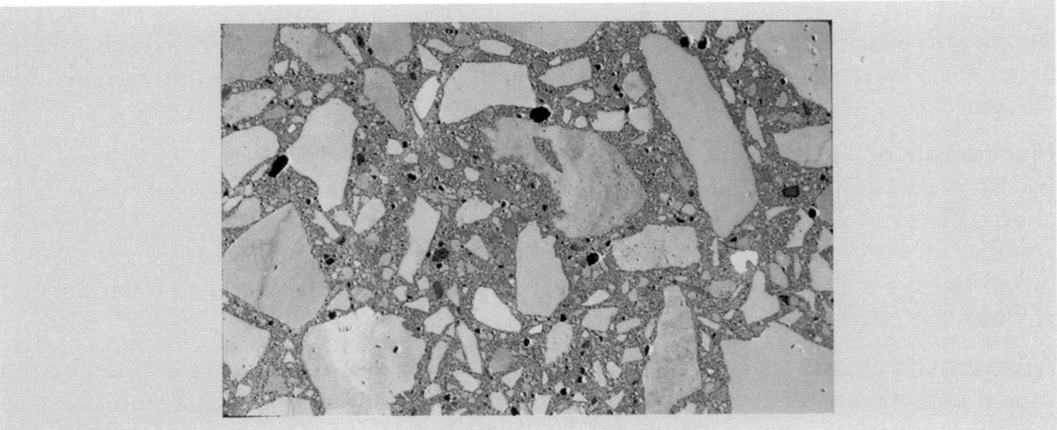

Figure 1-1
A cross section of hardened concrete. Note the good distribution of aggregate particles from the smallest to the $1^1/_4$-inch maximum size. (Courtesy of PCA)

Fresh Concrete. When concrete materials are first mixed together the concrete is said to be in a fresh condition. Fresh concrete is plastic—it has no fixed shape, and it changes shape easily. It can be manipulated and formed by means of molds.

The one property or characteristic of fresh concrete that concerns those on the job is the workability of the concrete. Workability is the ease with which concrete can be handled and placed with a minimum loss of homogeneity. The slump test is a good indicator of the potential workability or placeability of fresh concrete. See Figure 2-4.

Hardened Concrete. When the forms are removed from concrete after the concrete is several hours or a few days old, it is defined as *green concrete*. It still has a high moisture content and relatively low strength. The concrete can be damaged easily and must be supported as it cannot carry any load, not even its own weight.

After curing, the concrete matures and becomes hard. It is in this condition that the concrete has developed the required qualities of strength and durability, and can support a load. If it has been properly made, it will be free of cracks and other blemishes, with a surface that has a good appearance in accordance with the requirements of the exposure.

1.4. The Water-Cement Ratio

The amount of water to the amount of cement in concrete is called the water-cement ratio. Formerly expressed as the number of gallons of water for each bag of cement in the concrete, it is now more commonly expressed as a decimal number:

$$\text{Water} - \text{cement ratio} = W / C = \frac{\text{lb water}}{\text{lb cement}}$$

Note that the W/C can be expressed as the weights in a cubic foot, cubic yard, batch or day's production—in fact, any size unit of measurement.

Noting that a gallon of water weighs 8.33 pounds and that a bag of cement (United States) weighs 94 pounds, the pound of water per pound of cement ratio can be easily converted to the still common field usage—gallons of water per bag of cement:

$$W / C = \frac{\text{lb water}}{\text{lb cement}} (94 / 8.33) \frac{\text{gal. water}}{\text{bag cement}}$$

Example: W/C = 0.44 lb water/lb cement (94/8.33) = 5 gal. water/bag cement

Tests and field observations by many individuals have conclusively proved the importance of what has become known as the water-cement ratio law: the strength, durability and other desirable properties of concrete are inversely proportional to the water-cement ratio within the range of practical or usable mixes, all other factors remaining the same. However, because of the great number of aggregate sizes, shapes, types and qualities, variations in cements (even though they all meet the same specifications), different mixes, use of admixtures, and great diversity of environmental conditions of mixing and placing, this law should be looked upon as a guide rather than as an inflexible rule that applies without question to all concrete under all conditions.

The drying shrinkage of concrete is governed more by the unit water content of the concrete than by the water-cement ratio. See also Chapter 12. The relationship of water-cement ratio and several properties of the concrete are discussed in the following chapters. See Figure 1-2.

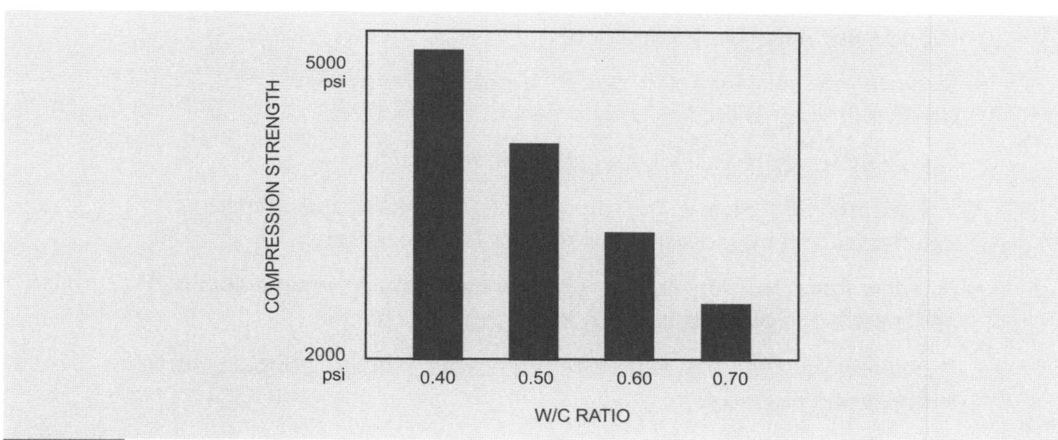

Figure 1-2
The descending columns show that the compressive strength decreases as the water-cement ratio goes up.

With the advent of other materials with cementing value, when used in concrete in combination with portland cement, such as fly ash, natural pozzolans, blast-furnace slag and silica fume, the traditional water-cement ratio has been renamed water-cementitious materials ratio in most written text and the ACI 318 Standard. *For brevity, unless required for technical clarity, the term water-cement ratio will be used (retained) within the text of this publication.* The reader should also refer to Chapter 12, Section 12.1.

1.5. The Role of Admixtures

It has long been known that certain compounds, when mixed into the concrete in small amounts, significantly alter the concrete in both the fresh and hardened state. Examples are adding calcium chloride to hasten setting and strength development, and chemical compounds to control the cement hydration. During the 1930s numerous compounds were proposed to reduce water requirement, entrain air, improve workability, accelerate, retard and otherwise improve or change the concrete. These admixtures, as they are called, are now well accepted in the industry when they are properly used. An admixture frequently provides the best way to achieve concrete with certain properties. On the other hand, concrete made without admixtures has always been acceptable. In today's market, greater than 90 percent of all concrete produced contains one or more admixtures to alter the properties of concrete to reflect construction and exposure needs, and to provide a more economical construction material.

The fundamentals of concrete, as discussed in Chapter 1, apply to all concrete, with or without an admixture. The introduction of an admixture modifies the concrete; hence, a discussion of admixture-concrete must consider the effect of the admixture. This in no way reduces or invalidates the basic concepts; it only warns us to be alert to admixtures and to recognize their effects, bad as well as good.

1.6. Good Durable Concrete

The concrete described as good, durable concrete, or quality concrete, is concrete that meets the structural and aesthetic requirements for the required life of the structure, at maximum economy. The properties this concrete must have are:

- *Workability* in the fresh condition;
- *Strength* in accordance with design, avoiding overstrength (uneconomical) as well as understrength (dangerous);
- *Durability* to resist attack by weather or substances;
- *Volume stability*—that is, minimum drying shrinkage and changes in volume caused by moisture and temperature variations;
- *Freedom from cracks* by reducing tendency to crack (volume stability) and installation of joints and crack-control devices;
- *Freedom from blemishes* such as rock pockets, scaling, popouts, surface softness and bug holes;
- *Watertightness* (where applicable);
- *Economy*; and
- *Good appearance*.

To obtain this quality concrete, one must start with good materials, properly proportioned, mixed and placed, with adequate inspecting and testing to verify the quality. To provide this kind of concrete requires responsible management at the time the project is conceived and throughout its life.

1.7. Distress and Failure of Concrete

Concrete and mortar are the only construction materials that are made on the site. Though subject to all the variations in weather, materials and methods, concrete under proper conditions of design, construction and control is nevertheless one of our most dependable and versatile construction materials, finding use in practically all types of construction, from the small family residence to high-rise buildings, highways, dams and other monumental works.

As in all materials, problems sometimes arise, and we may find concrete that has failed to perform as expected. The distress and failure sustained by concrete is no accident; failures do not just happen, they are caused. Somewhere along the line someone, through ignorance or carelessness, or in the misguided hope of saving a few cents, created or permitted a condition that led to the distress or failure of the concrete. The distress that we see being sustained by concrete is a symptom, or group of symptoms, indicating that there is something wrong someplace. In some cases, all we can do is remove or cover up the symptom, as we do when we take aspirin to stop a headache. In other cases, we can root out the basic trouble, thus eliminating the cause of the symptom, as we do when we have an appendix removed or treat any sickness.

So it is with concrete. Sometimes all we can do, for instance, is patch the cracks, thus covering up the symptom; on the other hand, we can avoid crack-inducing conditions or materials in the first place, thus preventing the basic trouble that shows up later as cracking.

Unfortunately, the cause of distress can rarely be traced to one single factor; distress usually results from several contributing causes. Broadly speaking, distress is caused by unsuitable materials, improper workmanship or the environment.

The first step in the treatment of concrete that shows damage or deterioration is to classify the damage. This will assist in diagnosis, as forces that produce each kind of damage can be generalized, thus narrowing the field of possible aggressors. After this classification, the cause may be determined. This process of diagnosis may be quite simple and almost automatic, or it may involve complicated and intensive detective work before the basic causes are found. Sometimes what seems to be the obvious cause is not the cause after all, or it may be one of several contributing to failure. The investigator should consider all facts that might be significant, such as determining the sources of materials used in construction, loading conditions, construction methods, design factors, conditions of exposure, presence of aggressive substances or forces, evidence of accidental damage (such as impact with vehicles) and foundation conditions.

Once the cause has been isolated, corrective measures are taken. Such measures include elimination of the cause, changes in the structure to enable it to withstand the destructive action, restoration of the damaged portions of the structure and protection to prevent further injury, abandonment of the structure and construction of a new structure to withstand the action, or a combination of these.

1.8. The Five Fundamentals [1.2]

The five fundamentals of concrete construction are investigation of the site, design of the structure, selection of materials and mix, workmanship in handling materials and concrete, and maintenance of the structure throughout its life. Consider how each of these five fundamentals helps create good, durable concrete.

(1) Investigation of the Site. Some sort of a site investigation is made for any structure, regardless of how insignificant it may be. This investigation is important for proper design of the structure and has a significant influence on selection of the materials and mix. An intelligent design cannot be made, nor can logical use be made of the available materials, without a thorough investigation of all features at the site. Proper investigation of the site means using a three-way approach that involves investigating the fitness of the location to suit the requirements of the structure, investigating the ability of the foundation to carry expected loads safely and investigating the existence of forces or substances that could attack the concrete.

In this investigation one is concerned with chemical or mechanical attack by outside agencies. Chemical attack is aggravated in the presence of water, principally because water is a vehicle for bringing substances into intimate contact with the concrete, to the extent of transporting them into the concrete through cracks, honeycomb or pores in the surface.

Cyclic forces of the weather may be destructive in two ways: first, by alternately expanding and contracting the concrete, whereby the concrete is stressed to the point where it cracks; and, secondly, by the entrance of destructive solutions into the cracks thus produced.

(2) Design of the Structure. Design of a structure must be adequate, accomplished by competent engineers in accordance with accepted safe practices. A structure cannot be designed by a cut-and-dried process that requires only the ability to read a table of values or take off a quantity from a graph. It requires intelligence and experience to design a structure adequately and to know that everything possible has been done to ensure the safety and economy of the design, based on results of the investigation. Failure or distress may range from small tension cracks to complete failure of the entire structure. There is not much that can be done for a structure that shows signs of distress caused by poor design other than makeshift expedients that will shore it up and provide additional support. Design should include a consideration of the capabilities of workers and machines so that formwork and other construction procedures are facilitated

(3) Selection of the Materials and Mix. Many cases of distress have been traced to faulty materials or improper mix proportions. In most cases, materials that are available have been used for making concrete for a number of years. If such is the case, histories of these materials and of the structures built with them should be examined. Information gained by such a study will reveal whether the materials under consideration are suitable. In any such investigation, one should keep in mind the fact that poor concrete can result even when good materials are used, if insufficient care is used in proportioning and handling the materials and concrete. In many instances, especially those concerning structures built many years ago, there is a lack of accurate information about the materials and methods used. Even so, the reputation of the local materials should be given careful consideration.

For large and important works, or in all cases if lacking historical data, suitability of materials must be determined by tests. These tests should not be used as an inflexible criterion for acceptance or rejection of the material, as they require sound engineering judgment in their interpretation and application.

Concrete mix proportioning has as its objective the production of concrete of maximum economy having sufficient workability, strength, durability and impermeability to meet the conditions of placing, exposure, loading and other requirements of the structure. Trial mixes, made under laboratory conditions and using the proposed materials, should always be made for large and important works. Trial mixes should also be made in those cases where inadequate information is available relative to existing materials. These mixes are subject to revision in the field as conditions require. Strength, water-cement ratio, maximum size of aggregate, and slump are usually specified, and the proportions of cement, aggregates, entrained air, and admixtures are determined by these trial mixes.

(4) Workmanship in Handling the Materials and Concrete. Probably, the greatest portion of concrete distress has been caused by improper methods of handling the ingredients, and the concrete itself after the ingredients have been combined in the batcher. Included in such minor failures are the many small discrepancies that are unsightly but not necessarily serious. It must be remembered, however, that serious difficulties frequently have small beginnings, and a small surface defect can develop into a major one.

This workmanship includes preparation of the aggregates, batching, mixing, transporting, placing and curing. The period of construction is the last opportunity for ensuring a well-built structure. Improvements to techniques and equipment are making it more and more difficult for poor construction practices to exist, but exist they do, and vigilance on the part of those charged with the responsibility for this phase of construction is necessary.

(5) Maintenance of the Structure Throughout its Life. This merely means what it says: inspection of the structure at reasonable intervals to determine whether unusual deterioration is taking place, along with adequate protection or repair to minimize the deterioration.

Application of the Five Fundamentals. Obviously, some structures, and certain parts of some structures, are more important than others when one considers the possible consequences of a failure or cost of repair and replacement. One can compare, for instance, the foundation and basement column of a high-rise concrete frame building with the driveway slab in the adjacent parking lot. Nevertheless, there is no excuse for sloppy and unsightly construction. Good, durable concrete is possible with the simplest equipment as well as with the most sophisticated plant facilities, and the five fundamentals can be applied in their proper small way to the small job, as well as in the thorough and elaborate detail expected for the big job.

In this text we will be concerned with the third fundamental, selection of the materials and mix, and the fourth fundamental, workmanship in handling the materials and concrete.

The Fresh Concrete

2.1 Significance of Workability

We have previously defined fresh concrete (Section 1.3) as exhibiting a transient condition of plasticity that permits the concrete to be molded and formed into its final configuration. When molding can be achieved with the least effort and minimum loss of uniformity and homogeneity, the concrete is workable.

Workability is probably the most important property of fresh concrete. It is a relative term, as any certain concrete mix may be considered workable under some placing conditions, but unworkable under others. Concrete for a thick airport pavement, for example, would not be suitable for a thin precast beam containing much reinforcing steel. This explains why workability is sometimes a subjective matter, meaning one thing to one person but something entirely different to another. See Figures 2-1, 2-2 and 2-3.

Figure 2-1
Roller compacted concrete (relatively nonplastic concrete of low slump) for dam construction. Note large cobbles in the concrete mix. (Courtesy of PCA)

Figure 2-2A
Concrete for slipform paving is of very stiff consistency. (Courtesy of PCA)

Figure 2-2B
Although of stiff consistency, pavement concrete is workable and can be successfully consolidated by the use of vibrators. No side forms are necessary. (Courtesy of PCA)

Figure 2-3
Concrete for smaller projects can be properly placed with a slump of between 3 and 4 inches. (Courtesy of PCA)

The workability of concrete depends on the properties and relative amounts of the materials in the concrete; that is, the amounts and characteristics of the fine aggregate, coarse aggregate, cement, water and admixture. On the other hand, the kind of structural element in which the concrete is being placed determines the amount of workability required. As stated above, pavements and massive structures of large cross section permit the use of relatively harsh, dry mixes; whereas small sections containing much reinforcing steel and embedded items require mixes with a high degree of workability. Stated another way, the available workability depends on the materials and how they are combined, whereas the workability required depends on the kind of structure and placing conditions.

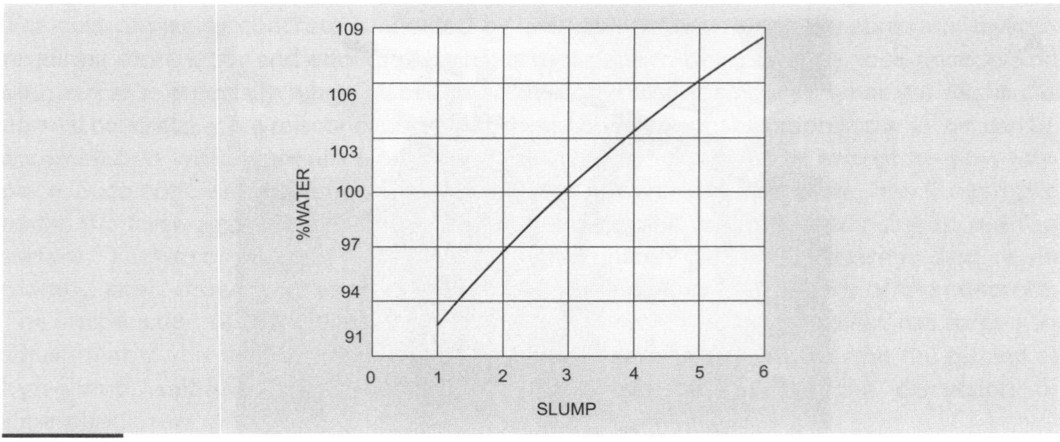

Figure 2-5
On an average, a change of 3 percent in total water per batch will change slump about 1 inch. If one assumes that the amount of water for a 3 -inch slump is 100 percent, then increasing the water by 3 percent will increase slump about 1 inch.

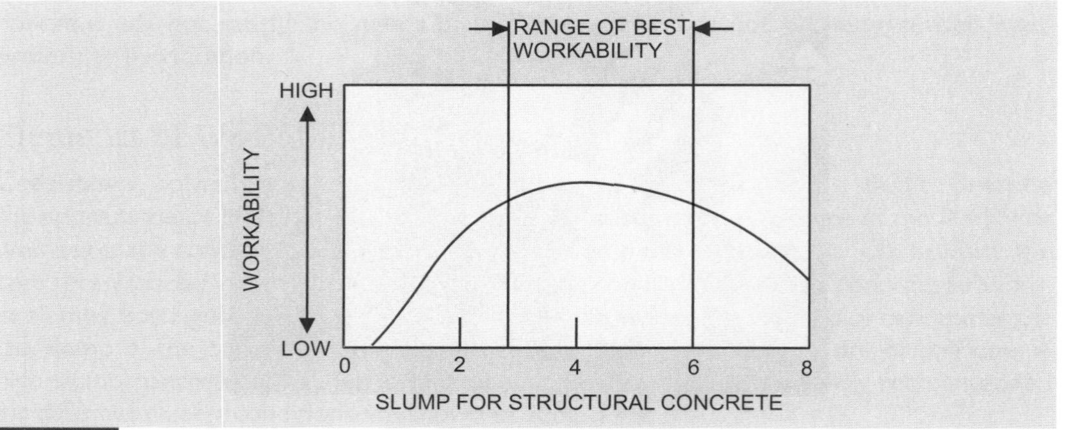

Figure 2-6
Workability is at a maximum in concrete of medium consistency, between 3-inch and 6-inch slump. Very dry and wet mixes are less workable.

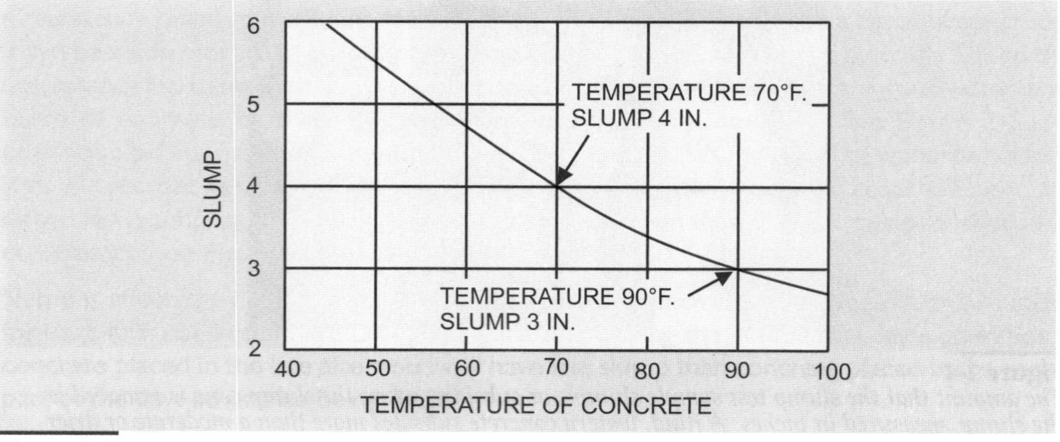

Figure 2-7
At the same water content, mixed at a higher temperature, concrete has less slump than the same concrete mixed at a lower temperature.

Cohesiveness is the element of workability that indicates whether the concrete is harsh, sticky or plastic. A good, plastic mixture is neither harsh nor sticky. It will not segregate easily. Cohesiveness is not a function of slump, as a very wet (high slump) concrete lacks plasticity. On the other hand, a low slump mix can have a high degree of plasticity.

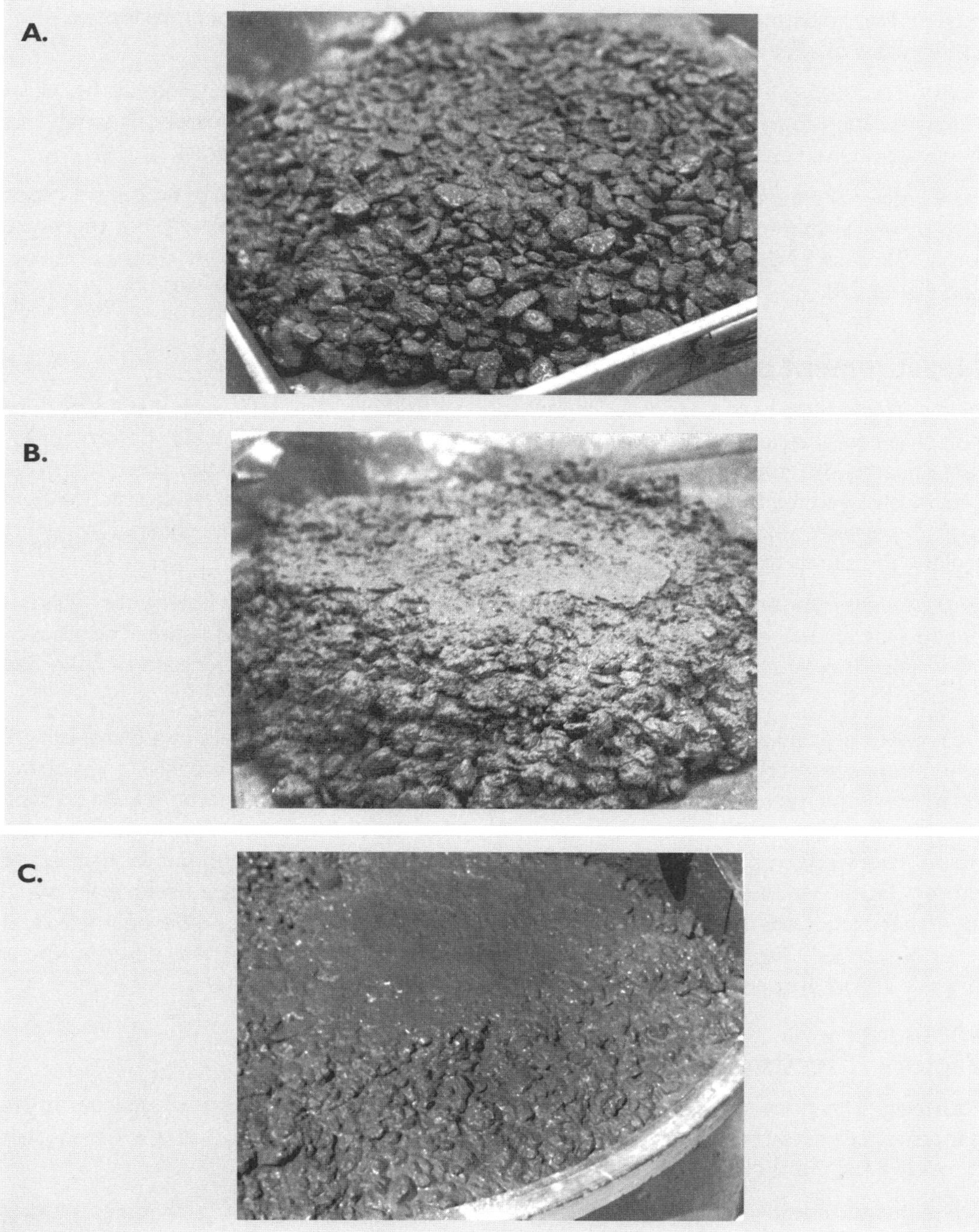

Figure 2-8
Workability is affected by the plasticity of the concrete. Coarse aggregate in the harsh mix, A, separates from the mortar. In the plastic mix, B, of good workability, the coarse aggregate is apparent in the mix, but it does not separate easily, and the surface can be troweled. The sticky, oversanded mix, C, lacks coarse aggregate and does not have the plasticity for good workability. All three mixes were made at about a 4-inch slump.

A harsh concrete lacks plasticity and cohesiveness. It segregates easily. Causes of harshness, other than high slump resulting from too much mix water, are low-cement content, lean mixes, coarse sand, a mix deficient in fines, rough and angular aggregates, or aggregate containing an excess of elongated or flat particles. Many times, the harshness of a mix can be reduced by the introduction of entrained air, or by adding cement, fine sand or fines to the mix. Proper adjustment must be made to the mix proportions to compensate for the air or other material added to the concrete.

Harsh, low-slump mixes are commonly used in pavements, mass concrete and precast concrete. Such concrete requires more vibration to consolidate it adequately, from which a better concrete results— a concrete with less tendency to shrink or crack.

Sticky mixes are those that possess a high cement content, or contain large amounts of rock dust or similar material. A fine sand might cause a sticky mix. Sticky mixes do not segregate easily. Because of their high water demands, they frequently develop shrinkage cracking. See Figure 2-8.

2.3. Measurement of Workability

Many devices have been proposed for measuring workability, most of them suitable for use only in a research laboratory. These devices measure fluidity, moldability, deformability, response to the application of work and many other qualities. So far, the slump test previously described, even with its limitations, has remained the best, most practical field test. Experience enables the technician to evaluate the concrete from observation while making the slump test.

Another test method for flow of fresh concrete involves the use of the flowmeter. This is a probe-type instrument that is inserted into the fresh concrete to a fixed depth. The amount of mortar flowing into openings in the instrument is reported as a measure of flow. See Section 13.4.

Equipment is available for installation on permanently mounted mixers in a plant, which, when connected to an indicator or recorder, gives a measure of the consistency of the concrete in the mixer. One type of meter measures the overturning moment acting on the mixer, a dry mix tending to accumulate at one end of the drum. Another type measures the power required to revolve the mixer, a dry mix requiring more power input to the electric motor. Other varieties are also available. Instruments of this kind are useful only when identical batches are repeatedly mixed, as they have to be calibrated for different mixes and sizes of batches. They are especially useful in plants for dams and other structures where the same kind of concrete is mixed day after day.

2.4. Factors Affecting Workability

Cement. The principal influence of cement on workability is through its amount in the concrete. Lean, low cement content mixes tend to be harsh and unworkable, whereas rich mixes are fat and sticky.

A fine-ground cement (Type III) increases the workability, especially of harsh mixes, making the concrete more cohesive and decreasing segregation tendencies. A coarse-ground cement reduces stickiness. These effects of fineness are relatively minor.

The chemical composition of the cement has no effect except in rare cases when there is unstable gypsum in the cement, the instability resulting from high temperature in the finish grinding mill. The effect in the concrete is called false set, in which the concrete appears to set or harden after only a few minutes. No heat is liberated, and the false set usually disappears upon prolonged mixing or remixing of the concrete. In ready-mixed concrete operations there is rarely any trouble with false set because the time lapse between introduction of the water and cement into the mixer and discharge of the concrete from the mixer at the job is of sufficient duration for the false set to disappear. If trouble is being experienced with false set, additional water should not be added to the batch. Instead, by waiting a few minutes, with or without mixing, the trouble will usually abate.

Other than its effect on workability, false set complicates and delays construction and may contribute to general lowering of the quality of the hardened concrete.

Contact of the cement and water at a high temperature is apt to cause a flash set. A flash set is an actual hydration of the cement, unlike false set, which is a temporary condition caused by hydration of calcium sulfate. Flash set can result from the use of hot mixing water during cold weather. Lumps of cement in the concrete commonly result, each lump consisting of a lump of dry cement covered with a layer of damp, partially hydrated cement, or a solid lump of partially hydrated cement. The simplest recourse is to change the batching sequence by introducing the hot water and aggregates into the mixer first. This cools the water, after which the cement can be admitted.

Aggregates. Grading and particle shape of the aggregates are factors that affect workability. Natural gravel consisting of rounded or subrounded particles is more desirable from a workability standpoint than crushed rock. A coarse aggregate containing a high percentage of flat or elongated pieces, either natural or crushed, leads to poor workability. If it is necessary to use a crushed or poorly shaped aggregate some improvement in workability is possible by using more sand in the mix, by increasing the cement content or by the addition of a powdered admixture, such as a pozzolan. More compactive effort in placing will be necessary.

Grading of coarse aggregate larger than $^3/_8$ inch is not as critical as minus $^3/_8$-inch material in affecting workability. The amount of pea gravel ($^3/_8$ inch by No. 4) (See Figure 2-9) is usually significant, especially if it contains a large amount of undersize, because the undersize will affect the sand grading.

Figure 2-9
Segregation or contamination of aggregate can result in grading variations that affect workability.

Variations in the grading of fine aggregate (sand) will cause variations in workability and water demand of the concrete and adversely affect the finish.

Aggregate gradings should conform to the requirements of ASTM C 33. The grading must not fluctuate between the low limit on one sieve and the high limit on an adjacent sieve.

Admixtures. Entrainment of the proper amount of air improves workability, especially in concrete made with poorly graded or irregular aggregates, and in lean mixes. Entrained air lessens the tendency of the concrete to segregate, slows the rate of bleeding and shortens the finishing time.

The plasticity of lean, harsh concrete can be improved by the addition of certain fine material, such as fly ash or other pozzolans, and rock dust.

Workability can sometimes be improved by the addition of certain admixtures that delay setting time or reduce water requirement.

OTHER CONCURRENT PROPERTIES

2.5. Segregation

Concrete is a heterogeneous mixture of several materials with widely different properties. Particle sizes range from cement particles a few microns in diameter to coarse aggregate as large as 6 inches, and specific gravity from less than two to more than three. Shapes and absorption values vary substantially. Because of these dissimilar properties of the several materials, forces are attempting to cause them to separate from each other. This separation is called segregation, usually manifested by the separation of the coarse aggregate from the mortar. Results of segregation in the hardened concrete are rock pockets (honeycomb), sand streaks, porous layers, scaling, laitance and bond failure at construction joints.

Certain mixtures have a tendency toward segregation. These are the harsh mixes—usually those that are too wet, sometimes very dry, or those that are undersanded. A well-proportioned cohesive mixture, in the plastic range of 1- to 4-inch slump, resists segregation, but any concrete will segregate if it is improperly handled through poorly designed equipment or improper procedures. Once segregation has occurred, succeeding handling of the concrete will not put it back together. Segregation must be avoided from the beginning and throughout the entire mixing, transporting and placing operations. Details of this control are discussed in Chapter 15.

2.6. Bleeding

During the period after concrete has been leveled off, surplus water in the concrete rises to the top surface of the concrete. Especially noticeable in flat slabs, this movement of water to the surface is called bleeding, or water gain, and is accompanied by a slight settlement of the solid particles (cement and aggregate) in the mix. Bleeding continues until the cement starts to set, until bridging develops between the aggregate particles, or until the solids reach their maximum consolidation. Mix proportions, sand grading, sand particle shape, amount of fines in the mix, cement fineness, water content of the concrete, admixtures, air content of the concrete, temperature and depth or thickness of the slab all influence the rate and total amount of bleeding. See Figure 2-10.

Figure 2-10
Bleed water on the surface of a slab must not be present when any finishing work is being done. (Courtesy of PCA)

Bleeding is influenced by the absorption of the subgrade or base on which the slab is placed. An absorbent sandy base will absorb part of the water from the concrete; hence, bleeding will be reduced somewhat. On the other hand, a slab placed on a nonabsorptive base (for example, a concrete floor placed on a plastic vapor barrier) will bleed more because all the free water must rise to the surface.

Detrimental effects of too much bleeding include delays in finishing; formation of a weak, nondurable surface if the concrete is finished when bleed water is present; formation of laitance; settlement of aggregate particles away from the underside of horizontal reinforcing bars or large aggregate particles, the voids thus formed becoming filled with water, causing a loss of bond with the steel or aggregate (See Figure 2-11); and increased porosity of the concrete because of the water channels that were formed.

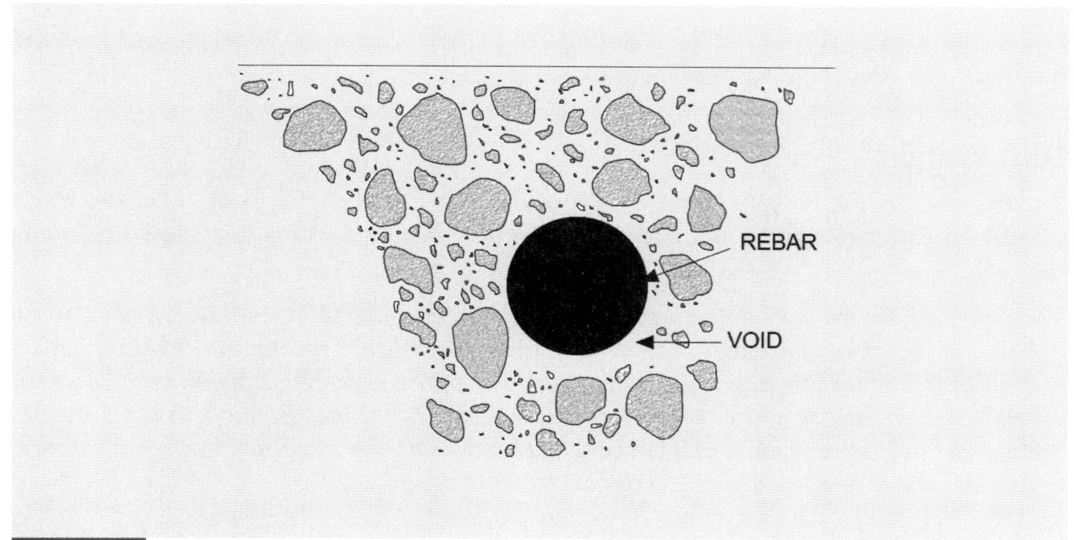

Figure 2-11
Bond with a reinforcing bar may be damaged by formation of water voids beneath the bar.

There are a number of expedients available that are more or less effective in reducing excessive bleeding. Air entrainment greatly reduces the rate, sometimes to the extent that water evaporates as fast as it reaches the surface. Well-graded sand produces concrete with a low bleeding rate. If a coarse sand must be used, it is beneficial to use a fine blending sand, mixing it with the normal sand in the aggregate processing plant, or batching it separately into the concrete. Other expedients that may be of value are to reduce the slump by reducing the amount of water in the batch, making the mix richer by increasing the amount of cement per batch, using a finer cement or adding a small amount of inert fines or pozzolan to the concrete.

If changes are made to batch proportions, the mix must be adjusted to compensate for the changes so as to maintain the proper strength and yield. Changes should be made only under proper authorization.

Once the concrete has been placed, there is little that can be done about excessive bleeding except to wait until the water disappears. Removal of the water can be expedited by dragging a length of hose across the concrete, thereby pulling the water off the edge of the slab.

Laitance. A light gray or nearly white substance sometimes appears on the top surface of a slab during or following consolidation and finishing. Called laitance, this layer consists of water, cement and fine sand or silt particles; it has no strength and is detrimental to the surface. On a horizontal construction joint or fill plane, laitance will destroy the bond between successive layers of concrete. If the joint is subject to hydrostatic pressure, water will probably pass through. On a floor or other exposed surface, it becomes a thin layer of inferior material that will scale and dust off after the floor is put in service. It contributes to hair cracking and checking.

Conditions and materials that are conducive to laitance are mixes containing more water than necessary, making them too wet for the placing conditions. Segregation during vibration causes the water and fines to rise to the surface. Floating and troweling too early and too much, or performing the finishing while bleed water is on the surface, will cause laitance. Excessive amounts of rock dust, silt, clay and similar materials in the concrete contribute to laitance.

Any of the measures that reduce bleeding will minimize laitance. Avoiding materials and methods that contribute to bleeding is essential.

2.7. Unit Weight

The density, or unit weight, is the weight of one cubic foot of the fresh concrete. Much valuable information can be obtained from the unit weight, and the test should be made regularly.

To make maximum use of the unit weight, it is desirable to know the weights of all ingredients used in one batch, day or any other measure. Further information can be computed if the specific gravities and proportions of the ingredients are known. Cement content, water content and air content of the concrete can be determined, as can the yield. Yield is defined as the volume of fresh concrete obtained from a known quantity of cement, aggregate, water and admixture divided by the unit weight.

A large discrepancy between the theoretical unit weight computed from the specific gravities and amounts of ingredients, and the measured fresh unit weight, indicates an error in specific gravity of the materials, batch proportioning, weighing or water content of the concrete. Such a discrepancy should be investigated immediately. An increase in water content of the concrete lowers the unit weight. An increase in cement content of the concrete raises the unit weight. These facts should be kept in mind when investigating problems in density or yield.

The fresh unit weight is a good indicator of what the density of the hardened concrete will be, and is especially critical in the control of lightweight and heavy concrete. As a matter of course, the importance of the fresh unit weight for normal-weight concrete should not be underestimated.

2.8. Air Content

All concrete contains a small amount of entrapped air in the range of 1 or 2 percent, but purposefully entrained air may be as high as 10 percent, depending on maximum size of aggregate. Determination of the air content of concrete in the field is significant only when air-entrained concrete is made. There are three methods commonly used: gravimetric, pressure and volumetric.

The gravimetric air content is determined from the fresh unit weight, by comparing the fresh unit weight with the theoretically computed unit weight of air-free concrete. The necessity for an accurate knowledge of specific gravities, moisture percentages and batch proportions limits the practical use of this method, except in the laboratory.

Boyle's law is applied in the pressure method, in which pressure, applied to the sample in a closed vessel, compresses the air in the concrete. Calibration of the apparatus permits direct reading of air content. The Washington type, which requires a small amount of water to fill the space over the sample, is very accurate because the water eliminates inaccuracies resulting from different quantities of air trapped in the instrument. No knowledge of the batch materials or properties is necessary.

Air content can be accurately determined by means of the volumetric meter, in which the difference in the height of a column of water before and after agitating the sample and water in the instrument indicates the air content. The volumetric method requires no knowledge of the batch material or properties and can be used in lightweight concrete. Tests are further discussed in Chapter 13.

A pocket size meter, called the Chace meter, is convenient for quick approximations of the air content and is valuable for determining if air contents are within specified limits. It is not as accurate as the other methods, and standard tests should be made as a check for record purposes.

The effect of entrained air in the fresh concrete, compared with similar concrete without entrained air, is improved workability, lowered unit weight and reduced rate of bleeding. The amount of air entrained is affected by many variables besides the amount of admixture used. The amount of air will rise if slump, water-cement ratio or percentage of sand in the mix is increased. A reduction in air content will follow an increase in temperature, mixing time, or amount of fines in the concrete, such as fineness or amount of cement, or fines in the sand. Thorough mixing, but not overmixing, is necessary. Worn blades in the mixer, overloading the mixer, or operating the mixer at the wrong speed may not provide vigorous enough mixing to ensure adequate entrainment of air.

The Strength of Concrete

The quality of concrete is judged largely on the strength of that concrete. Equipment and methods are continually being modernized, testing methods are improved, and means of analyzing and interpreting test data are becoming more sophisticated, yet we still rely on the strength of the same 6- by 12-inch cylinders, made on the job and tested in compression at 28 days age, as we did 90 years ago. Interestingly, the 2008 edition of the ACI 318 Standard (ACI 318-08) now specifically addresses the use of 4- by 8-inch cylinders for evaluation and acceptance of concrete (ACI 318 Section 5.6.2.4). See discussion on strength specimens in Chapter 13, Section 13.5.

3.1. The Importance of Strength

Obviously, the strength of any structure, or part of a structure, is important, the degree of importance depending on the location of the structural element under consideration. The first-floor columns in a high-rise building, for example, are more important structurally than a nonbearing wall. Loading is more critical, and a deficiency in strength can lead to expensive and difficult repairs or, at worst, a spectacular failure.

Strength is usually the basis for acceptance or rejection of the concrete in the structure. The specifications or code designate the strength (nearly always compressive) required of the concrete in the several parts of the structure. In those cases in which strength specimens fail to reach the required value, further testing of the concrete in place is usually specified. This may involve drilling cores from the structure or testing with certain nondestructive instruments that measure the hardness of the concrete.

Some specifications permit a small amount of noncompliance, provided it is not serious, and may penalize the contractor by deducting from the payments due for the faulty concrete. Statistical methods, now applied to the evaluation of tests as described in Chapter 26, lend a more realistic approach to the analysis of test results, enabling the engineer to recognize the normal variations in strength and to evaluate individual tests in their true perspective as they fit into the entire series of tests on the structure.

Strength is necessary when computing a proposed mix for concrete, as the contemplated mix proportions are based on the expected strength-making properties of the constituents.

3.2. Strength Level Required

The code and specifications state the strength that is required in the several parts of the structure. The required strength is a design consideration that is determined by the structural engineer and that must be attained and verified by properly evaluated test results as specified. Some designers specify concrete strengths of 5000 to 6000 psi, or even higher in certain structural elements. Specified strengths in the range of 15,000 to 20,000 psi have been produced for lower-floor columns in high-rise buildings. Very high strengths, understandably, require a very high level of quality control in their production and testing. Also, for economy in materials costs, the specified strength of very high-strength concrete is based on 56- or 90-day tests rather than on traditional 28-day test results. To give some idea of the strengths that might be required, Table 3.1 is included as information only. Remember that the plans and specifications govern.

Note that the *International Building Code* (IBC) (Section 1905.1.1) and the ACI 318 Standard (Section 5.1.1) indicate a minimum specified compressive strength of 2500 psi for structural concrete. Simply stated, no structural concrete can be specified with a strength less than 2500 psi.

Other properties of the concrete can be significant for concrete exposed to freeze-thaw conditions, sulfate exposure and chloride exposure (effects of chlorides on the corrosion of the reinforcing steel). Strength, however, remains the basis for judgment of the quality of concrete. Although not necessarily dependent on strength, other properties to improve concrete durability are related to the strength. Concrete that fails to develop the strength expected of it is probably deficient in other respects as well.

TABLE 3.1

STRENGTH REQUIREMENTS	
TYPE OR LOCATION OF CONCRETE CONSTRUCTION	SPECIFIED COMPRESSIVE STRENGTH, PSI
Concrete fill	Below 2000
Basement and foundation walls and slabs, walks, patios, steps and stairs	2500-3500
Driveways, garage and industrial floor slabs	3000-4000
Reinforced concrete beams, slabs, columns and walls	3000-7000
Precast and prestressed concrete	4000-7000
High-rise buildings (columns)	10,000-15,000

Note: For information purposes only; the plans and specifications give actual strength requirements for any job under consideration.

KINDS OF STRENGTH

Generally, when we speak of the strength of concrete, it is assumed that compressive strength is under consideration. There are, however, other strengths to consider besides compressive, depending on the loading applied to the concrete. Flexure or bending, tension, shear and torsion are applied under certain conditions and must be resisted by the concrete or by steel reinforcement in the concrete. Simple tests available for testing concrete in compression and in flexure are used regularly as control tests during construction. An indirect test for tension is available in the splitting tensile test, which can easily be applied to cylindrical specimens made on the job. Laboratory procedures can be used for studying shear and torsion applied to concrete; however, such tests are neither practical nor necessary for control, as the designer can evaluate such loadings in terms of compression, flexure or tension. See Figure 3-1.

3.3. Compressive Strength

Because concrete is an excellent material for resisting compressive loading, it is used in dams, foundations, columns, arches and tunnel linings where the principal loading is in compression.

Strength is usually determined by means of test cylinders made of fresh concrete on the job and tested in compression at various ages. The requirement is a certain strength at an age of 28 days or such earlier age as the concrete is to receive its full service load or maximum stress. Additional tests are frequently conducted at earlier ages to obtain advance information on the adequacy of strength development where age-strength relationships have been established for the materials and proportions used.

The size and shape of the strength test specimen affect the indicated strength. If we assume that 100 percent represents the compressive strength indicated by a standard 6- by 12-inch cylinder with a length/diameter (L/D) ratio of 2.0, then a 6-inch-diameter specimen 9 inches long will indicate 104 percent of the strength of the standard. Correction factors for test specimens with an L/D ratio less than 2.0 are given in the test methods for compressive strength (ASTM C 39 and ASTM C 42) for direct comparison with the standard specimen (Table 3.2.) For cylinders of different size but with the same L/D ratio, tests show that the apparent strength decreases as the diameter increases. See Figure 3-2. See also Chapter 13, Section 13.5.

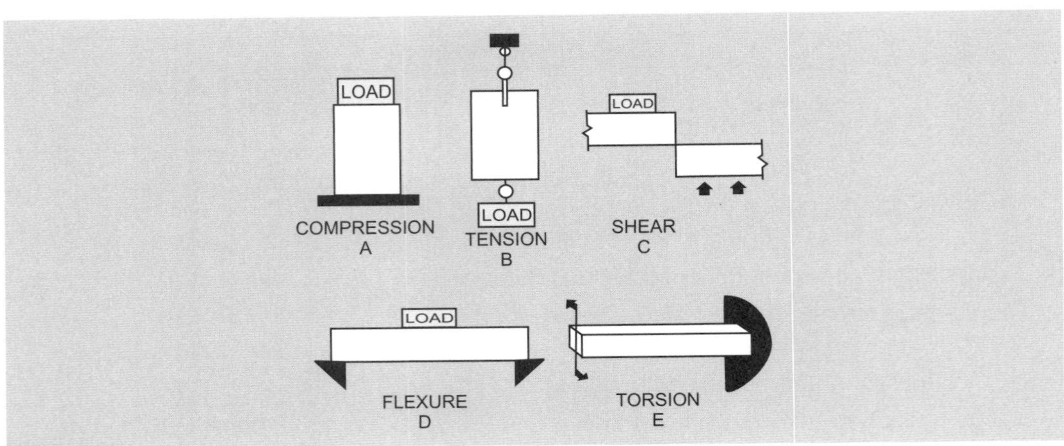

Figure 3-1
Concrete structures are subject to many kinds of loadings besides compressive. (A) Compression is a squeezing type of loading. (B) Tension is a pulling apart. (C) Shear is a cutting or sliding. (D) Flexure is a bending. (E) Torsion is a twisting.

TABLE 3.2

LENGTH DIVIDED BY DIAMETER	CORRECTION FACTOR
2.00	none
1.75	0.98
1.50	0.96
1.25	0.93
1.00	0.87

Example: A 6-inch core $8^1/_4$ inches long broke at 4020 psi
L/D = 8.25/6= 1.375
For: L/D of 1.375, the factor is 0.945.* Corrected strength is then: 4020 × 0.945 = 3800 psi.
*An example of interpolation.

	L/D RATIO FROM TABLE ABOVE	DIFFERENCE	CORRECTION FACTOR
Given value	1.50		0.96
		0.125	
Value to be determined	1.375	0.25	0.945
		0.125	
Given value	1.25		0.93

Note that the value to be determined lies halfway between given values; therefore, the correction factor is assumed to be halfway between values given.

3.4. Flexural Strength

Many structural components are subject to flexure, or bending. Pavements, slabs and beams are examples of elements that are loaded in flexure. An elementary example is a simple beam loaded at the center and supported at the ends. When this beam is loaded, the bottom fibers (below the neutral axis) are in tension and the upper fibers are in compression. Failure of the beam, if it is made of concrete, will be a tensile failure in the

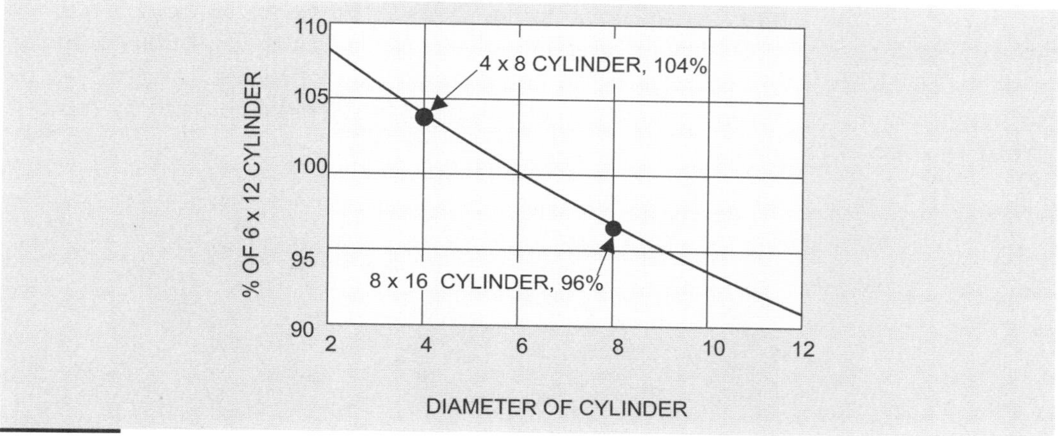

Figure 3-2
If we call the strength of a 6- by 12-inch cylinder 100 percent, then a 4- by 8-inch cylinder would indicate a strength about 4 percent higher (104 percent) for the same concrete, or an 8- by 16-inch cylinder would indicate only about 96 percent of the strength of the 6 by 12.

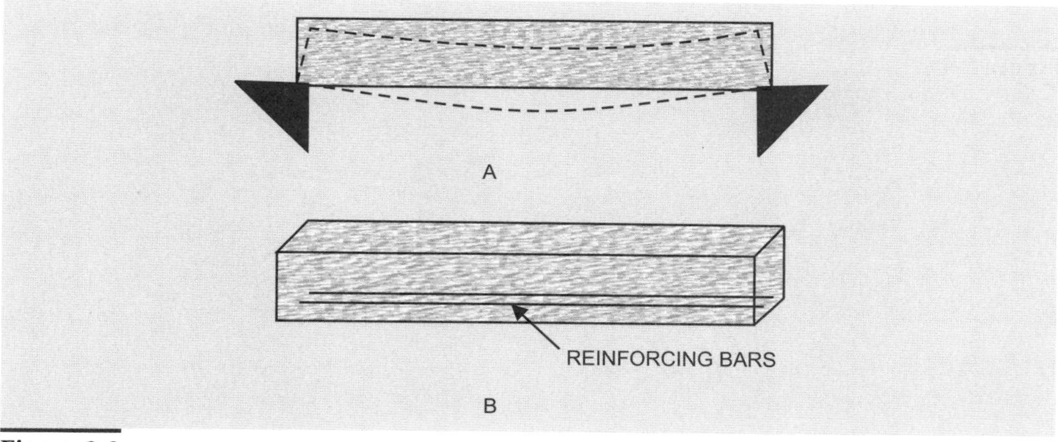

Figure 3-3
The bottom of a beam is in tension when the beam is loaded. Reinforcing bars are therefore put in the bottom of the beam to give it greater flexural strength.

lower fibers, as concrete is much weaker in tension than in compression. Now, if we insert some steel bars in the lower part of the beam (reinforced concrete), it will be able to support a much greater load because the steel bars, called reinforcing steel, have a high tensile strength. See Figure 3-3. Carrying this one step further, if the reinforcing steel is prestressed in tension (prestressed concrete), the beam can carry a still greater load.

The modulus of rupture is a measure of the flexural strength and is determined by testing a small beam, usually 6 by 6 inches in cross section, in bending. Usual practice is to test a simple beam by applying a concentrated load at each of the third points. See Figure 3-4.

Some agencies test the beams under one load at the center point, which usually indicates a higher strength than the third-point loading. Center-point loading is not usually used for 6-inch beams but is confined to smaller specimens.

3.5. Tensile Strength

There is no field test for direct determination of tension under axial loading. It is a difficult test to perform and the results are not reliable. There is, however, an indirect method called the splitting tensile test, in which a standard test cylinder is loaded in compression on its side. See Figure 3-5. By means of an equation, a value of tensile strength can be computed. Laboratory comparisons show that the tensile strength indicated by this test may be as much as 150 percent of the direct tensile strength.

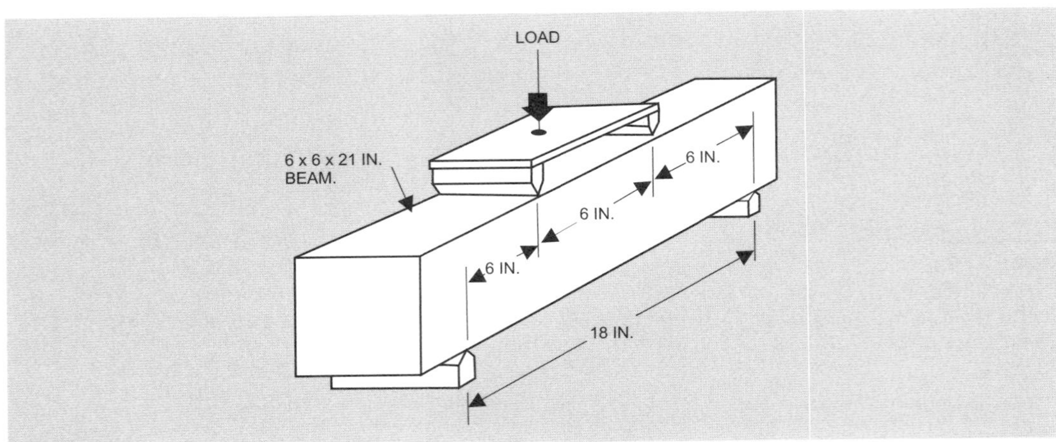

Figure 3-4
Testing a beam specimen in flexure.

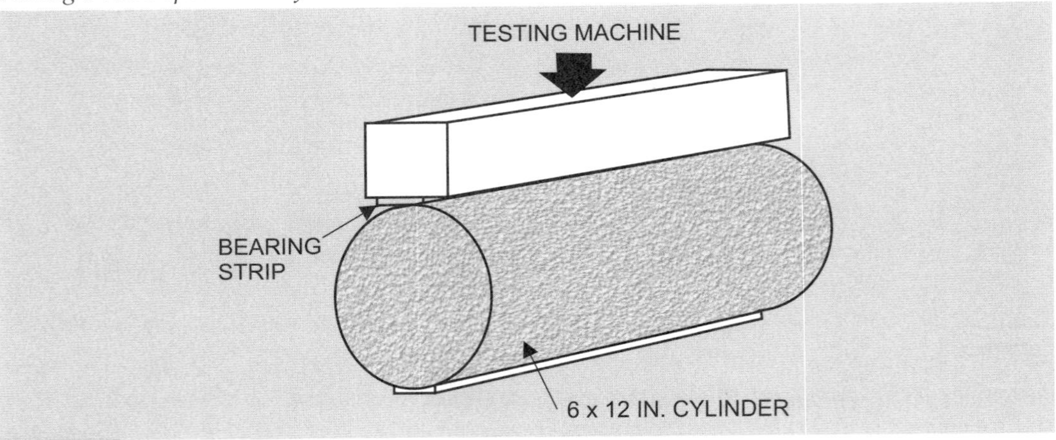

Figure 3-5
The splitting tensile test gives a good indication of the tensile strength of the concrete.

Concrete in the structure is rarely loaded in pure tension, the tensile stresses being in connection with flexure, torsion or a combination of loadings.

Awareness of the importance of tensile strength has increased, however, because of the significance of tension on the control of cracking. Research indicates that direct tension averages about 10 percent of the compressive, being about 7 or 8 percent for high-strength concrete (8000 to 10,000 psi compressive) and going as high as 11 or 12 percent for low-strength concrete in the range of 1000 psi compressive. See Figure 3-6.

3.6. Shear, Torsion and Combined Stresses

Shear is a loading in which a part of a member attempts to slide or shear along another part. See Figure 3-1. Because of the complexity of the action of forces in the concrete, it is not possible to make a direct determination of shear. Torsion, which is a twisting, is also complex and difficult to evaluate. When concrete fails, a combination of stresses causes the failure. Even a standard cylinder test, in which an axial compressive load is applied to the specimen, imposes shear and tension in parts of the specimen. Concrete in a structure is nearly always subjected to more than one type of stress—compressive, tensile, shearing—resulting from the application of various loads and moments on the members.

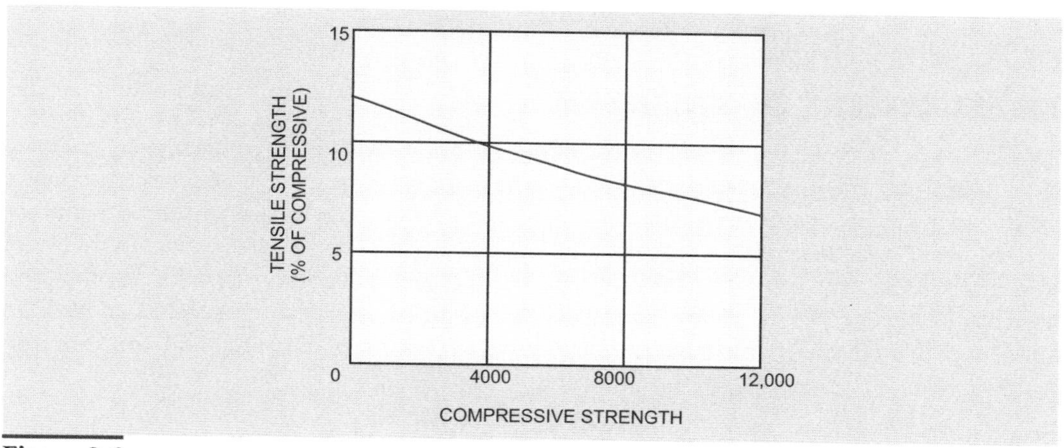

Figure 3-6
Tensile strength of concrete is about 12 percent of the compressive for low-strength concrete, going down to about 7 percent for high-strength concrete.

Theoretical and experimental relationships have been established that enable the engineer to relate forces and loads acting on members to compressive, flexural and splitting tensile values. By application of a suitable factor of safety, strength requirements for construction can then be specified.

3.7. Relationship of Test Strength to the Structure

In a later section of this book we will discuss how to test concrete and how to analyze test results. This analysis is necessary because of the natural variation of test results. This brings up two important questions: What is the relationship between the strength indicated by the test cylinders compared with the strength of the concrete in the structure? How are the variations in cylinder strengths reflected in the structure?

Consider the first question. Test specimens are made, cured and tested under certain standard conditions that are usually appreciably different from the conditions existing in the structure. Temperature and curing conditions can be vastly different. The value of the test specimens is that they give a measure of the strength potential and other properties of the concrete; they evaluate the materials and mix under certain standard conditions. If they indicate a low strength, then something is wrong with the materials or proportions. Actual strength of the concrete in the structure can be appreciably different. Besides temperature and curing, other variables are moisture content, size, shape, quality of consolidation, possible presence of defects such as rock pockets, restraint, and combinations of loading in the structure. It is because of these unknowns that a factor of safety must be considered by the structural engineer when the structure is designed.

In answer to the second question, the variations in cylinder strengths are not always reflected in the structure. If three specimens are made from one batch of concrete, under identical conditions throughout the test, there is no assurance that they will all fail at the same strength. In fact, the probability is that they will each break at a different strength. These are normal variations.

We do know that there are variations in the structure that are not caused by basic variations in the concrete itself. For example, when cores are taken from a column, the cores from the upper portion of the column invariably indicate lower strength than the cores from the bottom portion of the column. The reason for this is that the concrete near the bottom was compacted by the static hydraulic head of the concrete being worked above, yet there was no change in mix or materials.

MEASUREMENT OF STRENGTH

The strength of concrete can be determined by any one of four methods: by molding specimens from the fresh concrete on the job, by testing cores removed from the hardened concrete, by applying certain impact and rebound instruments to the hardened concrete and by sonic and electronic measurements applied to the hardened concrete in place. Specimens molded from fresh concrete on the job are universally used for control and acceptance testing. Tests by other methods are used for checking the results of molded specimens, especially in case of low strength indications or dispute, and for research involving existing structures.

Procedures for all strength tests are explained in detail in Chapter 13.

3.8. Job-Molded Specimens

For all concrete other than pavements, job-molded specimens are cylinders. The number of test cylinders required is given in ACI 318 Section 5.6.2, or in the project specifications. Test cylinders made to determine compliance with the specifications are made and cured under controlled conditions of temperature and humidity (ACI 318 Section 5.6.3). Test cylinders are sometimes made for curing under job or field conditions to determine time for removal of shoring or forms or when the structure can be put in service (ACI 318 Section 5.6.4).

3.9. Testing of Hardened Concrete

Samples sawed or cored from hardened concrete are not normally required, their use being confined to those cases in which some question or dispute has developed regarding the quality of the concrete as revealed by tests of molded specimens. Number, location, size and type of specimens are determined at the time sampling becomes necessary.

Coring and sawing specimens from hardened concrete are expensive expedients and should be adopted only as a last resort. Both procedures leave scars on the surface of the concrete that are difficult to eradicate, a condition that must be considered if the concrete is exposed to view. The possibility of structural damage, especially damage to reinforcement, cannot be ignored. If an approximation of the strength will suffice, one of the following described instruments can be used.

There are two simple instruments available for determining the strength of concrete in place. One of these, called a Swiss hammer (ASTM C 805), operates on the principle that the rebound of a spring-loaded steel plunger striking the surface of concrete is proportional to the strength of that concrete. It is a quick, nondestructive test that can be used for the determination of the approximate compressive strength of concrete in place, but it cannot be used to replace properly conducted cylinder or core tests. Testing is described in Chapter 13.

The second instrument is based on the principle that the penetration of a probe gauge into the concrete is inversely proportional to the compressive strength of the concrete. Called a Windsor probe (ASTM C 803), it uses a power-actuated device to drive the probe into the concrete. Accuracy is about the same as that of the Swiss hammer, but small indentations are left in the surface of the concrete, which might be unsightly in some exposed concrete.

In-place tests of concrete can be made with instruments that measure the velocity of a small mechanical pulse through the concrete. Known as the pulse velocity method, the apparatus consists of two vibration pickups (phonograph pickups), a hammer device to apply a blow to the concrete and an electronic circuit to measure the velocity of the sound generated by the hammer blow as the sound travels through the concrete from one pickup to the next. Considerable experience and expertise are required to interpret results correctly.

FACTORS AFFECTING STRENGTH

3.10. General Comments

When we ask what affects the strength of concrete, the answer is—just about everything. Among the factors are type, quality and amount of cement; quality, cleanness and grading of the aggregate; quality and amount of water; presence or lack of admixtures; methods followed in handling and placing the concrete; age of the concrete when placed in the forms; temperature; curing conditions; and age of the concrete when tested. Foreign materials may find their way into the concrete, thereby affecting the strength. Finally, the indicated strength of the test specimens may or may not actually represent the strength of the concrete in the structure. Table 3.3 shows some of the variables and their effects.

In using a table of this type, the first step is to eliminate the items that obviously do not apply, then to consider those that might be significant. In many cases, it will be found that there was more than one factor acting at the time. There may be one or more factors of high significance, or there may be several of minor significance that when acting together become highly significant.

In the following discussion we will assume that test specimens truly represent the concrete from which they had been sampled. One problem with strength tests is the time lapse between making the concrete and testing specimens. By the time strength results become available, it is too late to do anything about the concrete already placed, but the information at least provides a warning to avoid the troublesome materials or practices in future work.

Appreciable variations in strength of concrete result from the use of different brands intermittently or can even be caused by variations between shipments of cement from the same mill. Variations in raw materials, processing, age, fineness and temperature contribute to these variations. Undetected differences in cement types will affect strength, especially at early ages. If a batching plant has facilities for more than one type of cement, there is always the danger of using the wrong cement. There have been cases of accidental substitution of Type I for Type III. Also, accidental use of pozzolan instead of cement has occurred. Adequate inspection will minimize these incidents.

TABLE 3.3
SUMMARY OF COMPRESSIVE STRENGTH TEST VARIABLES [3.1]

BASIC CAUSE	ITEM	CAUSE OF VARIATION	PROBABLE OCCURRENCE	EFFECT ON STRENGTH
BATCHING AND MIXING				
Cement, material	1	Type and composition	With different brands	Considerable variation
	2	Manufacturing control	Any one brand	Can be considerable
	3	Age and condition	Always possible	Considerable variation
Water	4	Presence of salts	Infrequent	Not generally great
	5	Water/cement ratio	Dependent on control	Major effect
Sand, material	6	Chemically reactive	Common minor fault	Can be considerable
	7	Unsound particles	Infrequent	Not general
	8	Nonuniform properties	Infrequent	Not general
	9	Clean	Common minor fault	Not generally great
	10	Particle shape	Crusher and natural	Not within one type
	11	Grading	Always present	Through workability
Stone, material	12	Chemically reactive	Uncommon	Not appreciable
	13	Unsound particles	Depentent on source	Not generally great
	14	Nonuniform properties	With porous material	Not generally experienced
	15	Clean	Always possible	Can be considerable
	16	Particle shape	Crusher and natural	Not within one type
	17	Grading	Always present	Through workability
	18	Maximum size	With different mixes	Through workability
Temperature	19	Cement	Hot cement	Not appreciable
	20	Water	Extremes of climate	Not generally experienced
	21	Aggregates	Extremes of climate	Not generally experienced
Mix	22	Paste-aggregate change	Deliberate variations	Through workability
Cement Batching	23	Errors in weighing	Infrequent	Considerable
	24	Volumetric measurement	Not from central plant	Errors plus or minus 20 percent
Water measurement	25	Directly, added water	Where rely on judgment	Considerable if not measured
	26	Contained with sand	Most common	Considerable
	27	Sand bulking	Not from central plant	Can be considerable
	28	With coarse aggregate	Over period	Can be considerable
Sand measurement	29	Material changes, bulking	Not from central plant	Errors plus or minus 20 percent
Stone measurement	30	Material changes, operation	Where control limited	Not generally great
Mixing	31	Order of charging	Dependent on operator	Generally unimportant
	32	Priming mix	Occasional only	Can be considerable
	33	Mixer speed	With different plants	Not general
	34	Overcharging	Infrequent	Not general
	35	Time of mixing	Frequent	Variation can exceed 30 percent
TESTING				
Handling, sampling	36	Segregation	Chutes, transportation	Planes of weakness
	37	Constituent changes	Wherever retemper	Impossible to estimate
	38	Sampling	Different locations	Can be appreciable
	39	Bleeding	Mixes with water loss	Not generally great
Compaction	40	Hand tamping	Drier mixes	Considerable, exceed 50 percent
	41	Vibration	Over vibration	Segregation specimens
	42	Shock	Handling after setting	Damage creates weakness
	43	Particle orientation	Planes of weakness	Flat particles-40 percent
Size and shape	44	Wet screening	Mass concrete	Increase with screening
	45	Size of specimen	Nonstandard molds	Decrease strength with size
	46	Height/diameter ratio	Cylinders as overseas	Decrease as ratio increases
	47	Shape	Cube or cylinder	Cube strength greater
	48	Mold irregularities	Nonstandard molds	Nonaxial load
Curing	49	Drying out	Unlikely with good quality control	Great
	50	Moist curing	Not job curing	75 percent increase in 10 days
	51	Intial temperature	Freezing conditions	Considerable
	52	Temperature	Job curing in winter	60 percent variation possible
	53	Age	Compare at same age	Continuous increase
	54	Moisture content	When specimens dry	40 percent decrease
Capping	55	Plane ends	Dependent on laboratory	Concavity 30 percent, convexity 50 percent
	56	Capping material	Cylinders as overseas	Plaster of Paris-12 percent
	57	Axis of specimen	Technique problem	Not generally great
Testing machine	58	Bearing block	Dependent on laboratory	Can be considerable
	59	Centering	Dependent on laboratory	Can be appreciable
	60	Speed of loading	Dependent on laboratory	Can be considerable

3.11. Causes of Strength Variations

Strength test results falling below the anticipated values are the most common strength problem on the job. A high water-cement ratio from whatever cause decreases the strength of the concrete. Many factors besides the use of overly wet mixes with too much water can lead to a high water demand and consequent lowered strength.

Cement. The production of cement is done under close quality control, and cement is not usually the cause of strength problems. There have been, however, instances when the cement was the cause of the trouble. False set usually results in low and erratic strengths. False setting should be reported to the cement company immediately.

Old cement is rarely a problem with bulk cement in the modern batching plant except when it is allowed to accumulate on the sides of the silo if the silo is not drained every few months. Old cement containing lumps that can be broken with reasonable pressure between the fingers can be used, but hard lumps are apt to cause low strength. There have been reports of cement losing strength at rates ranging from one half of 1 percent to 5 percent per month when exposed to moist conditions that are due to improper storage.

Branches of the federal government and other large users of cement arrange for sampling and testing of cement at the mill and in this way keep informed of the quality. Every cement manufacturer will, upon request, furnish a certified mill test report with each shipment. This report identifies the shipment, shows the bin from which it was shipped and gives chemical and physical test results. These reports should form a part of the project records.

Aggregates. The quality of aggregates is probably more apparent in its effect on durability than in its effect on strength, but some properties of the aggregates, such as mineralogical composition, particle shape and graduation affect strength also. High-strength concrete is more responsive to physical properties of coarse aggregate than low-strength concrete. Surface texture, particle shape and elasticity are important in affecting both flexural and compressive strength. If an aggregate meets the requirements of ASTM C 33, it is satisfactory, as minor variations in the above properties can usually be compensated for by adjusting the mix proportions.

The effect of maximum size of aggregate (MSA) is variable. It is advantageous to use the largest MSA in low-strength concrete of the type usually used in mass concrete. When high strengths are required, a smaller MSA is desirable. See Figure 3-7. A large body of research by different agencies has led to the conclusion that there are three relationships between MSA cement content and compressive strength:

1. Maximum cement economy for low-strength concrete is achieved with a large MSA;
2. A rather wide range of MSA is suitable for medium-strength concrete at a consistent cement content; and
3. A higher strength for high-strength concrete is reached with a small MSA.

These relationships hold true for any kind of aggregate and age of concrete. It was also found that concrete with the smaller MSA has greater compressive strength than large MSA concrete at the same water-cement ratio, especially at water-cement ratios below 0.50. This information is shown in Figure 3-8.

Most of the deficiencies of aggregate that affect strength can be corrected by adequate and thorough processing of the raw material. Soft and friable particles, loam, clay, mica, coatings and organic material can all be removed by various steps in the manufacturing procedure.

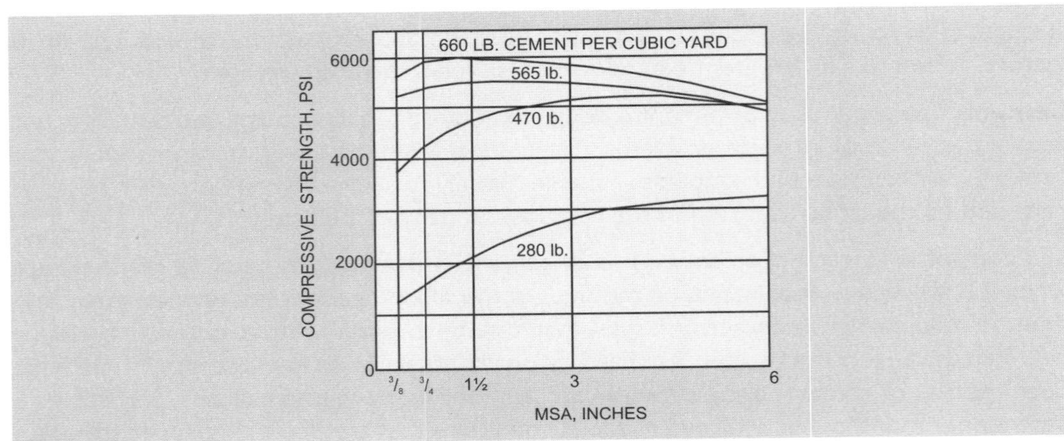

Figure 3-7
For high-strength concrete with a high cement content, it is advantageous to use a small MSA ($^3/_8$ to $^3/_4$ inch). For low- strength mass concrete, the largest practical MSA should be used.

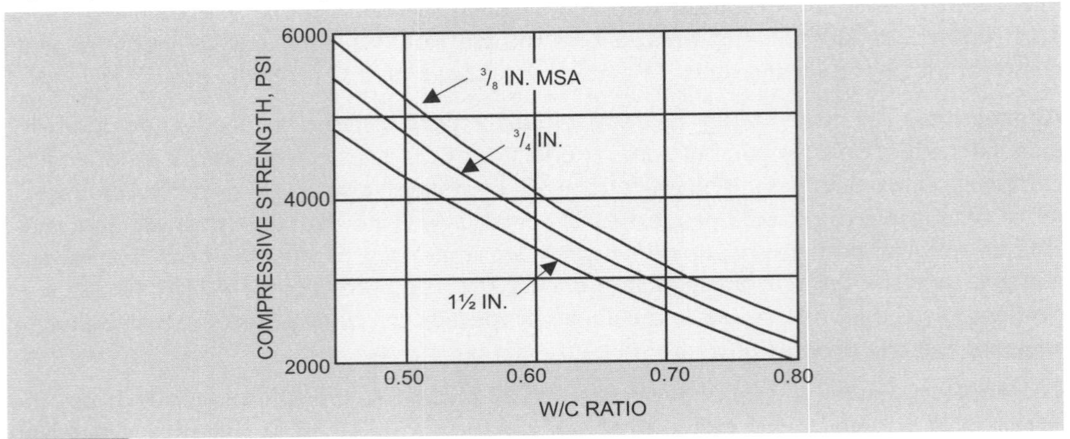

Figure 3-8
Small MSA concrete has higher strength than large MSA concrete at the same water-cement ratio.

Adherent coatings of clay, caliche or other materials prevent proper adhesion of the cement paste to the aggregate particles, resulting in lowered strength. Loose fines in small amounts are not harmful. Nearly all coarse aggregate, by the time it reaches the batcher, is coated with rock dust and other fine material. If the fines are loose and do not exceed about 1 to 1.5 percent of the coarse aggregate, no harm is done. Finely divided, uniformly distributed clay or silt in the sand is not harmful if it does not exceed 5 percent (3 percent for concrete subject to abrasion).

The full potential concrete strength will not be developed if the aggregate contains an appreciable amount of organic matter. Humus in surface soils, grass and roots require that aggregate deposits be stripped of the surface layer to a depth sufficient to eliminate the organic soil and roots.

Crushed stone, screened dry, is satisfactory as long as the stone is actually dry. The presence of varying amounts of moisture results in varying amounts of fines adhering to the stone, causing variations in strength of the concrete. Gravel usually contains moisture and fines in sufficient amounts to require wet screening. The scrubbing action of high-velocity water jets and movement of the aggregates removes all but the most stubborn coatings.

Certain aggregates of glacial origin contain organic materials that act as a foaming agent, entraining air in the concrete, thereby having a detrimental effect on concrete strength. Such aggregates should be avoided.

Some aggregates with a specific gravity less than 2.55 or absorption exceeding 1.5 percent may be deficient in potential strength. Unless such aggregate has a history of satisfactory use, tests should be made to determine how it will perform. Usual procedure is to compare it with an aggregate of known satisfactory performance.

Water from any municipal distribution system is satisfactory. Water from stagnant pools or swamps should be viewed with suspicion, especially if moss and algae are present. As little as one half of 1 percent of organic material in the mixing water causes a serious deficiency in strength. Impure water containing appreciable quantities of sulfates, chlorides or carbonates (mineral water) should not be used. Water from lakes, streams and ponds should be tested prior to use.

Mix Proportioning. The relative proportions of cement, water, aggregates and admixtures in the mix are fundamental factors affecting concrete strength. These are discussed fully in Chapter 12.

Making and Handling the Concrete. Inaccuracies in measuring batch quantities are a source of variations in concrete quality and therefore of strength variations. Volumetric measurement of solid ingredients should never be permitted, as variations of 40 percent or more in the strength can result. Fine aggregate is more critical than coarse aggregate in this respect, mainly because of variations in moisture content. See Figures 3-9 and 3-10.

Sources of batching errors are careless setting of weights on the scales, careless operation, material sticking in weigh hopper so that scales do not return to zero between batches, dirty or worn knife edges and fulcrums on scales, wrong allowance for suspense material in automatic plants, and uncompensated variations in moisture content of aggregates. Other strength variations may be caused by variations in sequence of charging materials into the

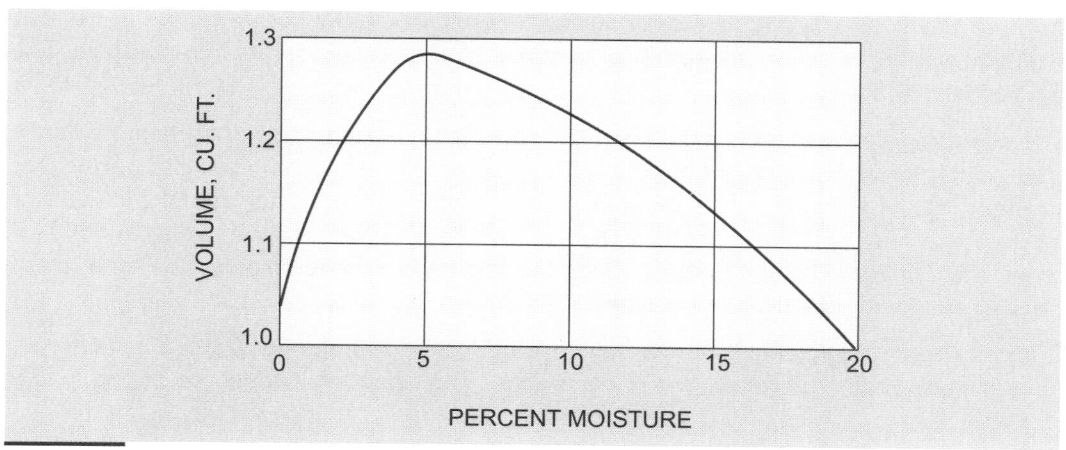

Figure 3-9
A serious objection to volume batching is the bulking of sand; that is, an increase in volume up to a maximum point, then a decrease in volume as the moisture content increases. Yet the actual weight of dry sand remains unchanged.

mixer, delays in mixing batches in which cement is in contact with wet aggregates, nonuniform or improper mixing time or speed, and delays in transporting or placing premixed or truck mixed concrete.

Figure 3-10
Footprint in pile of loose, damp sand, illustrating bulking effect of surface moisture. (Courtesy of PCA)

Handling and placing methods that lead to segregation of the concrete, and lack of thorough consolidation in the forms, cause variations that are reflected in the strength, although these variations will not be shown in test results.

Temperature and Curing. The rate of reaction between the cement and water varies over a wide range of concrete temperatures, proceeding slowly at low temperatures somewhat above the freezing point of water, and more rapidly at high temperatures somewhat below the boiling point of water. Concrete temperatures below 50°F are unfavorable for the development of early strength. Below 40°F the development of strength is greatly retarded. At freezing temperatures strength development is practically absent. There is some evidence, on the other hand, that curing at temperatures in excess of 150°F impairs the ultimate quality of the concrete.

A concrete temperature of 80°F or above during and immediately after placing will result in lower ultimate strength compared with concrete mixed and placed at 40°F to 80°F. Slow strength development and low-early strength result from low curing temperature, but strength will be better if the concrete is cured at a more moderate temperature. See Figures 3-11 and 3-12.

The effect of freezing depends on when the concrete is frozen. If it is frozen immediately after mixing or at any time while it is still in a plastic condition, it sustains a permanent strength loss of as much at 50 percent. If the freezing occurs after about 24 hours of moderate temperature, enabling the concrete to develop a compressive strength of 500 psi or so, the damage is probably slight. Once the concrete has been frozen, the duration of freezing seems to have little or no effect.

Curing of concrete accomplishes its objective by maintaining the concrete at a moderate temperature under constantly moist conditions. The effect of inadequate curing is low strength.

3.12. Apparent Low Strength

It sometimes happens that strength specimens do not truly represent the concrete from which they were made because of improper sampling or testing techniques. See Figure 3-13. Usually the result is low strength indications. If the concrete is sampled, made into specimens, and the specimens are handled and tested in accordance with ASTM C 172 (sampling), C 31 (making and curing) and C 39 (testing), the results should be representative. Any deviation in any of the steps will cause erroneous results.

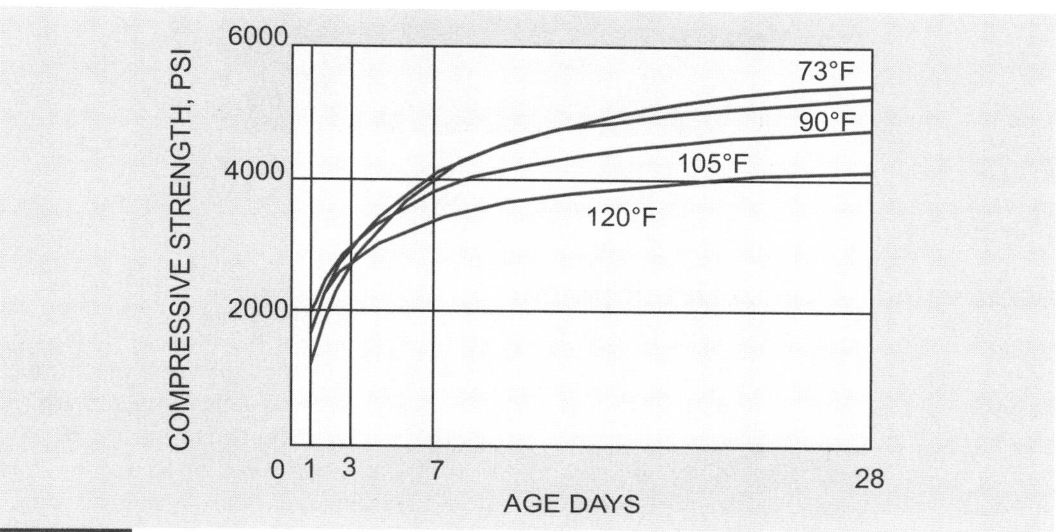

Figure 3-11
Concrete made and cured at high temperatures has high-early strength up to an age off four or five days, but beyond that time the strength is not as good as that for concrete made at 73°F. The strength difference continues beyond 28 days.

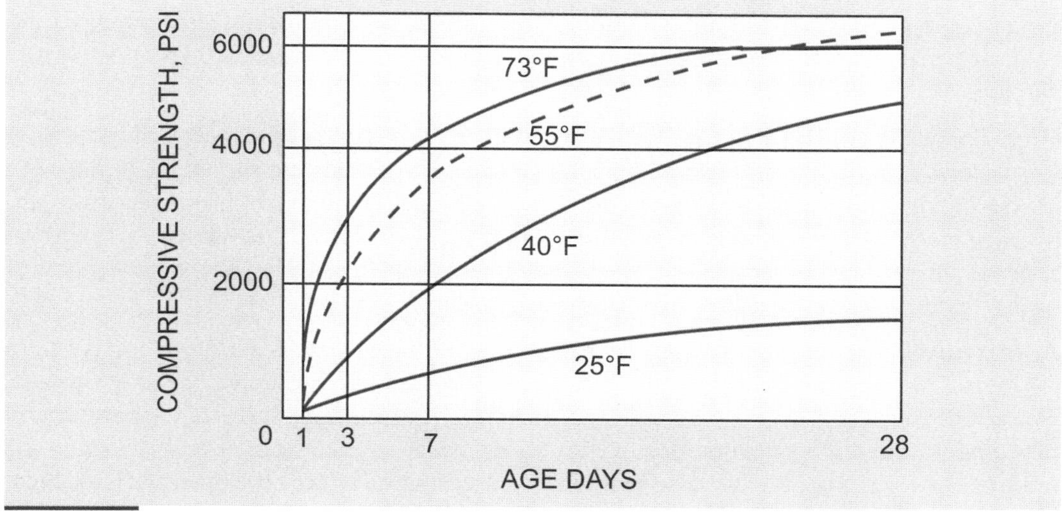

Figure 3-12
Concrete made and cured at low temperature does not develop strength as well as concrete made at 73°F. However, down to a temperature of 55°F the difference at 28 days is very small. The cooler concrete may actually have slightly more strength.

Figure 3-13
Improperly made strength specimens unfairly penalize good concrete because they are not representative.

3.13. Accelerated Strength Development

Rapid development of strength, especially in a few days or even hours, is frequently desirable. Emergency repairs of concrete must develop high strength in a matter of hours to permit use of the structure, or it may be necessary to put a new structure into service in a short time. During cold weather, high-early strength permits a shortened period of time during which the concrete must be protected from low temperatures or freezing. Rapid strength development is advantageous in precast and prestressed work, where a shorter operating cycle in the plant effects significant savings in cost by permitting early form removal and expeditious use of the castings, thus minimizing the number of forms required and reducing the storage area required for curing of units.

Five methods are commonly used, either singly or in combination, to accelerate the early strength of concrete. They are the use of early hardening or high-early-strength cement, the use of special rapid-setting cementing compounds, addition of an accelerator to the concrete mix, high-temperature curing and retention of the heat of hydration of the cement.

High-Early-Strength Cement. Modification of the compound composition of cement by increasing the percentage of tricalcium silicate (decreasing dicalcium silicate) together with finer grinding makes a cement that hydrates more rapidly, resulting in high-early strength. Known as Type III, this cement produces concrete with normal seven-day strength in approximately three days, and normal 28-day strength in seven days. Rapid generation of heat of hydration accompanies the early-strength gain.

Admixtures. The commonly used accelerator is calcium chloride added to the batch in solution. The effect on strength up to about seven days is similar to that of Type III cement. The effect of calcium chloride depends on the amount used, temperature, type and brand of cement and curing conditions. Calcium chloride should not be used when the temperature is above 60°F because of a possible detrimental effect on strength at later ages, and it should never be considered an antifreeze. It should not be used in prestressed concrete, and should be used with caution in reinforced concrete because of the potential for reinforcement corrosion. Nonchloride accelerators are commonly used in reinforced concrete to avoid corrosion.

Retention of Heat of Hydration. Because heat is liberated when cement hydrates, curing and strength development during cold weather can be accelerated if advantage is taken of this reaction. By insulating structural forms the heat generated is sometimes sufficient to protect the concrete from freezing until the necessary curing period has elapsed. The ACI report "Cold Weather Concreting" includes tables and charts showing insulation required for different temperature conditions.[3.2]

High-Temperature Curing. Central plant manufacture of precast concrete units lends itself particularly well to high temperature curing. Concrete masonry blocks, pipe, and precast and prestressed units of all types are universally cured by some sort of high temperature arrangement. By using rich, low-slump mixes containing over 700 pounds of Type III cement per cubic yard of concrete with a water reducing admixture and steam curing, compressive strength in excess of 4000 psi at 18 to 20 hours age is commonplace. See Table 3.4.

Both heat and moisture are necessary, usually supplied by maintaining the concrete in an enclosure charged with wet steam. See Chapter 17.

Rapid-Setting Cements. A number of compounds are available that set hard in a few minutes and develop several thousand psi strength in half an hour or less. Some are suitable only for emergency repair work. Some contain calcined gypsum (plaster of paris) and should not be used where they will be exposed to moisture. A few cement manufacturers make small quantities of a very rapid-setting portland cement. A mixture of portland cement and aluminous cement sets hard in a few minutes. Aluminous cement alone sets rapidly. Certain blended cements can achieve a compressive strength of 3000 psi in 6 hours or less.

TABLE 3.4
TYPICAL HIGH-EARLY-STRENGTH CONCRETE

CEMENT LB. PER CU. YD.	W/C RATIO	% AIR	COMPRESSIVE STRENGTH—PSI				
			14-HOUR	18-HOUR	3-DAY	7-DAY	28-DAY
750	0.34	2.6	6270	6730			8180
730	0.35	2.7	5740				7580
660	0.40	3.1	4500	4880		5680	
750	0.36	3.5			6530	7240	8940

Notes:
1. All concrete made with Type III cement, $^3/_4$-inch MSA, 1- to 2-inch slump.
2. Above values represent typical job values.
3. All specimens steam cured at 140°F to 160°F until broken, if less than 24 hours. Cylinders for later breaks then water cured at 70°F for balance of time.
4. Because of possible shrinkage and cracking, cement contents in excess of 800 pounds per cubic yard are not recommended.

3.14. Slow Strength Development

Concrete, after being placed, sometimes fails to harden properly, or fails to gain strength as rapidly as it should. Ultimately, such concrete may reach the normal strength expected for the mix and materials being used, although this is not always the case.

An overdosage of retarder in the concrete delays setting and early strength gain beyond the anticipated extent, although eventual strength appears to be unaffected, even by gross overages of two or three times the design amount of admixture. There have been many cases of concrete containing accidental double doses of retarder going into the forms. After

24 hours the affected concrete is so soft that it can be dug out with the hands, but after continued curing for a few days the concrete starts to develop strength, and at one month the affected areas are indistinguishable from the rest of the structure. Numerous readings taken with the impact hammer, cores cut from the concrete, and laboratory tests have verified this.

Errors and variabilities in formulation of admixtures are a potential cause of variations in concrete using these admixtures. For this reason, products of reliable manufacturers, with a satisfactory record of use, should always be used.

As mentioned earlier in this chapter, organic materials adversely affect concrete. The presence of sugar in any of its forms is detrimental and erratic in its effects. About 0.05 percent by weight of the cement in the concrete (less than one ounce per bag of cement) will completely retard the set for several hours, resulting in no strength at 24 hours and half of normal at three days, though it may cause a slight increase in strength at 28 days. Higher percentages can completely inhibit setting and hardening.

If concrete is placed cold and kept at a temperature near freezing (but not frozen) the hydration process will be slowed to such an extent that several hours can elapse before apparent setting begins, and strength gain will be very slow. Even at 28 days age, strength will be appreciably below normal. However, under proper sustained curing, especially if the temperature is raised to a minimum of 60°F, strength will increase. The sooner this curing is started, the sooner the concrete will develop adequate strength. No permanent damage is sustained by the concrete.

HIGH-STRENGTH CONCRETE (HSC)

The line of demarcation between normal structural concrete and high-strength concrete is about 6000 psi compressive strength. The increased need for reduced size and weight in structural elements, such as in columns of high-rise buildings (see Figure 3-14), and widespread use of precast and prestressed concrete have pinpointed the need for concrete with compressive strengths in excess of 10,000 psi. What we are concerned with in this section is not necessarily concrete with a high early strength, but concrete with a high ultimate strength. Under ordinary field conditions, the higher-strength concretes require a longer test age than the standard 28 days to develop the higher specified strengths. For

Figure 3-14
Reinforced concrete high-rise office building with 17,000 psi high-strength concrete in the columns. (Courtesy of PCA)

Figure 3-15
High-strength concrete testing.

example, the specified strength of concrete for the columns of a high-rise building may be indicated in the project plans and specifications as 10,000 psi at 56 days. Acceptance tests for the 10,000 psi concrete are at 56 days rather than the customary 28 days. The higher strength concretes also have a different strength breaking characteristic than that of the concrete strengths used in normal practice, as illustrated in Figure 3-15.

High-strength concrete is also commonly used in central casting yards of plants for the production of precast concrete products, where usual practice is concrete with slumps from 0 to 2 inches; very rich mixes of 700 to 800 pounds of cement per cubic yard; Type III cement; water-reducing admixtures; small MSA; sound, well-graded aggregate; extensive vibration and revibration; and high-temperature curing. All of these steps, except the very low slump and high-temperature curing, are available for structural concrete from a ready-mix plant. No new technology is required, but exercise of the best possible quality control measures is essential. The following discussion is concerned with ready-mixed structural concrete placed under field control conditions.

3.15. Selection of Materials and Mix

Cement should be selected on the basis of the best strength at 28 or 56 days. For this reason, there is probably no advantage in using Type III cement, as the advantage of Type III is in the strength prior to 28 days. There are differences between different brands of cement, and the best one should be selected.

Aggregates likewise should be selected on the basis of the best that is available. Irregularly shaped natural gravel or cube-shaped crushed rock with a rough and slightly porous surface will give the best bond with the cement paste. Bond with the sand is not so critical; hence, it is better to use a well-rounded sand because of better workability. See Chapter 8 for a discussion of aggregate properties. In the very rich mixes under consideration, the smaller MSA concrete gives the highest strength. See Figure 3-8. Some producers have found that $^{1}/_{2}$-inch or $^{5}/_{8}$-inch aggregate gives the best results. The grading of this coarse aggregate can vary over a wide range with little effect on strength as long as it remains within the limits of ASTM C 33. Sand must be well-graded and clean. The amount passing through the 100-mesh sieve should be minimum.

Laboratory trial mixes will reveal the strength-making properties of different aggregates. All properties of the aggregates—particle shape, surface texture, grading, composition—are critical.

Admixtures can be used to improve the concrete. There is no advantage in using an accelerator such as calcium chloride. Air entrainment is usually not required for most high-strength concrete, because the concrete is used where it is not likely to be exposed to freezing and thawing in a saturated condition. If it is exposed under these conditions, then the usual amount of air should be entrained, even at the expense of a loss of strength. The amount of air should be the minimum recommended for the MSA used. See ACI 318 Chapter 4, Table 4.4.1.

A water-reducing admixture or a superplasticizer can be used to advantage, and its use should be investigated. Tests should be conducted using the manufacturer's recommended dosage. The effect of any admixture varies with different cements, so tests should be made with the materials and mixes proposed for the work. Temperature affects the action of admixtures. The exact time that the admixture is introduced into the mixer is critical and requires accurate control.

Mineral admixtures, such as fly ash and especially silica fume, are used to achieve strengths between 8000 and 20,000 psi.

Mix proportions must be selected on the basis of trial mixes made in the laboratory, followed up with full-scale tests in the truck mixers that will be used on the job. Trial mixes must be proportioned with sufficient total water to give the necessary workability, usually obtainable with a 3- to 4-inch slump.

3.16. Handling and Quality Control

Tests should be made in the batch plant to obtain the optimum condition of charging the mixer and mixing. Central-mixed concrete is preferred; the so-called ribbon feed is preferred. Concrete should be kept as cool as possible because slump at higher temperatures is less and requires more water, which reduces strength.

High-strength concrete is often a sticky concrete. Water should not be added to the batches on the job except when, upon arrival at the jobsite, the slump of the concrete is less than that specified, and provided the water-cement ratio of the design mix is not exceeded. Close coordination between the contractor and ready-mix supplier is essential to avoid delay, and good communication must be maintained between the job and batch plant. Loads must be discharged promptly. Good concreting procedures are described elsewhere in this book, and they must be followed. Avoidance of segregation, application of good compaction and thorough curing are essential.

Quality control measures include, besides good construction practices, proper testing and evaluation of strength test results. See Chapters 13 and 26.

EARLY MEASUREMENT OF STRENGTH

The 28-day compressive strength test has become the standard of the concrete construction industry. The exact reason for this time period appears to be lost in the history of the industry. Probably, many years ago, it was thought that the strength at 28 days was nearly the ultimate strength of the cement. As a quality control measure, the 28-day strength is practically useless. By the time a month has passed since the placement of the concrete, there is nothing that can be done to improve low-strength concrete already in the structure. Additional stories have been erected in the building; a small structure may already be in use. To remove the defective concrete can be prohibitively expensive. For these reasons, many schemes have been advanced that would enable the concrete user to estimate 28-day strength from early-strength tests or other tests. These are discussed in Chapter 13.

EXPOSURE TO HIGH TEMPERATURE

3.17. Long-Time Exposure

Concrete is frequently used in locations where it is exposed for extended periods of time to high temperatures and low humidity. Chimneys, parts of furnaces and ovens, and incinerators are examples of the kinds of structures that might be involved. In most structures where high temperatures exist, only one face of the concrete is exposed to the heat, and the temperature decreases as it moves through the concrete. Therefore, only part of the concrete is affected by the temperature. The outer face will be considerably cooler than the heated face.

One effect of heating is to drive off the water that had combined with the cement during hydration. This drying, or dehydration as it is called, starts at the heated surface. Because the dehydrated concrete is a better heat insulator than normal concrete, the rate of dehydration is slowed considerably some distance from the surface.

Tests to determine the effect of heating have been made by several investigators with variable results, depending on size and kind of specimens, age of concrete, length of exposure, temperature, rate of cooling, mix proportions, type of aggregate and other factors. The average values shown in Table 3.5 give an indication of what might be expected. Individual laboratories reported losses between 0 and 40 percent at 500°F exposure, to as much as 50 percent to 95 percent at 1300°F.

In general, it appears that important load-bearing members with all surfaces exposed to the heat should not be continuously exposed to temperatures above 500°F. Members carrying light loads or with only one face exposed to the heat could probably withstand 900°F, although at this temperature spalling of the concrete becomes a hazard. Temperatures above 1000°F are definitely a hazard and require special treatment. Special refractory concretes designed and proportioned for high-temperature exposure are discussed in Chapter 22.

Concrete that will be exposed continuously to temperatures greater than 200°F should be laboratory tested to determine if the expected temperature will be detrimental.

TABLE 3.5
EFFECT OF HIGH TEMPERATURE ON COMPRESSIVE STRENGTH

	APPROXIMATE PERCENT LOSS IN STRENGTH AFTER EXPOSURE			
TEMPERATURE	500°F	800°F	1300°F	1700°F
Average % loss	0 to 40	50	30 to 96	97

Laboratory specimens were exposed on all sides for several hours to the temperature shown, then cooled and tested in compression.

3.18. Fire-Damaged Concrete

It sometimes becomes necessary to inspect and evaluate concrete in a building that has been damaged by fire to determine whether any parts of the building can be used. See Figure 3-16. These evaluations are difficult to make, requiring considerable experience to recognize the extent of the damage. If the surface of the concrete is soft and chalky, or if there has been spalling or scaling, the concrete has probably been damaged to the extent that load-carrying members can no longer function. Reinforcing steel in such affected concrete may have been damaged.

If the concrete can be chipped or picked to remove the damaged concrete down to the characteristic bright and crystalline appearance of good concrete, it might be possible to salvage the member. Swiss hammer readings and sonic measurements can give a good indication of the quality of the remaining concrete. Other laboratory tests are available that provide means to determine the maximum temperature suffered in various parts of the structure if samples of the concrete can be obtained within one or two days after the fire. Regardless of any tests that are made, the final decision is an engineering decision and should involve the structural engineer and the building department having jurisdiction over the building.

See also the discussion on high temperature in Chapter 4, Section 4.3.

Figure 3-16
Fire-damaged concrete resulting in spalling of the concrete that is due to fire exposure.

The Durability of Concrete

4.1. The Nature of Durability

When we build a structure of concrete, we expect it to last a long time; that is, we expect it to be durable. ACI Committee 201 in its report states "durability of concrete is defined as its ability to resist weathering action, chemical attack, abrasion, or any other process of deterioration. Durable concrete will retain its original form, quality, and serviceability when exposed to its environment."[4.1]

A discussion of concrete durability can hardly be made without a consideration of all the properties of concrete. Durability is intimately related to strength, shrinkage, water tightness and surface condition of the concrete; to structural design; and to materials and workmanship, as well as exposure conditions. In general, we may say that there are six factors that affect durability:

1. Characteristics of materials composing the concrete
2. Physical properties of the hardened concrete
3. Exposure conditions
4. Loads imposed on the structure
5. Practices used during construction
6. Structural design

Having defined durability, we are led to the second question: How can we measure the potential durability of any combination of materials under any given set of conditions? The complexity of this question was pointed out by Valenta, who stated the problem:

The importance of the durability of concrete increases with the development of cement production, scope of application of concrete, and the ever-increasing endeavor to learn more about the physical and structural mechanics of concrete. From the point of view of the majority of buildings and structures, particularly those of extraordinary technical and economic importance, the durability and the strength of concrete are of equal importance. In contradiction to strength, however, durability is a characteristic that is subject to a whole complex of laws and influences and that cannot be ascertained as simply as the strength of concrete; i.e., by means of a simple test. The strength of concrete, or any other of its physical or mechanical characteristics, is but a measure of its durability, which is necessarily influenced also by the problems of the actual determination of the strength on test specimens and its relation to the actual strength of the concrete in the structure.[4.2]

At present, methods of measuring durability are freezing and thawing tests of concrete specimens, tests to evaluate certain combinations of aggregates and cements, and indirect electronic tests. These are described in Chapter 13. All of these tests are valuable, and they give us a good indication of what to expect of the proposed concrete, provided that good construction practices are followed during construction.

Weathering. Any structure exposed to the elements for any length of time loses its new appearance, and concrete is no exception. This normal weathering may cause a slight roughening of the surface or slight rounding of edges, giving the structure a patina of age, but it is not harmful. The structure retains its good appearance and structural adequacy. The concrete is sound and strong. It is durable.

Accelerated weathering is manifested by cracking, spalling and disintegration of the concrete. Fragments of the concrete have a dull, chalky appearance; bond between the paste and aggregate is poor or nonexistent. Causes are many. Frost action causes expansion and cracking. Such weathering is cumulative, the rate of destruction increasing with every season. Attack by outside agencies, such as sulfates and acids, will hasten destruction. See Figure 4-1.

Figure 4-1
Freeze-thaw deterioration of nonair-entrained concrete.

Abnormal expansion and contraction of concrete cause stresses that are harmful. Such volume changes may be caused by changes in temperature or moisture, unsound aggregates such as shales and cherts, unsound cement (rarely), thermal differences in the materials, or reaction between aggregates and cement.

Sometimes there are differences in the durability of parts of a single structure that are difficult to account for. Among the factors contributing to the weathering damage shown in Figure 4-2 are the following:

1. Differences in surface area per unit volume, permitting easier access of the destructive agency to some parts of the structure
2. Differences in cement content, water-cement ratio, slump or other mix properties
3. Differences in methods of handling, finishing or curing
4. Differences in weather conditions when the concrete was placed

Figure 4-2
Small detailed portions of a structure have high surface area for each unit of volume, and the concrete may have had a high slump. Both factors make such small details more vulnerable to weathering damage.

THE AGENCIES OF DESTRUCTION

Concrete structures are in use today that are 50 to 100 years old. These structures are still performing satisfactorily and still present a good appearance. There is also concrete that has had to be repaired or replaced after only a few years of service. Why have some of these structures failed? What are the deficiencies that have marred the appearance, impaired the safety or reduced the serviceability of some of these structures?

When one examines a large number of concrete structures, the fact that stands out is the considerable difference in the appearance and usefulness of the concrete. In many cases it is quite difficult to explain why one structure, or even a part of a single structure, has shown serious deterioration while the adjacent concrete is still in excellent condition. Cracking, spalling, settlement and erosion are among the evidences of surrender to the agencies of destruction. These agencies are at work to some extent at all times and in all places. They may be classified into four general categories:

1. Deficiencies or weaknesses of the concrete itself, resulting from failure to follow the five fundamentals of good concrete construction, discussed in Chapter 1. Such weaknesses, although they may not themselves actually damage the concrete, can be the opening wedge for serious damage by actual attack. For example, low strength or shrinkage cracking could contribute to failure by attack by any of the forces in the three following categories.
2. Chemical or mechanical attack by outside agencies other than weather cycles.
3. Reaction between the constituents of the concrete itself.
4. Cyclic forces of the weather.

Briefly, such are the agencies of destruction. Steps necessary to protect the concrete from these attacking agencies seem obvious. Where there is danger of deficiencies in the concrete itself, care must be taken to provide the best concrete possible commensurate with requirements of the exposure. Where there are cyclic weather forces, good concrete must be provided, and it should be protected as much as possible. Where there is attack by outside agencies, good concrete is necessary, and the attacking forces must be neutralized. Where there is danger of reaction between the constituents, others must be provided, or they must be neutralized to make good concrete. In all cases, the starting point is good durable concrete.

Within the four categories listed above there are many conditions that can be harmful to the concrete, either individually or acting together. Some of the most important are:

1. A harmful reaction between certain minerals in the aggregates with the alkalies in the cement;
2. Exposure to sulfates in alkali soils;
3. Exposure to harmful substances in groundwater, seawater or in industrial processes;
4. Repeated cycles of freezing and thawing;
5. Repeated cycles of moisture or temperature changes above freezing;
6. Poor drainage in service; and
7. Inferior concrete resulting from inferior materials, high water-cement ratio, low cement content, high slump, inadequate compaction, segregation or lack of curing.

4.2. Deficiencies of the Concrete Itself

Concrete can be inherently nondurable if faulty materials were used, if the workmanship was inferior or if the mix proportions were not proper for the expected exposure conditions.

Faulty Aggregates. When deleterious materials find their way into the concrete, it is usually through the normal constituents of the concrete. In the case of aggregates, careful testing, including a petrographic examination, will reveal their presence. Natural aggregates of inferior or borderline quality can be improved by various beneficiation processes, such as heavy media separation, impact disintegration, jigging or elastic fractionation. Most aggregate producers can furnish laboratory reports covering their materials, showing whether deleterious materials are present in the finished material. Aggregates should conform to ASTM C 33.

Certain highly absorptive particles such as sandstone and chalk expand when they absorb moisture. Porous chert absorbs moisture that, upon freezing, exerts sufficient force to disrupt the concrete, causing a popout. See Figure 4-3. Cherts exist in gravel deposits and as lenses in limestone quarries in the north central states, especially northern Illinois and Indiana. Other contaminating substances sometimes found in the aggregates are silt, clay, mica, coal, shale, slate, humus and other organic matter, chemical salts, soft fragments, surface coatings, cemented particles, caliche and encrustations. In the Kings River in California, a black, soot-like coating was found on some of the gravel particles. Analysis showed it to be manganese and iron oxides, probably chemically harmless in concrete, but a possible source of inferior concrete because of its deleterious effect on bond between the paste and aggregate.

The chemical activity of certain reactive aggregates is discussed in Section 4.4.

Figure 4-3
A popout is the breaking away of a small fragment of concrete surface because of internal pressure that leaves a shallow, typically conical depression. Popouts are usually caused by (1) the swelling of certain aggregates as they absorb water, (2) the expansion of certain saturated aggregates as they freeze or (3) alkali-aggregate reactivity. (Courtesy of PCA)

The Cement. Rarely a direct cause of poor durability, the cement can nevertheless contribute to lack of durability if poor judgment is used in selecting the type of cement for certain exposure conditions. When concrete is to be exposed to sulfate soils or solutions, a Type V sulfate-resistant cement should be used in severe exposures, or a Type II in moderate exposures. Moderate sulfate- and high sulfate-resistant blended cements are also available.

Cement should be fresh and free of false setting tendencies. If it conforms with the requirements of ASTM C 150 and ASTM C 595, there should be no problem. By far, the largest volume of cement used is portland cement (ASTM C 150). Blended hydraulic cements (ASTM C 595 and ASTM C 1157) are available in some areas.

Water probably contributes as much to poor durability as anything else, not because of poor quality, but because of quantity. Harmful materials may enter the concrete through the mixing water. Normally, tap water is satisfactory, but sewage plants or manufacturing plants may contaminate streams or ponds with effluents containing harmful amounts of tannic acid, sugar, carbonic acid, sulfates or organic materials.

Admixtures. If admixtures are properly used, they can help make the concrete more durable. Improperly used, they can cause trouble. Admixtures from reputable manufacturers, used in accordance with the manufacturer's instructions, are satisfactory.

Inferior Workmanship. If we have selected our materials wisely, we should be well on our way to having our quality concrete. However, before we reach our goal, we have to put these materials together, mix them into concrete, transport the concrete to the forms, and there place, consolidate, finish and cure it. Much good concrete has been ruined at some point during its journey.

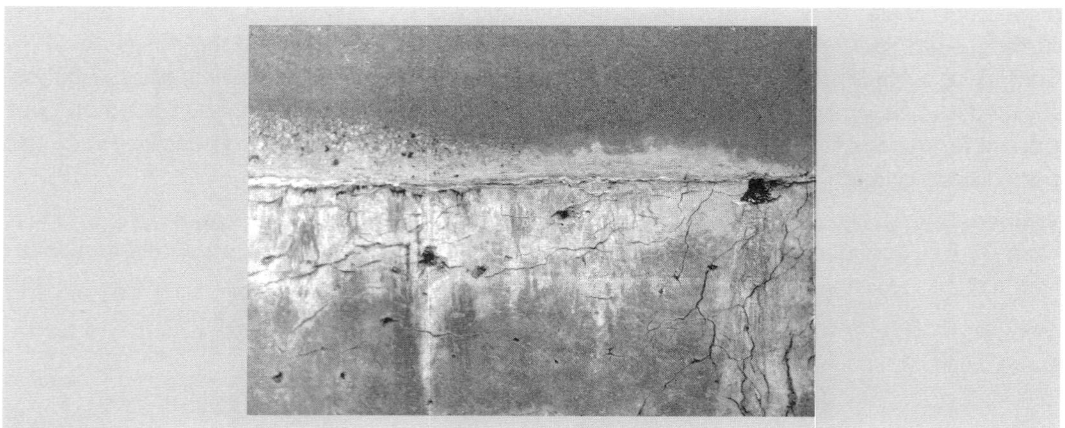

Figure 4-4
Shrinkage and cracking of weak concrete in the top of a lift of concrete at a construction joint, resulting from an accumulation of wet, soupy concrete.

Modern equipment and methods have steadily improved our ability to obtain good concrete. All persons concerned with concrete construction should be familiar with the equipment and methods available and should make sure that the best procedures are followed. One of the most common faults is the use of too much mixing water—in other words, high slump. Wet, high-slump mixes are vulnerable because of shrinkage cracking and permeability of the ensuing concrete and because wet mixes are more apt to segregate than more moderate ones. Segregation leads to rock pockets and weak, permeable layers. See Figure 4-4. The wrong equipment or methods can cause even good concrete to segregate. Weak, permeable layers will form where construction joints are not properly cleaned up. Finishing too soon, using water or dry cement to assist finishing, or over finishing all result in deterioration of the surface. Porous concrete or honeycomb result from insufficient consolidation. Deficiencies in curing are a constant source of distress.

Wrong Mix Proportions. The methods of proportioning (designing) concrete mixes described later in this book are based on producing the correct concrete to provide the required properties, including durability, in the hardened concrete. A high water-cement ratio, incorrect aggregate proportions, or improper application of an admixture can all lead to concrete of poor durability. Over sanded mixes, considered necessary for pumping by some operators, are a potential source of inferior concrete.

4.3. Chemical and Mechanical Attack

Most concrete, fortunately, is not subject to damage by aggressive substances or forces. For concrete that is to be exposed to attack, there are means available to prevent or minimize the deterioration. Good, impermeable concrete is resistant to many exposures. By wise selection of aggregates, proper mix proportions, care in handling and placing, correct finishing procedures and adequate curing, concrete can be made resistant to attack by many materials.

In severe exposures, concrete may need the additional protection provided by barriers of some kind. Construction of barriers is discussed in Chapter 23. The PCA concrete technology report "Effects of Substances on Concrete and Guide to Protective Treatments, (IS001.11)[4.3] includes a comprehensive list of materials that attack concrete, together with protective treatments. The report states:

The first line of defense against chemical attack is to use quality concrete with a maximum chemical resistance. This is enhanced by the application of protective treatments in severe environments to keep corrosive substances from contacting the concrete or to improve the chemical resistance of the concrete surface. Protective surface treatments are not infallible, as they can deteriorate or be damaged during or after construction, leaving the durability of the concrete element up to the chemical resistance of the concrete itself.

Proper maintenance—including regularly scheduled cleaning or sweeping, and immediate removal of spilled materials—is a simple way to maximize the useful service life of both coated and uncoated concrete surfaces.

Reference should be made to the PCA report for detailed information on all kinds of exposures. The following discussion touches only a few of the situations most likely to be encountered. Table 4.1 lists a number of common hazards and protective measures.

In virtually all cases of attack, water is a contributing factor, either water itself or water acting as the vehicle for aggressive materials. In the first instance, expansion and contraction of concrete, caused by alternate wetting and drying, contribute to weathering or failure, as do freezing and thawing of water within the concrete. Organic acids and other deteriorating substances may be carried into the concrete by water. Hence, there are instances when protective coatings are necessary. Whatever coating is selected, the anticipated life of the coating under the exposure and temperature conditions contemplated should be considered.

Attack by Substances. One weakness of concrete is its rather poor resistance to practically all acids, both organic and inorganic. Acids attack the concrete by dissolving the cement, thereby causing disintegration of the concrete surface, or internal damage if the acids can penetrate the concrete through cracks or other openings. Certain ores, coal or cinders stored near concrete are a source of mineral acids that will be leached out by rain or other water. Manufacturing plant wastes usually contain destructive chemicals. Mine drainage is usually acid and attacks concrete.

Except under unusual circumstances, petroleum oils do not attack concrete. One exception occurred a steel rolling mill where hot lubricating oil mixed with hot emulsifying water came in contact with concrete machine bases, causing eventual disintegration. Proper design or maintenance would prevent such an exposure.

Where oil is stored in concrete tanks, it is necessary to have good sound concrete, free of cracks to prevent loss of oil. Circumferential prestressing is of value in this case, or one of the commercially available coatings can be applied.

TABLE 4.1
PROTECTION OF CONCRETE FROM SUBSTANCES

EXPOSURE CONDITION OR MATERIAL	EFFECT	TREATMENT
Inorganic acids. such as sulfuric, hydrochloric, nitric, etc.	Active attack and disintegration	Thick bituminous mastic; vitrified tile with special cement; glass; lead: resin or rubber sheet; epoxy.
Organic acids: acetic (vinegar), lactic	Slow attack	Bituminous paint or enamel; phenolic resin varnish; chlorinated rubber; polyester, epoxy; neoprene.
Petroleum oils	Small loss of oil by penetration. No attack if no fatty oil additives are present.	Brick; tile; epoxy; polyester; phenolic; neoprene; vinyl.
Borax	Slow attack	Bituminous paint; chlorinated rubber; epoxy; vinyl; urethane; neoprene; sheet rubber.
Vegetable oils, such as olive, peanut, soybean, margarine, coconut, cottonseed. Lard, same as vegetable oils.	Slow attack	Fluosilicate; water glass; vinyl; polyester; urethane; epoxy; neoprene.
Sulfate salts present in alkali soils	Active attack on concrete of inadequate sulfate resistance.	Use Type V cement in severe exposure; Type II in moderate exposure; certain pozzolans.
Seawater	Moderate attack on concrete of inadequate sulfate resistance. Rusting of reinforcement.	Use Type II or Type V cement
Sewage	Usually not harmful, but hydrogen sulfide gas forms sulfurous acid, which attacks concrete.	Bitumen; epoxy; brick or tile; vinyl; polyester; neoprene.
Fertilizer	Fertilizers contain sulfates, nitrates and organic acids, all of which actively attack concrete.	Bitumen; brick or tile; vinyl; epoxy; magnesium or zinc fluosilicate; neoprene.
Bleaching solution (sodium hypochlorite)	Slow attack	Hypalon; vinyl; bituminous coatings; polyester, neoprene; sheet rubber.

Heat, blast and fuel spillage from jet aircraft cause scaling, cracking and crazing, especially of inferior concrete. Good dense concrete is usually resistant, and a properly constructed concrete slab should perform satisfactorily.

Calcium chloride is commonly used as a de-icing agent for pavements and similar areas. Ice and snow are melted with the direct application of a de-icing agent. This usage of de-icing salts is almost universal; only a few snowbelt states prohibit its use, even though the damaging effects are well documented. The effects of salts are similar to the effects of freezing and thawing, which are scaling, spalling and eventual failure. See Figure 4-5. These effects are minimized when the concrete contains the optimum percentage of entrained air.

Figure 4-5
Scaling of a concrete driveway exposed to de-icers. (Courtesy of PCA)

Small packages of de-icing agents containing ammonium nitrate and ammonium sulfate have appeared on the market from time to time for sale to motorists and householders. Because even weak solutions of these chemicals actively attack concrete, including air-entrained concrete, they should never be used on driveways or walks. The user should use only those de-icing agents known to contain only calcium chloride, sodium chloride or urea.

Food-processing plants impose special problems in the durability of concrete, presenting many types of corrosive organic compounds, as do tanning and fermentation plants. Exposure to organic acids and other compounds usually requires some kind of a protective coating. Examples of destructive compounds are fatty acids and blood in meat-processing plants, fruit acids in packing houses and canneries, and lactic acid in dairies. Hot lard oil actively attacks concrete. Long exposure to cleaning solutions can result in breakdown of surfaces. Solutions containing phosphoric acid and sodium hypochlorite can be quite destructive after repeated usage over a long period of time. Sometimes the attacking compounds may be entrained in the atmosphere or in steam or other vapors, attacking ceilings or other apparently safe concrete.

Concrete subject to such exposures should have a smooth, dense surface, adequately cured. Acid-proof coatings may be necessary, such as tile and certain proprietary compounds.

Coolers and freezers contain brine solutions that are likely to be harmful. Where the concrete is constantly wet, as in the bottom of a tank containing brine, there would be little effect on good dense air-entrained concrete containing about 560 pounds of cement per cubic yard at a water-cement ratio of 0.50 or less. The concrete should be permitted to dry out thoroughly after curing and before exposure. Two coats of raw linseed oil, brushed on hot, are said to seal the surface against penetration of brine. Other coatings may consist of epoxy resin, silicone solution or any of several synthetic materials.

A wash coat of cement and water of the consistency of heavy cream brushed on the interior of farm silos has been found to improve resistance of the concrete to attack by silage juices. Curing of the coating is important.

Organic acids resulting from the decay of vegetable or animal matter may be present in swamps. Drainage should be provided, if possible, to prevent contact of the water with the concrete.

In dealing with any case of potential or actual attack, there are two avenues of approach to the problem: either proper attention to produce resistant concrete will be the solution, or some sort of a barrier must be provided to prevent contact between the concrete and the aggressive material. Obviously, the first procedure can be applied only in the preliminary and construction stages. The second course of action should be similarly applied, although protective measures can sometimes be applied subsequent to construction.

Concrete can be made resistant or immune to attack by most substances if it is of high quality, i.e., made as described in this book. Low water-cement ratio, low slump, sound materials and proper construction practices can give us good, durable concrete.

See Reference 4.3 for more information.

Corrosion of Metals. Some metals, when embedded in or in contact with concrete, will corrode if moisture is present. Concrete in a dry environment will retain water for a long period of time after curing ceases. Relative humidity of the surrounding air and the dimensions of the concrete determine how long the moisture will remain in the concrete. Thin beams and slabs in the desert require several months to dry completely, even if there is no source to replenish the lost moisture, while massive sections can retain moisture for years. Where the concrete is exposed to water, it will never dry out.

In any case of corrosion of metal, the damage may take the form of disintegration and loss of the metal under attack, or disruption of the concrete because corrosion products occupy a larger volume than the metal that was corroded. See Figure 4-6. Usually, both reactions are significant. Another item of significance is that in most cases of corrosion investigated in the field and in the laboratory the presence of calcium chloride in the concrete aggravated the seriousness of the problem. See Figure 4-7.

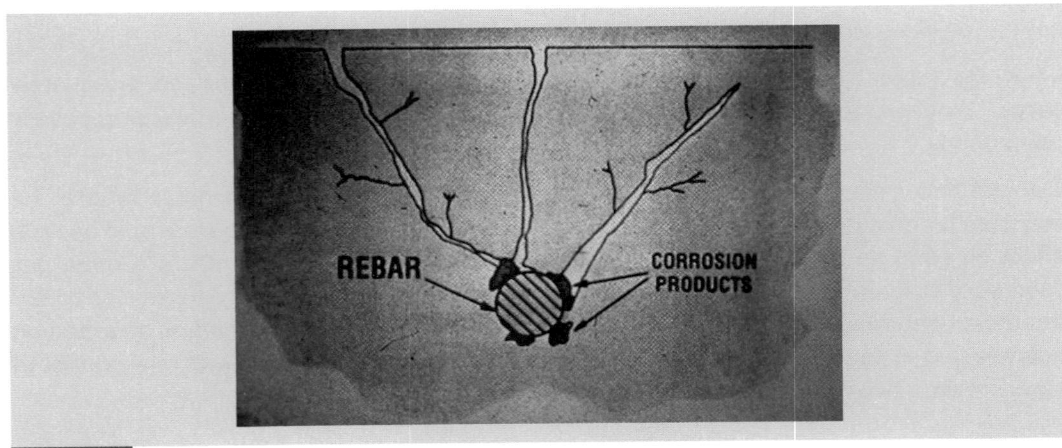

Figure 4-6
The expansion of corroding steel creates tensile stresses in the concrete, which cause cracking, delamination and spalling.(Courtesy of PCA)

Figure 4-7
The damage to this concrete parking structure resulted from chloride-induced (de-icing salts) corrosion of the steel reinforcement. (Courtesy of PCA)

On exposure to moist concrete, aluminum is attacked by caustic alkalies as long as moisture is present. Even though the concrete may dry out and the action cease, there is a possibility of further action if the concrete again becomes wet.

Lead will corrode in contact with green concrete but is not affected by hardened and dry concrete. If it is necessary to place lead in contact with fresh concrete, a sufficient extra thickness of lead should be provided to allow for initial attack. Bituminous or plastic coatings should be used on lead conduit or pipe embedded in concrete.

Copper has been used for many years as a grout and water stop in dams and other structures with complete success. Even when wet, copper alloys are immune to attack, although there have been cases of reaction in the presence of soluble chlorides.

Some corrosion of zinc will occur upon contact of zinc with moist concrete. The reaction is on the surface only and forms a dense film on the surface of the metal that protects the underlying metal from further attack. An exception occurs when calcium chloride used in the concrete intensifies the reaction, which may cause serious damage. This has occurred when concrete containing calcium chloride was placed against galvanized steel liners that were to remain in the structure. Corrosion penetrated through the galvanized sheets.

We are all familiar with the rusting of iron and steel when they are exposed to air and moisture. Corrosion of reinforcing steel becomes apparent when the concrete starts to crack and spall and rust stains disfigure the concrete. Water enters the concrete through cracks, honeycomb or other porosity. When it reaches the steel, corrosion starts. Because the corrosion product, rust, occupies a much larger space than the steel from which it came, a disruptive force is generated that causes more cracking and spalling in a cumulative spiral, which ultimately can cause failure of the structure. Prevention is accomplished by proper design and construction. Concrete should be dense, without honeycomb, cracks or other blemishes that can provide a means of ingress for water. Reinforcement should be covered by at least 3 inches of concrete in a seawater exposure, as described in Chapter 18.

Complicating the picture in some cases is the presence of what are called electrochemical couples. The moisture in concrete is highly alkaline and thus is an electrolyte; that is, it will conduct electricity. The presence of calcium chloride (used as a cement accelerator) increases the electrolytic activity. When two different metals are in contact with each other in an electrolyte, they form a couple in which small electric currents are set up. This is what happens in a battery. In concrete, for example, aluminum conduit can be in contact with steel (iron) reinforcement. The small electric current thus developed causes corrosion of the aluminum conduit. This is a greatly simplified explanation of the reaction that sometimes takes place. One outstanding case was spalling over aluminum conduit in the Washington Stadium that resulted in difficult and expensive repairs.[4.4]

In all cases of corrosion of metals, moisture is necessary. There is no reaction in dry concrete. In nearly all cases, the problem is intensified if calcium chloride was used in the concrete.

If danger of electrolytic action or stray electrical currents is expected, the risk of corrosion can be lessened by grounding reinforcing steel and other embedded metal parts at the time of their installation in the forms. Protective coatings on the metal also eliminate the problem.

High Temperature. We are concerned here not with high atmospheric temperatures, but instead with temperatures in furnaces, chimneys and in building fires. Concrete is frequently used for the structural components of furnaces, incinerators, foundry machinery, coke ovens and many other processing installations in which temperatures may exceed 2000°F. External parts of these machines can reach temperatures of 500°F or more.

The principal effect of such high temperatures is a reduction in strength. See Chapter 3. Indirectly, the strength loss could impair durability by weakening the load-carrying capacity of members, causing minor or major failure, depending on the degree of weakening.

Continuous exposure to temperatures in excess of about 900°F may result in spalling of the concrete. The choice of aggregate used in the concrete can influence the spalling resistance; limestone aggregate is better than siliceous aggregate, and crushed firebrick is said to provide improved resistance.

Concrete in chimneys is subject to acids resulting from buring fuel, especially from high-sulfur fuels. Frequently, moisture is present, either as rainwater or condensation from the flue gas. These hazards are in addition to the high-temperature menace and require good, durable concrete, together with careful supervision and inspection of construction. Linings, usually installed in chimneys built of concrete, may consist of special refractory concrete, hard-burned brick or firebrick. Selection of a liner is a design consideration and depends on type of fuel, kind of process, expected temperature and other factors. See Chapter 22.

Concrete in a building that has suffered a fire may come through practically unscathed or may be damaged to varying degrees, depending on the actual temperature (building fires can be over 1500°F), duration of the fire, possible thermal shock from water used to put out the fire, and the original quality of the concrete. See Figure 4-8. If the concrete appears sound, without cracking, spalling or deflection, it is probably safe to use. Nondestructive tests can be applied to determine approximate strength, elastic modulus and the presence of hidden cracks. Expert opinions, including that of the structural engineer, thorough examination of the structure and carefully performed tests are necessary when evaluating a fire-damaged structure.

Figure 4-8
When exposed to fire or unusually high temperatures, concrete can lose strength and stiffness. (Courtesy of PCA)

Structural and Accidental Damage. In a secondary way, the durability of concrete is affected by loadings, foundation failures and accidents. Any damage sustained by the concrete caused by such incidents can be the starting point for further damage by weathering or other mechanism.

Early indications of a foundation failure are the appearance of cracks in the concrete, settlement or heaving of all or part of the structure, differential movement of parts of the structure, or tilting. Further evidence is binding of doors and windows, walls out of plumb, and floors that are cracked and irregular. The most common cause of foundation failure is exceeding the safe bearing capacity of the soil, followed closely by failure to carry the footings deep enough. It is imperative that footings be carried below the frost line in those geographical areas where freezing of the soil is prevalent, or below the depth of influence of volume changes of the soil caused by wetting and drying. Some clay soils, known as *expansive soil*, possess the property of high volume change upon wetting or drying. The force exerted by swelling of such a soil upon wetting is tremendous and is sufficient to cause extensive damage to structures built in expansive soil areas. In particular, houses and other buildings erected on slabs on grade have sustained irreparable damage. If the soil was wet at the time of construction, subsequent drying can result in a loss of support because of shrinkage of the soil. A soil survey by a competent soils engineer will reveal any potential trouble areas, and should be required. See Figures 4-9 and 16-20.

Figure 4-9
Concrete slab on ground with severe damage that is due to expansive soil. (Courtesy of PCA)

Another case of foundation troubles is lowering of the water table subsequent to construction. Cases have been reported in which footings were supported on timber piles. Because the piles were completely and permanently submerged, there was no problem with decay. Pumping of groundwater in the vicinity lowered the water table, exposing the piles to a moist condition in the presence of air. Within a few years decay damaged the piles to the extent that they could no longer support the load, and the foundation failed. See Figure 4-10. Note that concrete piles are not subject to the decay that attacks wood piles.

Figure 4-10
Decay damaged wood piles could no longer support the load.

Sometimes excavation for a new building adjacent to an existing one removes the lateral support for the existing footings, causing slippage and settlement. Lateral support and shoring of the sides of the excavation can prevent such damage.

Overloading of part or all of the structure can cause deflections, failure of joints and connections, cracking, settlement and other unusual movements. Earthquake and blast damage can range from small cracks to complete failure. See Figure 4-10. Inspection of earthquake-damaged buildings has revealed misplaced reinforcing steel, bond failure at construction joints and other evidence of careless construction. Although we cannot say that good and proper construction would have prevented earthquake damage, we can certainly say that properly constructed structures will sustain less damage than shoddily built ones.

Figure 4-11
Extensive damage may result from a severe earthquake.

Soil Salts. The decomposition of rocks in nature results in the formation of salts, the most common being chlorides, carbonates, sulfates, and bicarbonates of magnesium, sodium, calcium and potassium. These salts, becoming part of the soil that results from rock decomposition, present no problem where there is sufficient rainfall and adequate natural drainage. In areas of slight rainfall, however, they frequently become concentrated in the soil, a condition that is aggravated by the application of irrigation water and the evaporation and transpiration of pure water, leaving the salts to accumulate in the soil. In the worst areas, the concentration is so high that the surface of the land is covered with a white deposit of alkali salts. These soils, sometimes called alkali soil in the United States, contain principally sodium and magnesium sulfates. The rate at which these sulfates attack any concrete structure depends on the type and concentration of sulfates in the soil and water, the quality of the concrete, the degree of wetting suffered by the concrete and the time length of exposure.

Where one side of a concrete slab is exposed to moisture, and evaporation takes place from the other side, we have a nearly ideal condition for sulfate deterioration, lending itself to increasing salt concentration and crystal growth in the concrete. See Figure 4-12.

Resistance of concrete to attack by sulfates is related to the water-cement ratio (amount of cement in the concrete) and the calculated amount of C_3A (tricalcium aluminate) in the cement, the resistance being enhanced for concrete with a lower water- cement ratio and for cement low in C_3A, as shown in Figure 4-13, based on Bureau of Reclamation data for exposure in 4.6 percent sodium sulfate. These are the most important parameters to be considered when evaluating the effect of any sulfate exposure. Of these, the C_3A content has the greatest influence. Other significant factors are the following:

- Air entrainment slightly improves sulfate resistance by reducing the water-cement ratio.
- The resistance of precast concrete, such as pipe, is improved by steam curing, either at atmospheric pressure or at elevated pressure in an autoclave.
- A drying period of three or four weeks after moist curing improves resistance.
- The use of calcium chloride as an accelerator reduces resistance.
- Concrete of high absorption is more vulnerable to attack than concrete of low absorption.
- Good workmanship in placing, finishing and curing, and producing smooth, dense surfaces free of honeycomb and cracks, is essential.
- A tested pozzolan is of value, especially if the cement contains more than 5 percent tricalcium aluminate.
- Fly ash and slag can improve sulfate resistance of concrete made with all types of cement.

4.4. Reactive Aggregates

Aggregates are usually considered inert; that is, they provide bulk for the concrete without entering into the chemical reactions that occur in concrete. There are, however, a few exceptions. One is what we have for a long time called the alkali-aggregate reaction, now more properly called the alkali-silica reaction. In this reaction, certain minerals and rocks react with the alkalies in concrete, causing an internal expansion in concrete that results in cracking and deterioration. The minerals usually associated with this reaction are opal, chalcedony, tridymite, cristobalite and certain zeolites; the rocks are glassy or cryptocrystalline rhyolites, dacites and andesites (including volcanic tuffs composed of these rocks) and cherts, both chalcedonic and opaline. Any aggregate containing a significant proportion of any of these materials should be considered potentially reactive.

Evidence of alkali-silica reaction consists of random or map cracking on a fairly large scale, the cracks opening up, in severe cases, over $^1/_2$-inch wide but seldom as deep as 18 inches. General deterioration of the concrete results. Parts of members may be shifted and control joints closed by the expansion of the concrete. If a piece of the affected concrete is broken off, it will appear dull and chalky, and individual aggregate particles, upon close examination, will show a thin layer of alteration or reactivity on their surfaces. Deterioration is progressive as long as moisture is present. See Figure 4-14. There is, however, a growing body of evidence now accumulating suggesting that an end point in the reaction may be reached within 25 or 30 years after the concrete was placed.

Indications that a proposed aggregate is reactive can be obtained by an examination of structures known to have been made with the proposed cement-aggregate combination. There are several laboratory methods for detecting potential reactivity, including a petrographic examination of the aggregate. These tests can provide valuable information when properly interpreted, but the service record of the materials must be relied upon.

Figure 4-12
Sulfate alkalies in the soil will attack concrete unless adequate resistance is provided by proper selection of materials, as shown by the failure of this concrete sidewalk. (Courtesy of PCA)

If a reactive aggregate must be used, there are several measures that can be taken to minimize the reaction. Most important is the use of a pozzolan. Fly ash, slag and silica fume are especially effective in controlling or preventing the reaction. Natural pozzolans, especially opaline chert, diatomaceous earth and some of the volcanic glasses and calcined or burnt clays are of significant benefit in controlling alkali-silica reactions. Blended cements aid in controlling the reaction, and air entrainment is said to be of some benefit. Low-alkali cements may also be available to help control the reaction. For a more in-depth discussion of the diagnosis and control of alkali-aggregate reactions in concrete, the reader is referred to references 4.5 and 4.6.

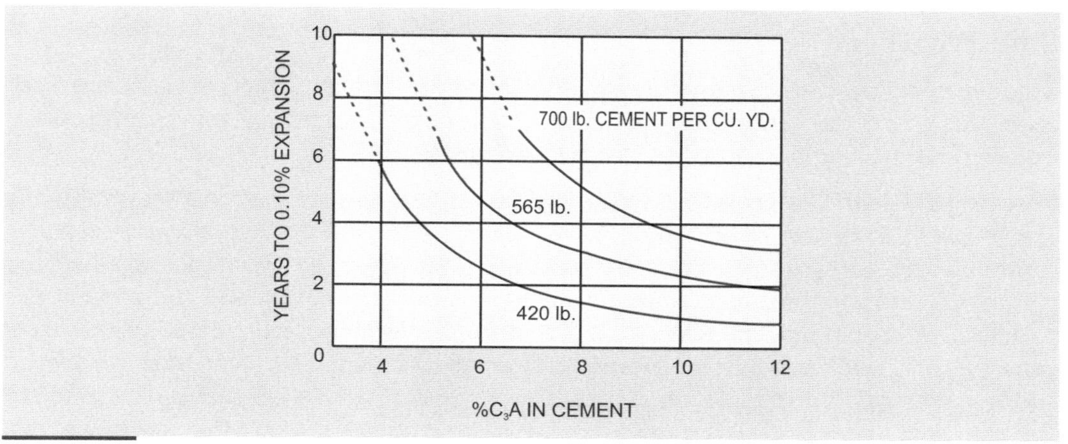

Figure 4-13
The composition of the cement and the water-cement ratio are the most important factors affecting resistance to sulfate attack.

Figure 4-14
Cracking, spalling and lateral offset caused by severe alkali-silica reactivity in wall. (Courtesy of PCA)

Chemical admixtures that control alkali-silica reaction are also available. The effectiveness of any method to control alkai-silica reactivity should be demonstrated by test. See footnotes. Certain fine-grained clay-like dolomitic limestones have been found to react with the cement alkalies. Most dolomitic limestones are satisfactory as aggregate, and those that are reactive are usually unsuitable for other reasons. They have been found in scattered localities in parts of Ontario, Canada and mid-western U.S. states. Limiting cement alkalies to less than 0.6 percent is not always effective; further reduction to 0.4 percent might be necessary. Dilution of the reactive rock with nonreactive aggregate is helpful.

In brief, aggregates available from most producers are inherently sound and clean, or they have been processed to remove unsatisfactory constituents. Potential sources of trouble may be physically unsound aggregate particles themselves that can be removed by processing; extraneous, deleterious material accompanying the aggregate that can be removed by processing; or chemically reactive aggregate that can be compensated for by rational selection of accompanying materials for use in the concrete.

4.5. Cold Weather

Cold weather can damage concrete in either of two ways: first, by freezing of the fresh concrete before the cement has achieved final set, and second, by repeated cycles of freezing and thawing over a period of months or years.

Freezing of Fresh Concrete. Only one episode of freezing while the concrete is still in the plastic state or during its initial hardening period may reduce durability, weathering resistance and strength by as much as one half. The length of the period during which the concrete is frozen is not important once the concrete has frozen, and the amount the temperature drops below freezing appears to have no effect. See Figure 4-15. Once frozen, concrete will never attain its full potential strength and durability, even after prolonged curing at reasonable temperatures. There is no material or admixture that can be added to the fresh concrete to lower its freezing temperature or to act as an antifreeze. How to mix and handle concrete during freezing weather is discussed in Chapter 19.

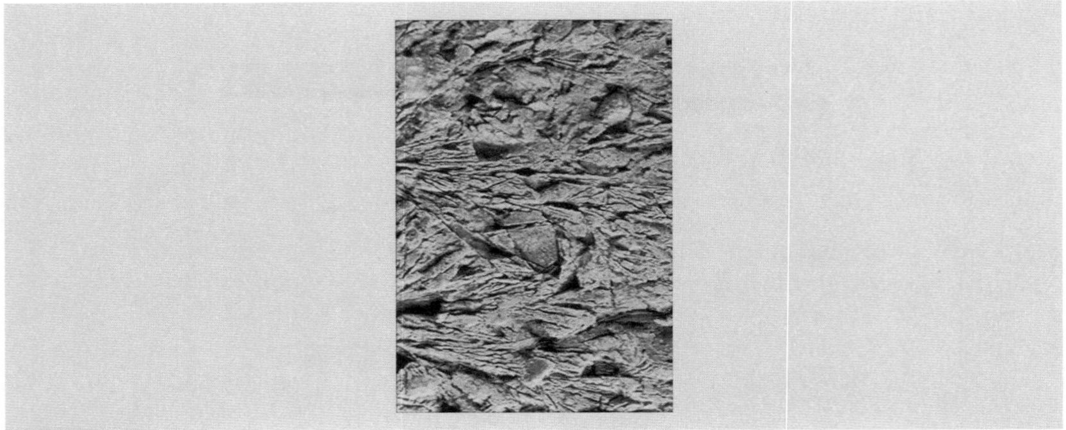

Figure 4-15
Frozen concrete. (Courtesy of PCA)

Frost Action. Frost, or cycles of freezing and thawing, damages hardened concrete by the freezing and consequent expansion of water in pores and openings in the concrete. Dry concrete, therefore, is little affected by such action. However, most concrete that is exposed to cold temperatures is also exposed to moisture or water and is therefore subject to attack.

Causes of poor frost resistance include poor design of construction joints; segregation of concrete while placing; leaky formwork and poor workmanship resulting in honeycomb and sand streaks; faulty cleanup of a joint surface before placing concrete against it; flat surfaces that allow puddles of water to collect on the surface of the concrete; insufficient or totally lacking drainage, permitting water to accumulate against the concrete; and cracks. The fineness and composition of the cement within the limits normally specified appear to have little or no effect.

To provide resistance against frost requires that good design principles be followed, taking care that proper and adequate drainage is provided. Horizontal construction joints should be avoided if possible. However, if such a joint is necessary, it should not be located near the water or ground line, but should be 2 or 3 feet above the ground line or high water line or the same distance below the ground or low water line. See Figure 4-15. Thorough cleanup of the previously placed concrete is essential. Concrete should contain 6 or 7 percent air for 3/4-inch or 1/2-inch maximum aggregate, and should be made of first class materials, carefully mixed and handled, with a water-cement ratio as low as possible. See Chapter 12.

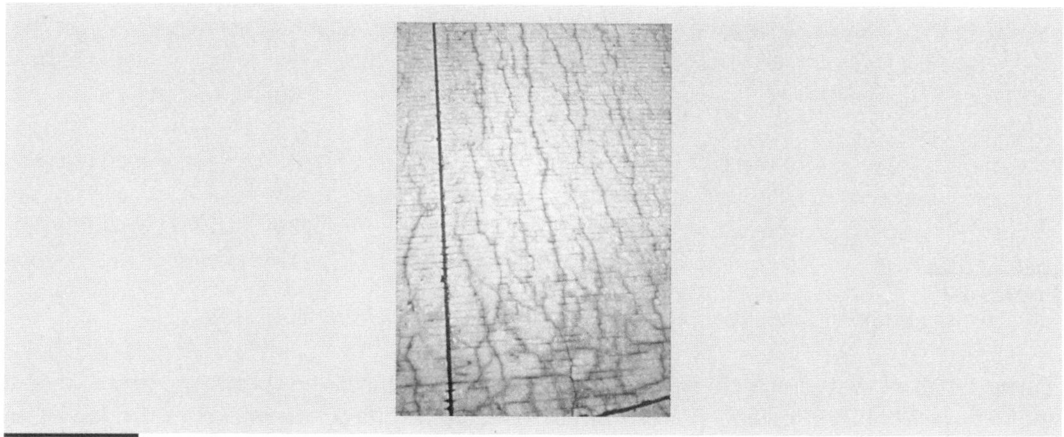

Figure 4-16
Frost damage (crumbling) at joints of a slab-on-ground and enlarged view of cracks. (Courtesy of PCA)

Structurally sound aggregate of low porosity should be used. Good construction practices should be followed throughout under proper supervision and inspection. Segregation, sand streaks and honeycomb must be avoided by careful placement of the concrete as near as possible in its final resting place, followed by thorough consolidation by means of internal vibrators. The objective is to produce good concrete with smooth, dense and impermeable surfaces.

MARINE ENVIRONMENT

A considerable portion of concrete placed every year is placed in a marine environment. Shore protective structures, breakwaters, seawalls and pilings are examples of the structural elements comprising this group. That entirely satisfactory concrete can be obtained is attested to by the continued good service of many of these structures that are 50 to 75 years old. Failures and distress, where they have occurred, can be traced to negligence in following good practices during construction of the facility.

4.6. Exposure Conditions

Broadly speaking, marine structures are exposed either to simple static conditions, as would exist on pilings supporting a harbor structure, or to dynamic conditions where they are exposed to waves and currents. The general exposure is usually a combination of each of these conditions, in which chemical as well as physical forces will be found. See Figure 4-17.

Waves and Currents. Changing economic conditions, increasing demands for harbor and beach facilities, and the need to preserve these facilities, make necessary the construction of shore and harbor protective systems. These systems consist of offshore breakwaters, groins, and seawalls or revetments.

Figure 4-17
Slow disintegration of concrete sheet piling exposed to seawater in the tidal zone.

Three types of waves are considered in designing structures: nonbreaking, breaking and broken waves. Nonbreaking waves, usually existing as diffraction waves in the lee of an obstruction, exert mainly hydrostatic pressure against structures, including deep-water structures. Many structures are also exposed to breaking waves. When a wave approaches shallow water near the shore or an obstruction, it starts to break and finally expends its energy on the beach or other obstruction. Waves that break directly upon a structure exert dynamic as well as hydrostatic pressure on the structure. Broken waves sometimes reach a structure that has been built above normal high water, especially during abnormally high tides and storms.

Because of the inherent inaccuracies in determining the loadings on structures exposed to waves, a safety factor larger than normally used is required. Practical consideration of the loadings requires evaluation of overturning moments, sliding and hydrostatic uplift. Stability of a wall will be endangered by water getting behind the wall by overtopping or other means, inducing lateral pressure tending to tip the wall forward, or eroding the foundation. The existence of littoral currents should be determined and their effect evaluated.

A littoral current is a current near the shore in a lake or ocean. Littoral drift is the material that moves along the shoreline under the influence of the littoral current and, to a lesser extent, of waves. Movement of sediments in this manner can cause serious erosion of the beach, necessitating the construction of groins, which are built perpendicular to the shoreline to trap the littoral drift, thus protecting the beach.

Tides. The principal effect of tides is a change in the zone of attack by changing the water surface elevation. Seawalls and bulkheads might be overtopped by waves during especially high tides. On the other hand, the stability of a structure might be threatened by the pressure of water remaining behind the structure during a low tide. Erosion and chemical attack are more severe in the portion of a structure alternately submerged and exposed by tides and waves.

Static Conditions. Concrete permanently submerged in seawater is exposed to possible deterioration resulting from rusting of reinforcement, abrasion or chemical action. However, as pointed out later, proper selection of materials and construction methods can virtually eliminate this problem.

Static loads, besides normal structural loads, might result from hydrostatic pressure of water, and loads imposed by back-fill against walls and bulkheads.

4.7. Nature of Attack

The nature of the attack and circumstances leading up to it can be classified as physical, chemical or, more commonly, a combination of both. See Figure 4-18.

Figure 4-18
Seawater attack on the pilings caused partial failure, which was aggravated by loading on the wharf, finally resulting in complete failure of the piles.

Physical. Damage to the concrete can occur as a result of abrasion, crystal growth in cracks and interstices, freezing and thawing, structural overloading, impact of moving objects, or wetting and drying.

Abrasion is the wearing away of concrete by sand and gravel (shingle) carried in suspension in a current of water. The current in some rivers is rapid enough to cause abrasion, and occasionally a littoral current will cause abrasion. Abrasion of concrete is most apt to occur just above the ground line, sometimes by material rolled or tumbled along the bottom.

In the tidal zone (the zone between low and high tide), the concrete is exposed to alternate wetting and drying. Evaporation of water from the concrete surface during low tide periods results in high salt concentrations that can lead to the growth of crystals in cracks and interstices of the concrete. The expansive force exerted by the crystal development can disrupt the concrete surface. A similar condition exists in the portion of the structure above high water exposed to wind-blown spray. Concrete can be located an appreciable distance away from the actual shoreline and still be exposed to spray. Beams and the underside of slabs of such over-water structures as wharves are potentially exposed to spray. Freezing and thawing exposure is especially destructive to concrete in a marine environment.

Structural distress results from loading a structure, or a portion of a structure, in excess of its capacity. An example is stacking heavy material on a deck, causing the slab or beam to deflect, resulting in localized failures that are characterized by hairline tension cracks. Such distress can result also from excessive stresses induced by unusually large storm waves, especially if they are breaking, and by sudden momentary loads caused by docking maneuvers of ships. Accidental impact of vehicles and other objects can cause localized failure.

Chemical. Probably the most common chemical distress is that resulting from rusting of reinforcing steel. Actually this is a combination of physical and chemical damage, as the water must first find its way to the steel through cracks and interstices in the concrete; then, because the iron rust has a greater volume than the original steel bar, the disruptive force of the rusting steel causes further cracking and spalling in a cumulative manner. Besides damaging the concrete, the rusting results in a loss of reinforcement, which could further compound the failure

Magnesium sulfate in seawater attacks some of the constituents of the cement paste, especially the aluminates. Chlorides promote rusting of steel. See Figure 4-19. Free carbon dioxide dissolved in water will leach free lime out of the concrete by reacting with the lime to make soluble calcium bicarbonate. If a high-alkali cement is used with an aggregate containing reactive silica minerals, excessive expansion of the concrete results in deterioration. This reaction is aggravated in the presence of moisture and is apt to be complicated with other reactions in the presence of seawater.

Figure 4-19
Damage resulting from chloride-induced corrosion of steel reinforcement. (Courtesy of PCA)

4.8. Resistant Construction

Protection of waterfront structures should commence when the structure is being designed. All too often the designer will carefully consider the structural and aesthetic aspects without recognizing the exposure conditions to which the structure will be subjected.

Chamfers and fillets not only improve the appearance of the structure but also serve a very useful purpose. By chamfering edges of members, the sharp arris, which is subject to spalling and chipping from moving objects, is avoided. See Figure 4-20. Fillets in re-entrant corners eliminate a possible source of cracking. Reinforcing steel should be well covered with sound concrete. Most agencies specify a coverage of 3 inches. If horizontal construction joints are necessary they should be located below low water or above high water.

Well-graded, first-class nonreactive aggregates conforming to ASTM C 33 should be used. Cement should be low in C_3A, either Type II or Type V. Consideration should be given to the use of an approved pozzolan with a record of satisfactory use in a similar exposure. Blended cements can also be used.

Concrete should contain an adequate amount of entrained air, with mixes proportioned in accordance with the procedures described in Chapter 12. The concrete should be workable, with slump and water-cement ratio as low as possible, containing at least 560 pounds of cement per cubic yard. Use of a water-reducing admixture will benefit the concrete by permitting the use of less mixing water for the same workability, also

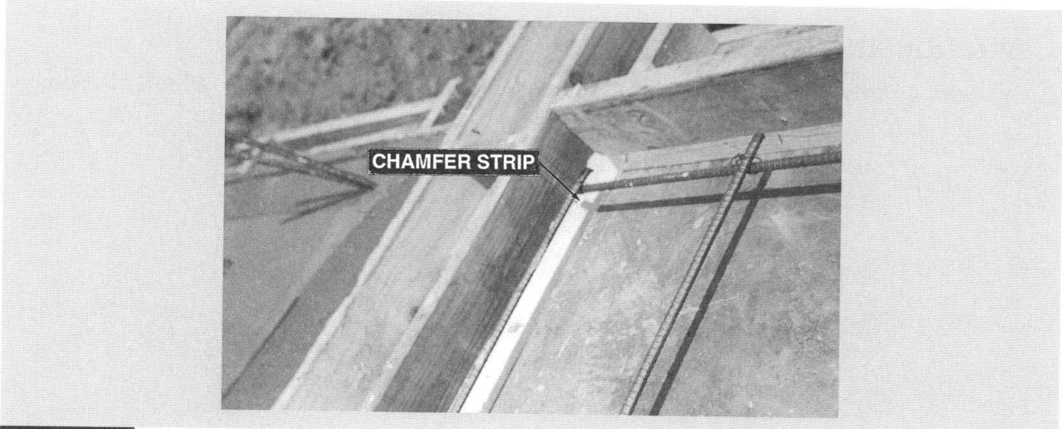

Figure 4-20
Chamfer strips installed in the form.

enhancing strength and durability. Calcium chloride admixtures should not be used.

Good workmanship during construction is essential. Concrete should be handled and placed in accordance with good practices. Special care is necessary during placing to avoid segregation, and the concrete should be thoroughly consolidated by means of vibration to preclude honeycomb and sand streaks. Construction joints should be cleaned up by means of a wet sandblast or equally effective means. Concrete should be cured for at least seven days, then permitted to dry out for as long as possible before exposure.

HYDRAULIC STRUCTURES

4.9. Exposure

Dams, canal linings, bridge piers and similar structures exposed to water, being partly wet and partly dry simultaneously or subject to cycles of wetting and drying, are subject to especially severe exposure conditions. The water may carry destructive quantities of acid, sulfides, sulfates or organic material. Hydraulic structures require good dense concrete with smooth surfaces; ample drainage; well-made joints, with water stops where movement is expected; good design, materials and workmanship; and entrained air. (Entrained air is not incompatible with dense concrete, as the latter implies well-consolidated concrete, without honeycomb or entrapped air.)

Movement of paving slabs or blocks on the face of embankments or reservoirs, sea walls or dams may be caused by hydraulic back pressure upon sudden lowering of the water level. Abrupt lowering of the water level might occur in the trough of a large wave. Prevention, in the case of continuous slab construction, consists of providing adequate porous drains on the back side of the slab and placement of weep holes through the concrete slabs. See

Figure 4-21. If the slope is paved with individual blocks or slabs, a space should be provided between individual blocks to enable the water to drain out between them.

Surfaces of hydraulic structures in cold climates may suffer damage near the water line by abrasion or adhesion of ice. Coatings seem to be of little value for protection, and the best defense is good, dense concrete with smooth and well-finished surfaces.

4.10. Cavitation and Erosion

Cavitation is likely to occur in conduits carrying water at high velocity where there is an

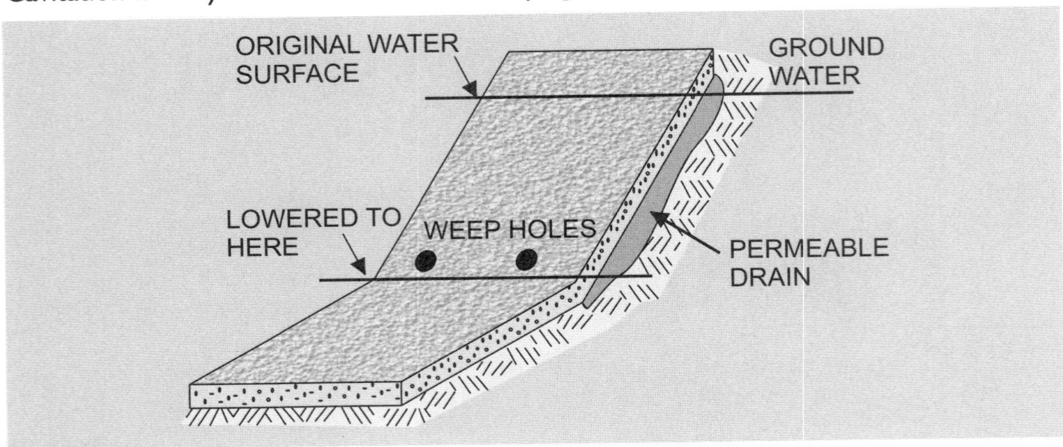

Figure 4-21
The pressure of groundwater is relieved through weep holes in the facing of the embankment.

abrupt divergence between the natural path of the water and the surface of the conduit, resulting in turbulence that creates localized areas of vacuum, causing particles of concrete to break loose. Such turbulence would exist at a sharp bend in the conduit or sudden change in cross section. Projections or depressions in the surface of the conduit will also cause cavitation. See Figure 4-22. For example, fins of mortar can originate in cracks in the forms when the concrete is placed. Clear water will cause cavitation if it is moving in excess of about 50 feet per second. A surface eroded by cavitation is rough, presenting a pock-marked or honeycombed appearance, in contrast to concrete that has been subjected to abrasion, which is usually quite smooth.

Proper design, providing streamlined flow of water, is important in preventing cavitation. Concrete surfaces should be dense and smooth without surface imperfections of any kind. Absorptive form lining or vacuum concrete will be of value, and unformed surfaces should be hard troweled. However, excessive troweling or working of the surface should be avoided. Properly proportioned mixes with high strength, low water-cement ratio (obtained by good practices, not merely more cement) and the optimum percentage of entrained air should be used. Even the best concrete will eventually show distress when exposed to cavitation; hence the importance of correct structural design and good construction practices. A heavy coat of resilient rubber is effective in controlling cavitation in an existing structure.

Erosion of concrete occurs when water carrying solid matter flows over concrete. Abrasion of the solids wears the concrete away. As in the case of cavitation, the best protection is proper design, maintenance and operation. Dense concrete of high strength offers some protection, but it too will be eroded under this exposure.

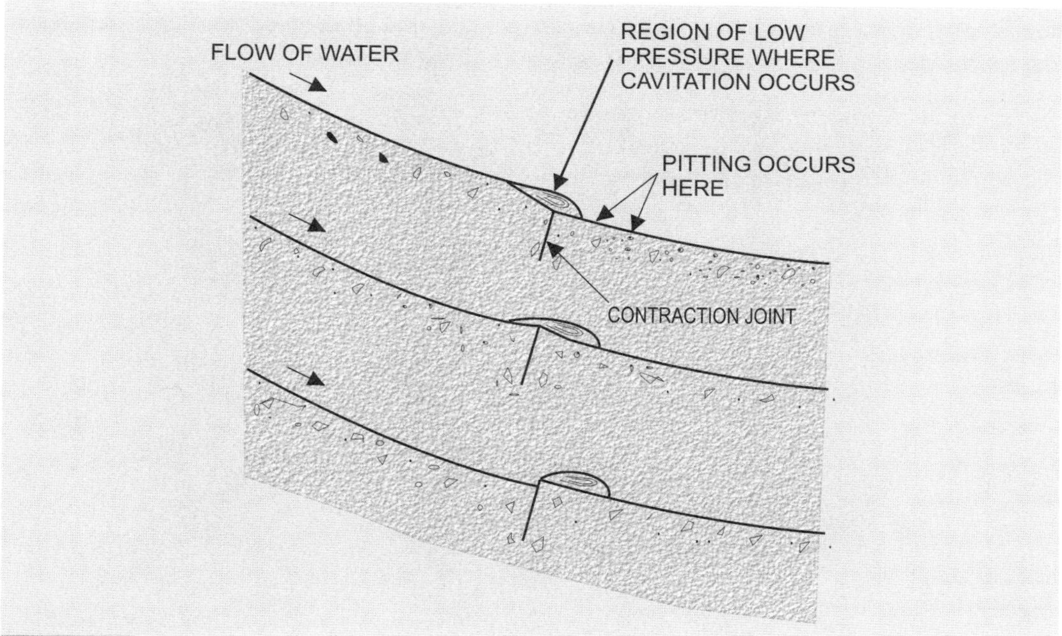

Figure 4-22
Water moving at high velocity will cause cavitation where there is a rough spot in the surface of the conduit.

There have been rare instances of attack of concrete by rock borers in tropical waters. It is doubtful if good concrete has ever sustained any damage from exposure of this nature.

SLABS ON GROUND

Slabs on ground (pavements, driveways, sidewalks and industrial floors) present many special problems in construction and maintenance that are due to their varied use and exposure conditions. Chapter 16 covers the construction of concrete slabs on ground, which, if properly performed, will do much to prevent the problems described below.

Poor construction and inferior materials are the principal causes of slab defects. These can be generalized as insufficient foundation support, inferior concrete, improper joint design and layout, improper or incomplete joint sealing, and soft concrete surface.

The best aggregate available should be used for slabs on ground. The concrete mix should have between 4 to 7 percent entrained air depending on exposure condition at a slump of 3 inches or less for concrete consolidated by mechanical vibration. The water-cement ratio should be as low as reasonably possible, not over 0.53 in a mild climate and not over 0.45 in a severe climate exposed to freezing and thawing.

4.11. Typical Problems

Scaling. Prior to the use of air entrainment, scaling was a serious problem in areas subject to freezing and thawing, especially when de-icing salts were used to melt snow and ice. See Figure 4-23. Air entrainment largely eliminated this problem. In addition to properly air-entrained concrete, freeze-thaw resistance can be significantly increased with the use of good quality aggregate, a low water-cement ratio, a minimum cement content, and proper finishing and curing techniques. The concrete should also be allowed to air dry at least 30 days after initial moist curing and prior to freeze-thaw exposure.

Figure 4-23
Scaled concrete surface resulting from lack of air entrainment, use of de-icers, and poor finishing and curing practices. (Courtesy of PCA)

Spalling at Joints. Poor timing in sawing joints may cause spalling. Other causes are carelessness in making construction joints (such as joints out of plumb) or the presence of old concrete or pebbles in the joint that prevent free movement of the joint when the concrete expands. Crushing of the concrete at joints is a similar phenomenon. See Figure 4-24.

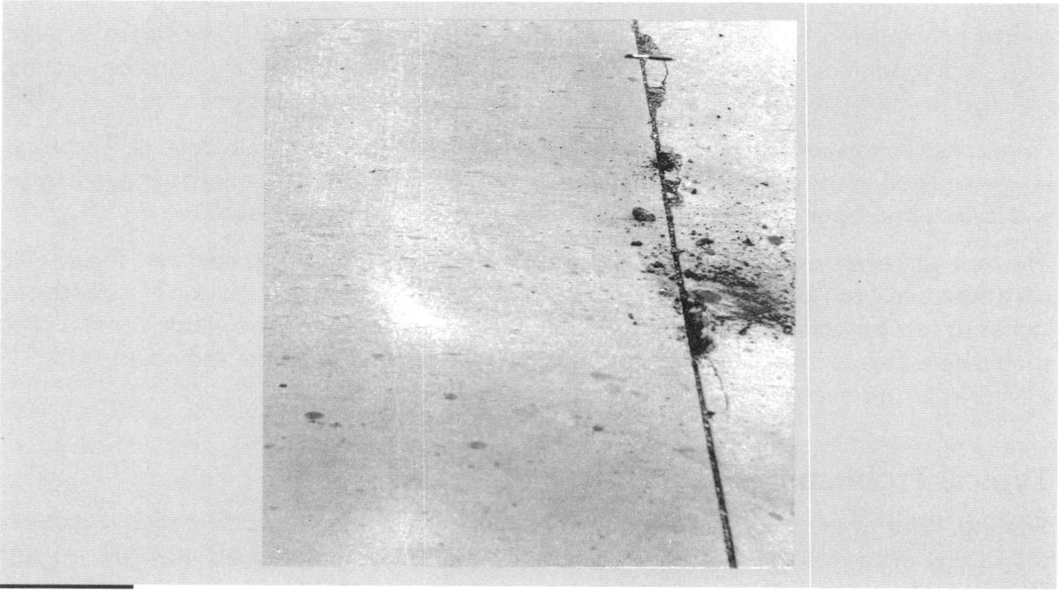

Figure 4-24
Spalling of a sawed contraction joint caused by dirt and foreign material in the joint that prevented movement. (Courtesy of PCA)

Subsidence. Failure of the foundation will permit the slab to subside or drop. This is especially prevalent at slab edges, where inadequate compaction of the sub-base within and adjacent to the slab edge, resulting from manual compaction, does not provide the required support. Subsidence or depression of a slab at a transverse joint may be caused by loads exceeding the capacity of the subgrade or, if installed, of the load transfer devices. The slab will break in an irregular pattern over a localized subgrade failure. Reinforcing steel will not prevent such failures, but it will hold the concrete together. See Figure 4-25.

Pumping. An action known as pumping may take place in the presence of moisture when a slab has been laid directly on a fine-grained, plastic, impervious soil. Poor drainage, even of a permeable sub-base, is a contributory cause. A slight vertical movement of the slab under traffic forces water and fine soil to the surface through cracks and joints. Evidence of pumping is especially noticeable immediately after a rainstorm and is characterized by the presence of fine soil and water adjacent to cracks and joints. When vehicles pass over the affected slabs, liquid may be observed spurting out of the crack. Continued pumping results in loss of foundation material and ultimate failure of the slab. Pumping can be prevented by placing the slab on permeable, granular sub-base and providing drainage away from the shoulders.

Figure 4-25
Failure of subgrade caused the pavement slab to subside on one side of the longitudinal joint. Dowels across the joint would have held the pavement in alignment.

When a slab is failing because of subsidence or pumping, mudjacking may save it. In mudjacking, a slurry of fine soil, cement and water is pumped under pressure through holes drilled through the slab. Under careful control of pressure and volume of slurry, voids beneath the slab can be filled and the slab brought back to grade. This work requires careful supervision and workmanship, as carelessness can result in cracked slabs or slabs raised too high. Only experienced workers should be permitted to do it.

Blowups. If a number of transverse joints become filled with solid material during cold weather, expansion of the concrete when the weather turns warm may cause the slab to buckle and crack. Contributory causes are inferior concrete and nonvertical joint surfaces. Joints should be kept free of pebbles and other foreign material, and should be properly sealed with an approved sealing or filling compound.

PREVENTION OF DETERIORATION

In each of the foregoing paragraphs, we have discussed specific measures that can be taken to prevent or minimize damage to concrete by the particular destructive medium under discussion. We have repeatedly stressed the importance of obtaining good durable concrete by the use of the right kind and amount of cement, sound aggregates, proper mix proportions, low water-cement ratio and total water approved methods of mixing, transporting, placing, consolidation and curing.

4.12. Air Entrainment

Throughout this chapter we have mentioned the importance of the proper amount of entrained air in the concrete as an aid to resistance of the concrete to deterioration. Any discussion of durability is not complete without mentioning the role of purposefully entrained air. Prior to 1940, use of air-entraining admixtures was virtually unheard of, but since the mid-1940s practically all exposed concrete, especially in severe climates where there are cycles of freezing and thawing, contains entrained air. Air entrainment is probably the most significant single factor contributing to durability of concrete as affected by freezing and thawing. See Figure 4-26.

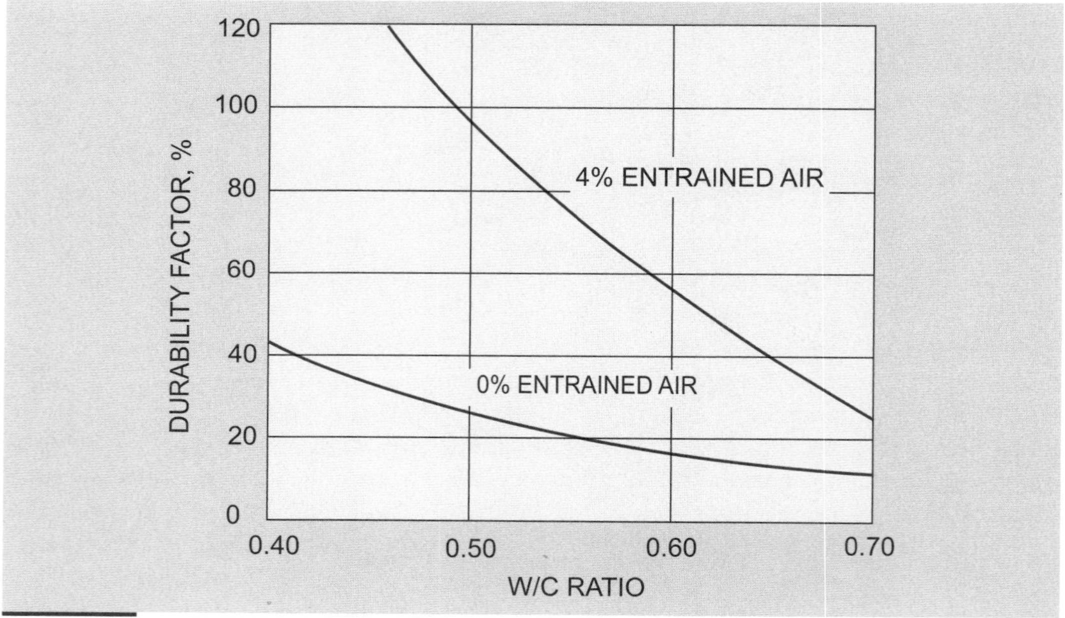

Figure 4-26
In both air-entrained and nonair-entrained concrete, the water-cement ratio has an important effect as well.

Addition of an air-entraining admixture to the mix results in the formation of innumerable microscopic spherical voids. These microscopic voids protect the hardened cement paste from the destructive action of freezing and thawing by absorbing or buffering the expansive force of freezing water in the water-saturated paste. Inasmuch as the air voids protect the paste only, concrete made with porous, unsound aggregates manifests poor resistance to freezing and thawing, whether or not it is air entrained.

Air entrainment is not a cure-all for whatever distress concrete may suffer. Entrained air does improve the durability and other characteristics of concrete, and its use should not be undervalued. However, it cannot take the place of good materials and quality construction. These compose the foundation of sound concrete construction.

Practically all concrete contains up to about 2 percent of entrapped air voids, by volume. These entrapped voids are much larger than the voids that are purposefully entrained in the concrete, and they are of no value in improving the concrete as the entrained air does.

Entrainment of air is accomplished by the addition of a small amount of a foaming agent that can be added as an admixture at the concrete batching plant. In some areas of the country, air-entraining cement is available that consists of portland cement with an air-entraining admixture interground at the cement plant.

With respect to the durability imparted to concrete by entrained air, two important points should be kept in mind:

1. Prolonged moist curing of the concrete, without an intervening drying-out period before exposure, tends to fill the voids with water and may actually reduce the potential durability originally imparted by the entrained air.
2. Concrete slabs on ground placed in the late fall (October, November) should not be subjected to de-icing salts during their first winter of exposure, as scaling may result.

4.13. ACI 318 Durability Requirements

The ACI 318 Standard addresses durability requirements for structural concrete in Chapter 4 in terms of exposure categories and classes depending on the severity of the exposure for structural concrete members. For concrete exposed to freezing and thawing conditions, air-entrained concrete must be used. For concrete exposed to sulfates in soil or water, sulfate resisting cement must be used, with a maximum water-cement ratio and a minimum specified concrete strength depending on the severity of the exposure condition. For concrete requiring low permeability (watertightness), a maximum water-cement ratio and a minimum concrete strength must be specified. For corrosion protection of reinforcement (chloride exposure), the concrete mixture materials (cement, aggregates, admixtures and water) must have a maximum amount of chlorides incorporated into the concrete. Depending on the severity of the chloride exposure, a maximum water-cement ratio and minimum concrete strength must also be specified. Conditions in structures where chlorides may be applied should be evaluated, such as parking structures where chlorides may be tracked in by vehicles, or in structures near seawater. For corrosion protection of the reinforcement, minimum cover is a critical consideration as well. For an in-depth discussion of the ACI 318 durability requirements, refer to Chapter 12.

Chapter 5
Volume Changes and Other Properties

Concrete, in common with other materials, is subject to changes in volume, either autogeneous or induced, depending on the environment and forces acting on it. To the practical person in the field, volume changes and thermal properties of concrete are of considerable importance, although at times the tendency is to consider them as being more in the province of the designer. However, they cannot be ignored in the field. Expansion and contraction are important to the extent that they affect dimensional stability and the formation of cracks. Creep or plastic flow may cause an undesirable change in distribution of stresses in the structure, and the thermal properties affect durability to the extent of their influence on expansion and contraction during temperature changes.

VOLUME CHANGES AND SHRINKAGE

5.1. Shrinkage

When we consider shrinkage of concrete, we find that we have two shrinkages: one occurs while the concrete is still in a plastic condition, and one occurs later after the concrete has hardened and begins to dry out. The basic cause of both is the same: loss of water from the concrete. If we can limit the amount of water in the concrete, and if we can control the rate of water loss, we will have come a long way in reducing shrinkage and consequent cracking. There are, of course, other factors to be considered. All of the following factors affect the amount of shrinkage:

- Total water in the concrete
- Water-cement ratio
- Quality of curing
- Condition of subgrade and forms
- Characteristics of the aggregate
- Grading of the aggregate
- Composition of the cement
- Size and shape of element being made
- Presence or absence of admixture
- Methods used in handling and placing concrete
- Weather conditions

An apparent shrinkage may be a loss of volume resulting from a loss of entrained air. Ordinarily, concrete loses some of its fresh volume because of normal shrinkage, and some (about 1 percent) because of loss of entrained air occasioned by handling and placing practices.

The properties of the cement have little effect on the amount of shrinkage, compared with the influence of job conditions and water content of the mixture. The amount of cement in the mix has a minor effect on shrinkage.

The most important requirement for minimizing shrinkage is that the total water per cubic yard be kept as low as possible. The use of high-slump mixes, which is probably the most common cause of abnormal shrinkage, should be avoided. (See Figure 5-1.) We must, however, be alert to the effect of the superplasticizers as described in Section 9.3.and 16.1.

It may be desirable to use a set-retarding and water-reducing admixture. Action of such an admixture not only reduces the total amount of water per cubic yard for equal workability but also delays hydration of the cement, thereby extending the period of plasticity so that the concrete will adjust better to early volume changes. However, use of a water reducer containing an accelerator to counteract the retarding action may result in more rather than less shrinkage, even though water content is reduced.

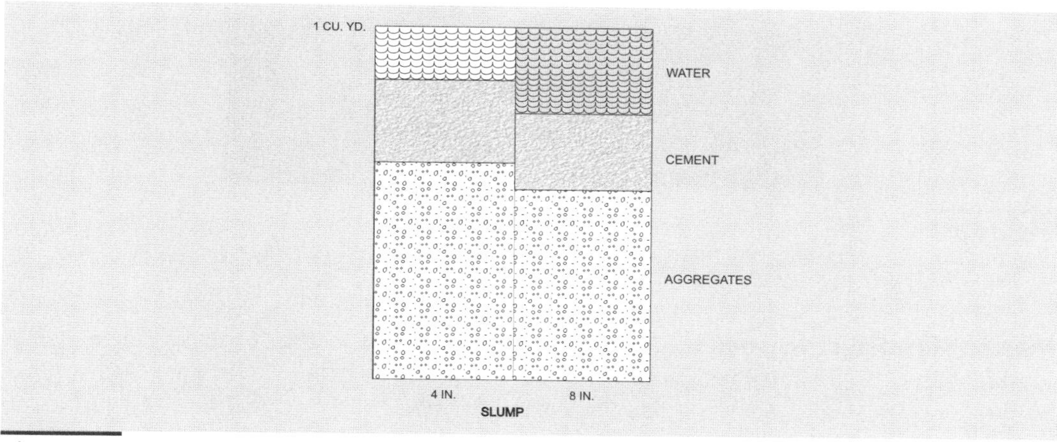

Figure 5-1
Because an 8-inch-slump concrete requires 35 pounds (almost 5 gallons) more water per cubic yard than a concrete with a slump of 4 inches, its shrinkage will be greater.

Proper selection of materials can reduce potential shrinkage by as much as half of that which might result from using shrinkage-reducing materials. Well-graded aggregates of the largest practical size consistent with size and design of the structural element being placed, containing an adequate amount of fines, should be used. Use the maximum amount of aggregate in the concrete consistent with workability and strength requirements. About 5 percent to 8 percent of the sand should pass the 100-mesh screen, and 15 percent to 30 percent should pass the No. 50. The sand should not contain more than 5 percent clay. Highly absorbent aggregates such as sandstone, shale and porous chert should not be used. Entrained air increases drying shrinkage, but because air entrainment permits the use of less mixing water, the net effect on shrinkage is negligible.

Obviously, the factor that lends itself most readily to control on the job is the amount of water that is introduced into the mixer. Hence the importance of placing concrete with the minimum practicable slump. Four-inch slump is adequate for any slab placement, whether it is driveway, floor or tilt-up. Narrow walls, beams and similar structural elements will require somewhat more slump. Concrete of the proper consistency will require the use of vibrators for consolidation in the forms.

Control of the rate and amount of water loss can be accomplished by dampening the subgrade and forms before placing the concrete and by the use of correct curing methods. Curing should be started immediately after finishing by covering the concrete with plastic sheeting, waterproof paper or sprayed-on membrane curing compound as described in Chapter 17. Two additional measures are of value during periods of dry, windy weather if the physical features of the job permit: shade the concrete from the direct rays of the sun (Figure 5-2) and erect windbreaks so the full force of the wind will not prevail over the surface of the concrete. Using fog nozzles (not spray nozzles) along the windward side of the concrete will raise the humidity over the area, thus lessening the rate of evaporation. See Figures 5-2 and 19-4.

Figure 5-2
Fog nozzles not only raise the humidity and lessen evaporation but also have a cooling effect when placed on the windward side.(Courtesy of PCA)

Plastic Shrinkage. As soon as concrete has been placed in the forms, it starts to lose water. Water can be absorbed by a dry subgrade, dry form lumber or dry aggregate; it can be lost through small cracks and openings in the formwork; or it can rise to the surface by bleeding and be lost by evaporation. Of these, evaporation accounts for the greatest loss of water. Loss of this water causes a decrease in volume of the concrete called plastic shrinkage. It is called plastic shrinkage because the shrinkage takes place while the concrete is still fresh, or plastic. The concrete has no strength, although it is beginning to assume some rigidity. Shrinkage occurs in the paste surrounding the aggregate particles; the aggregate itself does not undergo any change in volume. Hence, to control plastic shrinkage shrinkage of the paste must be controlled.

If the loss of water is reasonably slow, the concrete can adjust to the reduction in volume without difficulty, but a rapid loss of bleed water from the surface of the slab will introduce a tensile stress in the surface layer. Because the concrete has no strength, the tension causes cracks. These cracks, called plastic shrinkage cracks (Figure 5-3), appear on horizontal surfaces, developing rather suddenly about the time the water sheen disappears from the surface. See also Chapter 6.

Figure 5-3
Plastic shrinkage cracking. (Courtesy of PCA)

Under conditions of low humidity and drying wind, evaporation may be so rapid as to cause plastic shrinkage and cracking even before the concrete has been finished. The condition is aggravated if the concrete is placed on a dry, absorptive subgrade. One expedient that has been found to be helpful is the spray application of an extremely thin coating of an evaporation-inhibiting compound. The coating, sprayed on the concrete immediately following the first floating, prevents evaporation of water from the concrete but does not interfere with subsequent finishing procedures. If necessary, additional applications can be made. The material does not interfere with finishing or damage the concrete. It is a temporary expedient only and must be followed by normal good curing of the concrete.

Low humidity in the air and wind are the principal causes of high evaporation. Air temperature is also significant; the rate of evaporation can be high even at temperatures near freezing. The curves in Figure 5-4 can be used to estimate evaporation losses under different atmospheric conditions. [5.1]

The complaint is sometimes heard, especially in slab work, that everything on the job was the same from one day to the next, yet on the second day plastic shrinkage cracks occurred and on the first day it did not.

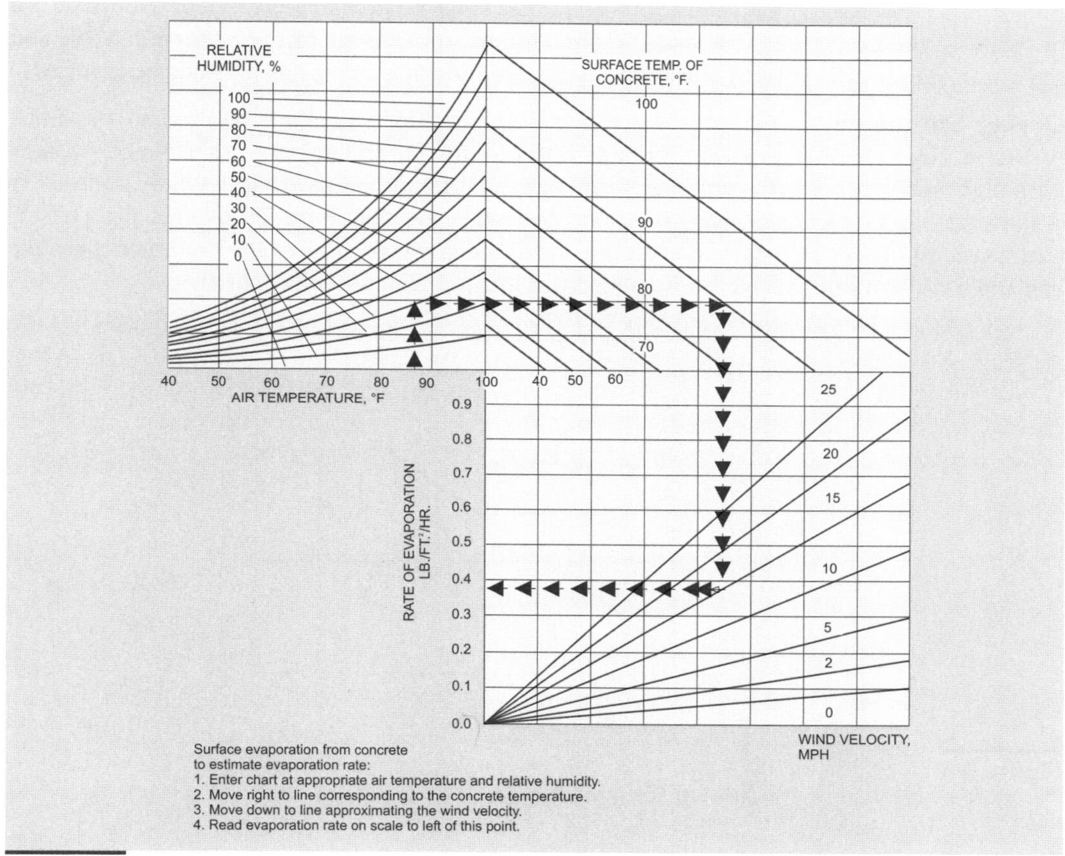

Figure 5-4
These convenient charts can be used to determine the amount of water evaporated from the surface. Assume that air temperature is 85°F, relative humidity is 30 percent, concrete temperature is 90°F, and wind velocity is 15 miles per hour. Enter the "air temperature" chart at 85°F, move straight up to 30 percent humidity, then across to the 90°F concrete temperature, down to 15 mph wind velocity, then left to edge of chart where we read 0.38 pound per square foot per hour of water evaporated, equal to about 4 $^1/_2$ gallons on a slab 10 feet square.

The point of course is that everything was not the same. Materials and methods might be the same, but the weather is not. An imperceptible change in wind velocity from 10 miles per hour to 15 miles per hour can cause half again as much evaporation. If the increase in wind velocity is accompanied by a drop of humidity from 40 percent to 20 percent the evaporation rate is nearly doubled. These changes in the weather, or even larger ones, can go unnoticed, but they can explain some of the differences in behavior of concrete.

See Section 22.11 (Plastic-Fiber-Reinforced Concrete) for a discussion on the use of plastic fibers as a deterrent to plastic shrinkage cracking.

Bleeding [5.2]**.** As soon as cement and water come together in a batch of concrete, complex physical and chemical activities commence. First, there is a reduction in the volume occupied by the water and cement. Then, after the concrete has been placed and consolidated—an action known as bleeding—sedimentation, or water gain occurs, as described in Chapter 2. A small amount of bleeding is not detrimental to the concrete; in fact, the settlement that causes bleeding actually results in a slightly stronger paste. Any plastic mix will bleed some, but when the bleeding gets to the point that a large amount of water collects on the surface, there may be trouble.

Movement of water through the concrete makes small flow channels that contribute to the porosity of the concrete. The water makes the topmost layer of the concrete weak and porous. Laitance is formed by the mixture of water and fines, delaying finishing operations.

Drying Shrinkage. Concrete loses moisture as it dries out, and this loss of moisture results in a loss of volume. The amount of water in the mix influences the shrinkage—mixes with high water content have higher shrinkage rates than mixes with low water content. It is the total water per cubic yard of concrete that is the dominant factor. See Figure 5-5. Composition of the aggregate, duration of initial moist curing, size of the concrete member and the environment in which the concrete is located all have their influence.

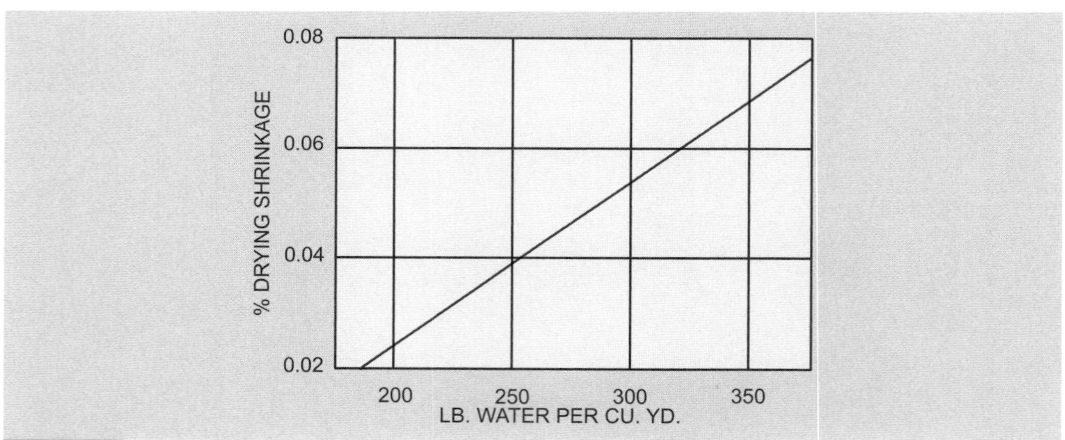

Figure 5-5
The more water there is in a batch of concrete, the more that concrete will shrink when it dries out.

Some admixtures affect drying shrinkage. The use of calcium chloride as an accelerator causes an increase in shrinkage. Some of the water-reducing admixtures increase drying shrinkage.

The mix proportions themselves have an insignificant effect on drying shrinkage, but through their effect on the water content of the concrete they become highly significant. Sand content of the mix should be the minimum necessary to provide the required workability, and the largest size of aggregate commensurate with structural requirements should be used. High sand content, small aggregate mixes considered necessary for pumping by some persons, should be avoided, as these can have a high shrinkage rate. As pointed out previously, concrete with a highmixing-water content will sustain more drying shrinkage than concrete with less water. Figure 5-5 shows that good concrete with about 300 pounds of water per cubic yard will have a drying shrinkage of about 0.05 percent ($^1/_{16}$ inch in 10 feet). The figure clearly shows the highly significant effect of water in the concrete. The drying shrinkage bears almost a straight-line relationship to the total water in the concrete; raising the water-cement ratio with no change in the cement content results in an increase in shrinkage.

With respect to curing, it appears that moist curing beyond seven days has only a small effect on shrinkage. Early curing, however, is important.

After the initial drying shrinkage has taken place, expansion resulting from subsequent rewetting will not return the concrete to its original dimensions. The initial drying shrinkage ranges from a low of about 0.02 percent for mixes with low water content made of good quality, properly graded aggregates to as high as 0.10 percent for high-slump mixes with inferior aggregates and high mortar content.

Because drying shrinkage implies a drying of the concrete, the environment of the concrete structure can assume importance. Temperature, relative humidity and air circulation all have an effect. Concrete in a damp or wet environment or that is subject to frequent wetting and drying will never reach the state of dryness of a member exposed to sun, wind and dry air.

The equivalents shown in Table 5.1 will be of assistance in comparing actual linear volume change measurements.

TABLE 5.1
LENGTH CHANGE EQUIVALENTS

MILLIONTHS	PERCENT	INCHES PER 10 FEET	
		DECIMAL	FRACTION
100	0.01	0.012	$^1/_{64}$
200	0.02	0.024	$^1/_{32}$
400	0.04	0.048	$^3/_{64}$
500	0.05	0.060	$^1/_{16}$
600	0.06	0.072	$^5/_{64}$
800	0.08	0.096	$^3/_{32}$
1000	0.10	0.120	$^1/_8$

NOTE: Fraction equivalents are approximate.

How to Limit Shrinkage. We have repeatedly stressed the importance of using the least amount of water that will give the necessary workability for the conditions in the forms. If the handling and placing equipment cannot manage such concrete, then the answer is to get equipment that can handle it. The answer is not more water and more slump. Other factors are as follows:

(1) Use good, workable mixes, properly proportioned with the largest amount of aggregate that is practical.

(2) Carefully schedule truck mixers to avoid delays in unloading.

(3) Thoroughly consolidate the concrete with vibrators.

(4) Cure the concrete properly.

(5) Intelligently use admixtures, avoiding those that might aggravate the shrinkage.

(6) Use the largest maximum size of aggregate to fit job conditions.

(7) Avoid aggregates that contain a large amount of clay.

(8) Use aggregates that have a low shrinkage when mixed in concrete.

(9) Maximize the coarse aggregate content.

(10) Follow good construction practices throughout.

Because shrinkage problems are intensified during hot weather, the suggestions in Chapter 19 should be observed when the air temperature climbs above 85°F.

It is sometimes desirable to estimate the amount that a concrete member will shrink in length when it dries out. If we know what the rate of drying shrinkage is for any certain concrete or mortar, the curves in Figure 5-6 will give us the change in length (shortening) of any length of member. For example, if we have a beam 70 feet long made of concrete that has a drying shrinkage of 0.04 percent, then that beam will shrink about $^3/_8$ inch when it changes from a saturated to a dry condition.

5.2. Volume Change

Volume change is the expansion and contraction of concrete that results from temperature changes or wetting and drying. These changes are reversible; that is, the cycles can be repeated and the concrete will expand or contract each time, depending on the force acting on it. A concrete unit will expand each time it gets warm and will contract or shorten when it cools if it is unrestrained. If it is restrained, the expansive or contractive force is there, but the concrete cannot move; consequently, the concrete may be damaged.

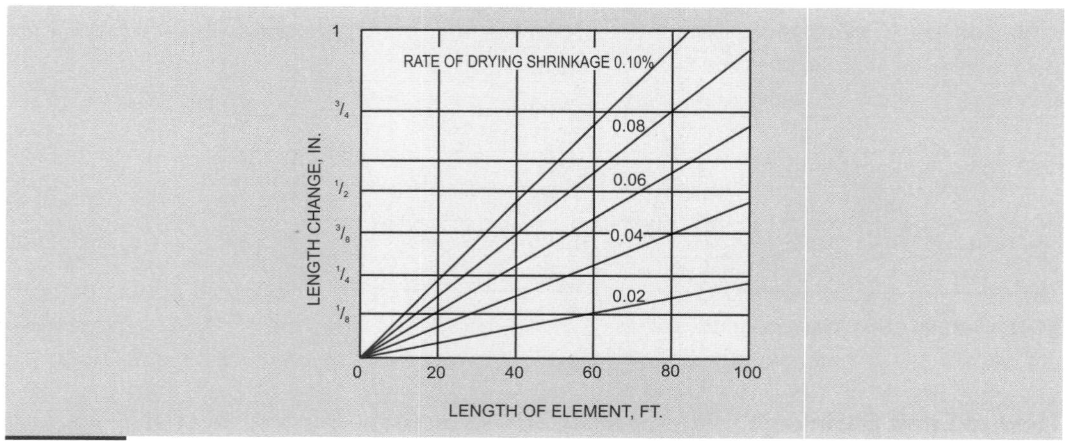

Figure 5-6
Drying shrinkage can be estimated by use of this chart. Normal concrete will have a shrinkage rate between 0.04 percent and 0.06 percent. Example: An 80-foot girder with a shrinkage rate of 0 .04 percent will shrink about $^3/_8$ inch.

If concrete is restrained, cracks form as a result of shrinkage or contraction combined with insufficient tensile strength. Expansion under restraint causes spalling at joints or excessive compressive stress. See Figure 5-7. Under the influence of moisture changes, concrete will withstand the compressive stress induced by wetting expansion but, if restrained, concrete will crack from tension resulting from shrinkage as it dries out.

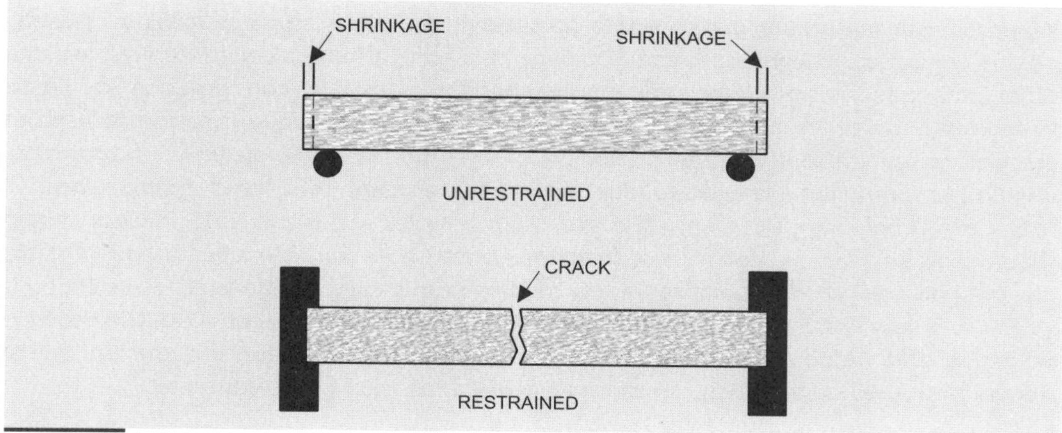

Figure 5-7
If concrete were free to move when it shrinks, shrinkage cracks would not be a problem. But concrete in the structure is restrained in some way; therefore, it cracks when it shrinks.

Volume changes caused by temperature fluctuations produce cracking and disintegration similar to that resulting from wetting and drying. The action, however, is somewhat different, as temperature effects involve the aggregate as well as the paste, while moisture changes normally involve only the paste.

5.3. The Role of Reinforcement

We have previously pointed out that concrete is relatively weak in tension compared to its compressive strength. Steel is strong in tension. We can now combine the two materials into what we call reinforced concrete to give us a structural material that can resist both compressive and tensile loads. This is discussed in Chapter 18.

Steel can also be used to aid in the control of shrinkage. Actually, the steel has a minor effect on the actual amount of shrinkage of a concrete member, but it does help control the shrinkage to the extent that the shrinkage stresses are distributed along the length of a bar of reinforcing steel. Consequently, there will be several very small cracks instead of one large one. The so- called temperature steel, a light reinforcement of wire mesh sometimes used in slabs, is an example.

5.4. Chemical Control of Shrinkage

Expansive cement is designed so that an expansive force is developed, which counteracts or compensates for normal shrinkage. This cement, used in shrinkage-compensating concrete, is discussed in Chapter 7. A shrinkage-controlling admixture is also available.

Nonshrinking concrete has a number of uses similar to ready-to-use nonshrink grout where larger clearances exist. It is not intended to be a structural concrete but is a specialized product for very specialized uses. Expansive cement or shrinkage-controlling admixtures should be used only under carefully controlled conditions and should follow the instructions of the manufacturer.

Also available is a magnesium phosphate concrete that is said to harden at low temperatures, expand slightly and develop as much as 5000 psi compressive strength in 24 hours.

When difficult placement makes a soft consistency necessary, such as filling a cavity by placing concrete through a restricted opening or placing around a congestion of steel and other embedded items where proper vibration is not possible, consideration should be given to the use of a superplasticizing admixture. Another technique occasionally used to prevent or minimize shrinkage of concrete is to add a small quantity of superfine, unpolished aluminum powder to the mix in the amount of 2 or 3 grams (about a teaspoonful) per bag of cement. The aluminum powder technique is not necessary and should not be used in normal construction, because of variable effects and possible lowering of strength. The painting variety of aluminum should not be used, as its action is slower and may not be complete by the time the cement starts to set. Also, the action is slower in cold weather. Tests should be made beforehand to determine the amount of aluminum to use, as too much can cause an actual expansion of the concrete.

In preparing a nonshrinking grout for use in such situations as placing under machinery, a superplasticizer may be effective, or the aluminum powder technique may be employed. In some cases preshrunk grout may be desirable. This is made by allowing the mixed grout to stand for an hour or two before use. No tempering water should be used, as this will destroy the effectiveness of the shrinkage control and weaken the grout.

5.5. Internally Induced Volume Changes

The activities we are concerned with in this section are those that result from chemical reactions within the concrete, resulting from the materials used in the concrete. Volume changes resulting from this activity are irreversible; that is, the reaction causes a damaging expansion to the concrete, but there is no subsequent contraction.

These reactions are discussed in Chapter 4, and will not be elaborated on here except to mention them as another group of activities that affect the volume of hardened concrete. Aside from the rather rare incidence of damage from foreign matter in the concrete, expansion most commonly results from reactions of certain aggregates with the alkalies in the cement. We say most commonly, but this does not imply that it is now a common occurrence; on the contrary, proper processing and use of materials now make these problems quite rare.

5.6. Measurement of Volume Changes

Direct measurement of length changes can be made on a structure in the field, but such measurements are awkward at best, although they are certainly revealing in the information they are capable of giving. Actual length of a long beam or wall can be made under different conditions of temperature or other influences, or the width of cracks and joins can be measured to give an indication of the length change of the unit.

If any requirement is stated in the specifications, it will usually be a limitation on drying shrinkage based on a laboratory method of test such as "Test Method for Length Change of Hardened Hydraulic-Cement Mortar and Concrete" (ASTM C 157). This is not an easy test to make, because of the close tolerances necessary in controlling the atmosphere in which the specimens are stored and the critical use of accurate length-change micrometer gauges. Test specimens of concrete are 3 inches by 3 inches by 11 inches. Length measurements are made at 24 hours, at the end of water curing at 28 days, then periodically for a period of up to a year in a drying atmosphere. See Figure 5-8.

It is difficult to relate shrinkage values made under fixed laboratory conditions to those that can be expected under the variable atmospheric conditions surrounding any structure in the field. Comparing unknown materials and mixes with both a field and laboratory history can be of value and gives an indication of probable behavior of the concrete in a structure. The least it can do is raise a warning when unusually high values are obtained in the laboratory for any proposed material or combination of materials.

Figure 5-8
Measuring the length of a 3-inch by 3-inch by 11-inch concrete bar to determine its shrinkage.

THERMAL PROPERTIES

5.7. Coefficient of Expansion

A common property of most materials is their ability to change in volume with changes in temperature; as the temperature rises, the volume of the material increases—the piece of material gets larger. When the temperature drops, the material contracts, or gets smaller. We can measure this temperature effect in the laboratory by taking a measured length of material, in this case a small concrete bar or beam about 12 or 18 inches long, and slowly heating it as we observe a micrometer gauge that measures the change in length of the bar. The unit of measurement is called the thermal linear coefficient of expansion, and is expressed as the amount, in inches, that a 1-inch long piece of concrete will increase in length for a 1-degree rise in temperature. The actual value varies slightly depending on the kind of aggregate in the concrete, cement and water contents, and other variables in the mix. On an average, the coefficient for concrete is generally considered to be 0.0000055 inch per degree Fahrenheit, sometimes shown as 5.5×10^{-6}. This, fortunately, is about the same as the coefficient for steel, which enables reinforced concrete to function.

The thermal coefficient is an inherent property of the concrete and is not subject to adjustment or control by the person on the job. It is, however, a very important property that must be taken into consideration when designing a structure. Provision must be made by means of joints of various types, by movable connections or other means to allow for the variation in length of structural members as the temperature varies. For example, a structural beam 80 feet long will increase in length by more than $^1/_4$ inch under a temperature rise of 50°F. (Daily temperature variations of 50°F or more are common in many areas.) Another important fact is that the force exerted by a concrete member expanding because of a temperature rise is equal to the force that would be required to stretch the member the same amount. That is why spalling, cracking, displacement and other damage arise when no provision is made to allow for movement.

The chart in Figure 5-9 can be used to estimate length changes of concrete elements of different lengths under different temperature changes. It can be used for estimating expansion with temperature rise or contraction with temperature drop.

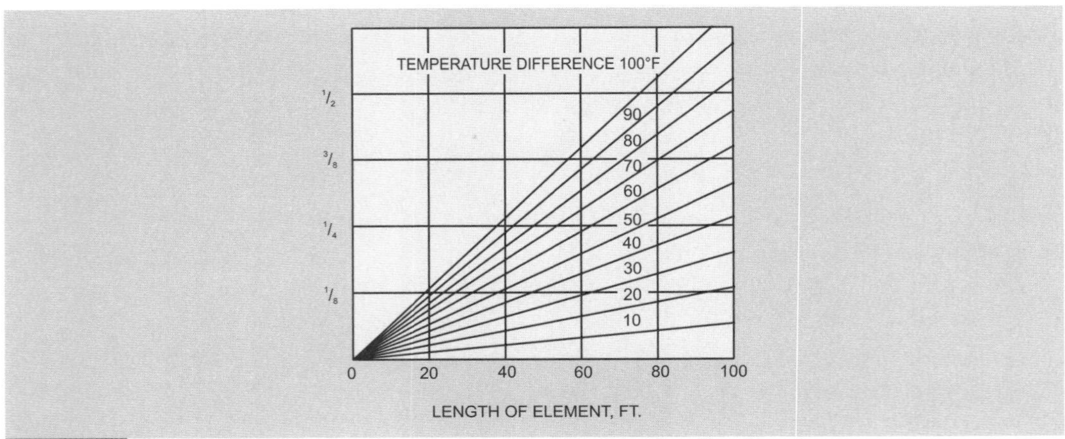

Figure 5-9
Chart for estimating length change of concrete with change in temperature. Example: A 75-foot girder changing in temperature from 40°F to 90°F (temperature difference = 50) will change in length about $^1/_4$ inch.

5.8. Conductivity, Specific Heat and Diffusivity

Different materials conduct heat, or allow heat to pass through, at different rates. The rate at which heat is conducted is called the conductivity, indicated by the letter k, and is defined as the amount of the heat the material will conduct through 1-inch thickness per unit of area. The unit used to measure heat, the British thermal unit (Btu), is the amount of heat required to raise the temperature of 1 pound of water 1 degree Fahrenheit. Some materials, such as aluminum and steel, have a high k value or conductivity, while heat insulators, such as rock wool and cork, have a low k value. Concrete has a fairly high k value—not as high as metals, but it will conduct considerable heat.

Conductivity is a function of the density of a material, and this applies to concrete. Other influences on conductivity are the kind of aggregate in the concrete, air content of the concrete and the amount of moisture in the concrete. Conductivity is lower in air-entrained concrete and increases as the moisture content of the concrete increases.

The conductivity of normal concrete, completely dry, is about 12 Btu-inches per square foot per degree-hour, going up to about 18 for saturated concrete. Dry lightweight concrete with a density of 100 pounds per cubic foot has a conductivity of about 4.

Use is made of the k value to determine the amount of heat that will be lost from a building during cold weather or the amount of heat that will enter a building during hot weather. From this information the engineer can calculate the requirements for air- conditioning equipment for cooling and heating the building. The conductivity is important in the solution of problems concerning condensation on walls and floors in buildings and is used in computations concerning heat movement in mass concrete.

The specific heat, S, of any material is the amount of heat required to change the temperature of 1 pound of the material by 1°F. By definition, the specific heat of water is one; that is, it takes 1 Btu to raise 1 pound (a pint) of water 1 degree in temperature. The specific heat of concrete is generally assumed to be between 0.20 and 0.22 Btu per pound per °F.

It is sometimes necessary to make computations regarding the flow or movement of heat through concrete. Problems of this nature arise when dealing with mass concrete in dams and similar structures, where the heat of hydration of the cement must be removed by some means, and we need to know the rate at which temperature changes take place within the mass. For these computations we need a value called the diffusivity, D, which is a measure of the rate of temperature change in concrete, expressed in units of square feet per hour. Diffusivity can be computed from the relationship.

$$D = \frac{\text{Thermal conductivity}}{\text{Specific heat} \times \text{density of concrete}}$$

If there is reason to believe that the thermal properties are causing trouble, laboratory tests will disclose whether this is true.

A high coefficient of expansion causes high stresses in the surface of concrete undergoing temperature changes, and a high-diffusivity gravel, compared with the mortar, causes differential volume changes, both of which adversely affect durability of the concrete. Over a given temperature range, a fast rate of temperature change is more damaging to the concrete than a slow change. These, however, are problems that are rarely encountered and need not be of serious concern to the person on the job.

ELASTIC PROPERTIES

5.9. Modulus of Elasticity

The word *modulus* merely means measure, so when we speak of the modulus of elasticity we are speaking about measuring elasticity of a substance.

Hardened concrete appears to be a rigid, unyielding material, and for most practical purposes it can be considered such. Materials, however, are not rigid, and concrete is no exception. If a sample of concrete, say a 6-inch by 12-inch test cylinder, is loaded in compression, the cylinder is shortened as the load is applied, and if the load is removed, the cylinder is returned to its original size and shape. See Figure 5-10. This property is called elasticity.

For most materials, the ratio of stress (load) to strain (deformation) is constant, and ratio is called the modulus of elasticity, known by the letter *E*:

$$E = \frac{\text{stress}}{\text{strain}}$$

or we can find *E* by dividing the load in pounds by the change in length, or deformation, in inches:

$$E = \frac{\text{load in pounds}}{\text{deformation at that load}}$$

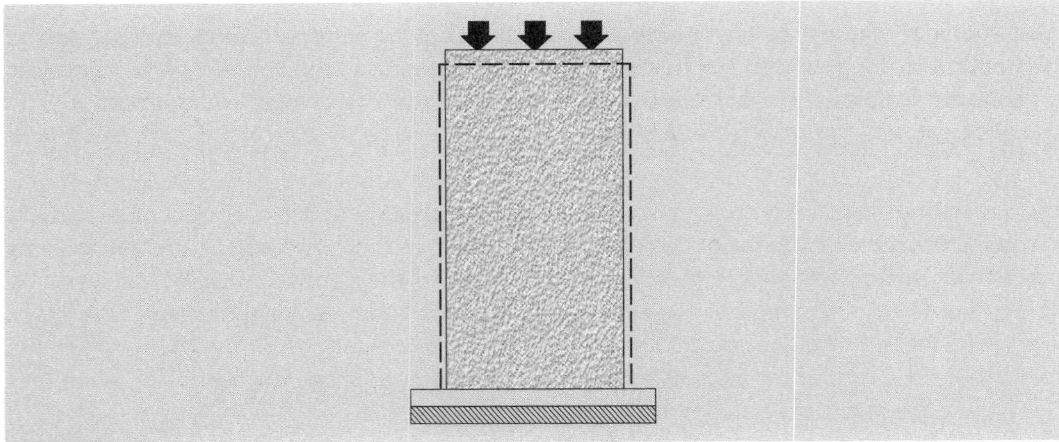

Figure 5-10
When a compressive load is applied to a concrete specimen, the specimen shortens, or compresses, and bulges slightly; however, it will resume its original shape when the load is removed.

Concrete is not a perfectly elastic material, the ratio of stress to strain decreasing as the load increases; that is, the stress-strain curve is slightly curved instead of a being a straight line. See Figure 5-10. However, within the range of usual working loads, the stress strain curve for thoroughly hardened concrete is nearly a straight line. The static modulus of elasticity is obtained by loading a specimen, usually a test cylinder, and observing the elastic deformation of the specimen under compression; that is, how much shorter it gets when it is loaded. The measurements are usually made at a load equal to about half of the estimated strength of the specimen.

In Figure 5-11 the estimated strength of the concrete was 5000 psi, so the deformation was measured at a load of 2500 psi. At this point the deformation was 0.00055 inch. Therefore *E* was; $\frac{2500}{0.00055} = 4{,}450{,}000$ psi.

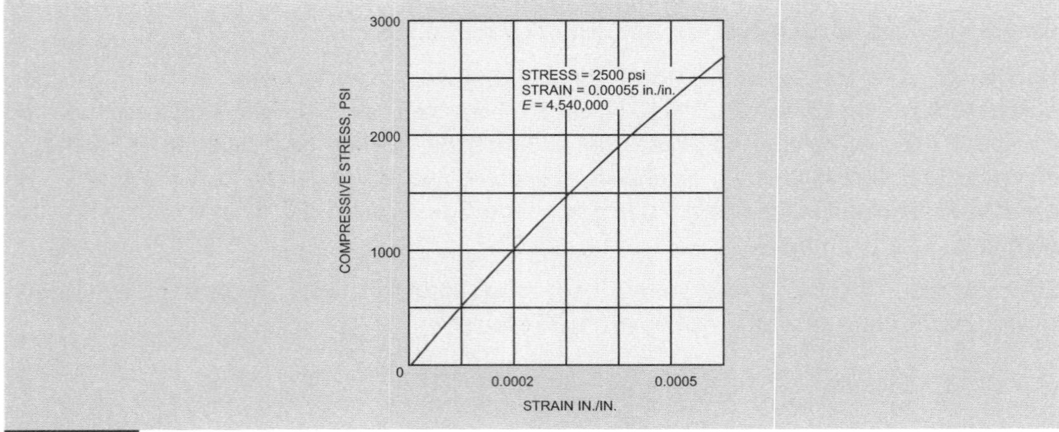

Figure 5-11
The stress-strain line for concrete is slightly curved. E is measured at a point about halfway to the estimated compressive strength, in this case, estimated to be 5000 psi (it was actually 4720 psi) . At the point of one half of 5000, or 2500 psi, the strain was 0.00055 inch.

Dynamic Modulus of Elasticity. Electronic equipment is available for determining the elastic modulus by measuring the velocity of sound waves through the specimen, or measuring the natural frequency of vibration of the specimen. These methods are specially useful in studying the deterioration of concrete, as specimens can be repeatedly tested without damaging them. A decrease in frequency of vibration or velocity of waves indicates a decrease in elastic modulus, which in turn reveals deterioration of the concrete.

High-Temperature Effects. Of serious importance when considering the effect of a fire on the concrete portions of a building is a significant reduction in the elastic modulus because of heating. Temperature, length of time the concrete is heated, age and quality of the concrete and other factors affect the actual reduction, but it is significant at 400°F and is more than half at 1000°F.

The modulus of elasticity is related to compressive strength, although not directly proportional. However, a high-strength concrete usually has a higher elastic modulus than a weak one. Cause of a low modulus may be a high water-cement ratio, an excessive percentage of entrained air, or any of the causes of low compressive strength.

5.10. Creep

Creep may be defined as a time-dependent deformation of concrete under sustained load, in contrast to elastic deformation that occurs immediately upon application of a load to the concrete and which disappears upon removal of the load. The terms creep and plastic flow describe the same phenomenon, with creep the preferred usage. The practical importance of creep is in the long-time deflection of flexural structural members. It is particularly significant in its effect on high-rise concrete buildings.

Tests by many investigators indicate that creep continues over many years, as long as the concrete is subject to stress. The rate of creep decreases rapidly at first and continues to decrease with time. One fourth of the total expected creep occurs within about two weeks of loading, and fully half of the creep develops within three months.

Creep appears to consist of two components: an irreversible flow or creep and a delayed elastic strain, which can be called recoverable creep. When the load is removed, the concrete will recover part of the creep deformation, but it will never fully return to its original dimensions.

With the increased use of prestressed concrete in modern construction, attention has been focused on creep as an important property of concrete, resulting in intensive study by many institutions and agencies. As results of these investigations become known, light will be shed on the practical measures that can be effected in the field to control the characteristics of concrete that affect creep. For example, we have found that high-pressure steam curing reduces creep potential significantly.

Within the range of normal concrete mixes, creep is proportional to the water-cement ratio and the amount of hardened cement paste. Hence, in those situations in which it is necessary to keep creep at a minimum value, such as in prestressed concrete, it is desirable to use the minimum cement content and minimum water-cement ratio that will produce concrete possessing other desired properties, including adequate strength. Aggregates should be well graded and of the largest practical size and should consist of types possessing high density and low absorption, with a high modulus of elasticity.

The effect of entrained air on creep is negligible. Therefore, an air-entraining agent may be used in low-creep concrete with confidence because of its other beneficial effects. The use of a water-reducing admixture may also be of value.

OTHER PROPERTIES

5.11. Watertightness

When we discuss the watertightness of concrete, we find that we are concerned with the permeability of the concrete on the one hand and the overall quality of the concrete on the other.

Permeability. The general definition of permeability is the property of a substance that allows the passage of fluids. With respect to concrete, it is the property that permits the passage of water through the concrete. Of course, we do not want our concrete to be permeable; we want it to be watertight.

Concrete contains a variety of pores, both in the paste and in the aggregate particles. The permeability of the concrete depends on the size, distribution and continuity of the pores, particularly the pores in the cement paste. Two important facts should be noted. First, the porosity of the paste is substantially affected by the water-cement ratio, directly in proportion to the water-cement ratio; second, the porosity decreases as the hydration of the cement continues. Concrete can be made virtually impermeable by following the rules of good concrete construction. Capillary flow of water through the porosity of the concrete (permeability) does not require a head of water (pressure), the flow resulting from a constant supply of moisture in contact with one surface of the concrete and evaporation from the other side. Capillary flow can proceed upward through a slab on grade to cause damage inside the building. Cases have been reported in residential construction in which water passing by capillarity through a thin concrete floor was in sufficient quantity to force the vinyl floor covering off the concrete, resulting in curling and loosening of the floor covering.

Laboratory tests to determine permeability can be made, but they are of little value for field application because so much depends on practices used in construction. Figure 5-12 is a typical curve that shows the effect of mixing water on the percolation of water.

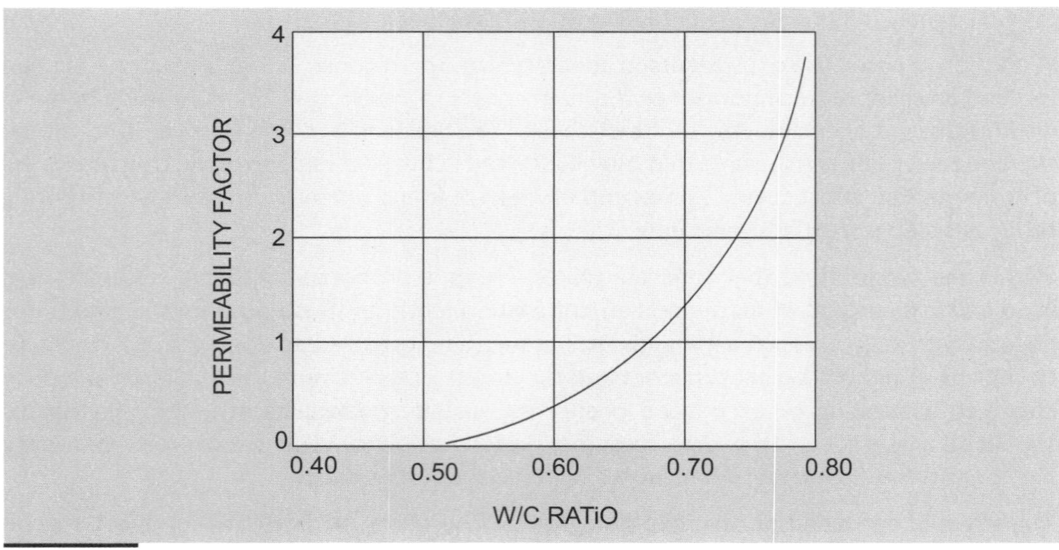

Figure 5-12
As the water-cement ratio goes up, the permeability of concrete also goes up. Tests were made in the laboratory on small disks of mortar.

Figure 5-13
This basement wall is sure to be a source of trouble in the future because water will pass through the uncompacted concrete. Placing the concrete in shallow layers properly vibrated would have prevented the rock pockets.

Flow through Openings. Another way in which water can pass through concrete is by flow through relatively large openings in the concrete, the water being under hydrostatic head and in contact with one surface of the concrete. Flow of this kind can be of considerable magnitude, depending on the head and size of the opening. Examples are rock pockets or poorly consolidated concrete and leakage through incompletely cleaned construction joints. See Figure 5-13.

Regardless of the way in which the water passes through the concrete, the results are not only unpleasant to look at but also can be damaging to the concrete. If aggressive solutions such as alkali water or water containing acid waste can penetrate the concrete, their damaging action can be multiplied manyfold.

How to Make Watertight Concrete. We are concerned here mainly with how to make the concrete itself watertight. Surface coatings for waterproofing and damproofing are discussed in Chapter 23.

The belief that using waterproofing admixtures or cement marked as "waterproofed" produces watertight concrete or mortar is not generally acknowledged by authorities in concrete technology. Studies made by both the Bureau of Reclamation and the Portland Cement Association indicate that close control of the water-cement ratio, proper placement of the mix, and proper curing afford the best guarantee of watertightness. Neither of these agencies recommends the use of "waterproofed" cement or waterproofing admixtures such as stearates, oleic acid, tallow, etc., to increase water resistance. The use of these additives does tend to make the dry cement powder water repellent. This water repellency necessitates longer mixing to obtain a uniformly plastic and workable mortar or concrete mix. Alkalies liberated during the hydration of the cement can cause these agents to deteriorate, and the hardened concrete eventually loses its water repellency.

Lack of watertightness can, in nearly every case, be traced to poor construction practice. For this reason, the best way to obtain watertight, impermeable concrete is to incorporate these properties into the structure when it is being built. This is accomplished by building the structure of good high-grade concrete. The following principles and precautions should be observed.

Figure 5-14
Leakage of water through construction joints in a dam constructed many years ago.

Construction joints are especially vulnerable, particularly the horizontal planes between lifts. Anything that weakens the bond on this surface will cause a leaky joint. See Figure 5-14. Among the causes are weak concrete on the top of a lift resulting from wet or over vibrated concrete, incomplete or totally lacking cleanup of the joint surface before placing the succeeding lift, laitance, rock pockets or honeycomb, and succeeding lift not vibrated enough. Cracks are also a source of leakage.

Use only sound, well-graded aggregates of flow porosity. Sand especially should consist of rounded particles instead of flat or angular ones.

Use a concrete that is plastic and workable, thoroughly mixed, with a water-cement ratio of less than 0.50 by weight in thin sections, or slightly higher in more massive sections. Avoid overwet mixes. Entrained air is beneficial by decreasing bleeding and interrupting the water channel structure within the concrete.

Handling and placing operations should be such as to avoid segregation and cold joints. Consolidate the concrete by means of vibration.

Use form ties in wall forms of the type that can be broken or removed below the concrete surface. Ordinary wire ties are not suitable. See Figure 5-15.

Finally, the concrete must be thoroughly cured. This consists of at least seven days of continuous wet curing or the application of a reliable curing compound.

The most important consideration is the application of sound construction practices, including good design, good materials, proper handing and adequate curing. The requirements for workable, durable, strong and crackless concrete are prerequisites for watertight concrete. The use of waterproofers or dampproofers, either integral or surface applied, cannot be accepted as compensation for poor workmanship, lean mixes or deficient materials.

Slab-on-grade floors in dwellings, stores, offices and for any occupancy that requires completely dry floors must be constructed with a vapor barrier. This is especially important for floors to be covered with vinyl tile or similar impermeable material. Many of the moisture problems associated with enclosed slabs on ground can be minimized or eliminated by using 4-inch granular sub-base (coarse gravel) to form a capillary break between the soil and the slab or by installing a vapor barrier, usually polyethylene sheeting or heavy asphalt roll roofing. See Figure 5-16. Special care is necessary to assure continuity of the sheet material at joints, at footings, around pipes and at similar points, and to avoid punching holes in it by tools or workers' shoes. A 2-inch layer of sand on the barrier ahead of concrete placing will protect the membrane and will absorb some bleeding water from the concrete. An integral admixture type of dampproofer is of no value in reducing transmission of moisture through a slab on the ground. Moisture transmission through a slab is proportional to the water-cement ratio of the concrete; hence the necessity of using the minimum amount of mixing water.

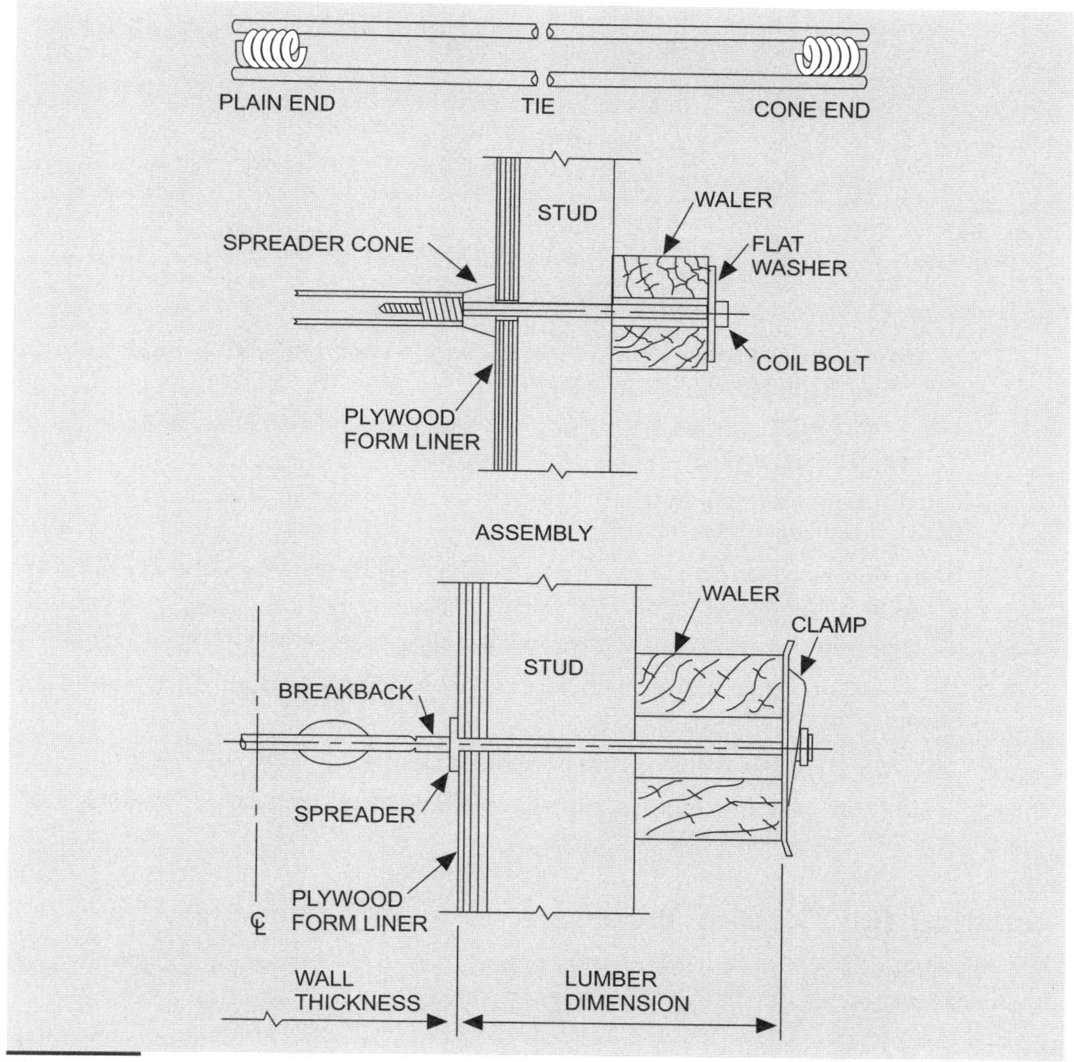

Figure 5-15
Some types of form ties use a spreader cone to serve as a spreader and to form a void in the surface of the concrete that can later be filled with mortar. The tie will break at the "breakback" when the form is removed.

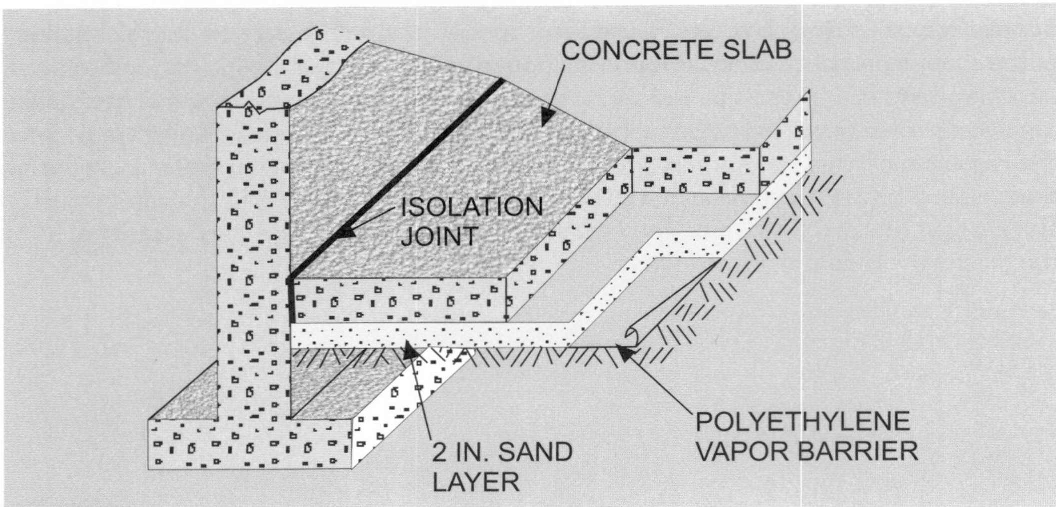

Figure 5-16
Section through a floor placed on a sand layer with a vapor barrier.

To summarize, the best way to obtain impermeable concrete is to:

1. Use a minimum amount of mixing water in the concrete, with a water-cement ratio not exceeding 0.50 by weight;
2. Use well-graded, sound aggregates in a mix containing at least 560 pounds of cement per cubic yard;
3. Entrain between 3 percent and 7 percent air, the larger amount in concrete with smaller size aggregate;
4. Keep the slump as low as possible for placing the concrete by vibration, not over 4 inches for slabs and 5 inches for structures;
5. Consolidate the concrete by vibration, avoiding segregation;
6. Clean all laitance and dirt off construction joints immediately before placing the next increment of concrete;
7. Use wall form ties that can be removed below the concrete surface; and
8. Cure the concrete by keeping it continuously wet for seven days or by application of a reliable curing compound.

5.12. Acoustical and Electrical Properties

The properties of concrete are not of great importance to the person on the job, but we should have some idea of what to expect of concrete.

Acoustics might be defined as the science of sound or the control of sound—in other words, noise control. Control of sound within a room requires that the walls, ceiling and floor be capable of absorbing noise instead of bouncing it back. A surface that is porous does not reflect sound; it absorbs it, so the room is quiet. The surface of concrete is hard and dense; it reflects about 98 percent of the sound that strikes it. For this reason, a room (say, a basement) with bare concrete floor, walls and ceiling is noisier than a room with carpet, drapes and a sound-absorbing acoustical ceiling. Lightweight structural concrete reflects about half as much noise as regular-weight concrete, so it is better for controlling sound within a room.

Another source of trouble is noise that passes through or is transmitted through a material. Because concrete is a dense material, it is a poor conductor of sound. Noise does not pass through a solid concrete wall as much as it does through a thin plaster wall. Lightweight concrete is not as good an insulator of sound passing through it as regular-weight concrete. The regular concrete would be better for keeping noise from passing from one room to the next.

Electrical properties, like acoustical properties, are of limited interest to us. In fact, they are practically not even considered when designing or building a concrete structure. An exception is the use of concrete for railroad ties, as the electrical properties can affect the signaling system. Dry concrete is a good insulator. Moist concrete has a fairly high resistance to the flow of electricity but not as good as dry concrete.

5.13. Fatigue

Fatigue means weariness or tiredness. Materials show fatigue after many cycles of application and release of load if the load each time is more than a certain amount. The number of applications of the load (stress) is many millions. Most materials have a fatigue limit—a stress level below which a specimen of the material will withstand an infinite, or unlimited, number of cycles of load application without failure.

There appears to have been considerable laboratory work done in investigating fatigue of concrete, but it is hard to draw any general conclusions except that more testing is required. The type of loading, whether plain reinforced or prestressed concrete, strength, temperature and moisture conditions, and many other factors influence fatigue of concrete. For example, shearing stresses in a beam may be more critical, as far as fatigue is concerned, than compressive stresses. Effect of reinforcing steel bond is varied, and if a beam cracks, subsequent loading usually causes static failure.

Whether concrete has a fatigue limit is subject to debate. However, a limit probably does exist. It depends upon design, method of loading, presence or lack of reinforcing steel, materials, effect of cracks and many other variables. We can acknowledge that fatigue exists, but beyond that it is of little concern.

Figure 5-17
Fresh unit weight of concrete is measured in a container of known volume. (Courtesy of PCA)

5.14. Yield

The yield of concrete is defined as the volume of concrete per batch of concrete and is found by dividing the actual weights of all materials (cement, aggregates, water) in the batch by the fresh unit weight (ASTM C 138). See Figure 5-17. Under good to excellent control conditions, the actual volume of the hardened concrete in place will be about 2 percent less than the fresh volume for air-entrained concrete, 1 to 1.5 percent less for nonair-entrained concrete. This normal loss of volume is caused by loss of part of the entrained air during handling and placing, decrease in the combined volume of water and cement during hydration, settlement, bleeding, drying shrinkage and compression of entrained air in the bottom of a lift. The apparent loss can be much higher if the forms bulge or spread. Small irregularities or inaccuracies of the magnitude of $^1/_8$ inch or $^1/_4$ inch in the construction of forms may cause discrepancies as high as 5 or 6 percent between the computed yardage and that measured in the forms. The error may be higher because of overrun on an irregular subgrade or failure to allow for wasted concrete. Errors in specific gravity of ingredients and in batch weights will cause errors in yield computations.

Discrepancies between computed yield at the batching plant and the volume of concrete measured in the forms are sometimes a source of disagreement between the ready-mix concrete producer and the contractor. A 4 percent error in 100 cubic yards of $60 ready-mixed concrete means $240 that someone will have to account for, one way or another. Therefore, it behooves all persons concerned to see that the many sources of error are minimized.

Cracks and Blemishes

Chapter 6

By the time the reader reaches this chapter he or she may have the feeling that we have been able to separate strength, durability, cracking and other properties of the concrete into separate compartments unrelated to each other. What concerns strength, for example, has nothing to do with cracking. This, however, is far from true. All properties of concrete are interrelated, and abnormal weakness in one will be reflected in another. For instance, in this chapter we will consider cracks and other blemishes. We cannot, however, avoid consideration of strength, durability, volume changes and so forth, because cracking and blemishes are a result or symptom of other deficiencies of concrete. These deficiencies may be the result of faulty design, the use of inferior materials, or improper construction practices.

CRACKING

It has been said that cracks in concrete cannot be completely prevented but that they can be controlled. Rare indeed is the concrete without at least a few shrinkage cracks of some kind. The reaction of most persons when they see cracked concrete is to assume that the concrete is defective. Certainly, cracks can be an eyesore, and occasionally they are an indication of weakness in the concrete. Cracks do not just happen; like accidents, they are caused. It is our purpose to find the causes, then see what we can do to prevent the cracks or at least minimize the cracks from marring our concrete.

6.1. Why Concrete Cracks

Concrete cracks because the tensile force pulling the concrete apart is larger than the tensile strength of the concrete at the time that load is applied. In other words, the concrete is not strong enough to resist. The load can vary from a very small stress that causes plastic cracking of fresh concrete to an applied load on a large, mature concrete structural member. Restraint is usually involved. If there were no restraint, all parts of the concrete member would be free to move, which would accommodate the stresses on the concrete, and there would be no cracking. This, however, is almost never possible in actual construction. There are many causes of cracking. Actual cracking usually depends on more than one factor.

Briefly, the causes of cracking are as follows:

1. Volumetric changes in the concrete resulting from changes in moisture content, temperature variations or internal reaction between components of the concrete;
2. Stress in the concrete as a result of loads applied to the concrete; and
3. Disintegration of the concrete by such outside forces such as freezing and thawing or aggressive solutions that first indicate their appearance by cracking.

Another way to classify cracks is to distinguish between cracks that occur in the fresh concrete before it hardens and those that develop in the hardened concrete. This is the classification we will follow.

For additional information on concrete slab surface defects, including causes, prevention and repair, the reader is referred to Reference 6.1.

CRACKS BEFORE HARDENING

In the previous chapter, we learned how concrete can shrink or lose volume while it is still soft or plastic because of a loss of mix water from the concrete. There are other causes of cracks at this time that we must also consider. Let us first give our attention to plastic shrinkage.

6.2. Plastic Shrinkage Cracking

Cracks in the fresh concrete, sometimes called preset, green or plastic cracks, are cracks that occur while the concrete is still in a plastic state, before the cement has set. They are very erratic in their appearance and occurrence, developing within one or two hours after the concrete has been placed, starting suddenly about the time the water sheen disappears from the surface. The cracks vary in width from fine hairlines to an eighth of an inch, and in length from an inch to several feet. See Figure 6-1. Depth is seldom more than 2 inches, although the cracks may extend through a thin slab. They may or may not be connected, and ordinarily do not extend to the edges of the affected member. Usually they have no definite pattern, although they may be generally at right angles to the long axis of the slab. They are entirely different in appearance from cracks that occur in hardened concrete.

Figure 6-1
Plastic shrinkage cracks caused by rapid loss of mix water while the concrete is still plastic. (Courtesy of PCA)

The latter are sharp-edged and clearly defined, sometimes breaking through aggregate particles, while plastic cracks follow around the aggregate particles and do not have the appearance of a clean break as do the cracks that form in the hardened concrete. They will frequently follow reinforcing bars or other embedded materials such as large aggregate particles. They are especially likely to occur in slabs, in the tops of walls and beams, and in footings on sandy soil. They are not progressive; that is, once the crack has formed, the stress in the concrete is relieved and no further cracking develops from this cause. These cracks are unsightly but are usually not dangerous, although they are a possible foothold for later trouble.

There are many factors involved in plastic shrinkage, and it is difficult to know in advance whether cracking might occur on any particular job. A large area of concrete, as on a floor or pavement, is much more liable to have plastic cracking than concrete in a form. The weather is very important. Concrete placed on a hot, windy day is more apt to crack than concrete placed in more moderate weather, especially if the concrete itself is hot. Methods that help to prevent or minimize plastic shrinkage cracking, in addition to the shrinkage control measures mentioned in Chapter 5, are as follows:

1. Keep the concrete temperature low. This might include sprinkling the aggregate stockpiles and using cool mixing water, even including ice during hot weather.
2. Keep mixing to a minimum. As soon as the required time or number of mixer drum revolutions is reached, stop the mixer or put it on agitation.
3. Keep the job moving so truck mixers do not stand around waiting to unload.

When cracks develop during finishing, the finisher can sometimes close and seal the surface by going over the cracked area with a float. Rapid evaporation is one of the most important factors in plastic cracking. The rapid evaporation (and consequent shrinkage) produces a tensile stress in the concrete and, because the concrete has no strength, it cracks. If the shrinkage can be delayed until such time as the concrete has some strength, the cracks may not form; if the concrete is slow in setting, the cracking might be worsened. This could happen if a retarding admixture is used in the concrete. The retarder keeps the concrete in a plastic condition longer, and therefore there is more opportunity for the cracks to develop. Admixtures containing calcium chloride should not be used during hot weather, as they sometimes contribute to cracking.

Shrinkage that results from segregation and bleeding causes settlement cracks in the tops of slabs, walls and beams. Well-graded aggregates and proper mix proportions including entrained air, with careful handling and placing, will minimize these cracks. High mixing, placing and curing temperatures are conducive to cracking.

The following case illustrates one example of plastic cracking. Pile sections 54 inches in outside diameter, 16 feet long, with a wall thickness of 4 $^1/_2$ inches were being made by the centrifugal spinning process. Twelve 1 $^1/_8$-inch-diameter longitudinal ducts for stressing wires were formed by means of steel rods inside rubber tubes attached to the end rings of the form. When the sections were removed from the spinning machine, it was observed that plastic cracks had developed on the interior of the sections, following along the ducts, that were due in part to slight movement of the rubber tubing during and immediately after spinning. Adjustment of the mix to provide a better grading served to lessen the frequency of the cracks.

In another plant producing the same type of piles, it was found that too-early removal of the rods and tubes caused longitudinal cracks inside the pile sections. Delaying removal of the rods for about 45 minutes eliminated most of the cracks.

6.3. Settlement and Movement

Early shrinkage is not the only cause of cracks in the fresh concrete. Sometimes shrinkage is blamed for cracks that may be due to other causes, such as settlement or movement of forms. Cracking may result from numerous causes, some of them obscure and difficult or impossible to identify. A perversity of concrete is that it sometimes cracks when cracks are not expected and fails to crack when cracks are expected.

Cracking of the concrete before hardening may result from movement of the forms, subgrade, reinforcing steel or embedded items; settlement of aggregate particles or reinforcing; false setting cement; or sagging or slippage of the concrete, especially on slopes. Cracks have been reported to occur even when the concrete surface is under water. Different sands and cements will affect their formation.

In Chapter 5 we saw how bleeding can result in settlement of the fresh concrete as the water bleeds to the surface of the concrete. The settlement of the concrete may be obstructed by reinforcing bars, items embedded in the concrete, or large aggregate particles, causing cracks in the fresh concrete over such obstructions. See Figure 6-2.

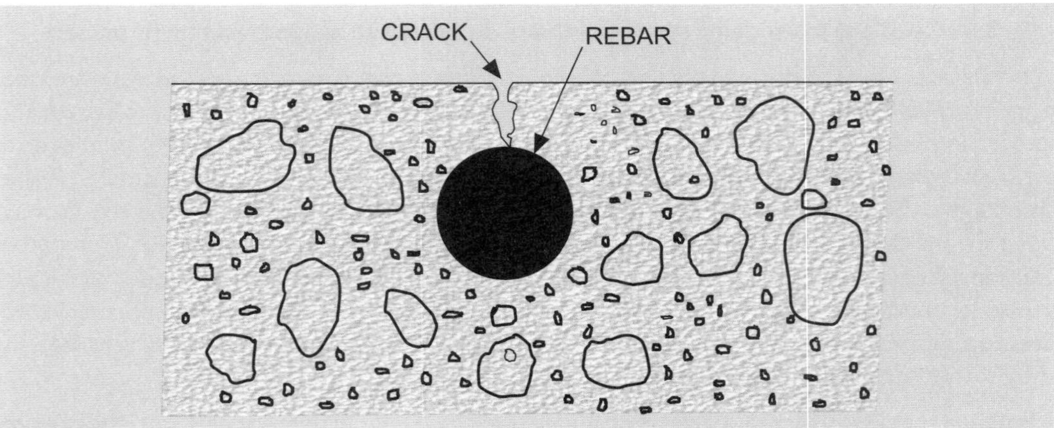

Figure 6-2
Settlement of the fresh concrete, especially of high- slump mixes, over a reinforcing bar too close to the top of a slab may cause a crack

Forms may slip or bulge if they are not adequately braced or if they are made of material that is not strong enough to hold the concrete. A subgrade that yields or settles when the concrete is placed on it will cause cracks. See Figure 6-3. Conditions that produce a yielding subgrade are lack of proper compaction of the soil, the presence of soft organic material (roots, sod, etc.), mud, or irregular surface that causes thin and thick areas in the slab.

Figure 6-3
Settlement of a soft subgrade or foundation will cause the concrete to crack.

Certain soils expand when they absorb moisture, and a slab laid on such a subgrade is apt to crack as the soil absorbs the moisture from the fresh concrete. Problem soils should be pointed out in the soils report, and proper measures taken to treat or remove such soils before concrete placing starts.

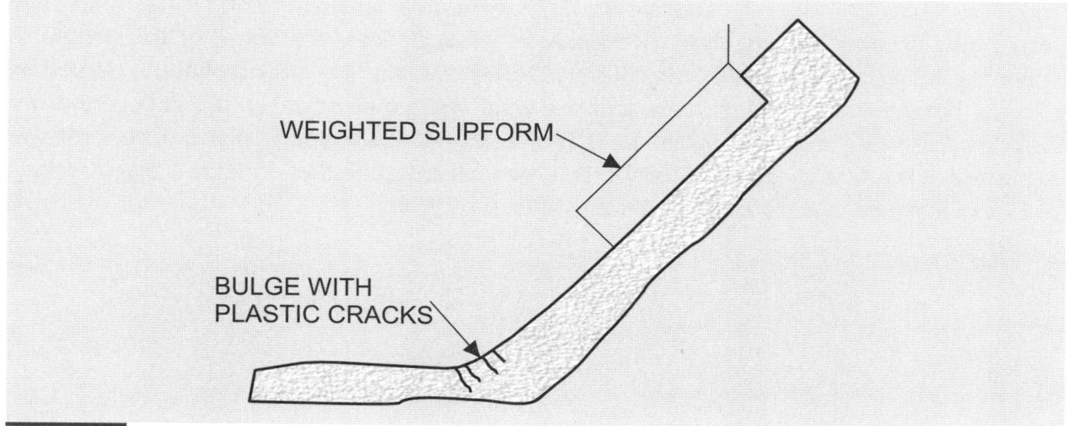

Figure 6-4
Bulging and cracking may develop near the lower end of the sloping slab, especially if the slump is too great. A short slip form pulled up the slope to confine the fresh concrete will be of some help.

Figure 6-5
Plastic shrinkage cracks will develop if reinforcing bars are too close to the surface.

When placing concrete on a slope without a top form, placing should begin at the bottom of the slope. See Figure 6-4. Even if the concrete is properly placed, it may bulge at the bottom if it had too high a slump.

Placing reinforcing bars too close to the top of a slab can cause a crack. See Figure 6-5.

Settlement cracks may occur at or near the soffit of a beam when the beam and column are placed monolithically (Figure 6-6), especially if overly wet or soft concrete is used, or they may develop at the top of a column or wall at the junction with a slab. The remedy in this case is to wait about two hours after placing the beam, wall or column before placing concrete in the slab, using concrete with the minimum slump possible.

CRACKS AFTER HARDENING

After the concrete has hardened (that is, after the hydration of the cement has progressed to the point that the concrete has some strength), there are many forces at work trying to damage the concrete. Drying shrinkage, alternations in temperature and moisture, chemical reactions within the concrete, aggressive environment, movement and settlement, freeing and thawing, overloading and accidents, are some of the problems concrete has to contend with. Many of these forces are out of the control of the person on the job, but we will consider them, and see what we can do to make the concrete more resistant to them. Basically what we need to do is make good, durable concrete as explained in the first chapters of this book. If we make that kind of concrete, then cracking will be eliminated in most cases and controlled in others.

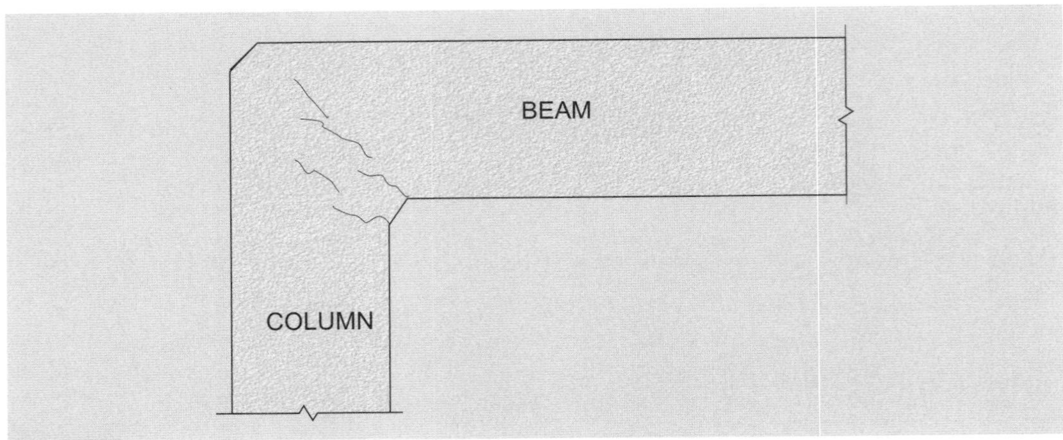

Figure 6-6
Placing the column and beam monolithically, without a delay at the soffit, may be the cause of cracking. Extra reinforcement helps to minimize the cracks.

6.4. Drying Shrinkage Cracks

Rapid loss of water from new concrete after it has hardened causes shrinkage stresses in the concrete. The longer the moisture loss can be delayed, the more strength the concrete will have for resisting the shrinkage. Delaying of the moisture loss is accomplished by curing the concrete.

If the concrete was unrestrained, cracks would not develop. However, all concrete is restrained in some manner—adjacent portions of the structure, such as the usual arrangement of slabs, columns and beams in a building, each restraining or interfering with free movement of adjacent members; the subgrade under a pavement or slab-on-grade; the presence of reinforcement—all of these prevent the concrete from contracting (shrinking) or expanding (swelling) freely. Consequently, the concrete cracks. See Figure 6-7. Drying shrinkage is affected more by the amount of water in the concrete (unit water content) than anything else. Other important factors are the composition of the aggregate and the duration and quality of curing applied to the concrete. Aggregate size, mix proportions and richness of mix (cement content) affect drying shrinkage because of their influence on the amount of water in the mix. See Section 5.1.

At a shrinkage of 0.05 percent, which is not an excessive amount for normal concrete, the concrete will shrink about $^1/_6$ inch in 10 feet of length, and if the concrete is restrained it will crack. For a strain (shrinkage) of 0.05 percent, a tensile stress of 1500 psi develops in the concrete, which is from three to five times the tensile strength of the concrete, depending on its age and the mix. The load (stress) greatly exceeds the strength, so the concrete cracks. Fiber reinforcement, however, may provide greater tensile strength. See Chapter 22.

Figure 6-7
Drying shrinkage cracks in slab-on-grade. See Section 16.2 discussion on the use of contraction joints in slabs on ground to control drying shrinkage cracking. (Courtesy of PCA)

6.5. Other Post-Hardening Cracks

Besides drying shrinkage, many other forces—some physical, some chemical—are attempting to cause cracks in the concrete. Regardless of the source of the tensile force, actual cracking occurs when the tensile force exceeds the tensile strength of the concrete, as previously pointed out. Higher compressive strength of the concrete is accompanied by higher tensile strength; hence, anything that improves or raises the compressive strength will produce concrete that can resist cracking more effectively.

General. After the concrete has hardened, it is subject to many forces that stress the concrete and cause cracking. Some of the forces, as discussed in the previous section, are due to temperature changes such as might occur in winter when concrete that has been kept warm is suddenly cooled, perhaps by the application of cold curing water. Changes may occur within the concrete itself: chemically, by such activities as reaction between aggregates and high-alkali cement, or chemical decomposition of deleterious aggregate particles; mechanically, as by expansion of aggregate particles caused by moisture absorption; or by freezing of water in pores of the aggregate. Loading may exceed design values, or the design may have been erroneous. Expansion or contraction joints may have been omitted or wrongly placed, as in the case of a low parapet wall that developed transverse cracks about a foot apart as a result of failure to provide weakened plane or dummy joints. Stress concentrations may occur at corners, changes in section or around openings. Impact may occur through accident.

Movement of part of a concrete building or other structure resulting from settlement will cause cracks in walls and slabs, or at external columns at the floor level. Sometimes a wall may crack because of settlement of the slab on which it is placed, the slab being relatively flexible compared to the wall. Moisture and temperature variations, or too much reinforcement congested in a small area, may cause cracks.

If settlement is unavoidable, heavy bases and columns should be constructed first and then, after settlement takes place, the remaining concrete in the comparatively thin walls may be placed. Adequacy of foundation material should be investigated beforehand to forestall such difficulties.

Diagonal cracks at corners of door and window openings can be controlled by the use of sufficient reinforcement. It is advantageous to place concrete to the top of such openings, then wait about two hours before continuing with the placement, permitting the concrete in the lower portion of the wall to shrink and settle. Concrete should be placed with the minimum slump practicable. A horizontal construction joint located at the level of the top of the opening will aid in preventing cracks. See Figure 6-8.

Structural and Accidental. Structural cracks can be defined as those cracks that come as a result of inadequate design, overloading or unanticipated movement of all or part of the structure.

The designer uses a factor of safety when designing the structure so the concrete will not be expected to carry a load that closely approaches the strength of the concrete. If a mistake is made in design, it is possible for the loads that develop in parts of the structure to cause cracking. An abrupt change in size of a section can be a source of cracking. Occasionally at a point of this kind there is apt to be a concentration or congestion of reinforcement that not only makes proper placing of the concrete difficult but also causes cracks to develop. A structural crack can develop as a result of misplacement of reinforcement. Location of the steel within the concrete is critical, and it must be placed exactly as shown on the plans.

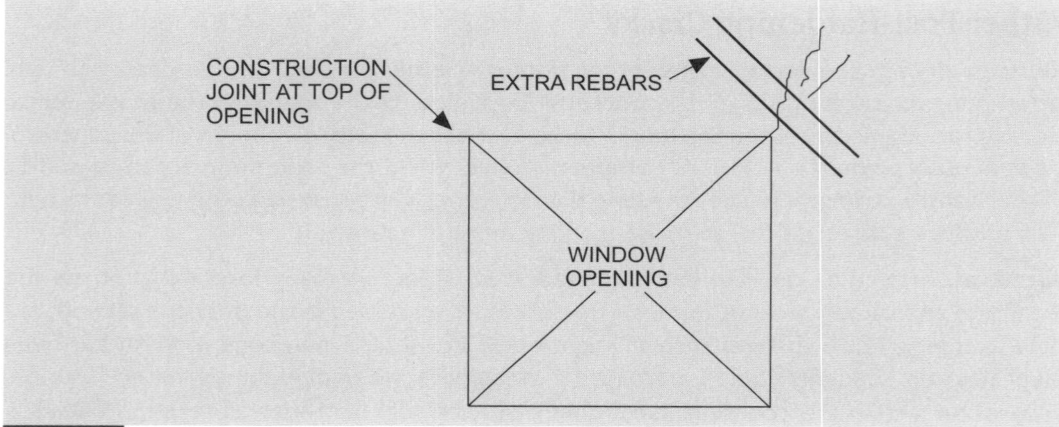

Figure 6-8
By delaying about two hours after placing concrete to the top of the opening before continuing with the wall above the opening, there is less likelihood that a crack will develop. Diagonal reinforcing bars placed as shown reduce cracking.

The only loading condition under the control of the person on the job is loading of parts of the structure during construction. Because the concrete is still green and does not have its full strength, it cannot be expected to carry heavy loads. Examples of poor practice are trucks, cranes or forklifts on recently placed slabs, and storing heavy bundles of reinforcing steel and other materials on floors and beams. Advice of the structural engineer should be obtained when considering such activities.

Movement of part of the structure can be a result of faulty design, overloading or foundation failure. Any of these situations can develop during construction. For this reason, the appearance of any crack should be investigated so as to determine its cause and provide a course of action. Early detection of a problem of this sort can frequently avert a serious failure.

Accidents during construction can cause cracks or more serious failure. Dropping loads, swinging crane loads against new construction, movement of shoring and falsework can all result in cracking of some kind. Acts of nature, such as high winds, earthquakes or accumulations of water from heavy rainstorms might overstress parts of the structure and cause cracks. Any unusual cracking should be brought to the attention of the structural engineer.

Reactive Aggregates. The deterioration caused by the reactions of certain aggregates in the concrete is discussed in Section 4.4. The cracking that is evidence of this type of reaction consists of random or map cracking on a fairly large scale—the cracks opening up, in severe cases, to over $^1/_2$ inch wide, but seldom as deep as 18 inches—resulting from abnormal expansion of the concrete, especially internal. See Figure 6-9. The cracks and voids are filled with a gelatinous or amorphous deposit, which also appears in cracks on the surface. The concrete, when broken, is lifeless and chalky, and individual aggregate particles, upon close examination, will be found to be altered or coated on the surface because of the reactivity. Deterioration is progressive and will continue as long as moisture is present. Cracking commences at any time from a few weeks to a year or more after the concrete was placed, the time depending on the percentage of reactive particles in the aggregate, alkali content of the cement, amount of moisture present in the environment, and temperature.

Figure 6-9
Deterioration of concrete from alkali-silica reaction. See also Figure 4-13. (Courtesy of PCA)

Rusting of Steel. Moisture can enter the concrete through small cracks, honeycomb or unsound construction joints and cause rusting of the reinforcing steel. This is especially serious in the presence of saltwater or certain contaminated atmospheres. Rusting is accompanied by an increase in volume with consequent disruption and cracking of the concrete. See Figure 6-10. Prevention is accomplished by proper design so that structural cracks will not form. Structures should be built of good dense concrete with at least 2 inches (3 inches in seawater) of cover over the reinforcing steel. Rock pockets and honeycomb can be prevented by good mix proportions and thorough consolidation by means of vibration. Construction joints must be properly cleaned. If rock pockets, honeycomb or sand streaks occur, they should be carefully patched. In some cases, waterproofing of the surface may be necessary. The measures that will limit or prevent cracks are described elsewhere in this book.

Temperature Effects. Sudden changes in temperature can stress the concrete and cause cracks. This is called thermal shock. Applying cool curing water to hot concrete surfaces

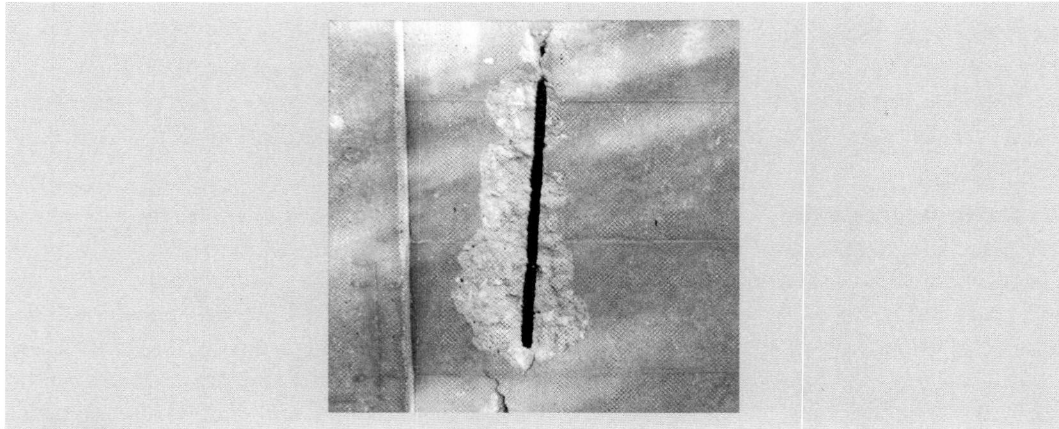

Figure 6-10
Rust on reinforcing placed too close to the surface, causing cracking and spalling of the concrete. (Courtesy of PCA)

produces a fine pattern cracking or crazing. This is especially prevalent on concrete in arid regions when the concrete is permitted to become dry and hot during intermittent applications of cool curing water. A similar condition exists when cold water is applied to freshly stripped warm concrete, particularly during cold weather. For best results, the temperature difference between the concrete and curing water should not exceed 25°F. If artificially heated concrete is suddenly exposed to very cold air by removal of the forms, crazing may result even without the application of water.

A case is reported in which vertical cracks about 8 feet apart appeared in a thin wall when the forms were removed. Cause of the cracks was traced to a combination of the use of excessively thick insulation on the forms for protection against freezing and restraint of the wall concrete by steel dowels and bond with the foundation concrete. Although the insulation was sufficient to protect the concrete to a temperature of 25°F below zero, actual temperatures never got below 25°F above zero. Temperature of the concrete, which at the time of placing was 75°F, was over 100°F when the forms were removed, whereas air temperature was 35°F to 40°F.

There have been cases in which anchor bolt holes, left exposed to weather, filled with water. Freezing of the water exerted sufficient force to crack the concrete. This happened on the pier cap of a bridge under construction.

We know that concrete shrinks when it cools and expands when it warms up. If the concrete is restrained, cracks will form. Shrinkage of a concrete member will stress the connections with adjacent members and cause distortion of the structure or cracking in the concrete. In a like manner, expansion can be a source of cracks. Although the expanding member itself will not crack, the change in length will affect and deform abutting construction. See Figure 6-11. Connections and supports can be designed to compensate for these variations in length.

Figure 6-11
The joint that had been provided in this structure was constructed with a lip at the top edge, which caused the concrete to break when the joint closed up.

Frost Action. Weathering is the action of frost, or cycles of freezing and thawing, and is manifested in its early stages by cracking. Weathering cracks develop in pavements, curbs, walls, railings and similar exposed concrete as many fine, closely spaced cracks more or less parallel to the edges of the affected members. See Figure 6-12. As they develop, they can become filled with a dark deposit of calcium carbonate and dirt. They show up in inferior concrete at any time, up to several years after construction, the period of time depending upon the quality of the concrete and the severity of the exposure. This type of cracking is progressive and, unless checked, will result in disintegration of the member. The concrete is low in strength, the matrix being dull and chalky. Cracks result basically from failure to provide quality concrete when the structure was built. Lack of entrained air contributes to the failure.

Figure 6-12
Cracks caused by weathering. The upper portion of this abutment has disintegrated, showing the ultimate failure first manifested by the cracks. See also Figure 4-1.

In some geographical areas, cycles of freezing and thawing occur almost daily. Air temperatures rise into the thirties and forties during the day (actual surface temperature of concrete exposed to the direct sun can be considerably higher), then plunge into the twenties or lower during the night. If there is moisture in the concrete, this is one of the most rigorous exposure conditions possible. In colder weather, when the concrete freezes and stays frozen all winter, the exposure condition is not nearly as severe. If the concrete is completely dry, damage is slight or absent.

Crazing or Hair Cracks. A pattern of fine cracks results from shrinkage of the surface of the concrete slab relative to the interior of the mass. This cracking consists of many small, shallow, random cracks in every direction, usually following a roughly hexagonal or octagonal pattern, usually connected and several inches apart but sometimes closer together. See Figure 6-13. Close examination reveals such cracking to a minor degree on many concrete surfaces, especially troweled surfaces. Cracking of this nature on mature concrete is especially noticeable after the concrete surface has been moistened with water, then allowed to dry. The water being retained in the cracks after the rest of the concrete has dried accentuates the cracks and makes them clearly visible.

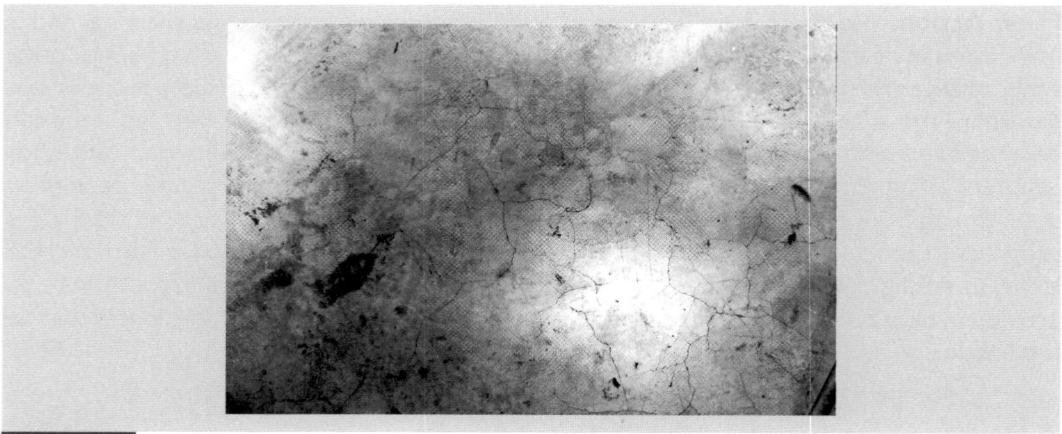

Figure 6-13
One example off fine craze cracking on a sidewalk. (Courtesy of PCA)

Figure 6-14
Spreading dry cement on a slab to dry up water standing on the slab is poor finishing practice and should never be permitted.

Minor cracking of this nature is not serious in itself, although it can lead to trouble in an especially severe exposure—for example, an ocean front structure exposed to freezing and thawing. Serious pattern cracking is caused by other activities and is discussed in other sections.

There are three general causes of crazing: improper and excessive finishing, rapid loss of moisture from the fresh concrete and temperature stresses. Irrespective of the basic cause, formation of cracks is predestined if overly fluid or rich mixes are permitted, and if the concrete is not properly cured.

Finishing faults include such practices as hastening finishing time by spreading dry cement on concrete that is too wet to trowel, or sprinkling water on concrete that has become too dry for proper finishing. These practices are the mark of an incompetent or lazy finisher and should never be permitted. See Figure 6-12. Overmanipulation during finishing, or troweling too soon after placing, causes a concentration of water and fines on the surface that creates a weak, soupy coating of laitance that will crack and peel. The amount of finishing necessary to produce the required surface should be kept to a minimum. Screeding, darbying and bullfloating may be done soon after placing, but floating and troweling should be delayed until the surface moisture disappears; that is, there should not be any bleed water on the surface.

SOME INDIVIDUAL PRACTICES

6.6. Slabs on Ground

Cracks that develop in slabs on ground usually owe their existence to careless construction practices or insufficient maintenance.

Transverse cracking and spalling near doweled joints in slabs will occur if joint dowels are carelessly placed. Dowels must be set parallel to the subgrade and centerline. See Figure 6-15. Devices for holding dowels prior to concrete placement must hold the dowels accurately in place to prevent dowel displacement when concrete is placed around them, yet be so designed as to permit subsequent slippage of the dowel in the hardened concrete. Dowels should be coated with a lubricant. Cracking at joints may also be caused by shifting of the slabs.

Random transverse cracking may be caused by subgrade failure, or frictional resistance of the subgrade to movement of the concrete resulting from volume change of the slab.

Longitudinal cracks result from differential settlement of the slab. This may occur as a result of softening of the subgrade by infiltration of water under the edge of the slab, the basic cause being poor drainage. Solid material entering a joint, or ice filling the joint, may also be the cause of a crack, especially if it begins at a transverse joint. If the filler in a preformed transverse joint does not reach the slab edge, leaving a concrete plug at the end of the joint, a crack will result. The same situation exists if a sawed joint does not extend all the way to the slab edge.

A corner crack is a short diagonal crack extending from a transverse joint to a longitudinal joint or to the slab edge. It is caused by solid material in the transverse joint, subgrade failure or weak concrete.

Figure 6-15a
Improperly aligned and supported dowels prior to placement of concrete. (Courtesy of PCA)

Figure 6-15b
Slab cracking along a line at the ends of the dowels. (Courtesy of PCA)

6.7. Mass Concrete

Concrete in structures consisting of a large amount of concrete in huge blocks or masses is called mass concrete. See Figure 6-16. Examples are dams, massive bridge piers and large footings. In these large massive structures, cracks may be caused by heat generation in the concrete as hydration takes place. The surface, being cooler, will be subject to tensile stress as the interior gets warmer. A large temperature difference in and near the surface of a dam, or a sudden cooling of the surface, will cause cracks to appear. If the interior temperature of the structure is still rising as a result of cement hydration, these surface cracks may enlarge and penetrate deeper into the structure. These are entering wedges for many physical and chemical activities such as seepage water and frost action.

Modified or low-heat cement should be used in concrete for massive structures of this kind, keeping the cement content as low as possible. By using pozzolans as a replacement for part of the cement, concrete for the interior of large dams may contain as little as 188 pounds (two bags) of cement per cubic yard. Entrained air is beneficial. The concrete should be as cool as possible at the time of placement (below 50°F at all seasons), use being made of such expedients as cooling the ingredients by sprinkling the coarse aggregate with cool water (taking care to permit the aggregate to drain before use), placing concrete during cool weather, using ice in the mixing water, early form removal and limiting the rate of placement so that the top of a lift is left exposed for as long as possible. In very large structures such as dams, pipe coils through which cool water is circulated are placed in each lift. Water circulation is started when concrete placing is started and continued until a satisfactory temperature is reached.

Figure 6-16
Mass concrete columns for San Francisco-Oakland Bay Bridge. Note the amount of reinforcement required in high seismic risk areas! (Courtesy of PCA)

The following methods or combinations of methods have been used for reducing the temperature rise in mass concrete and, consequently, the cracking:

- Use of low-heat cement.
- Minimum cement content.
- Use of pozzolanic material.
- Limitation on the rate of placement so that a greater part of the heat of hydration is lost from the top surface of the lift during construction.
- Placement of concrete during cold weather so that the heat of hydration will raise the temperature to, or only slightly above, the final temperature.
- Precooling concrete ingredients to reduce placing temperature of the concrete.
- Introduction of fine ice into the mix water.
- Early removal of forms.
- Use of steel forms to facilitate loss of excess heat from the surfaces.
- Artificial cooling, begun at the time or soon after the concrete is placed, not only reduces maximum temperature rise but also cools concrete to any desired temperature within a short time and permits grouting of contraction joints within a reasonable time after concrete placement.

6.8. Precast Concrete

Construction units that are made at some central manufacturing plant, then hauled to the site and installed in the structure, are called precast concrete. Concrete block and pipe can be called precast concrete, but what we are more concerned with here are wall panels, floor and roof slabs, girders, columns and other structural units. All of the instructions for making good concrete apply to precast as well as cast-in-place concrete, plus some special conditions that apply only to precast. See Figure 6-17. Refer to Chapter 20.

Figure 6-17
Precast wall units being lifted into place.

Cracking of precast concrete of any type can be minimized if units are designed properly, avoiding variable sections and providing adequate reinforcing. Molds should present a smooth, uncluttered surface to the concrete, with the necessary draft or taper to facilitate stripping of intricate details. Molds should be thoroughly cleaned and oiled after each use, using a form oil that is especially compounded for this use. Waste molds should be painted with shellac or other sealer.

Precast concrete is usually relatively rich concrete, and care is necessary to avoid shrinkage cracks. Water content of the mix should be as low as possible, and the units should be protected from drying out before curing has been completed. Castings should be removed from the molds as soon as possible. In the case of steam curing, this may be as soon as eight hours after making. For castings cured at normal atmospheric temperatures, a period of 16 to 24 hours is usually adequate. This is affected by the temperature, admixture, type of cement, richness of mix and kind of casting being made. Items such as packerhead pipe, tamped pipe and blocks, which use an earth-moist mix, are placed in the curing area and stripped immediately after casting. A fine fog should be used to prevent loss of moisture from the concrete until such time as the concrete is strong enough to withstand normal curing procedures. In certain cases calcium chloride may be used in the concrete to hasten the stripping time.

Sudden changes in temperature must be avoided, such as might occur by sprinkling cold water on castings recently removed from the steam chamber. Where additional curing is desired after the initial steaming, the castings can be returned to the steam room, or allowed to cool, after which water curing can be commenced. If the coating is not detrimental to subsequent processes, such as bonding to other concrete, painting or other treatment, membrane curing may be used. Castings should have a minimum of 12 hours of curing in wet steam between 130°F and 150°F commencing two to four hours after the concrete has been placed. They should be handled carefully at all times, especially during the period before final curing has been completed.

All cracking of precast units is not necessarily serious. Small surface cracks and hair cracks are not cause for rejection of units. They will frequently heal, especially in the presence of moisture. More serious cracking has to be evaluated in the light of the type of unit and exposure conditions. In many cases, a crack 0.01 inch in width would be acceptable, provided corrosion of the reinforcement is prevented by sealing the crack with slurry or epoxy resin, and provided that the unit is not weakened structurally by the crack.

More serious cracking would probably call for rejection of the unit. Seriousness may be judged either by the number of cracks in the unit, or by the width and extent of one crack. Basis for acceptance or rejection will be given in the job specifications.

6.9. Crack Control in Slabs on Ground

General jointing practices for slabs on ground are discussed in detail in Chapter 16. In this chapter we will consider joints in slabs from the viewpoint of crack control only. The objective of making joints in concrete slabs is to control cracking by providing for movement of the concrete as it expands and contracts. Both isolation joints and contraction joints are used to control cracking in floor slabs on ground, driveways and sidewalks.

Isolation joints (sometimes called expansion joints) should be provided whenever concrete abuts previously placed concrete, such as a previously placed slab, wall or footing. They permit both horizontal and vertical differential movements at abutting parts of a structure or slab. An isolation joint consists of premolded material $^1/_4$- to $^1/_2$-inch thick and as wide as the concrete is thick. Isolation joints are used also where a sidewalk meets a driveway, curb, another walk, building or any other structure that will interfere with free movement of the concrete.

Contraction joints (sometimes called control joints) provide for movement in the plane of the slab and induce cracking caused by drying and thermal shrinkage at preselected locations. If contraction joints are not used, or if they are too widely spaced, random cracks will occur when drying and thermal shrinkage produce tensile stresses in excess of the concrete's tensile strength. Transverse contraction joints in driveways and sidewalks should be made at intervals not greater than the width of the slab. Contraction joints in continuous floor slabs should divide large slab areas into relatively small rectangular panels. Panels should be as nearly square as practical. Contraction joints in slabs on ground are commonly made by a grooving tool that makes a groove in the freshly placed concrete, or by sawing a continuous slot in the top of the slab after the slab has hardened. The idea of a contraction joint is to decrease the slab section along the joint to provide a weakened plane where the concrete will crack when it contracts. Contraction joints, whether sawed, grooved or preformed, should extend into the slab to a depth of one fourth the slab thickness.

For further information on joint spacing and construction, refer to Chapter 16.

6.10. Crack Control in Foundation Walls

Not unlike slabs on ground, contraction joints should also be provided in plain concrete and lightly reinforced concrete foundation walls to control cracks. Properly spaced contraction joints will eliminate random shrinkage cracking and permit differential movements in the plane of the wall. Contraction joints in walls should be spaced not more than about 20 feet apart. In addition, contraction joints should be located where abrupt changes in thickness or height occur, and within 10 to 15 feet of corners. The thickness of the wall at a contraction joint should be reduced by at least 20 percent, preferably 25 percent. In lightly reinforced walls, half of the horizontal bars should be cut at the joint. Care must be taken to cut alternate bars precisely at the joint. At the corners of openings where contraction joints are located, extra diagonal or vertical and horizontal reinforcement should be provided to control cracking and keep the joint from opening wide.

Contraction joints in walls are made by attaching wood, metal or plastic strips to the inside faces of the formwork. The exterior side of the joint should be caulked with a chemically curing thermosetting joint sealant such as polysulfide, polyurethane or silicone that will remain flexible after placement. After the groove has been carefully caulked, a protective cover such as a felt strip 12 inches wide should be placed over the joint below grade. Some builders install a waterstop at contraction joint locations for added protection. Another method is to cut the contraction joints into the wall with a masonry saw. This should be done within a few hours after stripping the forms to prevent random cracking caused by drying shrinkage. With this method, a waterstop should be used.

Jointing details for crack control in foundation walls and slabs are summarized in Figure 6-18.

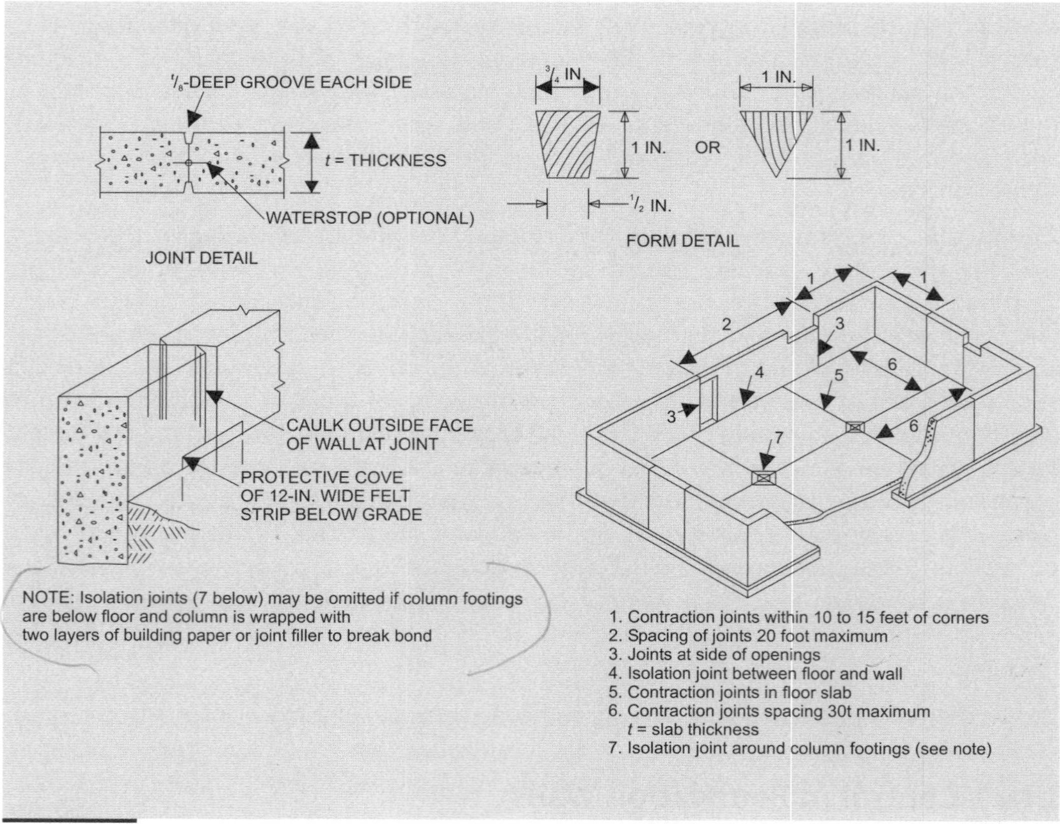

Figure 6-18
Location and spacing of isolation joints and contraction joints for foundation walls and slabs.

SURFACE BLEMISHES

Concrete may be cast against molds or forms of some kind, in which case the resulting surface, after removal of the mold or form, is called formed concrete. The surface of the concrete not cast against a form, which is worked with tools of some kind, is called (naturally) unformed concrete. The defects that can mar the surface of the formed concrete are different from the ones that occur on unformed concrete. In addition to this distinction, we must remember that there is more than one condition influencing the diagnosis and treatment of blemishes on the surface of the concrete. It is not always easy to determine the exact cause of imperfections and blemishes, as any one flaw may have several causes, and there may be as many or more possible methods of relief or repair.

Generally, defects may be classified into three groups, based on the method of their origin:

1. Blemishes that appear as a consequence of the use of inferior materials that cause abnormal activity within the mass of the hardened concrete. Examples are popouts and certain cracking.
2. Injury that appears as a result of outside forces attacking the concrete. Examples are sulfate or chemical attack.
3. Flaws resulting from poor construction practices. Examples are honeycomb, and unsatisfactory alignment of the concrete caused by inferior formwork.

6.11. Soft Surface

We expect the surface of concrete, either formed or unformed, to be hard and firm upon completion of curing. Anything less is unacceptable. Surface softness can occur on any kind of concrete surface. Dusting is evidence of a soft surface, and is defined as a surface condition that can be either:

1. A soil, chalky condition that occurs occasionally on formed concrete when the forms are removed, or
2. A condition of a concrete floor in which the wearing surface is so weak that the concrete wears off and dusts easily under any kind of traffic.

Formed Concrete. A soft surface on formed concrete can be identified by rubbing the hand over the surface. A soft surface will rub off or leave dust on the fingers.

One cause that has been identified a number of times is the use of new plywood form lining. On one job, a small panel was cast in the laboratory. One side was against plywood from stock and was known to be trouble free. The other side was cast against two pieces of plywood from the job; one was new, and one had had concrete cast against it. Otherwise, the job plywood samples were the same. All plywood was treated with job form oil four hours before the concrete was placed. When the form was stripped at 48 hours, the concrete cast against the laboratory plywood was hard; that cast against the job-used plywood was sound, with very slight evidence of dusting; that cast against the new plywood was dusty.

Experience has shown that the dusting surface will become hard if it can be kept moist for several days. Practical difficulties become apparent when this is attempted on the job, but it appears to be the only solution.

Job experience indicates that wood form lining is a potential cause of dusting whenever it is exposed to the weather for a long time before concrete is cast against it. Factory-treated plastic-coated plywood does not seem to give trouble, and plywood that has had concrete cast against it once gives no problem either. Cases have been traced to the use of too much form oil. Some types of hardboard or pressed board, when new, contain a constituent that causes a soft dusty surface when the board is used as a form liner, unless the board is sealed with shellac or a plastic coating.

New form lumber sometimes contains an excess of tannin, especially if the lumber has not been dried thoroughly. Sappy areas may be especially troublesome, even on dry lumber that has been oiled. These sappy areas should be painted with shellac, whitewash or a plastic form coating, because the tannin will cause a set failure on the surface of the concrete unless a barrier is provided.

When using a curing (sealing) compound, particular care should be taken to make sure that it is applied at the right time. If the concrete surface has become too dry before application of the compound, a soft surface is liable to result.

Unformed Concrete. Dusting as evidence of a soft surface is especially noticeable and offensive when it occurs on a concrete floor. In some manufacturing plants, the dust generated by traffic on the floor, even only foot traffic, can be hazardous to the equipment or product. Dusting can be avoided by proper attention to the use of good materials and correct construction methods described in this book. Good materials and good concrete, however, can be ruined by sloppy and careless finishing and curing. Instructions for finishing and curing are given in Chapters 16 and 17.

Carbon dioxide from unvented salamanders or heaters used for heating an enclosure during cold weather will react when it comes into contact with the surface of plastic concrete, impairing hydration and causing a soft surface. An immediate and complete sealing of the concrete by the application of a resin-based liquid curing compound effectively protects the concrete if it is impossible to vent the heaters outside the enclosure or use a source of heat that does not generate carbon dioxide. After about 24 hours the effect of carbon dioxide is negligible, or it may actually result in a harder surface. This does not mean that subsequent exposure to carbon dioxide of already softened concrete will restore the surface. It will not.

One of the most prevalent, and also the most needless, causes of dusting is incorrect finishing, including premature floating and troweling. Premature floating, in particular, mixes excess surface water with fines brought to the surface, producing a weak paste that has very little strength and frequently results in laitance. The surface water is bleed water, and there will be more of it if the slump is too high. The way to prevent this problem is to keep the slump at or below 4 inches and wait until the bleed water disappears before any troweling is attempted.

Lack of curing is another cause of soft surface. The concrete should not be permitted to dry out until after the specified curing period has elapsed.

Dirty aggregates can cause a soft surface. An excess of clay and silt mixes with the cement and fine sand on the surface and results in dusting.

See the discussion on floors in Section 6.14.

6.12. Pits and Voids

Small pits, bubbles or voids, sometimes called bugholes, often appear in formed concrete surfaces. See Figure 6-19. These voids are about half an inch, more or less, in diameter.

Occasionally, they are covered with a thin skin of dried paste that breaks away under slight pressure of the fingers, exposing a void that had previously been invisible. These voids may be the result of air voids, or small concentrations of free water. They are nearly impossible to prevent on vertical formed surfaces and are sure to occur on surfaces that are placed against forms that slope inward over the concrete.

There has been much controversy about the effect of entrained air on these defects; suffice it to say that they are no more numerous on air-entrained concrete than on plain concrete. Concrete placed against tight forms, such as steel or plywood, is bound to have at least a few of these small voids. They do not impair the structural integrity of the concrete, and in many instances their presence is not unsightly.

Prevention. Sticky, oversanded mixes should be avoided. Aggregates should be well graded with adequate, but not excessive, fines in the sand. Concrete should be of a plastic, workable consistency, neither too fluid nor too stiff, with a slump of 2 to 3 inches, or at most 5 inches, for any normal placement.

Observance of the following suggestions will help to minimize formation of these troublesome voids:

1. Do not place concrete rapidly in deep lifts. Place it in shallow layers, from a few inches to a foot in depth depending on the type of structure, taking care to vibrate each layer.
2. When placing against a form sloping inward over the concrete, vibrate each layer only enough for thorough consolidation. Slightly less vibration than is recommended for vertical surfaces should be adequate.
3. Sometimes a change in form oil will lessen the frequency of bubbles.
4. When placing against a vertical form, be sure to vibrate thoroughly, applying slightly more vibration than usual.
5. Vibrate the forms by means of properly designed form vibrators attached to the forms, permitting them to operate only on newly-filled forms, not on empty areas or on old concrete.
6. If the area is accessible, spade the concrete next to the form, using a $^{3}/_{8}$-inch rod bent into the shape of a stirrup about 6 inches across. Work this stirrup up and down briefly next to the form while vibration is in progress.
7. Make the forms of tongue-and-groove or shiplap lumber with small cracks through which air and water, but not mortar, can escape.
8. If the additional cost can be justified, use absorptive form lining or the vacuum process. However, these are expensive expedients and should be used only where an unblemished surface is imperative.

Filling of Voids. If, for the sake of appearance, it is considered necessary to fill the voids, the surface can be either sack-rubbed or stoned. These do nothing to improve the concrete structurally but, when properly done, do improve the appearance. Either treatment should be applied as soon as possible after the forms have been stripped. The greener the concrete when treated, the better chance there is that the treatment will be permanent. It is imperative that the work be done carefully; otherwise the surface will be soft and will peel off.

Figure 6-19
(A) The surface of concrete cast against plywood. Note the many small voids in the surface. (B) The surface of the wall shown in (A) after it had been sack rubbed, filling the voids.

In sack rubbing, the concrete surface should be moist but not wet when mortar, mixed to a barely damp consistency (earth moist), is spread over the concrete surface with a piece of burlap, rubber float or similar tool. Mortar should consist of one part cement to two or two and one-half parts sand passing the 16-mesh screen (ordinary window screen). Sufficient white cement should be used so the patches, when dry, will match the parent concrete. The concrete surface should be rubbed smooth with the burlap or float, making sure that the voids are filled flush with the surface. The mortar in the void should be neither raised above nor depressed below the concrete surface. Curing should be applied in the usual manner, either water curing or liquid membrane curing compound.

A stoned, or sand, finish is similar, except that the mortar is of a thick, creamy consistency. After the mortar has been spread, the surface is immediately rubbed with a carborundum stone, taking care to cover the entire area thoroughly. A thickness of mortar of about $^1/_{32}$ inch should be left to provide a proper finish over the area—as contrasted with sack-rubbing, which fills the voids only, leaving the surface unchanged. The final step is prompt and adequate curing.

It may be desirable to use one of the proprietary bonding agents for this application, after tests made to determine its suitability. These methods of patching should not be used for honeycomb or rock pockets, as such defects require removal and replacement of the unsound concrete. These procedures are described in Chapter 22.

Troweled Surface. Sometimes bubbles and blisters will appear in fresh concrete during troweling. In one case this was caused when the partially hardened base was sprayed with water and then dusted with dry cement. Apparently, water was trapped in the aggregate voids by the cement topping and was released when the troweling was done. This made small bubbles appear in the surface immediately after the troweling. Sprinkling water or cement on a surface to be finished is no way to prepare it and should never be permitted.

Power troweling too early can cause bubbles. Some operators have a tendency to tilt up the front edge of the blade to keep from picking up concrete on the blade, with the result that the concrete is troweled with the back edge of the blade. Air, trapped beneath the surface by this early troweling, forms into bubbles. If the concrete is still soft enough, it is sometimes possible to re-open the surface to let the air escape by using a wood float or a flat blade. The best remedy is to avoid the bubbles in the first place by not troweling too soon, and not tilting the blade so that troweling is done with the back edge.

6.13. Form-Related Defects

There are some surface problems that are related to formed surfaces only; that is, they do not happen on unformed concrete. Likewise, there are some problems that are confined to unformed surfaces only.

Rock Pockets. One defect that can happen on formed concrete is a rock pocket, or honeycomb. See Figure 6-20. When concrete is first placed in the form, it contains large amounts of entrapped air that cause large voids. The very nature of the concrete mixture does not lend itself to the automatic formation of a dense, homogeneous mass without the application of work to bring it all together and drive out the air. The picture is complicated if segregation of the concrete has occurred as a result of improper equipment or practices. See Chapter 15. In either case, the result is concrete that appears to consist of coarse aggregate coated with mortar but without enough mortar to fill the voids between the coarse aggregate particles. Of course, we can use the wet, high-slump, soupy mixes that flow easily without the expenditure of work, but we know that such mixes are completely unsatisfactory and lead to poor quality of the finished product. Besides, they are more apt to suffer from segregation than the more moderate mixes of reasonable slump. However, under proper supervision the use of a superplasticizing admixture does modify the concrete, so the high-slump mixes can be placed satisfactorily.

If the forms are not tight, a rock pocket can result from the loss of mortar or grout through leaks in the form, leaving behind the coarse aggregate without mortar to fill the voids.

Figure 6-20
This rock pocket occurred when the concrete was not thoroughly compacted by vibration or the mortar leaked through openings in the form.

Consolidation or compaction of the concrete in the forms is accomplished by the systematic and thorough application of vibration applied to each increment of concrete placed in the form.

To summarize, rock pockets are caused by the following:

1. Loss of cement grout or mortar from the concrete by leakage through the forms. Leakage of this type occurs wherever there are openings and can be minimized by making sure that individual boards and panels of lagging or sheathing fit accurately together. When a form is attached to old concrete, as is the case when constructing a wall in several lifts, the form must conform to the profile of the old concrete, without cracks. Knotholes and other holes should be plugged or covered.
2. Failure to consolidate the concrete thoroughly. Concrete should be consolidated by means of vibration, systematically and thoroughly applied to each increment of concrete placed in the form.
3. Segregation of the concrete, resulting from efforts to flow the concrete into place without adequate vibration, or resulting from the use of overly fluid mixes, or mixes that are lean and harsh, or mixes containing sand deficient in fines. Segregation also results from improper handling methods or equipment.

The only remedy for these defects, once they exist, is complete removal of all affected concrete down to good sound concrete, and replacement with new concrete or mortar.

Sticking to Forms. Almost as maddening as discovering honeycomb on a newly stripped concrete surface is to find that much of the surface concrete has pulled away by sticking to the form. Fortunately, this does not happen very often, and its prevention is comparatively simple.

Any form or mold used for containing fresh concrete until it hardens must be coated with a material that prevents bond between the concrete and the form. These form oils, or parting compounds, as they are called, may be specially compounded mineral oils, waxes, plastics or lacquers, or even job-mixed concoctions of fuel oil and lubricating oil. Correct use of any of the compounds commercially available will assure a satisfactory surface. Some plywoods that are factory treated with a plastic coating do not need application of form oil, although the plastic coating will perform better and last longer if a form oil is used.

Sometimes concrete will stick in localized areas to metal forms even though the form is apparently well coated with oil. Sticking may result if, in placing concrete, the concrete slides over the surface of the form, scraping the oil off the form.

In applying oil to any form, the surface must be clean before application of the oil. Spots of rust, dirt or old mortar are sure to cause trouble later when the form is stripped from the concrete. With respect to steel forms, care should be exercised in cleaning them. Although the metal must be clean, a too vigorous use of wire brushes, sand blast or abrasives— to the extent that bright metal is exposed—should be avoided. After cleaning and oiling forms with a nondrying oil, it is sometimes helpful if the forms can be exposed to warm sunshine for two or three days.

The use of galvanized sheet steel is not recommended for lining forms, because of the danger of excessive sticking.

Defects in Forms. The surface of the form is reflected in the surface of the concrete cast against the form. It is obvious that a smooth surface cannot be achieved with rough formwork. There are, of course, instances when rough lumber is used for lining forms to impart a rough texture to the concrete for aesthetic reasons, but these forms require careful workmanship the same as any other.

Unsatisfactory alignment of concrete surfaces results from poorly designed forms and slipshod construction. A form is a temporary structure that must be accurately built to carry heavy loads, not only the weight of the fresh concrete at 150 pounds per cubic foot, but also the weight of the workers and equipment necessary to construct the form and place the concrete.

The first consideration in designing formwork is safety. Failures of forms and shoring have resulted in spectacular construction accidents involving injury and death of workers and extensive property damage. Pressures on formwork can reach substantial values, as discussed in Chapter 11.

The material used for lining the form will bulge between studs under the pressure of the fresh concrete being vibrated if the studs are too far apart. See Figure 6-21. A lack of adequate ties will result in bulging as well, or the ties may be so overloaded that they give way and the whole form moves, causing a rough offset in the concrete. Proper spacing of studs and walers prevents bulging; adequate fastening, bracing and wedging restrain movement of the form under pressure of the concrete while it is being vibrated, and tight joints restrict leakage of mortar from the concrete.

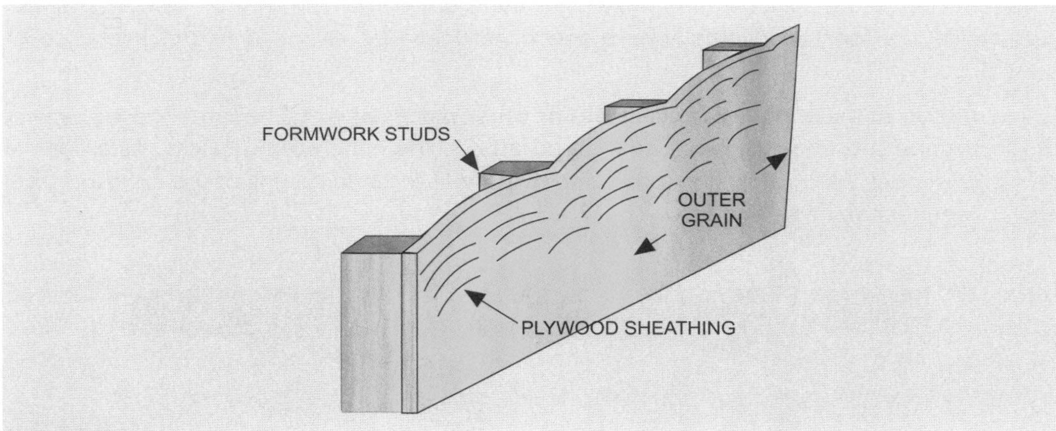

Figure 6-21
The proper spacing of studs and backing of form lining material will assure that the concrete surface be smooth. For maximum flexural strength, the outer grain of the plywood should be perpendicular to the studs; that is, it should span across the studs.

A blemish that sometimes occurs is an offset at a horizontal construction joint, where the form has moved a fraction of an inch at the bottom of the new lift. Not only is there an ugly offset in the concrete, but frequently the mortar leaks out and leaves a rock pocket at the bottom of the lift. The bottom edge of the form for the new lift of concrete should overlap the old concrete no more than 1 inch and must have sufficient ties so it can be drawn up tightly against the old concrete. This is illustrated in Chapter 11.

6.14. Floors and Similar Wearing Surfaces

Although some unformed surfaces comprise construction joints, or the tops of such structural elements as pier caps and walls, by far the greatest areas of unformed concrete are those comprising floors, pavements and walks. These surfaces may suffer cracking, crazing, peeling, dusting and other complaints if the concrete is not properly made and handled. Assuming that satisfactory materials were used, that the concrete was of the correct consistency and that mixing and placing operations were done in an approved manner, then the remaining operations are finishing and curing. It is here that much good concrete is ruined.

Because of the unique properties and problems associated with floors, this subject has been given special treatment in Chapter 16.

6.15. Stains and Discoloration

Blemishes of this kind result from a stain-inducing material that was inadvertently incorporated into the concrete when it was mixed, from material that was spilled or otherwise came in contact with the surface of the hardened concrete, or from material that contained an ingredient that discolored the concrete.

Certain iron pyrites will oxidize or rust when embedded near the surface of the concrete, causing a brown stain. If the aggregate contains pieces of pyrites, the suspected particles can be tested by immersing them in lime water. Reactive types of pyrites will produce a brown precipitate in a few minutes, whereas the nonreactive type remains stable, without the precipitate. Ironstones, or ferruginous concretions, will stain the concrete if they become embedded near the surface. In heavyweight concrete containing iron or steel aggregate, rust stains are apt to develop where metal particles are exposed in the presence of moisture.

Discoloration of the surface results if nails or other pieces of metal become lodged against the forms and subsequently become incorporated in the concrete. Careless placement of reinforcing steel, permitting it to touch the forms, will result in rusting of the exposed steel and consequent staining. See Figure 6-22.

Figure 6-22
Nails and pieces of wire, if lodged against the form when concrete is placed, will rust and discolor the concrete.

When calcium chloride is used as an accelerator, lumps may settle near the surface of the concrete, causing stains, unless the salt is added in solution to the concrete.

Form oils or coatings, if incorrectly used, sometimes cause stains. Form oils are especially formulated for specific uses (some for steam curing, etc.), and each material should be used for the purpose intended, as recommended by the manufacturer.

A difficulty that has occurred with certain plywoods is a pink discoloration of the surface of the concrete. This happens with the first usage of the plywood, but it does not happen upon subsequent usages. It is not a serious problem with concrete made with common gray cement, but it can be a problem with white concrete. Apparently, the discoloration is related to the glue or resin used in making the plywood. It is an erratic condition that cannot be foreseen. The pink color is difficult to remove, but it does fade out with time, sometimes taking several weeks.

Figure 6-23
Discoloration that is due to use of calcium chloride admixture in the concrete. (Courtesy of PCA).

Another discoloration, again not too serious on gray concrete but very unsightly on white concrete, is a tan or brown color that appears on surfaces cast against certain hardboard. This discoloration does not fade and is nearly impossible to remove. Great care in selection of forming materials should be exercised when using white concrete to avoid discoloration.

Dark discoloration on troweled surfaces is usually not a serious problem, but it can be a source of discontent to the owner, who will probably think that two different cements were used. Different cements can cause differences in color on either formed or unformed concrete. If ready-mixed concrete is being delivered from more than one plant, there might be different brands or types of cement.

One source of dark spots is the practice of sprinkling dry cement on the fresh concrete to speed up finishing. Dry cement should never be used to absorb water from the concrete to facilitate troweling. Dark spots, crazing and peeling will develop. Uneven pressure or variations in the blade angle in troweling can cause variations in color and texture. Lack of uniformity in curing is another cause.

Irregular dark areas developing in troweled slabs on grade have been a problem on some jobs. See Figure 6-23. The slab is structurally sound, and the finish is acceptable; however, irregular dark areas appear—much darker than the normal gray color—covering an area of several square inches to several square feet. When it occurs it is usually on smooth troweled driveways, patio slabs, tennis courts and similar exterior concrete.

The basic causes are the use of calcium chloride admixture, hard-troweled surfaces, inadequate curing, cement alkalies, changes in the concrete mix, and practices and finishing procedures that cause surface variations in water-cement ratio. Usually more than one of these factors is active in cases of severe discoloration.

One step that can be taken to minimize dark spots is to cure the concrete properly. This means keeping it wet for several days by ponding, spraying white pigmented curing compound, covering with wet sand or wet burlap. Tight plastic sheets or paper can be used, provided they are smooth with all joints and edges sealed against loss of moisture. Wrinkles in the sheet material may mar the surface, especially if calcium chloride was used in the mix.

Avoid the use of calcium chloride, especially if a hard-trowel finish is to be applied. Have adequate workers and equipment available to finish the concrete when it is ready. Remember that there is less open time when calcium chloride is used. Do not sprinkle dry cement on the surface being finished.

6.16. Efflorescence

Efflorescence is a deposit of crystalline salts on vertical surfaces of hardened concrete or masonry, brought from the interior of the mass by water and deposited on the surface by evaporation. Nearly all concrete and concrete products are more or less subject to efflorescence if moisture is available. Retaining walls are especially susceptible. See Figure 6-24. When water moves through cracks and porosity in the concrete, it brings to the surface the soluble calcium hydroxide that results from the reaction between cement and water. After evaporation of the water, the calcium hydroxide remaining on the surface reacts with carbon dioxide in the air, forming calcium carbonate, the familiar white crystalline deposit. Other rare forms are caused by sodium chloride or similar salts in the mixing water, organic matter in the mixing water or aggregates, high lime or gypsum in the cement, leaching of any water-soluble constituents, and materials that may be carried into and through the concrete by groundwater.

Figure 6-24
A heavy deposit of efflorescence. On this masonry retaining wall, the condition was worsened by transfer of salts from the soil through the concrete of the wall. (Courtesy of PCA)

Efflorescence is evidence of porosity, so anything that can be done to limit porosity will be helpful. Low total water per cubic yard of concrete, low slump, adequate consolidation of the concrete by vibration, tight construction joints and thorough curing are all essential.

Obviously, the best preventive is to keep moisture away from the concrete, where this is possible, by means of a vapor barrier, or by providing drains and weeps to carry water away from the back side of retaining walls and similar structures. See Figure 6-25. Well-proportioned mixes, thoroughly consolidated and cured, will produce watertight concrete and help prevent cracks through which water can pass. Construction joints should be avoided when possible, but if a joint is necessary it should be properly made.

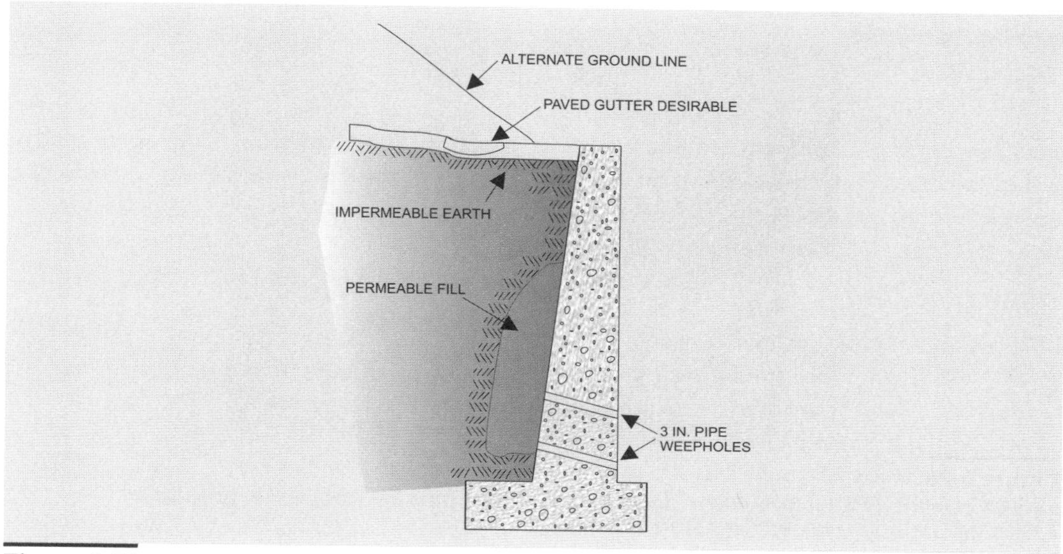

Figure 6-25
Water should not be permitted to collect behind a retaining wall. A permeable fill of granular coarse sand or fine gravel will intercept water that collects in the soil. The water then escapes through weep holes in the wall.

If efflorescence must be removed, it is suggested that attempts be made to wash it off with water. If this is not effective, it can be dissolved by dilute hydrochloric (muriatic) acid. The dilute acid is prepared by adding one part of acid to eight or ten parts of water. The surface to be treated is first moistened with water, then the dilute acid is brushed on, after which the surface is thoroughly washed with copious quantities of clear water. Because of its corrosive nature, the acid should be handled with great care. Workers should be provided with rubber protective clothing and goggles. The acid should not be permitted to come in contact with anything that might be damaged by it. A small inconspicuous area should be treated first to determine what effect the acid will have on the surface. The acid will etch the concrete surface and should not be permitted to remain on the concrete for more than a few seconds. Efflorescence that comes from salts within the concrete becomes less extensive with the passage of time. Each time it is cleaned off it will reappear within a short time, although each subsequent appearance is less severe.

Once in a great while a white efflorescence will appear on a slab on grade, caused by salts in the soil being carried by moisture passing through porosity in the concrete. This occurs in areas of soil with a high content of salts. It is practically impossible to eliminate because there is an almost endless supply of salt available to show up on the surface, no matter how many times the concrete is cleaned.

6.17. Laitance

Laitance occurs as a light gray or nearly white substance consisting of cement particles, water and the fine particles of silt and clay from the aggregates, appearing on the top surface of concrete during and immediately after consolidation. This layer of laitance has no strength and is especially undesirable on construction joints or fill planes, as its weakness prevents bond between the old, hardened concrete and the fresh concrete being placed in the succeeding lift. It has a high permeability and is a source of water leakage in hydraulic structures, retaining walls and basement walls. The presence of laitance on a floor results in a soft, dusting surface. See Figure 6-26. Layers of laitance deteriorate by weathering more rapidly than the sound concrete.

Figure 6-26
Laitance (sometimes referred to as "dusting") is evident by a fine powder that can be easily rubbed off the surface. (Courtesy of PCA)

The presence of excessive amounts of silt, clay, rock dust and similar materials in the aggregate increases the likelihood of laitance forming on horizontal surfaces of concrete containing these materials. Overly fluid mixes that segregate under vibration leave a layer of laitance on the surface. Excessive or too early floating and troweling, by bringing to the surface large quantities of water and fines, are conducive to the formation of laitance.

Low-water-content mixes are not inclined to accumulate laitance. Air entrainment reduces the bleeding capacity of concrete, which in turn reduces laitance.

Laitance that accumulates on the top of a lift that is to receive another lift or layer of concrete must be cleaned off before the new concrete is placed. See Chapter 15.

6.18. Scaling

The sloughing away or peeling of the surface in thin flakes is called scaling. It is one of the worst blemishes that can befall a horizontal concrete surface, yet control measures are available and are relatively easy to apply. See Figure 6-27.

The kind of scaling most commonly seen is a thin peeling of flatwork that starts several months after the concrete has been placed. The flaking can be as thick as $^1/_4$ inch or paper thin. Another kind of scaling occurs on nonair-entrained concrete caused by the application of de-icing salts. In both cases, the basic cause is failure to observe good construction practices.

Figure 6-27
Scaling of a slab, the result of a weak, nondurable surface where the surface mortar has peeled away, which usually exposes the coarse aggregate. (Courtesy of PCA).

The work procedures that cause the first kind of scaling are the same that cause a soft surface, dusting and laitance. They are as follows:

1. Finishing the concrete when there is free water on the surface. Mixing of this free water into a thin layer on top of the slab will cause segregation of the cement and aggregate fines, resulting in a thin layer of cement and fines underlaid by a layer of nearly clean sand that breaks the bond between the concrete and the thin surface layer. Subsequent drying causes shrinkage, then peeling.
2. Rapid drying of the surface because of a failure to cure the concrete.

Preventive measures have previously been cited: well-graded aggregates, minimum water content, finishing only when there is no free water on the surface, and prompt and thorough curing.

In geographical areas where freezing and thawing cycles are usual through the winter, it is common practice to spread calcium chloride salt on the surface of pavements, walks and driveways to melt the snow and ice. This salt is very destructive to concrete and will cause serious scaling of nonair-entrained concrete in a very short time. The use of entrained air in concrete provides the concrete with a remarkable resistance to such attack. Air entrainment should be used in any exposed concrete slab in a severe climate. In fact, it is a good idea to use air entrainment in all concrete exposed in such a climate.

6.19. Spalling

Spalling is a deeper surface defect than scaling, often appearing as circular or oval depressions on surfaces. See Figure 6-28. Spalls may be 1 inch or more in depth and 6 inches or more in diameter. Spalling is caused by pressure or expansion within the concrete. Spalls may occur over corroded reinforcing steel as illustrated in Figure 6-28. Spalling can also occur as elongated cavities along joints (joint spalls) caused by impact loads against improperly constructed joints.

Spalls can be avoided by properly designing the concrete element for the environment and anticipated service. The first line of defense against steel corrosion caused by chloride-ion ingress (use of de-icing salts) should be the use of a low-permeability concrete made with a water-cement ratio of 0.4 or less. Other means to reduce steel corrosion induced by chloride-ion ingress are use of epoxy-coated reinforcing steel, use of corrosion inhibiting admixtures and cathodic protection methods.

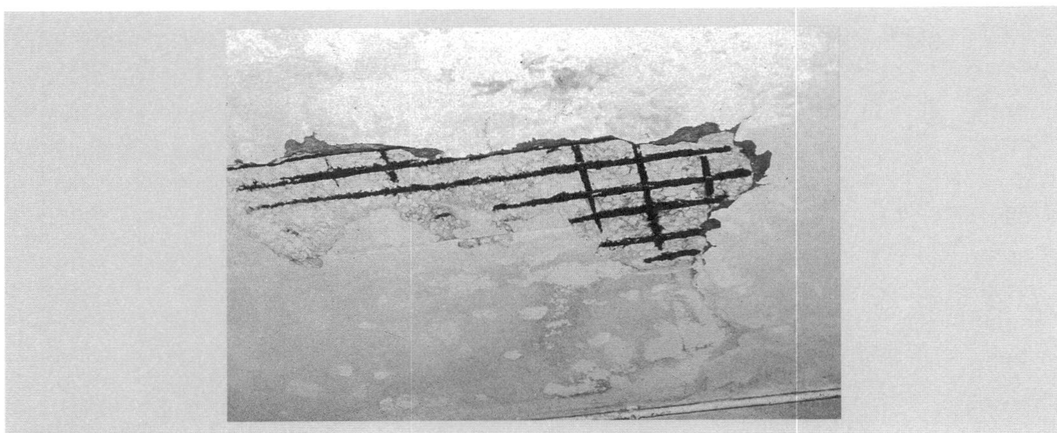

Figure 6-28
Spalling partly caused by the use of de-icing salt and the presence of reinforcing steel too close to the surface.

6.20. Popouts

A popout is the breaking away of a small piece of concrete in the shape of a cone, with the base on the surface of the concrete and the point at a particle of some kind, which expanded with sufficient force to break out the concrete. See Figure 6-29. Popouts are fairly common in flatwork, such as sidewalks. They occur in moderate as well as severe climates. They may develop any time after the concrete was placed, even a year or more later. Moisture is usually involved in the reaction, and freezing is sometimes present.

Figure 6-29
Popouts. In each case, there was a small particle that expanded and broke out the concrete.

Aggregates that might cause popouts are soft and lightweight particles of shale; soft, fine-grained limestone; and other light, porous materials. Porous cherts cause popouts where moisture in the chert freezes. Certain aggregates react with the alkalies in cement to cause popouts. On one job popouts were traced to a delayed reaction of calcined dolomite that had become incorporated into the concrete by way of the cement, which had been hauled in hopper-bottom cars without adequate cleaning after having been used for hauling magnesium compounds. More care in inspecting the cars before loading cement would have prevented this contamination. Cars, barges, trucks and other cement-handling equipment should be inspected and cleaned regularly. This is especially important if they have been used for handling other materials.

Another case of popouts was traced to pieces of glass in the coarse aggregate. Broken bottles had been thrown into the stockpiles from an adjoining soft drink bottling plant, and the reaction between the glass and high-alkali cement resulted in expansion of the glass fragments.

Cinder concrete may suffer from stains and popouts if the cinders are not aged sufficiently and if hard-burned free lime, free magnesia or calcium sulfate are present. The danger can be prevented if the crushed cinders are stored in a continuously wet stockpile for several weeks. Magnetic separation is desirable to remove tramp iron.

Rarely, slag concrete will suffer from popouts if particles of unburned flux from the smelting process find their way into the concrete. These particles are high in silica and magnesia.

The only cure for popouts is to prevent them in the first place, and the surest way to prevent them is to use aggregates free of deleterious particles. Some of the deleterious materials are difficult to identify, and some may enter the concrete by devious routes. Nevertheless, a petrographic examination performed by an expert in this field is the only way that one may be reasonably sure of identifying the constituents of the proposed aggregates and of knowing whether there are deleterious particles. Aggregates containing porous chert and other particles of low specific gravity can be beneficiated by heavy media separation, which removes the light particles by flotation. The use of low-alkali cement will frequently help. Often, popouts are the result of a combination of factors; hence the need to follow good construction practices in regard to the mix, water content, drainage, finishing and curing.

In a structure where appearance is especially important, popout cavities can be filled with an epoxy mortar consisting of epoxy resin and aggregate with enough cement to provide a satisfactory color match with the existing surface.

REPAIR OF DEFECTS

In spite of all the precautions and quality control measures that are taken in the production of concrete, there are occasions when concrete suffers damage and deterioration. Repair is greatly complicated if inferior concrete was placed originally, although protective measures, when properly applied, can be effective in limiting spread of the damage.

6.21. Diagnosis

The first step in repair is to diagnose the damage as described in Section 1.7. This includes determination of the cause of damage or deterioration and an evaluation of the extent of damage. Causes of deterioration are discussed in this chapter and in Chapter 4. In general, there may be inferior materials, poor design, faulty construction, chemical activity within the concrete, exposure to aggressive substances, long exposure to weathering, freezing and thawing action, or accidental damage. The next step is to determine the extent of the damage and the necessary repair measures. In some cases it may be more economical to remove a member rather than to attempt a repair.

When we consider patching and repair of concrete, we find that there are two reasons for making repairs: structural and cosmetic. Structural repairs are necessary to restore or maintain the structural load-carrying capacity of the structure. Cosmetic repairs are made for purely aesthetic purposes—to restore or maintain the appearance of the structure. Most repairs involve some features of both purposes and serve as well to prevent further damage to the concrete by weathering or other forces.

We can further classify repairs into those—usually of a minor patching character—that are the result of improper construction practices and those that are the result of damage to the concrete subsequent to construction.

One of the most difficult parts of the diagnosis is to determine the extent of repair required. The amount of concrete affected by construction faults, such as in a rock pocket, is relatively easy to determine, but the amount of concrete affected by attack on the hardened concrete—freezing and thawing, for example—can be difficult to establish. Nevertheless, one basic requirement of all methods of repair is that all defective or deteriorated concrete must be removed. If this removal is not done thoroughly, the repair will fail. See Figure 6-30.

Figure 6-30
When this wing wall was repaired, the apparently unsound concrete was removed and replaced with new concrete. The repair is now failing because all of the unsound concrete was not removed when the repair was made.

The final part of the diagnosis is the selection of materials and method of repair. This depends on the area and depth of the repair, whether structural valves have to be restored, the kind of deterioration and the appearance of the repair. New materials are constantly entering the market for which the user should carefully check independent laboratory tests and the history of use of the product to determine its suitability for the repair under consideration.

If appearance is important, care should be taken in the choice of materials and methods; otherwise the repair will be an unsightly eyesore, drawing attention to the patch rather than obscuring it. Architectural concrete surfaces that have had special treatment are particularly sensitive in this respect.

Methods of repair include dry pack, shotcrete or gunite, concrete replacement, preplaced aggregate, overlay, surface treatment, or the use of chemical repair material such as epoxy resin. Irrespective of the method selected, the repair must be completely bonded to sound concrete, free of cracks, compatible in appearance, impermeable and resistant to weathering or other attack.

Major repair work should be done only under the direction of an experienced engineer, as serious considerations encompassing not only the appearance but the safety of the structure are involved.

6.22. Preparation for Repair

Whatever the method of repair, the first step is removal of all unsound and disintegrated concrete. This requires rigorous and uncompromising removal of any concrete of doubtful soundness. The surface of the old concrete should be clean, and the fractured face should be bright and crystalline in appearance. A dull, dead-appearing surface indicates that more deteriorated concrete must be removed. Although good work can be done with hand tools, a faster and better job can be done with pneumatic chipping hammers. Light chipping guns are less likely than heavy jackhammers to cause damage to sound concrete. Edges are best made with a saw, making the cut at a slight angle so as to key the patch in place. Note, however, that edges for shotcrete should slope outward slightly to avoid inclusion of rebound. In using any saw or power tool, care must be exercised to avoid damage to reinforcement, inserts or sound concrete.

Reinforcing steel exposed during the concrete removal should be clean and rigidly fixed in place. For concrete or mortar patches, loose and scaley rust should removed; tightly bonded rust and bits of sound mortar tightly adhering to the steel need not be removed.

If the repair material is to consist of an epoxy resin base or similar chemical, the reinforcement must be cleaned to bright metal, which usually means sandblasting. If additional reinforcement is required within the repair area, it should be tied to existing steel or anchored by means of bolts set in the old concrete in expansion shields or epoxy resin.

After the unsound concrete has been removed, the area must be thoroughly cleaned. Areas on which mechanical tools were used should be blown clean with compressed air, then washed with a clean, high-pressure water jet. Final cleaning of the repair area should be done immediately before placing the patch to assure that the surface has not become soiled in the period between cleaning and application of the repair material.

Concrete and mortar patches should be applied to moist but not wet concrete. Old concrete should be kept moist for several hours before repair. This can be accomplished by covering the area with several layers of clean burlap or a curing blanket kept wet, wet sand, flooding or any method that will keep the concrete wet. Final cleanup is done after removal of the covering material. New, green concrete, if only a day or so old, probably contains enough moisture, but should be dampened just before placing the patch material. An air gun can be used to remove excess moisture.

Generally, epoxy resin compounds must be applied to completely dry concrete. However, there are some formulations that can be applied to wet surfaces. Advice of the manufacturer should be followed in either case.

6.23. Dry Pack

The dry pack method of repair has a distinct advantage in that it does not require any special equipment that is not available to all concrete finishers. However, it does require that the person making the repair be highly skilled in this particular art. If he or she is not, the results will be disappointing. A small portable gun is available, similar to a shotcrete gun, that appears to produce good dry pack patches.

In general, this method should be used for form bolt holes, deep but not large in area, rock pockets and any hole that is deep but not wide in area. When these conditions are satisfied, the dry pack method of repair is satisfactory because there is practically no shrinkage and the patch will develop high strength.

The dry-pack mortar should consist of one part cement to two and one-half parts of sand passing the 16-mesh screen (common window screen). Usually, it is necessary to use part white cement for a good color match. The mixture should contain just enough water so that it will form a ball when squeezed gently in the hand. Too much water will result in shrinkage and consequent loosening of the patch. Too little water will not make a sound patch.

If the patch is applied shortly after the form is removed from new concrete, it is not necessary to moisten the cavity. Old concrete should be saturated, then permitted to become surface dry before the cavity is filled.

After the cavity has been cleaned of all loose material, unsound concrete and oil, the mortar is tamped in layers about $^1/_2$-inch thick. Before the mortar is placed, a bonding grout should be brushed onto the surface. For good compaction and bond between layers, tamping should be done by hammering on a hardwood stick. The surface of the patch should be finished to match the existing concrete. However, even when matching a steel-formed surface, a steel trowel should never be used on the patch, as this leaves a dark surface that is impossible to remove. One method is to lay a piece of board over the patch and strike it several times with a hammer. The patch must be cured by the application of a membrane curing compound or water curing for three days.

6.24. Concrete Replacement

This method is used for repairs involving large and deep areas of more than a few square inches in area and 6 inches in depth. It is especially suitable for restoring large areas of hydraulic structures, where it has become necessary to remove concrete to an appreciable depth over a large area, and for replacement of old, deteriorated concrete. One important caution: the new concrete must be of a quality to resist the instrumentality that caused the original damage; otherwise it too will soon fail.

The concrete replacement method consists of replacing defective concrete with machine-mixed concrete of suitable consistency and proportions, which will bond with the base concrete. If the holes in the structure go entirely through a wall—or if the defect goes beyond the reinforcement, and in general if the quantity is large—then concrete replacement becomes necessary. Forming will usually be required for repairs in vertical surfaces.

Replacement is sometimes necessary for the repair of honeycomb in new work as well as for restoring old concrete. For new work the repairs should be made immediately after stripping the forms. Considerable concrete removal is always required for this type of repair. Excavation of affected areas should continue until there is no question that sound concrete has been reached.

Repairs in horizontal surfaces can be made without forms, except where replacement of corners and edges is required. Sloping and vertical surfaces require forms to confine the concrete.

Forms must be mortar-tight and must fit snugly to the old concrete to avoid loss of mortar or an offset around the perimeter of the patch.

The old concrete should be nearly dry when a $^1/_8$-inch coating of mortar, of the same proportions and water content as the replacement concrete, is applied by shotcrete, hand rubbed into the surface or dashed onto the surface. Any free water in the area should be blown out with oil-free compressed air before applying the mortar.

Concrete replacement follows immediately after the mortar coating. Concrete with a slump of 2 to 3 inches should be deposited in lifts, or layers, each layer consolidated by vibration using immersion vibrators if possible, otherwise by means of form vibrators, provided the forms are constructed to withstand form vibration. Concrete on the surface of a slab repair should be consolidated and finished to match the original surface.

Because of the relatively small amounts of concrete required, measuring and batching of the concrete might present a problem. In this case, consideration should be given to the use of pre-bagged mixed concrete, consisting of a proportioned mixture of cement, sand and coarse aggregate available in paper bags of several sides. Strength potential of the proposed material should be determined, especially if structural values are involved, and the color should be checked if color of the finished repair is important.

All repairs must be cured as described in Chapter 17.

6.25. Pneumatically Applied Mortar

Shotcrete is especially adaptable to thin areas of large extent, such as scaled areas. Preparation of the area and application of the shotcrete should be performed as described in Chapter 22.

6.26. Bonded Overlay

Scaled areas of pavements and similar flat slabs may be repaired by application of a thin overlay.

After removal of unsound concrete, the area to be patched is cleaned, preferably with a muriatic acid wash and thorough scrubbing and washing. After the surface becomes dry, a thin layer of bonding grout is spread, followed by air-entrained concrete containing coarse aggregate no larger than about one third of the thickness of the overlay. Slump should not exceed 1 inch. The concrete should be thoroughly tamped, then finished to match the original surface. Joints should be provided over joints in the original concrete.

The secret of the success of this method is to use concrete with no more than 1-inch slump applied over a thin grout layer on thoroughly clean and sound old concrete, thorough compaction by tamping or rolling, and proper curing. The temperature of the new concrete should be as near that of the underlying slab as possible for good bond. Minimum thickness of the overlay should be 1 inch. Improvement of the bond can be accomplished by the use of a bonding agent. See below.

A layered system can be made by the use of latex polymer modified concrete, fiber-reinforced concrete or internally sealed concrete. These highly specialized procedures utilizing unique materials and techniques should be performed by specialists in these fields. They are described in Chapter 22.

6.27. Repair of Cracks

Cracks in concrete are an indication of some condition or conditions adversely affecting the concrete. For this reason, repair and patching of cracks is a waste of time unless the condition that caused the crack is first corrected.

Probably there is little to be gained by patching cracking associated with weathering. Repair in this case consists of removing all the affected concrete down to sound concrete, as described under concrete replacement. Any concrete of doubtful soundness should be removed and the area patched in the best possible manner.

Before attempting to fill or repair any crack, the crack should be cleaned out. A high-pressure air or water jet may suffice for a fine crack. Larger cracks should be cleaned by the removal of spalled and loose concrete, old joint sealer and all foreign material. The use of picks or similar tools may be required.

There are two general guides to consider when repairing cracks: either the structural integrity of the concrete member must be restored, or merely filling the crack is all that is required. The choice of procedure is an engineering decision made after careful consideration of all the facts relating to the problem. If the cracking relieved the stresses that caused the crack or if other measures have been taken to relieve the stress, then a structural repair can be made.

Cracks can be filled with the same elastomeric materials that are used as joint fillers. Materials of this type retain their plasticity and permit slight movement of the concrete without breaking the seal, this being especially desirable to prevent passage of moisture through the crack. Resilient joint fillers may be either hot applied or cold applied. Materials for hot application consist of a mixture of asphalt, rubber and a filler, or similar materials, and are usually black in color. There are also asphaltic fillers for cold application, although the hot-applied filler is preferred.

Many kinds of sealants are available, most of them composed of polysulfide or similar plastics that retain their flexibility so that they will conform to the movement of the crack.

Mortar Filling. Large cracks can be filled with dry-tamped mortar. The crack can be well obscured, but strength across the crack is problematical at best. Mortar for this application consists of one part portland cement and two and one-half parts mortar sand passing the 16-mesh screen, mixed to a damp consistency such that a ball molded in the hands will stick together and hold its shape. In order that the repair matches the original concrete as closely as possible, white cement should be substituted for part of the normal gray cement. The mortar is tamped into the crack with sticks or similar blunt instruments and the surface struck off with a wood float. A steel trowel should not be used, as this will darken the patch. Upon completion of the repair, the mortar should be cured, either by the application of water or a liquid membrane-forming compound.

Bond of the mortar to the concrete is improved by the use of a portland cement bonding agent. There are several proprietary compounds available, including epoxy resin, which is applied to the concrete surfaces to be repaired, and a latex compound, which can be added to the patching mortar.

Epoxy Resins. By virtue of their high strength and adhesive properties, epoxy resins make a good repair where strength across the crack is necessary. Small cracks can be filled with epoxy resin, using a pressure process. For this method, holes are drilled about 2 feet apart in the concrete along the crack to a depth of about 1 inch. Zerk-type pressure fittings are sealed into these holes either by cementing them in place with epoxy resin or by inserting them in a soft metal sleeve, which is anchored by tapping the fitting with a hammer. The surface of the crack is then sealed with the epoxy adhesive, leaving small vent holes every 6 inches. After the adhesive on the surface has cured, the crack is filled by using a high-pressure hand-operated grease gun to force the epoxy adhesive into the crack through the openings. Epoxy for this application should contain no filler. Modifications of this method utilize other fittings such as one-way polyethylene valves, small tubing or pipe nipples. This procedure is useful on horizontal surfaces and can be used on sloping or vertical surfaces.

A low viscosity epoxy can be used for filling horizontal cracks by letting the resin penetrate the crack by gravity. Top of the crack should be veed out to a depth of about $^1/_4$ inch and the filler poured in. For cracks passing through the slab, the underneath side of the crack must first be sealed with a paste epoxy.

Large cracks can be filled with an epoxy mortar consisting of the epoxy adhesive mixed with mortar sand in the proportions of one part adhesive to three parts sand by volume. After first cleaning the crack of all loose concrete, dirt and other foreign material, the mortar is troweled into the crack. Complete filling is assured by working the mortar into the crack with a knife blade or similar instrument.

Some manufacturers formulate an epoxy resin that has been used as a resilient crack filler. Some of these can be used on damp surfaces.

6.28. Proprietary Materials

There are numerous crack fillers, patching compounds, bonding agents and other materials on the market, some of which are good products, but others of which are useless. In fact, they may be dangerous in that the users may be lulled into a false sense of security, not realizing their mistake until trouble develops again where they thought the problem was solved.

There are many good materials, and reliable manufacturers will be willing to give information on the composition of their products, where and when not to use them, how to use them and examples of jobs where they have been used. Among these are quick-setting cements that will set hard in a few minutes and compositions that can be used by stopping leaks through concrete under a hydraulic head.

A good quick-setting cement can be made by mixing common portland cement with calcium aluminate cement. Tests should be made with the materials at hand to determine exact proportions, but a hard set within a few minutes is possible. Aluminous cement alone sets quite fast. A mixture of cement and plaster of par- is makes a good temporary patching material that sets fast, but it cannot be used where it will be damp.

Bonding Agents. In many cases, the bond of the replacement or repair material to the parent concrete can be improved by the use of a material that improves the adhesion between the two materials. The cement grout coating previously mentioned in this chapter can be classified as a bonding agent. Chemical bonding agents are described in Chapter 10. Bonding agents are usually mixed with the cement grout and scrubbed into the old concrete immediately ahead of application of the patching mortar or concrete. In some cases, when patching spalls and nicks, a polymer bonding agent is mixed in the patching mortar or concrete. Some bonding agents will disintegrate in the presence of moisture. For this reason only those agents that are resistant to moisture should be used if there is any moisture present. Manufacturer's instructions should be followed.

Adhesives. Broken concrete can be joined by the use of epoxy resin adhesive. The piece to be bonded must be sound concrete such as that broken off by accidental impact. Precast elements, such as curbings, posts or traffic markers, and objects of metal may be bonded to concrete with epoxy.

Surfaces to be bonded must be sound and thoroughly cleaned, preferably by sandblasting or by washing with detergent followed by rinsing with water, then permitted to dry. The mixed adhesive is applied with a trowel or putty knife to both surfaces to be joined, then the piece to be bonded is pressed into place. If necessary because of the location, suitable supports or clamps should be provided until the epoxy cures, usually a matter of two to four hours, depending on the formulation of the adhesive and the ambient temperature.

Portland Cement

7.1. Raw Materials
7.2. Making the Cement
- Dry Process
- Wet Process
- Burning and Finishing

7.3. Composition
7.4. Types of Portland Cement
- Types I, II, III, IV and V
- Air-Entraining Portland Cements

7.5. Blended Hydraulic Cements
- Portland Blast-Furnace Slag Cement—Type IS
- Portland-Pozzolan Cement—Type IP and Type P
- Slag Cement—Type S
- Pozzolan-Modified Portland Cement—Type I (PM)
- Slag-Modified Portland Cement—Type I (SM)
- ASTM C 1157 Performance Specification for Hydraulic Cement

7.6. Other Cements
- Masonry Cement
- Mortar Cement
- White Portland Cement
- Plastic (Stucco) Cement
- Waterproof Portland Cement
- Expansive Cement

7.7. Other Nonportland Cements
- Aluminous
- Magnesite
- Rapid Setting

7.8. Properties and Characteristics
- Color
- Fineness
- Soundness
- Setting Time
- Compressive Strength
- Heat of Hydration
- Loss on Ignition
- Specific Gravity

7.9. Transporting and Conveying
7.10. Storage
- Bulk Cement
- Bagged Cement

Chapter 7

In Chapter 1 we saw how our present-day portland cement developed through the ages. Now we will turn our attention to the cement itself and learn how it is made and what it does. We take our cement pretty much for granted and do not give it much thought until some problem arises on the job that we feel is the fault of the cement.

Cement, although it is made of common earth materials, is a very complex chemical. Probably no other chemical is made in such large quantities, utilizing the largest moving machinery in industry and consuming huge quantities of coal, gas and oil for fuel. Cement is hydraulic in nature; that is, when mixed in the proper proportions with water, it will set and harden under water as well as in the air.

The powder we call cement is not actually a cement. It is only after it is mixed into a paste with water that the chemical process of hydration gives it a cementing property. Cement is ground so fine that a pound of it contains about 150 billion particles; 95 percent of these particles will pass through a sieve with 100,000 openings per square inch. See Figure 7-1.

Figure 7-1
Portland cement is a fine powder that when mixed with water becomes the glue that holds the aggregates together in concrete. (Courtesy of PCA)

Portland cements are manufactured according to ASTM C 150, "Standard Specification for Portland Cement" or ASTM C 1157 "Performance Specification for Hydraulic Cements." ASTM C 150 is discussed under Sections 7.1 – 7.4. ASTM C 1157 is discussed under Section 7.5.

PRECAUTION:

A quote from Reference 12.2 is pertinent for all personnel working with fresh concrete:

> When working with fresh concrete, care should be taken to avoid skin irritation or chemical burns (see warning statement in the box). Prolonged contact between fresh concrete and skin surfaces, eyes, and clothing may result in burns that are quite severe, including third-degree burns. Eyes and skin that come in contact with fresh concrete should be flushed thoroughly with clean water. If irritation persists, consult a physician. For deep burns or large affected skin areas, seek medical attention immediately.

Portland cement is alkaline in nature, so wet concrete and other cement mixtures are strongly basic. Strong bases—like strong acids—are harmful or caustic to skin.

WARNING . . . Contact with wet (unhardened) concrete, mortar, cement, or cement mixtures can cause SKIN IRRITATION, SEVERE CHEMICAL BURNS (THIRD-DEGREE), or SERIOUS EYE DAMAGE. Frequent exposure may be associated with irritant and/or allergic contact dermatitis. Wear water-proof gloves, a long-sleeved shirt, full-length trousers, and proper eye protection when working with these materials. If you have to stand in wet concrete, use waterproof boots that are high enough to keep concrete from flowing into them. Wash wet concrete, mortar, cement, or cement mixtures from your skin immediately. Flush eyes with clean water immediately after contact. Indirect contact through clothing can be as serious as direct contact, so promptly rinse out wet concrete, mortar, cement, or cement mixtures from clothing. Seek immediate medical attention if you have persistent or severe discomfort.[12.2]

7.1. Raw Materials

The raw feed is made up of common rocks, sand, silt and clay, carefully analyzed to give the proper mixture of basic chemicals from which the cement is made. The number of raw materials required at any one plant depends on the composition of these materials and the kind of cement being manufactured. See Table 7.1.

TABLE 7.1
RAW MATERIALS FOR CEMENT

CONSTITUENT OF THE CEMENT			
LIME	**SILICA**	**ALUMINA**	**IRON OXIDE**
Limestone	Sand	Clay	Iron ore
Chalk	Quartz	Shale	Mill scale
Marl		Slag	Blast furnace flue dust
Seashells			
Cement rock			
Marble			

Note: Each raw material may be the source of more than one constituent. For example, the limestone, although it is primarily a source of lime, might contain silica or alumina, or there may be iron present in the

Limestone constitutes the largest portion of the raw material. Large beds of hard limestone, from which the stone is quarried, exist throughout the world. See Figure 7-2. There are many kinds of limestone, but they all have one thing in common—they are made up principally of the compound called calcium carbonate which, when it is heated, becomes lime. Other sources of the carbonate rock, in addition to limestone, that have been used for making cement, are seashells, coral and chalk.

Cement contains many substances besides lime, and these come from other rocks or earth. Shale, slate, clay, sand, slag and iron ore are used. These materials contain compounds of aluminum, silicon and iron, which are necessary to make cement. They also contain unwanted compounds that must be kept at a minimum in the final product.

Preliminary investigations by cement manufacturer include surveys of prospective sources of raw materials to determine the quantity, kind and composition of materials in the deposit. This information enables the plant chemist to establish an approximate mix of raw materials to be used. Depending on the composition of the materials, the mix might be a simple combination of only two materials, or, more likely, a combination of three or four, sometimes as many as six.

Figure 7-2
Limestone Quarry. A primary raw material providing calcium for making cement. (Courtesy of PCA)

7.2. Making the Cement

The process of making cement is called pyroprocessing—that is, a process that requires heat. But before being heated, the raw materials require considerable preliminary preparation.

The first operation in the actual manufacture of cement is obtaining the raw material from the pit or quarry. See Figure 7-3. This usually requires blasting to break up the rock into pieces that can be handled. Picked up by power shovels, the stone is loaded into trucks, rail cars or conveyors that transport it to the primary crusher (see Figure 7-4), which crushes it to about 6 inches in diameter, then to a secondary crusher, a hammermill that reduces it to about $^3/_8$ inch in size. At this point, blending of the other raw materials with the limestone is accomplished, and the blend is conveyed to raw storage.

Samples are obtained at this point and immediately analyzed. In a modern cement plant (Figure 7-5) this sampling and testing gives the data that enables the plant chemist to develop the blending formula for the particular type of cement being produced. The blend

Figure 7-3
After the limestone has been broken by blasting, power shovels load it into trucks for transportation to the primary crusher.

Figure 7-4
Limestone fragments as large as 2 cubic yards in size are fed to the primary crusher for the first reduction in size.

consists of four or five parts of limestone to one part of other materials (clay, sand, etc.). The blended material is then stored in piles by means of belt conveyors, cranes or special stacking machines, and is later reclaimed by the same kind of machinery.

At this point in the process, there is a temporary divergence, depending on whether a dry or a wet process is being used. In the dry process, grinding and blending operations are done with dry materials. In the wet process, the grinding and blending operations are done

Figure 7-5
Arial view of cement plant. (Courtesy of PCA)

with the materials mixed with water in a slurry form. Steps in the manufacture of cement are illustrated in the flow chart in Figure 7-6. Although the operations of all cement plants are basically the same, no flow diagram can adequately illustrate all plants. Each cement plant is significantly different in layout, equipment and general appearance.

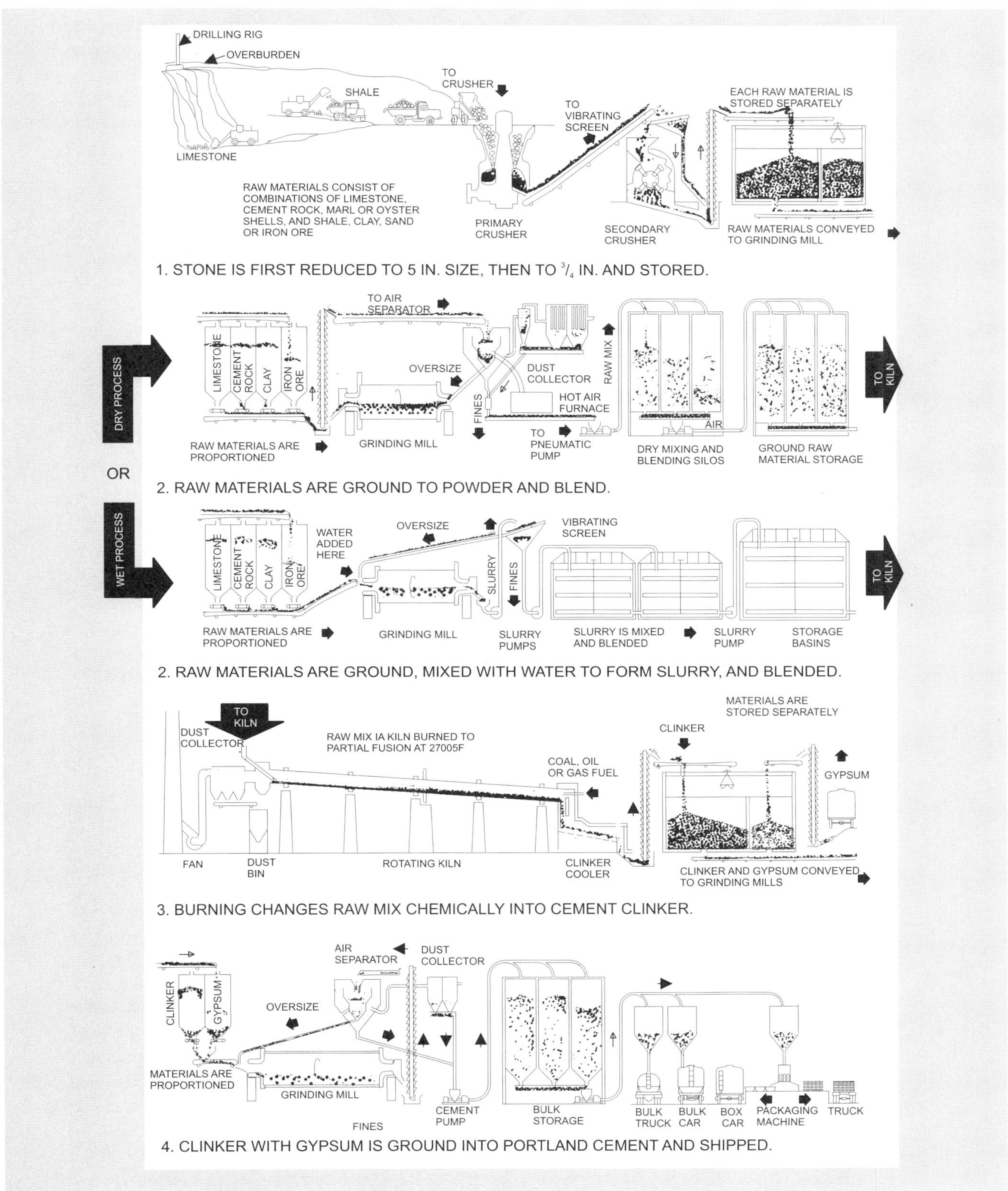

Figure 7-6
Flow of material through a cement manufacturing plant, showing both the dry process and wet process of preparing the kiln feed. (Traditional manufacture of portland cement.)

Dry Process. In the dry process, the material is now removed from the storage piles and stored in bins. Drying, if necessary, is in rotary driers, similar to horizontal kilns, except that the material is heated only enough to drive off the free moisture. From the storage bins or the drying kilns the material is delivered to the raw grinding department where it is reduced in size about 80 or 90 percent passing the 200-mesh screen. Grinding is accomplished in mills with a capacity of over 100 tons per hour per mill, usually operating in closed circuit with an air separator.

Dry-process mills may be tube mills, ball mills or compartment mills. See Figure 7-7. Sometimes the feed goes first through a ball mill, then through a tube mill for final grinding. A ball mill is a horizontal cylinder rotating at a speed of 15 to 20 rpm, containing a charge of steel balls 4 to 6 inches in diameter. As the cylinder revolves, the steel balls continuously cascade with the feed material, pulverizing the latter. A tube mill is similar to a ball mill except that it is more slender, contains smaller balls (between 1 and 2 inches in diameter) and accomplishes finer grinding. A compartment mill consists of two or three compartments separated by perforated steel plate bulkheads. Different sized balls in each compartment, graduated from large ones in the feed end to small ones in the last compartment, accomplish the same grinding as the ball- and tube-mill combination.

In any of these mills, a high-velocity airstream passes through the interior (hence the name air-swept), picking up the ground material and carrying it to the air separator that separates the fines from the coarse. Fines are conveyed to storage for kiln feed, and coarse oversize is returned to the mills for further grinding.

Figure 7-7
Ball and tube mills of various sizes are used for grinding the raw feed for the kiln. The same types of mills grind the clinker into cement. (Courtesy of PCA)

Wet Process. In the wet process, the raw feed is transferred from raw storage piles to the grinding mills, which are substantially the same as the ball, tube or compartment mills used for dry grinding. Water is introduced into the mill along with the feed. Mills are usually operated in closed circuit with some type of classifying equipment, such as hydroseparators, screws or rake classifiers, that separate the fines from the coarse. The fine slurry is then pumped to the thickeners, which remove much of the water, then into blending tanks or storage tanks, while the coarse material is returned to the mill for additional grinding. Storage tanks or basins are kept agitated to prevent settlement or segregation of the slurry before it is fed to the kilns.

Burning and Finishing. The temporary divergence of dry process and wet process paths ends when the kiln feed is put into storage. From now on there is only one process. Dry kiln feed or slurry is drawn from storage and fed into the kiln, a brick-lined steel cylinder that rotates at a speed of about one rpm. See Figures 7-8 and 7-9. Within the kiln, the feed as it advances does not flow in a straight line in the way water flows through an inclined pipe. Instead, the feed is raised partway up the arch as the kiln turns, then at a certain point gravity causes the mass to slide down the kiln shell. Thus, the feed takes a tumbling, cascading, zigzag course down the slightly inclined kiln. It is slowly heated to a temperature of about 2700°F as it is first dried. It then starts to melt or fuse into small lumps about the size of walnuts.

Figure 7-8
The two kilns shown are about 300 feet long and 18 feet in diameter. The red-hot clinker is discharged at the near end into travelling grate coolers where it is cooled to a safe handling temperature.

Figure 7-9
Artist's rendering of a cement kiln with portion of shell cut away to reveal the burner and flame inside. (Courtesy of PCA)

During the residence time of three to four hours in the kiln, the feed loses water and carbon dioxide gas, and new compounds are formed. This loss in weight as the kiln feed passes through the different burning stages in the kiln is known as *ignition loss*. For this reason it requires approximately 3000 pounds of kiln feed to produce one ton of clinker. Hence, we cannot say that a kiln will produce 500 tons of clinker, because the kiln was fed 500 tons of kiln feed in a given period of time.

Another important point is that composition of the kiln feed is not identical to the composition of a cement made from that feed. The chemist can determine these differences and thus can compute the feed composition required to give a certain cement. Ignition loss of the kiln feed is usually about 30 to 35 percent, whereas the cement has a loss of less than 2 percent. Because part of the alkalies evolve into the gas stream during clinkering and are removed from the kiln system with waste dust at the rear of the kiln, alkali content of the feed is appreciably higher than that of the cement.

The fused lumps, or clinkers as they are called, pass through a cooler where they are cooled to a temperature that permits them to be handled, then to storage or finish grinding. See Figure 7-10.

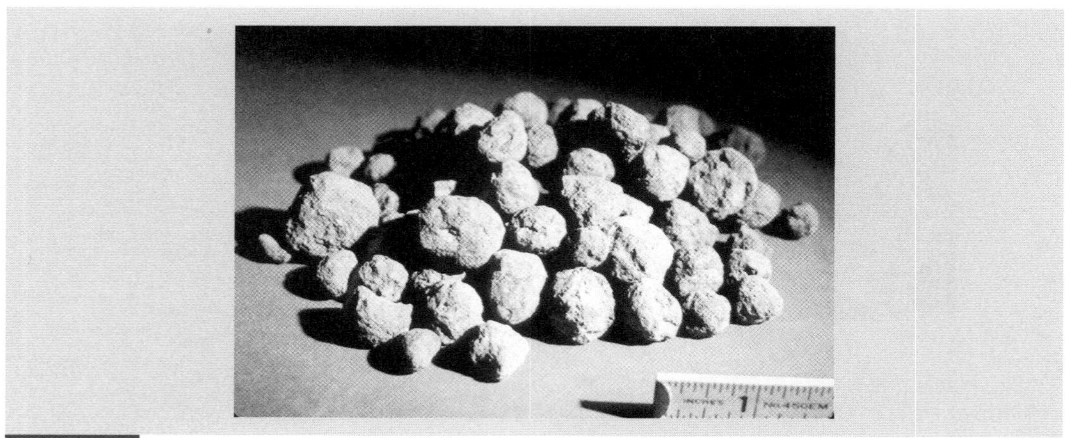

Figure 7-10
Cement clinkers are hard rock-like lumps that require only grinding to make portland cement. (Courtesy of PCA)

Clinkers can be stored for considerable periods of time. Thus, the manufacturer can make and stockpile clinker during periods of low demand to supply the requirements during periods of high construction activity.

Finish grinding mills are similar to the ball, tube and compartment mills used for raw grinding. During finish grinding, a small amount of gypsum, about 3 to 5 percent, is interground with the cement to control setting time of the cement when it is used. Other additives may be introduced at this time to facilitate grinding or to improve other properties of the cement. After grinding, the cement is conveyed to storage in silos or stock-houses and is now ready for shipment. Conveying can be accomplished by means of belt or screw conveyors, air-suspension troughs, or by pumping through pipes. Troughs and pipelines are most commonly used.

There are many brands of portland cement made by many producers, but they are all portland cement. In addition, there are several standard types of cement and a considerable number of specialized kinds of portland cement. All of them, however, are made from the same raw materials and are composed of the same compounds. The differences are in the relative proportions of the compounds and in the fineness. Some of the specialty cements have small amounts of additives to give the cement special properties for a certain limited usage.

7.3. Composition

A knowledge of the chemistry of cement is not necessary for those of us in the business of using the cement. We should, however, have a general idea of its composition.

Cement is composed of a number of compounds. These compounds are made by a reaction in the kiln. The analyst, in testing cement, is able to determine the percentage of these oxides in a sample of cement. From the oxide analysis the analyst computes the percentage of four compounds that compose almost all of the cement. The four primary compounds are tricalcium silicate (C_3S), dicalcium silicate (C_2S), tricalcium aluminate (C_3A) and tetracalcium aluminoferrite (C_4AF). There are other constituents, minor in amount, that have an important effect in the concrete.

Besides chemical tests, the analyst makes certain physical tests, including fineness, strength and setting time.

Cement particles, because of their small size, cannot be separated on sieves, so other means of measuring particle size are employed. The methods used are based on the assumption that the cement particles are spheres, and we can compute the specific surface of these spheres in a measured amount of cement. Size of the spheres is based on the Blaine air-permeability test (ASTM C 204). Table 7.2 is a typical analysis report of cement.

TABLE 7.2
ANALYSIS OF TYPE II CEMENT

CHEMICAL ANALYSIS		PHYSICAL PROPERTIES	
	PERCENT		
SiO_2	21.6	Fineness	
Al_2O_3	4.6	Blaine, sq cm/gm	3410
Fe_2O_3	2.8	Wagner, sq cm/gm	1920
CaO	62.8	Minus 325-mesh, %	95.4
MgO	3.9	Compressive Strength, psi	
SO_3	2.1	1-day cube	1300
Ign. loss	1.5	3-day cube	2550
Insol. Residue	0.1	7-day cube	3700
Free CaO	1.2	28-day cube	5650
Na_2O	0.41	Setting time, Hr & Min	
K_2O	0.24	Initial	3:25
Total Alkalies	0.57	Final	5:55
COMPOUND COMPOSITION			
C_3S	50	Air Content, %	10.4
C_2S	24	Autoclave Expansion, %	0.25
C_3A	7	Specific Gravity	3.15
C_4AF	8		

Note: This is a typical report that would be prepared by a cement plant chemist to show results of control tests.

7.4. Types of Portland Cement

The standard specification for portland cement is ASTM C 150. Table 7.3 shows typical properties of the five main types that are in common use. Modification of properties for each type is brought about by changing the compound composition and differences in grinding. The five types are described as follows:

TABLE 7.3
TYPICAL CHEMICAL AND COMPOUND COMPOSITION AND FINENESS OF CEMENTS

Type of portland cement	Range of chemical composition, %						Loss on ignition %	Na_2O eq	Range of potential compound composition, %				Blaine fineness m^2/kg
	SiO_2	Al_2O_3	Fe_2O_3	CaO	MgO	SO_3			C_3S	C_2S	C_3A	C_4AF	
I (min-max)	18.7-22.0	4.7-6.3	1.6-4.4	60.6-66.3	07-4.2	1.8-4.6	0.6-2.9	0.11-1.20	40-63	9-31	6-14	5-13	300-421
I (mean)	**20.5**	**5.4**	**2.6**	**63.9**	**2.1**	**3.0**	**1.4**	**0.61**	**54**	**18**	**10**	**8**	**369**
II (min-max)	20.0-23.2	3.4-5.5	2.4-4.8	60.2-65.9	06-4.8	2.1-4.0	0.0-3.1	0.05-1.12	37-68	6-32	2-8	7-15	318-480
II (mean)	**21.2**	**4.6**	**3.5**	**63.8**	**2.1**	**2.7**	**1.2**	**0.51**	**55**	**19**	**6**	**11**	**377**
III (min-max)	18.6-22.2	2.8-6.3	1.3-4.9	60.6-65.9	0.6-4.6	2.5-4.6	0.1-2.3	0.14-1.20	46-71	4-27	0-13	4-14	390-644
III (mean)	**20.6**	**4.9**	**2.8**	**63.4**	**2.2**	**3.5**	**1.3**	**0.56**	**55**	**17**	**9**	**8**	**548**
IV (min-max)	21.5-22.8	3.5-5.3	3.7-5.9	62.0-63.4	1.0-3.8	1.7-2.5	0.9-1.4	0.29-0.42	37-49	27-36	3-4	11-18	319-362
IV (mean)	**22.2**	**4.6**	**5.0**	**62.5**	**1.9**	**2.2**	**1.2**	**0.36**	**42**	**32**	**4**	**15**	**340**
V (min-max)	20.3-23.4	2.4-5.5	3.2-6.1	61.8-66.3	0.6-4.6	1.8-3.6	0.4-1.7	0.24-0.76	43-70	11-31	0-5	10-19	275-430
V (mean)	**21.9**	**3.9**	**4.2**	**63.8**	**2.2**	**2.3**	**1.0**	**0.48**	**54**	**22**	**4**	**13**	**373**

Type I. Common, regular or ordinary cement is a general purpose cement suitable for all uses when the special properties of the other types are not required. It is used in pavement and sidewalk construction, concrete buildings, bridges, railway structures, tanks and reservoirs, culverts, water pipe, masonry units and for all uses of cement or concrete not subject to such special conditions such as sulfate attack from soil or waters, or to objectionable temperature rise caused by the cement hydration. It will be supplied unless another type is specified.

Type II. Modified cement generates less heat and at a slower rate than Type I. It also has improved resistance to sulfate attack. It may be used in structures of considerable size such as large piers, heavy abutments and heavy retaining walls to minimize temperature rise, especially when the concrete is placed in warm weather. Type II cement may also be used where added precaution against moderate sulfate attack is important, as in drainage structures where sulfate concentrations in ground waters are higher than normal but are not unusually severe.

Type III. High-early-strength cement is used where high strengths are desired at early periods, usually a week or less. It is used when forms are to be moved as soon as possible or when concrete must be put into service quickly. In cold weather, the use of Type III cement allows reduction in the low-temperature protection period. It is also used when high strengths desired at early periods can be secured more satisfactorily or more economically than by using richer mixes of Type I cement. Concrete made with Type III cement has three-day compressive strength about equal to that made with Type I at seven days, and seven- day strength equal to Type I at 28 days. Ultimate strength is about the same as Type I.

Type IV. Low heat is a special cement, which generates less heat during hydration than Type II cement. Type IV cement develops strength at a much slower rate than Type I cement. It is used only in mass concrete such as large dams where the temperature rise resulting from heat generated during hardening is a critical factor. Type IV cement is available only on special order. It requires more curing than Type I.

Type V. Sulfate-resistant cement is a special cement used only in concrete exposed to severe sulfate action. It is used principally in soils or ground waters of high sulfate content. It gains strength more slowly than Type I cement. It is normally available from a few producers.

In addition to the five main types, three types of air-entraining portland cements are available:

Air-Entraining Portland Cements. Specifications for three types of air-entraining portland cement—Types IA, IIA and IIIA—are given in ASTM C 150. In these cements very small quantities of air-entraining materials are incorporated by intergrinding them with the clinker during manufacture. The air entraining portland cements produce concrete with improved resistance to freeze-thaw action and to scaling caused by chemicals applied for snow and ice removal. Note that air-entraining portland cements are available only in certain areas. Air-entrained concrete can also be produced by adding an air-entraining admixture directly to the concrete materials either before or during mixing. See Section 9.2.

7.5. Blended Hydraulic Cements

Blended hydraulic cements are used in concrete construction just like the various types of portland cement. The blended cements are produced by blending or intergrinding portland cement, ground granulated blast-furnace slag, fly ash and other pozzolans. See Figure 7-11.

Figure 7-11
Blended cements (ASTM C 595) use a combination of portland cement or clinker and gypsum blended or interground with pozzolans, slag or fly ash. (Courtesy of PCA)

Standard specification ASTM C 595 recognizes five classes of blended cement as follows:

Portland Blast-Furnace Slag Cement—Type IS. There are two types of cement, Type IS, portland blast-furnace slag cement, and Type IS-A, air-entraining portland blast-furnace slag cement. These cements can be used in general concrete construction when the special properties of other types are not required. There are also optional provisions for moderate heat of hydration (MH), or moderate sulfate resistance (MS), or both. The appropriate suffix may be added to the selected type designation, if desired. In these cements, granulated blast-furnace slag of selected quality is interground with portland cement clinker.

Type IS portland cement has a slower rate of strength development at early ages than Type I portland cement of comparable fineness. However, at 28 days the average strengths of concrete made with Type IS cement and Type I cement are approximately equal.

Portland-Pozzolan Cement—Type IP and Type P. These blended cements are manufactured by intergrinding portland cement clinker with a suitable pozzolan, by blending portland cement or portland blast-furnace slag cement and a pozzolan, or by a combination of intergrinding and blending. Type IP may be used for general construction, and Type P is used in construction where high early strengths are not required, as in massive structures such as piers, dams and large footings. Types IP and P may be designated as air-entraining, moderate sulfate resistant or with moderate heat of hydration by adding the suffixes A, MS or MH. For example, air-entraining Type IP cement with moderate sulfate resistance is designated as Type IP-A (MS). Type P may also be designated as low heat of hydration (LH).

Slag Cement—Type S. Slag cement is used in combination with portland cement for lower-strength concrete applications and in combination with hydrated lime in making masonry mortar. Air-entrainment may be designated in a slag cement by adding the suffix A. Types S and SA are not intended as principal cementing constituents of structural concrete.

Pozzolan—Modified Portland Cement-Type I (PM). This blended cement, for use in general concrete construction, is manufactured by combining portland cement or portland blast furnace slag cement and a finely ground pozzolan. The pozzolan content is less than 15 percent by weight of the finished cement. Air-entrainment, moderate sulfate resistance, or moderate heat of hydration may be designated in any combination by adding the suffixes A, MS or MH. For example, air-entraining Type I (PM) cement with moderate sulfate resistance is designated as Type I (PM)-A(MS).

Slag-Modified Portland Cement—Type I (SM). This blended cement is intended for general concrete construction. The slag component is less than 25 percent by weight of the finished cement. Type I (SM) cements may be designated with air entrainment, moderate sulfate resistance or moderate heat of hydration by adding the notations A, MS or MH.

The most common blended cements available are Types IP and IS. The United States uses a relatively small amount of blended cement compared to countries in Europe or Asia. However, this may change with consumer demands for products with specific properties, along with environmental and energy concerns.

ASTM C 1157 Performance Specification for Hydraulic Cement differs from ASTM C 150 and ASTM C 595 in that it does not establish specific chemical composition for the different types of cements. There are no restrictions as to the composition of ASTM C 1157 cements. The manufacturer can optimize ingredients, such as pozzolans and slags, to optimize for particular concrete properties. Individual constituents used to manufacture ASTM C 1157 cements must comply with the performance requirements specified in the standard. The standard also provides for several optional requirements, including one for cement with low reactivity to alkali-reactive aggregates.

7.6. Other Cements

In addition to the cements for which there are standard specifications, there are a number of special cements designed to meet specific conditions of usage and exposure. Some of these cements are of very limited usage and are available only from certain producers in certain areas. A user planning to specify any of these cements should check local availability.

Masonry Cement is a hydraulic cement for use in mortar for masonry construction. The primary ingredients used in the manufacture of masonry cement are portland cement, air-entraining agents and other additives that impart plasticity, workability and water retention to masonry mortars. Masonry cements must meet the requirements of ASTM C 91, which satisfies masonry cements such as Type N, Type S and Type M. Type N masonry cement is used in ASTM C 270 Type N and Type 0 mortars. Type S and Type M masonry cements are used in Type S and Type M mortars, respectively. In addition to mortar for masonry construction, masonry cements are used for parging and plaster (stucco). They must never be used for making concrete.

Mortar Cement is similar to masonry cement in use and function, consisting of a mixture of portland blended hydraulic cement and plasticizing materials, together with other materials introduced to enhance one or more properties such as setting time, workability, water retention and durability. Mortar cement must meet the requirements of ASTM C 1329, which requires lower air contents than masonry cement and includes a flexural bond strength requirement to enhance bond between the masonry mortar and the masonry units. Mortar cements are classified as Type N, S or M for use in ASTM C 270 Type 0, N, S and M mortar for masonry construction. Like masonry cement, mortar cement must never be used for making concrete.

White Portland Cement, which contains no iron, meets all specification requirements for Type I portland cement. It is pure white in color, and, unlike gray cement, its use makes possible an infinite range of tinted or colored concretes. Structurally it can be used exactly in the same manner as gray cement.

With the advent of modern forming techniques, architects and designers can, in many instances, obtain their structural and aesthetic requirements all in one forming operation. White cement can be used for any application where a white or true color surface is desired. Where economy demands, white can be used as an integral facing with gray. Typical uses are terrazzo floors, swimming pool plastering, manufacture of exterior stucco, precast concrete products, curtain walls, exposed aggregate applications, and many other decorative and structural treatments. With white cement, the selection of proper aggregates, especially light-colored sands, is very important. Tight formwork and careful placing are necessary. See Chapter 22.

Plastic (Stucco) Cement is a hydraulic cement, primarily used in portland cement-based plastering construction, consisting of a mixture of portland or blended hydraulic cement and plasticizing materials, together with other materials introduced to enhance one or more properties such as setting time, workability, water retention and durability. The term plastic refers to the ability of the cement to impart to the plaster a high degree of workability and the plaster to remain workable, or plastic, for a period of time. After initial application and floating on the wall, it can be reworked to obtain both densification and desired texture. Plastic cement is manufactured to ASTM C 1328. As indicated, it is manufactured primarily for portland cement plaster and stucco; however, its superior plasticity and workability also lend themselves very well to use in masonry mortar mixes.

Waterproof Portland Cement contains water-repellant additions such as stearates, oleates and tallow. Some degree of water repellency is imparted to concrete and mortar containing this cement, but alkalies liberated during hydration of the cement react with the waterproofing agent and tend to lessen its effectiveness.

Expansive Cement is a hydraulic cement that, when mixed with water, forms a paste that increases significantly in volume during and after setting and hardening. It should conform to ASTM C 845, Type K. It has been used to inhibit shrinkage of concrete, usually with regular portland cement, thus minimizing cracking. It appears to be especially advantageous in construction of water-retaining structures, as in sanitary engineering facilities where shrinkage cracking is significant and large placements are possible with fewer joints. It has also been used experimentally as "self-stressing" cement in prestressed concrete work. The concrete requires more mix water and is apt to suffer greater slump loss than concrete made with regular portland cement. It is not available from all cement producers.

7.7. Other Nonportland Cements

Aluminous. Also called calcium aluminate cement, high-alumina cement, or Lumnite, it contains calcium aluminates instead of calcium silicates. This is a refractory cement; that is, it is used for high-temperature exposures in ovens, furnaces and similar structures. It sets and develops strength faster than portland cement. Mortar or concrete containing a mixture of portland cement and aluminous cement will set hard in a few minutes. High-alumina cement should not be used for structural purposes.

Magnesite. Also called magnesium oxychloride cement or Sorel cement. Made by combining magnesium oxide with a saturated solution of magnesium chloride on the job. It has good strength and elastic properties. It is used for terrazzo because it is easily polished, and it is much used for lightweight floors in apartment buildings.

Rapid Setting. Some quick-setting proprietary cementitious compounds are available for repair work. Some of these are mixtures of portland and aluminous cements, portland cement and plaster of paris, or portland cement with a chemical accelerator or other additive. These materials all set and harden in a few minutes. Some can be used for sealing cracks and other damage against running water. For repair work to be exposed to water or to freezing and thawing, the user should check to determine whether the proposed material has a satisfactory history in a comparable exposure. Some of these materials deteriorate when exposed to water.

Some cement producers manufacture a regulated set cement that sets rapidly and apparently has a good durability record. Also on the market are a number of plastic compounds such as epoxy resin that are useful for special application. Repair materials are discussed further in Chapters 6 and 10.

7.8. Properties and Characteristics

Color. The color of cement is dependent on raw materials and is not an indication of quality. Fine-ground cement is normally lighter in color than a coarse cement of the same chemical composition.

Fineness. Greater cement fineness increases the rate at which cement hydrates and accelerates strength development. This contributes to the higher early strength of Type III portland cement. Finer cement fluffs more. Concrete made with fine cement may have more drying shrinkage.

Soundness. Soundness of a hardened cement paste is its ability to retain its volume after setting. Lack of soundness in cement is caused by free lime or excessive amounts of magnesia. Most specifications for cement limit the magnesia content and the autoclave expansion.

Setting Time. Time-of-setting tests are performed to determine whether a cement paste remains plastic long enough to permit normal placing of concrete without hampering finishing operations. The length of time that a concrete mixture remains plastic is generally more dependent upon the temperature and water content of the paste than upon the setting time of the cement.

Compressive Strength. Compressive strength of portland cement is the strength of standard 2-inch mortar cubes. These cubes are made and cured in a prescribed manner using a "standard sand." Strengths at various ages are indicative of the strength-producing characteristics of the cement but cannot be used to accurately predict concrete strengths because of the many variables in concrete mixtures. Comparative strengths of the five standard types are shown in Table 7.4.

TABLE 7.4
RELATIVE COMPRESSIVE-STRENGTH VALUES

TYPE OF CEMENT	AGE OF SPECIMENS			
	1 DAY	7 DAYS	28 DAYS	3 MONTHS
I Common or regular	100%	100%	100%	100%
II Modified	75	85	90	100
III High early strength	190	120	110	100
IV Low heat	55	55	75	100
V Sulfate resistant	65	75	85	100

Note: If we call the strength of regular cement 100 percent at any age, then the other cements will have the percentages shown. Example: If regular cement has 4000 psi at 28 days, then Type IV will have 75 percent of 4000 psi, or 3000 psi.

Heat of Hydration. Heat of hydration is the heat generated when cement and water react. In certain structures, such as massive dams, the rate and amount of heat generated are important. If this heat is not rapidly dissipated, a significant rise in temperature occurs. In mass concrete a rise in temperature may be undesirable insofar as it is accompanied by thermal expansion. Subsequent cooling of the hardened concrete to ambient temperature creates undesirable stresses and cracking. On the other hand, a rise in concrete temperature caused by heat of hydration is often beneficial in cold weather because it helps maintain favorable curing temperatures.

Loss on Ignition. Loss on ignition is determined by heating a cement sample of known weight to a full red heat. The weight loss of the sample is then determined. Normally this value does not exceed about 2 percent. Higher values usually indicate that the cement has prehydrated because of improper or prolonged storage.

Specific Gravity. Specific gravity of portland cement is generally about 3.15. Portland blast-furnace slag cement and pozzolan cement are about 2.90. The specific gravity of a cement is not an indication of quality; it is a value necessary for mix design calculations.

7.9. Transporting and Conveying

In-plant conveying of cement is accomplished with a great variety of equipment. Belt conveyors, bucket elevators, screws, troughs and pipelines are all used. Closed units are essential, not only to prevent a loss of cement, but because of rigid environmental pollution controls.

One system consists of a tank into which the cement flows by gravity from a truck, rail car or silo. The tank is then closed and air is admitted under pressure to aerate the cement and carry it out through a pipeline. The entire operation is automatic. In another system the cement flows by gravity into a short screw that forces the cement into a fluidizing chamber where it joins a stream of air that conveys the cement through a pipe. Pipelines can carry cement for hundreds of feet horizontally and can also move it vertically.

In the pneumatic trough conveyor, sometimes known as an air float or air slide, cement is carried in suspension in air and flows by gravity down the slightly sloping enclosed trough.

All of these systems require large volumes of dry, clean air, which requires that adequate dust collection facilities be provided.

Shipment of cement to the customer is made either in bulk or in 94-pound paper bags (1 cubic foot). Special trucks, rail cars, barges and ships are provided by the industry for movement of bulk cement. See Figures 7-12, 7-13 and 7-14. Users of large quantities of cement, such as ready-mixed concrete producers, heavy construction contractors and large precasting plants, specify bulk shipment, sometimes in their own vehicles.

Figure 7-12

Figure 7-13
Special rail car for transporting bulk cement. When the tank is pressurized with low-pressure air, the cement flows out through a hose attached to an outlet on the bottom of the tank.

There are two basic types of hauling equipment: bottom dumps and air pressure. Both truck-trailer combinations and rail cars are available in the two basic types.

Figure 7-14
Transporting bulk cement by ship tanker. (Courtesy of PCA)

In the bottom dump carrier, the cement flows by gravity out of a gate in the bottom of the vehicle. There may be more than one gate, with the bottom of the cement tank sloping toward the gate or gates. To unload, a boot on the receiving bin, which is located under the roadway or tracks, is attached to the gate opening on the carrier, the gate is opened, and the cement passes into the receiving bin from which it is conveyed to the storage silo.

Air vehicles can be pressurized; that is, they are designed to withstand the low pressure necessary to force the cement out. Many have their own air compressors. Unloading is accomplished by applying pressure to the cement tank to aerate the cement and force it out through a hose that is attached to the outlet of the tank. The hose leads to a pipe that delivers the cement to the silo. See Figure 7-19.

Cement for the small users, lumber yards, building material dealers and similar customers is furnished in bags which are usually grouped on pallets. See Figure 7-15. Practically all of the specialty cements are handled in bags.

Figure 7-15
A truck and trailer loaded with bagged cement on pallets. (Courtesy of PCA)

7.10. Storage

Cement, when protected from moisture and air, can be stored for months without deterioration. Moisture causes prehydration, and carbon dioxide causes carbonation, both of which result in gradual loss of desirable properties of the cement.

Cement, when used, should be free-flowing and completely free of lumps. Sometimes cement that has been in the lower bags of large piles for a considerable period assumes a "warehouse pack or set;" that is, it appears to be hardened. This condition can be minimized by piling cement for long storage not more than seven bags high. However, if warehouse set does develop, it can usually be broken up by rolling the sack on the floor, and the cement again becomes free-flowing and is suitable for use. The presence of lumps that cannot be pulverized readily in the hand, however, is indication that moisture has been absorbed. Such cement can be used for unimportant work by screening out the lumps.

Different brands and types of cement must be stored separately. Different brands should never be mixed, even though they are the same type, because of differences in color and other properties. Cement salvaged from spillage around the plant or broken bags should not be used.

Bulk Cement is usually stored in vertical silos or bins, although horizontal tanks are frequently used at temporary locations—for instance, at a batching plant for a major highway project. See Figure 7-16. A horizontal tank, sometimes referred to as a guppy, can be hauled over the highway as a semitrailer, then set up at the temporary plant and pipe connections made to receive hopper, pump and batching bin.

Figure 7-16
Horizontal storage tanks for bulk cement at a temporary transfer plant. Also shown is a semitrailer pressure vehicle.

When cement is stored in a vertical silo, there is a tendency for a hollow core to develop in the center of the silo when cement is withdrawn from the bottom. See Figure 7-17. When this happens, new cement, added to the cement already in the silo, will pass through to the discharge gate, bypassing the old cement already in the silo. For this reason it is a good idea to draw down and empty the silo every few months.

Sometimes cement will plug or arch and refuse to come out of the silo, leading the operator to believe that the silo is empty. See Figure 7-17. To prevent this, air jets can be installed near the bottom of the conical bottom to aerate or "fluff" the cement to make it free-flowing again. See Figure 7-18.

Silos, bins and all equipment for handling cement must be weathertight to prevent water from coming into contact with the cement. Gates, hatches, scuttles and manholes must be gasketed or sealed tightly. Pollution control laws strictly limit the amount of dust that is permitted to escape. See Figure 7-19.

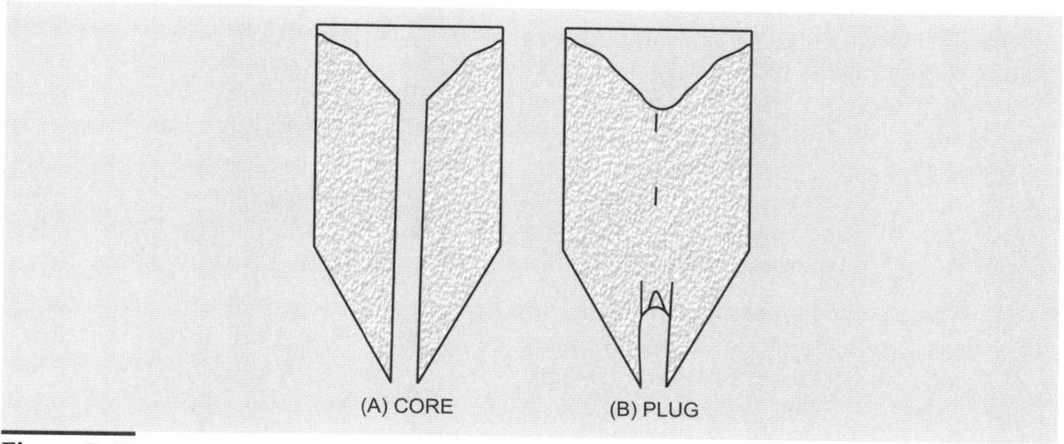

Figure 7-17
In Figure A hollow core has formed causing new cement to discharge first. In Figure B a plug has formed, preventing cement discharge.

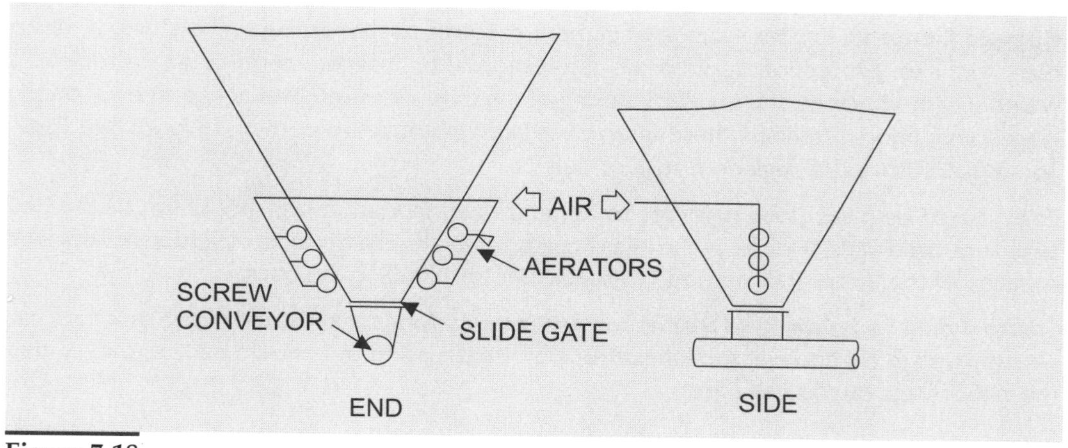

Figure 7-18
Air admitted near the bottom of a silo agitates and fluffs the cement so it will be free flowing.

Figure 7-19
Loading bulk cement into a truck semitrailer. The vehicle is standing on weigh scales, which enables the operator to cut off the flow of cement as the weight approaches the load limit.

Figure 7-20
An "air" truck and trailer. The tanks can be pressurized when unloading to force the cement out through a hose and pipeline.

Bagged Cement. Proper storage of cement requires the exclusion of moisture, which in turn requires protection against air circulation. For bagged cement this requires a weathertight structure and proper stacking of the bags. Bags can be stacked directly on the warehouse floor, provided there is no possibility of moisture coming through the floor. Some users store the bags on planks or pallets.

Bags should be piled close together to minimize air circulation but should not be stacked against outside walls. Height of the piles is determined by the physical limitations of the load capacity of the floor, and method or equipment for handling the bags.

Stacks should be arranged so that no moving or restacking is necessary and so that bags can be removed in chronological order of receipt. First in, first out should be the rule, so that the oldest bags can be used first.

For temporary storage on the jobsite where no storeroom is available, bags can be stored in the open if they are adequately protected. They should be on a tight platform well above the ground and should be covered with canvas or sheet plastic. Covering should cover the top of the pile and down all sides to the bottom and be well anchored so that it cannot be blown off by the wind. Edges of the platform must be protected to prevent rainwater from collecting and flowing under the bags. Some users place the first layer of bags on sheet plastic, then wrap the plastic entirely around the pile until it is completely enclosed. Folds in the plastic must be made so that they will not conduct water into the pile.

7.11 Supplementary Cementitious Materials

The practice of using other materials with cementing value in combination with portland cement, such as fly ash, pozzolans and silica fume, has been growing since the 1970s. Supplementary cementitious materials are added to concrete as part of the total required cementitious materials. They may be used in addition to or as a partial replacement for portland cement, depending on the properties of the materials and the desired effect on the concrete. Two or more of these supplementary materials can be combined to optimize the concrete properties. Supplementary cementitious materials are used in at least 60 percent of all ready-mixed concrete produced. Note that, with the advent of using other cementitious materials in combination with portland cement, the traditional water-cement ratio (w/c) has been renamed "water-cementitious materials ratio" (w/cm) in the ACI 318

Standard. For further discussion of this conceptual change see Sections 4.1 and 12.1. Three of the more commonly used supplementary materials are:

Fly Ash. Fly ash is the most widely used supplementary cementitious material in concrete mixtures. It is a by-product of the combustion of pulverized coal in electric power generating plants. See Figure 7-21.

Figure 7-21
Fly ash, a powder resembling cement, has been used in concrete as a partial replacement for portland cement since the 1930s. The color is usually gray, tan or white. (Courtesy of PCA)

There are two basic classes of fly ash: Class F is normally produced from burning anthracite or bituminous coal. This class has pozzolanic properties. Class C is normally produced from burning lignite or subbituminous coal. It has some cementitious properties, as well as pozzolanic properties.

Both classes have been successfully used. Lowered heat of hydration and better resistance to alkali-silica reaction and sulfate attack have been reported, as well as better economy depending on relative costs of cement, fly ash and other materials. Strength-producing properties can vary widely for different ashes. Setting time and control of entrained air may be affected slightly. Workability of the fresh concrete is improved, the mix water requirement is commonly lower than for comparable non-fly ash mixes, bleeding is less, and drying shrinkage is equal to control mixes. In any event, even though the ash complies with the requirements of ASTM C 618, laboratory tests or a service record of the material should be made. Compatibility of the proposed ash, cement and admixture should be investigated. Properly used, fly ash can be beneficial in high-strength concrete as well as in concrete of more moderate strength.

Pozzolans. As defined in ASTM C 618, a pozzolan is "a siliceous or siliceous and aluminous material, which in itself possesses little or no cementitious value but will, in finely divided form and in the presence of moisture, chemically react with calcium hydroxide at ordinary temperatures to form compounds possessing cementitious properties." The word pozzolan comes from the name of the town of Pozzuoli, Italy, situated near the source of volcanic ash used by the Romans in construction of many of their structures.

Pozzolans may be either natural or manufactured; some natural pozzolans are unprocessed; others are processed in some manner.

Natural Pozzolans. Volcanic tuff, volcanic ash, pumicite and obsidian are some of the common natural pozzolans. Pumicite usually requires no processing to prepare it for use. In Friant Dam the natural pumicite in the mass concrete was used directly from the deposit with no processing whatsoever. See Figure 7-22. This pumicite had over 95 percent passing the 325-mesh sieve. Other natural pozzolans are siliceous sedimentary rocks such as opaline chert and diatomaceous earth, the latter sometimes used without processing. The other natural pozzolans require grinding and size classification to reduce the material to the fine powder suitable for use.

Processed Natural Pozzolans. Calcined or burnt shales and clays, heated in rotary kilns, cooled, then crushed and ground to the required fineness, are a source of processed natural pozzolans.

Manufactured Pozzolans consist of crushed and ground blast furnace slag, fly ash and silica fume. Fly ash, sometimes called precipitator ash, is the fine combustion product resulting from burning certain types of pulverized coal in industrial furnaces. Principal sources are steam power plants. Fly ash consists of very fine spherical particles, which are carried out of the furnace in the flue gas and subsequently collected in precipitators.

Pozzolanic Action. Hydrated lime or calcium hydroxide is one of the products formed in the hydration of portland cement. This compound does not contribute to the strength of concrete. It is soluble in water and is removed by leaching. The pozzolan reacts freely with lime and will show some chemical reactivity when used with portland cement. The principal product of the reaction is a compound that has a relatively low solubility.

The chemical reaction between lime and pozzolans is not clearly understood. However, the generalities are well known and help to explain the physical effect of pozzolans in concrete. The pozzolan in itself is sound and insoluble, and appears to have no deleterious effect on the hardening of concrete.

Pozzolans usually improve workability. It has been observed that bleeding is eliminated or substantially reduced. Increased fines also reduce the tendency for concrete to segregate. The only detrimental effect of pozzolans on plastic concrete seems to be an increase in the amount of air-entraining agent required for the desired air content.

The principal effect of pozzolans on the hardened concrete is obtained by the reaction of pozzolan with the hydrated lime described above. This reaction increases the tensile and compressive strength, lowers the permeability, reduces leaching and improves the sulfate resistance of concrete.

An added benefit of some pozzolans is their effect in reducing the alkali-silica reaction in concrete. General effects of pozzolans are summarized in Table 7.5.

Use of Pozzolans. Pozzolans are used in large massive structure such as dams, where it is desirable to keep the heat of hydration as low as possible, in concrete exposed to seawater or sulfate attack, and as alkali-aggregate inhibitors. In those areas where the cost of a pozzolan is appreciably less than the cost of portland cement, proper use of pozzolan as a replacement for part of the cement results in a saving in cost per cubic yard of concrete. However, pozzolan should not be used without complete understanding of its character or without trials with the materials proposed for the project, including strength and durability tests of the concrete. The amount varies with the type of pozzolan, mix, exposure and other factors, ranging from 10 to 30 percent replacement of the cement.

A pozzolan is usually proportioned in the batch as a replacement for part of the cement. When used in this way, it reduces water permeability, especially for lean mixes, and improves resistance to aggressive solutions, such as seawater and sulfate or acid waters.

Effects of Pozzolans are summarized in Table 9.5. In general, heat generation in mass concrete is lessened because of the lower cement content.

Because pozzolans are usually used in lean mixes, there is an improvement in workability, with a reduction in segregation and bleeding tendencies. The effect on mixing water requirement is variable. Fly ash mixes require less water than plain concrete for the same consistency. Water requirements may be higher for some pozzolans, although the use of air entrainment tends to counterbalance this.

The effect on strength is quite variable, depending on the type of pozzolan used. It is usually improved for lean mixes. Strength gain is slow; hence, longer curing is necessary. With some pozzolans, strength after one year is higher than for plain concrete. Fly ash gives flexural and compressive strengths equal or superior to plain concrete after a year.

Most pozzolans have an inhibiting action on the alkali-silica reaction. Those high in opal are most effective.

Probably one of the most important uses of pozzolans is that they impart to concrete enhanced resistance to moderate sulfate attack, especially pozzolans high in silica. However, Type V cement should be used in a severe sulfate exposure. Whether to use Type V cement or a pozzolan depends upon economic considerations and the results of laboratory tests.

Figure 7-22
The natural pumicite used as a pozzolan in Friant Dam (California) required no processing to in preparing it for use. The dark- colored layer is a stratum of sandy pumicite that was stripped and wasted.

When a pozzolan is added to concrete as an addition to the cement, most of the above effects result, especially if the original concrete contained insufficient fines.

Silica Fume. Silica fume, also referred to as microsilica or condensed silica fume, is another material that is used in addition to portland cement. This gray or bluish-green-gray powdery product is a result of the reduction of high-purity quartz with coal in an electric arc furnace in the manufacture of silicon or ferrosilicon alloy. Silica fume is an extremely fine airborne material like fly ash, spherical in shape, about 100 times smaller than average

TABLE 7.5
GENERAL EFFECTS OF POZZOLANS

Cement quantity: Reduced.
Heat generation in mass concrete: Reduced.
Workability: Increased, especially for lean mixes. Segregation and bleeding tendencies reduced.
Alkali-silica reaction: Inhabited by most pozzolans.
Mixing water requirement: Fly ash mixes require less water than plain concrete for same consistency. Water requirement may be higher for some pozzolans, although the use of air entrainment tends to counterbalance this.
Durability: Effects vary.
Strength: Quite variable, depending on type of pozzolan used. Usually improved for lean mixes. Strength gain is slow; hence, longer curing is necessary. With some pozzolans, strength after one year is higher than for plain concrete. Fly ash gives flexural and compressive strengths equal or superior to plain concrete after a year.
Drying Shrinkage: Little effect.
Sulfate Resistance: Improved by most pozzolans, especially those high in silica.
De-icer scaling resistance: Slightly reduced.

cement particles. See Figure 7-23. The bulk density is about 16 to 19 pounds per cubic foot. Silica fume is sold in powder or liquid form. Silica fume for use in concrete must comply with the requirements of ASTM C 1240.

Silica fume as an admixture is used to provide a more impermeable concrete for applications such as parking deck slabs exposed to chlorides from de-icing salts. Silica fume is also a key ingredient in the production of very high-strength concrete (15,000 psi or greater). It is added as an additional cementitious material to the regular amount of portland cement, not as a partial substitute, at a dosage rate of 5 to 20 percent by weight of cement. Mix proportioning, production methods and the placing and curing procedures for silica fume concrete require a more concerted quality control effort than that for conventional concrete. It is imperative that the engineer, concrete supplier, contractor and inspector work as a team to ensure consistently high quality when silica fume concrete is specified.

Figure 7-23
Silica fume powder. (Courtesy of PCA)

Aggregates

8.1. Classification of Rocks
- Igneous Rocks
- Sedimentary Rocks
- Metamorphic Rocks

8.2. Sources of Aggregates
- Origin
- Preliminary Approval

8.3. Characteristics of Aggregates
- Soundness and Volume Stability
- Cleanness
- Hardness and Toughness
- Grading
- Particle Shape
- Texture
- Reactivity
- Specific Gravity
- Absorption
- Moisture Content
- Unit Weight and Voids

PROCESSING AGGREGATES

8.4. Sand and Gravel Deposits
- Coarse Aggregate
- Scrubbing
- Sand
- Aggregate Beneficiation

8.5. Quarries

8.6. Stockpiling

TESTING AGGREGATES

8.7. Sampling
- Quartering Samples

8.8. Testing

SPECIAL KINDS OF AGGREGATES

8.9. Blast Furnace Slag

8.10. Lightweight and Heavyweight Aggregates

Aggregates for concrete are sand, gravel, crushed stone, crushed slag, pumice and manufactured products. There are a few other specialty items, but they are of little concern to us. In this chapter we will consider sand, gravel, crushed stone and slag. Aggregates are frequently called a filler material because they occupy between 60 percent and 80 percent of the volume of ordinary concrete. See Figure 8-1. Because the aggregates occupy such a large part of the volume, their properties are very important in their effect on concrete.

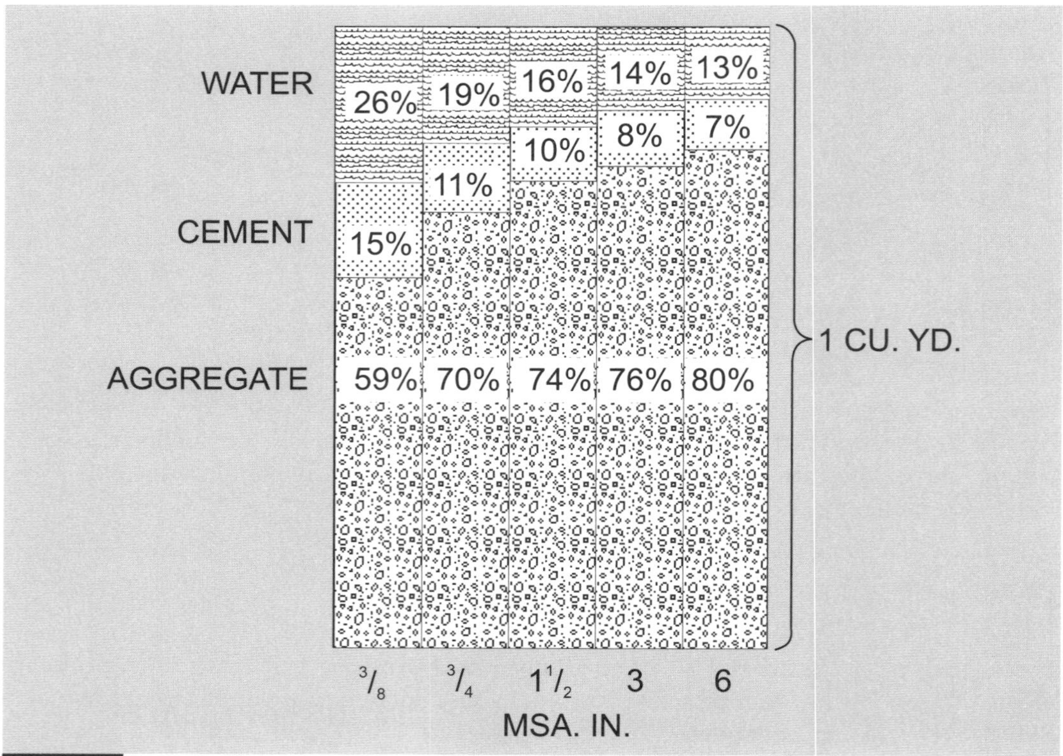

Figure 8-1
The aggregates occupy about three quarters of the solid volume of the concrete, the balance being cement, water and air voids. Each vertical column in the figure represents 1 cubic yard of concrete.

Aggregates are normally considered to be inert; that is, they are inactive. When they have been incorporated into the concrete, they do not enter into any chemical or physical reactions within the concrete. There are a few rock types and minerals, however, that are not inert and that under certain conditions will react in the hardened concrete to cause cracking, popouts and other disintegration. A knowledge and understanding of the characteristics of aggregates and how to determine these characteristics is accordingly essential if we are going to produce good and durable concrete.

8.1. Classification of Rocks

All natural aggregates owe their beginnings to the solid rock of the earth's crust, rock that consists of a mixture of different minerals. Over a period of many thousands of years, through the action of freezing and thawing, heating and cooling, wetting and drying, glaciers, streams and rivers, roots and chemicals, the solid rock is broken up into small pieces. Coarse pieces are called gravel. Fine particles are called sand, silt and clay.

The great masses of rock that compose the crust consist of three fundamental classes from which, through the processes of weathering described above, all gravels and soils emanate. The three classes are described here.

Igneous rocks were at one time intensely heated and in a liquid condition known as magma. Upon cooling, the magma solidified into crystalline bodies of varying sizes and shapes. Common igneous rocks are granite, diorite, gabbro, basalt and trap-rock (a variety of basalt). There are many less common varieties. The differences among the various kinds of igneous rocks are due to their mineralogical composition and to the different size of the individual crystals; granites are coarse grained and light in color, basalt is dark and fine grained. Igneous rocks, being hard and tough, are a good source of crushed rock. Sands and gravels of igneous origin, if the material has not weathered excessively, make good aggregate.

Volcanic scoria and pumice are used in some areas for lightweight aggregate.

Sedimentary rocks are secondary in their origin; the materials they comprise are the result of the weathering of some previously existing rock. Some sedimentary rocks are the result of mechanical transportation of sediments by streams into a body of water, where the sediments are deposited in layers. Other sedimentary rocks consist of materials dissolved by water circulating through rocks, carried to lakes or seas, where they are deposited by a chemical process. In the coarse-grained sedimentary rocks, the individual particles are cemented together by silica, iron oxide or calcium carbonate. Common sedimentary rocks are conglomerates and contain coarse pebbles, sandstone, shale, limestone and dolomite.

Sedimentary rocks range from poor to excellent as a source of aggregate. Hard and dense sandstones and limestones are good; shales are usually laminated and of poor quality; conglomerates are poor; dolomite, if hard and dense, is good.

Metamorphic rocks are rocks that have been metamorphosed, or changed, as a result of tremendous pressure, heat and chemical activity. All metamorphic rocks were at one time either igneous or sedimentary. They are usually banded or laminated. Types of metamorphosed rocks are gneiss, schist, slate, quartzite and marble. Their value as aggregates varies from poor to excellent depending on their hardness, density and freedom from laminations.

8.2. Sources of Aggregates

Origin. The natural aggregates that are used in concrete come from two sources: either solid bedrock or deposits of sand and gravel. Whether to use one or the other is a matter of availability and economics. Gravel deposits are usually preferred because they contain a supply of sand, and most users prefer gravel coarse aggregate over crushed rock. Deposits of good gravel are lacking in some areas, making necessary the importation of aggregates if no rock sources are nearby. Continued expansion of our cities results in many potential aggregate deposits being zoned and subdivided, thus reducing the available materials.

Quarries are exploited by blasting to make pieces that can be handled, then crushing and screening the pieces. The product, called crushed rock or stone, is angular with sharp edges. See Figure 8-2. In some plants, a manufactured sand is made by further crushing and grinding.

Figure 8-2
Crushed granitic rock is sharp and angular.

Gravel deposits may originate in several ways. These are stream and lake deposits, glacial deposits, alluvial fans, talus and windblown materials.

Windblown deposits consist of fine sand called blow sand. See Figure 8-3. These are apt to form in desert areas, although there are windblown deposits in more humid areas as well, along ocean fronts and near some large lakes, such as the south end of Lake Michigan. The sand is too fine to be used as concrete sand, but can be used as a blending sand to improve the grading of a coarse sand.

Figure 8-3
Fine blow sand, transported by the wind, is sometimes used as a blending sand with coarse or manufactured sand.

A talus, usually found at the bottom of a steep slope, consists of fragments broken off the solid rock above. These deposits are found in mountainous areas and are a very minor source of aggregate.

Glacial deposits exist throughout most of Canada and the northern midwest and northeastern U.S and are the remnants of glaciation that occurred during the ice age about 10,000 years ago. Kames, eskers and moraines are small hills, frequently long and narrow, containing a mixture of silt, sand and rocks, that are a source of sand and gravel. The material is a mixture of all sizes and many rock types, some of it of poor quality. With proper processing, good aggregates can be produced. Many aggregate producers in glaciated areas obtain their material from glacial deposits.

Probably the largest share of sand and gravel comes from water-transported deposits. There are two general kinds: stream and lake deposits, and outwash slopes or alluvial fans. Streams carry earth particles, the size depending on the velocity of the water. Floods and other changes affect the carrying capacity of the stream, which at times deposits material and at times cuts away material. The result are flood plains and terraces, sometimes of great depth, containing sand and gravel. The material is graded from fine to coarse and is quite well rounded; most of the unsound particles have been eliminated by the action of the stream. See Figures 8-4, 8-5 and 8-6. Numerous plants of all sizes operate on materials along water courses.

Figure 8-4
The high terrace rises above the flood plain. Both are potential sources of sand and gravel.

Figure 8-5
A close-up view of the terrace shown in Figure 8-4 showing the gravel exposed in the face.

Figure 8-6
A flood plain of a small river, which was later developed as a source of aggregate.

Alluvial fans are built up at the mouths of ravines and canyons along the base of mountains. The Piedmont Plain of the Atlantic coastal plain in the United States, which lies between the Appalachian highland and the coastal plain proper, is an example of an extensive series of alluvial fans. Materials may or may not be stratified and are usually somewhat angular. Some deposits are hundreds of feet thick and extend for miles along the base of a range of mountains. The alluvial fans at the base of the San Gabriel and San Bernardino Mountains in California have been worked for years by several large aggregate producers.

Preliminary Approval. One of the first steps at the start of a job is to obtain information about the aggregates. If preliminary aggregate tests are required, the engineer should have the laboratory obtain samples and make tests as soon as the contractor has designated the source. If no preliminary tests are required, the engineer should at least be sure that the proposed source is an established one, with a satisfactory record of use as regards durability, strength and other desirable qualities of concrete.

Large jobs in rural or isolated areas usually require that a new source of aggregates be developed. Prospecting for aggregate sources requires painstaking and thorough exploration of an area and should be done by experts familiar with the geological processes by which aggregate deposits are formed and the effects of aggregate characteristics on the properties of concrete.

Aerial reconnaissance, geophysical procedures and electrical and sonic methods are used extensively in making geological studies. Topographic and geological maps, available from state and federal agencies, are of considerable help in locating promising areas. Cut banks and bars along streams frequently provide information, as do excavations such as railroad and highway cuts. In the final analysis, if the project is large enough, test pits, core drill holes, cased holes and similar methods may be indicated.

Fortunately for us, most construction jobs, especially in urban areas, are supplied with concrete from established ready-mix plants, which in turn have well-established aggregate sources. These established sources have usually been well inspected and tested by various agencies, and reliable test results are available. The engineer should make sure that the test reports offered truly represent the material to be used and that there have been no changes in characteristics of the pit or quarry face being worked. There can be an appreciable difference in materials that come from different areas or strata in the deposit.

For the purpose of preliminary approval, the quality of the material is of greatest importance. Grading is of secondary importance, as it can be corrected by processing.

An inspection of structures known to have been made with the proposed materials, as described in Chapter 1, can be quite revealing. Regardless of whether the proposed aggregate meets all the code requirements, a careful examination of structures, performed by a competent engineer, will yield much valuable information. The structures examined, of course, must be subject to the same exposure conditions as the proposed structure, and the materials used must be the same as the ones proposed. In making such an inspection we must not be diverted by defects of the concrete that resulted from poor construction, such as form defects, excessive drying shrinkage caused by overwatered mixes, rock pockets and similar blemishes. We are concerned with the presence of popouts, spalling caused by unsound particles, soft spots, swelling and cracking caused by internal expansion, stains, unusually rapid weathering in freezing and thawing environment, and loss of bond between paste and aggregate particles.

The effect of the aggregates on the several properties of concrete are discussed in the chapters covering concrete properties. These should be reviewed at this time.

8.3. Characteristics of Aggregates

The quality of an aggregate is determined by visual examination and by physical and chemical tests. Seven properties designate the quality of an aggregate: it must be sound and stable, clean, hard and tough, well graded, suitably shaped, rough textured and nonreactive. These are the basic properties that apply to any aggregate. Listed in Table 8.1 are a number of characteristics and their effects on concrete. The characteristics and test methods listed in Table 8.1 are referenced in ASTM C 33. Reference 8.1 presents additional properties of concrete influenced by aggregate characteristics and corresponding test methods.

TABLE 8.1
CHARACTERISTICS OF AGGREGATES

CHARACTERISTIC	SIGNIFICANCE IN CONCRETE	TEST METHOD
Soundness	Strength, durability, appearance	ASTM C 88
Chemical stability	Alkali-silica reaction, popouts, disintegration, appearance	ASTM C 227, C 289, C 295, or C 342
Abrasion resistance	Wear resistance of floors, hardness	ASTM C 131 or C 535
Grading, or sieve analysis	Workability, density, economy, shrinkage	ASTM C 136
Maximum size of aggregate	Economy, shrinkage, density, strength	Inspection
Percentage of crushed particles	Workability, economy, strength	Separate and count particles
Particle shape	Workability, economy, shrinkage, strength	Inspection
Surface texture	Bond, strength, durability	Inspection
Specific gravity	Durability, density, needed for mix computations	ASTM C 127 or C 128
Absorption	Durability, needed for mix computations and control	ASTM C 127 or C 128
Moisture content	Needed for mix computations and control	ASTM C 70 or C 566 or other

Test methods listed in ASTM C 33 for evaluating aggregates are described below.

Soundness and Volume Stability. Determined by alternately soaking a sample in sodium or magnesium sulfate and drying for five cycles. The effect on the aggregate is similar to cycles of freezing and thawing, or wetting and drying. This test identifies aggregate particles that are highly absorptive, porous, easily cleavable, weak, or tend to swell when saturated. Certain shales and slate are examples of rock types that are apt to change volume with wetting and drying. Unsound aggregate produces unsound concrete—weak, nondurable, with cracking, popouts, spalling and poor appearance.

Unfortunately, some aggregates that pass the sulfate soundness test produce concrete having a low resistance to freezing and thawing. Conversely, some aggregates that fail the test make good concrete. For this reason, aggregates should not be accepted or rejected on the basis of only one test. Additional tests should be made, as described in this section, and the history of the material reviewed.

Cleanness. The presence of most contaminating substances can be found by visual examination of the material, and the amount can be evaluated by tests. There are a number of substances that are considered contaminating materials.

Specifications limit the amount of deleterious substances in the aggregate to a total of 4 or 5 percent by weight, with individual limits on each of the substances classified as deleterious. Soft and unsound fragments, clay lumps, coal, lignite, porous chert, shells, material finer than the No. 200 sieve, conglomerate and cemented particles are usually classified as deleterious. Other substances may be included, depending on local conditions. Table 8.2 shows the effects of some of the deleterious materials in concrete. Clay and silt (minus 200 mesh) are determined by washing a sample of the aggregate and decanting over a 200-mesh sieve. In concrete, excessive clay and silt form thin coatings on aggregate particles that interfere with bond of paste to aggregate, and also may increase mixing water requirement. Organic materials, such as humus in top soil, roots, grass, leaves and bits of wood delay setting and hardening of the cement and sometimes contribute to deterioration of the concrete. Coal, lignite and other lightweight materials, especially if they occur at or near the surface of the concrete, may disintegrate. Coatings and encrustations on aggregate particles interfere with bond. An opaline coating may be alkali-reactive. Caliche, a carbonate coating, occurs in some deposits in arid regions. Clay lumps, soft particles and lightweight particles may absorb some of the mixing water, break up during mixing, or result in soft spots in the concrete.

TABLE 8.2
CLEANNESS OF AGGREGATE

There are several kinds of deleterious or harmful materials that can get into the finished products. The method of detection and the effect on concrete are slightly different for each contaminant.

DELETERIOUS MATERIAL	SIGNIFICANCE IN CONCRETE	ASTM TEST METHOD
Clay and silt	Bond, durability, shrinkage, mix water requirement, strength	C 117
Organic impurities	Strength, durability, appearance	C 40 and C 87
Clay lumps	Appearance, mix water requirement, durability	C 142
Soft particles	Durability, appearance	C 235
Lightweight particles	Appearance, strength, durability	C 123

Results in concrete are low strength, unsoundness, poor durability, unsightly appearance, excessive shrinkage, popouts and stains.

Hardness and Toughness. This property is often used as a general index of aggregate quality and is especially important for concrete to be used in pavements and floors. As in the soundness test, there are cases in which the test does not accurately indicate the quality of the material, another reason for making the decision as to whether to use an aggregate for a certain exposure on the basis of tests of more than one property. Hardness and toughness are found by determining the abrasion resistance of the aggregate in a test known as the Los Angeles rattler test, which is a standard grinding test in a small ball mill. High abrasion loss indicates probable low-strength concrete and inferior resistance to the abrasion caused by traffic. See Figure 8-7. Specification limits for unsound and deleterious aggregates are listed in Tables 8.3 and 8.4.

Figure 8-7
The abrasion test of coarse aggregate is made in this standard laboratory ball mill. (Courtesy of PCA)

TABLE 8.3
LIMITS FOR DELETERIOUS SUBSTANCES IN FINE AGGREGATE FOR CONCRETE (ASTM C 33)

ITEM	MAXIMUM PERCENT BY WEIGHT OF TOTAL SAMPLE
Clay lumps and friable particles	3.0
Material finer than No. 200 sieve:	
Concrete subject to abrasion	3.0[1]
All other concrete	5.0[1]
Coal and lignite:	
Where surface appearance of concrete is of importance	0.5
All other concrete	1.0

1. In the case of manufactured sand, if the material finer than the No. 200 sieve consists of the dust of fracture, essentially free from clay or shale, these limits may be increased to 5 and 7 percent, respectively.

Grading. The property of aggregate most apt to change from hour to hour is the gradation, or distribution, of particle sizes on several specified sizes of sieves. The test for gradation is known variously as sieve analysis, grading analysis, mechanical analysis, gradation or simply grading. Aggregate grading requirements for fine and coarse aggregate for normal-weight concrete are listed in Table 8.5.

Sieve analyses are based on percents retained on or passing through square mesh sieves. The material can be sieved through each sieve individually, or, more commonly, put through all the specified sieves at one time by stacking the sieves on a mechanical shaker. See Figures 8-8, 8-9 and 8-10.

Figure 8-8
The gradation of both coarse and fine aggregates is measured with standard sieves. (a) Fine aggregate sieves. (b) Coarse aggregate sieves. (Courtesy of PCA)

Figure 8-9
Running a sieve test of fine aggregates in the field.(Courtesy of PCA)

Figure 8-10
Grading coarse aggregates in the laboratory. (Courtesy of PCA)

PASS	3/8 INCH	NO. 4	8	16	30	50	100
RETAIN	NO. 4	8	16	30	50	100	PAN
%	3	12	17	26	24	15	3

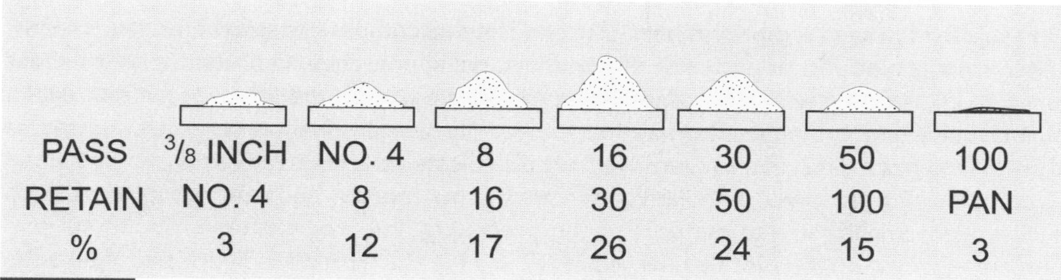

Figure 8-11
The sand sample has been separated into its component sizes with the amount of each size shown in each pan.

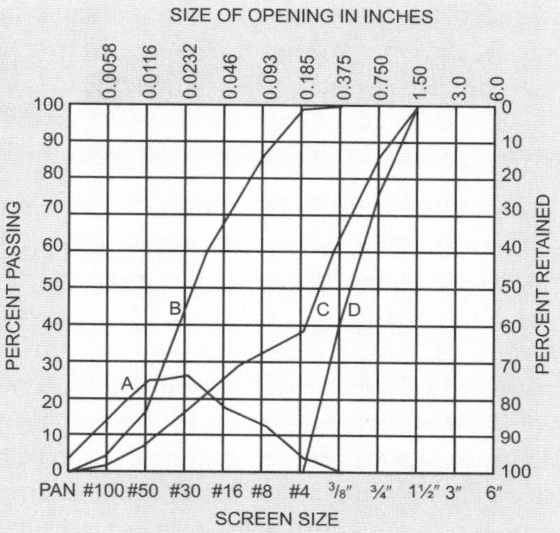

GRADING

SCREEN SIZE	% RET.	CUM. %	COMB. % RET.	COMB. % CUM.
6"				
3"				
1½"				0
¾"				15
⅜"	0	0		36
#4	3	3		62
FM				
#8	12	15		68
#16	17	32		74
#30	26	58		84
#50	24	82		93
#100	15	97		99
Pan	3			
FM	2.87			

Figure 8-12
Curve A shows the individual percents of each size of particles in the sand. Curve B shows the same sand grading plotted cumulatively. For comparison, Curve D shows a cumulative grading for 11/2-inch MSA coarse aggregate. The combined grading of the sand and coarse material is plotted in Curve C.

In Figure 8-11, a sand sample has been screened into its component sizes. This figure shows the individual amounts of each size of material passing one sieve and retained on the next smaller one, starting with the coarsest particles in the pan on the left— those that pass a $^3/_8$-inch sieve and are retained on a No. 4 sieve. These individual percents are plotted as Curve A in Figure 8-12. In this figure we have also plotted the cumulative percents retained (Curve B). To find the cumulative percents, you merely add the amount of each successively smaller size of material to the preceding one. For example, the cumulative percent retained on the No. 16 sieve is 3 + 12 + 17 = 32. This same figure also shows the cumulative grading curve for a $1^1/_2$-inch MSA mixture of fine and coarse aggregate for concrete, Curve C, consisting of the sand already shown, blended with the coarse aggregate shown in Curve D.

Plotting the sieve analysis graphically is of considerable value in visualizing the particle-size distribution of an aggregate, as shown in the figure. Data from coarse aggregate tests as well as from fine aggregate tests can be plotted in this manner. Some users compute and show individual percents passing one sieve and being retained on the next smaller sieve, whereas others make their analysis on the basis of cumulative percents passed or retained.

A useful number when studying aggregate gradations is the fineness modulus. The fineness modulus of an aggregate is a measure of its fineness. It is determined by adding together the cumulative percents retained on a specified series of sieves and dividing by 100. ASTM C 125 specifies No. 100, No. 50, No. 30, No. 16, No. 8, No. 4, $^3/_8$ inch, $^3/_4$ inch, $1^1/_2$ inch and larger, increasing in the ratio of 2 to 1. The fineness modulus (FM) does not provide information as to the grading of an aggregate, but when properly interpreted is useful in comparing the fineness of different aggregates. The FM does not tell the whole story, as aggregates with identical FMs can have different gradings, but in general a high FM indicates a coarse material and a small FM indicates a fine material. An example is given in Table 8-6. Although the FM can be applied to the entire aggregate in a mixture, most usage is found in its application to sand.

TABLE 8.4
LIMITS FOR DELETERIOUS SUBSTANCES AND PHYSICAL PROPERTY REQUIREMENTS OF COARSE AGGREGATE FOR CONCRETE (ASTM C 33)

TYPE OR LOCATION OF CONCRETE CONSTRUCTION	MAXIMUM ALLOWABLE, PERCENT						
	Clay Lumps and Friable Particles	Chert (Less than 2.40 sp gr SSD)[3]	Sum of Clay Lumps, Friable Particles and Chert (Less than 2.40 sp gr SSD)[3]	Material Finer than 75 Microns (No. 200 Sieve)[4]	Coal and Lignite	Abrasion[1]	Magnesium Sulfate Soundness (5 cycles)[2]
	MODERATE WEATHERING REGIONS[5]						
Footings, foundations, columns and beams not exposed to the weather, interior floor slabs to be given coverings	10.0	-	-	1.0	1.0	50	-
Interior floors without coverings	5.0	-	-	1.0	0.5	50	-
Foundation walls above grade, retaining walls, abutments, piers, girders and beams exposed to the weather	5.0	8.0	10.0	1.0	0.5	50	18
Pavements, bridge decks, driveways and curbs, walks, patios, garage floors, exposed floors and porches, or waterfront structures subject to frequent wetting	5.0	5.0	7.0	1.0	0.5	50	18
Exposed architectural concrete	3.0	3.0	5.0	1.0	0.5	50	18

1. Crushed air-cooled blast-furnace slag is excluded from the abrasion requirements. The compact unit weight of crushed air-cooled blast-furnace slag shall not be less than 70 lb/ ft.^{3} The grading of slag used in the unit weight test shall conform to the grading to be used in the concrete. Abrasion loss of gravel, crushed gravel or crushed stone shall be determined on the test size or sizes most nearly corresponding to the grading or gradings to be used in the concrete. When more than one grading is to be used, the limit on abrasion loss shall apply to each.
2. The allowable limits for soundness shall be 12% if sodium sulfate is used.
3. These limitations apply only to aggregates in which chert appears as an impurity. They are not applicable to gravels that are predominantly chert. Limitations on soundness of such aggregates must be based on service records in the environment in which they are used.
4. If the material finer than the 75-micron (No. 200) sieve is essentially free of clay or shale, this percentage may be increased to 1.5. A greater amount of material passing the 75-micron sieve may be permitted, provided the amount passing the 75-micron sieve in the fine aggregate (Table 1, ASTM C 33) is less than the specified maximum. In such case, the sum of the amounts finer than the 75-micron sieve from the separate fine and coarse aggregates shall not exceed the sum of the weighted maximum amounts permitted for the coarse plus fine aggregate.
5. For severe and negligible weathering regions, see Table 3, ASTM C 33.

TABLE 8.5
GRADING REQUIREMENTS FOR CONCRETE AGGREGATES (ASTM C 33)

TYPE OF AGGRE-GATE	GRADE OF AGGRE-GATE	AGGRE-GATE SIZE	PERCENTAGE (BY WEIGHT) PASSING SIEVE HAVING SQUARE OPENINGS														
			100 mm (4 in.)	90 mm (3½ in.)	75 mm (3 in.)	63 mm (2½ in.)	50 mm (2 in.)	37.5 min (1½ in.)	25.0 mm (1 in.)	19.0 mm (¾ in.)	12.5 mm (½ in.)	9.5 mm (⅜ in.)	4.75 mm (No. 4)	2.36 mm (No. 8)	1.18 mm (No. 16)	600µm (No. 30)	300µm (No. 50)	150µm (No. 100)
Regular Aggre-gates for Structural Concrete	Fine	-	-	-	-	-	-	-	-	-	-	100	95-100	80-100	50-85	25-60	10-30	2-10
	Coarse	3½"-1½"	100	90-100	-	25-60	-	0-15	-	05	-	-	-	-	-	-	-	-
		2½"1½"	-	-	100	90-100	35-70	0-15	-	05	-	-	-	-	-	-	-	-
		2"-1"	-	-	-	100	90-100	3570	0-15	-	0-5	-	-	-	-	-	-	-
		2"-No.4	-	-	-	100	95-100	-	35-70	-	10-30	-	0-5	-	-	-	-	-
		1½"-¾"	-	-	-	-	100	90-100	20-55	0-15	-	0-5	-	-	-	-	-	-
		1½"-No. 4	-	-	-	-	100	95-100	-	35-70	-	10-30	0-5	-	-	-	-	-
		1"-½"	-	-	-	-	-	100	90-100	20-55	0-10	0-5	-	-	-	-	-	-
		1"-⅜"	-	-	-	-	-	100	90-100	40-85	10-40	0-15	0-5	-	-	-	-	-
		1"-No.4	-	-	-	-	-	100	95-100	-	25-60	-	0-10	0-5	-	-	-	-
		¾"⅜"	-	-	-	-	-	-	100	90-100	20-55	0-15	0-5	-	-	-	-	-
		¾"-No.4	-	-	-	-	-	-	100	90-100	-	20-55	0-10	0-5	-	-	-	-
		½"-No.4	-	-	-	-	-	-	-	100	90-100	40-70	0-15	0-5	-	-	-	-
		⅜"-No.8	-	-	-	-	-	-	-	-	100	85-100	10-30	0-10	0-5	-	-	

TABLE 8.6
COMPARATIVE SAND GRADINGS PERCENTS RETAINED

SIEVE SIZE	SAND A		SAND B		C-33 ASTM LIMITS
	INDIVIDUAL	CUMULATIVE	INDIVIDUAL	CUMULATIVE	
4	0	0	0	0	0 to 5
8	12	12	11	11	0 to 20
16	20	32	15	26	15 to 50
30	24	56	32	58	40 to 75
50	24	80	26	84	70 to 90
100	15	95	12	96	90 to 98
pan	5	100	4	100	
Total		275		275	
FM	2.75		2.75		

NOTE: The fineness modulus (FM) is the same for these sands, but there is a significant difference in their gradings. Both individual and cumulative percents retained are shown.

A sand or coarse aggregate having a large excess or deficiency of any size fraction should be avoided. A smooth grading curve near the middle of the specified limits is the most desirable. Aggregate gradings that jump from a minimum on one size to the maximum on an adjacent size are unsatisfactory and will have a high percentage of voids or open spaces between particles.

Variations in grading cause a lack of uniformity of concrete from batch to batch, which makes control of the concrete difficult and causes difficulties in handling and placing the concrete.

The effect of maximum size of aggregate (MSA) is discussed in Section 3.11, as influencing strength. Usually, the small MSA mixes require more water than the large MSA mixes, as shown in Figure 12-7. For any given water-cement ratio the amount of cement required is less for large MSA mixes. Selection of the MSA depends upon its effect on strength, shrinkage, water demand and workability, and is influenced by the kind of structure, as discussed in Chapter 12.

Gap-graded aggregates are those that lack certain particle sizes. They have been successful in some specialized cases where no-slump concrete was consolidated by mechanical means. Gap-graded concrete tends to segregate easily and requires very close control of grading and water content.

Particle Shape. Aggregates are rounded, subrounded, sub angular or angular, ranging from well-rounded river gravel to crushed stone. See Figure 8-13. Shapes sometimes encountered are thin and elongated or flat and slabby. The main influence of particle shape is on workability of the fresh concrete. Angular aggregate makes concrete slightly stronger in flexure than rounded, but makes harsh concrete, sometimes requiring more sand, cement and water for workability. Particle shape has only a small effect on compressive strength of concrete. Flat and elongated pieces cause poor workability, requiring more cement and water to make workable concrete, but otherwise make good concrete. Generally, crushed and uncrushed coarse aggregate give about the same strength for the same cement content.

Texture. Some aggregate particles, usually fragments of individual minerals, have smooth or vitreous surfaces, and others are rough and coarse grained. Texture should not be

Figure 8-13
Sample A consists of good, well-rounded river gravel containing very few elongated or broken particles. Sample B consists predominantly of angular crushed particles. (Courtesy of PCA)

confused with particle shape, as well-rounded particles can have a rough texture. Roughness is desirable, as it provides better bond with the cement paste, thereby making concrete of better strength compared with smooth surfaced aggregate particles.

Reactivity. An aggregate that is chemically inert does not react with the cement and is not affected chemically by other influences in or on the concrete. A petrographic analysis identifies rock types so as to discover not only potentially reactive aggregates, but also coatings, lightweight particles, unstable materials, etc. Expansion tests like ASTM C 227, C 1260 and C 1293 can determine which aggregates are potentially reactive and confirm that pozzolans and blended cements are effective at controlling alkali-silica reactivity. A field service record is the best method to evaluate potential reactivity. See Section 4.4.

Specific Gravity. The ratio of the weight of a material to the weight of water is called the specific gravity. An average specific gravity for sand and gravel is 2.65, which means that the material is 2.65 times as heavy as water. Water weighs 62.4 pounds per cubic foot, so a solid cubic foot of stone weighs $2.65 \times 62.4 = 165.4$ pounds per cubic foot. This is called the density of the aggregate and is the weight of a solid cubic foot of the aggregate without any voids between the aggregate particles. See Figure 12-3.

The higher the specific gravity, the heavier the concrete. Low specific gravity (below 2.50) indicates a possibly porous, soft or highly absorptive aggregate of potentially doubtful quality, requiring additional tests to determine its suitability. Use of aggregate of low specific gravity in concrete sometimes results in low strength, popouts, scaling and poor durability. However, low specific gravity alone does not always indicate a poor quality aggregate, as many low specific gravity aggregates have given satisfactory service. Reference to the service record or laboratory tests shows whether the aggregate under consideration can be used to make concrete suitable for the expected exposure. It is necessary to know the bulk specific gravity for proportioning concrete mixes.

Absorption. The property of aggregate particles to absorb water into pores of the aggregate is called the absorption. Hard, dense stone such as granite may have absorption of only 0.2 percent, whereas absorption of a shale or porous chert is as high as 2 or 3 percent. Normally, absorption for sand should not exceed 1.5 percent, and not over 1 percent for coarse aggregate. High absorption indicates porous aggregate of low specific gravity, possibly leading to concrete of poor durability with scaling or popouts and with a probable high shrinkage rate. Aggregates of low specific gravity and high absorption should be investigated further by durability or abrasion tests, depending on expected exposure, and the service record should be studied. The amount of absorption must be known for proportioning and controlling mixes.

Moisture Content. One of four conditions can exist:

1. Oven dry, containing no moisture at all.
2. Air dry, containing less moisture than the aggregate capable of absorbing.
3. Saturated surface dry, containing only absorbed moisture, neither more nor less. Rarely attained except under laboratory conditions.
4. Moist or wet, containing free moisture on the surface in addition to absorbed moisture. Evaluation of moisture content is necessary for mix proportioning and field control. See Figure 8-14.

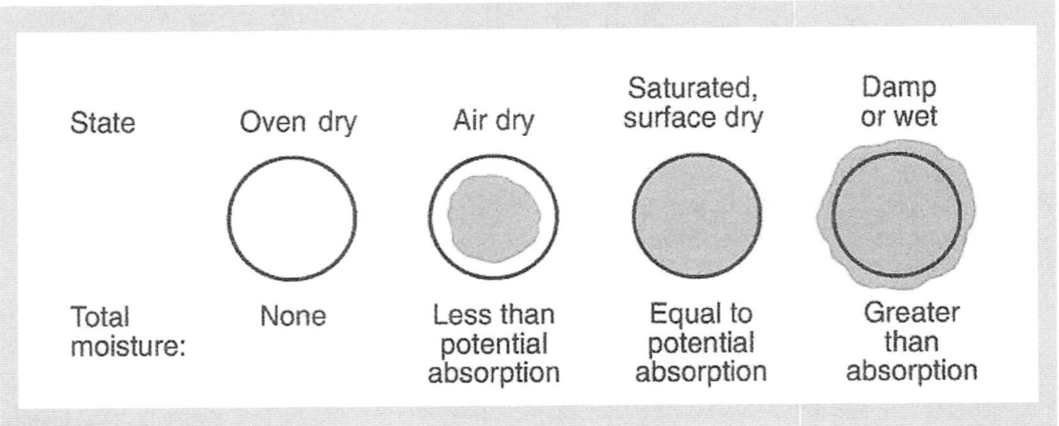

Figure 8-14
The amount of moisture in an aggregate particle can vary from completely dry to a wet condition with free moisture on the surface.

Sand is nearly always in condition 4 when it is batched. Coarse aggregate is usually in condition 2 or 4, rarely in condition 1 and (except for rare, temporary situations) never in condition 3.

Unit Weight and Voids. The weight per cubic foot is the weight of a cubic foot of the sand or coarse aggregate. It is frequently called the unit weight. Note that the unit weight is the weight of a cubic foot of the sand or coarse aggregate, whereas the density is the weight of a cubic foot of the solid rock. Both quantities are used when proportioning mixes. These are discussed further in Chapter 12.

The void content is a measure of the voids or spaces between the aggregate particles. A well-graded aggregate consisting of several sizes of material has a lower void content than a one-size aggregate. This is demonstrated in Figure 8-15, which shows that it took more water to fill the voids in the beakers with uniform but different size aggregates then the beaker filled with aggregates of both sizes. When different sizes are combined, the void content decreases. The amount of voids is affected by the particle shape as well as the grading, both in the sand and in the coarse aggregate. Rough angular particles have a higher void content than well-rounded particles. It has been estimated that an increase of 4 percentage points in the voids in sand will increase the water demand of the concrete by as much as 25 pounds per cubic yard, which in turn will reduce the compressive strength by 400 to 500 psi.

PROCESSING AGGREGATES

Aggregates, to most workers on the construction site, are sand and rock that are used in concrete. We rarely see them except as they come out of the ready-mix truck in the form of concrete. The previous pages in this chapter have described the processes by which nature prepares the materials that are used for aggregate and have briefly explained their properties. Now let us turn our attention to the processes by which these materials are removed from the ground and made into products that can be used by the construction industry.

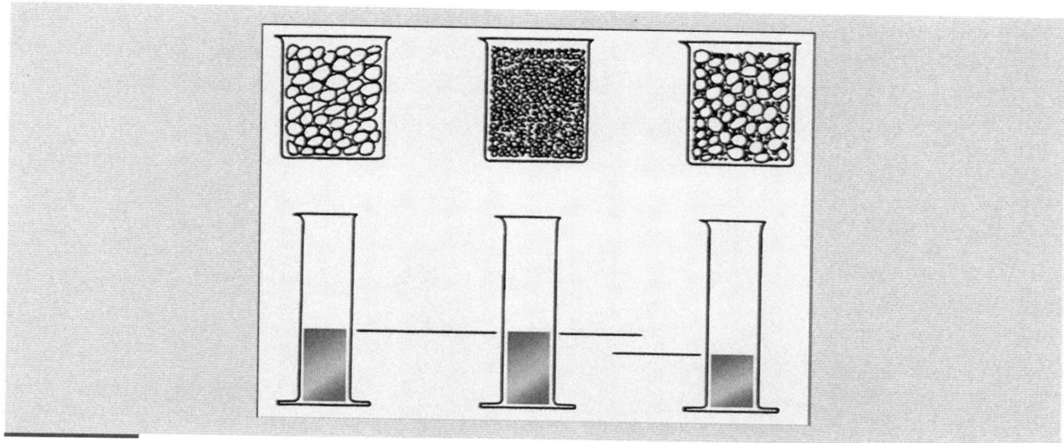

Figure 8-15
The level of liquid in the graduates (representing voids) is constant for aggregates of uniform but different size. When different sizes are combined, the void content decreases.

Screens may have square, rectangular or round openings. The comparison of equivalent sizes of screen cloth shown in Table 8.7 will be useful when comparing screen sizes. All tests are made on sieves with square openings.

TABLE 8.7
APPROXIMATE EQUIVALENT SQUARE AND ROUND OPENINGS OF STANDARD SIEVES

SQUARE HOLE	ROUND HOLE	SQUARE HOLE	ROUND HOLE
3	$3\frac{1}{2}$	$\frac{7}{8}$	1
$2\frac{1}{2}$	3	$\frac{3}{4}$	$\frac{7}{8}$
$2\frac{1}{8}$	$2\frac{1}{2}$	$\frac{5}{8}$	$\frac{3}{4}$
2	$2\frac{3}{4}$	$\frac{1}{2}$	$\frac{5}{8}$
$1\frac{3}{4}$	2	$\frac{3}{8}$	$\frac{1}{2}$
$1\frac{1}{2}$	$1\frac{3}{4}$	$\frac{5}{16}$	$\frac{3}{8}$
$1\frac{1}{4}$	$1\frac{1}{2}$	$\frac{1}{4}$	$\frac{5}{16}$
1	$1\frac{1}{4}$	$\frac{3}{16}$	$\frac{1}{4}$

NOTE: Measurements in inches.

8.4. Sand and Gravel Deposits

Natural sand and gravel cannot be dug out of the ground and used directly in concrete. It is true that there are rare deposits that are so clean that the material is clean enough to be used in concrete. Such deposits, however, usually have a poor ratio of sand to coarse aggregate and the grading of the material is variable. See Figure 8-16. Also, the material is generally not as clean as a casual examination would lead one to believe. Because of the lack of uniformity in source of supply, difference between grading and source and desired finish product, and the need to remove deleterious material, it is necessary to process pit-run aggregates to obtain an acceptable material in all respects. Unsatisfactory grading is corrected by crushing, screening, classifying and recombining; deleterious materials are removed by washing, scrubbing or beneficiation; and segregation and breakage are controlled by careful handling and avoidance of contamination.

Most deposits of sand and gravel are covered with a layer of soil in which vegetation of some sort is usually growing. The first job in developing the deposit is to remove all trees, shrubs and other vegetation, then stripping and wasting the surface soil to a depth sufficient to eliminate the topsoil. See Figure 8-17. Roots, sticks, grass and leaves must not be permitted to enter the processing plant.

Figure 8-16
The exposed cut face of sand and gravel in an open pit excavation operation. This bank is over 60 feet high.

Figure 8-17
A thin layer of dark-colored organic overburden is being stripped from the deposit to expose the underlying sand and gravel. The excavating equipment shown is the type commonly in use during 1941, but the operation is unchanged.

Evacuation of usable material is accomplished with shovels, draglines, cableway scrapers, front-end loaders and carryalls—the choice of equipment depending on the physical configuration of the deposit, depth of face excavated, presence or lack of groundwater, capacity of the plant and owner's personal preference. Excavated material is transported to the processing plant by means of trucks, conveyor belt or rail cars.

Sand and gravel should be processed wet. That is, screens should be provided with spray nozzles that direct high-velocity water jets onto the aggregate as it passes over the screens. The jiggling action of the aggregate on the screen, together with the scrubbing action of the water, removes all but the most stubborn coatings.

The amount of water required for washing aggregates varies widely, depending on the amounts of silt, clay or other material to be removed, size of plant and operating conditions. As a rough guide, each cubic yard of material produced per 10-hour day requires one gallon of water per minute. For example, a plant to produce 400 cubic yards of aggregate in a 10-hour day should be supplied with about 400 gallons of water per minute.

Disposal of wash water is a problem that has become acute because of the increasing concern with stream and lake pollution. Sometimes the water can be returned to an excavated portion of the pit, where it is permitted to remain long enough for the suspended solids to settle.

Passing the water through a sand filter usually removes most of the objectionable suspended matter. Reuse of the wash water may be desirable, provided care is taken to avoid a dangerous buildup of suspended or dissolved substances in the water that may do more harm than good.

Coarse Aggregate. One of the first steps in most plants is to remove or scalp the sand, including the recirculated fines from the crushers. This scalping may be done immediately before or after the primary crushing, after which the coarse material (larger than $^{3}/_{16}$ inch) is passed through various stages of crushing and screening.

Large jaw or gyratory (cone) crushers are used for the initial size reduction of oversize boulders; gyratory or cone crushers are suitable for intermediate sizes; and corrugated rolls are used for the final reduction. See Figure 8-18. Screens are usually of the vibrating type, either horizontal or sloping, single deck or multiple deck; although there are a few instances when revolving cylindrical screens are used, especially for scalping oversize material for crushing. Some plants keep crushed and uncrushed material separate, whereas others mix them. See Figures 8-19, 8-20 and 8-21.

Figure 8-18
Coarse material passing through the enclosed chutes enters the large gyratory crushers for secondary reduction.

Figure 8-19
Most aggregate classifying screens are vibratory, either horizontal or sloping, with water sprays to clean the aggregate.

Figure 8-20
A large sloping vibratory classifying screen heavily loaded with material.

Figure 8-21
The discharge end of a multiple-deck aggregate screen.

The variety of plant arrangements and flow sheets is almost as great as the number of deposits being processed. Each deposit and each usage imposes its own set of requirements on the processing arrangement. See Figure 8-22.

Scrubbing is required when adherent coatings of clay and silt cannot be removed from the aggregate by the usual washing and screening process.

A revolving scrubber, used before the screening operation, is a rotating cylinder with lifter flights that tumbles the material against a flow of water. It usually has a built-in screen to remove sand. It is sometimes called a blade mill or paddle mill.

A log washer is used on coarse aggregate after the sand has been removed. It consists of two slightly sloping shafts rotating against each other in a long trough or tank. These shafts have blades or paddles on them to cut or abrade the material. Using less water than a scrubber, a log washer is most effective with plastic clays that tend to ball.

Figure 8-22
A large aggregate plant, showing primary crushing tower on the left, secondary crushing tower in the center, and the main screening and classifying tower on the right. Main transfer and recirculating conveyors connect the three towers.

A screw washer is sometimes adequate if the aggregate is not too dirty. It consists of a long tub or tank full of water, in which one or two parallel screw flights turn. Material is fed into the lower end and is carried through the water up to the discharge end. Overflow water then removes the waste. Material must be sized first.

Sand or fine aggregate for concrete consists of aggregate particles passing a No. 4 ($^3/_{16}$-inch) screen. Ideally, it would be desirable to have all the sand conform to a grading similar to that shown in Figure 8-23.

We know, however, that such an ideal is practically impossible to attain and that it is not possible to hold the percentages of the several sizes to an exact value for any length of time. We can however set up an ideal grading, as shown in Figure 8-23, and we can make the sand to fit this gradation as nearly as possible. However, a tolerance must be provided; that is, a range from a minimum to a maximum value on each screen size. These limits, as they are called, then provide means to control the grading of the aggregate within a reasonable

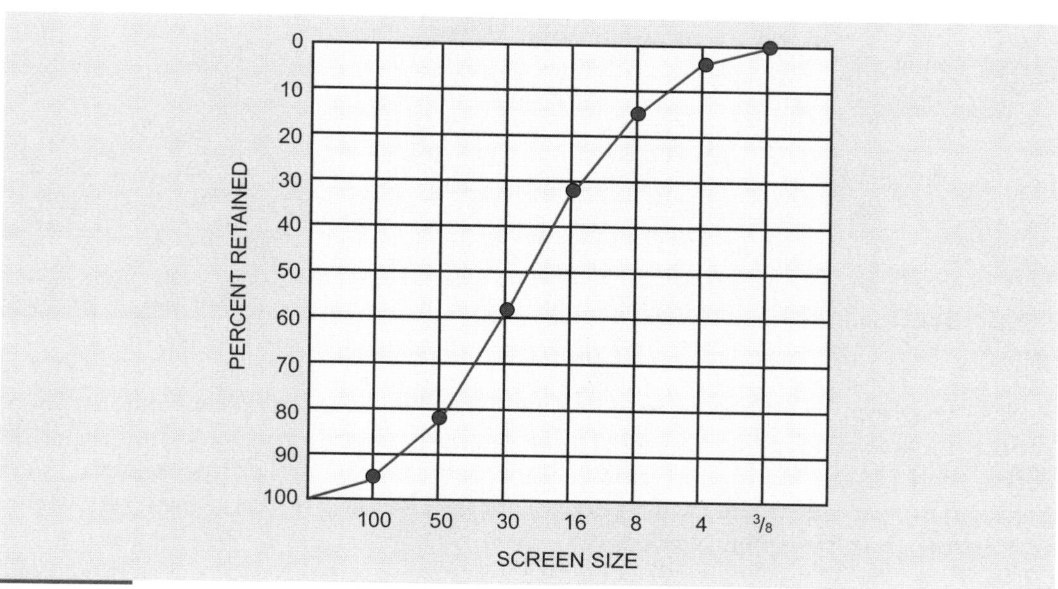

Figure 8-23
A sand grading curve for an ideal particle size distribution, plotted cumulatively.

variation. Figure 8-24 shows the limits applied to the sand grading just discussed. This curve is based on the grading requirements for fine aggregate in Table 8.8. As long as the plot of the test results for sand samples falls within these limits, we can say that, as far as grading is concerned, the sand is within specifications. If some of the values fall outside of the limits the sand is not in compliance, and steps must be taken in the production process to bring it back into specifications.

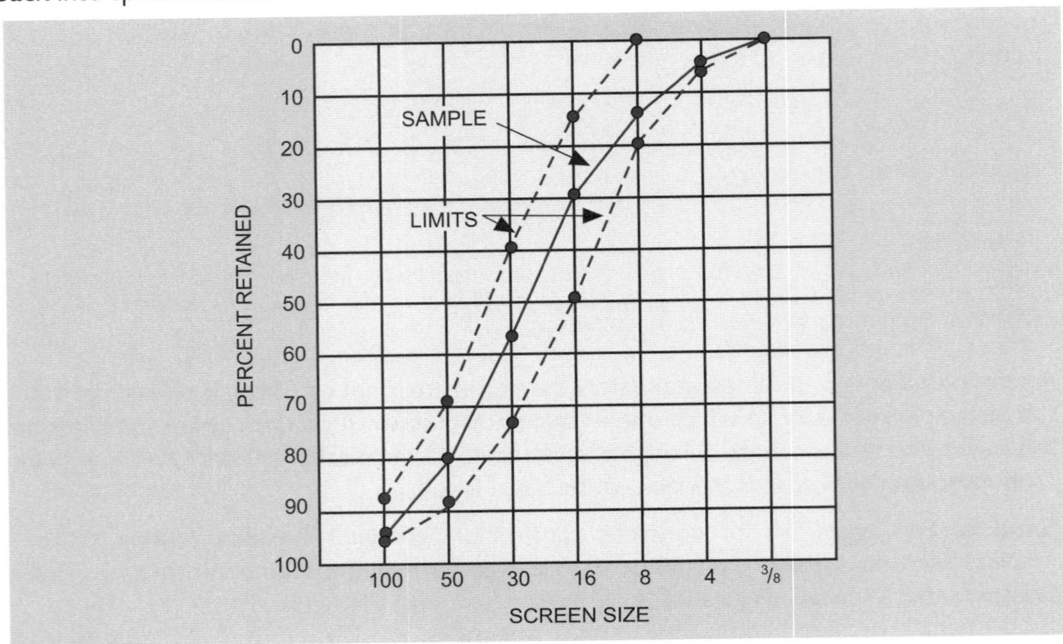

Figure 8-24
The sand grading curve from Figure 8-23 with maximum and minimum tolerances sketched on.

TABLE 8.8
GRADING REQUIREMENTS FOR FINE AGGREGATE
(See TABLE 8.5)

SIEVE	PERCENT PASSING	
	MIN.	MAX.
3/8-in.	100	
No. 4	95	100
8	80	100
16	50	85
0	25	60
50	10	0
100	2	10

In other chapters of this book we have discussed the effect of sand grading on the quality of concrete. We know, for example, that fine sand usually requires slightly more cement than coarse sand for the same water-cement ratio and slump. The principal effect of sand grading is on workability and finishing quality of the concrete. For this reason, very fine or very coarse sands should be avoided.

Even though tolerances are set up, there are times when the sand fails to conform to the specification; that is, somewhere along the grading curve there will be bulges outside the limits, indicating an excess or deficiency of some size or sizes of particles. Such material is classified as defective.

Pit-run sand usually does not conform to the specifications and must be processed. It can be called defective. Grading defects can be corrected by adding blending sand, by crushing a portion of the excess of large sizes, by removing a portion of the excess sizes or by a combination of these procedures.

If a correction of sand grading is to be accomplished by using a blending sand, two alternatives are possible. The blending sand can be fed into the aggregate plant feed and processed with the normal pit sand, or the blending sand can be fed onto a conveyor belt simultaneously with the coarse sand. Subsequent handling of the sand tends to mix it, as long as the sand is damp.

Sand, which should always be washed, may or may not pass through primary crushing, depending on the plant layout. Some plants scalp sand first and keep it entirely separate, whereas others process crusher fines along with the natural sand. Sand grading can be improved through the use of certain hydraulic machines called classifiers. Hydraulic separation rather than screening is best for sand sizes because water does not pass through the fine screen meshes as well as it does through meshes of $^1/_4$ inch or larger. Water actually reduces the efficiency of small-mesh screens.

Hydraulic classifiers depend on the fact that small particles settle more slowly than large ones when suspended in water. A screw classifier is similar to the screw washer for coarse aggregate and is used for removing excess clay, silt and fines. See Figure 8-25. Sand drags and rakes have a series of blades on a chain or reciprocating arm that move the sand up the sloping bottom of a tank full of water. Fines are carried off with the water. Other modern hydraulic sizers operate on the hindered settlement principle and are capable of classifying several different size grains of fairly uniform specific gravity. Hindered settlement is a process in which the material is fed into a vertically rising current of water. Feed and water velocity are so regulated so that the large particles settle and are removed from the bottom of the tank, while the small particles are carried by the water over weirs at the top.

Figure 8-25
A screw washer or classifier for sand. Sand and water enter through the large pipe near the center of the tank. Additional water is usually required. Washed sand is carried up the sloping tank and discharged at the high end (behind the platform). Water and waste fines pass out of the tank over adjustable weirs on the right.

Classifiers of this type may consist of one or more compartments, each compartment regulated to separate particles of different fineness.

Water jets installed in the bottom of a screw classifier can be used to vary the sand grading by varying the amount of water admitted.

After passing through the classifiers, the several sand fractions are recombined in the proper proportions to produce the required gradation.

In washing sand, uniformity should be required. The feed to the classifiers should be maintained at as constant a rate as possible. Water flow and weir elevations should be changed only as indicated by the cleanness and grading of the sand produced.

Aggregate Beneficiation. Many natural gravels include materials and rock particles that are not suitable for use in concrete and therefore require special processing to improve, or beneficiate, them. For example, some deposits in Michigan, Indiana and Illinois contain porous cherts of relatively low specific gravity. These particles are unsound and cause popouts and unsoundness in the concrete. Some river gravels in Arizona, California and other areas contain unsound and soft lightweight particles.

The term *beneficiation* usually refers to equipment and processes more complicated and more effective than the scrubber or log washer for improving materials of quite poor quality.

None of the processes is 100 percent effective, but they do make a significant improvement in the aggregate. The ones most commonly used are as follows:

1. Heavy media separation. In this process, coarse aggregate is passed through a tank containing a high-density solution, usually a suspension of pulverized magnetite and ferrosilicon in water. The suspension is regulated so that rock particles below a certain specific gravity, say 2.55, float in the suspension and are wasted. Heavy particles sink and are reclaimed. This process is effective for either hard or soft rocks and is effective in removing hard, porous chert. Subsequent washing of the gravel removes the suspension material, which is salvaged for reuse. Usually the separation is done in two or three stages.
2. Elastic fractionation. This process is based on the rule that hard particles, because of their elasticity, will bounce more than soft, friable ones. Separation is effected by dropping the gravel onto a sloping steel plate, so arranged that hard rocks rebound into receiving bins, and the soft material is taken off into waste. Usually the bounce material is recirculated, as separation is not thorough. This process is inefficient and rarely used.
3. Jigging. Based on the hindered settling principle, jigging is accomplished by passing the gravel through a tank in which water is subjected to vertical pulsations, either by means of compressed air jets or vibrating diaphragms. The pulsations float the light material if its specific gravity is appreciably lower than that of the heavier gravel.
4. Impact crusher. Sometimes referred to as a cage mill disintegrator. Gravel is fed through a chimney into a metal housing. Inside the housing a horizontal shaft equipped with vanes rotates at high speed. Soft particles are broken up when struck by the vanes or on impact with the housing. Subsequent wet screening removes the pulverized particles.

8.5. Quarries

Usually the stone to be quarried is covered with overburden of some kind, which has to be removed before drilling and blasting can be done. Hydraulic stripping, which has been practiced in some quarries, requires that drainage of the water and topsoil be away from the quarry, and that provision be made to prevent stream contamination. Usual practice is to use mechanical equipment for stripping. Shovels, draglines, bulldozers and scrapers can be used, depending on the amount of stripping and the configuration of the surface. If the overburden is not excessive, it can be permitted to fall into the quarry when the stone is blasted.

There are two general types of quarries, the bank quarry and the pit quarry. The bank quarry is located in sloping or hilly ground so the bottom of the quarry is about at the level of the surrounding country and rock is shot down to the crusher. The pit quarry is excavated below the level of the surrounding country, and the rock must be hoisted out of the pit.

All rock is blasted. The flow of material after blasting is similar to the flow through a gravel plant, except there is no sand to be processed. Most rock plants wash the material, and many make use of scrubbers and log washers.

The quality of rock in a quarry can be quite variable, especially the limestone. Lenses and pockets of soft rock, chert and clay are occasionally encountered. Igneous quarries are usually uniform in quality.

In areas where natural sand is scarce, stone sand is made by reducing stone to minus No. 4 size. Usually there is an excess of minus 100-mesh rock dust that has to be removed. This can be accomplished by washing.

8.6. Stockpiling

Finished aggregates must be stockpiled because the rate of production of each size is never the same as the rate of use. Stockpiles become necessary for storage of aggregates made during periods of slow demand, or for storage of sizes for which demand is temporarily low.

When aggregates are stockpiled on the ground, the ground should first be cleared of all vegetation and rubbish, then leveled. In removing aggregate from the pile, a layer of aggregate should be left on the ground so that the handling equipment will not pick up earth from the original ground. Thickness of this layer or pad depends upon the type of equipment used; a front end-loader may require a pad only 6 inches thick, while a 12-inch pad should be maintained if a clamshell is being used. Common practice is to spread rejected aggregate of the same size to be stockpiled over the stockpile area to provide a pad in advance of stockpiling operations.

Stockpiles will become contaminated unless sufficient area is provided for them, to prevent crowding and overlapping of piles. If there is not enough room in the area to provide a clear space between piles, separation should be achieved by means of stout partitions or bulkheads. See Figure 8-26.

When moving material into or out of stockpiles by means of a clamshell, the bucket should not be permitted to swing over a pile of one size aggregate while carrying a different size. For example, if gravel is being removed from a railroad car with a clamshell, the bucket should not be permitted to swing over the sand pile.

Material is removed from a stockpile by means of a clamshell, front end-loader or a conveyor belt operating in a tunnel beneath the pile. If two or more gates are provided to admit the material to the conveyor, good mixing is usually achieved. When loading out with a clamshell, good mixing can be achieved if the operator takes successive loads from different parts of the pile rather than removing from one low area where material is continually running down a slope. Use of a front end-loader is satisfactory if the machine is operated so that it takes vertical, or nearly vertical, slices through the pile rather than removing the aggregate from the periphery of the pile near the bottom, which causes the gravel to flow down the slope, resulting in segregation.

Figure 8-26
Well-separated stockpiles at an aggregate processing plant. Material is removed from the piles through gates under the pile feeding a tunnel conveyor. Several gates under each pile opened simultaneously minimize the segregation problem.

The greater the size range covered by the gravel in the pile, the greater the danger of harmful segregation. For example, aggregate graded from $^3/_{16}$ inch to $1\,^1/_2$ inches, when handled as one size of material, tends to segregate more than either the $^3/_4$ by $^3/_{16}$-inch size or the $1\,^1/_2$ by $^3/_4$-inch size.

Segregation can be minimized if stockpiles are built up in layers 3 to 4 feet thick. Stockpiles can be built up by discharging directly from trucks so that each individual dump is close to the adjacent ones; the material remains where it is dumped and is not permitted to roll down slopes. The trucks should not operate over the same part or layer repeatedly. Some protection of the aggregate is usually necessary to prevent contamination by dirt brought in on the wheels of trucks, especially during wet weather.

Stockpiles are frequently built up in high cone-shaped or wedge-shaped piles by discharging aggregates directly from stacking conveyors. Segregation is a serious problem resulting from material running down the sides of the pile. Some forms of stacking conveyors are articulated and movable, so that they can build the piles up in layers. Rock ladders minimize segregation and breakage of coarse aggregate, especially when the material is dropped off a high stacking conveyor. See Figure 8-27.

Sand, because it is usually handled in a moist condition, need not be subject to the above stockpiling limitations. However, if the sand is dry, it is very difficult to handle. The best remedy is to keep it moist. See Figure 8-28.

Figure 8-27
A rock ladder for 3-inch by $1^1/_2$-inch coarse aggregate is shown at the end of the stacking conveyor.

Figure 8-28
Moist sand stockpile built up from the discharge of a high stacking conveyor. Segregation of moist or damp sand is not a problem. The conveyor shown is capable of being moved.

Segregation of aggregate is minimized by observing these precautions: handle in closely graded sizes (that is, $^3/_{16}$-inch by $^3/_4$-inch, not $^3/_{16}$ by $1^1/_4$-inch, or similar sizes); handle and move as few times as possible; avoid high, cone-shaped piles; stockpile in layers; remove from stockpile in vertical slices, or use a tunnel conveyor with two or more gates under the pile; use rock ladders in piles and bins; drop material vertically into the bin, keeping bins full; use tall and thin bins, preferably circular in plan, with bottom sloping about 50 degrees from the horizontal.

TESTING AGGREGATES

Sampling and testing aggregates is usually done by quality control employees of the aggregate producer or inspectors at the aggregate processing plant. Check tests are frequently made at the concrete batching plant, especially when the job specifications state that the aggregate shall conform to certain standards when it is batched into the concrete.

Requirements are based on ASTM C 33, Standard Specifications for Concrete Aggregates. This specification lists a number of methods of sampling and testing for both fine and coarse aggregates to determine the properties that were briefly presented earlier in this chapter. For the technician on the job, the tests most likely to be performed at the batching plant are moisture content, sieve analysis, absorption and specific gravity, in that order of frequency.

The moisture content is usually fairly constant over any period of time. Changes are liable to occur when new materials are brought into the plant or after a heavy rainstorm. The coarse materials are usually dry, or nearly so, and it is rarely necessary to test them. Sand always contains some moisture.

The sand bin in a modern plant is fitted with a moisture meter that measures electrical resistance or conductance through the sand, the principle being that wet sand conducts electricity better than dry sand. See Figure 8-29. When properly calibrated for the particular plant and sand, these meters are quite accurate. A dial on the batching console gives the operator and inspector the percent moisture of the sand. In some plants, the moisture meter is connected electrically to the controls so that the amount of water introduced into the concrete and the weight of wet sand are automatically adjusted as the sand moisture content varies.

Figure 8-29
Sand moisture meter. (Courtesy of PCA)

8.7. Sampling

Truly representative samples are difficult to obtain, and the following paragraphs describe methods that will usually enable the technician to obtain samples as nearly representative as possible.

Considerable judgment is necessary in taking samples and drawing conclusions from the results of tests. Care should be exercised that samples are representative of the materials being tested. ASTM D 75 gives methods of sampling aggregates from conveyor belts, bins and stockpiles. These methods should be followed at all times.

Sampling from a stockpile is difficult, especially if the sample is for a sieve analysis, because of segregation in the pile and further segregation when the sample is removed. A sampling plan must be developed for different materials, plants and kinds of piles, requiring that several samples be taken from different parts of the pile, then combining them into one sample. A plan that has worked provides that the composite sample should be made up by taking one shovelful at the top of the pile, four at equally spaced points around the bottom, and four at random about halfway up the slope of the pile, all consisting of material from below the surface. If a short piece of board is held against the pile just above the point of sampling, unwanted material can be kept out of the sample. Individual sieve analyses are made if it is desired to determine whether the materials are segregated. If an analysis representative of the whole pile is desired, the samples are combined, then quartered to obtain a sample for testing.

Samples can be taken from a conveyor belt by stopping the belt and taking at least three portions and combining these to form a sample. ASTM D 75 requires that two templates shaped to fit across the belt be inserted and all material between the templates, including fines and dust, be removed. Sampling a moving belt should never be attempted, because it is dangerous and a representative sample cannot be obtained. See Figure 8-30.

The stream of aggregate emerging from a conveyor belt, bin, chute or other opening is not uniform in cross section with respect to particle size; hence the need for sampling the entire stream. The sampling bucket or container should not be permitted to overflow, as to do so may result in a sample with more fines than are present in the material being sampled. Most plants have some sort of special device by means of which a sample can be obtained. A sample should not be taken from the first material that is discharged by a bin or belt.

Figure 8-30
A sampling template of the kind specified in ASTM D 75, on a raw feed conveyor.

Representative sampling of trucks and railroad cars is extremely difficult and should be avoided if at all possible. It is necessary to dig into the load at several points, obtaining a rather large sample that has to be split down to testing size. Usually, a visual inspection of the car will give a general idea of the condition of the material in the car and can guide the sampler in determining where to take sufficient samples to be representative of the carload of material. Sand, if moist, presents no problem. Coarse aggregate segregates as it is loaded into the car, and this condition will be observed on the surface or by digging into the load.

Samples to be sent to the laboratory should be placed in clean cloth sample bags, plastic-lined burlap bags or other suitable containers. An identifying card or tag should be placed inside the bag or on the outside. The bag should be securely tied and sent immediately to the laboratory.

The size of the sample depends upon the tests to be run and the size of the largest particle. ASTM D 75 gives the weights of samples required. Generally, a sample of 75 to 100 pounds of coarse aggregate, or 50 pounds of sand, will be adequate.

Quartering Samples. Samples of aggregates, as obtained at the source or processing plant, are usually larger than is convenient for testing. Fine aggregate samples may be effectively reduced to test size by a sample splitter. See Figure 8-31. The quartering method should be used for reducing the size of coarse aggregate samples and can be used for reducing the size of fine aggregate samples if a sample splitter is not available.

Figure 8-31
A sample splitter divides the sample equally into two portions. Small gravel as well as sand can be processed on the one shown. (Courtesy of PCA)

In the quartering method, the sample is placed on a hard, clean surface where there will be neither loss of material nor accidental addition of foreign matter. In the field, a piece of canvas or plastic can be used. The sample is mixed thoroughly by turning the entire lot over three times with a shovel, beginning at one end and taking alternate shovels of the material the length of the pile. With the third or last turning, the entire sample is shoveled into a conical pile by depositing each shovelful on top of the preceding one. The conical pile is carefully flattened into a uniform thickness and diameter so that the material will not be transposed from one quarter to another. The flattened mass is then marked into quarters by two lines that intersect at right angles at the center of the pile. Two diagonally opposite quarters are removed and the cleared spaces brushed clean. The remaining material is mixed and quartered successively until the sample is reduced to the desired test size.

The practice of adding or removing small amounts of material to obtain a sample of a certain size or weight should be avoided, as these small increments are apt to affect the quality of the sample. Instead, the sample should be split to an amount reasonably close to the desired amount.

8.8. Testing

Tests for evaluation of the pit material and for quality control of the aggregates are made by the aggregate producer at the point of production. Once the quality of the material has been established, control of the production is accomplished by means of grading tests. Frequency of testing depends on the uniformity of the product and the rate of production. In a small efficient plant, one test per day might be sufficient for each material; in a large plant ten tests per day might be required. A sampling plan should be developed for each material in the plant, the objective being to perform sufficient tests to prevent any significant amount of defective aggregate from being produced. Methods of performing the tests are described in Chapter 24.

SPECIAL KINDS OF AGGREGATES

8.9. Blast Furnace Slag

When iron ore is processed to make iron, there is a large amount of rock material left over. This material is called slag. There are several varieties of slag, depending on the type of furnace from which slag comes. The one that is suitable for use as an aggregate for concrete is the slag from blast furnaces. The history of many structures going back more than 50 years and tests by various agencies demonstrate the satisfactory performance of air-cooled blast furnace slag aggregate.

Molten slag from the furnace is conveyed to pits or banks where it is cooled. After cooling, the slag is crushed, passed over magnetic separators to remove any bits of iron remaining, then is screened into appropriate sizes. The resulting product is hard, angular and slightly lighter in weight than natural aggregate.

Slag is not normally used for fine aggregate when natural sand is available, as it is not economical to crush it to sand sizes. ASTM C 33 specifies a minimum compact weight of 70 pounds per cubic foot for coarse slag. Other requirements of ASTM C 33 apply to slag exactly as they apply to natural aggregates.

The weight of concrete made with slag coarse aggregate depends on the weight of the slag and mix proportions and is usually in the range of 135 to 140 pounds per cubic foot. Strength, durability and other properties of slag concrete are comparable to those of concrete made with natural aggregates. Because of its rough and angular shape, slag makes concrete that is somewhat harsh. A slightly higher sand content usually solves this problem.

8.10. Lightweight and Heavyweight Aggregates

Aggregates for structural concrete weighing as little as 90 pounds per cubic foot may be either natural or artificial. Special low-density insulating concretes with density as low as 15 pounds per cubic foot are made with certain expanded minerals and foams. These are discussed in Chapter 21.

Heavy concrete for radiation shielding and counterweights may have a density of 300 to 400 pounds per cubic foot. These materials are covered in Chapter 21.

Water and Admixtures

9.1. Water

The reasons for using water in concrete are twofold: the water is required to react with the cement so that the concrete will get hard, and the water lubricates the fresh concrete to make it plastic and workable.

It requires slightly more than a quarter of a pound of water to hydrate a pound of cement. Expressed another way, a water-cement ratio of about 0.25 will furnish sufficient water to react with cement. Concrete with such a low water content, however, would be unmanageable and could not be placed, so we put in more water to make the concrete workable. This requires a water-cement ratio between 0.40 and 0.60 for normal structural concrete, giving us an amount of water in a cubic yard of concrete between 225 and 350 pounds, the exact amount depending on slump, materials, weather conditions and mix proportions.

The importance of the water-cement ratio has been previously pointed out in Chapter 1, where it was stated that the strength, durability and other desirable properties of the concrete go down as the water-cement ratio goes up. Against this, we have the increased fluidity and flowability of the concrete when more water is used. We know now that the high-slump, fluid concrete is not good concrete. For this reason the specification writer limits the water-cement ratio, total water and slump, and the worker on the job must see to it that the specification limits are not exceeded.

Cleanness. How dirty can water be and still be satisfactory for use in concrete? A measure sometimes used is that if the water is fit to drink it can be used in concrete. See Figure 9-1. In most cases this rule is good. It would permit the use of domestic water sources, most irrigation waters and some natural streams and lakes. It would also permit the use of carbonated spring water and so-called medicinal spring waters, but such waters would not be desirable in concrete. On the other hand, such a rule would prevent the use of water containing suspended clay or silt, although most specifications permit the use of water containing as much as 2000 parts per million of turbidity. Water of that turbidity contains about $^1/_4$ ounce of solids in a gallon—rather muddy water for drinking, but acceptable for mix water in concrete.

Figure 9-1
Water that is safe to drink is safe to use in concrete. (Courtesy of PCA)

When chemists speak of suspended or dissolved solids in water they use the term ppm, which means parts per million. One ppm means one part of solids in one million parts of water. The "part" can be a pound, gram or any unit of weight. For example, in water containing 2000 ppm of suspended material, we would have 2 pounds of solids in 1000 pounds of water. The letters TDS mean total dissolved solids. The suspended solids are found by filtering a sample of the water; the TDS are found by evaporating a sample of filtered water and weighing the residue.

There are many impurities that have a negligible effect on the quality of water for mixing, but others can have a significant effect, even in small amounts. The effect of some dissolved or suspended matters in the water is apt to be unpredictable, as the effect will differ with the different cements, the concentration of impurity in the water and the temperature. Sugar is a good example. As little as 300 ppm of sugar by weight of the cement (about $^1/_2$ ounce for 100 pounds of cement) will retard the setting appreciably, but will make the strength slightly higher. If we put in as much as $2^1/_2$ or 3 ounces of sugar per 100 pounds of cement, the concrete will probably undergo a very quick set, with low strength up to about seven days, with the possibility of higher strength at 28 days. In general, any sugar at all is undesirable because of objectionable effects on other properties of the concrete as well as erratic setting and variable strength.

Quality. The quality of water for use in concrete is seldom a problem in urban areas supplied by established ready-mixed concrete plants, because the water in most plants comes from the municipal supply. If the water comes from a private well, local health officials have probably tested the water, and records of these tests should be available. Usually the presence of harmful impurities will be revealed by the color, taste or odor of the water. If the water is reasonably clear and does not have a brackish, foul or salty taste or odor, it is usually satisfactory.

The source of water on a project in an isolated or remote location can sometimes be of doubtful quality and may require testing. Water from a stream or lake can be used, provided it is clean and otherwise conforms to the specification requirements. Stagnant or muddy pools, swamps and marshes should be avoided.

Natural sources are apt to vary appreciably between wet and dry seasons; hence, it may be necessary to obtain more than one sample during the progress of a job. The most serious danger is that the flow in a stream decreases to the extent that the concentration of organic material in the water becomes excessive.

Acceptance criteria for water to be used in concrete are given in ASTM C 1062. Water of questionable quality can be used for making concrete if concrete cylinders made with it have 7-day strengths equal to at least 90 percent of companion specimens made with drinkable or distilled water. In addition, tests should be made to ensure that impurities in the mixing water do not adversely shorten or extend the setting time of the cement.

Seawater. Whether or not to use seawater as mix water for concrete is a question that has been asked many times. The best advice is: Don't use it if it is at all possible to obtain fresh water. Seawater is generally suitable as mixing water for unreinforced concrete. Seawater contains about $3^1/_2$ percent salt, mainly chlorides, which act as an accelerator; the result is slightly higher early strength gain, but probably lower strength at 28 days or later. By adjusting the mix to lower the water-cement ratio, the required strength can be obtained.

Seawater should not be used in making reinforced concrete. The use of seawater as mix water greatly increases the risk of corrosion of the reinforcing steel, particularly in warm and humid environments. Failures of concrete structures have occurred, attributed in large part to corrosion of the reinforcing steel in concrete made with beach sand or seawater.

See Figure 9-2. The risk of corrosion, however, can be reduced if the water-cement ratio is kept below 0.45, the concrete contains adequate air entrainment, and the concrete is properly consolidated to produce a dense and impermeable concrete.

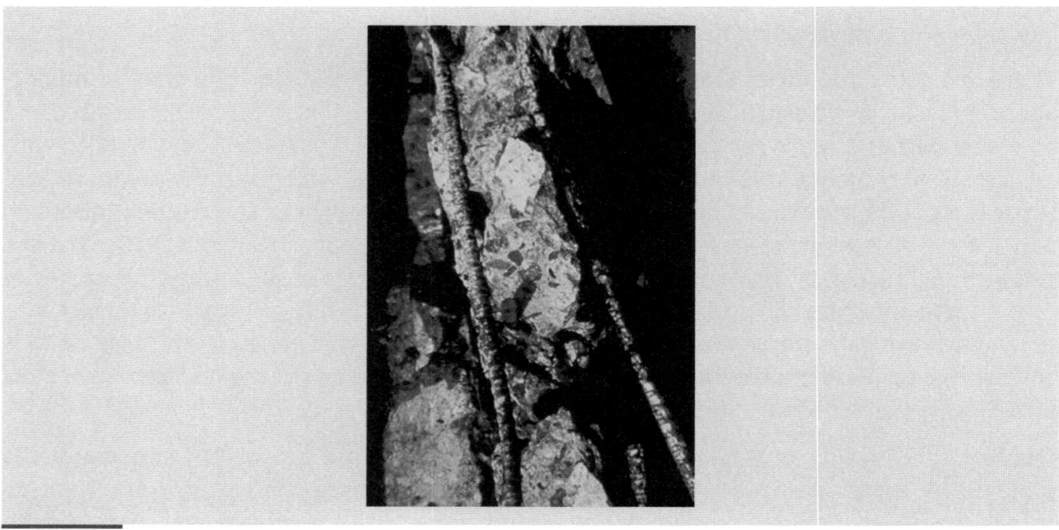

Figure 9-2
Complete destruction of seawall because of corrosion of steel reinforcement caused by use of seawater as mixing water in the concrete.

Seawater must not be used in making prestressed concrete, as it contributes to high-stress corrosion of the prestressing steel.

9.2. Admixtures

An admixture can be defined as any substance, other than cement, aggregate and water, that is added to a batch of fresh concrete for the purpose of altering any of the properties of the concrete either in the fresh condition or hardened. Two general classes of admixtures are chemical admixtures (ASTM C 494 and C 1017) and air-entraining agents (ASTM C 260).

ASTM C 494 chemical admixtures are described as follows:

WATER-REDUCING ADMIXTURE is an admixture that reduces the quantity of mixing water required to produce concrete of a given consistency.

RETARDING ADMIXTURE is an admixture that retards the setting of concrete.

ACCELERATING ADMIXTURE is an admixture that accelerates the setting and early strength development of concrete.

WATER-REDUCING AND RETARDING ADMIXTURE is an admixture that reduces the quantity of mixing water required to produce concrete of a given consistency and retards the setting of concrete.

WATER-REDUCING AND ACCELERATING ADMIXTURE is an admixture that reduces the quantity of mixing water required to produce concrete of a given consistency and accelerates the setting and early strength development of concrete.

WATER-REDUCING, HIGH-RANGE ADMIXTURE is an admixture that reduces the quantity of mixing water required to produce concrete of a given consistency by 12 percent or greater.

WATER-REDUCING, HIGH-RANGE AND RETARDING ADMIXTURE is an admixture that reduces the quantity of mixing water required to produce concrete to a given consistency by 12 percent or greater and retards the setting of the concrete.

PLASTICIZERS FOR FLOWING CONCRETE often called superplasticizers, are essentially high-range water reducers meeting ASTM C 1017; these admixtures are added to concrete with a low-to-normal slump and water-cement ratio to make high-slump flowing concrete.

When considering any proposed admixture, the user should know the type of admixture, trade name, manufacturer, local dealer or representative, and the length of time the product has been in use. The user should also determine the effect of the admixture on durability, permeability, strength (compressive, flexural, bond), drying shrinkage, time of set and, for pavement, resistance to salt scaling. Other properties of the concrete that may be affected are the amount of air entrained, workability, mixing water requirement, sulfate resistance, alkali-silica resistance and cracking tendency.

The user should know whether an accidental overdose of admixture can be tolerated without adverse effect; whether reliable, automatic dispensing equipment is available; and the type of service given by the distributor. Information of this nature is available from the manufacturer of the admixture and should include copies of reports from independent laboratories showing the results of tests. Products from well-established, reputable manufacturers are usually satisfactory.

The effect of an admixture may be variable, depending on brand, type and amount of cement in the mix, slump or water content of concrete, aggregate grading, length of mixing time, type of mixer, temperature and how and when it is introduced into the mix.

Some admixtures affect more than one property of the concrete, sometimes adversely. For this reason, any admixture must have a history of satisfactory use with the proposed materials, under the proposed conditions, or must be tested under these conditions.

The Need for Admixtures. The reason for using an admixture is to modify the properties of concrete so it will be more suitable for a certain usage. The modification might alter either the fresh concrete or the hardened concrete, or it might affect the concrete in both the fresh and hardened states. Among the effects sought by the use of admixtures are the acceleration or retardation of setting time, early strength development, reduction in water requirement, improved resistance to chemical attack and weathering, control of alkali-silica expansion, production of colored concrete and improvement in workability.

In general, concrete of satisfactory quality and lowest cost can best be provided by a properly proportioned mixture of aggregates, portland cement and water. Proper handling, placement and curing of this concrete will provide a long-lasting, well-performing concrete. However, admixtures can be used for the purpose of providing special properties not readily or economically attainable with local materials. When an admixture is considered for use, it should be tested with job-site materials under job conditions in advance of construction to determine its compatibility with the other materials and its ability under these conditions to produce the desired properties.

If we remember that an admixture is not a patent medicine to cure all ills but instead is a prescription to be administered under proper supervision and suitable conditions to produce a certain effect, then we are on safe ground and can use the selected admixture with beneficial results. When selecting an admixture, it is well to keep in mind three facts: First, some admixtures are a combination of materials intentionally selected to affect more than one property of the concrete. A water reducer, for example, might be combined with a retarder. Second, some admixtures, by their very nature, have side effects not related to their primary purpose, and the side effects might be different or even undesirable for certain combinations of materials used in the concrete. Third, an admixture cannot be used to cover up errors or carelessness in proportioning or using concrete.

Measurement. Admixtures can be measured by volume or by weight. Either method is acceptable as long as it meets certain requirements. When an admixture is used, it becomes an essential part of the concrete, and the dispensing of that admixture should be done under the same close control as applied to the other batch ingredients. Specifically, the dispensing system should have the following capabilities:

1. It must accurately measure the dose of admixture for any given batch of concrete. Most specifications permit a tolerance of 3 percent. Dispensers, provided by different manufacturers, accomplish measurement by weight, volumetric sight gauge, constant flow orifice and timer, or a positive displacement flow meter. See Figure 9-3.
2. The dispenser should deposit the measured dose of admixture into the mixer. Usually the admixture is dispensed into the water as the water flows into the mixer. The dispenser should be interlocked with the other batching apparatus so that failure of one batcher will interrupt all others. In a manual plant, the operator should be able to observe operation of the dispenser. Piping should be arranged so there are no dead spots where part of a dose can lodge.
3. The dispenser should be readily adjustable to dispense different amounts of admixture as required for different batches and conditions in the plant. It should be possible to draw off a measured dose to check the measuring accuracy.
4. The system should assure that the admixture is uniform in concentration from batch to batch. Some admixture solutions are unstable and must be constantly agitated to assure uniformity. Agitator and dispenser should be interlocked to prevent dispensing of incompletely mixed solution.

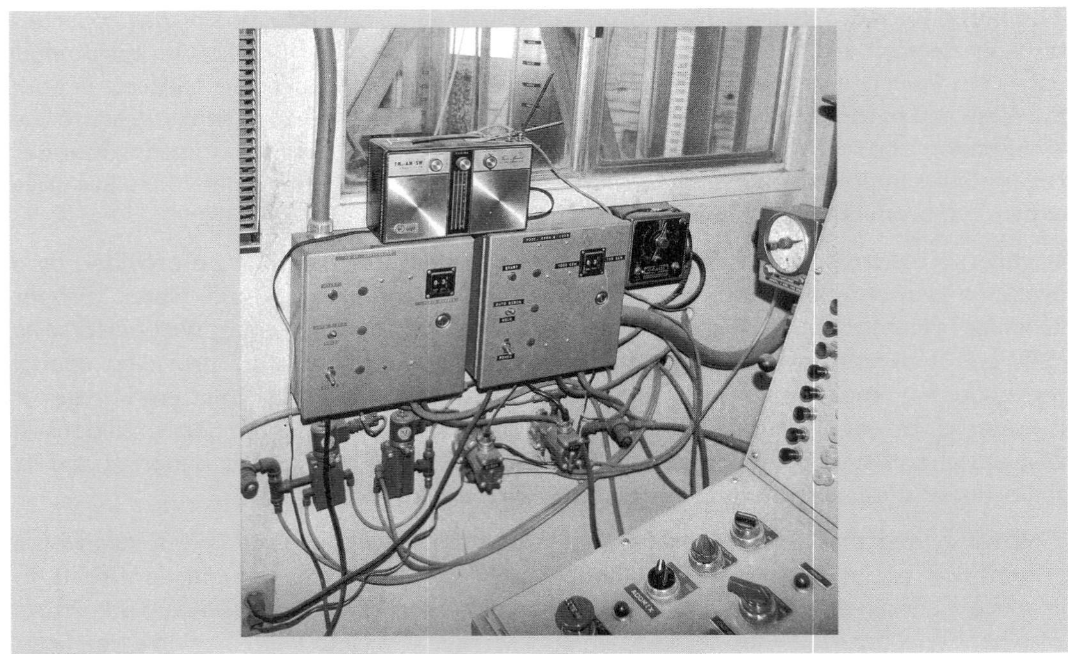

Figure 9-3
Various timers, pumps and sight gauges are used for measuring admixtures. Note the calibrated sight glasses through the window. These are used for checking the dosage of admixtures dispensed.

5. The system must be designed and installed so that all relevant parts are accessible for service and cleaning. Cleaning is usually accomplished by flushing with water.

A chemical admixture should never be dispensed into the concrete in a dry or powdered state, as it will rarely go into solution properly and will not be distributed completely throughout the batch of concrete. Instead, the admixture should be put into solution first and dispensed as a liquid. Coloring pigments, finely divided mineral admixtures and pozzolans are exempt from this restriction and can be introduced into the mixer in a dry condition, either in the form of prepackaged units or by weigh batching.

Liquids can be dispensed by weight or volume; most dispensers for liquids are of the volumetric type. Flow meters, similar to water meters, are very accurate when properly calibrated for the viscosity of the admixture being dispensed. A volumetric sight gauge, consisting of a positive displacement container, can become an automatic device when equipped with a float and electrical controls.

Manually operated sight gauge meters are also used. The timer-controlled orifice consists of a calibrated orifice that measures the amount of admixture by controlling the length of time a valve is open to permit the admixture to flow through the orifice. Inaccuracy results from restriction of the orifice by foreign matter and from variations in viscosity caused by temperature changes. Vents must be installed at appropriate points in the system to prevent air locks that can interfere with free flow.

Retarders, water reducers and air-entraining agents are usually supplied in liquid form and can be dispensed directly from the container. Others may be furnished as a concentrate that is diluted on the job. Some of them require continuous agitation to assure uniformity, especially the job-mixed solutions. Instructions of the manufacturer should be followed.

When using more than one admixture, they should not be intermixed prior to introduction into the mixer, unless the manufacturer states that it is permissible. Separation can be accomplished by admitting one of the admixtures with the mixing water in the usual manner and depositing the other in the measured batch of sand.

When the admixture is admitted to the mixer in the water, the time and rate of discharge must be regulated so that all the admixture is admitted before the flow of water ceases. The point in time at which some admixtures enter the mixer can be very critical. The amount of retardation, acceleration, air-entraining admixture or water required for the batch can vary significantly if the time of introduction is not uniform. Liquid admixtures should not come in contact with dry cement.

Liquid admixtures and the dispensing system must be protected from freezing. Heaters should be installed in storage tanks. Electrical immersion heaters are best, as they can be thermostatically controlled to prevent overheating.

Accelerators. Added to concrete, an accelerator speeds up setting time or increases the rate of early strength development, or does both. Benefits from this speedup are desired during cold weather to reduce pressure on the forms, permit early removal of forms, reduce the required period of curing and protection, permit early finishing, compensate for effect of low temperature on strength development and permit placing the structure in service sooner than would be possible without the accelerator. Calcium chloride is the one material that is commonly available for this purpose, and it is used to the virtual exclusion of everything else. It forms the basis for many brand-name admixtures.

Calcium chloride should be used in an amount not exceeding 2 percent by weight of the cement. In the generally mild winter temperatures prevalent in the south and southwest (except in mountain areas), 1 percent should be adequate.

Calcium chloride should be added to the batch in solution; to add it in flake or pellet form may result in stains or popouts where undissolved chloride concentrates in lumps. A good mixture is one in which one quart of solution contains one pound of chloride. This solution is measured into the mixing water as it flows into the mixer, by means of a mechanical dispensing device. Commercial calcium chloride for making the solution can be either the pellet form containing about 95 percent $CaCl_2$, or flake calcium chloride containing about 80 percent $CaCl_2$. The amount of water solution should be considered part of the mixing water when computing the water-cement ratio.

Concrete containing calcium chloride can have a compressive strength as much as 400 psi stronger at one day than plain concrete, 1000 psi stronger at three days and 900 psi stronger at seven days. The difference is still evident at 28 days, and sometimes even at one year. See Figure 9-4. Setting time is reduced by about a third to a half, workability of the fresh concrete is very slightly improved, early heat development is increased (but there is no increase in total heat), drying shrinkage is increased and bleeding is reduced. Calcium chloride should, however, be used with caution in reinforced concrete because of potential reinforcement corrosion; it also contributes to stress corrosion of prestressing steel and must not be used in prestressed concrete. It aggravates corrosion of galvanized metal and electrolytic corrosion of dissimilar metallic couples; hence, combinations of different metals, such as aluminum conduit and steel reinforcement, must not be present in concrete containing calcium chloride. Expansion of reactive aggregates with high-alkali cement is increased, and the resistance to sulfate attack is reduced. Resistance of concrete to abrasion and erosion is significantly improved, especially at early ages.

Calcium chloride should conform to the requirements of ASTM C 494. Because is absorbs water from the air, it should be kept in tightly sealed containers at all times. It should not be used if it becomes sticky or caked.

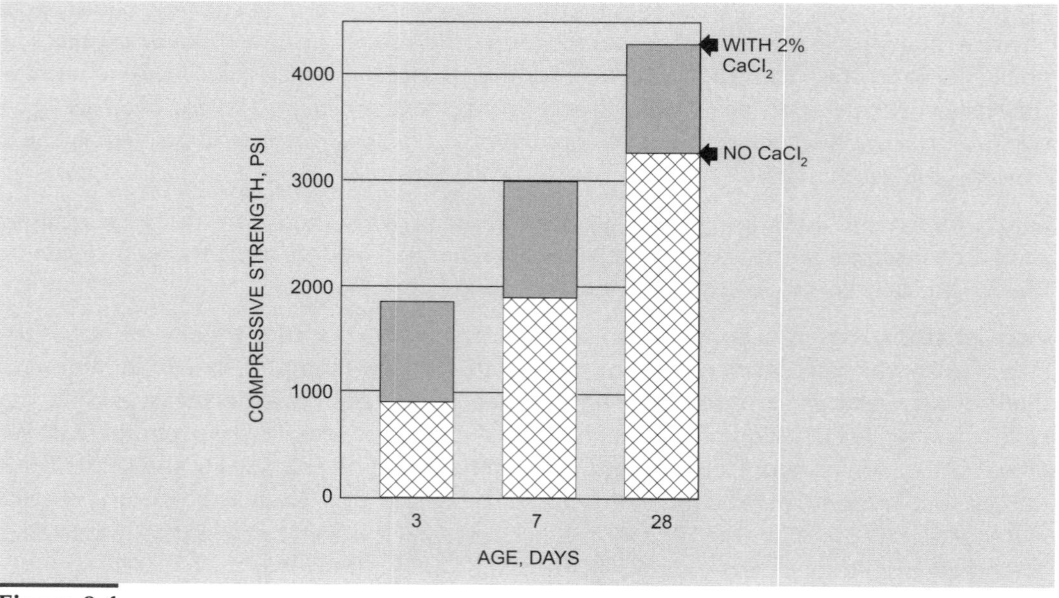

Figure 9-4
When concrete specimens were made at a temperature of 50°F, then cured at 40°F after three days, the specimens containing 2 percent calcium chloride had twice the strength of plain concrete at three days age, and almost 30 percent greater strength at 28 days.

The effect of calcium chloride depends on the amount used, type and brand of cement, temperature and curing conditions. It is more effective in rich mixes and usually increases the slump for the same amount of total water. It will precipitate most air-entraining agents unless added to the batch separately. Separate addition can be accomplished by permitting the chloride to flow in with the mixing water, introducing the air entrainer into the batch with the sand.

The chloride ion of the calcium chloride is the culprit that leads to problems with corrosion. For this reason, other accelerating chemicals have been researched in the hope of finding a cheap, efficient, noncorrosive accelerator. Formulations of calcium nitrite, calcium formate and formic acid offer the best potential and their use is rapidly expanding.

There are a number of other chemicals that accelerate the hydration of cement, but few are used to any extent, because of the difficulty in handling them, cost or erratic results. With some of these materials, almost infinitesimal changes in the amount have a pronounced effect. Triethanolamine is an accelerator, sometimes used in formulating a water-reducing admixture to overcome the retarding tendency of the water reducer, or for use in prestressed concrete. Other accelerators are sodium carbonate, calcium oxychloride, soluble silicates and fluosilicates, and mixtures of aluminous and portland cement. Very rapid-setting cements are discussed in Chapter 7.

Water Reducers. An admixture of this kind, as its name implies, is used for the purpose of lowering the mixing water requirement, at the same time providing equal or superior workability. Many of the retarding admixtures also reduce the water requirement (for the same slump or consistency), and some entrain a small amount of air. Water reducers are also used to reduce cement content while maintaining water-cement ratio; however, this usage of water reducers can have the adverse effect of decreasing durability and increasing finishing and workability problems.

Chemicals commonly used for water reduction are the lignosulfonates (calcium, sodium or ammonium) and salts of hydroxy-carboxylic acids.

Inasmuch as strength increases as the water-cement ratio goes down, other things being the same, and durability is higher for low water-cement ratio concrete, it may easily be seen that it is advantageous to keep the water-cement ratio as low as possible, within reasonable limits. See Figures 9-5 and 9-6.

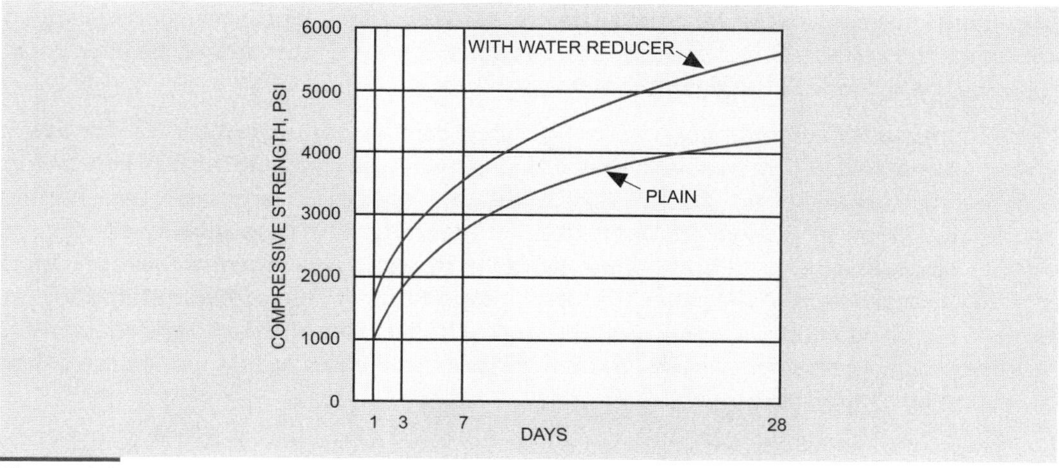

Figure 9-5
When the mix is properly adjusted to maintain the same slump and cement while reducing water, the mix containing the water-reducing admixture has higher strength at all ages than plain concrete.

It sometimes happens that the combination of materials in use on a job just will not produce concrete of adequate workability and consistency without exceeding the specified water-cement ratio. A situation of this nature calls for the use of a water reducer.

Concrete of very high strength, as that specified for prestressed concrete, is more easily attained, usually with less cement, if a water reducer is used in the concrete.

A water reducer should be used only after field tests have demonstrated its effect on mix proportions, entrained air and other properties of the concrete. Trials should be made with varying amounts of admixture, starting with a relatively small amount. The amount used should be the minimum to give the desired results, taking into account the varying effects of temperature, mix proportions and placing conditions.

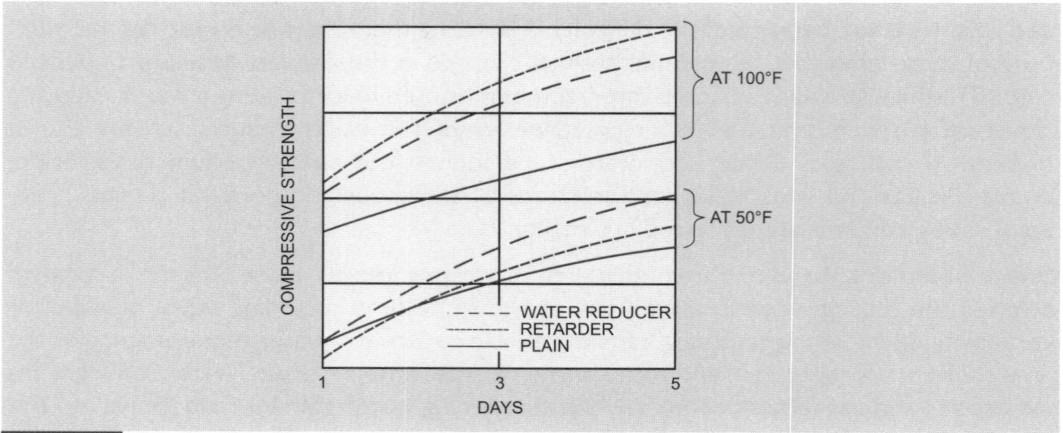

Figure 9-6
The early strength advantage of concrete containing a water reducer is more evident when the temperature is high than it when it is low. This figure also shows that concrete containing a retarder gains strength after the initial period of retardation, even at 50°F. The effect is greater at the higher temperature.

Retarders. A retarder is an admixture that slows the chemical process of hydration so that the concrete remains plastic and workable for a longer time than concrete without the retarder. However, once the cement starts to set, strength gain should be at the normal rate. See Figure 9-4. Retarders are used to delay the set of cement during difficult placements that require the concrete to be in a plastic condition longer than normal, and to overcome the acceleration of set during hot weather. Rarely, a retarder may be effective in reducing the tendency of a cement to false set.

Many chemicals have a retarding action on portland cement, some of which are extremely erratic and unreliable. Inorganic compounds exhibiting these characteristics include boron compounds (borax, boric acid, calcium borate), sodium bicarbonate and certain phosphates. Commonly used retarders are either metallic salts of lignosulfonic acid, such as calcium lignosulfonate, or salts of organic hydroxycarboxylic acid. Many of these products are refined byproducts of the paper-manufacturing industry. These admixtures, in addition to their retarding action, act as water reducers. Modification of the material during the manufacturing process permits the producer to make an admixture that will feature either the retarding or the water-reducing property.

Evaluation of a retarder is accomplished by testing in accordance with ASTM C 403. This method specifies the use of a calibrated needle that is pressed into a sample of mortar sieved out of the concrete. See Figure 9-7. When the hardening process has progressed to the point where the penetration resistance is 500 psi, the concrete is said to have reached the vibration limit and can no longer be made plastic by revibration. Beyond this point, a delayed second layer of concrete would not become monolithic with the layer below, and a cold joint will probably result. The vibration limit is the point in the hydration process that it is desirable to postpone by means of a retarder.

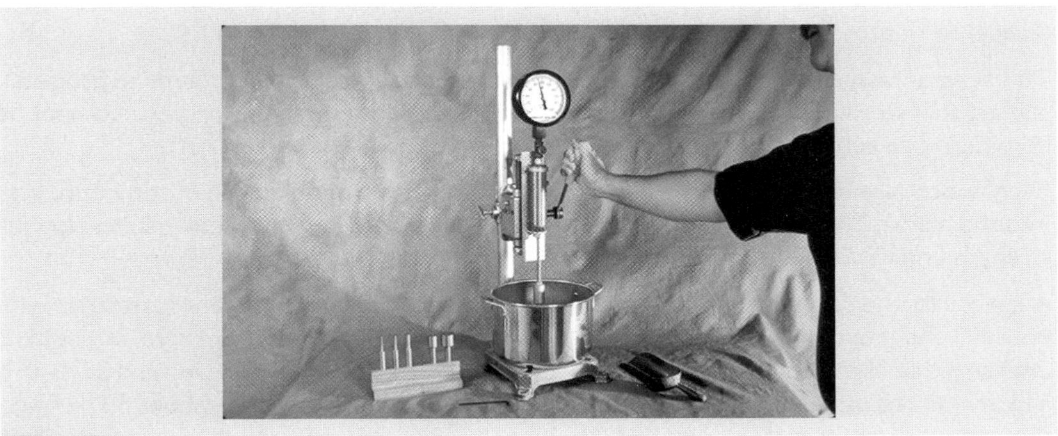

Figure 9-7
Time of setting equipment (ASTM C 403) to measure the resistance of plastic concrete to the penetration of a needle. (Courtesy of PCA)

As hydration continues, the concrete begins to gain some strength. When the penetration resistance reaches 4000 psi, the concrete has a compressive strength of about 100 psi. The period of time elapsed between the vibration limit and 100 psi strength should be about the same for retarded concrete as for the unretarded concrete. See Figure 9-8. This test evaluates a retarder as such; but, when using a retarder, trial mixes should be made in the field, using the equipment that will be used for mixing on the job. This is because small laboratory batches may not give the same results as job batches.

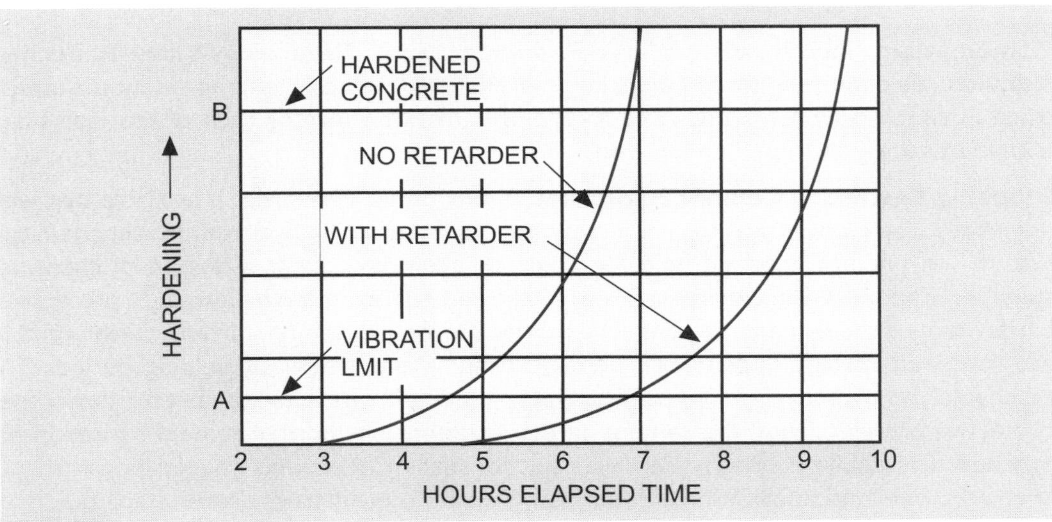

Figure 9-8
At point A the penetration resistance, 500 psi, shows that the concrete is no longer plastic. For plain concrete this occurred after an elapsed time of about 4 $^1/_2$ hours. Use of the retarder delayed this condition $2^1/_2$ hours, to an elapsed time of 7 hours. Once the vibration limit had been passed, both concretes developed strength at the normal rate.

Temperature is important, not only because the admixture is affected by temperature, but also because the amount of retardation desired depends upon temperature. In the summer when air temperatures are high, cement sets or hydrates more rapidly than it does during cool weather. For this reason, it is often desirable to use a retarder during hot weather to lengthen the time the concrete is plastic and workable. This retardation is especially desirable when placing conditions are such that the concrete has to be placed slowly in small increments. In ordinary concrete placements no retardation is necessary when the temperature is below 75°F.

The several manufacturers of retarders furnish automatic dispensers for their products, and a dispenser should always be used. A laborer with a coffee can or pop bottle cannot be considered an automatic dispenser, although this procedure has been used.

Even under careful control, there may be occasions when a double dose of admixture gets into the mixer. This need not be cause for alarm, as the affected concrete will develop full strength under adequate curing.

A lignosulfonate retarder reduces water requirement by 5 to 10 percent compared with nonretarded concrete and entrains a small amount of air. Compressive strength is unchanged or slightly higher up to three days and appreciably higher at seven and 28 days. If a hydroxylated carboxylic acid is used in an amount to retard the set by about 30 percent, the water requirement is reduced about 5 to 8 percent, and no air is entrained. Compressive strength will be slightly lower up to 24 hours but appreciably higher after three days.

Flexural strength of retarded concrete is somewhat increased over that of unretarded concrete, but not to the same extent as compressive strength. Usually, a retarder has no effect on the rate of slump loss, even though the setting time has been delayed. There have been instances in which the slump loss was faster with a retarder than without.

Effects of these admixtures vary with different cements, temperature, water-cement ratios, mix proportions and job conditions. For this reason, tests should be made with job materials and, insofar as possible, under job conditions, to determine the actual results of using any certain admixture.

Combination Accelerating-Water-Reducing and Retarding-Water-Reducing Admixtures combining these functions are frequently marketed and have found a useful place in concrete construction. The foregoing remarks regarding each of the individual admixtures apply to these combinations.

Stopping/Restarting Cement Hydration. A new breed of retarder is available that will essentially stop (and restart) the cement hydration process. The two-component chemical system will put hydration on hold for hours or even days. First, a dosage of chemical stabilizer is added to the concrete; this stops the cement hydration by forming a protective barrier around the cement particles. The barrier prevents the cement from achieving initial set. Required dosage depends on cement content and length of holding period. To reactivate the hydration, a chemical activator is added. The activator breaks down the protective barrier around the cement particles and permits normal cement hydration to proceed. The stabilizer system was developed for ready-mix producer applications such as overnight/weekend stabilization of returned plastic concrete in truck drums, stabilization of freshly batched concrete for long hauls, stabilization of leftover concrete from pump lines in the concrete hopper, and other applications where, for whatever reason, the cement hydration process must be stopped.

Air-Entraining Agents. Admixtures of this type (commonly abbreviated AEA) are now generally accepted for use in nearly all concrete, especially in areas where the concrete is exposed to freezing and thawing. Their use is recommended for all exposed concrete. The types commonly available are organic salts of sulfonated hydrocarbons and salts of sulfonated lignin. Synthetic detergents, petroleum acid salts, fatty and resinous acids and salts of wood resins are also sources of air-entraining agents.

Air-entraining agents should conform to the requirements of ASTM C 260, and the manufacturer should furnish proper certification to this effect.

The benefits of air entrainment have been recognized for many years, especially in regard to the great improvement in durability of concrete exposed to freezing and thawing conditions. See Figure 4-26. There are other benefits too, especially to fresh concrete, as well as a few drawbacks that should be considered. However, the benefits outweigh the disadvantages, even when the concrete is not exposed to freezing and thawing. See Table 9.1.

By improving workability of the fresh concrete, entrained air permits the use of harsh and poorly graded aggregate, reduces bleeding and reduces segregation tendencies. Handling and placing of the concrete are facilitated, and finishing of slabs can be accomplished sooner than with plain concrete. The reduction in permeability of air-entrained concrete compared with plain concrete makes the hardened concrete more resistant to the passage of moisture.

At constant water-cement ratio, compressive strength of concrete is reduced by 4 to 6 percent for each percent of air entrained, but by taking advantage of the increase in workability to make a reduction in water the strength loss can be partially compensated. Drying shrinkage of concrete is increased as the amount of entrained air increases. However, by keeping slump constant and again taking advantage of the reduction in water, which is about 2 to 4 percent, net shrinkage of the concrete is not appreciably increased. See Figure 9-9.

TABLE 9.1
AIR-ENTRAINED CONCRETE

As compared with similar, nonair-entrained concrete, air-entrained concrete has the following properties.

1. Greatly improved resistance to weathering damage from cycles of freezing and thawing.
2. Greatly improved resistance of pavements to scaling by de-icing salts.
3. Greatly improved workability of the fresh concrete.
4. At the same water-cement ratio, compressive strength of the concrete is reduced by 4 to 6 percent for each of entrained air. However, strength of lean mixes may be increased.

5. Resistance to attack by most chemicals is increased slightly.
6. Elasticity and erosion resistance follow strength.
7. Drying shrinkage increases as amount of air increases, but this is offset by lower water requirement. Hence, negligible effect.
8. Increased watertightness.
9. Minor reduction in susceptibility to alkali-aggregate reaction.
10. Unit weight of concrete reduced in direct proportion to the amount of entrained air.
11. Reduced rate of bleeding.

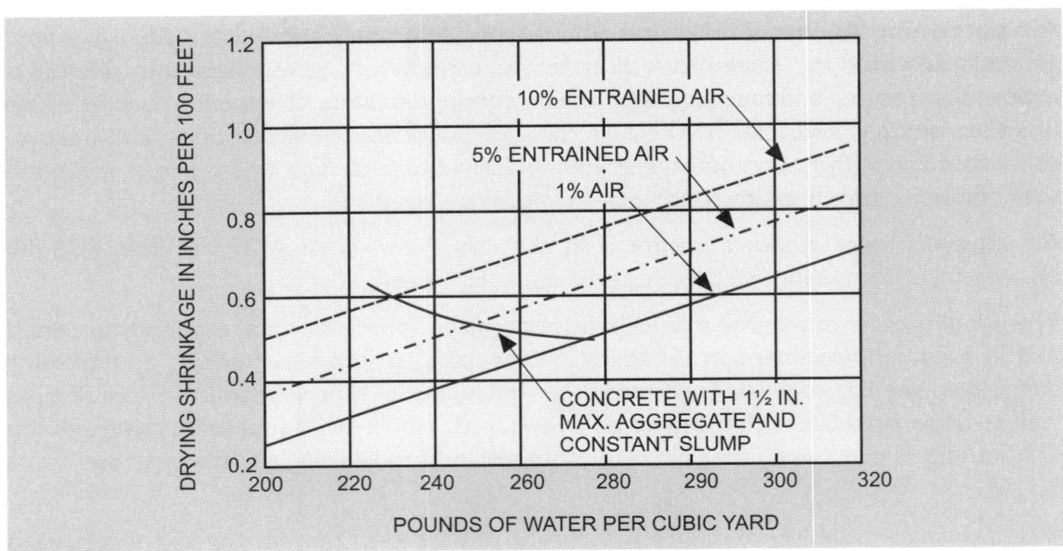

Figure 9-9
Although the drying shrinkage potential of concrete increases as the amount of entrained air is increased, the actual shrinkage is affected only slightly because of the lessened water requirement. This is illustrated by the curve for the $1^1/_2$-inch MSA mix, which shows the same shrinkage at 5 percent air as at 1 percent for constant slump.

If the concrete mixer is overloaded, the blades worn or the mixer operated at the wrong speed, the batch will not be mixed vigorously enough to make efficient use of the air-entraining agent. Different waters may affect the amount of entrained air.

Air-entraining agents are not permanently damaged by freezing, but stratification in the tank will result, requiring energetic remixing. During cold weather when using calcium chloride in the concrete, the air-entraining agent and chloride must be admitted to the batch in the mixer separately, as the chloride tends to prevent entrainment of air. One expedient is to introduce the chloride in the mixing water, and the air-entraining agent in the sand.

In certain parts of Canada, Washington and the Midwest, some sands contain organic material that acts as a foaming agent, entraining air in the concrete. This entrained air is unstable in the fresh concrete, and the remaining voids are poorly spaced and of the wrong size to be of any value in improving durability. When this condition is encountered, a minute amount of isopropyl alcohol will act as a defoaming agent. Sometimes an excess of air-entraining agent is effective in removing this "organic air," as it is sometimes called.

The air-entraining agent is usually introduced into the concrete as a separate ingredient, added at the batch proportioning plant by means of a dispensing device. In some areas it is interground with the cement, in which case the cement is known as air-entraining cement. Best control of the amount of air is obtainable when the AEA is batched separately, as the amount can be varied in accordance with results of tests made on the concrete. However, for the job being done under minimum or partial inspection it is best to use air-entraining cement if available, as this will ensure at least a minimum amount of air in every batch. Additional AEA can be added to each batch at the proportioning plant, if required to maintain the proper percentage of air in the concrete. The ACI 318 required air contents for concrete exposed to freezing and thawing conditions are shown in Table 9.2. Exposure Class F1 (moderate exposure) is an environment in which the concrete is exposed to freeze-thaw conditions but will not be continually wet nor exposed to water for long periods before freezing, and will not be in contact with de-icers. Exposure Class F2 and F3 (severe exposure) is an environment in which the concrete is exposed to wet freeze-thaw conditions or to de-icing salts for snow and ice removal. Project specifications should allow the air content of the delivered concrete to be within (-1.5) and (+1.5) percentage points of the Table 9.2 target values.

TABLE 9.2
TOTAL AIR CONTENT FOR CONCRETE EXPOSED TO CYCLES OF FREEZING AND THAWING (BASED ON ACI 318 TABLE 4.4.1)

MAXIMUM SIZE AGGREGATE, IN.	TOTAL PERCENT AIR	
	EXPOSURE CLASS F1	EXPOSURE CLASS F2 AND F3
$^{3}/_{8}$	6	7.5
$^{1}/_{2}$	5.5	7
$^{3}/_{4}$	5	6
1	4.5	6
$1^{1}/_{2}$	4.5	5.5

Special lightweight concrete with as high as 85 percent entrained air is discussed in Chapter 21.

The amount of air is affected by many variables other than the amount of agent used, as shown in Table 9.3.

TABLE 9.3
INFLUENCES ON AIR CONTENT
The amount of air is affected by many variables other than the amount of agent used. An increase in any of the following has the indicated effect on the amount of entrained air:

INCREASE:	% AIR CHANGE
Slump	Increased
W/C ratio	Increased
Percent Sand	Increased
Fines in Sand	Decreased
Cement Content	Decreased
Temperature	Decreased
Mixing Time	Decreased
Cement Fineness	Decreased

Increasing the slump (amount of water per batch) or percent of sand will usually increase the amount of entrained air. A rise in temperature or lengthening of mixing time results in less entrained air. Content of air is lessened with an increase in the amount of fines in the concrete, such as would result from more fines in the sand, increased cement content or finer cement.

There are three methods of determining the air content of fresh concrete: the gravimetric, the pressure and the volumetric, each of them covered by an ASTM standard test method. The Chase air indicator, a thimble-sized device that can be used for rapid approximations of the air content, does not meet ASTM requirements, but it is small, easy to use and handy for quick checks on the mix. Testing is covered in Chapter 13.

Dampproofing and Permeability-Reducing Admixtures. In nearly all cases, leakage of water through a concrete structure can be traced to poor construction practices such as rock pockets or honeycomb, sand streaks, poor cleanup of construction joints, cracks and lean, soupy mixes.

Good materials and workmanship are the first consideration in any structure and are essential for watertight concrete. The use of waterproofers and dampproofers, either integral or surface applied, should not be considered compensation for poor workmanship, lean mixes or deficient materials. Many of the admixtures called waterproofing or dampproofing improve the quality of the concrete because of some beneficial effect such as entrainment of air or improvement of workability.

Powdered admixtures, either inert or pozzolanic, may be of value as a waterproofer in lean mixes or normal mixes lacking fines. They are of no value in rich mixes, or normal mixes with adequate fines. In fact, they may be detrimental in the latter.

Calcium chloride is sometimes used for improving watertightness, but its value for this purpose is questionable. Some proprietary compounds sold as waterproofers consist of little more than a calcium chloride solution.

Stearates reduce absorption and retard capillary action but are of little or no value if the water is under pressure. Butyl stearate is sometimes used in an amount not exceeding 1 percent by weight of the cement. It should be added to the batch as an emulsion for proper distribution throughout the concrete. Butyl stearate, in the recommended amount, has no serious effects on strength.

Related materials are the soaps, salts of fatty acids such as ammonium or calcium oleate or stearate, that act primarily as water repellants. These are foaming agents and may cause an increase in permeability if the water is under pressure. If used, they should not exceed 0.2 percent by weight of the cement. In water curing concrete containing a water repellent, the concrete should not be permitted to dry out until curing has been completed, as the concrete cannot readily be wetted once it has dried. Colloidal asphalt solutions are sometimes proposed for waterproofing.

Workability agents, by improving the ease of placing and consolidation of concrete, tend to promote watertightness. Examples are water-reducing agents and air-entraining agents.

Numerous proprietary compounds are on the market, some of which are of practically no value. The user should be certain of what is in a compound or admixture before using it in concrete, as some materials may have a detrimental effect on strength, durability or other properties of the concrete.

Bonding Agents. Some bonding agents are applied to the surfaces to be bonded, whereas others are used as admixtures. Bonding admixtures are made from natural or synthetic rubber, or polymers such as polyvinyl chloride, polyvinyl acetate, acrylics and butadiene-styrene copolymer. Some of these latexes are reemulsifiable when exposed to moisture and are apt to soften with age, thus damaging the bond and strength of the repair. Only the nonreemulsifiable materials, stabilized to inhibit coagulation in the presence of cement, should be used. See Chapter 10.

Alkali-Silica Inhibitors. Some pozzolans are very beneficial in reducing the severity of the reaction.

Antifreeze Compounds. There is no material that can be put into a batch of concrete to lower the freezing point of the fresh concrete without seriously damaging the concrete. Concrete work can be done in subfreezing weather, as described in Chapter 19, when adequate precautions are taken to protect the concrete. These precautions, however, do not include the use of any magic admixture that will prevent freezing of the fresh concrete.

Coloring Pigments are classified as admixtures. Coloring admixtures should meet the following requirements:

1. Colorfastness when exposed to sunlight.
2. Chemical stability in the presence of alkalinity produced in the set cement.
3. No adverse effect on setting time or strength development of the concrete.
4. Stability of color in steamed or autoclaved concrete products during exposure to the conditions in the autoclave.

Trial panels should be made to determine the exact color. Color of cement and aggregates, curing and finishing all affect the color. Selection of aggregate is especially important when white cement is used. Many manufacturers package colors in units, each unit containing the correct amount of pigment, plasticizer and other ingredients to color a cubic yard of concrete containing 560 pounds of cement, or a similar volume of concrete. Only pure mineral pigments should be used, as shown in Table 9.4.

TABLE 9.4
COLORING ADMIXTURES

Grays and Black	Black iron oxide Mineral black Carbon black
Browns and Reds	Red iron oxide Brown iron oxide Raw and burnt umber
Blue	Cobalt blue Ultramarine blue Phthalocyanine blue
Green	Chromium oxide Phthalocyanine green
Ivory, cream or buff	Yellow iron oxide
White	White cement

Workability Agents. Any material in the concrete mix that improves workability can be considered a workability agent. In all cases in which the aggregate shape or grading causes poor workability, entrained air has a beneficial effect. Entrained air reduces the tendency of the concrete to segregate, lowers the bleeding rate and expedites finishing. The addition of fines increases the plasticity of lean, harsh mixes, thereby improving workability. Examples of fines are pozzolans, fly ash, rock dust, colloidal silica, diatomaceous earth and bentonite. The water reducers and retarders can be classified as workability agents.

Expansion-Producing Admixtures. These admixtures compensate for drying shrinkage of the concrete and are usually incorporated in expansive cement but could be used as an admixture. Expansive cements are a mixture of portland cement, sulfoaluminous cement and blast furnace slag. These are discussed in Chapter 7.

Gas-Forming Agents. Shrinkage and bleeding of fresh concrete is apt to result in voids under forms, embedded items or machinery—causing a loss of bond and reduction in watertightness—when it is necessary to place fluid mortar or concrete in difficult, restricted areas. This shrinkage can be prevented by the proper use of a material that reacts with the cement to produce an expansive gas. Unpolished aluminum powder is commonly used for this purpose in an amount equal to about one teaspoonful to a sack of cement. See Section 5.4, Chemical Control of Shrinkage.

Finely Divided Mineral Admixtures. There are four kinds of materials in this classification. Based on their chemical and physical properties, they are classified as cementitious materials, pozzolans, pozzolanic and cementitious materials, and inert fines. Inert fines are listed under workability agents. They are used in lean, harsh mixes to improve workability, reduce bleeding and increase strength. They should not be used in mixes that have adequate fines.

Cementitious materials are natural cement, hydraulic lime, mixtures of blast furnace slag and lime, and granulated blast furnace slag. Besides providing fines in the mix, these materials provide some cementing value.

Plasticizers for Flowing Concrete.[9.1] Since the mid-1970s a group of admixtures known as super- plasticizers has become available. Briefly, a superplasticizer can be used in concrete:

(a) As a water reducer to give concrete a very low water- cement ratio but normal workability and high strength.

(b) To provide normal workability and strength at a reduced cement content but normal water-cement ratio.

(c) As a plasticizer to produce very workable concrete; that is, a flowing, self-leveling concrete with a high slump and high compressive and flexural strength. See

Figure 9-10
Flowable concrete with high-range water reducers. (Courtesy of PCA)

Figure 9-10.

These plasticizers are essentially high-range water reducers meeting ASTM C 1017.

Although these materials are relatively new, their advantages are becoming well documented. Much laboratory and field research has been done and more is being accomplished. When superplasticizers first became available, the principal objection to their use was the rapid loss of slump. There was an immediate gain of several inches slump when the admixture was first added to the concrete, followed by a rapid loss of slump in which the concrete reverted back to its original slump within 30 or 60 minutes. See Figure 9-11. This rapid slump loss posed no serious problem in a precasting plant; but for ready-mixed concrete it became necessary to introduce the admixture into the mixer at the job site. Now, extended-life superplasticizers that do not have rapid slump loss are available for ready-mix batching at the plant.

Dosage of the admixture is relatively high, running between $^1/_2$ pound to 3 pounds per bag of cement. This makes these admixtures appreciably more expensive than conventional ones.

Trial mixes should be made in the laboratory prior to use of these admixtures on the job. Compared with concrete without the superplasticizer, there is normally no change necessary in mix proportions. Slightly more fines may be desirable for the flowing, self-leveling mixes, and the air-entraining admixture may need adjusting. Segregation, bleeding and setting time are not significantly affected. Strength is usually better than that of plain concrete because of the lowered total water content and water-cement ratio. Freeze-thaw durability is usually not affected, and resistance to sulfate attack is not changed. Bond to the reinforcing steel may be improved and corrosion of the reinforcement is unaffected. Mixes containing fly ash appear to be affected about the same as those without. In general, the results are affected by the kind and dosage of admixture, type of cement and temperature. See Section 16.1 for word of caution on use of superplasticized concrete for slabs on ground construction.

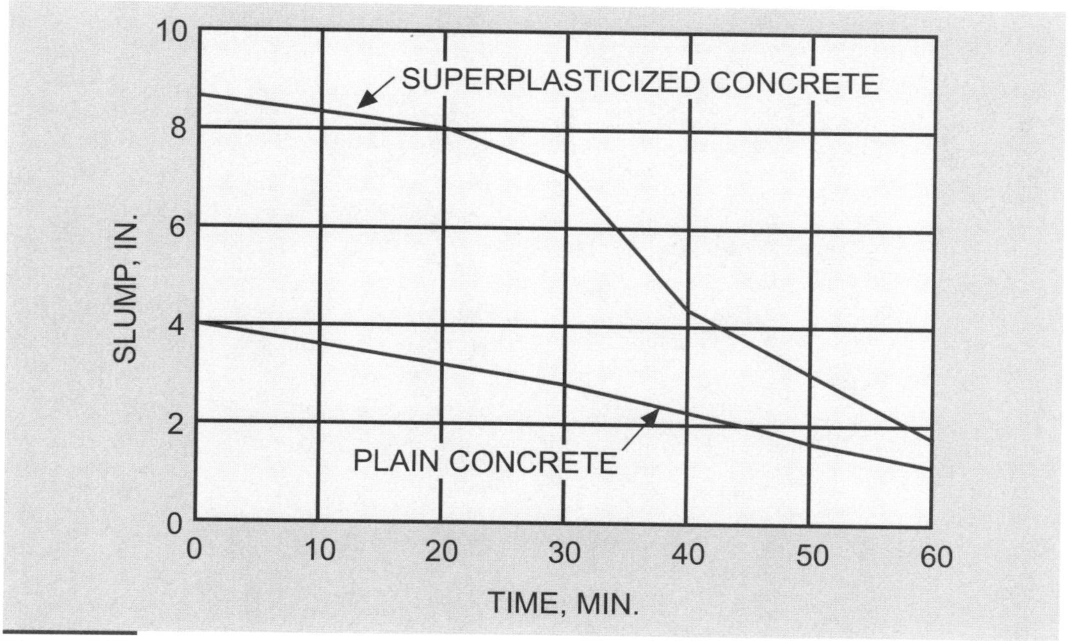

Figure 9-11
The immediate increase in slump when the superplasticizer is added to the concrete is lost after about an hour.

Accessory Materials

10.1. Field-molded Sealants
- Mastics
- Hot-applied Thermoplastics
- Chemically Curing Thermosetting
- Solvent-Release Thermosetting
- Rigid Materials

10.2. Preformed Sealants

10.3. Epoxy Resin

10.4. Bonding Agents and Adhesives

10.5. Surface Coatings
- Paint
- Waterproofing

10.6. Patching Compounds

10.7. Surface Retarders

Joints in concrete are necessary to allow for changes in volume of structural units, to permit movement of adjacent components of a structure or to facilitate construction. See Chapter 16. Irrespective of the kind of joint (or crack), nearly all of them must be filled or sealed to assure satisfactory performance of the structure. The materials used for filling joints or cracks are known as joint fillers or sealants, or caulking compounds. They may be field-molded materials that are poured, gunned or troweled into position, or they may be rigid or flexible preformed materials. See Figure 10-1.

10.1. Field-Molded Sealants

Sealants of this kind consist of oil-base mastics and elastomeric materials such as polysulfides, urethane, epoxy, neoprene, asphalt and rubber-asphalts. Some are applied hot, some cold.

Mastics are composed of thick liquids such as low-melting point asphalts containing asbestos or other filler. They can be used where only small joint movements are expected. They harden with time, thus reducing their serviceability.

Hot-Applied Thermoplastics of the solvent or emulsion type set either by the release of a solvent or breaking of an emulsion upon exposure to air. They can be applied at normal temperature or by heating to not over 120°F. Among these materials are acrylic, vinyl and modified butyl types in several colors. They can be used only in joints with a small movement.

Chemically Curing Thermosetting sealants are either one- or two-component liquid systems that become solid by a chemical reaction. They include polysulfide, silicone, urethane and epoxy-base materials. They can be used in many situations for both horizontal and vertical joints with considerable movement. Although somewhat expensive, they have a long service life.

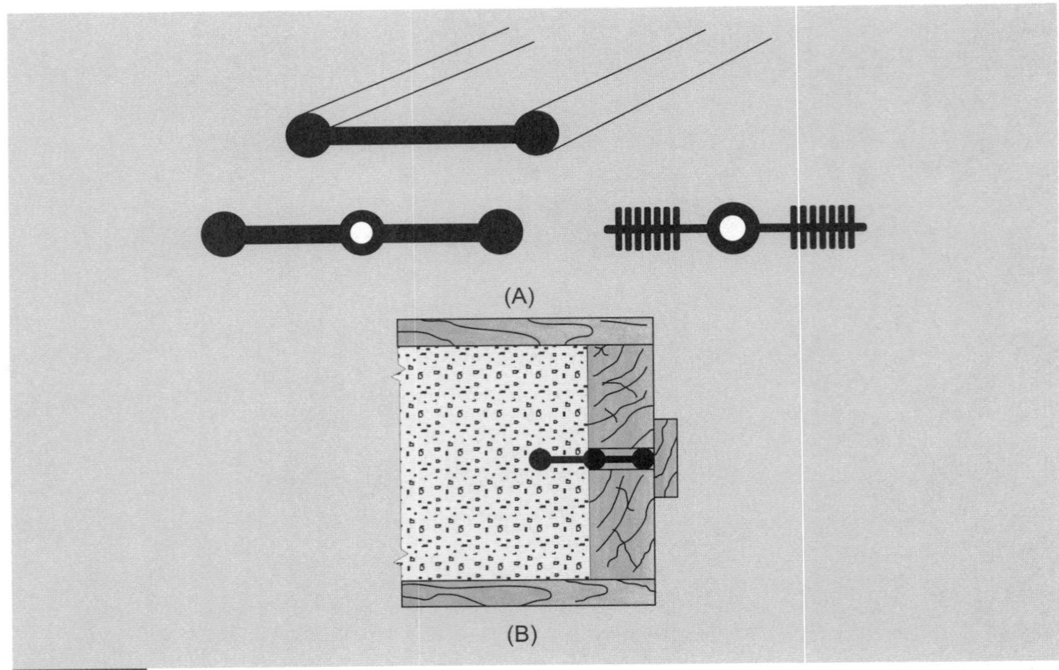

Figure 10-1
Flexible rubber or plastic waterstops come in a variety of shapes, three of which are shown in A. One typical method of installing a waterstop in the form is shown in B.

Solvent-Release Thermosetting sealants cure by release of a solvent. Certain polyethylene, butyl and neoprene materials are included. They are suitable for horizontal and vertical joints in buildings having small movements.

Rigid Materials such as lead, sulfur and modified epoxy resin can be used in joints where no movement is expected.

10.2. Preformed Sealants

Premolded sealants may be embedded in the concrete, called a waterstop, or installed by compressing the sealant into a joint slot, called a compression gasket. Another type is a rigid porous strip impregnated with asphalt or similar materials, which is used to form an isolation joint.

Fillers for construction and isolation joints are usually preformed strips of the required dimensions to fill the joint space. They may consist of various felts and fibers saturated and bonded with bituminous binder (usually asphalt), self-expanding cork, sponge rubber or certain synthetic plastic materials. Sometimes strips of selected cypress or redwood are used.

Waterstops are used for the purpose of preventing the flow of water through a joint in the concrete. Rigid waterstops are usually made of sheet copper, although steel and lead have been used. Flexible waterstops are frequently used instead of rigid ones and are necessary if joint movement is expected. Natural and synthetic rubbers and polyvinyl chloride are widely used. See Figure 10-1.

The plans and specifications for the project will detail all joints, including the kind of sealing and filling materials to use. Many joint fillers are proprietary materials, and the instructions of the manufacturer should be followed.

10.3. Epoxy Resin

There are literally hundreds of compounds that are known as epoxy resins. ASTM C 881, a performance specification based on end use of the material with concrete, provides requirements for three types, three grades and three classes of epoxy resin systems, depending on end use, flow characteristics and temperature exposure conditions. In general, there are four broad areas in which epoxy resins are used with concrete: bonding hardened concrete, steel, wood, brick and other materials to hardened concrete; bonding plastic (fresh) concrete to hardened concrete; producing a skid-resistant surface on concrete; and filling cracks and performing other repairs.

Epoxy resins are designed for a wide range of temperature conditions, surfaces and applications. They will adhere to many materials, including concrete, plaster, wood, metals and most plastics. Some are formulated to stick to wet surfaces. They will not adhere to waxed or greased surfaces. Properly formulated epoxy systems exhibit only minor shrinkage, or none at all, upon curing. Toluene is sometimes used for thinning the resin; however, its use is not recommended, because of the danger of increasing curing shrinkage and weakening of the material.

In the use of any epoxy system the recommendations and instructions of the manufacturer should be followed. The ASTM standards cover specifications for the material only.

Epoxy resin is sold under a multitude of trade names. The material consists of two components: the basic resin and a curing agent, which are packaged so that a full can of one component is mixed on the site with a full can of the second component in small batches that can be used up in two or three hours or less, depending on the pot life. Pot life is the period of time during which the material can be used after it has been mixed. It is different for different formulations and decreases with rising temperature. Epoxies have a relatively short pot life. They become hard and unworkable in a matter of a few minutes to two hours or so, depending on the formulation and temperature. For this reason, small amounts should be mixed at a time. Tools and mixing equipment should be cleaned before the material starts to harden by scrubbing with toluene. Because of their toxic nature, epoxies should be used in a well-ventilated place, and contact with the skin should be avoided.

10.4. Bonding Agents and Adhesives

Polyvinyl acetate (PVA), which improves the bond of concrete to old concrete, is usually supplied as an emulsion or dispersion and is used in varying amounts, depending on the application. Other advantages besides improvements in bond strength are crack resistance, better strength, elasticity and wear resistance. Some polyvinyl acetate latexes are reemulsifiable when exposed to moisture and are apt to soften, thus damaging the bond. For this reason, only those PVA bonding agents known to contain an ingredient designed to render the dried film resistant to moisture should be used.

A copolymer latex of butadiene or styrene may be used for a bonding agent, provided it is suitably stabilized to inhibit coagulation in the presence of cement and contains a moisture protection ingredient.

Formulations of epoxy resin and polysulfide liquid polymer are available as bonding agents. Bonding agents are also used in the form of an admixture. See Chapter 9.

Bonding agents can, in a general sense, be classed as adhesives, as they promote the bond (adhesion) of new concrete to old. In its strict sense, however, the term adhesive means a substance that is used to join hardened concrete or other solid material to hardened concrete. Epoxy resin is extensively used for this purpose.

10.5. Surface Coatings

Color in concrete is usually obtained by the proper selection of cement and aggregates, and through the use of coloring admixtures described in Chapter 9. There are also a number of chemical stains available. These are proprietary compounds and normally present no problem in their selection.

Paint can be applied to concrete to make it watertight, to improve its durability, to decorate or for a combination of these reasons. See Chapters 17 and 22.

Paints consisting of a portland cement base can be applied to new, damp concrete. Commercially available paints should be used, but in an emergency a white portland cement paint can be made with the following proportions:

White portland cement	20 pounds
Hydrated lime	4 1/2 pounds
Pigment (titanium dioxide or zinc sulfide)	3/4 pound

Calcium chloride	3/4 pound
Aluminum or calcium stearate	1/4 pound

The ingredients should be mixed dry, screened through a piece of ordinary window screen, then mixed with water. Use only sufficient water to make a thick paste; mix thoroughly, breaking up the lumps; then add the rest of the water.

For the job-proportioned paint, the amount of water will be about 3 quarts for 10 pounds of the dry material. The manufacturer's instructions should be followed regarding the amount of water to use and method of mixing.

Other paints include oil-base paints consisting of opaque pigments suspended in a vehicle of drying oils and thinner (linseed oil paints); resin-emulsion paints consisting of water-reducible pigment paste in an emulsified oil-extended resin, usually glycerol phthalate; synthetic rubber paints, either the emulsified synthetic-rubber resin type or the rubber-solution type; and other resin or plastic types of paints.

Numerous surfacing materials are used for protecting concrete from attack by aggressive substances. A few of them are listed in Table 4.1.

Waterproofing and dampproofing of concrete are sometimes accomplished by means of coatings on the concrete.

Solutions of methyl and ethylsilicone resins dissolved in toluene make effective water repellents when applied to the surface of concrete. Epoxy resins may also be used. Phenolic resin varnish, microcrystalline wax, neoprene and coal-tar cutback are also effective. These materials should be applied to cured and dry concrete.

Powdered iron preparations, consisting of mixtures of powdered iron and cement, usually with an oxidizer such as ammonium chloride, are effective. These preparations range from thin coatings to a stiff consistency and are applied with a brush, as paint. Caution should be exercised in their use if they are to be exposed to the weather, as they may cause rust stains.

There are numerous proprietary materials on the market; many of them based on the foregoing materials. The user should make sure that any proposed material has a satisfactory service record and should use it strictly in accordance with the manufacturer's instructions.

10.6. Patching Compounds

Many compounds proposed for use as rapid-setting cements designed for quick repair of damaged concrete are available. Some of these compounds consist of portland cement with dehydrated gypsum (plaster of paris). The plaster of paris causes a set in a few minutes, which makes the material desirable for quick patching of small blemishes and damaged areas. The use of this material, however, should be discouraged, as it has poor weathering resistance, and is apt to fail in the presence of moisture. Good compounds are available, and the supplier should be requested to furnish independent laboratory reports showing the satisfactory results of the use of a proposed material.

A mixture of portland cement and aluminous cement sets in a few minutes and develops strength rapidly. See Chapter 7.

10.7. Surface Retarders

One method of exposing the aggregate on the surface of architectural panels is to retard setting of a very thin layer of cement on the surface of the concrete. Subsequent washing of the surface removes the unhydrated cement, revealing the aggregate. The retarder can be applied to the form for formed concrete, or directly to the surface of the concrete if the surface is unformed.

The retarder for this application has the same formulation as the admixture but is a more concentrated solution.

Formwork

11.1. Formwork Requirements
11.2. Wood for Forms
- Plywood
- Hardboard

11.3. Other Forming Materials
- Fiberglass
- Plastic and Rubber Liners
- Steel
- Paper and Cardboard
- Waste Molds

11.4. Form Accessories
- Form Clamp
- Snap Tie
- Coil Tie
- She-Bolt
- Inserts

11.5. Form Oils and Compounds
11.6. Falsework and Shoring
- Permanent Shores
- Reshores

11.7. Slipforms
11.8. Precast Concrete Forms
11.9. Prefabricated Forms
11.10. Use of Forms
11.11. Removal of Forms and Shoring

In this chapter the reader will learn why forms are necessary, how the forms are made, what they are made of and how they are used. The design of concrete forms can almost be considered a branch of engineering in its own right, combining as it does features of civil, structural and mechanical engineering.

11.1. Formwork Requirements

Forms are necessary to confine the fresh concrete and mold it to the required shape. The simplest form is a hole dug in the ground, into which concrete is placed for a foundation or footing to support part of a structure. More elaborate forms are designed for special construction techniques. Forms, although their function is temporary, are engineered structures. They must be designed and built to impart the specified surface to the concrete in the most economical manner; to resist high loadings, both moving and stationary; to resist the hydraulic head of a heavy fluid weighing 145 pounds per cubic foot; to withstand the effects of vibration of the concrete; to confine the concrete without shifting or permitting loss of part of the concrete or mortar; and to lend themselves to rapid construction, erection, stripping and transporting. Above all, the forms must be constructed with an adequate factor of safety to prevent the possibility of failure or collapse. Some common construction deficiencies that lead to form failure are the following: [11.1]

(a) Inadequate diagonal bracing of shores.

(b) Inadequate lateral and diagonal bracing and poor splicing of double-tier shores or multiple-story shores.

(c) Failure to control rate of placing concrete vertically without regard to changes in setting times.

(d) Failure to regulate properly the rate and sequence of placing concrete horizontally to avoid unbalanced loadings on the formwork.

(e) Unstable soil or inadequate bearing under mudsills, sometimes caused by wash water from the forms; in no case should mudsills or spread footings rest on frozen ground.

(f) Failure to inspect formwork during and after concrete placement to detect abnormal deflections or other signs of imminent failure, which could be corrected.

(g) Insufficient nailing, bolting or fastening.

(h) Failure to provide adequate support for lateral pressures on formwork.

(i) Shoring not plumb and thus inducing lateral loading as well as reducing vertical load capacity.

(j) Locking devices on metal shoring not locked, inoperative or missing.

(k) Vibration from adjacent moving loads or load carriers.

(l) Inadequately tightened or secured form ties or wedges.

(m) Form damage in excavation by reason of embankment failure.

(n) Loosening of reshores under floors below.

(o) Premature removal of supports, especially under cantilevered sections.

(p) Connection of shores to joists, stringers or wales that are inadequate to resist uplifts or torsion at joints.

(q) Failure to comply with manufacturer's recommendations for standard components.

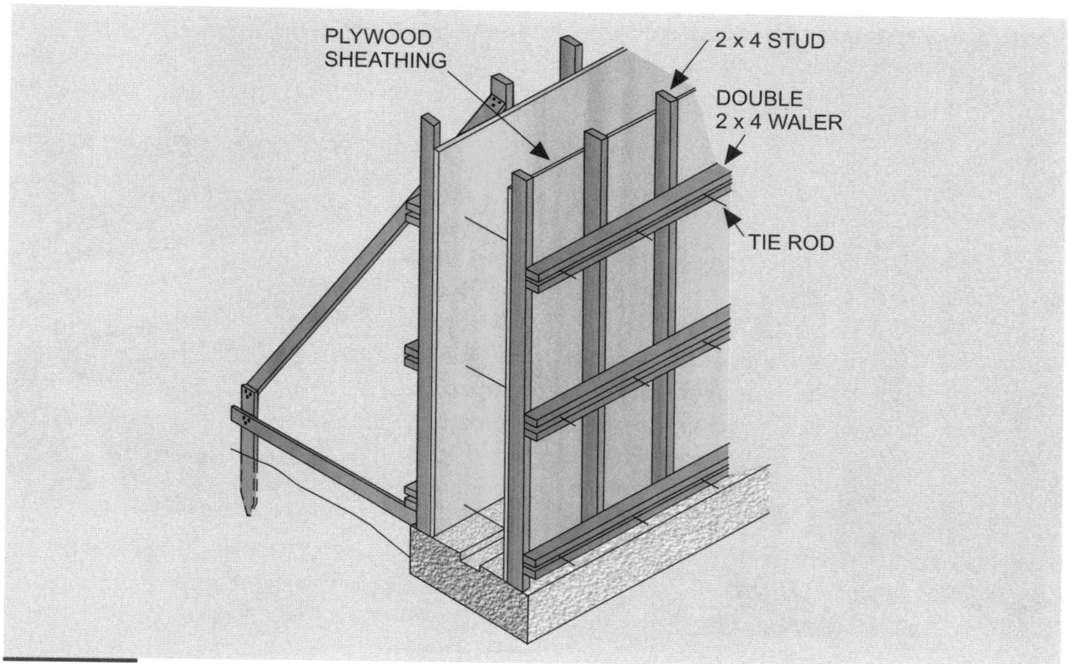

Figure 11-1
Basic elements of a simple wall form erected on a concrete footing. Reinforcement for the wall is not shown. The tie bars, which are designed to act as spreaders also, will be anchored by different devices on the outside of the walers.

Forms are made of wood, metal, cardboard, fiberglass, plaster, plastic, insulated concrete forms (ICF), Styroform deck forms, concrete or combinations of two or more of these materials. By far the greatest proportion of forms are made of wood, mainly because of its general availability and ease of fabrication. A section of a simple wall form is shown in Figure 11-1. See Figures 11-2 and 11-3.

The first consideration in designing formwork is safety. Failures of forms and shoring have resulted in spectacular construction accidents involving injury and death to workers and extensive property damage.

The concrete surface reflects the form surface. It is quite obvious that a smooth surface cannot result from rough formwork. Of course, there are instances when rough form lumber is used to impart a rough texture to architectural concrete, but these forms require as careful quality of construction as any other formwork. See Figure 11-4.

Unsatisfactory alignment of concrete surfaces results from poorly designed forms and slipshod construction. Occasionally, misalignment of the concrete is caused by movement of the form while the concrete is being placed and vibrated. Movement of forms during concrete placement can be prevented by attention to important details. Proper spacing of studs and walers prevents bulging; adequate fastening, bracing and wedging restrain movement of the form under pressure of the concrete while it is being vibrated; and tight joints restrict leakage of mortar from the concrete.

Forms should be constructed and maintained so that the finished concrete will be true to line and grade, and of the dimensions and shape shown on the plans, keeping in mind that tolerances in dimensions refer to the dimensions of the concrete in the structure, not the formwork. Forms should be constructed so that they can be easily removed without damage to the green concrete.

Figure 11-2
Foundation wall formwork.

Figure 11-3
Formwork for isolated column footing with rebars and column base plate and anchor bolts secured in place ready for concrete placement.

Figure 11-4
An example of an attractive texture obtained by the use of rough lumber form liners.

Forms should be mortar tight, sufficiently rigid to prevent distortion caused by pressure of the concrete and other loads incidental to construction, and constructed and maintained to prevent warping and opening of the joints that are due to shrinkage of the form material. Molding or chamfer strips should be placed in the corners of forms to produce beveled edges on permanently exposed concrete surfaces. See Figure 11-5. Interior angles on such surfaces and edges at formed joints usually do not require beveling. Chamfer strips may be job-cut strips of wood, plastic or rubber. The plans should govern relative to the use of such items.

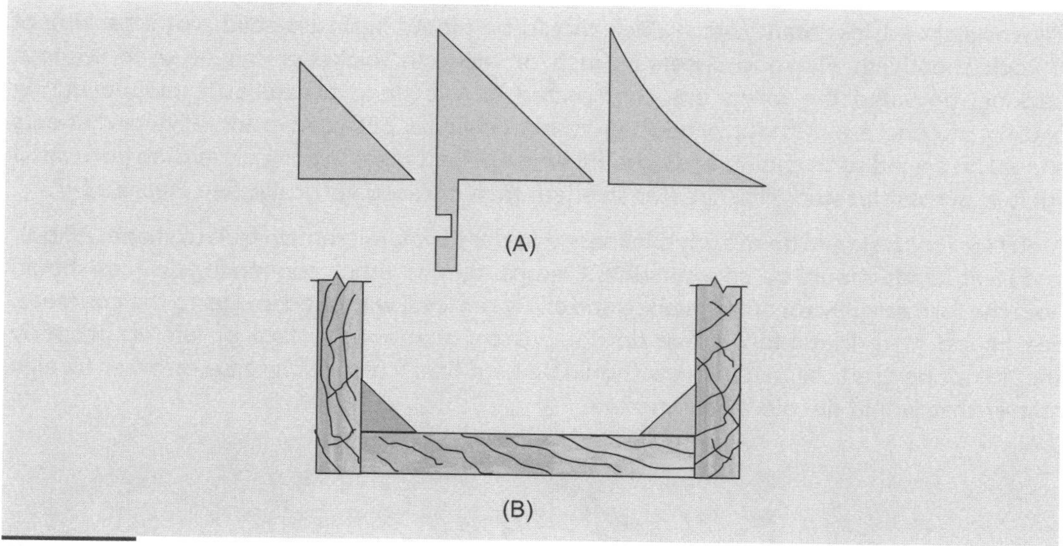

Figure 11-5
Cross sections of some chamfer strips made of flexible polyvinyl chloride are shown in A. Installation of two kinds is shown in B. See also Figure 4-20.

All lumber in contact with concrete should be free of knotholes, loose knots, cracks, splits, warps or any other defects that would mar the appearance of the finished structure. Lumber having defects affecting its strength should not be used.

Wood lining for forms should be of such kind and quality, or should be so treated or coated, that there will be no chemical deterioration or discoloration of the formed concrete surface. The type and condition of form lining and the construction of the forms should be such that form surfaces will be even and uniform. The wood or metal lining should be placed so the joint marks on the concrete surfaces will be in general alignment, both horizontally and vertically.

Form panels, either of wood or metal, should be constructed and assembled to result in tight joints between the panels. Panel joints should match in general alignment the joints of the lining or sheathing. See Figure 11-6.

Figure 11-6
Foundation wall forms secured in place. Note the tight joint between each panel.

Plywood sheets less than $^5/_8$-inch thick should be placed against a solid wood backing of $^3/_4$-inch sheathing. Plywood sheets $^5/_8$ inch or more in thickness can be used without backing, provided the forms are constructed to withstand pressure developed during placing of concrete without producing visible waviness between studs. Plywood sheets should be placed so that joints are tight. Plywood placed with the long dimension horizontal (that is, across the studs) has greater strength than if placed vertically. See Figure 11-7.

Metal for forms should be of such thickness that the forms will remain true to shape. All bolt and rivet heads should be countersunk. Clamps, pins or other connecting devices should hold the forms rigidly together in place and allow removal without damage to the concrete. See Figure 11-8. Metal forms that do not present a smooth surface or line up properly should not be used. All metal forms should be kept free from rust, grease or other foreign matter that would discolor the concrete.

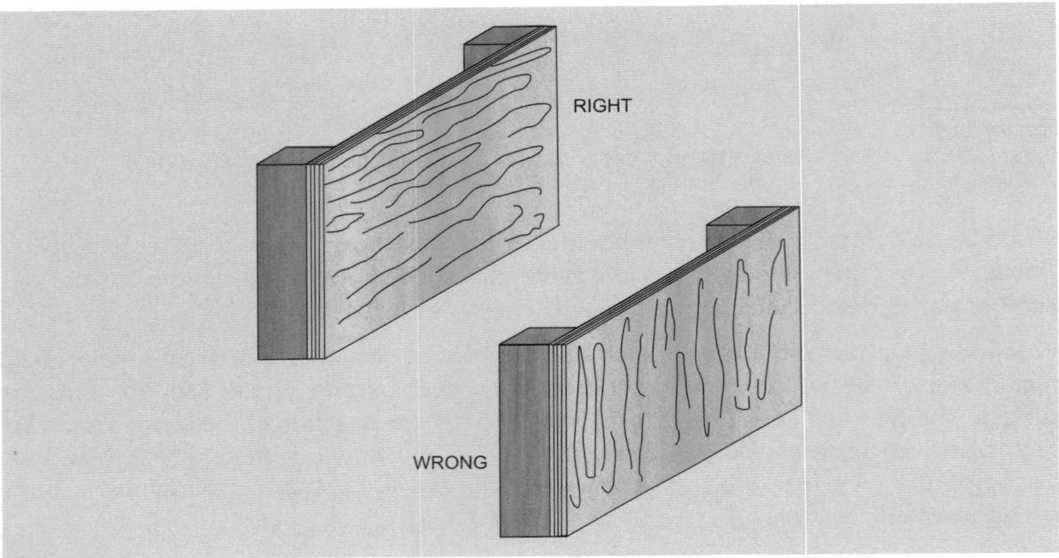

Figure 11-7
Plywood has greater strength and will deflect less if the grain of the outside plies runs across the studs, rather than parallel with them. See also Figure 6-19.

Figure 11-8
A simple clamping device made of $^3/_4$-inch plywood holds the side forms temporarily without any bolt or tie within the concrete.

Forms should be securely tied together with approved rods, and braced in a substantial and unyielding manner. In general, tie rods should be designed to act as struts or spreaders also. Wood struts should not be permitted to remain in the concrete.

For concrete surfaces that will be permanently exposed to view, metal ties or anchorages within the forms should terminate beneath the formed face of the concrete. The ties should be constructed so that removal of the ends or end fasteners can be accomplished without causing appreciable spalling at the faces of the concrete.

Devices that, when removed, will leave an opening entirely through the concrete are usually not permitted. Wire ties may be used when both sides of the concrete will be covered with backfill or otherwise not permanently exposed. Wire ties, if used, should be cut flush with the concrete surface.

In placing a successive lift of concrete on previously placed and hardened concrete, the horizontal joint between the two lifts is often a source of disfigurement. No methods of minimizing this are shown. The grade strip shown in Figure 11-9 should be set accurately with its bottom edge about 1/2 inch below the finished elevation desired for the lift. Grooves shown in Figure 11-10 should be straight and continuous across the structure. Their location should be planned beforehand to give a pleasing appearance to the completed structure. Form anchorages should be provided about 4 inches below the top of the lift. When the form is set for the succeeding lift, the sheathing should overlap the previous concrete by about 1 inch and should be drawn up snug by means of the anchorage in the concrete below and by ties close to the bottom of the new lift. Proper observance of these precautions will assure a neat-appearing joint in the structure.

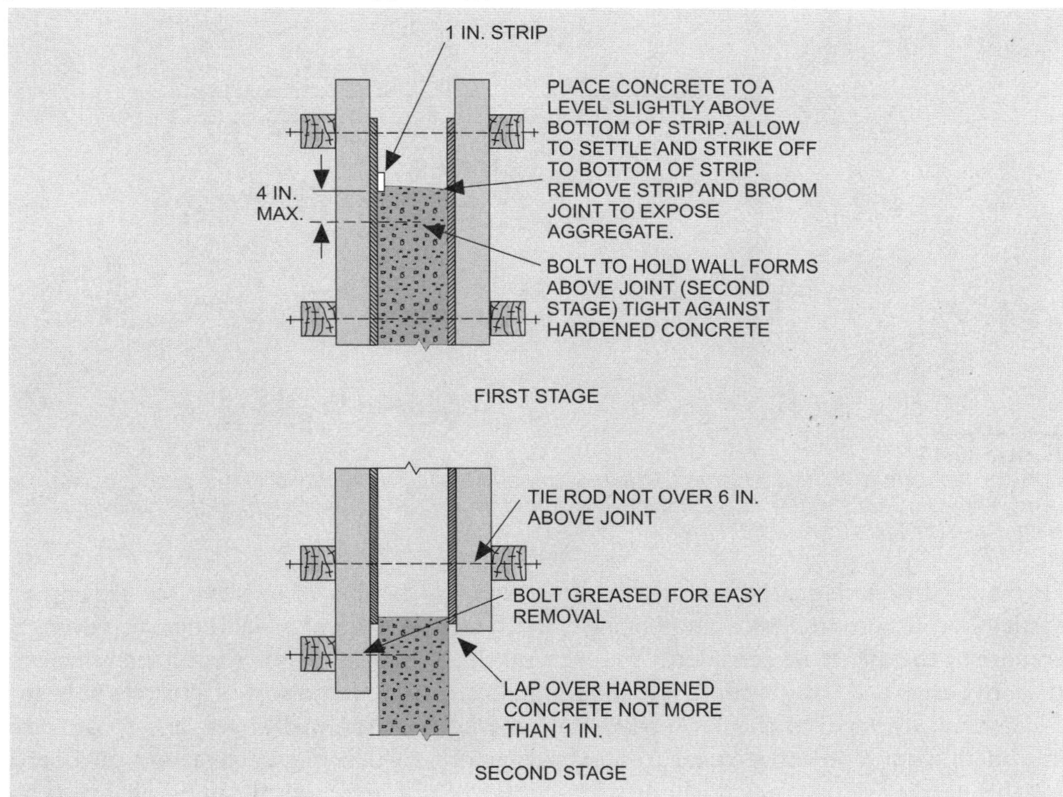

Figure 11-9
The use of a grade strip will assure a straight line at a horizontal construction joint. After the top of the lift has been cleaned, there may be a small recess or offset where the grade strip was removed. This recess may be filled with mortar when the succeeding lift of concrete is placed.

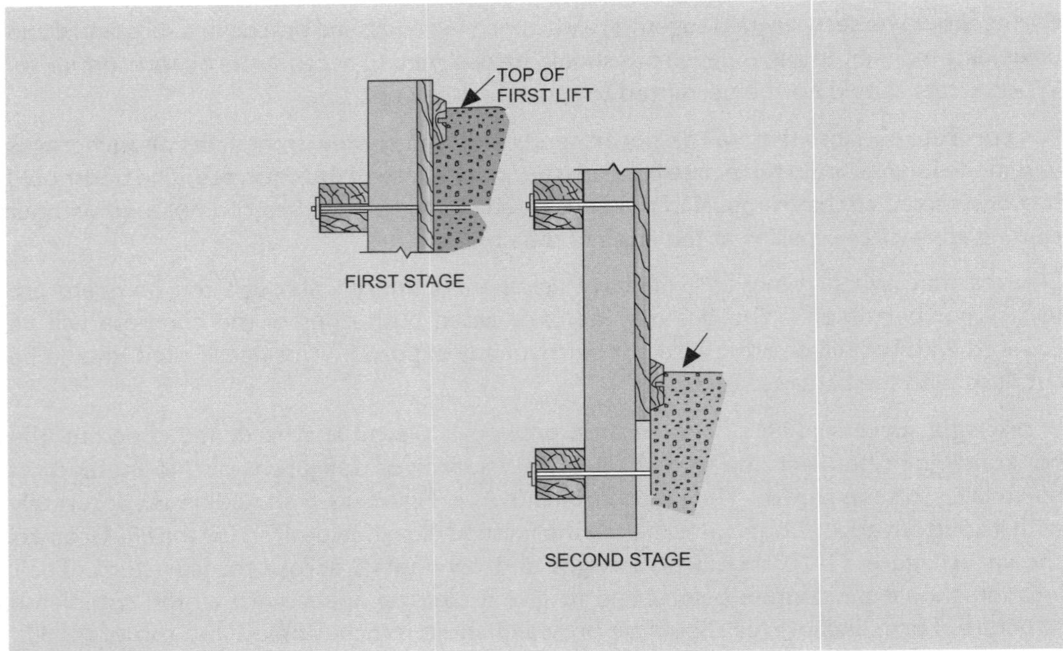

Figure 11-10
A rustication groove is formed in the concrete at the joint by the use of a trapezoidal-shaped strip of wood attached to the form. The use of double-headed nails permits the form to be removed while the strip remains in place for the next lift.

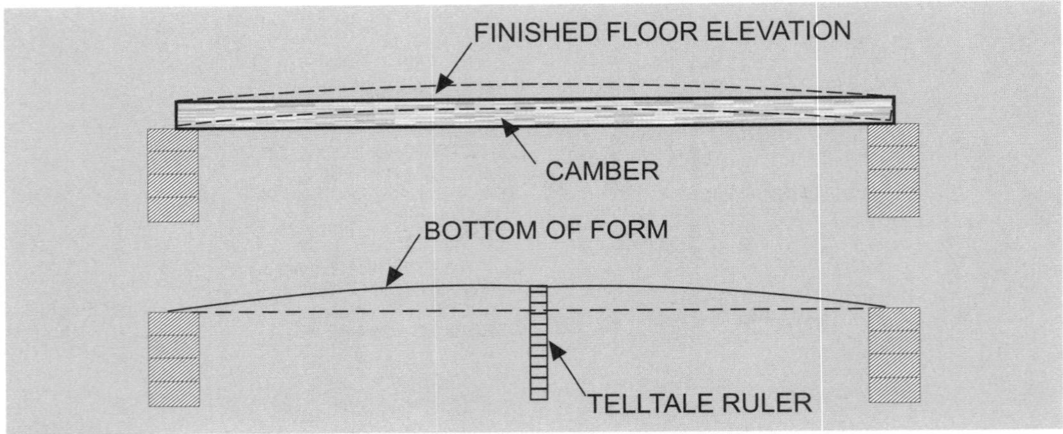

Figure 11-11
Because the weight of the fresh concrete may cause slight sagging of the form, the form is usually cambered, or raised slightly in the center. The deflection caused by the weight of the concrete will then cause the form to straighten out.

Forms for suspended slabs and beams are frequently cambered to allow for sagging or settlement, a common allowance being $^1/_4$ inch per 16 feet of span. This, however, is something that has to be considered for each individual form, depending upon many factors of form design and construction. It is good practice to install some sort of a telltale beneath a slab or beam form to check on settlement during concrete placement, and to provide information for corrective measures. A satisfactory telltale might be a vertical board suitably marked or a ruler attached to the center of the span, which can be observed by means of a surveyor's level. See Figure 11-11.

Most forms must be treated with a form oil or parting compound (form release agents) that will prevent the form from sticking to the concrete.

Normally the choice of form materials and design of the forms is in the hands of the contractor, although there are occasions when the specifications state a requirement for a certain surface or effect on the concrete and may specify a special forming material.

11.2. Wood for Forms

Softwoods are generally used for form lumber, the species depending on the kind that is available in the local market. The most common varieties are Douglas fir, southern pine, western hemlock and eastern spruce. Other species may be used in limited amounts in certain areas.

Lumber for forms consists of standard sizes, either rough or dressed. Because of the loadings imposed on forms, a good grade of lumber should be used; the quality known as construction grade is usually specified. Shoring or falsework should be of select grade. Partially seasoned wood, known in the trade as shipping-dry lumber has been found to be the most stable. Green lumber will warp and crack, whereas kiln-dried lumber will swell excessively when it becomes wet.

Plywood is universally used for sheathing or lining forms because it gives smooth concrete surfaces that require a minimum of hand finishing and because the relatively large sheets are easy and economical to use. Boards are sometimes used for sheathing when the architect requires the imprint of the boards on the surface of the concrete.

Plywood is a fabricated wood product consisting of an odd number of plies or layers of wood placed crosswise of each other and glued together. See Figure 11-12. The grain of alternate plies being at right angles to each other gives the plywood strength in both directions as well as minimum warping and splitting. Plywood is furnished in either interior or exterior type, with the exterior type preferred for forms because it is made with waterproof glue. Standard sizes of plywood are 48 inches wide with various lengths, a 96-inch length being common. Thickness ranges from $^{3}/_{16}$ inch to $1^{1}/_{8}$ inches. The thin

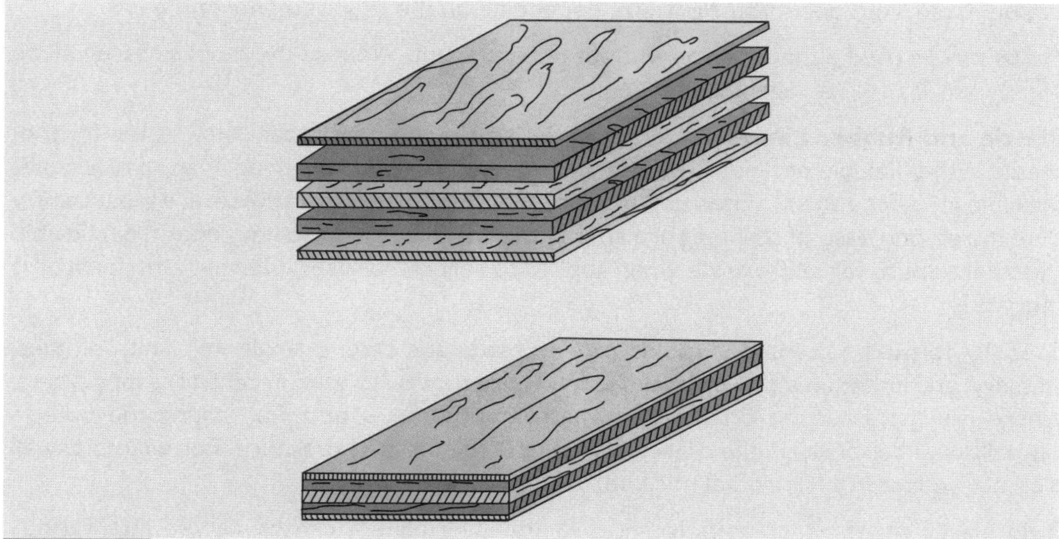

Figure 11-12
The construction of plywood. A is a blown-up view, showing how the grain of the thin veneer plies is alternated. The finished plywood is shown in B.

plywood comprises three plies, with more plies (always an odd number) in the thicker boards. Plywood for concrete forms is usually factory treated with a form oil or parting compound. With proper care, plywood can be reused many times.

Coated plywood is frequently used. Called overlaid or plastic coated, it is ordinary exterior plywood with resin-impregnated fiber facing material fused on one or both sides. The overlay hides the grain of the wood, resulting in a very smooth surface. Overlaid plywood can be used without a release agent, but a light application before each use will be beneficial in lengthening its useful life. More than 100 reuses are commonly realized. Two grades are available, high density and medium density, the difference being in the density of the surfacing material.

Hardboard or fiberboard is made of wood particles that have been impregnated, pressed and baked. When used as a form liner, hardboard must be backed with sheathing of some kind, as it does not have the strength that plywood has. Tempered hardboard, which is preferred for formwork, has been impregnated with drying oil or other material that makes the board less absorbent and improves its strength.

11.3. Other Forming Materials

Fiberglass. The material popularly known as fiberglass is finding increasing usage as a forming material. Correctly known as glass fiber reinforced plastic, it is especially suitable for repetitive production of complicated shapes, particularly precast elements. The initial cost of the mock-up is high, but the cost can be spread over many production units. Some users in precasting plants have reported over 200 uses of one mold with practically no evidence of wear.

The first step in making a fiberglass mold is to make a mockup or master pattern in exactly the configuration of the finished concrete member. The master pattern can be of wood, plaster, or any material that lends itself to the design to be achieved. After coating the master pattern with a bond-breaking wax, layers of polyester resin and glass fabric are applied until the required thickness is reached. Suitable reinforcement and backing must be provided so that the mold will not be deformed when it is handled. Any number of molds can be made from one master pattern, depending on the requirements of the job.

Molds can be used either with or without release agent. Wear of the mold surface will be slightly less if a release agent is used.

Plastic and Rubber Liners. Materials of this nature are those that come in the form of sheets, either flexible or rigid, to be attached to the solid form sheathing. Many patterns are available in "off the shelf" materials such as rubber floor mats. Repetitive use, a great variety of textures, and ease of stripping are among the advantages of such material. Thin flexible sheets are apt to sag and wrinkle when applied to vertical surfaces. Normally, no form oil is required.

Specially formed soft rubber molds can be made for casting small wall units or tiles, statuary, and imitation stone or brick facing. These molds are used in central casting plants, where the mold can be oriented in whatever position is best for casting the objects regardless of the position the objects will take in the finished structure. For example, wall tiles can be cast in a horizontal position.

Rigid plastic sheets, such as polyurethane insulating panels, can be carved into various designs and used as a single-use liner for forms.

Steel. Fabricated and rolled steel shapes are used as supporting members on forms, especially where many reuses are expected, and for load conditions in which it is impractical to use timber. Light-pressed steel shapes are used in constructing prefabricated forms, which may be lined with sheet steel, plywood or hardboard. Pressed steel pans and domes, placed with steel or timber supports, form the underside of waffle slabs and pan joists. See Figure 11-13. Corrugated steel sheets serve as permanent bottom forms for decks; that is, they remain in place after the concrete has been placed. Smooth sheet steel can be used for facing forms, and fabricated steel forms are extensively used in the precast and prestressed concrete industries.

Figure 11-13
Waffle slab formwork utilizing wood soffit boards.

Paper and Cardboard. Multiple layers of heavy paper are bonded together and impregnated with waxes and resins to become a water-repellent cardboard, usually found in the shape of cylindrical single-use molds for columns. Thickness may be as much as $^1/_2$ inch. Void formers and forms for pan joists and waffle slabs are composed of heavy corrugated paper similar to the material used in boxes and cartons. See Figure 11-14 and Table 11.1.

Figure 11-14
Heavy fiber forms, shown in A, are used for forming circular columns, B.

TABLE 11.1
SIZES AND WEIGHTS OF SONOTUBE FIBER FORMS

REGULAR (STANDARD LENGTH-18~)			
INSIDE DIAMETER (IN INCHES)	WT PER FT (IN POUNDS)	INSIDE DIAMETER (IN INCHES)	WT PER FT (IN POUNDS)
8	1.7	28	10.4
10	2.3	30	11.8
12	2.8	32	12.6
14	3.6	34	13.4
16	4.8	36	14.5
18	5.4	38	16.9
20	7.5	40	17.8
22	8.2	42	18.6
24	8.9	44	20.5
26	9.6	48	23.5

Waste Molds. Complicated and ornamental surfaces can be cast in plaster waste molds. A waste mold is made of casting plaster reinforced with fiber and supported on a wood framework. See Figure 11-15. The plaster can be either cast against a master mold or carved. Carving should be done while the plaster is still moist from the mixing water, as it is easier to carve when it is in this condition.

Great care is necessary in fitting a waste mold into the formwork. It should be well supported but should not support any of the formwork that is erected above. Joints within the mold and between the mold and the formwork should be carefully filled or pointed to eliminate joint marks in the finished concrete. Waste molds should be sized with shellac or lacquer when they are made, and coated with a form oil or parting compound just prior to placing of the concrete. When the form is later stripped from the hardened concrete, the waste mold is removed piecemeal and destroyed or wasted.

Figure 11-15
Configurations that are difficult to form in other materials conveniently lend themselves to plaster waste molds.

11.4. Form Accessories

Hardware for forms consists of fasteners, ties, spreaders, braces, hole covers, rustication strips, chamfer strips, void formers and inserts.

The most common fastener is the nail in various sizes. Double-headed nails are frequently used, as they can be withdrawn with minimum effort and damage to the structure. See Figure 11-16.

Many varieties of ties are available. Some of them serve as wall-form spacers as well as their primary function of holding the forms together. Most form ties are expendable, or partly expendable; that is, all or part of the tie remains in the concrete. Some of the commonly used ones are as follows:

Form Clamp. This consists of a smooth round rod with clamps attached to each end of the rod where it extends beyond the walers. Set screws hold the clamps in place. It provides no spacer action. Some specifications prohibit the use of a clamping device (such as this) that uses a straight rod extending through the wall. The rod can be pulled out of the wall with special tools or cut off flush with the surface of the concrete.

Figure 11-16
Double-headed nails can be driven so that they hold tightly but can be easily removed to facilitate form stripping.

Snap Tie. Especially common in wall forming, the snap tie is made of a single piece of wire cut to length and headed on each end to hold the snap-tie clamp. It is notched or weakened at a point where it can be broken inside the wall after the concrete hardens. When the form is stripped, the tie is bent parallel with the concrete surface and then turned to break it at the notch. A flattened portion at the center of the tie prevents the rest of the rod from turning.

Coil Tie. Helical wire coils are welded to opposite ends of steel rods (struts) to form the expendable portion of the tie. There may be either two or four struts, depending on the working load. A coil bolt is threaded through a washer and the form into the open end of the helical coil.

She-Bolt. Two nut washers, two waler rods and a central tie rod comprise a she-bolt assembly for a wall form. Usually cone nut spacers are included. She-bolts are of greater capacity than the previously described ties and are frequently used in construction of heavy forms for mass concrete.

Inserts are of many designs. They are attached to the form in such a way that they remain in the concrete when the form is stripped. They provide for anchorage of brick or stone veneers, pipe hangers, suspended ceilings, ductwork and any building hardware or components that must be firmly attached to the concrete.

11.5. Form Oils and Compounds

Most forms must be treated with a form oil or parting compound to prevent adhesion of the form to the concrete. Exceptions are the overlaid plywoods and smooth fiberglass molds. Even these, however, will have a longer life if they are treated. Numerous oils, shellacs, waxes, lacquers and plastic coatings are available, and the choice of one or another is largely a matter of personal opinion. However, in choosing any form coating, the conditions of use must be considered. The material should be formulated for the particular usage and form material intended. The form coating should not interfere with subsequent curing, painting, plastering or other surface treatments of the concrete, nor should it stain the concrete or cause permanent softening or interfere with the subsequent application of adhesive for floor tile or other floor or wall covering. Instructions of the manufacturer should be followed, as some coatings are designed for certain usages and will not necessarily be satisfactory for others.

Compounds can be broadly classified as those that provide a barrier between the form and the concrete and those that are chemically active. Barriers can be further broken down into a group of oily liquids that must be applied shortly before the concrete is placed and that must be in a moist or wet condition on the form when the concrete is placed against the form, and another group of barriers composed of those lacquers and plastic coatings that are applied in a fluid condition but must be permitted to dry before the concrete is placed.

Chemically active coatings are applied as a liquid and are usually permitted to dry before the concrete is placed. A constituent in the coating temporarily prevents hydration of a very thin layer of cement adjacent to the form, thus preventing adhesion. After the form has been stripped, further curing of the concrete permits the surface to harden.

11.6. Falsework and Shoring

Some structures require support for the formwork and green concrete other than the actual forms themselves. Such a temporary supporting structure is called falsework, shoring or centering. It may consist merely of a few simple post shores, or it may be a complicated engineered structure supporting the forms for an arch or dome.

Vertical shoring under a beam or slab form can be one of two kinds:

Permanent Shores, in which the form and shoring are designed so that the post shores will be undisturbed when the forms are stripped. This usually means leaving a small strip or section of the form in place directly over each post until the post is removed after the concrete attains the required strength.

Reshores, in which the form and post shores are removed simultaneously, each individual post being immediately replaced by a new post wedged in place to support the concrete.

Falsework should be designed and built to carry the full load of the forms, including walkways and platforms, the weight of the fresh concrete (assumed to weigh 145 pcf for normal concrete) and an additional live load caused by workers and construction equipment.

The foundation for shoring is of great importance, insofar as settlement must be minimized. In some cases, such as centering for an arch bridge, the falsework can be supported on piling. Mud sills are usually acceptable if they are of adequate area and are carefully placed on firm earth. Supports should never be placed on frozen or unstable ground. In a multiple-story building, supports for successive floors should be placed directly over those below. Frequently, the use of jacks or wedges is required to take up any settlement in the falsework, either before or during concrete placement.

There are several kinds of fabricated shoring components made up of wood or metal posts, jacks, adjusting devices, horizontal shores or adjustable beams, scaffold shores and joists, all of which offer a wide range of products to fit almost any form-supporting condition. See Figure 11-17.

Figure 11-17
This formwork system utilizes wood joists, stringers and shores.

External bracing is required on some forms for architectural concrete when the presence of patches or plugs remaining on the surface of the concrete, as a result of using through-the-form spreaders and ties, is unacceptable.

11.7. Slipforms

A slipform, or sliding form, is a movable form that is raised vertically as the concrete is placed. A form of this kind is called a vertical slipform and is used on buildings and silos. See Figure 11-18. Horizontal slipforms are used for highway pavements, curb and gutter, and canal linings. See Chapters 16 and 22.

Figure 11-18
Self-climbing slipform for "Burj Khalifa Tower" . . .the tallest man-made structure ever built at 2684 ft.

11.8. Precast Concrete Forms

In this category we find precast wall panels, precast deck slabs, precast column enclosures and other items made of precast concrete. These units are usually of special architectural design and are more easily made in a casting yard where the finish can be achieved under closer control than on the job. After they have been cured, they are set in place on the structure, and backup structural concrete is placed. The precast unit thus serves as a form that confines the fresh concrete, then remains in place and becomes a permanent part of the building.

It is essential that the fresh concrete bond to the precast form. Bonding is accomplished by grooving or roughening the back side of the precast unit. Grooving or roughening can be achieved by the proper selection of the form liner, or it can be done with cutting tools, sandblasting or acid etching. Bond can also be made by means of metal anchors. One end of the anchor is embedded in the precast concrete, and the other end is embedded in the cast-in-place backup concrete. Usual construction provides a combination of both methods.

11.9. Prefabricated Forms

Prefabricated forms, made by several manufacturers, can be divided into two groups. First are the modular forms that can be used for many applications. The user adapts the formwork to make use of the modular dimensions of standard off-the-shelf units. The second group of prefabricated forms consists of those units that are manufactured for one certain job. Dimensions of the units are based on job requirements. Prefabricated forms are frequently rented by the contractor.

A form unit usually consists of a plywood, hardboard or sheet steel replaceable panel that is held in a frame of light pressed steel shapes. Patented locking devices join the units together, and spacer ties are included as part of the package.

Some units are self-supporting and can be joined together and stacked to form large areas. Others require studs or walers to support them.

Column forms are made of sheet steel, cardboard or fiber board. Single-use forms are made of paperboard tubing that can be stripped by making one or two vertical cuts with a saw or special knife, then peeling the paperboard off the concrete. Sheet steel is frequently used for reusable forms. Square and rectangular columns can be formed with sheets of plywood held in place by special column clamps. See Figure 11-19.

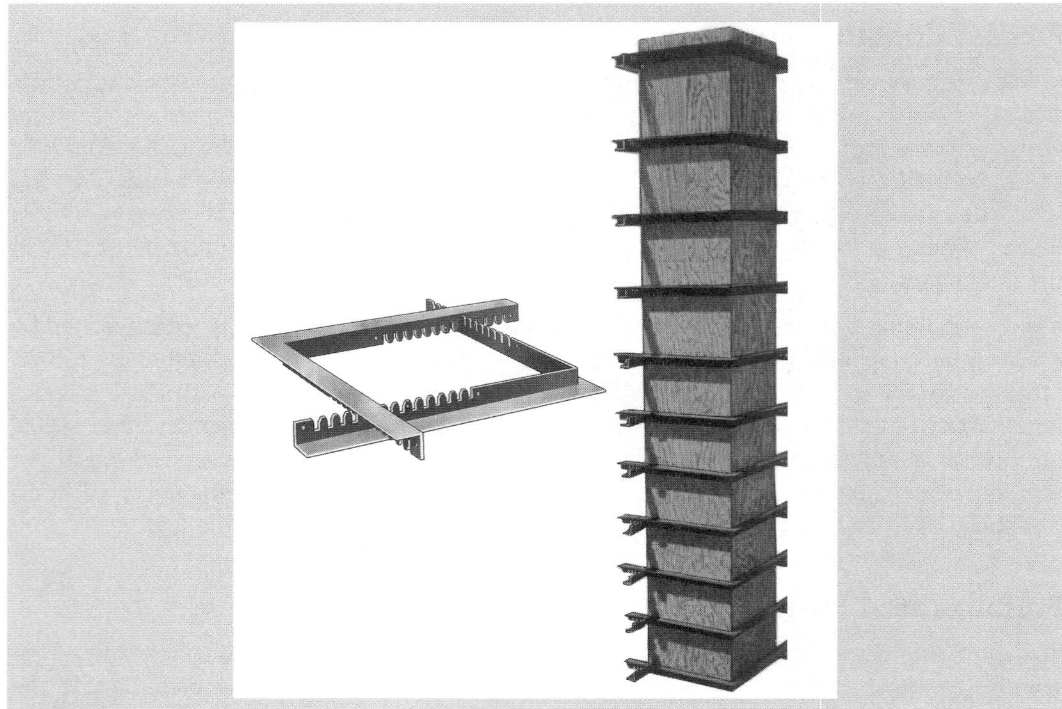

Figure 11-19
Patented clamps of this type simplify forming of square and rectangular columns. (Sonoco Products Company.)

Prefabricated forms usually require the field application of some sort of form oil, with the exception of paperboard tubes that are factory treated with a parting compound.

11.10. Use of Forms

Dimensions of forms in place should be carefully checked before concrete is placed. Immediately prior to placing concrete, any warping or bulging should be corrected and all dirt, sawdust, shavings or other debris removed. In narrow walls, where the bottom of the forms is otherwise inaccessible, the lower boards or panels should be left loose on the back side so that extraneous material can be removed just prior to placing concrete; the board or panel can then be replaced.

Whenever possible, the inspector should get down into the forms to make an inspection. Although the forms are sometimes so congested and tight that an inspector cannot get in; it is nevertheless difficult to inspect a tall form from the top only.

Runways for buggies and workers should not be laid on the reinforcing steel mat unless special high supports are inserted under the top layer of bars to keep them from being pushed down. When using metal chairs to support the reinforcing bars on a wood or plywood bottom form, the chairs should be checked to see that they do not punch into the wood, thus lowering the steel from its intended position. Chairs are available that overcome this punching tendency.

It is sometimes necessary to install a construction joint with the reinforcement extending across the joint. To avoid bending and possibly damaging the steel, the bulkhead form should be split along the line of the reinforcing bars so that it can easily be removed in two parts without disturbing the steel in the hardened concrete. See Figure 11-20.

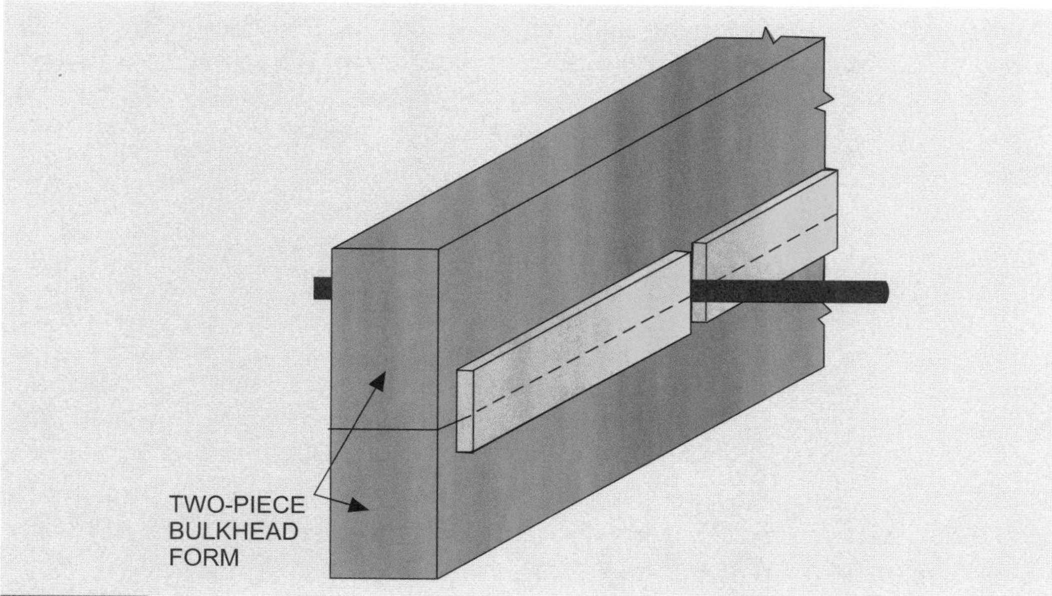

Figure 11-20
If the header board is split along the line of the reinforcing bars, the bulkhead can be removed without bending or otherwise damaging the bars that extend across the construction joint.

During concrete placing, it may be necessary to have form watchers observe the forms for evidence of settling, bulging, slipping or movement of any kind. Telltale indicators are frequently used, as discussed in Section 11.1.

11.11. Removal of Forms and Shoring

The care exercised in handling and using a form will be reflected in the serviceability of that form. With few exceptions, forms are designed and built for reuse. This is particularly true of prefabricated forms. Single-use job-built forms are built of components that can be reused. Plywood can be used several times; coated plywoods can be used many times. Timber for studs, walers, shoring and other structural parts of the form can be salvaged and reused. When the forms are carefully stripped (removed from the concrete), not only is the concrete surface undamaged, but the form itself is preserved for additional use. After a form has been stripped, it should first have all dirt and mortar removed; it is then coated with an appropriate parting compound. Metal scrapers and wire brushes should be used with care, as they can scratch and roughen plywood, and overuse on steel is apt to clean the steel down to bright metal, which is undesirable. A small block of wood makes a good scraper. During removal and cleaning of forms, buckets or other containers should be handy to receive the many items of small hardware that are reusable. Solvents are available for cleaning these items. If the forms are not to be reset immediately, they should be carefully stacked to prevent warping.

The length of time that forms and falsework must be left in place depends on the type of structural element under consideration and the strength development of the concrete. Vertical side forms on walls, beams and columns can usually be removed on the day following placing of the concrete. It may be necessary to retain shoring under slabs and beams for three weeks or even longer. Normally, the specifications are quite explicit on the strength required of the concrete before support can be removed. The building code requirements for removal of forms are stated in ACI 318 Section 6.2.

Proportioning the Concrete Mixture

Characteristics and properties of concrete depend on the materials used in making the concrete, the proportions of these materials and the care used in making and handling the concrete. In this chapter we will learn how to proportion the materials into a concrete mixture (frequently called a mix) and how to adjust the mix to maintain the required quality. Basic requirements for the concrete, established by the type of structure and exposure conditions, are adequate workability of the fresh concrete, strength in accordance with design loading of the proposed structure and durability to enable the structure to endure for a long period of time. In addition to all of this, maximum economy is necessary. See Figure 12-1. The requirements for the concrete are established by the specifications, either directly, by stating actual values of selected properties (a definite strength, for example); by reference to certain industry standards; or, more commonly, by a combination of both methods.

The ingredients for concrete should be selected to make the most economical use of available materials that will produce concrete of the required quality. Basic relationships have been established that provide guides in approaching optimum combinations, but final proportions should be established by trial mixes and adjustment in the field.

In properly made concrete, each particle of aggregate, no matter how large or small, is completely surrounded by paste, and all spaces between aggregate particles are completely filled with the paste. The aggregates are considered inert materials, whereas the paste is the cementing medium that binds the aggregate particles into a solid mass. In some cases, an admixture is added for the purpose of modifying or improving the concrete. When sources of ingredients, type of cement and quantity of admixture remain the same, the quantity of cement, the grading of the aggregates and the consistency of the concrete can be varied over a wide range without materially affecting strength, provided the quality of cement paste, as determined by the water-cement ratio, is maintained constant. When sources of ingredients vary, concrete strengths may vary appreciably even though the water-cement ratio is held constant.

If we have selected our materials wisely, then we are well on the way to having quality concrete, possessing the correct degree of workability, strength, durability, volume stability, freedom from cracks and blemishes, watertightness and economy.

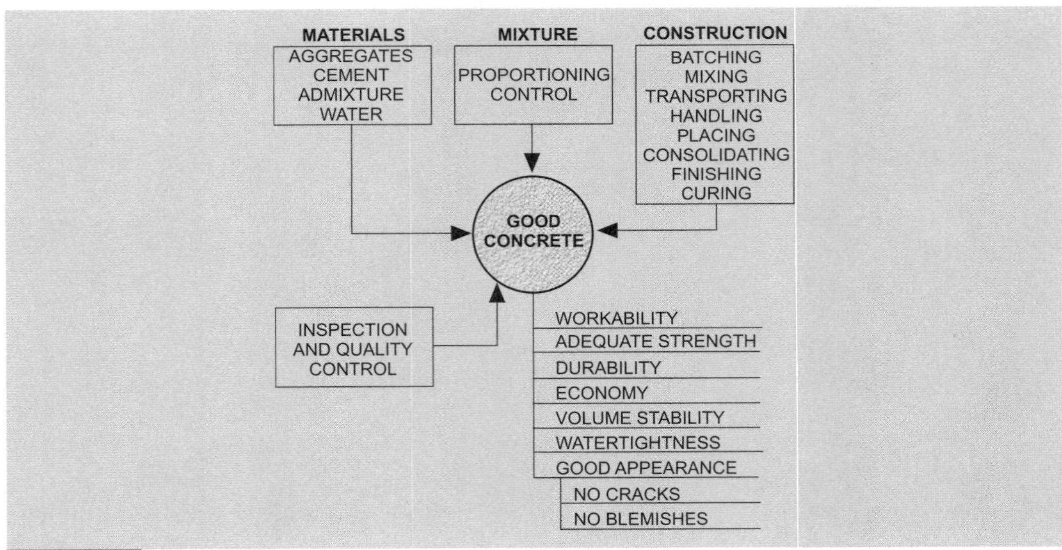

Figure 12-1
Good materials, well-proportioned mixes and quality construction produce concrete with the necessary properties.

Before we reach our goal, however, we have to put these materials together, mix them into concrete, transport the concrete to the jobsite, and there place, consolidate, finish and cure it. This journey starts at the concrete proportioning and mixing plant and is not complete until the completed structure has been turned over to the owner. The first step is to proportion the mixtures to make best use of the materials to give the concrete the required properties.

There are a number of different ways to design or proportion concrete mixtures, some of which are rather complex and cumbersome, and others that are based on a degree of refinement that is never attained under field conditions. The methods described in this chapter for selecting proportions for normal-weight concrete are based on values gained from experience with many materials and the report of ACI Committee 211, with modifications to suit the special requirements of the code. Those persons desiring to pursue this subject further should have a copy of the ACI 211 report.[12.1] Design and Control of Concrete Mixtures by the Portland Cement Association[12.2] is also an excellent technical reference reflecting industry practice on proportioning concrete mixtures.

12.1. Codes and Specifications

Mixes should be proportioned to make the best use of available materials and to produce concrete possessing suitable workability, which will, upon hardening, develop the required degree of strength, durability and other properties. The properties of the aggregates and cement have a pronounced effect upon the workability and quality of hardened concrete; hence, any standard method of proportioning mixes must, of necessity, be based upon average conditions and furthermore must be adjustable to make it fit specific conditions and materials.

Nearly all project specifications for concrete specify a strength, usually requiring that the average strength shall be a certain percentage or amount higher than the specified compressive strength f'_c. Specification requirements may also include a minimum cement content, maximum water-cement ratio, range in slump and range in entrained air content.

With the increasing use of other materials with cementing value in combination with portland cement in the production of today's concrete, the traditional water-cement ratio has been renamed water-cementitious materials ratio in most written text and code and specification documents, including ACI 318 Chapter 4. This change in naming of the basic parameter affecting the quality of concrete clarifies that other cementitious materials permitted by the code are additive to the basic cementing ingredient in concrete (portland cement) to satisfy the water-cementitious materials ratio limitations for the special exposure conditions addressed in ACI 318 Chapter 4. For the calculation of water-cementitious materials ratios, as limited by the code, cementitious materials may include the following:

- portland cement (ASTM C 150);
- blended hydraulic cement (ASTM C 595);
- expansive hydraulic cement (ASTM C 845); either by themselves or in combination with:
- fly ash (ASTM C 618);
- raw or calcinated natural pozzolans (ASTM C 989); or
- silica fume (ASTM C 1240).

For brevity, unless required for clarity, the term "water-cement ratio" is used within the text of this publication.

ACI 318 Durability Requirements. ACI 318 Section 5.2 states the general requirements for the quality of concrete that must be attained on the job. Specifically, it provides that the concrete shall conform with certain strength requirements; it shall possess "workability and consistency to permit concrete to be worked readily into forms and around reinforcement under conditions of placement to be employed without segregation or excessive bleeding"; and shall provide "resistance to special exposures as required for applicable exposure categories of Chapter 4."

The 2008 edition of ACI 318 (ACI 318-08) introduced new exposure categories and classes with applicable durability requirements for concrete. Strength and water-cement ratio requirements for various exposure conditions are shown in Tables 12.1 and 12.2. The water-cement ratio selected for mix design must be the lowest value required to meet anticipated exposure conditions. Table 12.1 also shows minimum strength requirements for various exposure conditions. For concrete exposed to sulfates, sulfate resisting cements must also be specified as indicated in Table 12.2. When durability does not control, the water-cement ratio should be selected on the basis of structural requirements. The following three examples illustrate the code durability requirements for potentially hazardous concrete exposure conditions.

1. If the concrete will be exposed to freezing and thawing, it must contain entrained air (see Table 12.3). Also, if the concrete is exposed to freezing and thawing in a moist condition, the water-cement ratio is limited to 0.45 with a corresponding minimum specified strength of 4500 psi. See Table 12.1.
2. Concrete containing steel reinforcement and exposed to de-icing salts must have a water-cement ratio not exceeding 0.40 with a corresponding minimum specified strength of 5000 psi. See Table 12.1.
3. Concrete that will be exposed to seawater must be made with Type II cement, with a water-cement ratio not exceeding 0.50 with a corresponding minimum specified strength of 5000 psi. See Table 12.2.

It is noteworthy that both a maximum water-cement ratio and a minimum strength are required for concrete durability. This dual requirement is to help ensure that the W/C ratio required for durability will actually be obtained in the field. It is difficult to accurately determine the W/C of the concrete during production and especially as a jobsite control. The minimum strength requirement for durability effectively transfers quality from a W/C limitation to a more easily controlled specified compressive strength. A minimum strength requirement for concrete durability will also provide some control on the indiscriminate use of some of the lesser quality portland cement replacements permitted by the code to satisfy the W/C limitations for durability.

The code also limits the amount of portland cement replacement for concrete exposed to de-icing salts. See ACI 318 Section 4.4.2. Fly ash is limited to 25 percent of the total weight of required cementitious materials. For example, if a parking deck slab is to be exposed to de-icing salts, Table 12.1 indicates a maximum water-cement ratio of 0.40 with a minimum specified strength of 5000 psi. If the mix design requires 288 pounds of water to produce an air-entrained concrete mix to a specified slump, the required total weight of cementitious materials is 288/0.40 = 720 pounds. The 720 pounds of cementitious materials may be all portland cement or a combination. If the local ready-mixed producer decides to use some fly ash as a portland cement replacement to provide a more economical mix, the maximum amount of fly ash is limited to 0.25(720) = 180 pounds maintaining the same required W/C = 288/(540 + 180) = 0.40.

TABLE 12.1
DURABILITY REQUIREMENTS FOR VARIOUS EXPOSURE CONDITIONS

Exposure Category	Exposure Condition	Maximum Water-cement Ratio (w/c)	Minimum Design Strength, f'_c (psi)
P0 P0 C0 C1	Concrete protected from exposure to freezing & thawing, deicing salts, or aggressive substances	Select water-cement ratio based on strength, workability, and finishing needs	2500*
P1	Concrete intended to have low permeability (watertightness)	0.50	4000
F1 F2 F3	Concrete exposed to freezing & thawing in a moist condition or deicing salts	0.45	4500
C2	For corrosion protection of reinforcement for concrete exposed to chlorides from deicing salts, salt water, brackish water, seawater, or spray from these sources	0.40	5000

Adopted from Reference 12.2. ACI 318-08 introduced four exposure categories for determining durability requirements for concrete: (1) F-Freezing & Thawing; (2) S-Sulfates; (3) P-Permeability; and (4) C-Corrosion Protection of Reinforcement. Increasing numerical values represent increasing severe exposure conditions.
*See ACI 318 Section 5.1.1

TABLE 12.2
TYPES OF CEMENT REQUIRED FOR CONCRETE EXPOSED TO SULFATES IN SOIL OR WATER

Sulfate Exposure Class		Sulfate (SO_4) in soil (% by mass)	Sulfate (SO_4) in water, (ppm)	Cement Type (ASTM C 150)	Maximum Water-Cement Ratio (w/c)	Minimum Design Strength, f'_c (psi)
S0	Negligible	Less than 0.10	Less than 150	No special type required	-	2500**
S1	Moderate*	0.10 to 0.20	150 to 1500	II	0.50	4000
S2	Severe	0.20 to 2.00	1500 to 10,000	V	0.45	4500
S3	Very severe	Over 2.00	Over 10,000	V	0.40	5000

TABLE 12.3
TOTAL AIR CONTENT FOR FROST RESISTANT CONCRETE

Maximum aggregate size (MSA) (In.)	Air content (percent)	
	Exposure F1	Exposure F2 & F3
$^3/_8$	6	7.5
$^1/_2$	5.5	7
$^3/_4$	5	6
1	4.5	6

Adopted from Reference 12.2

Water-Cement Ratio. Water-cement ratio to be used on the job must be the lower of the one determined based on strength and the one based on durability.

The code permits alternate methods of arriving at the mix proportions for any proposed job, all described in Chapter 5 of ACI 318. First is a statistical method, using existing data from field experience with the proposed materials. Fortunately, in practically all communities of any size, established sources of aggregates and ready-mixed concrete can be found. Most of these producers have a record of the use of their materials. Under these circumstances the code provides that a statistical analysis of strength tests can be used as a basis for establishment of the mix proportions. Statistical methods, explained in Chapter 26, require at least 15 tests of the materials.

If enough test results are not available for a statistical analysis of field tests, then it is necessary to make laboratory trial batches to determine the concrete proportions. A curve must be made showing relationship between compressive strength and water-cement ratio (or cement content) at a minimum of three points on the curve representing batches with strengths above and below that required. See Figure 12-2. Each point on the curve represents the average of at least three specimens tested at the designated age, usually 28 days. Strength values should be similar to those that would be required if results of field tests were available.

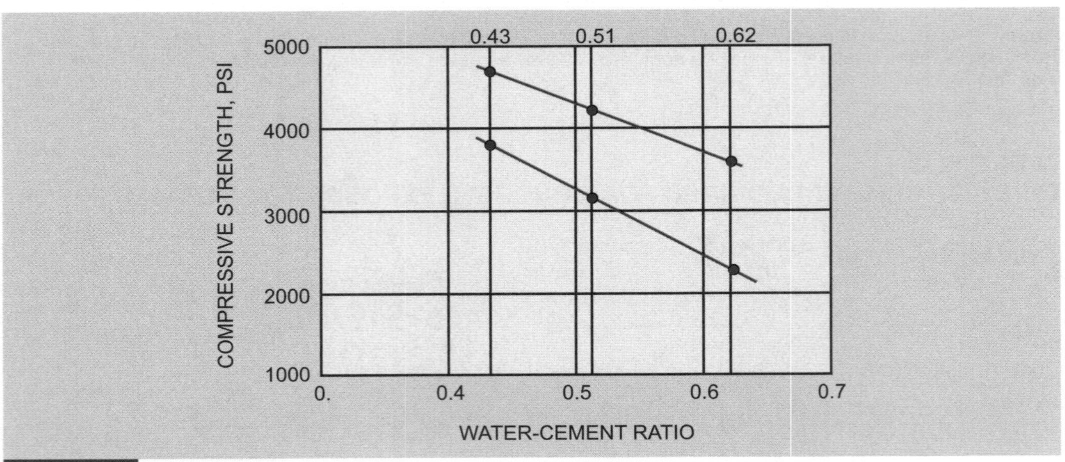

Figure 12-2
Typical strength curves from trial mix data for a required 28-day strength of 4000 psi.

If adequate data from both field tests and trial mixtures cannot be obtained, as a third alternative to establishing the concrete mix for a proposed job, the target strength for selection of mixture proportions may be based on a required overdesign. See ACI 318 Section 5.4. If, for example, the specified strength is 3000 psi, the strength used as the basis for selecting concrete mixture proportions (water-cement ratio) must be based on 3000 + 1200 = 4200 psi. In the interest of economy of materials, the use of this option for mix proportioning should be limited to relatively small jobs where the added cost of trial mix data is not warranted.

12.2. Properties of Materials

Any method of mix proportioning requires that certain properties of the materials be known. The following definitions, as they apply to cement and concrete materials, should be of assistance.

Specific Gravity. The ratio of the weight of a solid piece of aggregate or cement of one cubic foot volume to the weight of one cubic foot of water is the specific gravity. Water weighs about 62.4 pounds per cubic foot at normal temperatures and has a specific gravity of 1.0. The specific gravity of cement is usually assumed to be 3.15. Thus, the weight of cement is 3.15 times heavier than water. The specific gravity of normal aggregates is between 2.50 and 2.85 and is determined by tests.

Bulk Specific Gravity, the value used in mix proportioning, is used for aggregates because aggregate particles are not solid. They contain a small amount of small pores. In the method of mix proportioning used in this book, the bulk specific gravity based on saturated, surface-dry condition (SSD) is used. The ACI method uses the specific gravity based on the dry weight.

For field control, it is simpler to use the specific gravity based on a saturated, surface-dry condition. Aggregates are nearly always batched in a wet or damp condition, and the use of the saturated, surface-dry specific gravity eliminates the necessity of considering aggregate absorption when making adjustments to the mix under operating conditions.

Density. The density of a material is the weight of one cubic foot of the solid material, including any voids within the material. It is the bulk specific gravity multiplied by 62.4, the density of water. For cement, the density is 3.15(62.4) = 196 lb/cu ft. Density is measured in pounds per cubic foot.

Voids. There are two kinds: tiny openings or spaces inside the aggregate particles (these are the pores we consider when computing bulk specific gravity) and large voids that exist between the aggregate particles.

Unit Weight. When aggregate is placed in a container, the container contains aggregate particles plus large voids between the particles. The unit weight of aggregate is the weight of one cubic foot of the aggregate particles. The unit weight of cement is the weight of one cubic foot of dry cement. It is 94 pounds per cubic foot.

These terms are further discussed in Section 8.3. Figure 12-3 shows that a cubic foot of aggregate consists (for example) of 103 pounds of aggregate particles with voids, or air spaces, between the pieces of aggregate. If we were to compress the aggregate into a solid piece of stone, we would still have 103 pounds of stone, but the volume would be only 0.62 cubic feet.

This is the solid volume of the stone, sometimes called the absolute volume. In a similar manner we see that a sack of cement contains many small voids between the cement particles, and the solid volume of the sack of cement is actually 0.48 cubic feet. See Figure 12-4. Densities of the materials normally used in concrete are shown in Figure 12-5. Density of the aggregate shown is the average value.

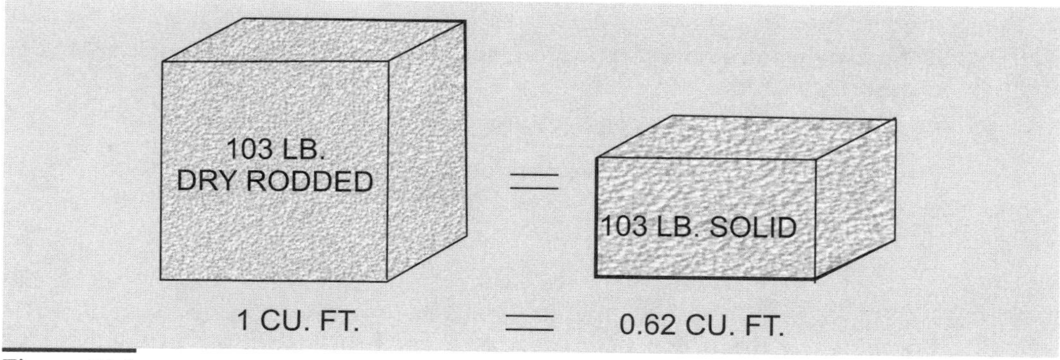

Figure 12-3
When aggregate is placed in a container, there are void spaces between the particles. If the aggregate could be compressed so there were no voids, the cubic foot of loose aggregate would then occupy only 0.62 cubic foot, although the weight is unchanged.

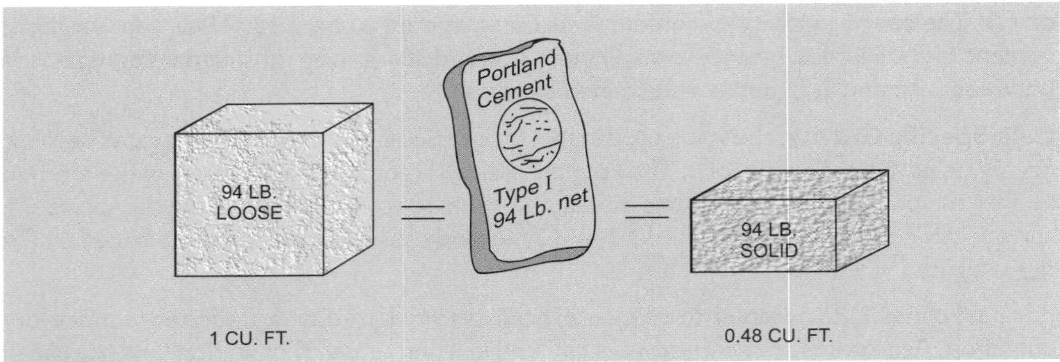

Figure 12-4
One bag (94 pounds) of cement has a loose volume of 1 cubic foot, but when compressed to a voidless solid the 94 pounds of cement occupies only 0.48 cubic foot.

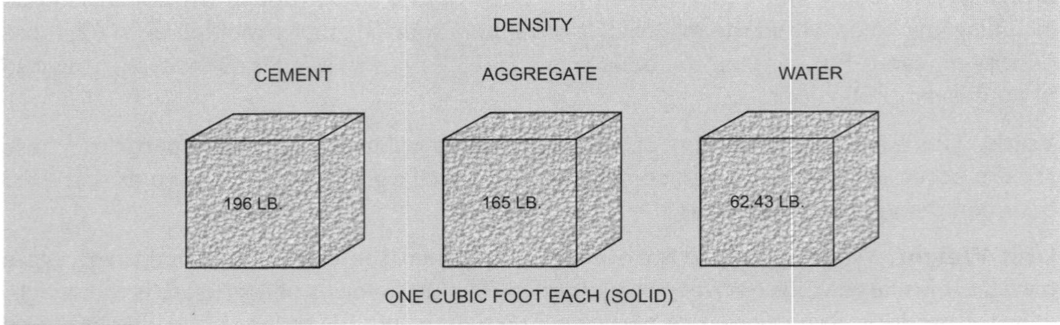

Figure 12-5
Average densities or weights per cubic foot of solid material.

12.3. Selection of Mix Characteristics

When information based on tests or a history of use is not available, it becomes necessary to make estimates of mix proportions by following a logical series of steps by means of which proportions for a trial batch can be established. This method would be used for establishing a beginning mix.

Briefly, the steps are as follows:

1. Select the maximum size of aggregate from the specifications or based on the job condition;
2. Determine water-cement ratio from specified limits, strength requirements or exposure condition;
3. Select slump from specified limits or based on construction condition;
4. Estimate the total water requirement per cubic yard;
5. Estimate air content;
6. Compute cement content;
7. Compute aggregate content;
8. Compute trial mix proportions based on Steps 1 through 7, correcting them for aggregate absorption and moisture content; and
9. Adjust the mix proportions under field conditions.

Certain requirements for the concrete will be given by the job specifications or the code. Among the items that might be specified are compressive strength, water-cement ratio, slump, cement content and maximum size of aggregate. The specifications will state whether admixtures, including air-entraining agents, can be used. Other information will be developed from tests of the aggregates. Among the items in this category are bulk specific gravities, absorptions, moisture content, sieve analyses of fine and coarse aggregates and the unit weight of the coarse aggregate. These properties are discussed in Chapter 8. Other information needed for proportioning the mix includes brand and type of cement. In the following example, it is assumed that Type I portland cement will be used.

The sequence of determining the several properties of the concrete, when they are not specified, depends on the method used for proportioning the concrete.

Maximum Size of Aggregate. Normally the maximum size of coarse aggregate (MSA) is specified. If not, it can be determined from the size of the structural element and the amount of reinforcing steel. The MSA should not exceed one third the depth of a slab, three fourths of the minimum clear spacing between reinforcing bars or between reinforcing bars and forms, or one fifth of the narrowest dimension between sides of forms, whichever applies to the structure under consideration. (See Figure 12-6.)

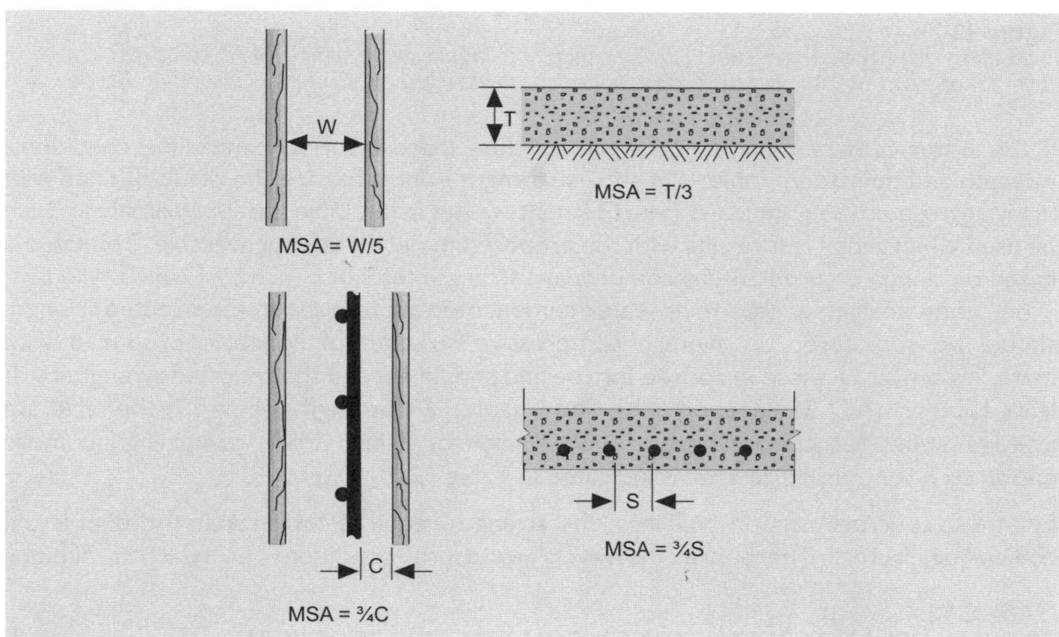

Figure 12-6
The MSA depends on the size of structural element and spacing of reinforcing steel. See ACI 318 Section 3.3.2.

As a practical measure, only one MSA, based on the dimensions that will give the smallest MSA, is normally used throughout the job, especially on the smaller job. Considerable yardages of different concretes sometimes justify different aggregate sizes for different parts of the structure. Common practice is to use aggregate of ³/₄-inch or 1-inch maximum size in practically all structural concrete. Lower-strength concrete in massive footings and similar portions of the structure can be made with 1¹/₂-inch or 3-inch MSA. For high-strength concrete, best results can usually be obtained with the smaller MSA. See Chapter 3. Although the small-aggregate mixes require more water per cubic yard (see Figure 12-7), they are always made with relatively high cement contents; hence, the water-cement ratio is lower, which results in higher strength.

Water-Cement Ratio. Because most of the desired properties of the concrete depend on the quality of the cement paste, selection of the correct water-cement ratio is an important step. Usually, a limiting water-cement ratio is stated in the specifications, and the mix proportioning becomes one of selecting other characteristics of the mix to maintain the water-cement ratio below the specified limit.

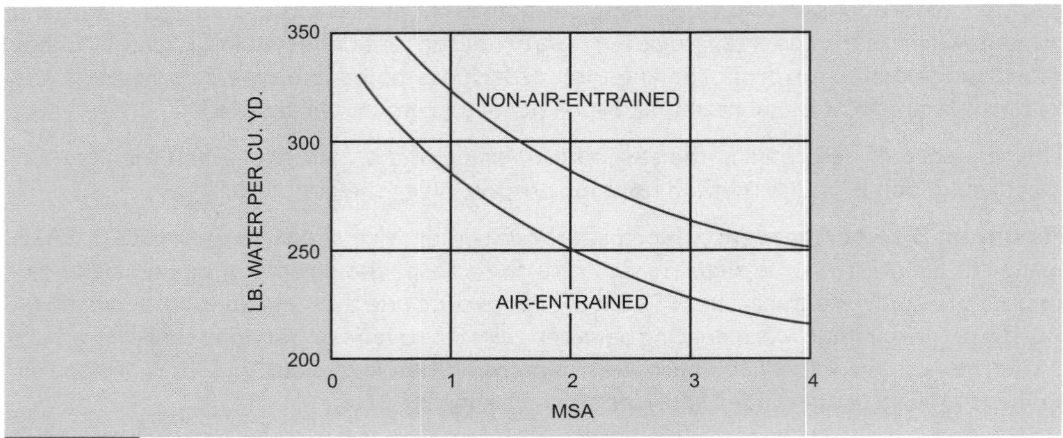

Figure 12-7
Total water requirement per cubic yard of concrete is higher for the small MSA concrete. Strength of small MSA concrete is high; however, because cement content is also high, the

If the water-cement ratio is not specified, then there are two controlling conditions: strength and durability. Table 12.4 shows strength values that are the result of tests with many aggregates using standard Type I cement. Values in this table are approximate and can be used when actual test results with the proposed materials are not available. This table is based on standard cured 28-day compressive strength tests of 6-inch by 12-inch cylinders. When using strength to determine water-cement ratio in this manner, the selected strength should be considered an average compressive strength of concrete produced with materials similar to those proposed for use and should exceed the specified strength by at least 1200 psi (ACI 318 Section 5.4.). See Chapter 26. Strengths shown in the table are average values for concrete made with $^3/_4$-inch or 1-inch MSA, containing no more entrained or entrapped air than is indicated in Table 12.7

For most concrete used in buildings, the strength required for structural design is the controlling factor. There are, however, exposure conditions in which a limiting

TABLE 12.4
TYPICAL RELATIONSHIP BETWEEN COMPRESSIVE STRENGTH AND WATER-CEMENT RATIO

COMPRESSIVE STRENGTH, PSI	WATER-CEMENT RATIO BY WEIGHT	
	NONAIR-ENTRAINED CONCRETE	AIR-ENTRAINED CONCRETE
2000	0.82	0.74
3000	0.68	0.59
4000	0.57	0.48
5000	0.48	0.40
6000	0.41	0.32
7000	0.33	—

water-cement ratio and corresponding minimum strength is required for durability. Section 12.1 discusses the code limitations on water-cement ratios and minimum strengths based on exposure. When exposure conditions must be considered, the concrete mix proportions must be based on the lower of the water-cement ratio required for strength and that required for exposure. The specified strength of the concrete must also be based on the higher of that required for structural design and that required for exposure.

Slump. Normally, the specifications designate the maximum slump, or range of slump, for concrete in various parts of the structure. As a guide, Table 12.5 can be used if no slumps have been specified.

Slump is a measure of the consistency of the concrete and should be the lowest "maximum" that will permit efficient placing of the concrete. Values in the table are for concrete consolidated by vibration.

TABLE 12.5
RECOMMENDED MAXIMUM SLUMP FOR VARIOUS TYPES OF CONSTRUCTION

CONCRETE CONSTRUCTION	MAXIMUM SLUMP, IN.
Reinforced foundation walls and footings	3
Plain footings, caissons and substructure walls	3
Beams and reinforced walls	4
Building columns	4
Slabs-on-ground	3
Mass concrete	3

Adapted from Reference 12.2
Plasticizers can effectively provide higher slumps.

Water Content. The total amount of mixing water per cubic yard of concrete for the required slump depends on the maximum size and grading of the aggregate, as well as the shape and texture of the aggregate particles. It is not significantly affected by the cement content, but it is affected by the temperature, as shown in Figure 12-8. The use of air-entraining and water-reducing admixtures can have a significant effect on the water requirement.

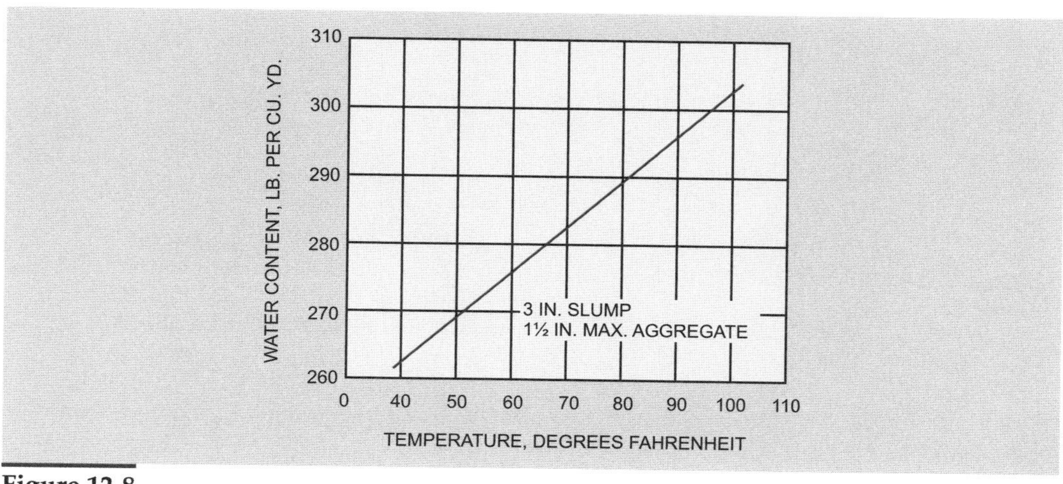

Figure 12-8
A temperature difference of 10°F can affect the slump by almost 1 inch, a fact that should be considered when proportioning the mix. See also Figure 2-7.

The water contents shown in Table 12.7 are based on reasonably well-shaped rounded or subangular coarse aggregates of average grading, with natural sand with a fineness module of about 2.75, in mixes with a water-cement ratio of about 0.55. Adjustments for other conditions can be made in accordance with those shown in the table. This table also shows recommended percents of sand and air contents.

Air Content. Normal concrete without entrained air contains a small amount of entrapped air that should be included in the computations. Entrapped air in nonair-entrained concrete, and total air in air-entrained concrete, can be estimated from Table 12.7.

Cement Content. The starting cement content is determined from selected water-cement ratio and water content:

$$\text{weight cement per cubic yard} = \frac{\text{weight water per cubic yard}}{\text{water-cement ratio}}$$

If a minimum cement content is specified, then the starting cement content should be the larger of either the one specified or the one computed as shown above.

Aggregate Content, Percentage Method. Total volumes of water, cement and air are now computed. The solid volume of any material is its weight divided by its density, the density being the specific gravity of the material multiplied by 62.4, which is the weight per cubic foot (density) of water. This computation can be simplified by the use of Table 12.6. The solid volume of aggregate per cubic yard of concrete equals 27 (the number of cubic feet in a cubic yard) minus the sum of the volumes of water, air and cement. The volume of sand can now be computed from Table 12.7, which shows sand as a percentage of the total aggregate volume. Coarse aggregate then is the remaining volume. Although angular aggregate will probably require more sand to render the mix workable, the percentage of sand should be kept to a minimum because raising the sand percentage to make a more workable mix increases the water requirement, as shown in Figure 12-9.

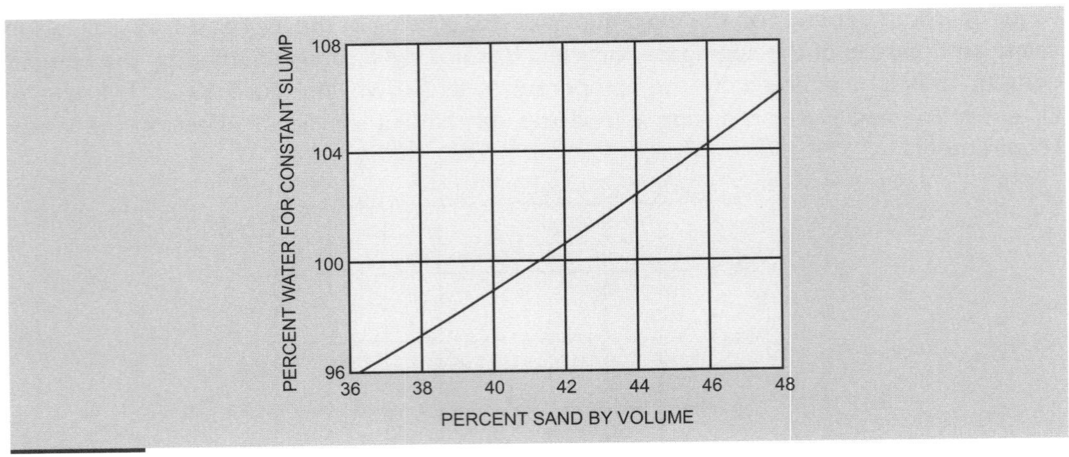

Figure 12-9
Each increase of two percentage points in sand volume increases water requirement by almost 2 percent if slump is held constant.

Weights of fine and coarse aggregates are obtained by multiplying volume times density, from Table 12.6.

Example. Normal-weight concrete is to be placed in a structure consisting of slabs and formed walls and columns. Unreinforced slabs are 5 inches thick; the smallest structural elements are 8-inch walls with No. 6 reinforcing bars spaced at 8 inches—2 inches clear of the forms. The structure is not in a severe exposure, and air entrainment is not specified.

Materials are assumed to have been investigated for quality, availability, etc. A satisfactory history of use in the area, based on a careful inspection and evaluation of existing structures known to have been constructed with the proposed materials, may serve as the basis for acceptance. Laboratory tests may also serve as a basis for evaluation.

Unless otherwise specified by job specifications, materials should conform to the appropriate ASTM standards.

The amount of preliminary investigative work done depends on the size and importance of the project and the service conditions to which it will be exposed. However, it should be remembered that a small yardage of concrete does not necessarily mean that it is unimportant.

Methods of testing aggregates, cement and concrete are described in Chapter 13.

Items specified:

Average compressive strength, allowing for safety margin to keep the number of low tests within specified limits: 4000 psi

Maximum water-cement ratio: 0.60

Slump: Maximum 4 inches

Specific gravity of cement: 3.15

Properties of aggregates, determined by tests:

Coarse aggregate

Bulk specific gravity, SSD basis: 2.67

Percent absorption: 0.2 percent

Moisture content at time of use: dry

Dry rodded unit weight: 99 pcf

Fine aggregate

Bulk SSD specific gravity: 2.64

Percent absorption: 1.0 percent

Moisture content: 5.5 percent

Fineness module: 2.75

Selection of MSA:

The contractor proposes to use a 1-inch MSA pump mix. This is smaller than any of the MSA limitations from Figure 12-6, so 1-inch MSA is acceptable.

Water-cement ratio:

For a strength of 4000 psi, Table 12.4 gives us a water-cement ratio of 0.57 for nonair-entrained concrete. This is lower than the specification maximum, so 0.57 governs.

TABLE 12.6
DENSITY RELATIONSHIPS

SPECIFIC GRAVITY (S)	DENSITY (D)
2.45	152.88
2.46	153.50
2.47	154.13
2.48	154.75
2.49	155.38
2.50	156.00
2.51	156.62
2.52	157.25
2.53	157.87
2.54	158.50
2.55	159.12
2.56	159.74
2.57	160.37
2.58	160.99
2.59	161.62
2.60	162.24
2.61	162.86
2.62	163.49
2.63	164.11
2.64	164.74
2.65	165.36
2.66	165.98
2.67	166.61
2.68	167.23
2.69	167.87
2.70	168.48
2.71	169.10
2.72	169.73
2.73	170.35
2.74	170.98
2.75	171.60
3.10	193.44
3.15	196.56
3.20	199.68

D = 62.4 S
W = DV
Weight of water is 62.4 pcf

NOTE:
S = specific gravity
D = density, pcf
W = weight, pounds
V = volume, cubic feet
V = W divided by D

TABLE 12.7
APPROXIMATE MIXING WATER, SAND AND AIR CONTENTS

	NONAIR-ENTRAINED CONCRETE			AIR-ENTRAINED CONCRETE		
MSA	LB WATER PER CU YD	VOL. SAND % OF TOTAL AGGREGATE	% AIR CONTENT (ENTRAPPED AIR)	LB WATER PER CU YD	VOL. SAND % OF TOTAL AGGREGATE	% AIR CONTENT* (ENTRAINED AIR)
3/8	385	61	3.0	340	59	6.0 to 7.5
1/2	365	53	2.5	325	50	5.5 to 7.0
3/4	340	45	2.0	305	42	5.0 to 6.0
1	325	41	1.5	295	37	4.5 to 6.0
1 1/2	300	36	1.0	275	33	4.5 to 5.5
2	285	33	0.5	265	30	4.0 to 5.0
3	245	31	0.3	225	28	3.5 to 4.5

Above quantities based on the following:
Slump: 3 to 4 inches
Sand Fineness Module: Approximately 2.75
Well-shaped angular coarse aggregates graded within limits of ASTM C33

Slump:

Specified at 4 inches maximum.

Water content:

From Table 12.7 for a 3-inch to 4-inch slump, 1-inch MSA, total water per cubic yard of concrete is 325 pounds.

Air content:

From Table 12.7 the amount of entrapped air for 1-inch MSA nonair-entrained concrete is 1.5 percent.

Cement content:

$$cc = \frac{\text{weight water per cubic yard}}{\text{water-cement ratio}} = \frac{325}{0.57} = 570 \text{ pounds per cubic yard}$$

Sand percentage:

From Table 12.7 for 1-inch MSA concrete, sand equals 41 percent of solid volume of aggregate.

Coarse aggregate:

The remaining volume is coarse aggregate. Computation of the mix is shown in the accompanying Table 12.8. Batch weights for aggregates and water must be corrected for aggregate moisture and absorption. The coarse aggregate is dry; therefore, it will absorb from the mix water,

$0.002 \times 1816 = 4$ pounds water

The sand contains 5.5 percent total moisture, but the sand absorption is 1.0 percent, so the net free water to be contributed to the mix water is 4.5 percent,

$0.045 \times 1249 = 56$ pounds

TABLE 12.8
TRIAL MIX COMPUTATIONS

		WEIGHT LB PER CU YD	SOLID VOLUME CU FT PER CU YD
Cement		570	
Table 12.6 for Sp. Gr. 3.15:			
	570/196.56 =		2.90
Water		325	
	325/62.4 =		5.21
Air			
	0.015 x 27 =	_____	0.41
TOTALS		895	8.52
Volume of aggregate, cu ft			
	27 - 8.52 =		18.48
Sand			
Volume	0.41 × 18.48 =		7.58
Table 12.6 for Sp. Gr. 2.64:			
	7.58 × 164.74 =	1249	
Coarse aggregate			
Volume	18.48 - 7.58 =		10.90
Table 12.6 for Sp. Gr. 2.67:			
	10.91 × 166.61 =	1816	_____
TOTALS		3960	27.00
Theoretical unit weight of concrete = 3960/27 = 146.7 pcf			

Adjusted mix weights (pounds):

Cement		570
Water	325 + 4 - 56 =	273
Sand	1249 + 56 =	1305
Coarse aggregate	1816 - 4 =	1812
Total		3960

Note that the total weight per cubic yard remains the same.

To mix a small batch in the laboratory, the original mix can be proportioned for whatever size batch is convenient. Assume that a batch of about 3 cubic feet (450 pounds) is required. The original adjusted mix weights are proportioned in the ratio:

$$\frac{450}{3960} = 0.114$$

	Cubic Yard	Batch
Cement	570 x 0.114 =	65.0 pounds
Water	273 x 0.114 =	31.1
Sand	1305 x 0.114 =	148.7
Coarse aggregate	1812 x 0.114 =	206.8
	3960	451.6 pounds

Exact quantities of cement and aggregates are batched by weight; water may be weighed or measured by volume in a graduated container. The aggregates are weighed cumulatively, beginning with the smallest size, on a scale accurate to 0.03 pound, or $^1/_2$ ounce. A container of sufficient size to hold either a half batch or a full batch is placed on the scales, tare correction is made for the weight of the container, and the aggregates are weighed separately. Aggregates should be at room temperature (65° to 75°F) before being mixed. In batching water, a quantity greater than computed requirement is measured. Admixture (if any) is added to a portion of this water, and this portion of water is the first part to be added to the batch in the mixer. Water from the weighed container is then added until the concrete has the required slump. After the concrete reaches this slump, water remaining is measured, this amount is subtracted from the total measured, and the net water that has been added is computed.

In the example,

Estimated water requirement (pounds)	= 31.1
Total weight of water plus container	= 45.0

An estimated 10 or 11 pounds of water is poured from this container into a bucket, the total admixture is added to the bucket, and this mixture is poured into the materials already in the mixer. More water is added from the weighed container to obtain the 4-inch slump.

Weight of container plus water (pounds)	= 45.0
Final weight of container plus remaining water	= 11.3
Total water used for 4-inch slump (pounds)	= 33.7

Capacity of the mixer should equal or slightly exceed batch size. Before a trial batch is placed in the mixer, the mixer should be primed by mixing a small partial batch (one half of normal batch, or less) of approximately the same composition as the trial batch, or by mixing a small amount of mortar of the same composition as mortar of the batch. This priming batch is discharged and wasted, leaving a coating of mortar on the mixer interior. Place dry ingredients in the mixer and add about two thirds of expected water, including all the admixture. Allow it to mix about one-half minute, then add clean water from the remaining measured amount until the desired slump is attained, as judged by appearance of the concrete in the mixer.

Mix the batch for a total of three minutes. Dump the concrete into a watertight and nonabsorptive receptacle of such size and shape that the concrete can be turned over with a shovel to eliminate segregation. A clean, damp concrete floor or wheel-barrow is satisfactory.

Immediately after the concrete is dumped from the mixer, tests for temperature slump, unit weight, yield and air content (for air-entrained concrete) are made, then cylinders are cast for strength tests. Concrete used in determining air content by the pressure method should be wasted. Concrete used in unit weight tests and slump can be recombined with the concrete remaining in the pan for making six 6-inch by 12-inch cylinders.

The trial mix in the example required more water for the specified slump than was originally estimated; therefore, the mix must be adjusted to compensate for the additional water. The actual water used in the trial batch was 33.7 pounds, which equals

$$\frac{33.7}{0.114} = 295 \text{ pounds per cubic yard}$$

which is 22 pounds more than the estimate. Total water per cubic yard is thus increased by 22 pounds to 347 pounds, and the cement must be increased to maintain the water-cement ratio:

$$\text{New cement weight} = \frac{347}{0.57} = 608 \text{ pounds}$$

The new adjusted mix is shown in the accompanying "Adjusted Mix" Table 12.9-A. Aggregate and water weights must again be corrected for absorption and free moisture, as before:

Coarse aggregate

$0.002 \times 1763 = 4$ pounds absorption

Sand

$0.045 \times 1212 = 54$ pounds free moisture

Adjusted mix weights (pounds)

Cement		608
Water	347 + 4 - 54 =	297
Sand	1212 + 54 =	1266
Coarse aggregate	1763 - 4 =	1759
Total		3930

When setting scales in the batching plant under operating conditions, the proportions should be rounded off to the nearest 5 pounds of cement and water and 10 pounds of aggregate. Computation, however, should be carried out to the nearest pound to avoid cumulative errors.

At this point, characteristics of the proposed materials have been well enough established to permit making a group of mixes to determine the effect of water-cement ratio on the concrete strength. A series of mixes is made at different cement contents and constant slump. Total water per cubic yard will be almost constant for these mixes. Information from each mix includes mix number, total water per cubic yard, pounds of cement per cubic yard, yield, slump, unit weight, percent air, fine-aggregate content, coarse-aggregate content and notes about the workability of the batch. Subsequently, results of strength tests are included.

To comply with the requirements of the trial mix method, the code requires that three mixes be made to establish three points on the strength versus water-cement ratio curve. Figure 12-2 is a typical curve for 4000 psi concrete developed in this manner after the first trial mix had been adjusted and rerun. In this case, the water-cement ratios for the two additional trial batches, after adjustments, were 0.43 and 0.62. As soon as the strength tests became available, it was possible to plot the curves shown in Figure 12-2.

All batches should be tested for fresh unit weight, slump, air content, temperature and compressive strength as described in Chapter 13.

All cylinders should be cured in accordance with ASTM C 192. Two cylinders will be broken at 7 days, two at 14 days and two at 28 days.

Cylinder numbers, markings, transmittal forms and reports should indicate clearly that the cylinders were cast from trial mixes.

Final adjustments of the proposed mix should be made under full-batch field-operating conditions at the beginning of concreting operations.

As an alternative procedure, instead of mixing small batches in the laboratory, it is sometimes desirable to mix full-size batches in the ready-mix concrete plant, provided the proper materials are available as required by the specifications and job conditions, and following the same procedures that were described for the small laboratory batch. Corrections must be made for moisture in the aggregates. A series of mixes should be made so that a curve similar to Figure 12-2 can be made.

The foregoing procedure should always be followed when proportioning mixes using materials from a new source not previously used for concrete. On the other hand, in urban areas where ready-mixed concrete plants have been established, it is possible to use existing mixes as a basis for new mixes for any specific job. Usually, the commercial mixes commonly in use will have to be adjusted in some way to make them acceptable for a particular job, but the existence of these mixes and the knowledge of their characteristics and properties greatly simplify the work of proportioning new mixes. A pair of curves similar to those in Figure 12-2 should be prepared. Curves of this type can be prepared for the commercial mixes from a ready-mix plant and will be useful in selecting mixes for those jobs for which commercial mixes are suitable. Different curves are necessary for different maximum sizes of aggregate in the mixes and for different admixtures.

TABLE 12.9-A
ADJUSTED MIX

		WEIGHT	VOLUME
Cement		608	
	608/196.56 =		3.09
Water		347	
	347/62 =		5.56
Air		——	0.41
TOTALS		955	9.06
Volume of aggregate			
	27 = 9.06 =		17.94
Sand			
	0.41 x 17.94 =		7.36
	7.36 x 164.74 =	1212	
Coarse aggregate			
	17.94 = 7.36 =		10.58
	10.58 x 166.61 =	1763	——
TOTALS		3930	27.00

TABLE 12.9-B
FLY ASH MIX

Adjust the mix in Table 12.9-A for fly ash.

1. Decrease cement by 80 lb.
2. Add 100 lb fly ash.
3. Reduce mix water content 5 percent.

		WEIGHT	VOLUME
Cement	608 - 80 =	528	2.69
	528/196.56 =		
Water	347 x 0,95 =	330	5.29
	330/62.4 =		
Fly ash		100	
	100/(2.30 x 62.4) =		0.70
		______	0.41
TOTALS		958	9.09
Volume of aggregate			
	27 - 9.06 =		17.91
Sand			
	0.41 x 17,94 =		7.34
	7.36 x 164.74 =	1209	
Coarse aggregate			
	17.94 - 7.36 =		10.57
	10.58 x 166.61 =	1761	______
TOTALS		3928	27.00

12.4. The ACI Method

In the ACI Method, properties of the materials are estimated or determined in the same manner as previously described. In addition, it is necessary to know the dry-rodded unit weight of the coarse aggregate, fineness module of the sand and weight of the concrete in pounds per cubic yard. Aggregates are proportioned on a dry basis rather than on the saturated, surface-dry condition. The procedure is the same up to the selection of aggregate content.

After determining the amounts of water, air and cement in the manner previously described, the next step is to compute the amount of coarse aggregate using Table 12.10. The volume of dryrodded coarse aggregate for a cubic yard of concrete is equal to the value from Table 12.10 multiplied by 27. By multiplying this volume by the dry-rodded weight per cubic foot of the coarse aggregate, we obtain the dry weight of coarse aggregate per cubic yard of concrete.

TABLE 12.10
VOLUME OF COARSE AGGREGATE PER UNIT OF VOLUME OF CONCRETE

MAXIMUM SIZE OF AGGREGATE (IN.)	VOLUME OF DRY-RODDED COARSE AGGREGATE* PER UNIT VOLUME OF CONCRETE FOR DIFFERENT FINENESS MODULI OF SAND			
	2.40	2.60	2.80	3.00
3/8	0.50	0.48	0.46	0.44
1/2	0.59	0.57	0.55	0.53
3/4	0.66	0.64	0.62	0.60
1	0.71	0.69	0.67	0.65
1 1/2	0.75	0.73	0.71	0.69
2	0.78	0.76	0.74	0.72

*Volumes are based on aggregates in oven-dry-rodded condition as described in ASTM C 29.
These volumes are selected from empirical relationships to produce concrete with a degree of workability suitable for usual reinforced construction. For less workable concrete, they may be increased about 10 percent. For more workable concrete, such as may sometimes be required when placement is to be by pumping, they may be reduced up to 10 percent.
(Adapted from Reference 12.1)

TABLE 12.11
FIRST ESTIMATE OF WEIGHT OF FRESH CONCRETE

MAXIMUM SIZE OF AGGREGATE (IN.)	FIRST ESTIMATE OF CONCRETE WEIGHT, LB PER CU YD*	
	NONAIR-ENTRAINED CONCRETE	AIR-ENTRAINED CONCRETE
3/8	3840	3710
1/2	3890	3760
3/4	3960	3840
1	4010	3850
1 1/2	4070	3910
2	4120	3950

*Values calculated for concrete of medium richness (550 pounds of cement per cubic yard) and medium slump with aggregate specific gravity of 2.7. Water requirements based on values for 3- to 4-inch slump. !f desired, the estimated weight may be refined as follows if necessary information is available: for each 10 lb difference in mixing water values for 3- to 4-inch slump, correct the weight per cubic yard 15 pounds in the opposite direction; for each 100-pound difference in cement content from 550 pounds, correct the weight per cubic yard 15 pounds in the same direction; for each 0.1 by which aggregate specific gravity deviates from 2.7, correct the concrete weight 100 pounds in the same direction. For air-entrained concrete, air content for a severe exposure (ACI 318 Table 4.4.1) was used. The weight can be increased 1 percent for each percent reduction in air content.
(Adapted from Reference 12.1)

To obtain the amount of sand, it is necessary to know the weight of the concrete per cubic yard. If this is not known from experience, it can be determined from Table 12.11. Adding together the weights of water, cement and coarse aggregate computed as above and substracting from the weight per cubic yard of the concrete gives the weight of dry sand per cubic yard of concrete.

Alternatively, the amounts can be computed on a volumetric basis. The total volume of water, cement, air and coarse aggregate is subtracted from 27 to give the volume of sand. The volume of any material is its weight divided by its density from Table 12.6.

Because this method is based on dry aggregates, the computed weights of aggregates will be slightly different from those computed in the previous example. The computed weight of coarse aggregate is the dry weight, so no adjustment of coarse aggregate weight is necessary, but the water content must be increased by an amount to compensate for the coarse aggregate absorption. The sand weight will have to be increased by an amount equal to its total moisture content, and the water weight reduced by the net free water in the sand. Further adjustments and tests are made as previously described.

Example. Weight and volume of water, cement and air, determined as previously described.

Weight	Volume
895	8.52

From Table 12.10, for 1-inch MSA and sand fineness module of 2.75,

Volume = 0.68 × 27 = 18.36 cubic feet aggregate.

Note that this volume is loose aggregate, not solid volume.

Weight = volume × dry-rodded weight

= 18.36 × 99 = 1818 pounds

From Table 12.7, solid volume coarse aggregate,

Volume = 1818/164.74 = 11.03

Total Volume = 19.55

Volume of sand

27 - 19.55 = 7.45

Weight of sand, Table 12.7

7.45 × 164.74 = 1227

Batch proportions are now:

	Weight	Volume
Cement	570	2.90
Water	325	5.21
Air		0.41
Sand	1227	7.45
Coarse aggregate	1818	11.03
Total	3940	27.00

The slight differences in results between the two methods are inconsequential. In fact, differences in aggregate weights of 50 pounds would be acceptable.

Trial mixes and adjustments are now made as previously described.

12.5. Admixtures

Air Entrainment. For proportioning air-entrained mixes, the procedure is the same as for nonair-entrained concrete except that the amount of entrained air should be the percentage shown in Table 12.7. Note that the weight of water per cubic yard is less and the percentage of sand is less for the air-entrained mixes. The amount of air entrainer should be recommended by the manufacturer.

The effects of air entrainment and the factors affecting it are discussed in Chapter 9.

Superplasticizers. The introduction of the high-range water reducers, or superplasticizers, has presented another variable in proportioning mixes. In general, the use of these admixtures as water reducers does not require a significant change in mix proportions, other than a water reduction possibly as high as 30 percent. Additional fine material may be necessary for the flowing consistencies. Fly ash appears to be effective. Some of the admixtures cause some reduction in entrained air. Laboratory trial mixes are necessary to develop satisfactory mixes that will produce the required concrete. The manufacturer will furnish necessary data on the particular admixture under consideration and should assist in preparing mixes.

Fly Ash. With the introduction of fly ash into the concrete, special considerations arise that require attention. The possible benefits and problems from the use of fly ash are discussed in Section 9.5. There are differences in fly ashes, and the producer should assist the user in the proper proportioning of mixes containing any certain fly ash. One producer makes these recommendations as a starter when using their ash to adjust, for example, a mix of the type shown in Table 12.9-A (per cubic yard):

Reduce cement content 80 pounds

Add 100 pounds fly ash (specific gravity 2.30)

Reduce mix water content 5 percent

The revised mix is shown in Table 12.9-B. Note that the aggregate proportions are virtually unchanged. It may be desirable to decrease sand percentage slightly if the mix tends to be sticky. All admixtures can be used in fly ash mixes. Air entrainment is necessary, however, as in all mixes, for freeze-thaw resistance and other durability exposures. The superplasticizers can be used successfully with fly ash.

Other Admixtures. The use, effects and handling of admixtures are discussed in Chapter 9. Most admixtures affect the water content and slump of the concrete; therefore, their effects must be taken into consideration when proportioning mixes contain them. The manufacturer's data should be considered when any admixture is to be used.

12.6. Yield

The yield is the volume of concrete produced per batch. It is the weight of all materials batched, including water and admixture, divided by the fresh unit weight of the concrete determined by test. The unit weight and yield of freshly mixed concrete are determined in accordance with ASTM C 138. See Figure 12-10. It may not be exactly equal to the

Figure 12-10
Fresh concrete is measured in a container of known volume to determine unit weight. (Courtesy of PCA)

designed size of the batch because of variations in water content, air content, specific gravity of materials or amount of consolidation of sample in the container, or because of poor sample selection.

The relative yield is the actual yield divided by the designed size of the batch. A relative yield greater than 1.0 indicates that an excess of concrete is being produced; that is, the plant is over-yielding. A relative yield smaller than 1.0 indicates that the batches are short; the plant is under-yielding, or producing less concrete than the mix design calls for. Relative yield should be between 0.993 and 1.007 for good control (a variation of about 0.2 cubic foot in a cubic-yard batch).

12.7. The Small Concrete Job

For a very small job for which small batches of concrete will be mixed on the job there are two alternatives:

1. Use ready-bagged concrete mix. Each bag contains a premeasured amount of cement, sand and coarse aggregate weighing either about 60 pounds or 100 pounds. Adding water makes it ready for use. Bagged concrete is available from most lumber yards and building material suppliers. Compressive strength at 28 days is in the 3000 to 4000 psi range if care is taken to keep the slump within reasonable limits.
2. Use bagged cement (94 pounds per bag) and select a mix from Table 12.12. If only a few cubic feet of concrete are required it can be mixed in a wheelbarrow or mortar box. It is better, however, to rent a small mixer. See Figure 12-11.

Figure 12-11
Jobsite portable mixer (6 cu ft capacity) for small jobs. (Courtesy of PCA)

The proportions in Table 12.12 are only a guide and may need adjustments to obtain a workable mix with locally available aggregates. Where commercial sand and gravel are available, these materials should be used. Allow 15 percent extra for waste.

Bank run sand and gravel are sometimes used, but their use should be discouraged because of variations in and lack of control of grading and quality. However, if conditions are such that this material must be used (in an isolated location, for example) then it should be screened over a piece of $^1/_4$-inch hardware cloth, and the large stones that are too large for the structure removed. The fine and coarse materials should be measured in the correct proportions shown in Table 12.12, making sure the material is free of grass, roots, clay, excessive silt and similar contaminants. Material containing particles of decomposed granite or soft shale should not be used.

Weight measurement of aggregates is best if scales are available. After measuring the aggregates, the correct amount of cement is added and the dry batch mixed, then water is added. The air-entraining agent, if used, should be mixed with part of this water. Add water and mix vigorously until the concrete is of a plastic consistency, that is, when it is plastic and cohesive. A shovelful dumped into concrete in the wheelbarrow will blend into the concrete but will stand up slightly. The concrete should not be so soupy that it runs like water because such concrete will not have good strength or durability and will probably crack.

On all but the smallest and most isolated jobs, it is best to use ready-mixed concrete, if the quantity amounts to the minimum load normally delivered. The order should be placed with the supplier on the day before the concrete is needed. The concrete supplier will require the following information:

1. Where and when to deliver the concrete.
2. The amount of concrete in cubic yards, including an allowance of about 10 percent more than computed to the actual lines of the forms, to allow for waste, uneven subgrade, etc.
3. The mix. Specify the cement content in bags or pounds per cubic yard and a maximum coarse aggregate size of 1 inch. Specify a slump of 4 to 5 inches.
4. If the job will be exposed to freezing and thawing, or exposed to seawater, specify a minimum cement content of six bags, or 560 pounds, per cubic yard and 6 percent air entrainment for durability.

In some areas, building material dealers have facilities for preparing small amounts of concrete and will furnish a small trailer to be towed by an automobile for delivering the concrete. Another good source for small amounts of concrete is the presacked dry mix that requires only the addition of water and mixing prior to placement. Sacked concrete is available from most building material dealers and lumber yards.

TABLE 12.12
PROPORTIONS BY WEIGHT TO MAKE ONE CUBIC FOOT OF CONCRETE FOR SMALL JOBS

MAXIMUM-SIZE COARSE AGGREGATE, INCHES	AIR-ENTRAINED CONCRETE				NON-AIR-ENTRAINED CONCRETE			
	CEMENT, LB	WET FINE AGGREGATE, LB	WET COARSE AGGREGATE, LB[1]	WATER, LB	CEMENT, LB	WET FINE AGGREGATE, LB	WET COARSE AGGREGATE, LB[1]	WATER, LB
3/8	29	53	46	10	29	59	46	11
1/2	27	46	55	10	27	53	55	11
3/4	25	42	65	10	25	47	65	10
1	24	39	70	9	24	45	70	10
1 1/2	23	38	75	9	23	43	75	9

1. If crushed stone is used, decrease coarse aggregate by 3 pounds and increase fine aggregate by 3 pounds. (Adapted from Reference 12.2)

No-Slump Concrete

In some applications, especially in precast and prestressed work, concrete of very dry consistency is used, the slump being between zero and 1 inch. Water content can be reduced to less than the amount that will give a zero slump in some cases. Special equipment is available for measuring the consistency of these extremely dry mixes, because in this range of mixes the slump test as a measure of consistency is of no value. However, the normal procedures for mix proportioning can, in general, be applied. The principal difference is in the amount of water required per cubic yard of concrete. If the amount of water for a mix with a slump of 3 to 4 inches is considered 100 percent, then the amount of water for these very dry mixes is between about 78 percent for extremely dry mixes and 93 percent for stiff, plastic mixes, according to the ACI method of proportioning these mixes.[12.3] Selection of MSA, water-content ratio, air content and aggregate quantities follows conventional methods.

Admixtures can be used for essentially the same purposes as in conventional concrete, although tests may indicate a need for a somewhat different dosage in the no-slump concrete. Fly ash has been found to be beneficial in providing plasticity.

12.9. Gap-Graded Concrete

Most concrete contains aggregates continuously graded from the finest material passing the 100-mesh screen to the maximum size of coarse aggregate suitable for the structure. In some situations claims have been made that "gap graded" aggregate produces better concrete. A gap grading is one in which some sizes of the aggregate are not used. A typical aggregate might consist of sand with $^3/_4$-inch coarse aggregate, without any pea gravel size. Gap gradings are sometimes used for exposed-aggregate architectural concrete where special surface effects are desired. Gap-graded mixes are more sensitive than conventional concrete to variations in aggregate grading and water content, and to the inclusion of undesirable particles. They require close control and supervision. Advantages claimed are lower water-cement ratio, leaner mixes, lower slump and some improvement to properties of the hardened concrete.

The amount of cement, sand, water and air in the mix is usually about 50 percent of the total volume, more being required for crushed coarse aggregate. The one-size coarse aggregate composes about 70 percent of the total aggregate volume, and cement content is about 600 pounds per cubic yard. Slump is less than 3 inches.

Testing and Controlling the Concrete

13.1. Mix Adjustments

In Chapter 12 procedures were given for adjusting the batch weights for absorption and free moisture in the aggregates. In addition to this, there are several other methods or rules of thumb that are useful when adjusting mixes. A laboratory mix will rarely be entirely suitable under field conditions. Adjustments to the amount of water or admixtures are commonly required. Lean mixes require more sand than rich mixes. Once the optimum amount of sand has been found for a certain mix, the water-cement ratio and cement content can be adjusted by interchanging equal solid volumes of sand and cement. This will affect the strength but will have little effect on slump and water content. If cement content is increased by 100 pounds, the sand should be decreased by 75 pounds; coarse aggregate content and batch size remain the same. Other suggestions include:

0.10 increase in sand FM	Increase sand percent by 1/2 percentage point
1-inch increase in slump	Increase water 3 percent
1 percent increase in sand	Increase water 1 percent content
1 percent increase in air content	Decrease water 3 percent Decrease sand 0.5 to 1 percent

An increase of approximately 3 percent in total water will increase slump by about 1 inch (see Figure 2-5) and reduce compressive strength about 150 psi. Adjusting mix proportions to compensate for difference in volume of water is most easily accomplished by increasing or decreasing aggregate volume by a quantity equal to the decrease or increase in water volume, keeping cement and air volumes constant. Making an adjustment in this manner introduces a slight error in the amount of water required in the new batch, but the error is not significant if estimated water requirement of the first trial batch was reasonably close to actual. Changing the ratio of fine to coarse aggregate affects the amount of water for constant slump. (See Figure 12-9.)

Under field conditions, fluctuations in temperature, aggregate gradation and other factors cause changes in the amount of water required for constant slump; hence, variations of water-cement ratio of ± 0.02 by weight are considered normal and do not require that the mix be adjusted. However, if the water-cement ratio is consistently high, an adjustment should be made to bring it down. If it is consistently low, it may be raised if strength of concrete is adequate. Steps that aid in reducing the water-cement ratio include reducing the percentage of sand in the mix, improving sand grading, using a larger-size coarse aggregate, using a water-reducing admixture or using an air-entraining agent. Improvement in handling and placing procedures that permit the use of lower-slump concrete should be considered.

Any of the above changes works equally well in the opposite direction.

The coarse aggregate volume should be maintained constant because it simplifies computations and operations in the batching plant.

Changing the cement content of nonair-entrained mixes at the same slump has no appreciable effect on the total amount of water per cubic yard.

Sometimes, during concreting operations, the technician may receive complaints of harshness of the concrete. If there has been an actual change in workability of the concrete, it is nearly always caused by a change in aggregate grading, either the fine or the coarse, or both. Correction may be made by adjusting the aggregate percentages. For example, a change in the sand-fineness modulus can be partially offset by adjusting the sand percentage as previously described. Changes in coarse-aggregate grading may affect the percentage of voids and therefore the rodded unit weight of the aggregate, which is reflected in a change in the amount of sand required.

Adequate control of the concrete requires more than the application of these simple measures. Control of the concrete is accomplished by observing the concrete, taking samples and making tests, in this way providing a true picture of the quality and characteristics of the particular concrete under consideration.

13.2. Sampling and Testing

A sample is a small part of a stock or quantity of a material. The material sampled may be the amount produced in an hour or day, a certain number of units, a continuous flow, a truck or carload, a batch or any other reasonable unit of quantity. Sampling procedures are described in the applicable standards, but some sort of a sampling plan should be developed. It may be one sample for a certain quantity, such as cubic yards, tons, batches or loads. It may be based upon time intervals.

The basic requirement of any sampling procedure is to obtain a sample that is truly representative of the material being sampled. Sampling methods are based on the mathematics of probability, and a sampling plan can be developed for any set of conditions. However, it is not necessary for the field person to make a statistical analysis whenever he or she wants to sample any given material. Instead, standards have been established for this purpose by the American Society for Testing and Materials (ASTM) and other organizations, based on the appropriate statistical background. Reference to the ASTM or code designation for any material provides information on the sampling method to be adopted.

If all is properly done, the sample represents the material from which it was taken and tests reveal the quality of the sample. If, however, the sample does not properly represent the material, then the test results are in error, and misleading information will be obtained.

"One test is worth a thousand opinions." So goes the old saying. While it is no doubt somewhat of an exaggeration, the truth is that a test properly performed does give factual information that cannot be disputed. To be properly performed, a test must:

1. Be made on a sample that is representative of the material under consideration.
2. Follow standard methods of sampling and testing that have been approved by the industry and by the agency requiring the test.
3. Make use of properly calibrated and standardized apparatus designated for the selected testing method.
4. Be performed by qualified personnel experienced in testing and familiar with the material under test.
5. Be interpreted by a person fully acquainted with the material and the test method, capable of evaluating the test and able to interpret the results within the context of the project requirements.

Testing of concrete and concrete materials is important and necessary. This has been demonstrated many times, especially when one reads of the spectacular structural failures caused by inferior materials. But there are other failures too, not so spectacular, but nevertheless serious and costly to the owner. Examples are popouts, cracking and unsoundness of concrete resulting from the use of aggregates containing inferior constituents— constituents that would be revealed by proper tests, properly performed.

Adequate quality control of materials, including laboratory and field tests, informs the engineer, architect, producer, contractor and owner of the properties of materials proposed for construction and serves as a guide to the producer in maintaining the products within specification limits.

The Need for Tests. Why are tests necessary? The reason for testing is that concrete, and all the materials that go into concrete, have variable properties. No two batches of concrete are ever exactly alike, any more than any two supposedly identical automobiles are exactly alike. Tests are made to find information about the properties of materials. Tests do two things: they reveal the quality of a product when it was sampled, and they show (when there are several tests) how uniform the product is.

We govern the quality of construction materials by means of control tests made at the producing plant and at the jobsite. In the case of concrete these tests are made to assure that the materials are uniform, to determine the properties of the plastic concrete, and to provide specimens for strength tests of the concrete at later ages.

Throughout the following discussion we keep repeating that conclusions should be based on the results of more than one test. This is true not so much because a test might be wrong, but because there are so many small variables that can affect the concrete and the testing method. It we take, say, three tests, we can see if one test is out of line. Furthermore, the average of these tests is more apt to be accurate than just one test result alone.

Testing is a precision operation. The person making a test must follow an exact procedure that has been standardized and documented, keeping within well-established limits in all steps constituting the test. Completion of the test enables the tester to report results in objective numerical values. See Figure 13-1.

A TEST is

Precise

Standardized

Documented

TEST RESULTS are

Numerical

Objective

TESTS can

Control
- Measure properties of product
- Assure reliability and uniformity

Evaluate
- Measure suitability and performance

Verify
- Check previous tests or option

Figure 13-1
A properly performed test is a precise measure of a certain property of a material.

One important fact should remain constantly in mind: An improperly made test is worse than no test at all. Because important decisions are frequently based on the results of tests, the significance of the test result is easily apparent. The specified method must be rigidly adhered to. When the standard says 25 strokes of a $^5/_8$-inch tamping rod, it means exactly 25 strokes, not 24 or 26, and it means the $^5/_8$-inch tamping rod described in the standard, not a piece of reinforcing steel or a stick of wood. Strict and undeviating observance of the specified procedures will enable the technician to achieve accuracy and reliability.

Nonstandard Tests. Some agencies occasionally develop special test procedures that are particularly applicable to some special requirement of that agency. When this occurs, a detailed instruction is usually developed by the agency. Other cases of nonstandard tests may result from the development of new equipment and procedures. These are the result of research by manufacturers and other groups.

Experiments have been made in the use of rapid determinations of the water content of fresh concrete, moisture and cement content of hardened concrete, and similar procedures. These, when standardized and accepted by the industry, will eventually lead to our being able to control the quality of concrete at the time it is manufactured instead of waiting for the results of long-time tests.

13.3. Application of Tests to Concrete

Two Kinds of Tests. When we consider the application of tests to concrete we find that there are two kinds of tests pertaining to the concrete. The first group includes the tests that are applied to the fresh concrete for the purpose of controlling or verifying the quality of the concrete. Tests in this category include tests for slump or consistency, air content, unit weight, cement content, mix analysis and yield. Also, specimens for strength tests are made from the fresh concrete. These are the tests that are normally made by the inspector or the producer's quality control technician.

The second group of tests comprises those tests that are made on samples of the hardened concrete, not including strength tests of specimens that were molded from the fresh concrete. The purpose of these tests is to either verify or refute the quality of the hardened concrete, usually as the result of unsatisfactory results obtained on samples of the fresh concrete. Embraced within this group are strength tests made on the surface of the hardened concrete; strength tests on samples drilled, sawed or otherwise removed from the structure; mix analysis of the hardened concrete, including air content; and several other complex tests that require very specialized equipment and techniques.

Testing Standards. Testing procedures based on standards of the American Society for Testing and Materials (ASTM) have been developed by committees representing all segments of the concrete construction industry. They are continually reviewed and revised. Many of the standards for concrete were originally issued 70 or more years ago but are kept abreast of developments and requirements of the industry through the continued program of review and revision.

There are several standards with which the concrete technician and inspector must be thoroughly familiar because these standards will be used nearly every day. The seven basic field tests on freshly mixed concrete are the following:

ASTM NUMBER	TITLE
C 172	Sampling Freshly Mixed Concrete
C 143	Test for Slump of Portland Cement Concrete
C 1064	Test for Temperature of Portland Cement Concrete
C 231	Test for Air Content of Freshly Mixed Concrete by the Pressure Method
C 173	Test for Air Content of Freshly Mixed Concrete by the Volumetric Method
C 138	Test for Unit Weight, Yield and Air Content (Gravimetric) of Concrete
C 31	Making and Curing Concrete Test Specimens in the Field

It is essential that field technicians and inspectors have proven skills in performing these seven standard field tests on freshly mixed concrete. It is equally important that everyone involved with the construction project have confidence in the technician's skills. Knowledge and ability to perform the seven standard field tests can be demonstrated by certification as an ACI Concrete Field Testing Technician, Grade I. This field testing certification is administered by local ACI chapters nationwide. The national certification for reinforced concrete special inspector administered by the International Code Council (ICC) requires ACI Concrete Field Testing Technician certification as part of the requirements for ICC certification for Reinforced Concrete Special Inspector. See Section 25.9.

Other test methods that should be known to the concrete technician/inspector include the following:

C 42	Obtaining and Testing Drilled Cores and Sawed Beams of Concrete
C 803	Test Method for Penetration Resistance of Hardened Concrete
C 805	Test Method for Rebound Number of Hardened Concrete
C 900	Test Method for Pullout Strength of Hardened Concrete

THE FRESH CONCRETE

13.4. Testing Freshly Mixed Concrete

The purpose of this section is to assist the technician in making the specified tests by explaining some of the procedures and equipment. It is not intended as a complete treatment of each testing method. The technician must refer to the specified method for the correct way to make the test and should have available a copy of the appropriate ASTM standard. It is noteworthy that, where time has an effect on test results, the ASTM testing

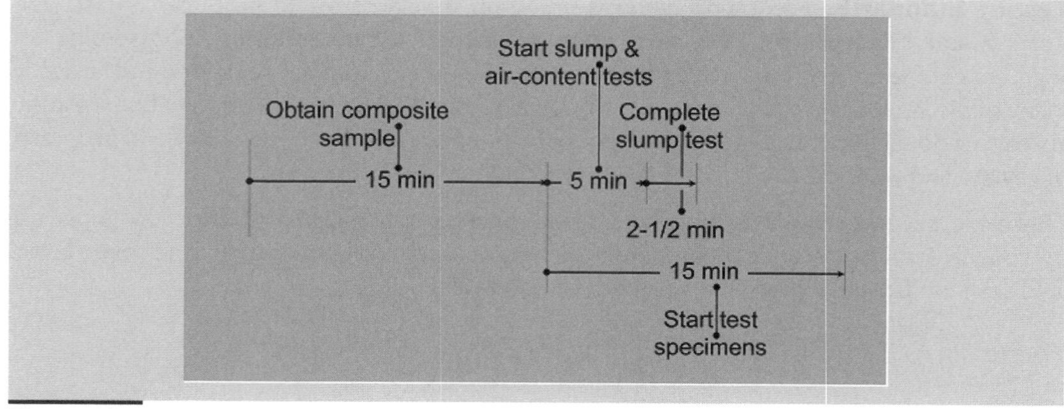

Figure 13-2
Time limits for field testing. . .when "time" has an effect on test results

procedures have time limits for conducting the tests, as shown in Figure 13-2. It is certainly not the intent that the technician carry a stop-watch. The technician should, however, be aware of proper use of time in conducting a test within reasonable time limits.

Sampling. Samples should be taken from predetermined batches in accordance with a logical sampling plan. Samples can be taken from batches of low or high slump to check such batches and for special information, but sampling only such batches will be misleading and will not provide proper quality control information.

ASTM C 172 requires a sample of at least 1 cubic foot for strength tests but permits smaller samples for slump or air content tests.

Obtaining samples truly representative of the concrete to be tested is of great importance and is often given insufficient attention. In order to minimize sampling errors, the sample should be obtained from the middle portion of a batch representative of those being used. A sample being taken from a stream discharging from a mixer should cut across the entire cross section of the stream and should be taken at two or more regular intervals throughout the discharge of the batch, or the stream should be diverted completely into a container avoiding both the first and final portions of the discharge. See Figure 13-3. The sample should be thoroughly remixed in a wheelbarrow, large pan or other nonabsorptive surface. If only a slump test is desired to check on the consistency of the batch in the mixer, a small sample can be taken as soon as the first part of the batch has been discharged. The batch from a paver should be dumped and a sample collected from five different places in the pile, taking care to avoid taking part of previous batches. When sampling from a truck mixer, the entire stream of concrete should be taken, and the discharge gate on the mixer should not be choked down. Gate should be fully open with rate of flow regulated by rotational speed of the mixer drum.

Figure 13-3
Sampling fresh concrete. Note the ineffective placement (location) of the wire mesh.

Normally, samples should not be taken from the forms, because segregation, bleed water and other factors are apt to make the sample nonrepresentative. Occasionally, samples can be taken from the structure, but these would be for special purposes, not for routine control purposes. Sometimes sampling will be specified at discharge of a pump.

A sample from which strength tests are to be made should be carried to the place where the specimens are to be made. Do not move the specimens for at least 12 hours.

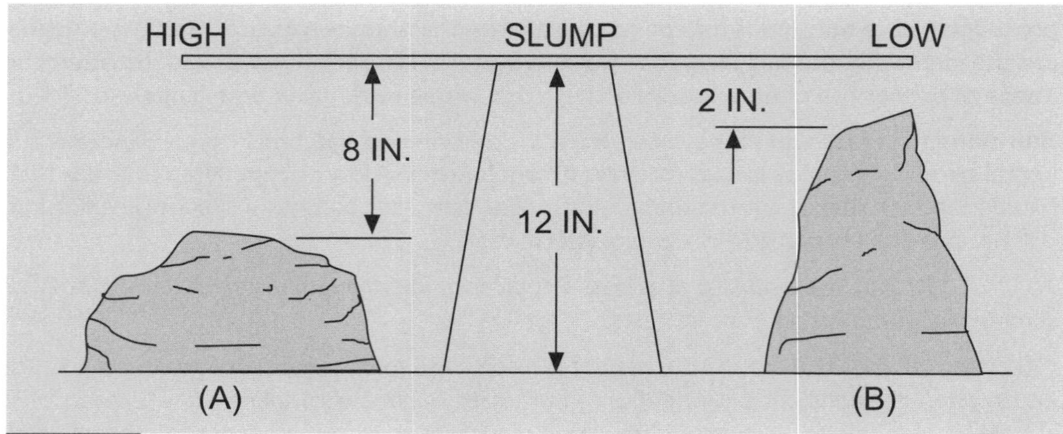

Figure 13-4
When we speak of "high slump" or "low slump" we are referring to the number of inches the concrete specimen slumps or subsides when the cone is removed. The concrete at A is described as a high-slump, wet, fluid or soft concrete; that at B is a low-slump, dry or stiff concrete.

The sample must be processed promptly. A delay of 10 or 15 minutes, especially during hot weather, can reduce the results of a slump test by one half and can adversely affect other test results.

Slump Test. Fluidity, softness or wetness of a batch of concrete is indicated by its consistency, which is determined by the slump test, ASTM C 143. Slump is measured in inches, a low slump indicating a stiff or dry consistency, and a high slump indicating a soft or wet consistency. See Figure 13-4.

The slump cone is a sheet metal cone, open at both ends, 12 inches high with the base 8 inches in diameter and the top 4 inches in diameter, provided with foot pieces and handles. See Figure 13-5. Some equipment suppliers furnish a cone and a base that are clamped together, the clamping arrangement being such that it can be released without disturbing the mold.

Figure 13-5
Making the slump test. Note that the technician is holding the cone down by standing on the footpieces while he fills the cone and rods the concrete. (Courtesy of PCA)

Slump is determined by measuring the vertical subsidence of the sample upon removal of the cone. Measurement is made by placing the slump cone beside the slumped sample and laying the tamping rod across the cone, extending over the specimen, then measuring from the bottom of the rod to the slumped concrete at a point over the original center of the base of the specimen. This measurement was shown in Figure 2-4. After the slump measurement is completed, the side of the slumped concrete should be tapped gently with the tamping rod. The behavior of the concrete under this treatment is a valuable indication of its cohesiveness, workability and placeability. A well-proportioned, workable mix will slump gradually to lower elevations and retain its original identity, while a harsh mix will crumble, segregate and fall apart.

Slump specimens that break or slough off laterally give incorrect results and should be remade with a fresh sample. A slump test that indicates noncompliance with the specified limits should be immediately checked by at least one additional test.

Procedure for the slump test is as follows:

1. Obtain a representative sample of freshly mixed concrete from the truck mixer (two 40-pound bucketsfull).
2. Moisten the inside of the cone and place it on a flat, moist, nonabsorbent surface at least 1 foot by 2 feet (plank, piece of heavy plywood, concrete slab, etc.). Surface must be firm and level. Hold slump cone in place by standing on foot pieces.
3. Fill cone one-third full of concrete and rod this layer exactly 25 times with the tamping rod. Distribute rodding evenly. Use standard steel tamping rod, 5/8-inch diameter by 24 inches long with one end rounded to a hemispherical tip. Do not use a piece of resteel. Depth of the concrete in the cone should be about 2 1/2 inches (one-third volume).
4. Fill cone with second layer until two-thirds full and rod this layer 25 times with the rod penetrating into but not through the first layer. Rod evenly over entire area. Depth of the concrete should now be about 6 inches (two-thirds volume). See Figure 13-6.
5. Overfill the cone slightly and rod top layer 25 times with the rod penetrating into but not through the second layer. Rod evenly over entire area.
6. Use the tamping rod to scrape off excess concrete from top of cone and remove concrete from around bottom of cone.
7. Slowly lift the cone vertically. Avoid jarring or bumping the concrete.

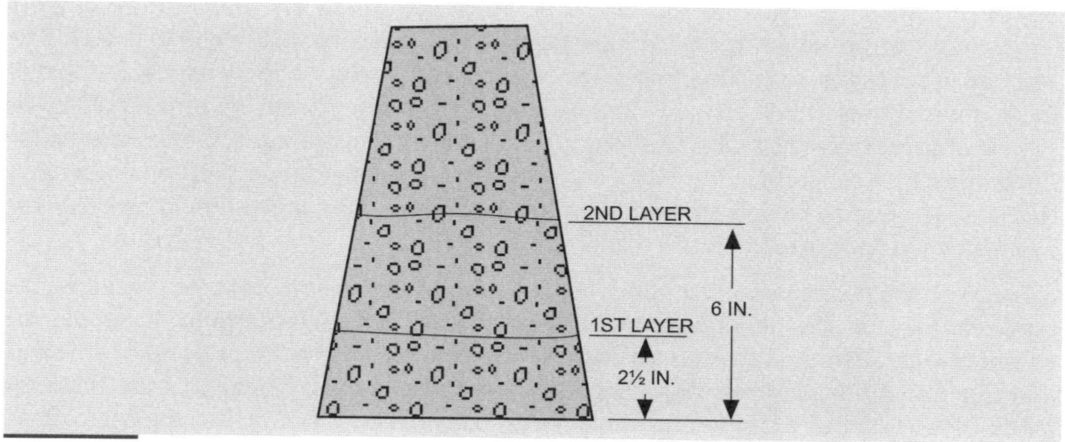

Figure 13-6
Each layer of concrete in the cone should be about one third of the volume of the cone.

8. Set slump cone on surface next to but not touching slumped concrete, and lay the tamping rod across top of cone. Measure amount of slump to nearest $^1/_4$ inch from bottom of tamping rod to a point over the original center of the base of the specimen.
9. Gently tap the slumped concrete and note whether it subsides further or crumbles. A harsh, unworkable specimen will tend to crumble or segregate when tapped lightly with the slump rod. A plastic, workable mix will stick together and subside unbroken. Passing a trowel over the sample in the pan gives an indication of the cohesiveness of the mix, as shown by the effort required to smooth the surface. Concrete used in the slump test can be remixed with the remaining concrete for making strength specimens.

Flowmeter. The flowmeter is a new and updated electronic version of an original K-Slump tester, developed over 20 years ago to measure consistency and workability of fresh concrete as an alternative to the more time-consuming slump test. The flow- meter (see Figure 13-7) is a hollow, pointed plastic tube, perforated with rectangular slots, with a built-in liquid crystal display (LCD) in the handle. When inserted to a fixed depth into the concrete, mortar from the concrete flows through the slots. After pushing the tube into the concrete, the operator pushes a button on the handle to start the timer. In 40 seconds the LCD reads zero and the operator lowers a plunger until it touches the mortar surface inside the tube. Contact is indicated by a black dot appearing at the lower corner of the LCD. A flow reading then appears as an LCD readout. The flow number is related to the amount of mortar flowing into the tube and correlates with standard slump test results. A temperature sensor in the tube permits the operator to also measure concrete temperature by pressing a button and observing the LCD.

The device works equally well on fresh concrete with or without a superplasticizer, giving results in just over one minute. The manufacturer does not, however, recommend using it on concrete with a standard slump less than 2 inches. The flowmeter is a newly adopted ASTM standard test method (ASTM C 1362) to determine the flow of freshly mixed concrete, either in the field or in the laboratory.

Temperature of Fresh Concrete. Because of the effect of temperature on the properties of the concrete, such as slump, air content, water demand, strength and durability, it is a good idea to check the temperature of the concrete whenever a slump test is made. Additional checks of the temperature should be made during hot or cold weather when the concrete temperature may be approaching the specified limits.

ASTM C 1064 prescribes the standard method of measuring the temperature of fresh concrete. An armored thermometer (see Figure 13-8) graduated in 2°F divisions should be inserted in the concrete and permitted to remain until the reading becomes stable. Two or three observations should be made for each test. The thermometer can be inserted in the sample of concrete while slump or strength samples are being prepared, or the observation can be made on concrete in the form, depending on whether the specification limits apply to the concrete as mixed or in the forms. In any event, temperature should be observed whenever strength specimens are made.

Sometimes it is desirable to compute the probable temperature of concrete based on the temperatures of the ingredients. For example, in the summertime it would be advantageous to know if the temperature of the concrete can be kept below a specified maximum or if the water or aggregates will have to be cooled. Because of the relatively minor effect of cement temperature and the difficulty of cooling cement in the batch plant, no consideration is given here to any attempt to cool it. To change the temperature of average concrete by 1°F requires a temperature change of 9°F in the cement, but only $3\frac{1}{2}$°F in the water or $1\frac{1}{2}$°F in the aggregate.

The following equation gives a close approximation of the concrete temperature, based on the temperature of its ingredients:

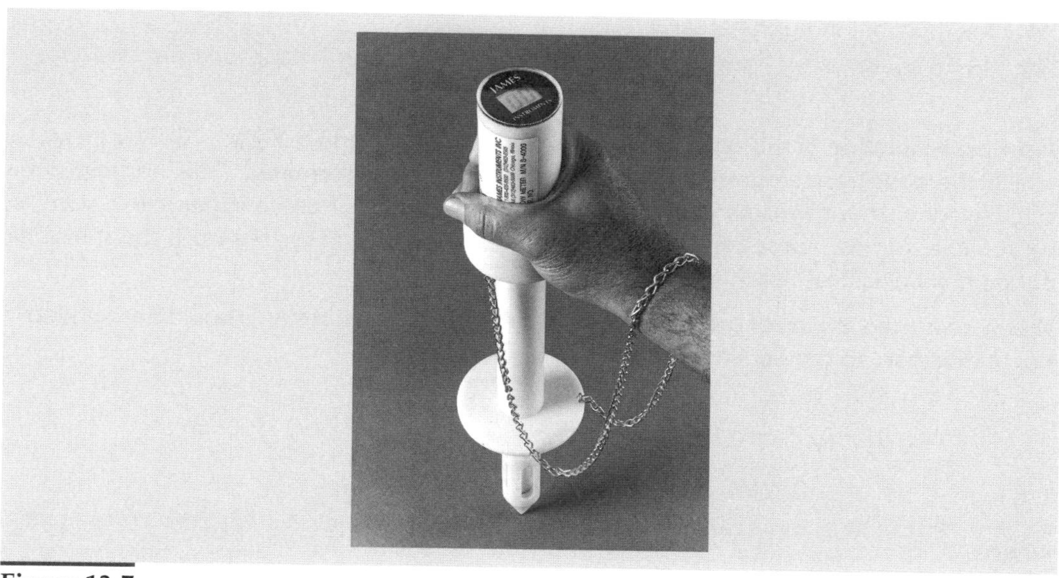

Figure 13-7
The flowmeter, "Slurpy," gives flow values that correlate with standard slump-test results. A temperature sensor also permits the operator to measure concrete temperature electronically.

Figure 13-8
A small dial thermometer with an armored shaft (a) can be inserted in the concrete (b) to determine the temperature.

$$T = \frac{0.2(T_a W_a + T_c W_c) + T_f W_f + T_m W_m - 144 W_i}{0.2(W_a + W_c) + W_f + W_m}$$

In which

0.02 = assumed specific heat of dry materials

T = temperature of concrete

and

Weight Symbol, Lb	Temperature Symbol, °F
W_a Aggregates (surface dry)	T_a
W_c Cement	T_c
W_f Free moisture in aggregates	T_f
W_m Mixing water	T_m

A simple adaptation of this equation is shown in the nomograph in Figure 13-9 by means of which the concrete temperature can be approximately determined. This nomograph is suitable for use with mixes containing about 500 pounds of cement per cubic yard, at reasonable slumps. These calculations do not allow for heat absorbed from the sun or air during mixing or heat resulting from the work of mixing.

When ice is used as part of the mixing water, the term "minus $144W_i$" should be included in the numerator, so the equation now reads:

$$T = \frac{0.2(T_a W_a + T_c W_c) + T_f W_f + T_m W_m - 144W_i}{0.2(W_a + W_c) + W_f + W_m}$$

where

W_i = weight of ice,

and

W_m = weight of mixing water plus ice.

Air Content. Limits and tolerances for entrained air for each type of concrete are given in the specifications. If air-entrained concrete is specified, the air content of fresh concrete should be determined on the first batch for the day and each time cylinders are cast or whenever there are significant changes in the weather or other conditions during the day. During hot or cold weather concrete placing, frequent checks should be made, particularly when concrete ingredients are heated.

There are two types of air meters in regular use, the pressure-type air meter (ASTM C 231) and the volumetric air meter (ASTM C 173), the latter sometimes called a roll-a-meter. The pressure air meter (see Figure 13-10) relates pressure to volume. Air meters of this type are calibrated to read air content directly when a predetermined pressure is applied. The applied pressure compresses the air within the concrete sample, including the air in the pores of the aggregates. For this reason, air content tests by the pressure method are not suitable for determining the air content of concretes made with lightweight aggregates. Aggregate correction factors that compensate for air trapped in the aggregates are relatively constant and, though small, should be subtracted from the gauge reading to obtain the correct air content.

The volumetric air meter consists of a two-part container with a graduated neck. See Figure 13-11. The percentage of air is read directly on the graduations on the instrument. Air content of lightweight concrete made with lightweight aggregate, slag or any aggregate of high porosity must be determined with the volumetric apparatus, which can also be used on normal-weight concrete. To make the test, the sample is placed in the vessel, lid attached, and the vessel filled with water to the mark on a scale. The meter is then rolled until all air in the sample has been displaced with water, the meter placed in an upright

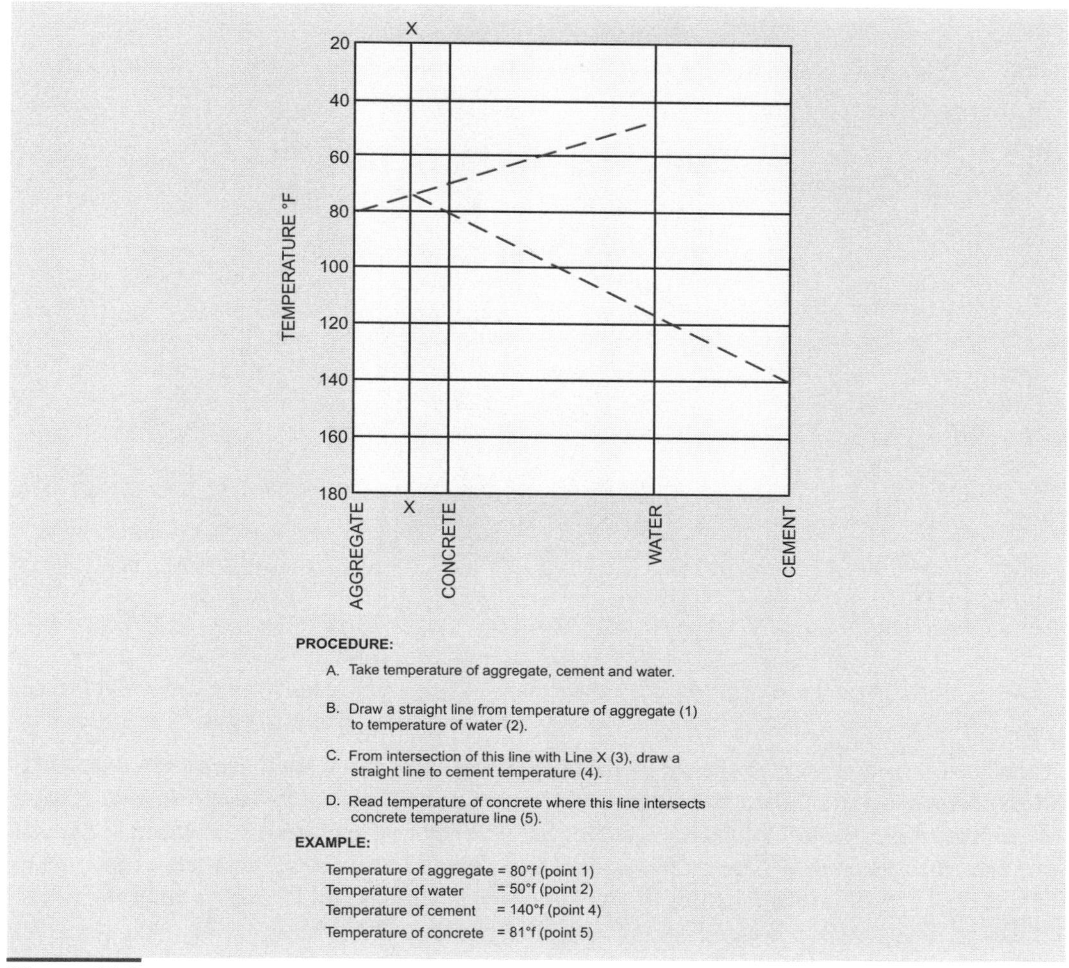

Figure 13-9
This nomograph can be used for finding approximate temperature of concrete, applied to normal mixes containing about 450 to 600 pounds of cement per cubic yard, based on temperature of ingredients.

Figure 13-10
Pressure type air meter (ASTM C 231). (Courtesy of PCA)

Figure 13-11
The volumetric air meter (roll-a-meter) (ASTM C 173). (Courtesy of PCA)

position, and the percent of air read on the scale. Because of the bulkiness and weight of the "roll-a-meter", the device is not considered to be very user-friendly by field technicians.

A quick and easy check of the air content is made with the Chace air meter (AASHTO T199), which is a small glass vial or tube with a glass stem. See Figure 13-12. A small sample of mortar (about a thimbleful passing a No. 10 sieve) from the concrete is placed in the vial. The tube is then filled with isopropyl alcohol and shaken to remove the air from the mortar. The drop in level of the alcohol in the tube gives a measure of the air content of the concrete. This method is rapid and simple to perform and can be used for making a quick estimate of the air content. The average of three tests should be used. This test is not a substitute for the more accurate pressure and volumetric methods, and the latter should be used for control tests.

The manufacturers of air meters provide detailed instructions for calibrating and using their apparatus, and these instructions should be followed. In addition to such instructions, the technician should be governed by ASTM C 231 for normal weight concrete, or ASTM C 173 for lightweight concrete.

All equipment must be calibrated to determine volumes and weights of the apparatus, and the pressure devices must be corrected for compression of air within the interstices of the aggregate particles. The correction is small and reasonably constant for any one aggregate. The test is not difficult to make and must be made for different aggregates. It cannot be ignored. ASTM C 231 explains the procedure in detail, and the manufacturer's instructions explain the procedure as applied to their own products. An aggregate correction factor is not required for the volumetric apparatus.

One of the main sources of error, if the meter has been calibrated correctly, is in leveling the concrete in the bowl. Leveling should be done carefully, using a straightedge with a sawing motion to remove the excess concrete, being especially careful to remove mortar that rises behind the straight edge. The edges of the bowl, where the lid gasket fits, should be wiped clean with a rag and the lid carefully seated in place. All valves should be operated in accordance with the manufacturer's instructions.

Figure 13-12
The Chace air meter.

Concrete used in the air meter (in which water is used to fill the container) should not be used for slump tests or for making strength specimens.

A gravimetric method to determine air content of concrete, ASTM C 138, requires accurate knowledge of aggregate specific gravities, batch weights of the particular batch sampled, and absolute or solid volumes, and is usually a laboratory procedure. See following discussion on unit weight and yield. The procedure of ASTM C 138 should be followed.

Let

W_l = total weight of all materials in the batch, pounds

V = total absolute volume of all materials in the batch, cubic feet

T = theoretical weight of concrete on air-free basis, pound per cubic foot

then

$$T = \frac{W_1}{V}$$

Also

A = Air content of concrete, percent

W = fresh unit weight of concrete, pounds per cubic foot

Y = yield, volume of concrete produced per batch

$$= \frac{W_l}{W}$$

then

$$A = 100\frac{T - W}{100T}$$

or

$$A = 100\frac{Y - V}{Y}$$

To find V, determine absolute volume of each material from the relationship

$$\text{Absolute volume} = \frac{\text{weight batched}}{6.24 \times \text{specific gravity}}$$

See Table 12.5. Bulk specific gravity and weight are based on saturated, surface dry condition of aggregates. Specific gravity of cement is 3.15.

Unit Weight and Yield. The unit weight (or density) of any material is the weight of 1 cubic foot of the material. Specifications do not ordinarily impose any limits on the unit weight of concrete, except for lightweight or heavy concretes for which close regulation of the weight is essential for proper control. Complete quality control of normal concrete, however, is not possible without regular determinations of the fresh unit weight of the concrete, following the method of ASTM C 138. The unit weight is required for computing the volume of concrete produced and for gravimetric air content, as well as determination of the yield and actual cement factor. It is necessary to know the specific gravity of the ingredients and mix proportions to make these computations.

The ASTM method specifies that the measuring vessel shall have a capacity of either $^1/_2$ or 1 cubic foot, depending upon the maximum size of aggregate in the concrete being tested. See Figure 13-13. However, if no unit weight measure is available, the calibrated bottom part of a pressure air meter makes a good substitute for the routine checking of unit weight. This is accomplished by filling the air meter bowl with concrete in the normal manner, striking off and weighing. Having predetermined the exact volume of the air meter bowl by the method described in ASTM C 29, it is a simple calculation to find the concrete unit weight, which equals:

$$\frac{(\text{weight of container and concrete}) - (\text{weight of container})}{\text{volume of container}}$$

After weighing the bowl and concrete, the top of the meter is put in place and the air

Figure 13-13
Unit weight of fresh concrete (ASTM C 138). 0.5 cu ft container commonly used with aggregates up to 2 in. (Courtesy of PCA)

content found in the usual way.

The fresh unit weight gives a good indication of the unit weight that can be expected of the hardened and dried concrete. Large MSA normally produces concrete with a higher unit weight. An abnormally low unit weight indicates either a high air content or excessive amount of mix water. A high cement content results in a high unit weight.

In the following computations, let

W = fresh unit weight of concrete, lb per cu ft

W_l = total weight of all materials in batch, lb

Y = yield, volume of concrete produced per batch

Y_d = *design, volume of batch of concrete, cu yd*

R_y = *relative yield*

$$Y = \frac{W_l}{W}$$

$$R_y = \frac{Y}{Y_d}$$

Also

N = actual cement content, pounds per cubic yd

N_t = weight of cement in the batch, pounds

$$N = \frac{N_l}{Y}$$

Example. In a certain plant, the total weight of cement, added water, damp sand, damp coarse aggregate and admixtures for a 5-cubic-yard load was 19,860 pounds, including 2750 pounds of cement. The batch was designed to produce 5 cubic yards.

Then

W_l = 19,860 pounds,

Y_d = 5 cubic yards, or 135 cubic feet, and

N_r = 2750 pounds = 550 pounds per cubic yard.

Using the bowl of a pressure air meter, the technician obtained the following:

Volume of container = 0.28 cubic foot, obtained as described in ASTM C 29.

Weight of container plus concrete sample = 55.3 pounds.

Weight of container = 14.5 pounds.

$$W = \frac{55.3 - 14.5}{0.28} = \frac{40.8}{0.28} = 146.0 \text{ pcf}$$

$$Y = \frac{W_l}{W} = \frac{19{,}860}{146} = 136.2 \text{ cu ft} = \frac{136.2}{27} = 5.05 \text{ cu yds}$$

$$R_y = \frac{Y}{Y_d} = \frac{136.2}{135} = 1.01$$

This indicates that the plant is over-yielding slightly (about 1 percent).

$$N = \frac{N_t}{Y} = \frac{2750}{5.05} = 545 \text{ pounds per cubic yard}$$

If additional tests consistently indicate that the plant is over-yielding, it might be well to decrease the batch size slightly. In this case, the yield is 136.2 – 135.0 = 1.2 cubic feet per load excess.

$$1.2 \times 146 = 175 \text{ pounds per batch}$$

By reducing the aggregate weight by 175 pounds per batch, the yield can be corrected. This correction should be made only if the design water content is being used and air content is normal, and only if further tests show consistent over-yielding.

Density of concrete can also be found by measuring and weighing compressive strength specimens. This should be done as soon as the specimens have been stripped from the molds, before they dry out.

Mixer Performance Test. Some agencies require that mixer performance or efficiency tests be performed. These tests, sometimes called tests of the uniformity of concrete, may include slump, unit weight, air content, strength, cement content, water content and unit weight of air-free mortar. Not all of these items are required by every agency. Evaluation is based on a comparison of results of tests performed on two samples of concrete, one from the first part of the batch and one from the last part. In some cases, evaluation may be done on samples from successive batches. Specifications for each individual job should be consulted to determine what, if anything, is required. Suggested requirements for uniformity are given in Table 13.1, which is based on the requirements of ASTM C 94 for ready-mixed concrete.

The results of several hundred tests made on large stationary mixers (4 cubic yards and larger) showed that variation in the unit weight of mortars and variations in water-cement, sand-cement and water-fines ratios reflect the adequacy of mixing.

After any arbitrarily selected mixing time, samples of concrete are obtained by one of two methods: (1) Stop the mixer and remove samples at approximately equal distances from the front and back of the mixer or (2) while the mixer is discharging take one sample after

TABLE 13.1
REQUIREMENTS FOR UNIFORMITY OF CONCRETE

TEST	REQUIREMENT, EXPRESSED AS MAXIMUM PERMISSIBLE DIFFERENCE IN RESULTS OF TESTS OF SAMPLES TAKEN FROM TWO LOCATIONS IN THE CONCRETE BATCH
Weight per cubic foot calculated to an air-free basis, pounds per cubic foot	1.0
Air content, volume percent of concrete	1.0
Slump: If average slump is 4 inches or less If average slump is 4 to 6 inches	 1.0 1.5
Coarse aggregate content, portion by weight of each sample retained on No. 4 sieve, percent	6.0
Unit weight of air-free mortar based on average for all comparative samples tested, percent	1.6
Average compressive strength at 7 days for each sample, based on average strength of all comparative test specimens, percent	7.5

The maximum permissible difference in results is the difference between the test values for the samples taken at two locations in the batch. (Adapted from Table A1.1 of ASTM C 94.)

about 15 percent of the batch has discharged, and one after about 85 percent has discharged. Do not sample the very first or very last part of the batch, and be sure to sample the entire cross section of the stream of concrete being discharged. The following procedure is recommended in ASTM C 94:

> *12.5.1 Sampling for Uniformity of Concrete Produced in Truck Mixers—The concrete shall be discharged at the normal operating rate for the mixer being tested, with care being exercised not to obstruct or retard the discharge by an incompletely opened gate or seal. Separate samples, each consisting of approximately 2 ft^3 shall be taken after discharge of approximately 15% and 85% of the load. These samples shall be obtained within an elapsed time of not more than 15 min. The samples shall be secured in accordance with C 172, but shall be kept separate to represent specific points in the batch rather than combined to form a composite sample. Between samples, where necessary to maintain slump, the mixer shall be turned in mixing direction at agitating speed. During sampling the receptacle shall receive the full discharge of the chute. Sufficient personnel must be available to perform the required tests promptly. Segregation during sampling and handling must be avoided. Each sample shall be remixed the minimum amount to ensure uniformity before specimens are molded for a particular test.*

Each sample is then tested for slump, and (using the air meter) for unit weight and percent air. Following standard procedures, 6- by 12-inch cylinders for strength tests can be made at this time. Three specimens should be made for each test.

13.5 Compressive Strength

The compressive strength of concrete, discussed in Chapter 3, is one of the most easily determined properties of concrete. Properly made, a strength specimen provides much valuable information about the concrete. Improperly made, the test results can lead to misleading conclusions, which can have serious repercussions. The specimens shown in Figure 3-12 are obviously poorly made, and their test results would be valueless. However, seemingly sound specimens can have serious defects as well. Nonrepresentative sampling, careless handling and improper curing can cause serious deficiencies that would not be apparent on the specimen but still affect test results.

Strength Specimens. A noteworthy revision has been adopted in the 2008 edition of ACI 318 (ACI 318-08): Both the 6-inch by 12-inch and the 4-inch by 8-inch test cylinders are now permitted for evaluation and acceptance of concrete. (See ACI 318 Section 5.6.2.4.) The smaller 4-inch by 8-inch cylinders are common for acceptance of very-high strength concrete (greater than 10,000 psi) and are the common size for plants and production of precast and prestressed concrete products. The smaller size is now code recognized for all concrete strength testing. Note that the 4-inch by 8-inch cylinder is easier to cast, requires less sample, weighs considerably less than the 6-inch by 12-inch cylinder (easier to handle) and requires less moist-curing space. In addition, the smaller cross-sectional area allows higher compressive strengths to be tested by smaller load capacity testing machines.

According to Section 5.6.2.4. a single-strength test result is the average of the strengths of at least two 6 by 12-inch cylinders or at least three 4-inch by 8-inch cylinders made from the same sample of concrete. The commentary states "testing three 4-inch by 8-inch cylinders preserves the confidence level of the average strength because 4-inch by 8-inch cylinders tend to have approximately 20 percent higher within-test variability than 6-inch by 12-inch cylinders." Interestingly, the difference in indicated strength between the two cylinder sizes appears insignificant, as illustrated in Figure 13-14. (See also Figure 3-2.) In any case, maybe the day has arrived when most evaluation and acceptance of concrete will be done using the smaller, more convenient, 4-inch by 8-inch cylinders. A special note: if the 4-inch by 8-inch cylinders are being used, the diameter of the tamping rod should be $^3/_8$ inch, not the customary $^5/_8$ inch.

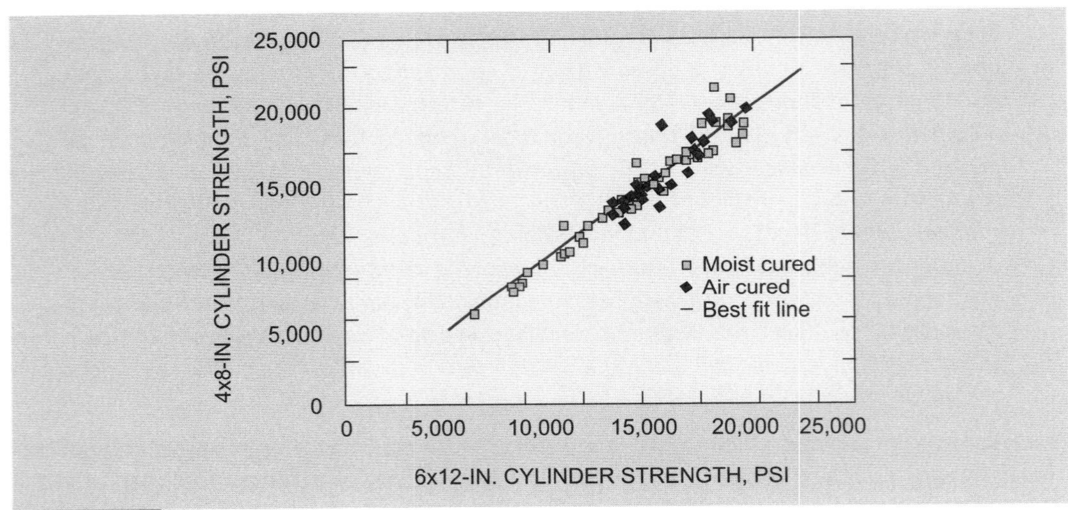

Figure 13-14
Comparison of 4-in. x 8-in. and 6-in. x 12-in. cylinder strengths. (Courtesy of PCA)

Frequency of Testing. Generally, the number of specimens is designated in the job specifications. Each mix should be sampled each day it is used, and every major structural unit of the building should be represented by at least one set of specimens. Specifications will require a set of cylinders for a designated maximum yardage each day, or a set of cylinders for a certain maximum slab or wall area cast per day. The number of specimens in a set depends on the type and size of the structure, volume of concrete and the use to be made of the results. A single specimen should never be depended on for reliable results.

The reader is referred to ACI 318 Section 5.6.2 regarding frequency of testing:

5.6.2. Frequency of testing

5.6.2.1. Minimum number of strength tests <u>per day</u>

- Once per day
- Once for each 150 cu yds placed
- Once for each 5000 sq ft of surface area placed

5.6.2.2. Minimum number of strength tests <u>per project</u>

- Five strength tests from five (5) randomly selected batches or from each batch if fewer than five batches

5.6.2.3. Total quantity less than 50 cu yds

- Strength tests may be waived by building official

Making Cylinders. In making test specimens, the concrete should be taken from batches spread throughout the period of concreting operations, not from just one or two batches. The inspector or technician must make the specimens. He or she may be assisted by a laborer in obtaining the concrete sample and transporting it to the point of casting (see

Figure 13-15), but it is the technician's responsibility to make the specimens and perform tests. All testing equipment should be clean, without hardened or dry concrete adhering to it, and free of oil or other dirt. A long-handled scrub brush should be a part of every set of equipment. As soon as a group of tests has been completed, the equipment should be scrubbed clean, using the brush and plenty of water.

Figure 13-15
Sampling fresh concrete (ASTM C172). (Courtesy of PCA)

The method for making specimens is described in ASTM C 31 for specimens made in the field, or ASTM C 192 for specimens made in the laboratory. The designated method must be carefully followed.

Molds for cylinders are of steel, cast iron, plastic, coated cardboard or tin cans conforming to ASTM C 470. See Figure 13-16. Cardboard molds are coated with paraffin or other waterproofing material.

Figure 13-16
Commonly used plastic cylinder mold. The molds must be covered to prevent loss of moisture.

Molds should be clean and watertight. Metal molds should be oiled very lightly.

Molds should be placed on a smooth, level, firm surface before filling. For 6-inch by 12-inch cylinders, the molds should be filled one-third full and rodded exactly 25 times with a $^5/_8$-inch-diameter tamping rod. The second and third layers are similarly placed. In rodding the first layer, the rod should not come in contact with the bottom of the mold. In rodding subsequent layers the rod should just penetrate into the layer below. See Figure 13-17. For the 4-inch by 8-inch cylinders, the molds should be filled in two layers of approximately equal depth and rodded with a $^3/_8$-inch-diameter tamping rod. See ASTM C 31 Tables 1 and 3.

The top layer should have a slight excess of concrete, which is struck off with a trowel after rodding has been completed. Excess finishing or working of the top should be avoided. Surface of the concrete should be perpendicular to the sides of the mold.

If the specimen is to be consolidated by vibration, both the 6-inch by 12-inch and 4-inch by 8-inch molds are consolidated in two layers, using a vibrating element approximately 1 inch in diameter. For the 6-inch by 12-inch cylinders, the vibrating element is inserted two times in each layer for three or four seconds per insertion. The duration of vibration is longer for stiff, unworkable mixes than it is for fluid or wet mixes. The vibrator should not touch the mold. When vibrating the top layer, the vibrator should penetrate the lower layer about 1 inch. Mortar should not run over the top of the mold when the vibrator is inserted in the top layer. After vibration has been completed, concrete should be added and worked into the specimen with a trowel, after which the top is struck off. For the 4-inch by 8-inch cylinders, the vibrating element is inserted once in each layer. See ASTM C 31 Table 4.

External vibration is sometimes used, especially for stiff concrete with a very low slump. Care must be taken to avoid over-vibration, as this will cause segregation of the concrete in the mold. Usual practice is to finish the surface of the specimen after vibration has been completed.

Figure 13-17
Consolidating the concrete in the molds by rodding. (Courtesy of PCA)

The specimens should be properly identified. Scratching a number on the side of the mold or in the fresh concrete on the top of the cylinder, if carefully done, is usually satisfactory. Whatever system is used, it must be waterproof, nonfading and permanent. Some felt tip pens will meet these requirements. Another method is to use a paper tag with a wire attached. (A cloth tag is apt to disintegrate in the presence of moisture.) Write on the tag, with indelible pencil or ball-point pen, the specimen number, job identification and date made; then attach the tag by embedding the wire in the fresh concrete at the edge of the specimen.

Tops of cylinders should be covered with glass plates, sheet plastic, wet burlap or similar material to prevent drying out. A plastic refrigerator bowl cover with an elastic band makes a good cover.

Specimens should immediately be placed where they are out of the way of construction activities and where they can be kept between 60°F and 80°F and in a moist environment preventing any loss of moisture for up to 48 hours. It is good practice to keep them in a field office where they can be kept warm in winter and cool in summer. Under no circumstances should specimens be moved before 12 hours, as they can easily be damaged.

The test cylinders are usually kept in a specially constructed storage box to provide the necessary moisture and temperature control. (See Figure 13-18.) If this box is out of doors during cold weather, it should be insulated. A source of heat, such as an electric light bulb or lighted lantern, should be provided in the box. A thermostatically controlled heater of the type used in chicken incubators has been found to be quite useful. Overheating and drying of the specimens must be avoided. The box should be kept in a shady location during warm weather. Evaporation of water from sand, burlap or other porous surfaces helps to keep the temperature down.

Figure 13-18
Placing cylinders in curing box. (Courtesy of PCA)

Portable concrete cylinder curing boxes are available to maintain a precise curing test environment, as illustrated in Figure 13-19. The unit shown contains a heat pump that both heats and cools automatically so that a constant temperature is maintained within the unit in any environment. An LED display constantly monitors temperature. Such manufactured units are lightweight and easily transportable to and from the job site.

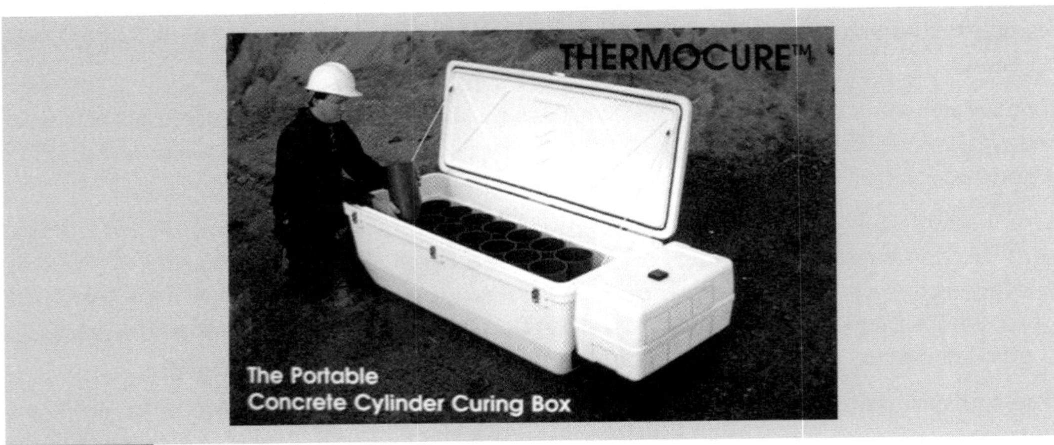

Figure 13-19
A portable concrete cylinder curing box.

After the initial curing period in the field, usually one day, the specimens are transported to the laboratory, where the molds are stripped and the specimens placed in 73°F moist curing. At this time, the identifying data should be marked on the specimen with graphite lumber crayon or with a felt tip pen using black waterproof ink. If it is necessary to transport them by car or truck they should be placed in boxes padded with sawdust, plastic foam or other material that will protect them from jostling and bumping. Specimens should never be hauled loose in the trunk of a car or in the back of a pickup truck. See Figure 13-20. They should not be permitted to dry out and must be protected from freezing. Specimens should never be shipped by public carrier unless they are well padded in stout boxes.

Figure 13-20
How not to transport specimens to the laboratory!

Jobsite Curing. Many times, specimens are stored at the jobsite for several days before they are shipped to the laboratory. This practice is harmless as long as the specimens are stored under conditions of moisture and temperatures that meet the requirements of ASTM C 31. Wet sand, burlap or water tanks are used.

The danger of jobsite storage is that the specimens might not be properly cured. If the cylinders are exposed to the wrong temperature, are frozen or are permitted to dry out, they will show serious deficiencies in strength. On one job, over 300 test cylinders were made during late summer to check 4000 psi concrete being delivered. Some of the cylinders were sent to the laboratory promptly after one day; others remained on the job for up to six days. Average strength of the cylinders job cured for one day was 4360 psi. Average strength declined steadily for extended job curing, and the six-day job-cured cylinders averaged only 3800 psi.

Specimens are sometimes cured under job conditions to determine when forms can be removed or a structure put into service. Field-cured specimens of this kind should be removed from the molds after 24 hours and stored as nearly as practicable under the same conditions as the portion of the structure they represent. Before they are to be tested they may be stored (dry) in the laboratory, but for no more than one week.

Note also that according to ACI 318 Section 5.6.4 strength test of cylinders cured under field conditions may be required by the building official to check the adequacy of curing and protection of concrete in the structure.

Testing in the Laboratory. In the laboratory, specimens are stored under moist conditions at 73.4 ± 3°F until it is time to cap them for breaking. Storage may be in a 100-percent relative humidity room or in tanks containing a saturated lime water solution. (See Figure 13-21.)

Figure 13-21
Moist storage room. Temperature and humidity are controlled within the tolerances specified by the standard specifications. (Courtesy of PCA)

Inasmuch as the ends of cylinders are never exactly smooth and at right angles to the axis, it is necessary that they be prepared before being placed in the testing machine. An exception is the bottom of a cylinder cast in a heavy steel or iron mold, the mold being fitted with a heavy machined plate firmly and tightly attached to the mold.

Some laboratories, especially research laboratories, grind or cap the ends, using special jigs to assure accuracy. The most common method is to cap the cylinder with a fluid or mastic material, which, upon hardening in contact with a machined metal plate, forms a smooth and strong bearing surface. Neat portland cement, quick-setting cement and plaster of paris have been used for this purpose. Present practice is to use a thermosetting compound that develops its strength in a matter of a half hour or so. Mixtures of powdered sulfur with an inert filler such as fly ash or fireclay are satisfactory, as are a number of proprietary compounds. Special high-strength compound is available for use in testing specimens with strength exceeding 6000 psi. A mixture of three parts sulfur to one part fireclay, by weight, heated to 350°F – 400°F is satisfactory. It should be kept in mind that it is not the purpose of capping to cover up poor quality in making cylinders.

The following precautions should be observed:

1. Cylinder ends that are extremely rough or crooked should be smoothed somewhat with a coarse horseshoe rasp or similar tool.
2. Cylinders should be at room temperature.
3. Capping material must be at the proper temperature—neither too cold nor too hot. The wrong temperature will cause spongy caps.
4. A capping jig is necessary to assure smooth and square caps. See Figures 13-22 and 13-23.
5. The caps should be as thin as possible, perpendicular to the cylinder axis, smooth and parallel with each other.
6. The cap should be permitted to cure for at least 30 minutes before the specimen is tested.
7. It is not necessary to oil the cylinders before capping, although there is no objection to a very thin oiling with a light oil.
8. There is no objection to salvaging a portion of the used caps, provided that most of the compound in the heating pot is new material, and provided that the material has not been overheated.

Figure 13-22
A 6-in. x 12-in. test cylinder being capped in the capping jig. The jig holds the cylinder in the vertical position until the capping compound hardens, usually less than 30 seconds. (Courtesy of PCA)

Figure 13-23
A capped cylinder ready for testing. (Courtesy of PCA)

9. After the cap is affixed, it should be tapped lightly with the handle of a screwdriver or putty knife to locate bubbles. If bubbles are found, the cap should be removed and a new one applied.

ASTM C 1231 describes the use of unbonded neoprene caps that are not adhered or bonded to the ends of the specimen. This method of capping uses a disk-shaped neoprene pad that is approximately the diameter of the specimen. The pad is placed in a cylindrical steel retainer. A cap is placed on one or both ends of the cylinder. See Figure 13-24. The specimen is then tested in the same manner as bonded capped specimens in accordance with ASTM C 39.

Figure 13-24
Unbonded caps are often the preferred capping method because they allow immediate testing at the same moisture condition at which the cylinders were cured. (Courtesy of PCA)

Testing of specimens is covered in ASTM C 39. The specimen should be centered on the platen of the testing machine and the bearing block carefully brought into contact with the specimen. The load is applied at a rate of about 2000 psi per minute. It is permissible to apply the load at a slightly faster rate up to about half of the expected breaking load. The ASTM method requires that the cylinder be tested in a damp condition. See Figures 13-25, 13-26 and 13-27.

Figure 13-25
A 6-inch by 12-inch specimen in position in the testing machine, ready to be tested in compression. (Courtesy of PCA)

Figure 13-26
The compressive load is applied at a rate of about 2000 psi per minute. (Courtesy of PCA)

Figure 13-27
Test specimen after testing. (Courtesy of PCA)

It is difficult to look at a cylinder that broke at an unusually low strength and try to second guess what was wrong with it. There are, however, certain indications that sometimes give a clue as to what was wrong. A cylinder that dried out too soon will usually have a variation of color on the broken surface when the specimen is tested. Segregation, with a concentration of large aggregate toward the bottom of the specimen (as cast) indicates possibly poor sampling, or the use of a square-end tamping rod. Voids in the interior may result from failure to tap the side of the mold when it is filled, especially when testing low-slump concrete.

Methods of Rapid Strength Measurement

Average strength curves can be used if tests are made at three or seven days. The curve shown in Figure 13-28 is based on the average of many tests. Obviously, actual conditions on any one job might be significantly different, and the curves can be used for rough approximations only. An equation, known as the Bureau of Standards formula, was advanced several years ago, giving similar results. The equation is

$$S_{28} = S_7 + 30\sqrt{S_7}$$

in which

S_{28} = 28-day compressive strength, and

S_7 = 7-day compressive strength.

The seven-day strength is not much better than the 28-day strength for control, so we are not much better off.

Many procedures have been suggested, most of them relying on heat to accelerate hydration. All of them have the drawback that they can give only an approximation of the 28-day strength.

ASTM C 684 has standardized four procedures of accelerated strength testing. In Procedures A, B and C, specimens are normally 6-inch by 12-inch cylinders made in the usual manner in metal molds with tight lids conforming to Specification C 470. Paper molds

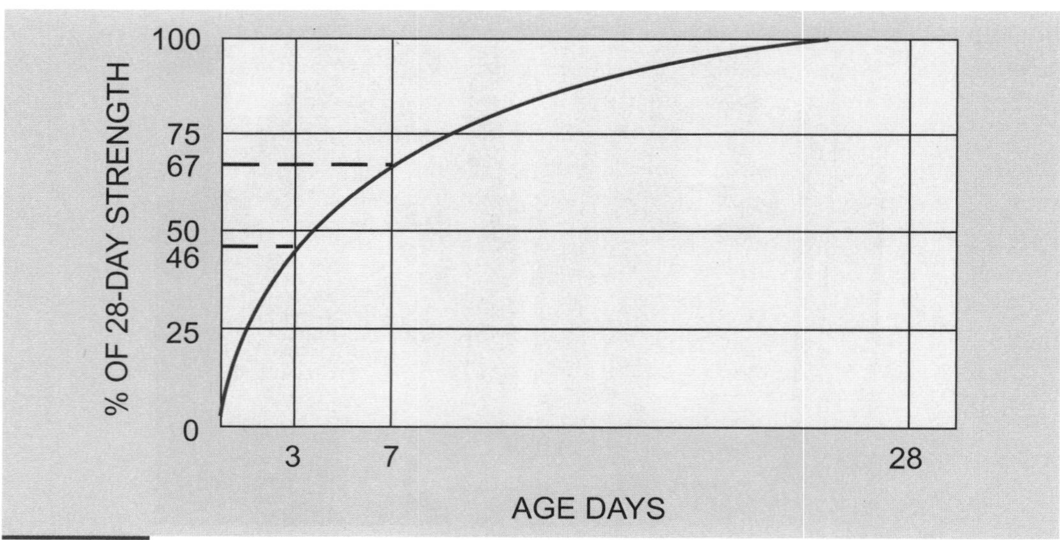

Figure 13-28
Normal concrete develops about 50 percent of its 28-day strength at 3 days and 70 percent at 7 days under standard curing at 73°F and 100 percent relative humidity.

are excluded. For Procedure A – Warm Water Method, the specimen immediately after casting is placed in water at 95°F for 24 hours and then tested. Results can be obtained 25 hours after casting. The procedure is relatively simple, but strength gain is not as high as in Procedure B – Boiling Water Method, in which the specimens are standard cured for 23 hours and then placed in boiling water for $3^1/_2$ hours.

In Procedure C – Autogenous Curing Method, specimens are put into insulated containers immediately after casting and held for 48 hours, at which time they are removed and tested.

Procedure D – High Pressure and Temperature Method, involves simultaneous application of elevated temperature and pressure to the concrete, using special containers. The specimens are cured at an accelerated temperature of 300°F with a 5-hour age of testing,

The accelerated curing procedures provide, at the earliest practical time, an indication of the potential strength of a specific concrete mixture. These procedures also provide information on the variability of the production process for use in quality control.

Maturity Methods. When inspectors or lab technicians measure and report or look at strength versus age data (7-day versus 28-day) for concrete cylinders, they are using maturity. Simply stated, maturity relates time (and temperature) to strength. Maturity methods for predicting strength gain of freshly placed concrete have become increasingly popular in recent years. ASTM C 1074 provides maturity functions to estimate the strength gain of in-place concrete. The basis for maturity methods is that the strength of concrete is directly related to curing time and temperature. The benefit of maturity methods is that in-place concrete strength can be predicted from simple measurements of concrete temperature over time. The maturity concept is based on the principle that each concrete mixture has a unique strength versus time-temperature relationship. Thus, a specific concrete mix will have the same strength at a given maturity (time and temperature) no matter what conditions (time or temperature) occur before measurement. Knowing the time and temperature of the concrete allows the strength of the concrete to be estimated. Before maturity of job-placed concrete can be determined, a maturity curve must be developed in the laboratory that correlates the strength gain for a specific concrete mix to both time and temperature. After a maturity curve is established for a particular concrete mixture, the concrete strength can be estimated from a measured time-temperature record of the in-place concrete. Maturity can be determined with commercially available maturity equipment consisting of temperature sensors (thermocouples) embedded in the concrete, and data recorders. (See Figures 13-29 and 13-30.) The temperature sensors must be embedded in the freshly placed concrete, e.g., at the exterior surface of the placement. The maturity method can be very effective when making decisions about form removal, post-tensioning, sawing joints, opening slab work to traffic and controlling heat curing in precast plants. To date, maturity methods are not recognized as an acceptable method for final evaluation of concrete in the ACI 318 standard.

Figure 13-29
Installation of temperature sensors in large reinforcement cages. (Courtesy of PCA)

Figure 13-30
Maturity meter with datalogger. (Courtesy of PCA)

All of the early-strength test methods must be standardized using the materials and mixes currently in use on the job. Type and brand of cement, aggregates and use of admixtures all have a significant effect. Any equation or relationship would have to be developed for each set of materials. None of these methods is acceptable as a method of control at this time, although there is great interest in the subject, and much research being done both in the laboratory and in the field.

How much better it would be if we could control the concrete at the time it is made. The methodology is available. The problem is in obtaining acceptance by the industry. Part of the procedure would be tests of the materials. Cement is always tested at the mill where it is made. A few relatively simple tests, such as fineness and setting time, can be made on samples at the job to check whether the proper cement was received. Aggregate tests, as previously described, would be part of the procedure. Adequate inspection of batching, mixing, placing and curing would be necessary, together with tests of concrete for unit weight and slump. Methods of analyzing the constituents of fresh concrete are available. Nuclear and atomic tracer methods of analysis are under investigation in many laboratories. No doubt the day will come when we will know the quality of the concrete when it is made, instead of waiting a month for the results of tests. Perhaps our preoccupation with 28-day strength has caused us to belittle or ignore many methods of test and control that are available to give us the answers when they are of the greatest value.

13.7. Other Strength Tests

Flexural Strength. On some work, especially pavements, flexural strength is more significant than compressive strength. (See Figure 3-1.) Flexural strength is determined on small beams that are loaded in the center (ASTM C 293) or loaded at the third points (ASTM C 78). Usual practice is to test beams 6 inches by 6 inches in cross section by loading at the third points. This is shown in Figure 3-4.

Beams are made in accordance with the procedure described in ASTM C 31 for field specimens or ASTM C 192 for laboratory specimens. Length of 6-inch beams is at least 20 inches to allow for a slight overhang at the supports.

For beams to be compacted by rodding, place concrete in two layers with each layer rodded 60 times (for a 20-inch length). Spade each layer along sides and ends with a trowel or similar tool. If the concrete is to be vibrated, place and vibrate it in one layer. After compacting the concrete, strike off the top of the specimen with a straightedge and finish with a wood float. Do not overwork the surface. Proper curing is very important.

Drying of the top surface during the early life of the specimen is particularly damaging.

Splitting Tensile Strength. The increased emphasis on control of cracking of concrete has resulted in more interest in the tensile strength of concrete. The test, ASTM C 496, is particularly designed for testing lightweight structural concrete specimens. See ACI 318 Section 5.1.4.

The specimen for this test is a standard test cylinder (4-inch by 8-inch or 6-inch by 12-inch), the same as the compressive strength specimen. The test is made by applying a compressive load on the side of the specimen, which causes the specimen to split lengthwise. (See Figure 13-31.) A formula is provided in ASTM C 496 for computing the splitting tensile strength in pounds per square inch. Normal precautions applied to compressive specimens should be observed. The test setup is shown in Figure 3-5.

Figure 13-31
Splitting tensile strength test (ASTM C 496). (Courtesy of PCA)

HARDENED CONCRETE

There are several ASTM standards for sampling hardened concrete and testing hardened concrete for strength, percent air, specific gravity and cement content. Because these tests require laboratory facilities and equipment, they are usually performed only in case of question regarding the quality of concrete already placed.

Another group of tests includes petrographic examination, elastic properties, bond and certain dynamic tests that can be classified as research testing.

To the person in the field, most of these tests are of limited interest because of their complexity and the elaborate laboratory equipment required. We are, however, concerned with sampling the hardened concrete and some of the in-place strength tests. This discussion is therefore confined to sampling and coring concrete, testing cores and the use of the Swiss hammer and Windsor probe for in-place compressive strength testing.

13.8. Sampling Hardened Concrete

The most common method of sampling hardened concrete is to extract cores from the structure using a diamond drill. (See Figure 13-32.) Drill bits are available in several diameters. The diameter of the core depends on the thickness of the concrete member being sampled, maximum size of aggregate in the concrete, and presence of reinforcing steel or other embedded material that might be damaged by coring.

The core drill should be well anchored and in good operating condition. Any looseness in the machine, loose bearings or similar conditions may cause the bit to wobble, which might damage the core.

ASTM C 42 covers the removal of cores and sawed beams from hardened concrete. Samples for strength tests should be representative of the concrete placed, as far as can be determined. Obviously, abnormal areas, joints, rock pockets, fill planes and edges should be avoided. If the location of batches or loads of concrete can be identified in the structure, the core should be taken from near the center of the batch. Diameter of the core must be

Figure 13-32
Coring machine (core drill) removing core from slab on-ground. (Courtesy of PCA)

at least twice the nominal size of the largest aggregate in the concrete; three times would be better. Ratio of length to diameter, after the core has been capped for compression testing, should be about 2. A core with a length less than 95 percent of its diameter should not be used for compressive strength testing.

Cores are sometimes taken for purposes other than strength tests. It may be desirable to check the depth of honeycomb or surface deterioration, to check the thickness of a slab (or location of slab reinforcement) or to examine a construction joint. In such cases, location of the cores depends on the surface appearance of the concrete and the best approach to the affected area.

Samples of concrete broken off a structure are sometimes sent to the laboratory for analysis or testing. Selection of the sample, of course, should be made in such a manner that the sample truly represents the exact portion of the concrete it is desired to investigate. A single piece, or several pieces, with a total volume of about 1 cubic foot should be obtained if at all possible for concrete with a 1-inch or $1\,^1/_4$-inch MSA.

Each sample should be marked with proper identifying information and placed in a separate bag or box. Samples that are liable to break or disintegrate should be well padded or otherwise supported to prevent further damage to the concrete. If the moisture content is important, the sample should be immediately placed in a waterproof plastic bag. Remember, however, that most coring and cutting machines use water for lubrication and cooling; therefore, the moisture content of samples obtained with these machines is not significant.

Regardless of what further testing is to be performed, it is a good idea to obtain the specific gravity and density of the concrete; however, samples for strength tests should not be boiled before testing for strength.

13.9. Strength Tests

Some specifications require that cores be saturated before compression testing, whereas others state that the core should be in a moisture condition similar to that of the structure, which usually means dry. ASTM C 42 recommends moisture conditioning by soaking the cores in lime water for at least 40 hours.

Quite frequently, the length of cores is less than twice the diameter, the desired value. If such is the case, a correction factor can be applied to the indicated strength to give a corrected strength; see Table 13.2. Although there appears to be some question about the advisability of making such a correction for dry cores, slightly more accurate strength values will usually be obtained if it is applied. The correction should always be made for cores tested in a moist condition. Actually, for L/D ratios of 1.5 and higher, the size of the correction is less than the normal experimental error of sampling and testing plus the normal variations in the uniformity of the material being tested.

TABLE 13.2
CONVERSION FACTORS TO BE APPLIED TO INDICATED STRENGTHS OF NONSTANDARD-LENGTH COMPRESSIVE-STRENGTH SPECIMENS

LENGTH/ DIAMETER	CORRECTION FACTOR
1.94 – 2.10	1.00
1.75	0.98
1.50	0.96
1.25	0.93
1.00	0.87

Example: A 4-inch-diameter core, 6 inches long, broke at 4000 psi.
L/D ratio is 6/4 = 1.5. Corrected strength is 4000 x 0.96 = 3840 psi.

Cores should be dressed by sawing or tooling the ends so that the ends are fairly smooth and perpendicular to the axis of the core. The reason for this dressing is to remove projections and to make the ends at right angles to the side of the core within the tolerances shown in ASTM C 42. After the preliminary dressing, the cores should be capped as described in the section on testing of cylinders, then again measured for overall length, including the caps, after which they can be tested in compression.

Computations:

1. The end of a core $= \dfrac{(\text{average diameter})^2}{4} \times 3.1416$
2. Compressive strength $= \dfrac{\text{total load applied to specimen}}{\text{end area}}$
3. Corrected strength = strength computed in Item 2 above multiplied by correction factor from Table 13.2 if length of core is not two times diameter.

EXAMPLE

1. Measurements of diameter were 3.1, 3.0 and 3.0 inches

$$\text{Average diameter} = \frac{3.1 + 3.0 + 3.0}{3} = 3.03\ \text{inches}$$

2. $\text{End area} = \frac{(3.03)^2}{4} \times 3.1416 = 7.23\ \text{square inches}$
3. Measured length after capping, 5.0 inches
4. $\text{L/D ratio} = \frac{5.0}{3.03} = 1.65$. From Table 13.2

   Correction factor = 0.98
5. Total load applied by testing machine, 39,600 pounds
6. $\text{Compressive strength} = \frac{39{,}600}{7.23} = 5480\ \text{psi}$
7. Corrected strength = 5480 × 0.98 = 5370 psi

It is difficult to interpret the relationship between the strength of standard molded cylinders and cores drilled from the structure. In general, most investigators report that the strength of cores is lower than the strength of standard cylinders tested at identical ages. There are, however, many exceptions to this, hence the difficulty in interpreting results of core tests and applying the results to the structure being tested. Besides the difficulty in interpreting results, other objections are the cost of core drilling and the resulting scarring or damage to the concrete. In the final analysis, in the hands of an expert, cores can be a valuable means of assisting in the evaluation of concrete.

In-Place Strength Tests. Various devices have been proposed to be inserted in the concrete of the structure and subsequently pulled out of the hardened concrete for determining the in-place strength of the concrete. In-place (usually called in situ) tests of concrete can be accomplished by either of two methods. One method makes use of an insert embedded in the fresh concrete to be pulled out of the hardened concrete at a later date (ASTM C 900). The other method makes use of a device that is inserted into a hole drilled into the hardened concrete, to be subsequently pulled out as well. In both methods the pullout force is a measure of the strength of the concrete.

Currently, ASTM C 873 provides a test method for determining the compressive strength of concrete cylinders molded in place using a cylinder mold assembly consisting of a mold

Figure 13-33
Concrete cylinders cast in place in cylinder molds provide a means for determining the in-place compressive strength of concrete. (Courtesy of PCA)

and a tubular support member fastened within the concrete formwork prior to placement of the concrete. See Figure 13-33. The cylinder mold is filled in the normal course of concrete placement. The specimen is then cured in place in the same manner as the rest of the concrete. The specimen in the "cured-in-place" condition is removed from its in-place location immediately prior to de-molding, capping, and testing. The "cast-in-place" cylinder strength relates to the strength of concrete in the structure on account of the similarity of curing conditions, insofar as the cylinder is cured within the slab. The cast-in-place cylinders can be used in concrete that is 5 to 12 inches in depth.

13.10. Nondestructive Strength Tests

The use of impact and similar instruments for in-place strength testing of concrete was discussed in Section 3.9. Shown in Figure 13-34, the Swiss hammer (ASTM C 805) is an instrument of this type. Easy to use, this instrument provides a rapid means of obtaining an

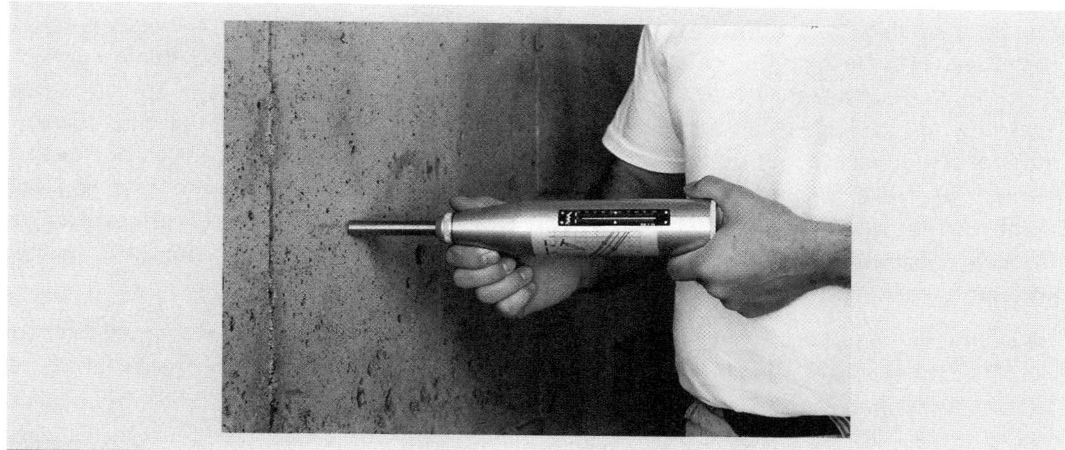

Figure 13-34
The Swiss hammer being used for determining the compressive strength of hardened concrete. (Courtesy of PCA)

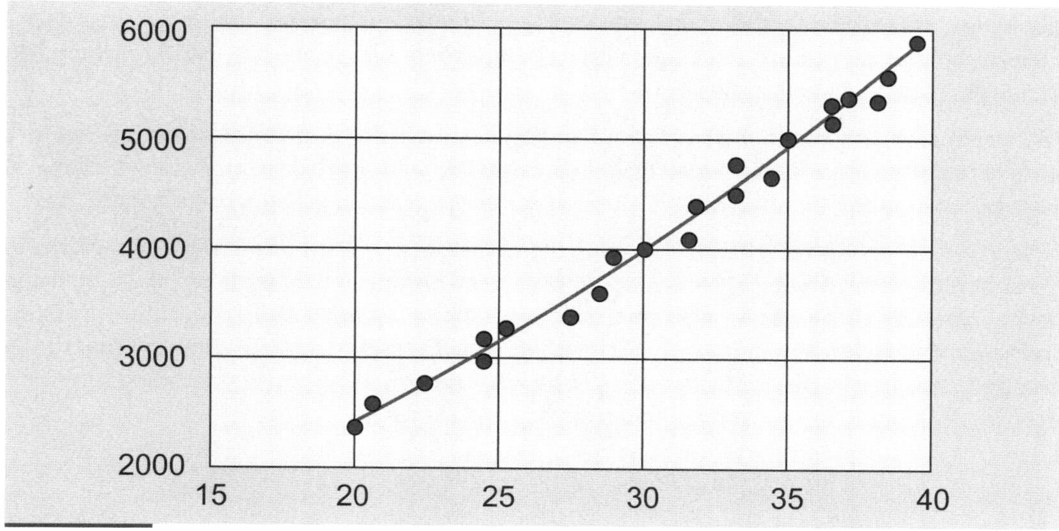

Figure 13-35
A typical calibration curve for a Swiss hammer.

approximation of the strength of concrete in the structure, its accuracy being about plus or minus 10 percent to 20 percent, depending on how well it is calibrated.

The instrument should be calibrated on concrete composed of the same materials as the concrete to be tested. To calibrate, place a capped 6-inch by 12-inch cylinder in a testing machine and apply a compressive load of at least 300 psi or 10 percent of the expected strength. With the cylinder under this load, take about 15 readings on the middle two thirds of the cylinder with the test hammer, then test the specimen in compression in accordance with standard procedure. Normal practice is to disregard the highest and lowest values and base the average on 10 or 12 individual readings. After a number of cylinders of different strengths have been tested in this manner, an average curve can be drawn showing compressive strengths against instrument readings. See Figure 13-35. There is a calibration chart accompanying the instrument, which can be used for rough approximations, but more accurate results will be achieved if the instrument is properly calibrated for the specific concrete being tested.

Accuracy of the test hammer is sufficient to detect wide variations of strength within a structure, but the hammer cannot be used as a substitute for properly made strength specimens. Accuracy is affected by the surface and moisture conditions of the concrete, presence of large aggregate particles or voids at point of impact, age of the concrete and mass of the concrete unit under test. Concrete sections less than 4 inches thick must be backed up with support of some kind. Accompanying the instrument is a small carborundum stone that should be used to remove the surface skin on smooth surfaces, to smooth rough surfaces, remove carbonation or reveal surface voids that will give an erroneous reading.

Usual practice is to take 15 readings (it requires only a minute or two) and average the best 10 of the 15 readings. These 10 readings should show a deviation of no more than about plus or minus $3^1/_2$ on the impact scale.

The direction of impact affects the readings. The instrument can be used horizontally, vertically up, vertically down or inclined. The calibration chart supplied with the instrument has separate curves for the different directions of impact. A job-calibrated instrument can be used for impacting in directions other than the direction in which it was calibrated by applying a correction to the observed readings of the same amount shown on the factory calibration chart. Rebound readings with the instrument horizontal are usually higher than readings taken in a vertical position.

Concrete less than seven days old is apt to register low rebound values, whereas old, dry concrete (older than 90 days) will have higher values than normal. Also, readings taken on wet concrete will be as much as 20 percent lower than readings taken on dry concrete.

The Windsor probe (ASTM C 803) is a hardness tester similar to the Swiss hammer and gives similar results. See Figure 13-36. It measures hardness to a greater depth than the Swiss hammer, but making the test is a little more complicated, and small holes or indentations are left in the surface of the concrete. Testing is accomplished by measuring the penetration of a powder-actuated probe into the concrete, the action being similar to that of powder-driven fastening tools (Ramset and similar).

Figure 13-36
The Windsor-probe technique for determining the compressive strength of hardened concrete. (Courtesy of PCA)

13.11. Other Tests

Laboratory procedures previously mentioned for determination of the cement content of hardened concrete, density, voids, air content and absorption are sometimes resorted to, especially when there is any question regarding the quality of the concrete in the structure. Properly interpreted when the properties of the aggregates, cement and admixtures are known, and the mix design is available, the results of these tests enable the engineer to deduce considerable further information regarding the concrete.

Tests of this nature are frequently performed on broken pieces of concrete, either fragments broken out of the structure, or pieces of cylinders or cores that had been tested in compression. Sampling for these special tests should be done as previously outlined in this chapter.

13.12. Future Tests

Some of the ASTM standard field tests in long use to control the quality of concrete have a simple defect. By the time the tests indicate that there is a problem, the hardened concrete is already in place, and the most common corrective action is no corrective action: the concrete is used as is. New and updated testing methods and devices are continuously being developed that will measure fresh concrete properties quickly enough to allow quality control personnel to rapidly detect changes in the concrete. Such rapid detection ability would permit timely mix adjustments that can reduce the amount of poor-quality in-place concrete. Test methods and procedures of the future share a common desired feature: they must provide results faster than conventional tests that measure similar properties. Shorter testing time is an essential feature to allow technicians to test concrete more frequently, enabling them to spot mix changes before they become a problem. The flowmeter discussed in Section 13.4 is an example of a faster field test to measure consistency and workability of fresh concrete.

Batching and Mixing the Concrete

14.1. Handling Materials
- Aggregates
- Cement
- Water
- Admixtures
- Superplasticizers
- Pozzolans

14.2. Batching Controls and Systems

14.3. Auxiliary Plant Equipment
- Recorder
- Moisture Meter
- Consistency Meter

14.4. Batch Proportioning
- Accuracy of Measurement
- Calibrating and Checking Equipment
- Batching

14.5. Mixers
- Concrete Plant Mixer
- Nontilting Mixer
- Tilting Mixer
- Vertical Shaft Mixer
- Horizontal Shaft Mixer
- Operation and Control

READY-MIXED CONCRETE

14.6. History

147. Equipment
- Mobile Batcher Mixed Concrete (Continuous Mixing)

14.8. Operation and Control
- Mixing Speed and Time
- Control of Water
- Load Tickets
- Additional Information
- Prolonged Mixing
- Delayed Mixing
- Disposal of Waste
- Recycled Water in Concrete

14.9. Responsibilities
- Producer Responsibilities
- Contractor Responsibilities
- Joint Responsibilities

Concrete is a perishable material. For this reason it must be used within a few hours after the materials are first joined together in the proportioning or batching plant. Batching and mixing equipment ranges in size and complexity from simple wheelbarrow scales supplying a small portable mixer to complicated automatic stationary plants with a capacity of several hundred cubic yards of concrete per hour. With the greatly increased usage of ready-mixed concrete, the small portable layouts have practically disappeared from the construction scene. Nearly every town and city has at least one ready-mixed concrete plant, some of which, however, offer virtually nothing in the way of control.

Control is not a function of size. Accurate control of concrete quality is possible with wheelbarrow scales and a three-bag mixer under careful operation and qualified inspection. On the other hand, the most modern automatic equipment is of no avail in the hands of sloppy operators and careless inspectors.

14.1. Handling Materials

Aggregates. When batched into concrete, the aggregate should be at a reasonably uniform moisture content and grading, and free of contaminating materials. It must meet the requirements of ASTM C 33. An aggregate failing to meet Standard C 33 can be used if tests or actual service show that satisfactory concrete can be produced, as authorized by the building official.

Specific requirements of ASTM C 33 cover gradation, physical properties and limitations on deleterious substances and are included in Chapter 8. The user should refer to the applicable standard for definite requirements.

Fine aggregate (the material nominally all passing the No. 4 sieve) is required to be natural sand, manufactured sand or a combination of these two. See Figure 14-1. Manufactured sand is the product resulting from crushing and grinding rock and stone, and is normally somewhat angular and harsh.

Figure 14-1
Fine aggregate (sand). (Courtesy of PCA)

Coarse aggregate should consist of crushed stone, gravel, air-cooled blast furnace slag or a combination of these materials.

Figure 14-2
Coarse aggregate (crushed stone). (Courtesy of PCA)

Most coarse aggregate from gravel sources contains some crushed material resulting from crushing of oversize rocks and boulders in the deposit.

Of special importance are the precautions that must be taken in those geographical areas where reactive aggregates are found. This problem is discussed in Section 4.4.

Unless otherwise approved by the building official, the maximum size of coarse aggregate will be determined by clearances in the forms, shown in Figure 12-6. These clearances are one-fifth of the narrowest dimension between sides of forms, one-third of the depth of slabs, or three-fourths of the minimum clear spacing between reinforcing bars, bundles of bars, prestressing strands or post-tensioning ducts. Tests have demonstrated that the space between reinforcement and side of forms is not significant if the MSA conforms to these limitations.

Ordinarily, aggregates conform to the grading requirements of ASTM C 33 without too much difficulty. In some cases, the producer might install finish screens at the batch plant, which will lessen the need for exact compliance with the grading requirements when the aggregates are delivered. The reason for finish screening is to remove excessive undersize material from the aggregates. Coarse aggregates, in addition to being washed and crushed, are screened into several sizes so that the concrete mixes can be proportioned properly. However, by the time these separated sizes reach the batching plant they are apt to contain a large amount of undersize caused by numerous handling operations, or they may be seriously segregated. Finish screening of coarse aggregates at the batching plant largely eliminates the accumulations of undersize and reduces segregation. See Figure 14-3.

When aggregates are delivered to the plant by truck, there are potential sources of trouble. In some plants the material is dumped at the base of a stockpile and cast into the pile with a clamshell. Trucks are apt to carry mud and clay into the stockpile area, or the crane operator may get careless and pick up earth with the aggregate, thus contaminating the pile.

In case the aggregates are placed in a stockpile on the ground, a considerable amount of material at the bottom of the stockpile will, in all probability, be rendered unfit for use because of its becoming mixed with undesirable foreign matter. If aggregates are being taken from the bottom of the pile, unusual care must be exercised to avoid the inclusion of objectionable material. See Figures 14-4 and 14-5.

Sometimes a truck driver or crane operator will place the wrong material in a pile or bin. The only recourse is to remove the offending material, which may require emptying the bin. Such drastic measures will be avoided if th batching plant is equipped with a finish screen that removes the undesirable material.

Figure 14-3
Unwanted fine material in the coarse aggregate results from segregation, abrasion and contamination in handling.

Figure 14-4
Properly cleaned up stockpiles present a neat appearance.

When aggregates of different types or sizes are placed in adjoining compartments of the same storage bin, the partition between the two should be built to a height sufficient to prevent material from flowing from one compartment into the other. Partitions should be tight and free from holes through which fine materials might leak from one bin to another.

Figure 14-5
Waste and contamination result from careless operations of the clamshell.

Aggregates, when batched, usually have varying moisture content, depending on time-length of storage, weather conditions and other factors. Unless correction is made for these variations, water content of the concrete will vary, with consequent fluctuations in slump and strength. Moisture changes in the aggregate frequently result from the practice of charging the plant feed alternately from wet and dry portions of a stockpile, or alternately from a relatively dry pile and cars or barges containing wet sand. Operators of end loaders and cranes should be instructed to avoid charging any one material into the plant from more than one source.

Aggregates should not be used direct from the processing plant, but should be permitted to drain for at least 24 hours. Shipment by rail can be assumed to allow sufficient time for drainage. If it is necessary to moisten coarse aggregate, the operation should be completed several hours before the material is required. Moistening can be accomplished by setting sprinklers on the stockpiles. Attempting to moisten aggregates in plant bins can lead to problems with control of the excess water and slump of the concrete.

Stockpiling of aggregates on the ground at the batching plant should be avoided, if at all possible. If, however, it is necessary, then the recommendations in Section 8.6 should be observed.

Cement. Handling and storage of portland cement is discussed in Sections 7.8 and 7.9. Many batch plants have facilities for more than one type of cement, in which case each type should be stored in a compartment or silo separate from other types. Each bin from which cement is batched should have separate facilities for handling the cement to the weigh batcher, including gate and conveyor.

Cement shipments should be checked before unloading, especially if the plant is using more than one type or brand, and care exercised to ensure that cement is placed in the correct silo or bin. Dust abatement equipment must be operating properly.

Except for rare instances when small amounts of a specialty cement is required, bagged cement is seldom used.

Differences in type or brand of cement can usually be noted by comparing the color of small samples of the cements.

Water. Storage and handling of water practically never presents any problems, the water normally being piped into the plant from a municipal distribution system of some kind. In those plants in which water is stored in tanks, normal precautions to prevent contamination should be observed. Fluctuations in line pressure may affect the accuracy of certain measuring devices, making necessary the installation of a pressure regulator.

When ice is used for cooling during hot weather, provision must be made to include the ice as part of the mixing water. Ice should be in the form of fine chips or slush ice. Measurement of water in a weigh batcher or volumetric tank usually permits measurement and introduction of the ice with the water. If water is measured through a volumetric meter or in a closed tank, other means must be found to measure the ice. This may take the form of an auxiliary tank or scale.

Admixtures. When received from the manufacturer, an admixture may be in one of three conditions: a ready-to-use solution, a concentrated liquid that must be diluted with water before use, or dry powder or granules that must be dissolved in water before use. Long-time storage of some liquid admixtures may result in segregation or stratification of the material in the barrel or tank. The admixture is not harmed by this separation and can be completely restored by agitating. Small pneumatic or electric mixers that can be inserted in a drum or tank do a good job of mixing. Admixture solutions that have been frozen likewise will stratify but can be similarly restored.

Dry admixtures should be protected from contamination and moisture during storage. Calcium chloride, easily absorbing moisture from apparently dry air, becomes gummy and lumpy and should not be used in this condition.

The manufacturer's instructions regarding storage, mixing and use of any admixture should be observed.

Superplasticizers. Because of the rapid slump loss when using some of these admixtures, it is necessary to introduce them into the mixer immediately before discharge of the concrete into the receiving equipment. This is feasible in most precasting plants where the mixer is located adjacent to the point of usage. In the case of the jobsite mixer discharging directly into the bucket, conveyor or other direct-handling equipment, there is no problem. When using ready-mixed concrete, however, this means adding the admixture after the truck arrives at the construction site, unless an extended-life admixture is used, which permits addition at the batch plant. One solution to short-life superplasticizers is to attach a small tank to the ready-mix truck. The correct dosage of admixture, having been admitted into the tank when the truck was at the batch plant, can now be discharged into the mixer after arrival at the construction site. A minimum number of revolutions of the drum will mix the admixture into the concrete.

Pozzolans. Materials in this classification (fly ash, for example) should be handled in the same manner as cement. Because of their extreme fineness, pozzolans require positive shutoff devices to prevent the flow of pozzolan after the shutoff point.

Pozzolan in bags is used in many plants because the infrequent requirement for the material makes bulk handling facilities uneconomical. Concrete batches in this case should be of such size that the required amount of pozzolan in the batch is a multiple of full sacks. Partial bags should not be used unless the pozzolan is weighed.

14.2. Batching Controls and Systems

Control systems range from manually controlled individual batchers that depend entirely on the operator's visual observation of a scale or volumetric indicator, to fully automatic systems that are actuated by a single starting signal and stop automatically when the designated weight of each material has been reached. Some of the equipment is shown in Figures 14-6 and 14-7. Also available are computer-actuated systems of high accuracy with digital readout for immediate information of the batcher operator and printout for permanent record.

Figure 14-6
Control room for batching equipment in a typical ready mixed concrete plant. (Courtesy of PCA)

Figure 14-7
Control panel for modern computer-based automation system for ready mix concrete production. (Courtesy of PCA)

The Concrete Plant Manufacturers Bureau defines batching controls and systems as follows: [14.1]

Batching controls. The part of the batching equipment that provides the means for controlling the batching device for an individual material. It may be mechanical, hydraulic, pneumatic, electrical, etc., or a combination of these means. A hatching system is a combination of batching controls necessary to proportion the ingredients for concrete. Batching controls or systems are so located with respect to the batching equipment being controlled that visual monitoring for accuracy, calibration of controls and manual batching can be accomplished.

Manual Controls. Manual controls exist when the batching devices are actuated manually, with the accuracy of the batching operation being dependent on the operator's visual observation of a scale or volumetric indicator. The batching devices may be actuated by hand or by pneumatic, hydraulic or electrical power assists.

Semi-Automatic Batcher Controls. When actuated by one or more starting mechanisms, a semi-automatic batcher control starts the weighing operation of each material and stops automatically when the designated weight of each material has been reached.

Semi-Automatic Interlocked Batcher Controls. When actuated by one or more starting mechanisms, a semi-automatic batcher control starts the weighing operation of each material and stops automatically when the designated weight of each material has been reached, interlocked in such a manner that the discharge device cannot be actuated until the indicated material is within the batching tolerance.

Automatic Batcher Controls. When actuated by a single starting signal, an automatic batcher control starts the weighing operation of each material and stops automatically when the designated weight of each material has been reached, interlocked in such a manner that:

(1) The charging device cannot be actuated until the scale has returned to zero balance;

(2) The charging device cannot be actuated if the discharge device is open;

(3) The discharge device cannot be actuated if the charging device is open; and

(4) The discharge device cannot be actuated until the indicated material is within the batching tolerances.

For cumulative batchers with tare compensated controls that treat the start of the weighing of each ingredient as zero, interlocked sequential controls are provided, and the batching tolerances apply to the required weight of each individual material being batched. For cumulative batchers without tare compensated controls, interlocked sequential controls are provided, and the batching tolerances apply to the required cumulative weight of material as batched.

Automatic Volumetric Controls for water or admixtures, when actuated by a single starting signal, start the batching operation and stop automatically when the designated volume has been reached. The batching control includes visual means of observing either the quantity set or the quantity batched, and an indication of the completion of the batching operation.

A Manual Batching System consists of the required combination of individual manual batcher controls.

A Partially Automatic Batching System consists of the required combination of batching controls, at least one of which is for controlling the cement and aggregates, either semi-automatically or automatically.

A Semi-Automatic Batching System consists of the required combination of semi-automatic interlocked batching controls or of semi-automatic interlocked and automatic batching controls.

An Automatic Batching System consists of the required combination of automatic batching controls. All batching equipment in the system for batching ingredients by weight are activated by a single starting mechanism. A separate starting mechanism may be provided for volumetric batching of water and/or admixtures not batched at the time of weighing. Each automatic batcher is designed to return to zero tolerance, and each volumetric device resets to start or signals empty before being recharged. The discharge of any ingredient from the system does not start until all batching controls have been cleared of the previous batch, the scales returned to zero tolerance and volumetric devices reset to start or signal empty. The discharge into the mixer of any scale does not start until all weighed ingredients have been batched in that scale. Provisions are made for adjusting the sequential discharge of the batchers or measuring devices and the rate of discharge of materials.

14.3. Auxiliary Plant Equipment

Most modern plants are equipped with recorders that make a printed record of each batch, a moisture meter that measures and indicates the amount of free moisture in the sand in the batcher, and an indicator that gives a measurement of the consistency of the concrete in the mixer (in stationary plants).

Recorder. The function of a recorder is to make a permanent record of plant operations. The code does not require the use of a recorder, but many specifying agencies require them, and many ready-mix plants use them as a part of their normal control and record system.

The Concrete Plant Manufacturers Bureau defines batching recorders as follows: [14.1]

Batching recorders produce a record of the batch weights or volume of each material being batched, a batch identification or a batch count, day, month, year, time of day and a register of empty balance. A batching recorder may be either graphic or digital.

A Graphic Recorder is an instrument that scribes a line on a graphic chart simultaneously with the indication of the scale as the materials are being weighed. Each scale may have its own recorder, or a series of scales may simultaneously record on a single graphic chart.

Figure 14-8
An electronic moisture meter. When properly calibrated, it can be an invaluable aid in controlling slump and water-cement ratio.

A Digital Recorder is an instrument that prints the weight or volume of a material or materials. The recording of each material is done after each material is batched, or after the total materials for a mix have been batched. Each measuring device may have its own recorder, or a series of measuring devices may record on the same tape or ticket.

A Digital Batch Documentation Recorder records the required information for each material in a total batch, identifying each material used along with a mix formula identification, size of the total batch or load, and an identification of the production facility. Where certain required information is unchanged from batch to batch, it may be preprinted, stamped or written on the record. The load may be identified by a batch count number, a ticket serial number or both.

A Digital Concrete Certification Recorder produces at least two tickets of a batch or load, which, in addition to the required information, includes the percent of sand moisture compensation, identification of the purchaser; identification of the job or project, and/or the particular placement location of the concrete. Space for identification of the delivery vehicle (truck number), driver's signature, purchaser's signature (representative receiving the concrete) and the amount of water added at the jobsite is also provided.

Moisture Meter. Many permanent and semipermanent plants are equipped with moisture meters. See Figure 14-8. An instrument of this kind consists of two electrodes (sometimes the steel batcher side is one electrode) in the sand weigh-batcher connected electrically to an indicating device at the operating console by which the operator and inspector are informed of the percentage of moisture in the sand being batched so that proper allowances can be made. One type of meter is connected to a recorder so that a permanent record of moisture variations is made. This recorder also serves to count the number of sand batches weighed out. More complex installations include compensating devices by means of which the amount of free water in the sand is automatically offset on the water batcher so that the correct amount of total water is introduced into the mixer, while at the same time the sand weight is corrected to maintain the correct dry weight of sand.

Electrodes should be maintained in a clean condition and all parts occasionally. The instrument can be calibrated as follows:

1. Check the zero reading with no sand in the batcher.
2. Load the batcher in the usual manner, and note the meter reading.
3. Take a sample of sand (about one pound), making sure that it is identical with that surrounding the electrodes.
4. Determine the moisture content of the sand sample by drying.
5. Adjust the meter to correct for any discrepancy between meter reading and moisture content corrected for absorbed moisture.

Actual adjustments to the scales and meters should be made by the plant personnel. The inspector should refrain from operating or adjusting any of the plant equipment.

Consistency Meter. Instruments for measuring the consistency of the concrete in the mixer (stationary mixer or truck mixer) are also available. See Figure 14-9. Sometimes called slump meters, one is a wattmeter that measures the power required to turn the mixer (a dry batch requiring more power than a wet one). Another measures the overturning moment acting on a tilting mixer (a dry batch having a tendency to concentrate in one end of the mixer, rather than level out as a wetter batch would do). Others have been experimentally used, one being so designed that it would automatically introduce water into the mixer if the slump were too low, a procedure that could lead to excessive, uncontrolled water in the batch of concrete. Any of these instruments requires calibration for different mixes and different sizes of batches.

Figure 14-9
Truck mixer slump meter.

14.4. Batch Proportioning

A wide variety of batching and weighing equipment is available, which may be assembled into many combinations subject to one basic requirement: the cement must be weighed separately from the other materials. Some old manual plants weigh the cement first on cumulative scales on which the aggregates are weighed; however, this is not a desirable method and no longer complies with specifications. A satisfactory arrangement is to suspend the cement weigh-batcher in the center of, but independent from, the aggregate weigh-batcher or hopper. Cement is then weighed first on its own independent scale, followed by the aggregates on their scales. Such an arrangement lends itself well to the ribbon feed method of charging the mixer.

Batching of aggregates may be cumulative, weighing each aggregate in turn on one weigh-batcher; or individual, weighing each aggregate separately in its own weighbatcher. Scales may either be beam or dial type for either weighing method.

If the water is weighed, separate scales should be provided for it. Water in most plants is measured volumetrically, usually by means of a meter or, occasionally, in a calibrated tank. Water batchers on pavers and portable mixers are of the volumetric overflow or siphon type.

Admixtures that are handled in solution can be measured either by weight or volume. See Figure 14-10. One type of dispenser for liquid admixtures controls the amount measured by means of a timer that regulates the period of time during which a small pump forces the material through a calibrated orifice. Others depend upon a variable-stroke piston, overflow or siphon tanks, and other devices. Dry powders should be weighed. The code permits fly ash and other pozzolans to be batched cumulatively with the cement, provided the cement is weighed first.

Figure 14-10
Liquid admixture dispenser provides accurate volumetric measurement of admixtures. (Courtesy of PCA)

Accuracy of Measurement. Tolerances in measurement of materials are given in Section 9 of ASTM C 94 for ready-mixed concrete. These tolerances apply to batching for all types of plants unless modified by the job specifications. Under special circumstances the purchaser may permit measurement of cement in whole standard bags. Batch weights of aggregates should be based on dry weights corrected for moisture content. The tolerances for water apply to the amount of water measured, not to the total water in the batch of concrete.

Variations in water pressure may result in variable amounts of water batched in some types of volumetric measuring devices, as the batcher may not fill completely when the pressure is low. If such is the case, a storage tank at the plant or mixer, feeding into the batcher, serves to equalize the pressure, in this way yielding uniform batches of water.

A source of slump variations is sometimes a careless or inexperienced operator who fails to operate the weighing and batching equipment properly. Variations in batch weights of the ingredients result, and these variations are reflected in variable yield, consistency and strength of the concrete. Another source of trouble is equipment that is worn out or out of adjustment. A leaky valve on the water batcher can result in wide fluctuations in the amount of water batched, especially if there are variable periods of time between batches.

The sequence of charging materials into the mixer affects the efficiency of mixing, and therefore affects the quality and uniformity of the concrete being produced. A batching sequence should be established for every plant that produces concrete of the best and most uniform quality. Water should precede, accompany and follow the solid ingredients into the mixer. The so- called "ribbon feed," by which the cement and all sizes of aggregate are fed into the mixer simultaneously, promotes thorough blending of all materials in the mixer and is thus conducive to efficient mixing.

The time and method of charging admixtures has a significant effect on the efficiency of any admixture. The admixture should be charged at the same point in the mixer charging sequence for every batch. Liquids can be metered into the mixing water as it flows into the mixer, and dry admixtures can be ribboned with the aggregate. When more than one admixture is being used, they must be kept separate until they are in the mixer.

Calibrating and Checking Equipment. Scales and batchers should occasionally be checked for accuracy. In most jurisdictions, the building official requires that the sealer of weights and measures inspect and calibrate the scales at stated intervals. When a calibration is made, it should include the entire weighing system, using dead weights, and not merely a calibration of the scale head or beam. Readings on scale dial or beam, operator's control dial if in a different location, and recorder should be taken at zero balance and at several points to the capacity of the scale.

Scales are inherently rugged and accurate devices. With proper maintenance, they will consistently give good service. Scales and batching equipment should be kept clean and in adjustment at all times. Binding of dull or dirty knife edges and fulcrums causes serious weighing errors. Tare weights may vary if materials hang up or stick in the weigh-batcher. Dial scales should be checked daily to ascertain that the hand returns to zero when the batcher is empty. Beam scales should balance at no load when only the tare beam is "in."

The method and equipment needed to check scales will vary somewhat depending on the size and complexity of the weighing and recording system. In general, the following procedure is suggested, subject to modification depending on the kind of plant. Once a plant has been checked and adjusted, it should accurately produce many cubic yards of concrete as long as it is maintained in good condition. Occasional short-cut checks can then be made to assure accuracy.

If the plant is equipped with an autographic recorder that makes a record of the batches, the recorder should be checked at the same time as the scales. The procedure is as follows:

1. See that batching hopper is empty and clean.
2. See that hoppers, fulcrums, knife edges and all moving parts are free—no binding, no rubbing, no friction. Knife-edge bearings must be centered, and knife edges must be sharp.
3. Balance scales on zero. Be sure recorder, if any, indicates zero.
4. Place test weights on the hangers, recording the scale reading and recorder reading at each 500-pound increment. Scales and recorders both must be calibrated together with all equipment in operating condition. Do not disconnect the recorder. Allowance should be made for dunnage, such as planks and hangers to support the weights.
5. Usually the number of test weights will not total the maximum weight desired. In this case, remove all weights and dunnage, balancing the scales and recorder on zero; fill the batcher until the scales indicate the maximum reached under item 4 above; then attach hangers for weights and proceed as under item 4.
6. Workers should be cautioned to place and remove weights carefully so as to cause as little extraneous movement of the scales and recorder as possible.

Water-measuring devices should be calibrated carefully and kept in good operating condition at all times. Before attempting to calibrate a water batcher, the valves and other mechanisms should be checked and any necessary repairs made. The water system on a modern mixer or plant is so designed that measured water can be drawn off through a special outlet in the discharge line. Calibration is accomplished by passing any given amount of water through the meter or batcher, drawing it off through the special outlet and catching the measured water in a 50-gallon drum or other container, which is then weighed on a platform scale. Conversion to gallons, if necessary, is done by dividing the weight by 8.33.

The indicator on the measuring tank or meter should be set at a reading slightly below any anticipated amount to be used and the water discharged into the barrel, where it is weighed. The indicator should be advanced about 25 pounds, the tank refilled, and the water again discharged and weighed. This should be repeated until the capacity of the batcher is reached. Accuracy of measurement should be within the specified tolerance.

Batching. Manual batching, that is, opening and closing of gates by hand-operated levers and controls, with cutoff controlled by the operator's observation of a scale dial or indicator, is somewhat slower than automatic or semiautomatic batching. There is, however, a control of charging and discharging rates that is not possible with any of the automatic systems. The operator can regulate the flow of material from a full-open position of the gate to a cracked open condition, in which it is barely open for slow feed of the material. Automatic gates usually have a dribble-feed setting, in which slow feed is accomplished by automatic rapid opening and closing of the gate. This would be done in the final stage of weighing, when the weight in the hopper approaches the set weight.

In some modern high-speed plants, the speed with which materials are weighed—especially in a cumulative batcher—makes it necessary to make an allowance for suspense material at the end of each weighing. See Figure 14-11. As the charging gate on one aggregate material automatically closes when the set amount of material has passed into the weigh hopper, the gate on another aggregate simultaneously opens, and the second aggregate starts to pass into the weigh hopper. However, a small amount of the first aggregate, which had passed through the gate but had not yet reached the hopper, should be included in the weight. This is the suspense material, and it is necessary to make allowance for it when setting the batch weights on the scales. This allowance is not necessary if the scales are permitted to come to rest after each increment of weighing.

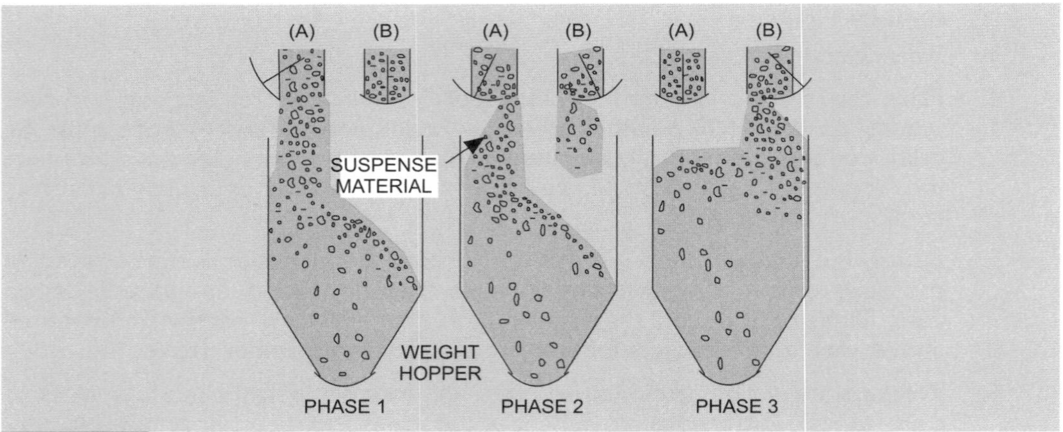

Figure 14-11
In phase 1, Gate A is open and Gate B is closed. Aggregate A is passing into the weigh hopper. Phase 2: Gate A closes and Gate B opens simultaneously, but part of Aggregate A is still in suspense, not yet in hopper. Phase 3: Gate A is closed and Gate B is open, Aggregate B now being weighed.

The scales, when set at the required weight, should be accurately balanced by the load for each batch. Any overload of aggregates in excess of the allowable tolerance should be removed. In weighing cement, care should be taken that all of the cement weighed out for each batch is completely discharged. When a dial scale is used, the operator and inspector should make frequent observations to make sure that the indicator returns to the zero reading upon discharge of the hopper. When a beam scale is used, frequent observations should be made to make sure that the tare beam is in balance upon discharge of the hopper when the weigh beams are locked out. Failure to show a zero balance indicates that material is hanging up in the hopper. Unless otherwise provided for in the specifications, the combined error of weighing and batching each material should not exceed the values shown in ASTM C 94.

Transfer of batches to the mixer should be as direct as possible. If conveyor belts are used, care must be exercised in design to avoid flat spots in transfer points, where parts of batches can hang up. One of three conditions may exist:

1. The mixers and batchers are all located in one plant, the batchers feeding the mixer directly through short conveyor belts, holding hoppers or other direct means.
2. Dry batches are hauled by truck from the batch plant to the mixer where water and admixtures are added.
3. Dry batches, or dry batches and water, are dumped into truck mixers.

14.5. Mixers

Mixers may be either portable or stationary, tilting or nontilting, in a large range of sizes from small 3-cubic-foot laboratory models to those with a capacity of 15 cubic yards. The choice between tilting and nontilting is largely a matter of personal preference, as either type is capable of mixing concrete efficiently and thoroughly.

The Concrete Plant Manufacturers Bureau defines the several varieties of mixers as follows:

Concrete Plant Mixer—A machine used to combine cementitious materials, water, aggregates and other ingredients to produce concrete in a batch, and usually operated in a fixed plant location while mixing concrete.

Nontilting Mixer—A rotating drum mixer that charges, mixes and discharges with the drum axis horizontal.

Tilting Mixer—A rotating drum mixer that discharges by tilting the drum about a fixed or movable horizontal axis at right angles to the drum axis. The drum axis may be horizontal or inclined from the horizontal while charging and mixing.

Vertical Shaft Mixer—A mixer with an essentially level floor and cylindrical or annular mixing compartment, with one or more vertical rotating shafts to which blades or paddles are attached. The mixing compartment may be stationary or rotate about a vertical axis.

Horizontal Shaft Mixer—A mixer with a stationary or rotatable cylindrical mixing compartment with the axis of the cylinder horizontal and one or more rotating horizontal shafts to which mixing plates are attached. [14.2]

The nontilter has a cylindrical drum and is usually charged or loaded by means of a skip or hopper at one end of the drum, with the mixed concrete discharged by means of a swinging discharge chute at the other end. See Figure 14-12.

Tilting mixers are found in many stationary or permanent plants. The tilter has a bowl-shaped or conical drum and may be charged at either the front or back, depending on the manufacture and installation.

Figure 14-12
A nontilting mixer. The direction of rotation of the drum is reversed to discharge the batch.

Discharge of the mixed concrete is accomplished by tilting the front end downward. See Figure 14-13.

Figure 14-13
Central mixing in a stationary mixer of the tilting drum type with delivery by truck agitator to the jobsite. (Courtesy of PCA).

The horizontal shaft mixer, usually called a pug mill or mortar mixer, is frequently installed in plants manufacturing masonry block, pipe and similar items in which very stiff, nonplastic concrete is used.

Most mixers will tolerate a 10 percent overload. It is permissible to load mixers to this capacity if the manufacturer's nameplate guarantees that the mixer will handle a batch of this size and mixer performance tests demonstrate that the concrete is properly mixed. See Section 13.2.

Operation and Control. A review of the previously mentioned general factors leads to consideration of specific elements that must be evaluated and controlled on any job. Homogeneity and uniformity of the concrete depend on the following:

- Design and condition of the mixer drum, blading and discharge spout,

- The degree of blending achieved by the method of charging the mixer,
- Size of the batch in relation to the gross volume of the mixer drum,
- The speed of rotation and number of revolutions of the mixer drum during charging,
- Speed of rotation and number of revolutions of the mixer drum during mixing, and
- Timing and rate of addition of admixture.

Sometimes cement balls in the concrete result from poor distribution of the cement when it comes in contact with water in the mixer. Specific causes are feeding cement or water too fast, batch exceeding mixer capacity, worn or improper blading, or holding the unmixed batch in the mixer for several minutes with the mixer not rotating. Once the cement comes in contact with damp aggregates or water, the batches should be mixed and used as expeditiously as possible. A maximum delay not exceeding two hours may be permitted, but the specifications should be consulted for specific information. In the case of dry batches hauled in batch trucks or held in a holding hopper, extra cement must be added to the batch if the cement is in contact with the moist aggregate for more than two hours. See the discussion in the following section on ready-mixed concrete.

If cement enters the mixer drum first, it is apt to stick to the interior of the mixer. This trouble is obviated by leading a portion of the water ahead of the aggregates and cement, with the aggregates leading the cement slightly.

To promote thorough mixing inside the mixer drum, the blading should be designed to move the concrete from one end of the drum to the other, with many crossings of paths; however, concrete should not be permitted to drop from near the top of the drum. Mixers for mass concrete containing large aggregate do not require as much blading as mixers for small aggregate concrete. In complete mixing results from too short a mixing time, worn or improper blades, interior of mixer encrusted with old concrete, speed of rotation too fast or too slow, size of batch exceeding the mixer capacity or improper batching sequence.

A minimum mixing time is usually stated in the specifications. Time starts when all materials, except the last of the water, are in the mixer, and continues for the periods shown in Table 14.1. Overmixing should be avoided, because the grinding action causes objectionable fines in the mix, resulting in a requirement for more water. There may also be some loss of entrained air. During hot weather, overmixing is especially objectionable because of the heating effect.

TABLE 14.1
SUGGESTED MINIMUM MIXING TIME

CAPACITY OF MIXER (CU YD)	MIXING TIME, MINUTES
2 or less	$1\frac{1}{2}$
3	2
4	$2\frac{1}{2}$
5	$2\frac{3}{4}$
6 or larger	3

The mixer should be equipped with a timing device that automatically starts timing at the specified time and prevents discharging the mixer until the set time period has elapsed. The timer should be adjustable and locked. To check the mixing time, the timer should be set on the selected interval while the mixer is operating under load. Time should start when the skip reaches the top of its movement or the last of the aggregate and cement are in the mixer, and continue until the discharge gate opens. In the case of a dual-drum mixer, the second batch discharged after the start of timing will be the one being timed.

Some agencies specify a mixer performance, or mortar efficiency test, to provide information for adjustments to the mixing time. This test compares concrete samples from two or more parts of the batch on the basis of unit weight of mortar and percentage of coarse aggregate. Mixing time may be increased or decreased, depending on results of this analysis. The test is described in Section 13.2.

READY-MIXED CONCRETE

Ready-mixed concrete is concrete that is mixed in a central batch plant and hauled to the jobsite in either agitating or nonagitating truck hauling units, or concrete that is batched into a truck mixer and mixed en route to the site, or a combination of partial mixing in the central plant and final mixing en route in a truck mixer. These methods are known respectively as central-mixed concrete, truck-mixed concrete and shrink-mixed concrete.

All of the general factors affecting batching and mixing concrete apply to ready-mixed concrete as well as to job-mixed concrete. In addition, there are certain factors that are especially significant in a ready-mix operation. These include, in general, the following:

- Method and sequence of charging dry materials into the mixer,
- Handling of supplementary water added during transit or at the jobsite,
- Disposition of mixer wash water,
- Effect of time of haul on quality of concrete,
- Control and assurance of uniformity within batch and from batch to batch,
- Dispatching of loads to fit placing schedule without delays, and
- Time lapse between charging the mixer, mixing and discharging.

All of these factors and many others can successfully be controlled, as attested by the greatest share of structural concrete in building construction being the product of the ready-mixed concrete industry.

14.6. History

Today, the ready-mixed concrete industry accounts for about half of the portland cement consumed, with several thousand plants engaged in this activity. Growth of the industry has been steady since the first pioneer ventured forth with a horse-drawn rig back in 1909. We have records of a number of operators central mixing concrete and hauling it to the job- site in the years prior to World War I. During the 1920s the economics and convenience of ready-mixed concrete were becoming apparent throughout the construction industry. It soon became apparent that the advantages of ready-mixed concrete were the convenience and savings for the contractor as well as uniformity of the product. Storing aggregates and cement on the site, with the consequent congestion of the construction area and waste of materials, is no longer a problem when ready-mixed concrete is used.

Nearly all of these early operations were central mixed, with the concrete hauled in dump bodies of some sort. Some small use was made of truck mixers, a typical unit consisting of a small stationary mixer mounted on a truck chassis, which led to the development of a specially designed mixer drum driven by the truck engine through a power takeoff. Between 1930 and 1940, several manufacturers started producing truck mixers, a common size being a 3- or 4-cubic-yard unit.

It was during this period that many agencies became aware of the problems associated with some of the machines, and considerable research and testing went into the development of equipment and techniques to improve the industry. A large share of the credit for these improvements must go to the National Ready Mixed Concrete Association, which was organized in 1930 as a nonprofit trade association of ready-mixed concrete producers. One of the earliest technical papers dealing with ready-mixed concrete was "Organization of Central Mixing Plant," which appeared in the Journal of the American Concrete Institute in 1925.[14.3] In 1930, ACI published a symposium of papers on central mixing plants, including one that discussed the new trend toward mixing in transit.[14.4] In 1937 the high-discharge mixer appeared and made the horizontal units obsolete. See Figure 14-14.

Figure 14-14
One of the first (1937) high-discharge truck mixers, a 4-cubic-yard unit. (Courtesy of PCA)

Acceptance and use of ready-mixed concrete became widespread during and after World War II with the development of improved methods and equipment. Units are now designed for transit mixing loads of 12 and 15 cubic yards, and even larger.

Figure 14-15
A modern 15-cubic-yard-high discharge revolving drum truck mixer.

14.7. Equipment

There are three types of units that are used either as truck mixers or agitators: the horizontal-axis revolving drum type; the inclined-axis, high-discharge revolving drum type (see Figures 14-15 and 14-16); and the open-top revolving-blade or paddle type (see Figure 14-17). Standards for operation of these mixers are covered in ASTM C 94. The same machine can be used for either truck mixing in transit or hauling central-mixed concrete. When used for hauling only, the machine is called an agitator, the only difference being that, when used as an agitator, the mixer drum or paddle shaft is rotated at a much slower speed than when used as a mixer. Also, an agitator can handle a batch half again as large as a batch to be mixed. Practically all ready-mixed concrete today is transit mixed in high discharge revolving drum machines.

Figure 14-16
A 10-cubic-yard truck mixer with the mixer discharge over the cab of the truck. This greatly facilitates positioning the truck for discharging the concrete. (Courtesy of PCA)

Figure 14-17
Truck agitators are used for hauling central-mixed concrete. A word of caution: When central mixed concrete is transported to the jobsite in an agitator unit, no water can be added at the jobsite. The agitator is not designed to operate at mixing speed to remix the concrete after addition of water! Note the proper placement of the welded wire reinforcement. (Courtesy of PCA)

Another kind of hauling unit consists of a truck-mounted dump body (popularly called a bathtub because of the rounded contour of the corners). This is a nonagitating type of unit in which discharge is accomplished by raising the truck body, the concrete sliding out through a gate in the rear of the container. See Figure 14-18.

Figure 14-18
Concrete can be mixed in a stationary mixer and then hauled to the jobsite in a nonagitating truck. (Courtesy of PCA)

Each truck mixer should have attached to it in a prominent place a metal plate on which is stamped its rated capacity in cubic yards as guaranteed by the manufacturer, and the recommended speed for mixing and agitating. The mixer should be provided with a batch meter and locking device capable of preventing discharge of the concrete prior to the required number of revolutions of the drum, or with a suitable revolution counter for indcating the amount of mixing. An accurate water meter, having indicating dials and totalizer, should be located between the supply tank and mixer.

For concrete to be properly mixed in a truck mixer, the gross volume of the mixer drum must be considerably greater than the capacity of the drum for mixing. This is necessary so that there will be sufficient excess volume to permit the several ingredients to be loaded into the mixer and thoroughly mixed. The mixer manufacturers recognize this reality when they designate the capacity of their machines. Standards of the industry fix the maximum capacities in terms of a percentage of the gross volume.

Adequate maintenance of the mixer is highly significant in achieving a good mixing within the specified number of revolutions. Mechanical maintenance is, of course, obvious and necessary. The interior of the mixer drum requires inspection and maintenance also. Concrete must not be permitted to accumulate on the mixer shell or around the blades. The blades themselves should be replaced when they show excessive wear. Water entry nozzles should not be plugged with mortar or concrete. The user's manual will explain these situations and their correction.

Mobile Batcher Mixed Concrete (Continuous Mixing). Mobile batcher mixers are special trucks that batch by volume and continuously mix concrete as the dry materials, water and admixtures are continuously fed into the mixer. The concrete must conform to ASTM C 685.

One make of this type of plant can be mounted on a truck or trailer, or set up in a stationary location. It consists of separate bins and tanks for aggregates, cement and water, which can be loaded at a central plant, the truck driven to the site, and the concrete mixed as required for the job. Materials are metered at a specified engine speed through a constant cement discharge, adjustable aggregate bin openings and adjustable water gauge setting, which are calibrated to continuously discharge the proper ratio of materials in accordance with preselected openings based on a conventional weight-basis mix design. Mixing is accomplished in a continuous auger flight with the discharge end higher than the receiving end.

Figure 14-19
A continuous volumetric batching and mixing unit for supplying small amounts of concrete. The mobile batcher measures materials by volume and continuously mixes concrete as the dry materials, water and admixtures are fed into a mixing trough at the rear of the unit. (Courtesy of PCA)

One of the principal advantages is the portability of the unit and the ease with which small amounts of concrete can be mixed and distributed. Control of the concrete is simplified because the concrete is constantly in view in the mixing trough. Figure 14-16 shows one of the units in operation.

14.8. Operation and Control

The method of batching and introducing the cement and aggregates into the mixer drum has a very important influence on the efficiency of mixing. If some degree of blending of the materials can be accomplished prior to or during charging of the mixer, better mixing can be assured. From this standpoint, the so-called ribbon loading method is best. See Figure 14-20. In this method, all materials are fed into the mixer simultaneously. There are obvious difficulties with certain plant arrangements that make such a system impractical, but even so, a certain degree of ribbon feeding can be accomplished. In a plant in which all aggregates are weighed cumulatively in one weigh hopper, the mixer charging sequence is limited to what can be done with the relative sequential positions of aggregates, cement and water.

Figure 14-20
Ready-mixed concrete plant with a conveyor belt system (ribbon-loading method) of charging the truck mixers.

In most transit-mixed operations, most of the water is introduced into the mixer at the proportioning plant, the balance being added from a supply on the truck at the jobsite. Ideally, the plant water should lead, accompany and follow the solid materials. More water should be admitted to the hub end of the mixer than is added at the discharge end. Coarse aggregate should lead the sand and cement slightly. Admixture should be introduced with the water, preferably after the first portion of water has entered the mixer. Introduction of the batch into the mixer is normally a continuous process. There are, however, situations in which this is not so. For example, in some plants it is necessary to charge the mixer in more than one dump of material because the mixer has a capacity greater than the capacity of the batcher. An 8-cubic-yard mixer would require two dumps of material when being loaded from a batch plant with a 4-cubic-yard capacity. Figure 14-21 shows graphically an idealized charging sequence.

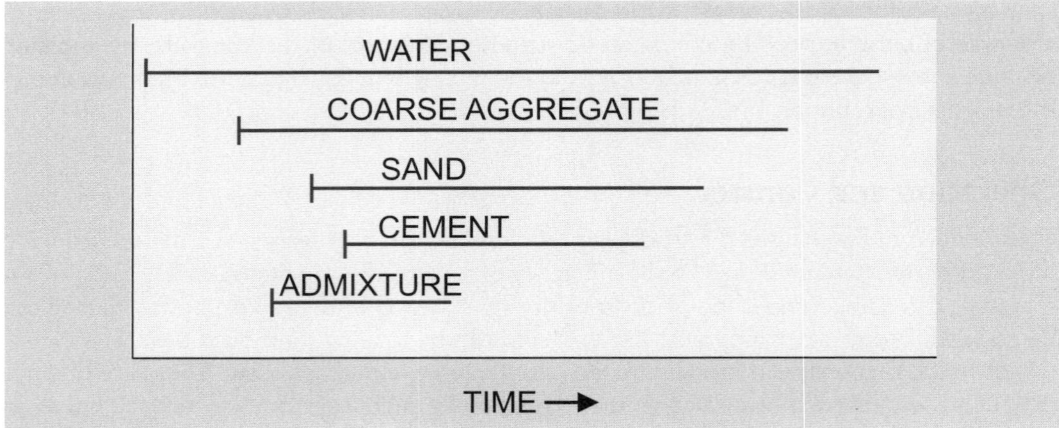

Figure 14-21
The best charging sequence is one in which the water leads and follows all solid materials, with the coarse aggregate leading and following the cement. Admixtures should always enter the mixer at the same point in the sequence.

Speed of the mixer drum while charging is also important. Most tests indicate that a slow drum speed is desirable.

When concrete is central mixed, the truck mixer carrying such concrete, acting as an agitator, is set on the slow agitating speed for the trip to the jobsite. Just before discharging the batch, the mixer is operated at mixing speed for a few revolutions. Some mixers agitate and mix in a rotational direction opposite to their discharge rotation.

Concrete has a tendency to lose consistency or stiffen when over-mixed, owing to generation of heat and the grinding action within the mixer. This is more likely to be a problem during hot weather. For this reason, the importance of uniformity in number of revolutions of the drum or paddles is readily apparent. The code places an upper limit on the total number of revolutions. The revolution counter, previously mentioned, provides the necessary information for control.

Careless and improper discharge of concrete from the truck mixer is a cause of segregation. The ideal discharge is at full discharge opening and full drum speed without interruption. This ideal, however, is not always possible or practical under existing job conditions. Usually the discharge is intermittent, as into buckets or buggies, and is also at a restricted rate. In any event, the discharge gate should be fully open during discharge, and the rate controlled by varying the mixer speed. Trying to force the concrete out through a partially open or restricted discharge opening results in segregation, especially toward the end of the batch, as it tends to hold back the coarse aggregate and permit the mortar to pass through. This is especially troublesome with high- slump mixes and those mixes containing larger-size coarse aggregate. The mixer should always be run at mixing speed for a few revolutions just before discharging if it has been on agitating speed.

Mixing Speed and Time. ASTM C 94 for truck-mixed concrete provides for making mixer performance tests to determine the amount of mixing necessary to completely mix the concrete in 70 to 100 revolutions of the drum. When satisfactory mixing is found in one mixer, it can be assumed that satisfactory mixing will be accomplished in other mixers of substantially the same design and condition of blades. Additional revolutions of the mixer beyond the number required for adequate mixing should be at agitating speed.

For stationary mixers in a central-mix plant, mixing time is based on performance tests, or on a scale of one minute for mixers of 1-cubic-yard capacity, plus 15 seconds for each cubic yard of capacity over one.

Truck mixers should be operated at the speed recommended by the manufacturer, which, in some cases, will be found to be as high as 18 rpm. Experience has shown that best mixing is obtained with drum speeds between 12 and 18 rpm, although slower speeds also can give good results in some mixers. Speeds greater than 18 rpm are likely to be inefficient because of the influence of centrifugal force. Regardless of the speed (within reasonable limits) the principal influence on uniformity of the concrete, as far as rotation of the mixer is concerned, is the total number of revolutions at mixing speed rather than the speed of rotation.

Agitating speed should be just fast enough to prevent the concrete from settling and stratifying in the mixer drum. Manufacturers recommend a speed between 2 and 6 rpm, the slower speed usually being adequate.

Control of Water. The total water in a batch of concrete includes free water on the aggregates, water contained in admixtures, ice when used during hot weather and water added to the batch. ASTM C 94 requires that added water shall be measured to an accuracy of 1 percent of the required total mixing water, and water from all sources in the batch shall be measured to an accuracy of plus or minus 3 percent of the specified total.

Job specifications may vary somewhat in their requirements for control of the water. Most specifications refer to ASTM C 94 requirements. Some specifications have required that all supplementary water (that is, water in addition to the water contained in the aggregate) be added at the plant, with no water carried on the truck, or that all water be added from an outside source at the site of the structure. Fortunately, these restrictions are rarely encountered anymore; practically all job specifications now accept the code requirements without modification.

Wash water is necessary for cleaning out the mixer drum occasionally. Methods of controlling wash water are sometimes a source of controversy on the job. If specifications permit it to be carried on the mixer truck, the water should be in a special tank or compartment, entirely separate from the mixing water.

Wash water is sometimes retained in the mixer drum for use as part of the mixing water for the succeeding batch. Technically, there is no serious objection to such a procedure, and it is permitted by the code, provided the water can be accurately measured. There are practical limitations, however, and measurement of this water may not be possible. If control can be set up as a batch plant function, as suggested by some operators, there is usually no problem.

A modern truck mixer carries adequate water for normal operations and is equipped with a meter for measuring the water. The water system is designed so that water can be admitted under pressure at the hub end of the mixer drum, through a hollow shaft. Admitting the water in this manner provides for optimum distribution of the water through the batch in the drum. Satisfactory distribution of water by spraying the water into the spout of the mixer with a hose is difficult, if not impossible.

It is common practice to introduce only part of the total supplemental water at the batch plant, and then to add sufficient water after the truck arrives at the jobsite to bring the concrete up to the designated slump. This procedure is entirely acceptable, provided the total water in the batch does not exceed the maximum allowable water-cement ratio, and provided that it is done under adequate supervision and control. Whenever supplementary water is added to the concrete, the mixer should be operated at mixing speed for at least 30 revolutions after all the water has been admitted.

Occasionally a supervisor or other person will request that extra water be added to a batch, either before discharge or more frequently during discharge of the concrete, because of a supposed difficulty in placing or finishing. Such abuse should not be permitted. Also, adding water to the concrete after placement, as shown in Figure 14-22, is strictly forbidden. Any water that is introduced into the mixer after it leaves the batch plant should be noted by the truck driver on the delivery ticket. In any event, the specified water-cement ratio or slump should not be exceeded.

Figure 14-22
"Retemporing" concrete!

Load Tickets. Each load of concrete is accompanied by a load ticket. See Figure 14-23. The code requires that such a ticket be provided. As a minimum, the following information is required:

1. Name of ready-mix batch plant
2. Serial number of ticket
3. Date and truck number
4. Name of contractor
5. Specific designation of job (name and location)
6. Specific class or designation of concrete in conformance with that employed in job specifications
7. Amount of concrete (in cubic yards)
8. Time loaded or first mixing of cement and aggregates

Additional Information. Additional information designated and required by the job specifications shall be furnished upon request. Such information may include the following:

1. Reading of revolution counter at first addition of water
2. Signature or initials of ready-mix representative
3. Type and brand of cement
4. Amount of cement

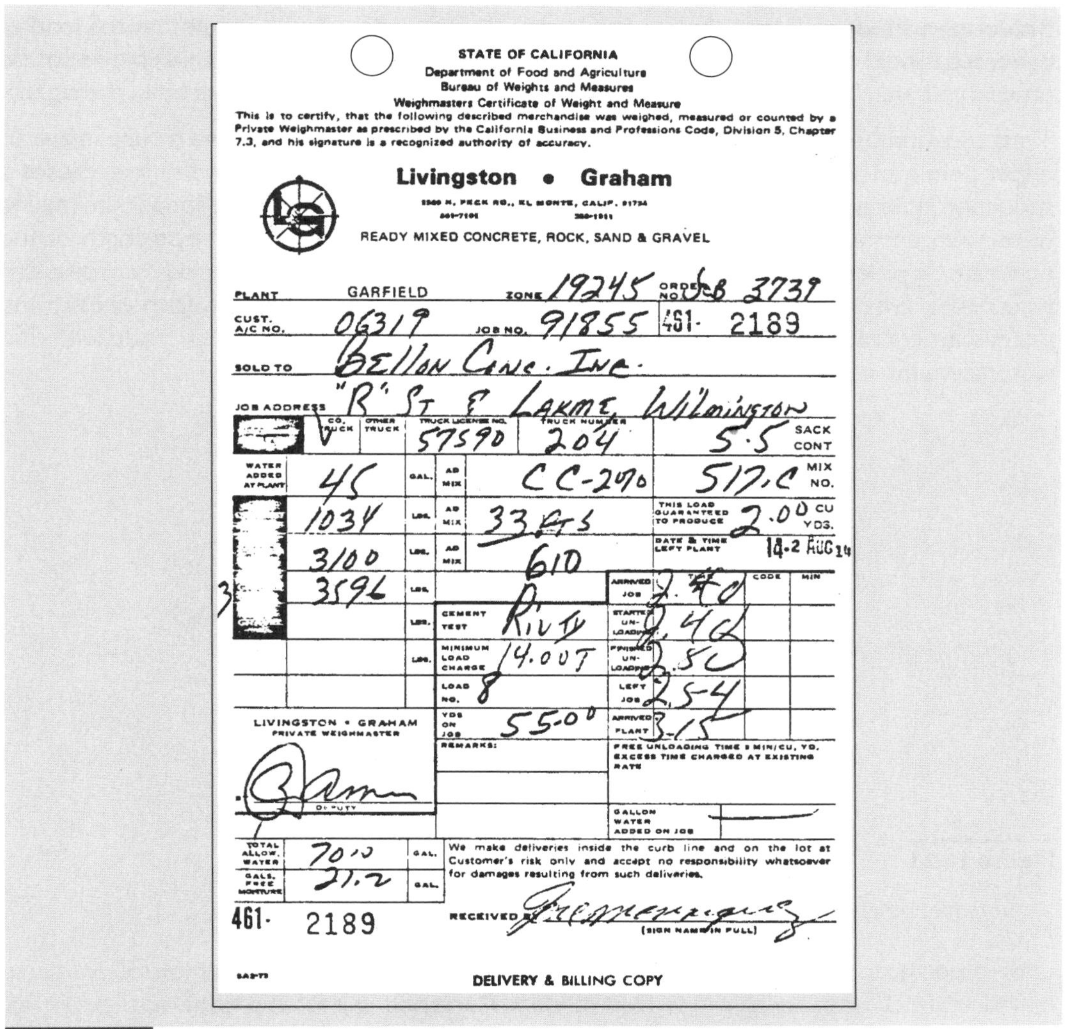

STATE OF CALIFORNIA
Department of Food and Agriculture
Bureau of Weights and Measures
Weighmasters Certificate of Weight and Measure

This is to certify, that the following described merchandise was weighed, measured or counted by a Private Weighmaster as prescribed by the California Business and Professions Code, Division 5, Chapter 7.3, and his signature is a recognized authority of accuracy.

Livingston • Graham

READY MIXED CONCRETE, ROCK, SAND & GRAVEL

PLANT GARFIELD ZONE 19245 ORDER NO. 3739
CUST. A/C NO. 06319 JOB NO. 91855 461- 2189
SOLD TO Bellon Const. Inc.
JOB ADDRESS "R" St & Lakme, Wilmington

CO. TRUCK	OTHER TRUCK	TRUCK LICENSE NO.	TRUCK NUMBER	SACK CONT
✓		57590	204	5.5

WATER ADDED AT PLANT	45	GAL.	AD MIX	CC-270	517.C	MIX NO.
	1034	LBS.	AD MIX	33 Ots	THIS LOAD GUARANTEED TO PRODUCE 2.00	CU YDS.
	3100	LBS.	AD MIX	610	DATE & TIME LEFT PLANT	14-2 AUG 14
	3596	LBS.				

		TIME	CODE	MIN
CEMENT TEST	Rivty	ARRIVED JOB 2.40		
MINIMUM LOAD CHARGE	14.00T	STARTED UNLOADING 2.46		
LOAD NO.	8	FINISHED UNLOADING 2.50		
YDS ON JOB	55.00	LEFT JOB 2.54		
REMARKS:		ARRIVED PLANT 3.15		

LIVINGSTON • GRAHAM PRIVATE WEIGHMASTER

DEPUTY

FREE UNLOADING TIME 3 MIN/CU. YD. EXCESS TIME CHARGED AT EXISTING RATE

GALLON WATER ADDED ON JOB

TOTAL ALLOW. WATER	70.0	GAL.
GALS. FREE MOISTURE	21.2	GAL.

We make deliveries inside the curb line and on the lot at Customer's risk only and accept no responsibility whatsoever for damages resulting from such deliveries.

RECEIVED BY (SIGN NAME IN FULL)

461- 2189

DELIVERY & BILLING COPY

Figure 14-23
A good example of the kind of load ticket that should accompany every load of ready-mixed concrete.

5. Total water content by producer (or water-cement ratio)
6. Water added by receiver of concrete and his or her initials
7. Admixtures and amount of same
8. Maximum size of aggregate
9. Weights of fine and coarse aggregate
10. Indication that all ingredients are as previously certified or approved. It is especially important that the additions of water to the load be well detailed as to amount, time and who requested it. A common procedure is for the ticket to be in duplicate, with one copy given to the purchaser and one copy returned to the plant with the truck driver after the load and additional water have been acknowledged by the contractor's supervisor.

Prolonged Mixing. Occasionally a truck mixer is delayed, with the result that the load of concrete is held beyond the specified time limit. If the specifications are flexible as to the time limit, it may be possible to use the load, provided the concrete has not been damaged.

Tests by a number of investigators have conclusively proved that long-time mixing, up to an elapsed time of several hours, with the addition of water to maintain the slump, causes a reduction in strength and other desirable properties of the concrete. The longer the mixing time, with periodic additions of water to maintain slump, the lower the strength of the concrete. In practically all of the tests in which the concrete was retempered by addition of water to maintain slump, the strength of the concrete bore the same relationship to the water-cement ratio after each addition of water as if the concrete had been made with that water-cement ratio in the first place. (See Figure 14-24.)

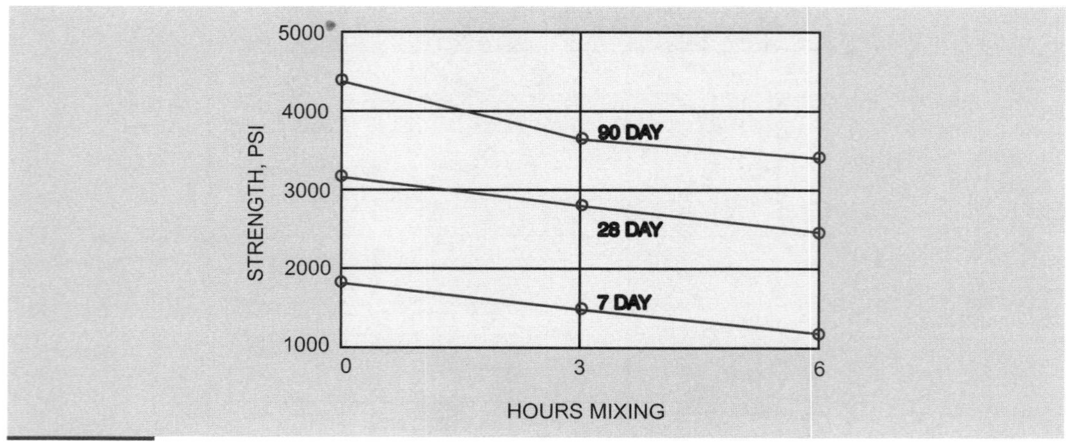

Figure 14-24
Long-time mixing of the concrete with water added to maintain the slump causes a loss of strength and other properties.

Long-time mixing, either continuous or intermittent, without the addition of water to maintain the slump, results in a loss of slump or stiffening of the concrete. At an air temperature of 60°F, concrete with a slump of 4 to 5 inches immediately after mixing will lose as much as 2 inches of slump after a 90-minute haul; at 80°F the slump loss will be 3 inches or more. Results of one series of tests are shown in Figure 14-25. Long-time mixing without regauging the mix with additional water to maintain the slump usually results in slightly higher compressive strength of the concrete as long as the concrete remains workable enough to be cast into cylinders. Depending on temperature, materials and job conditions, probably about three hours is a practical limit.

The air content in air-entrained concrete decreases as the amount of mixing is extended, as shown in Table 14.2. Concrete was continuously mixed in a truck mixer with samples taken as shown. After 300 revolutions, a small amount of water was added, which increased slump slightly; however, air content continued to fall with additional mixing.

In general, delays have an unfavorable effect on mixed concrete. The only possible exception is long-time mixing, either continuous or intermittent, without addition of water. The concrete will stiffen or lose slump, but the strength may be slightly higher. Temperature of the concrete will rise and air content drop. Workability of the concrete will limit the tolerable length of the delay time. Flexural strength and durability are affected in much the same way as compressive strength.

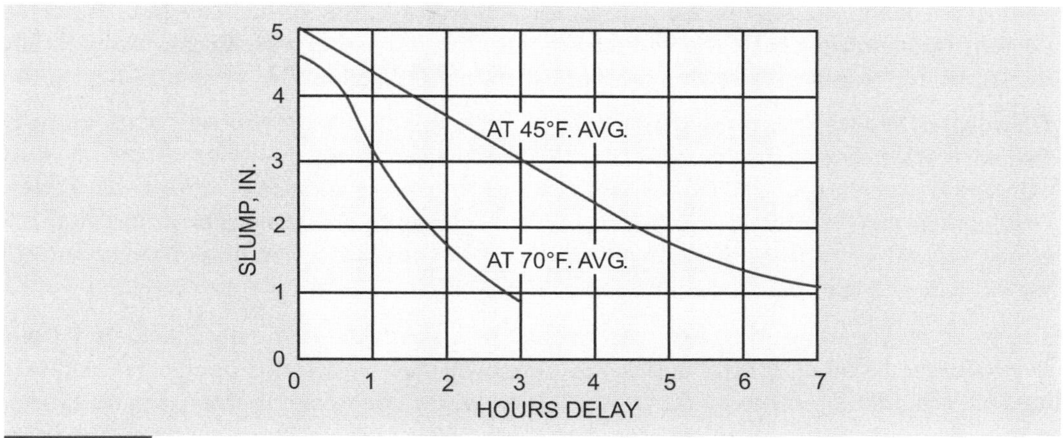

Figure 14-25
Loss of slump is more rapid at higher air temperatures if water is not added to the concrete.

TABLE 14.2
EFFECT OF LONG-TIME MIXING ON AIR CONTENT

NO. OF REVOLUTIONS	INCHES SLUMP	PERCENT AIR
60	3.1	3.2
100	3.4	3.1
200	1.2	2.7
300	0.8	2.5
Water added to partially restore slump		
320	1.8	2.1

Good construction practices, as well as specifications, will always impose a time limit on the length of time cement can be exposed to moisture in a batcher or mixer. A period of one and one-half hours is generally agreed upon in the industry as a practical limit; however, this limit has to be flexible as permitted or required by job conditions. Temperature, kind of cement, use of admixtures and conditions of exposure all influence the allowable limits.

Delayed Mixing. A situation sometimes arises when, because of a long haul, it appears desirable to batch the aggregates into the mixer, perhaps with part of the water, and introduce the cement last so that it will lie on top of the other materials. After a long delay, even as long as several hours, the truck is driven to the job-site. The balance of water can be added before the truck starts out, in which case the load is mixed in transit, or the water can be added and mixing done after arrival at the job. Some additional mixing may be required to achieve a homogeneous batch of concrete, and a small amount of extra cement should be included to allow for loss of cement by hydration of the part of it in contact with wet surfaces during the delay. For a delay of two or three hours, 6 percent more cement should be added, then 3 percent for each additional hour. Total delay should not exceed seven or eight hours, less during hot weather.

This procedure is contrary to a strict interpretation of the ASTM mixing requirements. From a quality standpoint, however, when done under close supervision and control, it is probable that acceptable concrete will be made.

With the advent of a new breed of retarder admixture to essentially stop and restart the cement hydration process, the concern of a long haul has been technically solved. See discussion of "stopping/restarting cement hydration" in Section 9.2.

Disposal of Waste. In any ready-mix operation there is always the problem of disposing of the wash water and waste material resulting from cleaning out the mixer drums. In many plants, the procedure is to introduce wash water into the mixer, rotate the drum for a few revolutions, then discharge into a sump. Water passes to a settling tank or pond and is either wasted or reused for washing or mixing. Sumps and tanks must be cleaned out periodically and the soiled waste hauled to the dump.

The cost of disposing of waste, the economic loss of possible reusable material and environmental pollution restrictions now imposed have led some operators to devise means of reclaiming some of the waste. Using waste concrete for making patio blocks, stepping stones and similar items helps, but the market for these items is limited, and there is still the problem of disposing wash-out material. This problem has been solved at many plants by installing screws or screens that separate the reusable sand and gravel, passing the water on to a settling sump where the water remains for several hours or overnight. After settling, the clean water is pumped into a clean-water storage tank from which it can again be used for washing out trucks. There appears to be a ready market for the reclaimed aggregate, and the only waste is the sludge that must be removed from the sump.

Recycled Water in Concrete. The ready mixed industry is faced with the challenge of managing about 3 percent to 5 percent of its estimated annual production of 400 million cubic yards as returned concrete. In addition, about 80,000 truck mixers are washed out, using about 200 to 400 gallons of water each, on a daily basis. With increasingly more restrictive environmental regulations on discharge of water from batch plants, the industry is being forced to look at recycling some of the process water generated at ready mixed concrete plants.

ASTM C 94 Specifications for Ready Mixed Concrete permits the use of recycled wash water as mixing water in concrete. Recycled water is primarily a mixture of water, cementitious materials and aggregate fines resulting from processing returned concrete, and could include truck wash water. In most batch plants, the recycled water is passed through settling ponds where the solids settle out, leaving a clear water. In some plants, the recycled water from a reclaimer unit is kept agitated to keep the solids in suspension (See Figure 14-26) for use as a portion of the batch water in concrete. The C 94 specification permits a maximum solids content of the wash water of 50,000 parts per million, or 5 percent of the total mixing water. This amounts to about 15 lb/yd^3 solids in a typical concrete mixture. Research has shown that using recycled wash water with solid contents within ASTM C 94 limits, as the total mixing water, has no significant effects on fresh and hardened concrete properties. As more restrictive environmental regulations are established, the ready mixed concrete industry will be forced towards zero-discharge production of ready mixed concrete.

14.9. Responsibilities

In any job being supplied with ready-mixed concrete, there is a division of responsibilities between the ready-mix supplier, contractor, inspector, testing agency and engineer or architect. The division in responsibilities between the concrete producer and the contractor needs to be clearly defined. When this is done, quality control, testing and inspection features are facilitated. The division of the two areas of responsibility is marked by the transfer of the concrete from the producer's conveyance into the contractor's

Figure 14-26
Recycled water and reclaimed aggregate at a ready-mixed concrete plant.

handling equipment. The National Ready Mixed Concrete Association and the Associated General Contractors of America have published a joint Statement of Responsibilities in this respect. It is quoted below:

> Concrete as supplied for construction is a perishable product, requiring the informed cooperation of both the ready- mixed concrete producer and the contractor to assure its satisfactory performance. It is the purpose of this statement to outline the functions of those two parties and to define their respective areas of responsibility. It is assumed that requirements for the concrete are properly covered in the specifications, and that it is not the function of the contractor or producer to overcome deficiencies or oversights in design of the structure.
>
> Concrete construction involves two phases, which correspond to the areas of responsibility of the producer and contractor. The production phase belongs to the ready-mixed concrete operator and encompasses the operations of securing satisfactory ingredient materials, combining them in proper proportions, mixing them into a homogeneous product and delivering the mixture into the hands of the contractor in such condition that it can be satisfactorily placed. The second phase, under the purview of the contractor, includes the movement of the concrete into the forms, its proper placement and consolidation, the finishing operations, and adequate protection and curing to assure developing the quality potential. Transfer of the concrete from the producer's delivery conveyance into the handling equipment of the contractor marks the division between the two areas of responsibility. Specific functions and responsibilities of the two parties are described below.

Producer Responsibilities

The ready-mixed concrete producer shall:

1. Use material—aggregates, cement, water and admixtures, if required—meeting the requirements of the specifications and capable of producing concrete of the required quality.
2. Provide adequate personnel and equipment to assure continuous production at a rate to meet the needs of the work. The equipment shall conform to and be operated within requirements of the specifications for such features as accuracy of measurement, rate and amount of mixing and volume rating.

3. Proportion and batch all concrete to meet specification limits. Depending upon the nature of the specifications, the limits to be met may include one or a combination of the following:
   a. Quantity of cement per unit volume of concrete
   b. Ratio quantity of mixing water to quantity of cement.
   c. Consistency, usually measured as slump
   d. Air content
   e. Specific ratios or quantities of ingredients
   f. Strength (Note: The producer's responsibility for strength extends only to measurements by recognized methods for evaluating quality of the concrete as delivered.)
4. Cooperate with inspection services by making all facilities and operations conveniently accessible for examination and securing of test samples.

Contractor Responsibilities

The contractor shall:

1. Provide the producer in advance with all information necessary to establish mixtures and costs, including: limitations on materials, proportions, strength and consistency; location and nature of project; quantity of concrete required; rate and method of placement; anticipated unusual conditions.
2. Organize concrete placement to permit advance scheduling of deliveries and prompt discharge after delivery.
3. Perform all operations of handling, placement, consolidation, protection and curing in conformance with specification requirements to assure adequate quality of the end product.
4. Cooperate in facilitating inspection and, where required, engage competent personnel for sampling and testing the concrete.

Joint Responsibilities

Except for larger projects where specialized inspection is required, concreting operations are not usually under continuous surveillance. The producer and contractor must see that satisfactory control is maintained.

It is particularly important to avoid the use of excessive mixing water. The contractor's forces should be informed of the dangers and be firmly disciplined in avoiding high slump. At the same time, the producer should discourage the use of high-slump concrete by requiring that extra water, over that needed to produce the specified slump, be noted and signed for. Authoritative cooperation on the part of the contractor is needed to see that this procedure is followed.

Conformance with good practices in controlling the amount of mixing and the elapsed time between batching and discharge requires attention from both contractor and producer. No difficulty is encountered if ordering and delivery schedules are consistent with the capacity of placement facilities.

All parties have an important stake in seeing that concrete testing is properly done. Particularly in the case of strength, errors in measurement can lead to unjustified concern over quality, resulting in costly retesting, delays or replacement. Violations of standard sampling and testing practices should not be tolerated by either the ready-mixed concrete producer or the contractor. [14.5]

Handling and Placing the Concrete

The phase of the work called placing concrete can be divided into three operations: preparation to receive the concrete, conveying and placing the concrete in the forms, and care of the concrete after it has been placed. In this chapter we will be concerned with preparation, conveying and placing.

15.1. Preparation for Placing

Foundations. Preparation of surfaces against or upon which concrete is to be placed depends upon the type of foundation material and requirements of the structure. Excavation for foundations should extend into sound, undisturbed soil or rock. If earth is overexcavated, the overexcavated portion must be back-filled with select material and compacted to the specified density, or filled with low-strength concrete (see Section 22-17 Controlled low-strength material). Rock surfaces should be clean and sound. If free water is present, it should be blown out with air jets or otherwise removed. In some cases it may be necessary to provide a sump (outside the form area) into which the water drains, for removal by means of a pump.

Foundations should be free of frost and ice when concrete is placed. During dry seasons, earth should be moist but not muddy. Concrete should not be placed in running water, although underwater placement is permissible in still water if the tremie method is used. See Chapter 22.

Steel shells for cast-in-place piles should be inspected by lowering a light into the shell. In the same manner, shafts for caissons should be inspected. Compact television equipment especially designed for the purpose permits an inspection of the entire shaft to be made from the surface.

Construction Joints. In most structures, it is not possible to place all of the required concrete at one time. The volume of concrete, structural limitations and time limitations make it necessary to discontinue placing concrete at some point, to be resumed later after the concrete has hardened. The plane of separation between the placements is called a construction joint. A construction joint may be horizontal, as between lifts in a wall or column, or it may be vertical, as one formed by a header board in a slab or beam. Locations of construction joints are shown on the plans. If because of a stoppage of the work for several hours or longer it becomes necessary to install a joint, the joint should be made in a plane normal to the main reinforcing bars and in a region of minimum shear. Minimum shear in simply supported slabs or girders is at or near the center of the span. The design engineer should be consulted with respect to location of construction joints, as a wrongly located joint can seriously alter the load- carrying capacities of the structure.

ACI 318 Section 6.4 provides as follows:

Construction joints must be made and located so as not to impair the strength of the structure. Where a joint is to be made, the surface of the concrete must be cleaned and laitance removed. Immediately before new concrete is placed, all construction joints must be wetted and standing water removed.

Construction joints in floors must be located within the middle third of spans of slabs, beams and girders. Joints in girders must be offset a minimum distance of two times the width of intersecting beams.

Beams, girders or slabs supported by columns or walls must not be cast until the concrete in the vertical support members is no longer plastic or semifluid. (See Figure 15-22.) Beams, girders, haunches, drop panels and capitals must be placed monolithically as part of a slab system, unless otherwise shown in design drawings or specifications.

Formed construction joints should be avoided whenever possible, as they are planes of weakness and potential sources of leakage of water through the structure. Waterstops should be installed if leakage of water through the joint must be prevented.

After the first run of concrete at a vertical construction joint has gained sufficient strength, the form or header board should be removed and the concrete washed with a jet of water of sufficient volume and velocity to wash the cement paste from the surface. This washing should be done at such a time that the aggregate will not be loosened. Horizontal joints should be similarly washed.

Roughness is not essential to a good joint. In fact, a better joint is obtained if the surface of the old concrete is regular and smooth, avoiding large pieces of aggregate protruding above the surface, or depressions such as footprints. The ideal cleanup exposes particles of fine aggregate but does not cut deep enough to loosen the coarse aggregate.

Reinforcing steel is normally continuous across a construction joint, or dowels are provided. Some designs call for shear keys in the joint. Shear keys, formed in the run of concrete composing the lower portion of a horizontal joint, may be formed by inserting and subsequently removing beveled strips or blocks of wood that were saturated with water prior to insertion. Keys in a vertical joint may be formed by affixing beveled blocks or trapezoids to the forms. (See Figure 15-1.)

A satisfactory horizontal joint is not difficult to obtain if good, low-slump concrete has been used, and if the concrete has not been excessively worked. High-slump mixes, overvibration (especially near the top of a lift), too much use of a jitterbug or other tools that bring mortar to the surface, excessive job traffic—all tend to bring water and fines to the surface, which results in a weak layer of laitance that must be removed.

When good concrete is used and proper placing techniques are followed, a minimum of cleanup is required. Cleaning with an air-water jet or wire brushing can be done while the concrete is still soft enough so that the laitance and scum can be removed without loosening the aggregate. The cleaned surface must be protected until the next lift of concrete is placed, otherwise it will be so dirty that any benefit from the cleanup is lost. A covering of damp sand will protect the surface as well as provide curing. An inferior surface will require more intensive cleaning to remove the accumulation of laitance and other unsuitable surface materials.

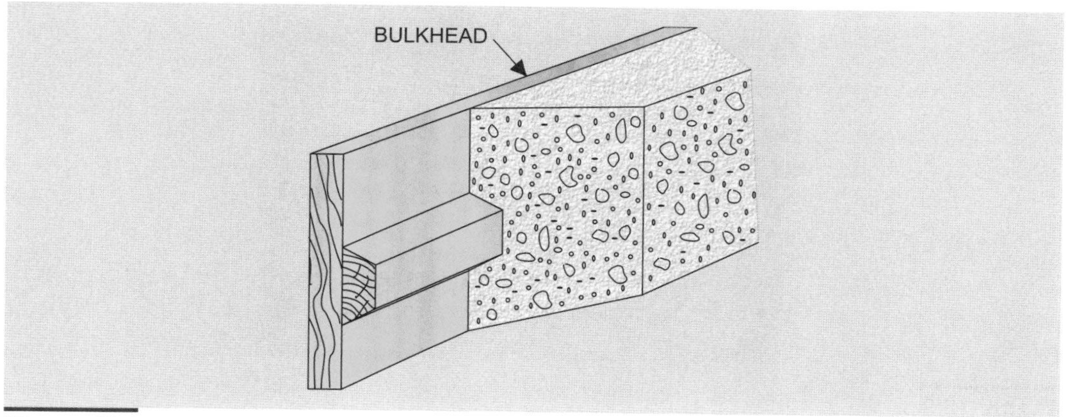

Figure 15-1
Trapezoidal-shaped blocks can be inserted in the formwork to make a key in the joint. See also Section 16.2 for proper sizing of keyed construction joints.

Immediately prior to placing the second run of concrete, the joint surface is thoroughly washed to remove all sand, mortar, laitance, nails and other foreign materials. The concrete can then be placed. Because of the need for this final cleanup, it is sometimes more economical to forego the preliminary cleanup and use a wet sandblast just prior to placing the concrete.

Forms. Chapter 11 covers formwork and the use of forms in considerable detail. Forms should be clean, tight and properly braced. The form lining should impart the desired surface appearance to the hardened concrete and should be coated with an appropriate oil or parting compound. Form oil should not be applied so thickly that it runs down and collects in the bottom of the form or on a construction joint. Particular care must be taken to avoid splashing or dripping form coatings on reinforcing steel, prestressing strand, construction joints or on any other surface where bond with the concrete is required. Wood forms should be moistened with water so that they will not absorb water from the concrete, which causes the wood to swell.

Openings in the forms must be available for cleanup, placing and vibrating the concrete, and inspection of the formwork and concrete placing. Plugs and coverings must be available for closing the openings at the proper time without causing a delay in placing the concrete. Grade strips must be in place if a horizontal construction joint is to be made.

Wall form ties are usually designed to serve also as spreaders and remain in the concrete. If wood spreaders are used, vigilance is necessary to make sure that they are removed just ahead of the concrete as it rises in the form. A stout wire attached to the spreader and leading to the top of the form will be of help in removing the spreader. Many specifications now prohibit the use of wood spreaders.

Wooden boxes or blocks are sometimes used to make block-outs, which form openings and voids in the concrete. (See Figure 15-2.) They should be well soaked with water ahead of concrete placing so that they will not absorb water from the concrete; otherwise they may expand and crack the green concrete.

At least one worker, usually a carpenter, is delegated to act as a form watcher while concrete is being placed. That worker's responsibility is to observe the forms for any movement, check bracing and wedges, and in general detect any movement that requires corrective action.

Figure 15-2
A wood blockout for an opening in a tilt-up panel. Wood should be well soaked with water before concrete is placed around it. (Courtesy of PCA)

Reinforcing Steel. At the time reinforcing steel is embedded in the concrete it should be free of dirt, paint, oil, grease or other foreign substances. A thin coating of rust or mill scale is not detrimental, provided it adheres tightly to the steel. Dried mortar splashed on the steel ahead of the concrete's being placed should be removed by wire brushing. If the mortar cannot be removed by vigorous wire brushing, it is probably safe to leave it on the steel. Note that vigorous brushing is required.

All reinforcing steel should be accurately placed and, during the placing of concrete, held firmly in position. Distances from the forms should be maintained by means of chairs, ties, hangers or other approved supports.

Embedded Items. Most concrete structures have objects and fixtures embedded in them. Among these items are castings for manholes and catch basins, anchor bolts, pipes and conduits, inserts of various types and instruments. With few exceptions these are fixed in place prior to concrete placement by being attached to the forms or to the reinforcing steel. Adjustments to the steel location to accommodate these items should be made only as shown on the plans. Nothing should be inserted into the concrete without first considering the effect of the insertion on the strength of the structural member.

Many precast concrete products are manufactured in such a way that it is difficult, if not impossible, to position reinforcement that protrudes from the concrete before the concrete is placed. According to ACI 318 Section 16.7, such items as dowels or inserts can be placed while the concrete is plastic if proper precautions are taken:

When approved by the engineer, embedded items . . . that either protrude from concrete or remain exposed for inspection may be embedded while concrete is in a plastic state, provided:

Embedded items are not required to be hooked or tied to reinforcement within plastic concrete,

Embedded items are maintained in correct position while concrete remains plastic, and

Concrete is properly consolidated around embedded items.

This exception to the general rule that all reinforcement and embedded items be placed and supported before concrete is placed does not apply to reinforcement that is completely embedded. The inspector should be satisfied that the embedded items are properly placed and anchored and that the concrete is properly compacted around the items.

Final Inspection. Performance of all of the inspection discussed in this section requires that immediately prior to concrete placement a final inspection should be made, covering all features of foundations, forms, steel and embedded items, cleanup, etc.

At this time the inspector and the contractor's supervisor should make sure that all plant and equipment are ready to go. The correct quality of concrete should be ordered and the ready-mix supplier prepared to furnish concrete at the required rate. Transporting equipment, such as pumps, cranes, batch trucks, buckets, conveyors and helicopters should be standing by and should be capable of handling the concrete at the required rate without segregation and at the specified slump; sufficient vibrators, with extra standby units, should be on hand; curing materials should be available. Special protective facilities for hot weather, freezing temperatures or rain, as the season dictates, should be available. If the work is expected to continue after nightfall, sufficient lights should be in readiness. Once these details have been prepared ahead of time, then all systems are go and placing of the concrete can proceed. See also Chapter 25.

15.2. Conveying Concrete

After the concrete has been discharged from the truck mixer, it must be placed in the forms. For most structures—because of long distances, difference in elevation or obstructions that block movement of the truck—some means must be provided for moving the concrete from the mixer chute into the forms.

Conveying equipment must be selected with care. Not only must cost per cubic yard and cubic yards per hour be considered, but also the layout of the site and characteristics of the concrete to be handled. Maximum size of aggregate, slump and sand content, whether normal weight or lightweight, must all be evaluated and equipment selected that will efficiently handle the kind of concrete specified. Essential requirements of any system of moving concrete from the mixer to the forms are to minimize segregation, prevent loss of part of the concrete and avoid excessive loss of consistency (slump). Among the methods used are direct discharge from mixer into the forms, crane and buckets, pumps, conveyors, buggies, wheelbarrows, pneumatic placers, small rail cars or a combination of two or more of these methods. Helicopters have been employed for transporting equipment and concrete buckets into especially isolated or difficult sites. See Figure 15-3.

The method used depends on the size of the job, adequacy of space, availability of equipment and other factors. The job should be planned so that the ready-mix trucks can get as close to the actual site as possible. Unused equipment and piles of materials should be out of the way to give the trucks more freedom to maneuver. The method and equipment used should not impose a restriction on the slump of the concrete being placed, as the slump is determined by the specifications for the structure.

Figure 15-3
A helicopter can be used for transporting concrete to a site that is difficult to reach with any other conveying system. The bucket used is a special lightweight bucket.(Courtesy of PCA)

Figure 15-4
When the truck can get fairly close to the forms, concrete can be discharged directly into the form. Extension chutes carried on the truck widen the placing area. Note the height of discharge on this inclined-axis mixer. (Courtesy of PCA)

Direct Discharge. Some jobs are so laid out that the ready-mix truck can discharge directly into the forms or on the subgrade. Extra lengths of chute are carried on the truck that provide a placing radius of about 15 feet from the truck. (See Figure 15-4.) This method minimizes the amount of handling of the concrete, but long extensions of the chute at a flat angle are likely to result in demands for more mix water so that the concrete will flow down the chute. Such demands, of course, should be resisted as long as the concrete is at the specified slump.

When the truck can move alongside a slab form the extension chutes can enable coverage of a relatively large swath as the truck moves along. The flat angle of the chute, however, makes it necessary to assist the concrete to move down the chute, unless the truck is elevated slightly above the slab or a high-slump concrete is used. (See Figure 15-5.)

Figure 15-5
Directing the chute and helping the concrete flow. (Courtesy of PCA)

Concrete can be discharged directly into any structural form as long as the mixer discharge is sufficiently high above the top of the form. How high depends on the kind and size of mixer and how close to the form the truck can be spotted; usually a height of about 4 feet is sufficient.

Chutes. Some sort of chuting arrangement enters into many concrete distributing systems for moving concrete to a lower level. Every truck mixer has two or three short lengths of chute that enable the truck to discharge concrete over an extended area because of the high discharge nature of the mixer.

Chutes should be of rounded cross section, made of metal or lined with metal, smoothed to prevent the concrete from sticking, and of the proper slope so that concrete of the required slump will slide, not flow, fast enough to keep the chute clean but so fast as to cause segregation. Stiff mixes require a steeper slope than wet or fluid mixes, and as long as there is no segregation or separation any reasonable slope can be tolerated. A slope of about 1:3 is about as flat as can be used, and slopes steeper than 1:2 are apt to give trouble with segregation. (See Figure 15-6.)

Figure 15-6
Truck discharge of concrete into continuous footing using several lengths of chute. Note the semicircular cross section of the chutes. (Courtesy of PCA)

Regardless of the slope, there is always some segregation at the discharge of the chute. For this reason, end control should be provided so that the concrete will drop vertically without segregation from the end of the chute. Two or more sections of metal drop chute "elephant trunk" will serve to control end segregation. (See Figure 15-7.) A mere baffle is not adequate. Care should be exercised to prevent water used for flushing the chute from entering the forms

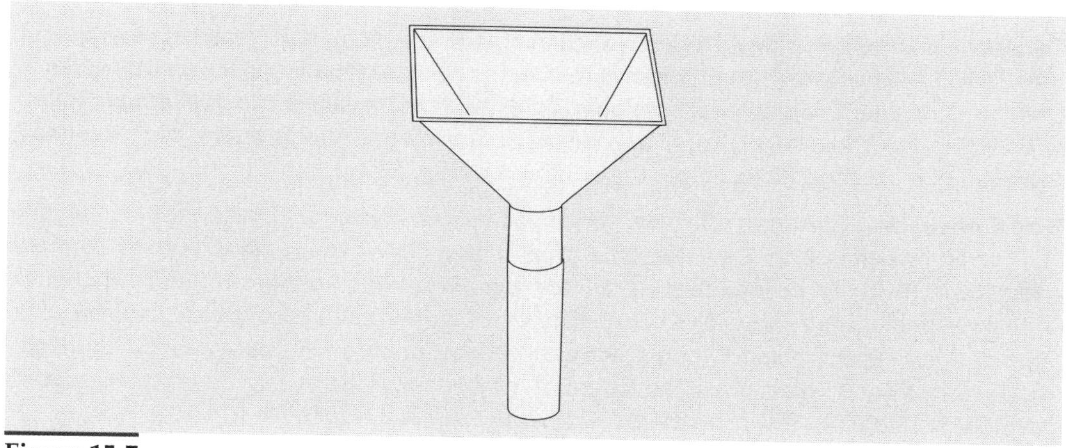

Figure 15-7
A collection hopper and drop chute confines the concrete and prevents segregation.

Buckets, properly designed and operated, are an excellent means of conveying concrete. Capacities range from less than 1 cubic yard for structural use to 12 cubic yards for mass concrete. Each bucket should have a capacity of at least one batch of concrete as mixed. This requirement is waived when truck mixers are used. Buckets can be handled by cranes, derricks, trucks, rail cars, helicopters or cableways. Moving by cranes is the usual method of handling buckets in building construction.

Cylindrical, bottom-discharge buckets are used in most structural applications. (See Figure 15-8.) Size of the bucket, depending on the capacity of hoisting or other handling equipment and the job conditions, is frequently less than 1 cubic yard but may be as high as 2 or 3 cubic yards. Buckets should have nonjamming gates, and the discharge rate should be controllable so that small increments of concrete can be discharged at a time. Gates are usually manually operated except on large mass concrete buckets that have pneumatic or hydraulic controls.

Best results with any bucket are obtained when the discharge gate is symmetrically located in the bottom of the bucket. Less segregation of the concrete results, and there is less tendency of a suspended bucket to kick sideways when concrete is discharged. Common practice when dumping concrete from the bucket directly into a narrow form is to attach a rubber boot or sleeve to the bottom of the bucket to direct the flow of concrete.

Figure 15-8
The workhorse of concrete buckets, the 1-cubic-yard general purpose bottom-dump bucket. (Courtesy of PCA).

Care should be taken to avoid shaking and jarring the concrete, as this causes segregation, especially if relatively high-slump concrete is being handled. Hauling buckets on trucks or cars for a considerable distance, especially if the buckets are subject to appreciable jolting or bumping, can cause segregation. In addition to detrimental effects on the concrete itself, segregation makes it difficult to get the concrete out of the bucket.

Belt Conveyors. Concrete is frequently moved by belt conveyor. One system, consisting of a conveyor belt system mounted on a truck mixer, allows the ready-mix truck to place concrete as much as 12 feet above grade or any place within about 40 feet of the truck discharge spout. (See Figure 15-9.) Other conveyor systems consist of portable conveyors, a series of conveyor flights for long-reach applications, or spreading conveyors with means of side discharge at points along the length of the conveyor. Belts can move concrete long distances horizontally and, to some extent, vertically. (See Figure 15-10.)

Figure 15-9
A belt conveyor mounted on a truck mixer places concrete up to about 40 feet of the truck discharge spout. (Courtesy of PCA)

Figure 15-10
Conveyor belt mounted on a crane. Multiple conveyor belts and crane-mounted belts can be used in series to place concrete rapidly over long distances. Proper chute connections between belts must be used to avoid aggregate segregation. (Courtesy of PCA)

Two problems associated with belts are segregation of concrete and loss of concrete and mortar that adhere to the belt. Segregation can be controlled by the use of discharge control as discussed under "chutes," and by the use of hoppers to feed the concrete onto the belt in a continuous stream or ribbon. Feed hoppers should be constructed in such a way that there are no dead areas where concrete might hang up. Loss of mortar on the return belt can be prevented by placing a rubber or other suitable scraper at the discharge end of the belt. Scraped mortar should be fed into the concrete-receiving hopper. Long conveyors, or series of conveyors, should be covered when ambient conditions make it necessary to protect the concrete from rain, hot sun, wind or other unfavorable climatic conditions.

Belts are not large, a 16-inch width being common. Troughing of the belt is accomplished either by the use of troughing rolls or by a continuous pan over which the belt passes, the advantage of the pan being less segregation of the concrete. Individual flights, commonly about 50 or 60 feet long, can be joined together to permit the moving of concrete over long distances. Transfer of concrete from one conveyor to another is accomplished by a swivel arrangement that supports the conveyor ends and guides the concrete from one belt to the next. The swivel enables the receiving conveyor to be swung through an arc to cover a large placing area. In addition, the end conveyor telescopes under the preceding flight to provide still more maneuverability. (See Figure 15-11.) The maximum angle for elevating concrete, about 30 degrees, is reduced when high-slump mixes are transported. Also, the belt capacity, which is usually in the range of 50 or 60 cubic yards per hour, with rates as high as 100 cubic yards per hour reported, is curtailed when wetter mixes are used. Power for operating the belt is supplied either by a gasoline engine through a hydraulic drive or by an electric motor.

Figure 15-11
Placing concrete by belt conveyor. A metal hopper and dropchute (elephant trunk) prevents the concrete from segregating as it leaves the belt. (Courtesy of PCA)

Wheelbarrows and Buggies. Pneumatic-tired wheelbarrows can be used for moving small amounts of concrete for short distances. (See Figure 15-12.) About 200 feet is the maximum horizontal distance for a wheelbarrow. One person with a wheelbarrow can move a maximum of about $1^1/_2$ cubic yards of concrete per hour.

Figure 15-12
Transporting and handling concrete by wheelbarrow. (Courtesy of PCA)

Two-wheel buggies or carts may be either manually driven or power driven. Hand-operated carts can carry about 6 or 8 cubic feet each, with a maximum haul of about 200 feet. One person and cart can move a maximum of 5 cubic yards of concrete per hour under favorable conditions. A power-driven cart has a capacity of up to $^1/_2$ cubic yard and can move a maximum of 20 cubic yards of concrete per hour on a moderate length of haul. Maximum haul should not exceed 1000 feet. (See Figure 15-13.)

Figure 15-13
Power buggies can move all types of concrete over short distances. (Courtesy of PCA)

Runways for wheelbarrows and buggies should be rigid, smooth and level. (See Figure 15-14.) They are best supported by the forms, although sometimes they may be at least partially supported on the reinforcing steel, provided the steel is secured in such a way that it will not be displaced by the weight and impact of the moving carts.

Other Methods. On many jobs it has been found convenient to discharge the truck mixer into a hopper from which the concrete can be loaded into carts, buckets or other conveying equipment. It may be necessary to elevate the truck on a ramp to provide sufficient headroom. Employing a surge hopper in this manner provides some flexibility of operation and occasionally permits the truck to unload without delay.

Small side-dump cars operating on portable sections of monorail track have been used for transporting concrete about the job. They are limited to horizontal, or nearly horizontal, movement. End-dump buckets transported by forklift have been used.

Hoists and elevator towers are used in high-rise building construction. The ready-mix truck can discharge into the hoist bucket directly or into a receiving hopper. At the upper end, the hoist bucket usually charges another hopper from which the concrete is taken by wheelbarrows or carts for distribution about the floor. (See Figures 15-14 and 15-15.)

Figure 15-14
Concrete being elevated by bucket and crane to upper floors of high-rise building construction. (Courtesy of PCA)

Figure 15-15
The tower crane and bucket can easily handle concrete for tall-building construction. (Courtesy of PCA)

Segregation. One of the most important considerations in handling concrete, as has been pointed out several times in this chapter, is the avoidance of segregation, or separation of the coarse aggregate from the mortar.

Concrete is not a homogeneous material and is subject to forces attempting to separate the component materials. This is discussed in Chapter 2. Separation should be prevented before it happens, rather than an attempt being made to correct it afterward. Concrete should drop vertically, regardless of the type of equipment it is coming out of or going into. Concrete should not be dropped through reinforcing steel or other objects that tend to separate it, nor should it be directed against the forms. Figure 15-16 illustrates several methods of minimizing segregation when discharging concrete from a belt or chute. If such end controls are employed, there should be little trouble with segregation in transporting concrete.

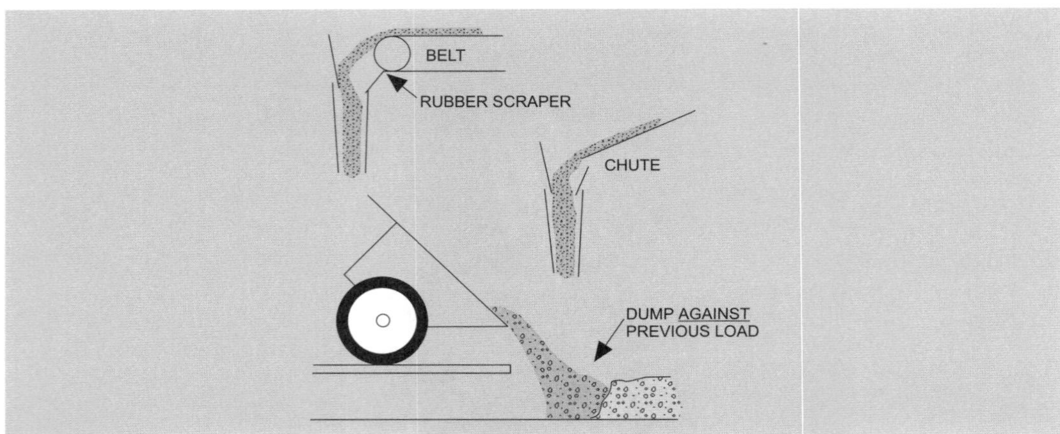

Figure 15-16
Various expedients can be employed to minimize segregation of concrete.

15.3. Pumping Concrete

Transporting concrete through a pipeline has been employed since the 1930s using heavy pumps that can handle concrete containing aggregate as large as $2^1/_2$ or 3 inches. These heavy pumps are especially useful on bridge, dam and similar heavy construction projects. In building construction, great use is made of the small-line pumps.

Conditions that lend themselves especially well to pumping exist on those sites and areas where access is limited and the site is crowded with materials and equipment, such as many city building sites. Nearly all pumps, large and small, can pump vertically, thus lending themselves very well to high-rise building construction. A pump takes up little space and can be located any place that the ready-mix truck can reach. Conveying hose and pipe are easily placed out of the way and take little room. In locations difficult to reach with the ready-mix truck, such as on a steep hillside, a pump can easily move the concrete over obstructions that would be exceedingly difficult for the trucks to overcome.

Equipment. A large variety of pumping equipment is available that is suitable for almost every concreting job. Initially, the so-called small line pump evolved from grout and plaster pumps. Improvements have been continuous, with larger sizes, greater pressures and more capacity. Present pumps are available with capacities in excess of 150 cubic yards per hour and capable of pumping as much as 100 feet vertically and 3000 feet horizontally. Pressure on the concrete may reach 2500 psi. Placing booms can reach over 100 feet. Many of these rigs are small and portable enough to be mounted on a trailer that can be towed behind a pickup truck. See Figure 15-17. Others are truck mounted. See Figure 15-18. Capacities depend on size of pump, length of horizontal distance or vertical lift, slump, mix proportions and maximum size of aggregate.

Figure 15-17
Small pumps like this one are popular for small jobs (Courtesy of PCA)

Figure 15-18
Truck-mounted pump and boom can conveniently move concrete vertically or horizontally to the desired location. (Courtesy of PCA)

There are several makes of piston pumps, either hydraulically or mechanically driven, most of them with two pistons alternating on the power stroke. (See Figure 15-19.) The large (6-inch to 10-inch diameter) low-velocity pistons force the concrete through reductions to the pipe or hose, which may be from 2 to 4 inches in diameter. Concrete from the ready-mix truck is deposited in the holding or loading hopper leading directly to the loading chamber, passing through valves into the unloading chamber where the piston forces it into the pipe or hose for delivery to the forms.

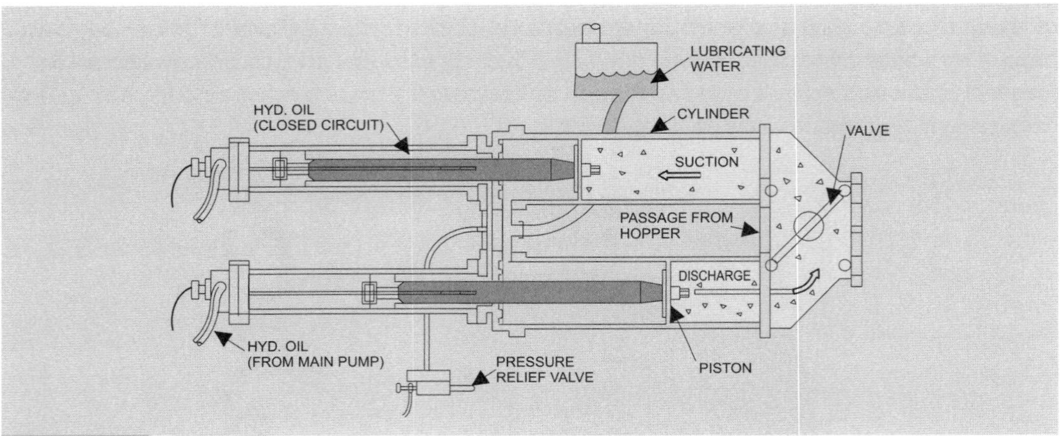

Figure 15-19
Typical twin-cylinder piston-type concrete pump. (Courtesy of PCA)

Another type of pump is the pneumatic type in which the concrete is carried through the pipe by air pressure in a manner similar to shotcrete; however, discharge is at low velocity. (See Figure 15-20.)

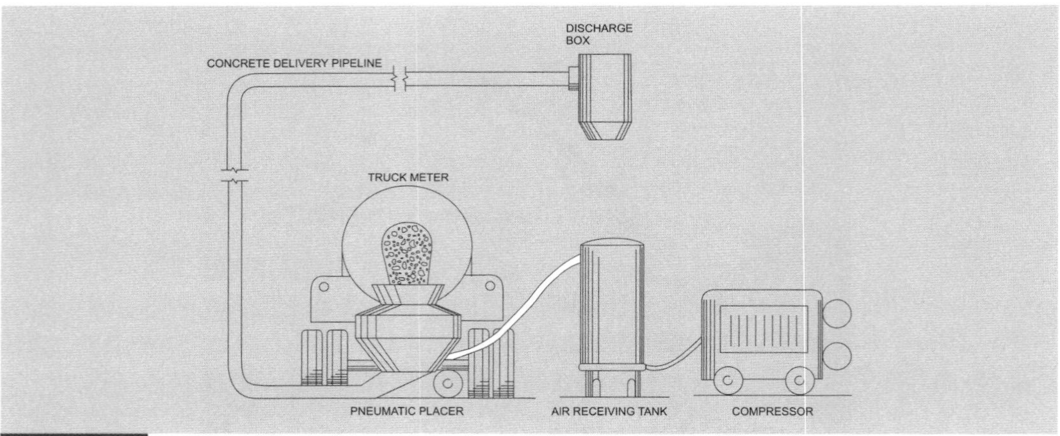

Figure 15-20
Typical pneumatic concrete pumping system. (Courtesy of PCA)

Delivery lines from the pump can be either rigid pipe or hose. Pipe is usually of steel. Some usage has been made of aluminum pipe, but it has been found that a reaction between the concrete and the aluminum causes the formation of hydrogen gas, which seriously reduces the strength of the concrete. For this reason aluminum pipe should not be used. Flexible conduit and hose, although they develop more resistance to pumping and for this reason should be kept at a minimum on any piping layout, provide flexibility in distributing the concrete. Rubber hose should not be inserted adjacent to the pump, especially in a long line. Change in direction is best accomplished with long radius curves rather than elbows. Minimum pumping resistance will be achieved with a maximum of straight runs of pipe, the use of pipe free from dents and encrusted concrete, a minimum use of bends and hose, and the largest practical diameter of pipe. Boom extensions are available that aid in getting the conduit over obstructions to reach the forms.

Materials. The size, shape, grading and proportions of aggregates are all important in obtaining a pumpable concrete. Some operators suggest that the maximum size of coarse aggregate should be no larger than about 40 percent of the conduit diameter. If we follow this rule, then conduits and maximum aggregate sizes would be about as shown in Table 15.1.

Rounded or subrounded aggregates make better mixes for pumping than aggregates containing a large proportion of crushed material, although the latter can be used satisfactorily. If it becomes necessary to use crushed rock coarse aggregate, every measure should be taken to assure maximum plasticity of the concrete as described in the section on mixes.

TABLE 15.1
RELATIONSHIP OF CONDUIT SIZE TO MSA

PIPE OR HOSE I.D., INCHES	MAXIMUM AGGREGATE SIZE, INCHES
$1^1/_2$	$^5/_8$
2	$^3/_4$
$2^1/_2$	1
3	$1^1/_4$
4	$1^1/_2$

Grading of the aggregates should conform to the requirements of the code or specifications under which the work is being performed. Sand must contain adequate fines, with 15 to 20 percent passing the No. 50 screen and at least 3 percent passing the 100-mesh. In 1-inch or $1^1/_2$-inch mixes, the total aggregate should contain about 10 to 15 percent pea gravel.

Poorly graded aggregates are apt to result in concrete that is difficult to pump, the friction of such concrete in the conduit offering so much resistance that maximum pumping efficiency is not achieved. Skip or gap gradings are not suitable.

Any of the usual types of portland cement can be used in concrete to be pumped.

Good construction practices govern the use of admixtures, and neither special limitations nor tolerances need to be applied as far as pumping is concerned. During hot weather a set retarder may be desirable if concrete is being pumped a long distance.

Any admixture that improves workability of the concrete will improve pumpability. Water reducers, air entrainers and fine mineral admixtures can be of advantage in the same way that they improve any concrete mixture, and their usage should be governed by the same rules and standards. Fly ash is generally beneficial. Use of a superplasticizer can significantly reduce line pressure and pumping resistance. The use of special proprietary pumping aids may improve pumping by reducing friction and bleeding in the piping system.

Mixes. Concrete for pumping must be plastic and workable. Many people have therefore felt that a very high percentage of sand, as much as 65 percent of total aggregate for a 1-inch maximum aggregate concrete, is necessary. However, pumps have been improved to the point where they can handle mixes with a higher proportion of coarse aggregate. Although a slightly oversanded mix is more pumpable, all else remaining equal, excessive oversanding is not considered necessary. The percentage of sand in the mix should be based on the void content of the coarse aggregate, a properly proportioned mixture containing enough sand to fill the voids in the coarse aggregate and enough paste to coat the aggregate particles. If this condition is achieved, a pumpable concrete can be realized even with relatively poorly graded coarse aggregate. However, the better the overall grading, the better the pumping results will be.

If harsh, angular coarse aggregate must be used, it is especially important to make sure that both the fine and coarse are well graded, as previously described. A somewhat higher cement content and percentage of sand than normal will be helpful, and the maximum size of coarse aggregate should be about 3/4 inch. Entrainment of about 4 percent air will be beneficial. Some improvement in workability without excessive water may result from the use of a water-reducing admixture if this can be done under careful technical control.

A plastic, workable mix with a slump of about 3 to 6 inches is best. If the slump is too low, the friction of pumping is higher, resulting in less volume of concrete per unit of pump horse-power. However, segregation tendencies are lessened. Mixes that are too wet, on the other hand, are more likely to segregate, a condition that is liable to result in line blocks or plugs. Entrained air is sometimes desirable to enhance workability, especially with poorly graded or angular aggregates. Cement content is usually around 500 pounds per cubic yard, with a lower limit in the neighborhood of 420 pounds per cubic yard. Richer mixes can be pumped successfully. Pumps can handle 2-inch aggregate mixes, but 3/4-inch or 1-inch mixes are commonly used. (See Table 15.2 and Figure 15-21.) In general, pumping difficulties that are caused by the concrete mix result from attempting to pump concrete with too much or too little slump; harsh, unworkable mixes caused by poorly graded or angular aggregates; concrete containing porous, highly absorptive aggregate, or aggregate too large for the machine or conduit.

TABLE 15.2
CONCRETE MIXES THAT HAVE BEEN PUMPED

MAXIMUM AGGREGATE SIZE	3/8 IN.	3/4 IN.	1 IN.	1 1/2 IN.	1 1/2 IN.
Percentage of sand	60	43	45	39	40
Cement pounds per cubic yard	565	525	525	510	565
Slump, inches	4	3 to 4	5	3 to 4	3 to 4

At these slumps, water contents will normally produce concrete with well over 3000 psi compressive strength at 28 days.

Figure 15-21
Pumped concrete to a structural floor. Concrete is delivered in a continuous stream at low velocity.

A loss of slump during pumping is normal and should be taken into consideration when proportioning the concrete mixture. A loss of $^1/_2$ inch per 100 feet of conduit is not unusual, the amount depending upon ambient temperature, length of line, pressure used to move the concrete and moisture content of aggregate at time of mixing. The loss is greater for hose than for pipe and might be as much as $^3/_4$ inch per 100 feet.

Pumping. Before starting to pump concrete, the conduit should be primed by pumping a batch of mortar through the line to lubricate it. A rule of thumb is to pump 5 gallons of mortar for each 50 feet of 4-inch hose, using smaller amounts for smaller sizes of hose or pipe. Concrete is dumped into the pump hopper before the last of the mortar disappears into the pump loading chamber, pumped at slow speed until concrete comes out the end of the delivery hose, then sped up to normal pumping speed.

Concrete should be kept in the pump hopper at all times, which makes necessary the careful dispatching and spacing of ready-mix trucks. From the standpoint of the pumper, it is better to have a ready-mix truck stand by for a few minutes than to shut down the pump to wait for the concrete. However, having ready-mix trucks standing by waiting to discharge their concrete is neither efficient nor desirable. Nevertheless, once pumping has started it should not be interrupted, as concrete standing idle in the line for any length of time might cause a line block.

Other causes of line blocks are slump too high; harsh, unworkable mix resulting from poor aggregate grading; mix too dry or undersanded; bleeding of the concrete; a long line exposed to the hot sun; improper adjustment of the pump valves; dirty and dented pipe sections; or a kinked hose. When a block occurs in a hose, it can sometimes be located by walking back on the hose from the nozzle end and then dislodging by rapping the hose. It is dangerous to attempt to dislodge a block by applying more pressure at the pump, as this will only worsen the condition and perhaps damage the equipment. Most pumps are designed so that the direction of pumping can be reversed, thereby drawing the concrete out of the line and sometimes relieving the block.

During an extended delay at the placing end, it is good practice to run the pump for a few strokes every few minutes, even though the concrete has to be wasted, in order to avoid a block in the line. This is especially necessary in hot weather.

Concrete can be pumped upward, but downhill pumping has a few special problems because the concrete is apt to separate or segregate in the pipe unless there is resistance to pump against. Resistance can be provided by a valve at the discharge end that can be adjusted to restrict the flow of concrete, or by inclining the final lengths of pipe or hose upward.

If it is necessary to run the line up and over an obstruction, then down a considerable distance, it may be advisable to install an air-release valve at the highest point of the line to prevent an accumulation of air that could result in a line block.

Upon completion of pumping, the pump and line are cleaned out by pumping clean water. To accomplish this, place large sponge or rubber flush plug is placed in the feed end of the conduit. (In the case of the vacuum-roller machine, the sponge is inserted in the pump suction line.) After being washed out, the hopper is filled with water and the pump started, forcing the remaining concrete out of the conduit, followed by the sponge or plug and water. One type of pump has a separate wash-out pump that supplies water to force the plug through the conduit. Hoppers, heads, cylinders and all parts must be cleaned to remove adhering concrete and mortar.

Pumping Lightweight Concrete. The main problem in pumping lightweight concrete is the loss of plasticity of the concrete resulting from absorption of water by the porous aggregate particles when the concrete undergoes pressure in the line. Some operators report a slump loss of about 3 inches between the time the concrete is dumped in the pump hopper and when it comes out the hose. This has made necessary a slump of 7 inches or more at the pump. Such abnormally absorbed water subsequently escapes from the aggregate upon release of pressure after the concrete has been placed, resulting in undesirable excessive bleeding. Success in pumping can sometimes be at least partially assured if natural sand can be used for the fine aggregate, and if the coarse aggregate is prewetted before batching. In areas where they are available, it has been reported that the use of coated coarse aggregate obviates the high-absorption problem.

If exposure or other considerations do not present restrictions, the use of coarse aggregate that has been treated by a vacuum absorption process has been found to give good results. This process fills the pores of the aggregate with water by treatment at the manufacturing plant; thus, there is no absorption of water during pumping to cause a loss of plasticity of the concrete.

15.4. Placing Concrete

The correct term to describe the act of depositing and consolidating concrete in forms is placing. Concrete is not "poured," although this expression is commonly accepted in our everyday language on the job. The use of this word owes its beginning to the days when sloppy, wet concrete really was poured and permitted to flow into place.

The first uses of concrete were in dams, heavy foundations and other relatively massive structures in which an earth-moist concrete was compacted in place with much hand labor. Later, when reinforced concrete was introduced, the narrow forms containing reinforcing steel required a wet, plastic consistency, and concrete passed into the era of very fluid mixes that were poured in place. However, research pointed out the faults and shortcomings of high water contents. In about 1930, the introduction of high-frequency vibration for consolidating concrete presented a practical way of handling relatively dry and unworkable mixes. Today, practically all concrete, regardless of its application, is consolidated by vibration, and the need for wet mixes no longer exists.

It should be kept in mind that the slump of concrete is prescribed by specifications and is not to be determined by the condition or type of construction equipment. If the equipment will not handle concrete of the specified maximum slump, then the correction lies in the direction of proper equipment, not wetter mixes.

Depositing Concrete in Forms. A basic rule is that concrete should be deposited as nearly as possible in its final location. It should be dropped as nearly as possible vertically.

Figure 15-16 shows correct and incorrect methods of placing concrete under various conditions. As long as the concrete is deposited near its final position without undue segregation, any method is acceptable. Drop chutes, or elephant trunks, are usually necessary to guide the concrete through congestions of reinforcing steel or other items. A crane bucket, moving along the top of a wall form, is satisfactory as long as the concrete is confined in a drop chute attached to the bucket. A hopper or series of hoppers (see Figure 15-7), with drop chute attached, are commonly used when placing concrete in a wall form. Any reasonable height of unconfined fall is acceptable, provided the form is clear and open so that the fall is vertical and segregation does not occur. In practice, it is usually necessary to limit free fall to a few feet because of obstructions in the form. As a practical safety measure, specifications usually limit the height of free fall.

Placing should be rapid enough to cover encrustation of mortar on steel or forms before the mortar dries. If the encrustation is on areas within the lift to be completed within a few hours it need not be removed. In using elephant trunks, the workers should be cautioned to avoid pushing the bottom section at a considerable angle from the vertical. The bottom section should be vertical, although upper sections may be at an angle to facilitate placing of the concrete.

Concrete in walls, footings, beams or any other structural components of appreciable height should be placed in horizontal layers not exceeding about 18 inches in depth, unless another thickness is specified, starting at the ends or corners of the forms and working toward the center. The first layer on a rock foundation or construction joint should be preceded by a layer of mortar, not over 1/2 inch thick, well broomed into the surface, or by a layer of concrete containing one-half the amount of coarse aggregate in the regular mix, spread to a thickness of at least 2 inches. A satisfactory mortar is made by omitting the coarse aggregate from a batch of concrete, using just enough water to provide a soft, mushy consistency. Any admixture used in the concrete should also be included in the mortar batch. Subsequent layers, continuing to the full height of the structure, should be placed and consolidated before the underlying layer has hardened. As the top of the placement is approached a drier concrete consistency should be used.

In some work, such as beneath openings in walls, it may be necessary to move concrete a short distance horizontally. Such horizontal movement should be kept to a minimum.

In placing concrete in high, thin walls or similar structural units, it is common practice to provide ports or windows in the forms. If possible, these windows should be made on a surface that will not be exposed to view in the finished structure, such as the back side of a wingwall on a highway structure. When the level of the fresh concrete within the structure approaches the window, the hole should be closed as tightly and neatly as possible. Because of the danger of segregation resulting from a high velocity stream of concrete entering the form at an angle, and because of the surface blemishes usually resulting in the area where the hole was closed, it is best to avoid use of these ports, if at all possible, or provide a collecting hopper outside the opening.

Concrete in cast-in-place piles and deep caisson footings must, of necessity, be dropped a considerable distance. Placement should be as nearly continuous as possible, as consolidation in the lower portion of the footing depends upon the impact of succeeding increments of concrete. A plastic consistency of about 4-inch slump should be adequate, although a particularly deep or narrow form may necessitate a slump as high as 6 inches.

In placing a slab, batches of concrete should be placed against or toward the preceding ones, not away from them. Batches should not be dumped in separate, individual piles. If the slab is on a slope, placing should start at the lower end of the slope.

In a structure consisting of monolithic columns, beams and slabs, concrete should be placed to the top of the columns or beams, then allowed to set for two or three hours, depending on the weather, for settlement to take place. The slab may then be placed. (See Figure 15-22.) If this procedure is not followed, cracks are liable to form where the slab joins the beam or the beam joins the column. (See Figures 6-6 and 6-7.)

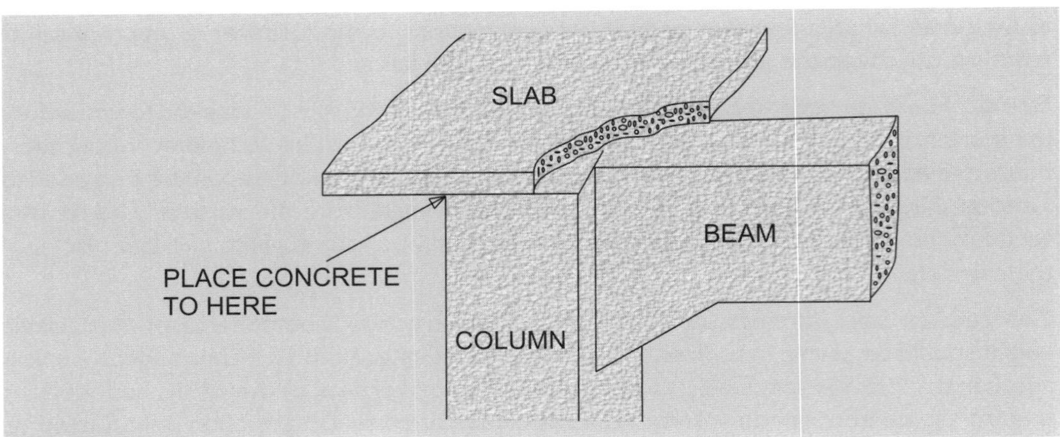

Figure 15-22
A delay in placing concrete where structural members join can help to reduce the incidence of "subsidence" cracking.

Waste molds, or other molds for intricate architectural details, should be protected from accumulations of mortar or other damage while the underlying concrete is being placed. One method of accomplishing this is to cover the inside of the mold with canvas, polyethylene or similar material, which is removed when the level of the concrete in the form reaches the mold.

Mass concrete in dams and similar massive structures is usually placed in lifts of 5-foot or $7^1/_2$-foot depth, each lift consisting of several layers. To avoid cold joints, these layers are carried across the form in a series of steps, the first step being the mortar, then step 2 and so on until the final step at the top. Concrete contains cobbles as large as 6 inches in diameter, has a low cement content and is placed at a slump of 1 or 2 inches or even less. Segregation is an ever-present problem. Buckets, which may have a capacity of as much as 12 cubic yards, should have full-opening bottoms that discharge the concrete vertically and quickly.

Placing Concrete During Rain. Under rain conditions, trouble can be forestalled by proper preparation and planning. A supply of protective coverings should be available nearby. Placing of concrete should not be commenced during a rainstorm but should be delayed until there is a reasonable assurance that the placement can be completed before the rain starts. If the rain starts while concrete placing is in progress and the work must be continued because a construction joint cannot be made, no damage will be done during a light drizzle, provided the following precautions are taken:

1. Place the concrete at a slightly lower slump.
2. Dry up puddles of water collected on the foundation or old concrete in the joint before new concrete is placed.
3. Cover the working area with tarps or tents, and keep them in place until the concrete has set.
4. Keep the surface of the new concrete on a slight slope so water will run off.
5. Avoid working the surface of the new concrete. After the concrete reaches grade, a slight slope should be provided, if feasible, for drainage.
6. If the rain is so heavy that it is not possible to dry up the puddles or keep the rain from washing the surface, work should be discontinued. The inspector should consult his or her supervisor regarding the placing of bulkheads and dowels and making a joint.

During thunder showers of short duration, it is frequently possible to cover the forms with a temporary cover and suspend work until the storm passes.

Special Placing Methods. There are a number of occasions that present unique problems in handling, placing and consolidating concrete. Among these are slip forms, underwater placing, preplaced aggregate concrete and several others, all of which are discussed in Chapter 22.

In some cases it becomes necessary to modify good concreting practices. When this is done, it is done knowingly and carefully, with safeguards and controls to compensate for the modifications. Modification of good practices should not be construed as license to eliminate all controls.

15.5. Consolidation

Once the concrete has been deposited in the forms, it has to be compacted or consolidated, to make it into a solid, uniform mass, without voids, rock pockets or sand streaks. Many years ago, consolidation was accomplished by laborers wielding a variety of spades, tampers and similar tools. Now nearly all concrete is consolidated with high-frequency vibrators. Vibrators come in many sizes and kinds but all can be grouped as either internal immersion vibrators or external vibrators, depending on whether they operate immersed in the concrete or apply their vibration externally.

Equipment. Internal spud vibrators to be immersed in the concrete, operating at speeds of 5500 to 15,000 vpm (vibrations per minute), range in size from 3/4-inch diameter to 6 1/2-inch diameter with head lengths from 10 to 28 inches. (See Figure 15-23.) The larger ones for mass concrete operate in the lower range of frequency; many require two workers to handle them. Most vibrators are either electric or air-powered, and some are driven by small gasoline engines. The motor may be in the head of the vibrating unit, or the vibrating unit may be connected to the motor by means of a flexible shaft. This classification includes gang-mounted and tube vibrators used on paving machines, as well as vibrators for structural concrete. Vibrators used in building construction include those from 3/4-inch to 3 1/2-inch diameter, with speeds from 8000 to 15,000 vpm. Vibration is accomplished by the rotation of an eccentrically loaded shaft. Some manufacturers furnish a frequency converter to change normal 60-cycle electric current into a high-frequency current, the better to operate the vibrator at these high speeds.

Figure 15-23
Different types of spud vibrators. (Courtesy of PCA)

External vibrators include form vibrators that are attached to the forms and vibrate the concrete by vibrating the forms, surface vibrators that apply vibration to the surface of a slab and vibrating tables used in precasting plants.

Form vibrators operating at a minimum speed as low as 1000 vpm are attached to the exterior of the mold or form. They are used in locations where it is not possible to use internal vibrators, such as in tunnel linings or heavily congested forms; they are also used for making pipe, masonry units and many other types of precast concrete. Pneumatically driven units develop vibration by the rotation of an eccentric weight. The speed can be varied by changing the volume of air supplied. One type, consisting of a loose weight inside a circular housing, is known as the police whistle because of its construction. Electric vibrators are either the rotary type or electromagnetic type, the latter consisting of a heavy armature that vibrates at synchronous speed (3600 vpm on 60-cycle current).

The second group of external vibrators includes surface, pan or screed vibrators that operate on the surface of a floor, slab or pavement. (See Figures 15-24 and 15-25.) Minimum frequency should be 3000 vpm. Consolidation of the concrete in thin slabs is accomplished by drawing a vibrating screed or strike-off unit slowly over the surface. Slabs up to 8 inches thick can be consolidated adequately. Thicker slabs require additional internal vibration. Screed lengths as long as 40 feet have been used. In addition to consolidating the concrete in the slab, the unit strikes off the surface and prepares it for final finishing. Pans, grids and rollers equipped with vibrators of some kind fit into this category.

Figure 15-24
This vibrating screed is being used to straight-edge the concrete to proper grade. The edge forms are actually the screeds. (Courtesy of PCA)

Figure 15-25
Laser screed leveling an industrial floor slab. Laser screeds are used primarily for placing high-quality, flat, interior concrete floors. (Courtesy of PCA)

A third group covers the table vibrators that are used in precasting plants. These are not normally encountered on construction sites. They accomplish their compaction either by means of external vibrators attached to the table or by means of a very slow, high-amplitude vibration induced by raising and dropping the table on which the form for the concrete unit is attached.

The frequency of vibration can be determined by the use of a vibrating reed tachometer. The instrument can be held against the vibrator while the latter is immersed in concrete, or it can be held on the formwork nearby.

Operation. Concrete should be consolidated to the maximum practicable density, so that it is free of pockets of coarse aggregate and entrapped air and closes snugly against all surfaces of forms and embedded materials. Vibrators should be applied to the concrete immediately after it is deposited. In consolidating each layer of concrete, the vibrator should be operated in a nearly vertical position (see Figure 15-26), and the vibrating head should penetrate slightly into and revibrate the concrete in the upper portion of the underlying layers. (See Figure 15-27.)

Application of vibrators should be at points uniformly spaced, close enough together to ensure complete consolidation (usually not more than twice the radius over which the vibration is visibly effective), and of sufficient duration to thoroughly consolidate the concrete, ordinarily 5 to 15 seconds per insertion. Vibrators should be used sparingly or not at all in wet, high-slump concrete. Harsh, low-slump concrete requires more vibration than concrete of a moderate slump. Vibrators should not be dragged through the concrete nor should they be used for the purpose of moving the concrete about in the forms.

The vibrating head should not be held against the reinforcement. The head should be withdrawn from the concrete slowly and immediately reinserted. Allowing the vibrator to run for a long period of time with the head out of the concrete can damage it. The running vibrator should not come in contact with the form, as a rough spot may be caused on the surface of the concrete.

Vibrators cannot recombine concrete that has already segregated. If segregation has occurred because of improper handling techniques, concentrations of rocks should be broken up by shoveling the rocks onto areas of concrete containing a sufficiency of sand.

Completion of vibration is indicated when the surface of the concrete takes on a flattened glistening appearance, the rise of entrapped air bubbles ceases, the coarse aggregate blends into the surface but does not disappear, and the vibrator, after an initial slowdown when first inserted into the concrete, resumes its normal speed.

Overvibration can and does occur, especially when overly wet mixes are being placed. The correction is to reduce the slump first, then adjust the amount of vibration. If concrete has been overvibrated, the coarse aggregate will have sunk below the surface, and the surface may have a frothy appearance.

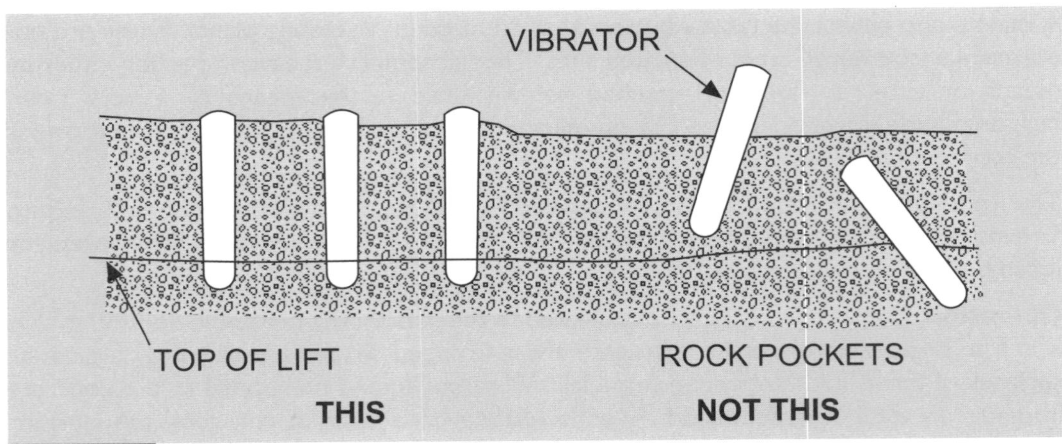

Figure 15-26
The vibrator should be as nearly vertical as possible when it is immersed in the concrete.

Figure 15-27
Vibrating concrete with a spud vibrator. (Courtesy of PCA)

Revibration occurs when the vibrator, in consolidating a layer of concrete, penetrates into the layer below to weld or unite the two layers. The intentional revibration of previously placed concrete occurs when a running vibrator will sink of its own weight into the concrete. The general effect of revibration is improved strength, durability and appearance. It tends to reduce water pockets under horizontal reinforcing steel and reduces the appearance of settlement cracks. In moderation, the net effect of revibration is beneficial. It must, however, be done by experienced operators under adequate supervision. (See Table 15.3.)

When consolidating concrete against intricate form faces for architectural work where the surface must be as flawless as possible, vibration should be supplemented by spading or rodding along the forms, especially in corners and angles. Placing the concrete in thin layers is also beneficial.

Surface voids on the concrete and how to minimize them are discussed in Section 6.12.

Form vibrators are occasionally used in structural applications but should not be depended on to consolidate concrete if internal spud vibrators can penetrate to the area in question. This may take a little ingenuity in forming and placing but is well worth the extra effort. The configuration of some structures requires that some of the concrete be flowed in place, in which case form vibrators can be of considerable value.

The use of surface vibrators is discussed in Chapter 16.

Guidelines for vibrating concrete are summarized in Table 15.3.

TABLE 15.3
GUIDELINES FOR VIBRATING CONCRETE

Proper Vibration	1. Causes concrete within "field of action" to act like a liquid 2. Increases strength and bond between concrete and rebar and decreases permeability 3. Decreases cold joints, honeycombing, excessive entrapped air and segregation
How to Vibrate	1. Insert vibrator vertically, allowing it to penetrate under its own weight and vibrating action (of its own accord) to bottom of lift and at least 6 inches into previous lift 2. Allow vibrator to contact bottom of form and tilt slightly (to effectively vibrate bottom of first lift) 3. Hold at bottom of lift for 5 to 15 seconds 4. Lift vibrator up at about same rate as downward rate
Spacing Tips	1. Watch concrete to determine vibrator's "field of action" 2. High-powered vibrators and high-slump concretes have larger "fields of action" 3. Insert vibrator so "fields of action" overlap 4. Rule-of-thumb . . . "field of action" is roughly eight times vibrator head diameter
Stop Vibrating when	1. Concrete surface takes on a sheen 2. Large air bubbles no longer escape 3. Vibrator changes pitch or tone
Vibrating Don'ts	1. Don't use a vibrator to move concrete horizontally 2. Don't force or push a vibrator into concrete. Let it penetrate under its own weight and vibrating action
Revibration Produces Better Concrete	OK to revibrate if vibrator can sink into concrete under its own weight!

Slabs on Ground

One of the most important properties of any slab is a hard, wear-resistant surface. Of course, the hardness of a floor in a dwelling to be covered with carpet or other floor covering need not be of the same quality as a floor in a warehouse where it is subject to heavy trucks and abrasion. Nevertheless, any slab requires a durable surface, free of cracks, one that is smooth, in a true plane, and which meets the aesthetic and structural requirements of the structure. This surface is dependent on the quality of the concrete, the care used in placing the concrete, the use of proper finishing techniques and, finally, the curing. In this chapter we will discuss how to construct the slab up to the time finishing commences. Finishing is described in Chapter 17.

16.1. The Slab on Ground

Concrete slabs in this category include sidewalks, driveways and floors of all kinds, from light-duty residential floors to heavy-duty industrial floors. Street, highway and airport pavements are not included. All slabs must start with proper preparation of the site.

Subgrade. Side forms are set to elevation by the use of string lines fastened to stakes or other firm supports at the required elevation, or by measurement from existing structures. In either case the area on which the slab is to be placed is dug out to the depth necessary to provide the specified slab thickness.

The subgrade must be prepared by first removing all organic material such as grass, roots, etc., and removal of any soft or yielding soils, replacing such materials with a select granular soil. Pipe trenches and excavations must be backfilled and compacted. Backfill over the area, if required, should consist of a base of granular soil compacted in layers about 6 inches thick. This granular base should be permeable; hence, fine sands or silty soils should be avoided. Plastic clay should not be used. The subgrade should give uniform support to the concrete. If additional fill is required to bring the subgrade to the required elevation, it should be made of granular material compacted in layers about 4 or 5 inches thick. (See Figure 16-1.)

Figure 16-1
A concrete slab on ground is only as good as its subbase! Before placing any slab on grade, proper compaction of the subbase is essential, preferably by a vibratory compactor. Hand tamping should only be used along edges and in isolated areas where the compactor does not fit. (Courtesy of PCA)

The fill should extend about one foot beyond the edge of the sidewalk or driveway with a gentle slope to carry rainwater away.

Drainage. Elaborate provision for drainage of the subgrade is usually not necessary for sidewalks, driveways, patio slabs and similar areas. Floors may require special preparation. Good drainage is essential to any floor. Some specifications require that the finished floor elevation be a minimum distance above the final grade after landscaping; 6 or 8 inches is sometimes specified. Ground around the building should slope away from the building. In wet areas, special drains may be required. In industrial sites, good drainage usually is available and ordinarily should present no problem in constructing the floor. Perimeter drains and underdrains are nearly always provided in such areas.

A subgrade for any type of slab, if it is water soaked much of the time, should be covered with a permeable layer of sand, or sand and gravel, to a depth of at least 6 inches. A permeable base of this kind should be drained so that water will not accumulate.

Forms. Edge forms must be set accurately to line and grade. Edge forms can be of 2- by 4-inch or 2- by 6-inch lumber set on edge and nailed to stakes about 4 feet apart, with a stake at every joint in the form to hold both ends in alignment. The top of the stake should be below the top of the form board so that it will not interfere with striking off the concrete. All form stakes should be vertical and outside of the area to receive concrete. (See Figure 16-2.)

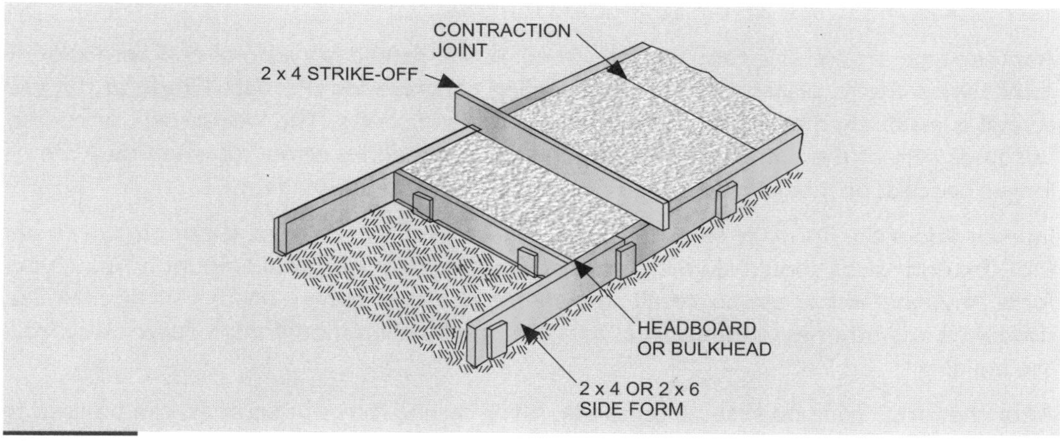

Figure 16-2
Steps in the construction of a sidewalk or similar slab are shown. Bulkheads are installed whenever concrete placing is interrupted long enough for the concrete to set.

Quarter-inch plywood with the outside grain vertical when in place can be used for forming curves. Stakes must be closer together on curves to hold the form in place. It is better to use a few extra stakes than to skimp.

When setting forms for a narrow walk or driveway it is sometimes more convenient if one side form can be carefully set to line and grade. The other side can then be set by means of a tape and carpenter's level. Attach the level to the edge of a length of 2- by 4-inch or 1- by 6-inch lumber slightly longer than the slab width. To provide the necessary cross slope, a block of wood is attached under one end of the board so that when the level indicates that it is horizontal one side form will actually be lower by an amount equal to the thickness of the block. (See Figure 16-3.)

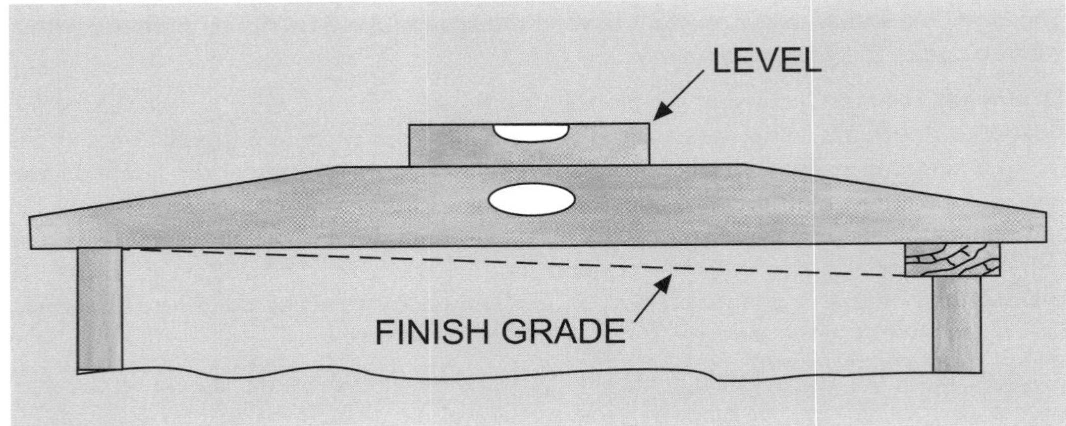

Figure 16-3
By inserting a block of the required thickness under one end of the straightedge, the correct elevation for the side form is established when the level bubble is centered.

Forms for slabs or floors of large area can be set in the same way but, because of the large area, it is necessary to provide intermediate guides to assure the correct grade or elevation of the slab. The guides are called screeds. A screed can be a 2-inch timber, pipe or other shape that can be accurately supported at the required elevation. After the concrete has been placed and struck off, the screeds are removed.

Another type of screed is called a wet screed. A wet screed is a strip of concrete about 6 inches wide that is placed just ahead of placing concrete for the slab. Grade of the wet screed is established by straightedging between grade stakes. The wet screed, of course, becomes part of the complete slab. Grade stakes should be removed when they are no longer needed, or they can be driven down flush with the subgrade.

Interior floors are normally level. If a slope is necessary for drainage, it can be $^1/_8$ inch per foot. Exterior slabs should always be sloped for drainage; the usual amount is $^1/_4$ inch per foot. Anything less is apt to result in ponding or "bird baths" on the surface. Walks, driveways and other exterior surfaces adjacent to a building should slope down away from the building.

After the forms have been set, a template, riding on the forms or screeds, can be used to provide a smooth, uniform subgrade at the proper depth to produce the designed thickness of slab.

Reinforcement. Many slabs constructed without reinforcement perform in a completely satisfactory manner; hence, reinforcement is rarely used. Reinforcement is occasionally specified for a floor to be exposed to particularly heavy loading. If reinforcement is used, it should be supported on chairs or mortar blocks that will hold the steel in place during construction until it is firmly embedded in the concrete. Metal chairs should be provided with sand plates or otherwise designed so they will not penetrate into the subgrade. See also Chapter 18.

Light reinforcement is sometimes specified for floor slabs of large area if heating ducts are embedded in the concrete, or for certain soil conditions.

Vapor Barrier. An interior floor should be built on a vapor barrier if there is any likelihood of moisture passing through the concrete to damage floor coverings, furnishings or equipment inside the building. Most specifications require a vapor barrier of some sort.

(See Figures 16-4 and 16-5.) An acceptable construction, shown in Figure 5-15, includes a sheet of 4-mil polyethylene sheeting overlaid with 2 to 4 inches of sand to protect the membrane and absorb some of the mix water out of the concrete. If the subgrade is gravelly with many rocks it may be desirable to place a layer of sand on the sub- grade before laying the polyethylene to protect the latter. Workers should be cautioned to use care to avoid punching holes in the vapor barrier. The barrier should be smooth and uniformly supported, lapped 6 inches at joints, and carefully fitted around pipes and other service openings, using tape or adhesive to hold it in place. When properly installed, a vapor barrier of this kind will meet the requirements of some codes for a termite barrier also. A vapor barrier is important even in desert areas, as irrigation of shrubbery around the building will cause moisture to be present under the slab. This moisture can pass through the slab by capillarity and cause damage in the interior of the building.

Figure 16-4
A vapor barrier is usually placed under all concrete floors on ground that are likely to receive an impermeable floor finish, or used for any purpose where the passage of water vapor through the floor is undesirable. (Courtesy of PCA)

Figure 16-5
Vapor retarder being placed in staggered pattern for hardwood flooring for gym. (Courtesy of PCA)

For certain noncritical occupancies, where there is no danger of the water table ever coming within several feet of the slab, a fill of coarse sand or pea gravel 4 to 5 inches thick beneath the slab will usually suffice. As long as a small amount of capillary moisture passing through the slab can be tolerated, this construction is satisfactory. The site must be well drained.

The Concrete Mixture. Whether to use air entrainment depends on local practices. All concrete to be exposed to freezing and thawing should contain entrained air, and unexposed concrete is improved by the entrainment of an appropriate amount of air. Mixes should be based on the concepts discussed in Chapter 12. The suggestions in Table 16.1, adapted from "Guide for Concrete Floor and Slab Construction (ACI 302.1R),"[16.1] can be used for proper construction of concrete floors and slabs on ground. Anyone involved in concrete floor and slab construction should have a copy of the ACI 302.1R document. The guide discusses how to produce good quality floors and slabs for various classes of service, emphasizing site preparation, concreting materials, concrete mixture proportions, concreting, workmanship and curing.

Concrete for exterior walks and driveways should contain at least 520 to 560 pounds of cement per cubic yard and should have a slump between 2 and 4 inches. A well-proportioned mix with this cement content and at this slump will have a good water-cement ratio. Concrete to be machine finished should have a slump at the lower end of the suggested ranges. Slump for floor concrete to be placed over a vapor barrier should be at the minimum also.

Placing the Concrete. General requirements for placing concrete are discussed in Chapter 15.

Figure 16-6
Dampening the subgrade, yet keeping if free of standing water, will lessen drying of the concrete and reduce problems from hot weather conditions. (Courtesy of PCA)

Figure 16-7
Once the concrete is discharged from the chute, come-alongs or square-ended shovels are used to quickly spread and level the concrete in the forms. This allows screeding to occur more readily. (Courtesy of PCA)

The subgrade should be saturated one day before the concrete is to be placed and should be damp at the time the concrete is placed. (See Figure 16-6.) There should be no wet or muddy spots. After the screeds, bulkheads and joint materials have been installed, the concrete is placed, beginning at a bulkhead or edge form and working toward the center. Concrete should be placed as close to its final position as possible.

The first operation is spreading the concrete, which is done with short-handle, square-end shovels or special hoes (come-alongs). (See Figure 16-7.) Rakes should not be used, as they cause segregation. If the concrete is fluid enough to flatten and spread out when it is dumped on the subgrade, it is too wet. After it has been spread, the concrete is vibrated to compact it. At this time, the level of the concrete should be slightly above the screeds.

Concrete is best consolidated by vibration, then struck off or screeded to the required grade by means of the working straightedge, which removes excess concrete by a sawing motion as it is advanced along the forms or screeds. (See Figure 16-8.) If the surface of the slab is to be crowned (that is, to be higher in the center so water will drain to both edges), the straightedge can be curved to fit the crown. A small amount of concrete should be pushed ahead of the straightedge to fill in low spots. Sometimes a vibrator is mounted on the straightedge, helping to consolidate the concrete as well as bringing it to the required elevation. (See Figure 16-8.) Low spots behind the strike-off should be filled by shoveled concrete, then struck off again. The work of striking off must be completed as rapidly as possible, before the appearance of bleed water on the surface.

Figure 16-8
Using the working straightedge, the workers are leveling the concrete between the side forms. (Courtesy of PCA)

Figure 16-9
Vibratory screed. (Courtesy of PCA)

TABLE 16.1
(Adapted from ACI 302.1R)
A—CONCRETE FLOOR CLASSIFICATIONS

CLASS	ANTICIPATED TYPE OF TRAFFIC	USE	SPECIAL CONSIDERATIONS	FINAL FINISH
1	Light foot	Residential surfaces; mainly with floor coverings	Grade for drainage; level slabs suitable for applied coverings; curing	Single troweling
2	Foot	Offices and churches; usually with floor covering	Surface tolerance (including elevated slabs); nonslip aggregate in specific areas	Single troweling; nonslip finish where required
		Decorative	Colored mineral aggregate; hardener or exposed aggregate; artistic joint layout	As required
3	Foot and pneumatic wheels	Exterior walks, driveways, garage floors, sidewalks	Grade for drainage; proper air content; curing	Float, trowel or broom finish
4	Foot and light vehicular traffic	Institutional and commercial	Level slab suitable for applied coverings; nonslip aggregate for specific areas and curing	Normal steel trowel finish
5	Industrial vehicular traffic—pneum atic wheels	Light-duty industrial floors for manufacturing, processing and warehousing	Good uniform subgrade; surface tolerance; joint layout; abrasion resistance; curing	Hard steel trowel finish
6	Industrial vehicular traffic—hard wheels	Industrial floors subject to heavy traffic; may be subject to impact loads	Good uniform subgrade; surface tolerance; joint layout; load transfer; abrasion resistance; curing	Special metallic or mineral aggregate; repeated hard steel troweling
7	Industrial vehicular traffic—hard wheels	Bonded two-course floors subject to heavy traffic and impact	Base slab—good uniform subgrade; reinforcement; joint layout; level surface; curingTopping—composed of well-graded all-mineral or all-metallic aggregate; mineral or metallic aggregate applied to high-strength plain topping to toughen; surface tolerance; curing	Clean-textured surface suitable for subsequent bonded topping
8	As in Class 4, 5 or 6	Unbonded toppings—freezer floors on insulation, on old floors or where construction schedule dictates	Bond breaker on old surface; mesh reinforcement; minimum thickness 3 in.; abrasion resistance and curing	Hard steel trowel finish
9	Superflat or critical surface tolerance required. Special materials-handl ing vehicles or robotics requiringspecifi c tolerances	Narrow-aisle, high-bay warehouses; television studios	Varying concrete quality requirements. Shake-on hardeners cannot be used unless special application and great care are employed. Proper joint arrangement	Special finishing techniques required

B—MINIMUM CEMENT CONTENTS*

MAXIMUM SIZE OF COARSE AGGREGATE, IN.	MINIMUM CEMENT CONTENT, LBS PER CU YD
$1\frac{1}{2}$	470
1	520
$\frac{3}{4}$	540
$\frac{1}{2}$	590
$\frac{3}{8}$	610

*These mixes are specifically for normal weight aggregate.
Different mixes may be needed for lightweight aggregate concrete.
Cement quantities may need to be greater in order to satisfy requirements of finishability and/or resistance to freezing and thawing, and to de-icing salts.

C—AIR CONTENTS FOR NORMAL-WEIGHT CONCRETES

MAXIMUM SIZE OF COARSE AGGREGATE, IN.	TYPICAL AIR CONTENTS OF NON-AIRENTRAINED CONCRETES	RECOMMENDED AVERAGE AIR CONTENT FOR AIR-ENTRAINED CONCRETES,PERCENT		
		Mild Exposure	Moderate Exposure	Severe Exposure
$\frac{3}{8}$	3.0	4.5	6.0	7.5
$\frac{1}{2}$	2.5	4.0	5.5	7.0
$\frac{3}{4}$	2.0	3.5	5.0	6.0
1	1.5	3.0	4.5	6.0
$1\frac{1}{2}$	1.0	2.5	4.5	5.5

Tolerances: For average air of 6 percent or greater, + 2 percent; for average air less than 6 percent, + $1\frac{1}{2}$ percent.

D—COMPRESSIVE STRENGTH AND SLUMP FOR EACH CLASS OF CONCRETE FLOOR*

FLOOR CLASS	28-DAY COMPRESSIVE STRENGTH, PSI	SLUMP, IN.
1	3000	5
2	3500	5
3	3500	5
4	4000	5
5	4000	4
6	4500	4
7 Base	3500	4
8 Topping	5000 – 8000	2
9 Superflat	4000 or higher	5

*On Class 2 through 9 floors, compressive strength of concrete slab before allowing construction traffic should be at least 1800 psi.
Maximum slump measured at point of placement, e.g., at the discharge end of the pump line if concrete is pumped.
A high-range water-reducing admixture (superplasticizer) meeting the requirements of ASTM C 494 (Type F or G) or a combination of admixtures meeting requirements of ASTM C 494 (Type A, C, D, E, F or G) may be used to increase slump level, providing the resulting mix conforms to all other requirements.

Immediately following the strike-off, the final compacting work on the soft concrete is accomplished by the use of either a darby or a bullfloat. (See Figures 16-10 and 16-11.) Only one of these tools should be used, as they both do the same thing. This operation fills the surface voids and removes the ridges and rough spots left by the straightedge. The darby produces a surface to a closer tolerance than the bullfloat, but the latter is easier to use on large areas. After this treatment, which must be completed before the bleed water appears on the surface, the concrete is left alone until it has stiffened enough for finishing.

Figure 16-10
The darby is used to smooth the surface of the concrete after the strike-off. It is also used in tight places where the bullfloat cannot reach. (Courtesy of PCA)

Figure 16-11
On large clear areas, the bullfloat can be used for the initial smoothing. (Courtesy of PCA)

On some jobs the tamper or jitterbug is used after the strike-off and before the darby or bullfloat. (See Figure 16-12.) The objective is to force the large aggregate particles slightly below the surface to facilitate the steps that follow. This can be accomplished on very dry mixes with a slump of 1 inch or less. The problem with the jitterbug is that when it is used on the more plastic mixes, even high-slump mixes, it makes a layer of weak, soupy mortar on the surface. It is best not to use the tamper on concrete with a slump of much over 1 inch.

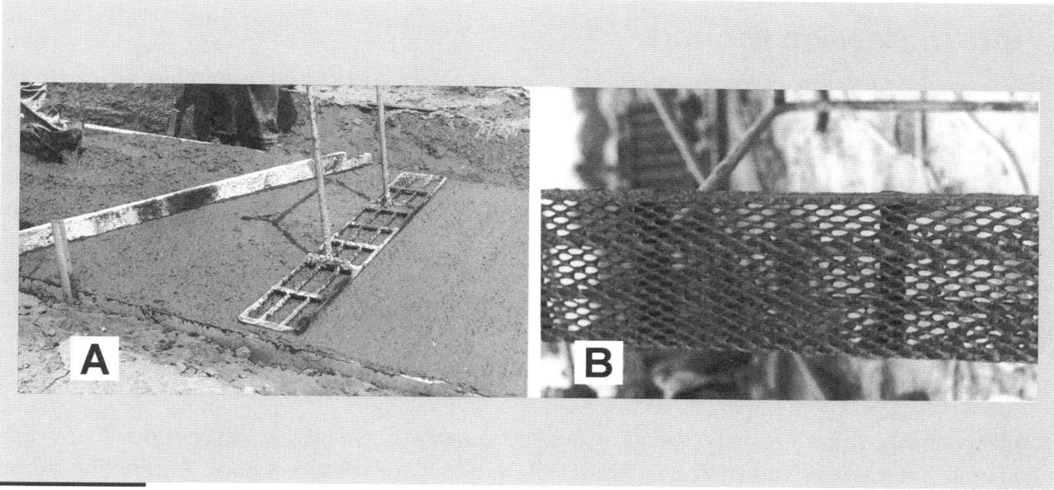

Figure 16-12
The tamper, sometimes called a jitterbug. Figure A shows the tamper in use on a sidewalk. A detail of the tamping face is shown at B.

Superplasticized Concrete. When a superplasticizing admixture is used to produce a self-leveling mixture with a slump of 8 to 10 inches some modification of placing procedures becomes necessary. This very workable concrete flows into place and can be brought to a smooth surface quickly with the application of significantly less labor than concrete without the admixture. The resulting slab is remarkably flat, requiring very little smoothing to prepare it for the finishing operations. (See Figure 16-13.)

Overworking must be avoided. Finishing is accomplished in the usual manner. A word of caution in the use of superplasticizers in concrete for floor and slab construction: the high slump super- plasticized concrete will have about the same drying shrinkage and potential shrinkage cracking as regular high-slump concrete.[16.2] Some specifications will state "4-inch maximum slump with or without admixtures" to guard against excessive shrinkage potential (cracking) of flatwork.

Figure 16-13
Flowable "superplasticized" concrete with a high slump is easily placed. Note that the concrete truck should not drive on the reinforcement mats! (Courtesy of PCA)

16.2. Joints in Slabs on Ground

The primary function of most joints in concrete is to control or minimize cracking by permitting shrinkage and other volume changes, or to permit relative movement of adjacent portions of the structure. There are three broad types of joints in slabs on ground: isolation joints, contraction joints and construction joints. Isolation joints permit slabs to move (very slightly) vertically and horizontally. Contraction joints prevent vertical movement but permit slight horizontal movements. Construction joints are stopping places for a day's work. Construction joints are frequently designed and constructed to provide continuity by bond of the concrete across the joint. Construction joints can be constructed to function as contraction joints.

See also Section 6.9 for discussion of crack control in slabs on ground.

Construction Joint. At any location where concrete placing is discontinued for any reason, it is necessary to install a bulkhead and make a construction joint. A joint should be made at the end of each day's work. On large slabs, a predetermined layout is followed in locating the joints. A bulkhead form can be of metal or wood and must be set accurately with its top edge at the finish grade of the slab. Common practice is to install a load- transfer device consisting of a key formed in the joint, or smooth dowels. Deformed reinforcing steel tie bars are required if bond across the joint is required. (See Figure 16-14.) The

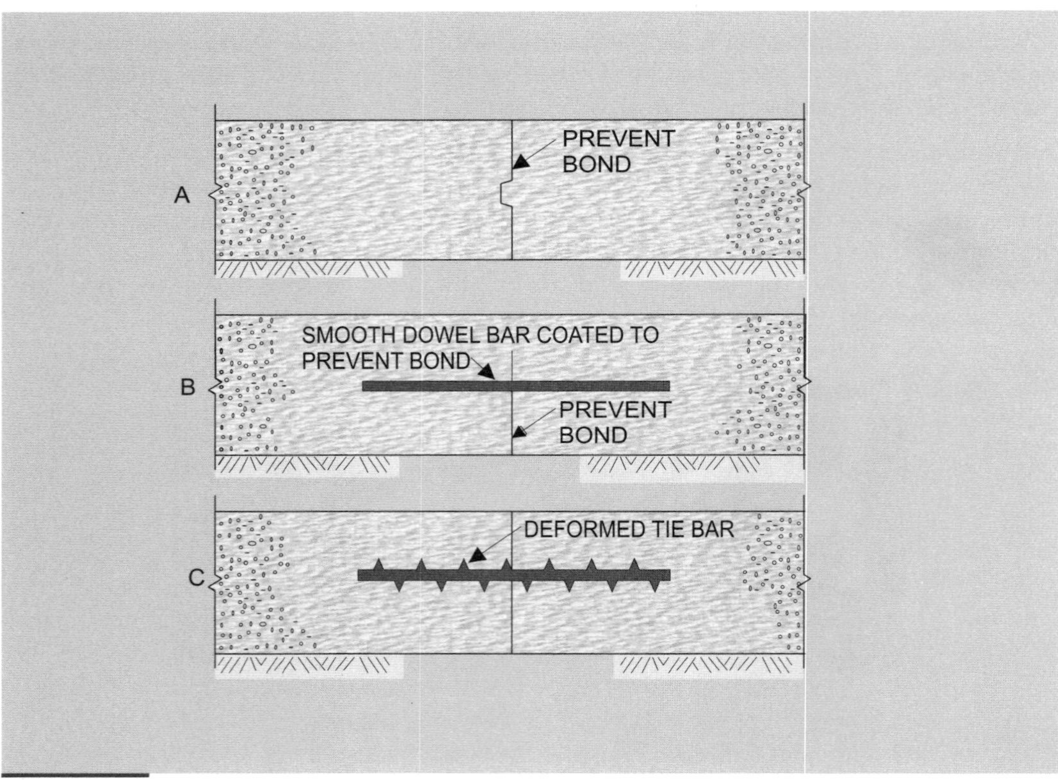

Figure 16-14
Three construction joints are shown. A: The keyed joint can be made by inserting the block shown in Figure 15-1. The surface of the first run of concrete should be painted with curing compound or similar material to prevent bond of the second run. B: Smooth dowels permit horizontal movement at the joint but prevent vertical displacement. There should be no bond between the concrete surfaces. C: Deformed reinforcing tie bars provide continuity across the joint. The two concrete surfaces should be permitted to bond. Joints A and B also act as contraction joints.

object of using a load transfer device is to aid in transferring any applied load across the joint and to prevent vertical displacement of the slabs at the joint. Special care is required in placing concrete to assure that the concrete closes in tightly against the bulkhead and keying devices. Smooth dowels must be rigidly held in place so they will not be displaced. A dowel that is not perpendicular to the bulkhead is apt to cause a crack in the concrete later on. (See Section 6.6 and Figure 6-15.) Construction joints should be located not less than 5 feet from any other joint to which they are parallel[16.3] It is sometimes possible to locate the construction joint so it can also function as a contraction joint. For the keyed construction joint illustrated in Figure 15-1, the sizing of the keyway is important to prevent a void in the keyway when the second run of concrete is placed. (See Figures 16-15 and 16-16.)

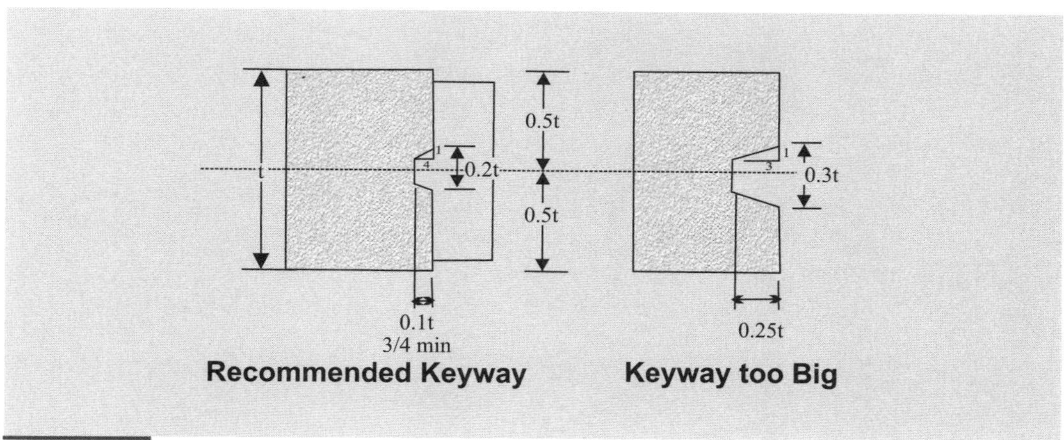

Figure 16-15
Proper sizing of keyway for keyed construction joint.

Figure 16-16
Keyway too big resulting in keyway not completely filled with concrete.

Contraction Joint. The drying shrinkage of the concrete in a large slab will cause random cracks in the slab unless means are provided to relieve the stress induced in the concrete by shrinkage. This relief is furnished by contraction joints (sometimes called control joints) made in the concrete at regular intervals.

There are several ways to make contraction joints. In one method a small groove, made with a grooving tool, is cut in the fresh concrete just before floating. (See Figure 16-17.) This reduces the slab thickness along the line of the groove, thereby making a weakened plane in the slab because the thickness of concrete under the groove is less than at any other place in the slab. Consequently, when the concrete shrinks it will crack in this weakened plane (See Figure 16-18) rather than at random locations throughout the slab as illustrated in Figure 6-7.

Figure 16-17
A contraction joint cut in the fresh concrete with a grooving tool on a bullfloat. (Courtesy of PCA)

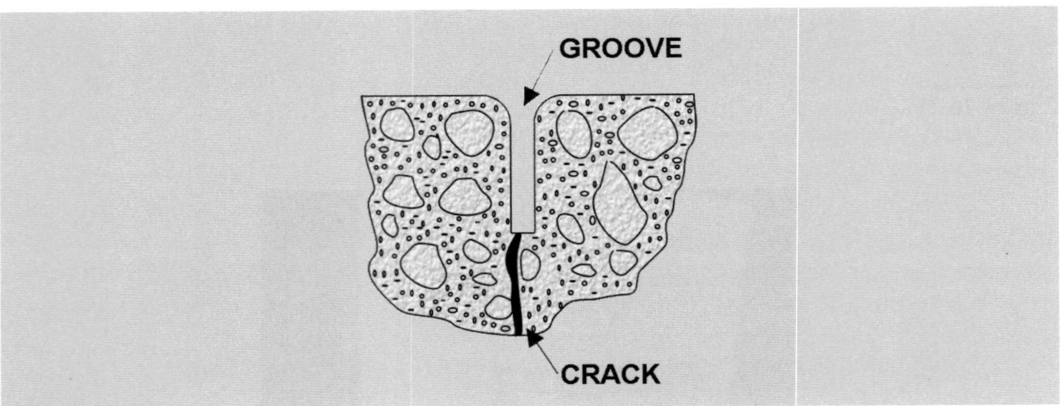

Figure 16-18
Because the thickness of the concrete is a minimum where the groove is cut, the slab, when it contracts, will crack along the joint as illustrated in Figure 16-19.

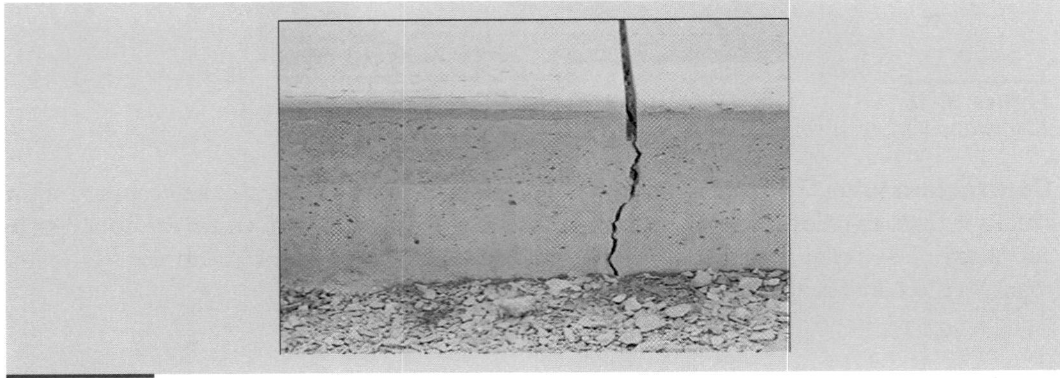

Figure 16-19
Controlled shrinkage cracking. (Courtesy of PCA)

In another method, a forming strip of wood, plastic or metal is embedded in the fresh concrete and subsequently removed after the concrete takes an initial set, leaving a groove in the concrete. (See Figure 16-20.) Also available is a shaped strip of plastic that is embedded permanently in the concrete. There are a number of materials on the market that produce a good, straight groove. On the other hand, a poorly designed or carelessly placed strip can result in an irregular groove, with the material displaced laterally or buried in the concrete, or no groove at all, causing expensive delay and repairs to remove the displaced material from the hardened concrete and install an acceptable groove.

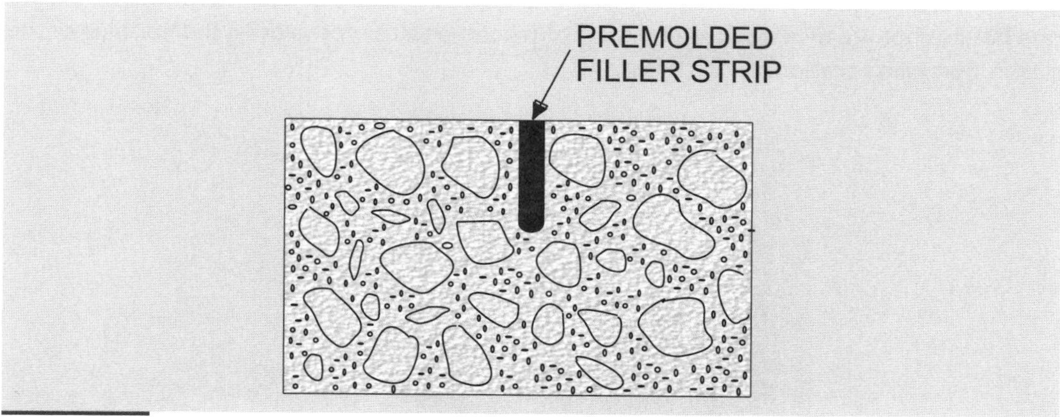

Figure 16-20
A forming strip can be inserted in the fresh concrete to make a weakened plane. The strip is usually removed, although occasionally the strip may serve as the joint filler.

On jobs of considerable extent (continuous floor slabs), the joint can be cut with a saw after the concrete has hardened. (See Figure 16-21.) Power-driven saws are available, fitted with either abrasive or diamond blades. Timing of the sawing is critical; if the sawing is done too early, pieces of aggregate will be loosened and the concrete will ravel, resulting in a rough and irregular groove. If the sawing is delayed, it will be more difficult and cracks may have developed before the sawing is completed. Experienced operators can judge when the concrete is ready for sawing; usually it is between 4 and 12 hours after the concrete has been placed, depending on temperature, the mix and other factors.

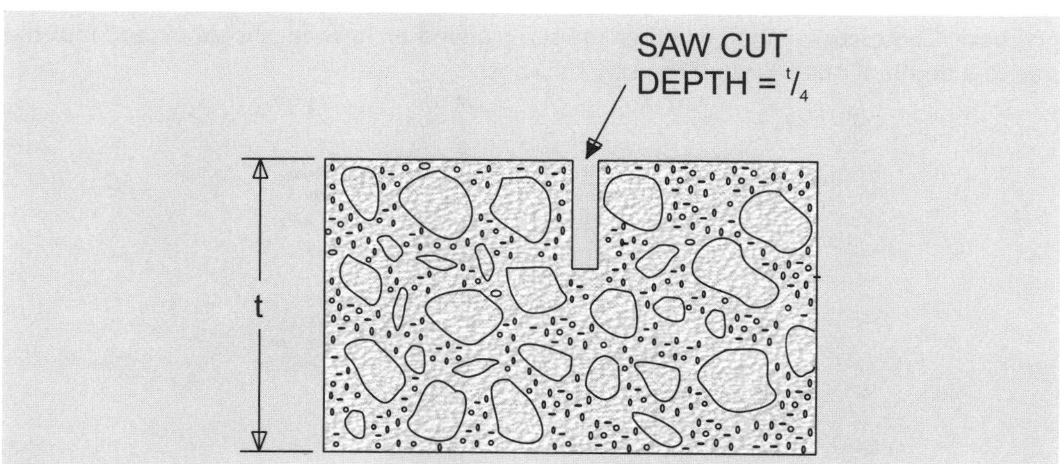

Figure 16-21
The sawed joint performs the same function as the grooved or formed joints. These contraction joints are sometimes called "dummy joints."

In contrast to the conventional wet-cutting process, another method of saw cutting joints uses a special type of dry-cut saw (see Figure 16-22) that can cut joints in concrete soon after placement. The saw cuts resulting from this "early-entry dry-cut" process are not as deep as those produced using the conventional wet-cut process ($1\,^1/_4$ inch maximum). The timing of the early-entry process has proven very beneficial in controlling random cracking and finishing the saw-cut operations before major stress builds, and the concrete seeks its own relief via a random crack. This dry-cut method is used in the initial set drying stage rather than the permanent or (green) set stage of the concrete, which is often too late for crack control. Typically, the waiting period for the early-entry dry-cut saws will vary from one hour in hot weather to four hours in cold weather after completing the finishing of the slab in that joint location.

Figure 16-22
Using the early-entry dry-cut method, workers can saw joints in floor slabs immediately after finishing. (Courtesy of PCA)

Maximum spacing of contraction joints in continuous floor slabs on ground is a function primarily of slab thickness. Unless reliable data indicate that more widely spaced joints are feasible, contraction joints spaced at intervals not exceeding 30 times the slab thickness, in both directions, should be used for well-proportioned concrete with aggregates having normal shrinkage characteristics. Contraction joints should divide a large slab area into relatively small rectangular panels. (See Figure 16-23.) Panels should be as nearly square as practical. Contraction joints, whether sawed, grooved or formed, should extend into the slab to a depth of one fourth of the slab thickness.

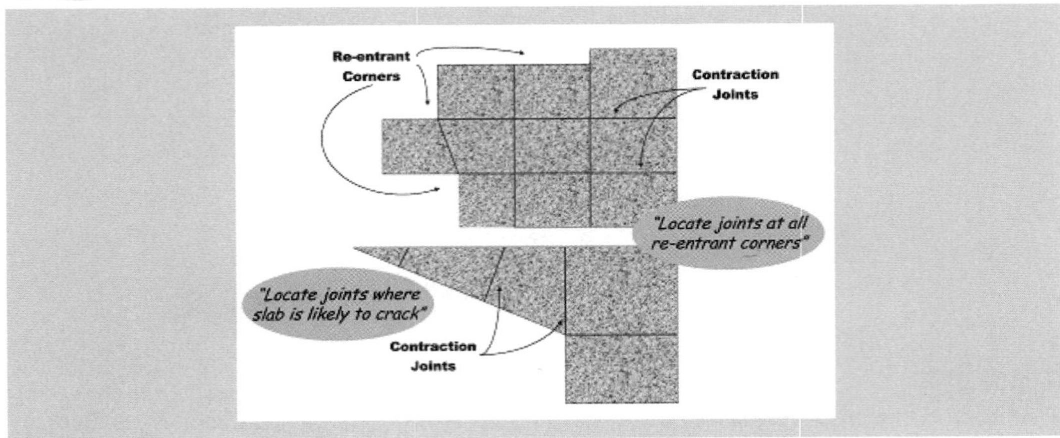

Figure 16-23
Location of contraction joints in slabs on ground.

In grooves made in fresh concrete, the edges should be slightly rounded and care must be taken to make sure that the groove is free of projections of concrete or aggregate particles. Joints made by sawing or a forming strip are filled with a mastic material to prevent foreign material from entering the joint and to seal against entrance of water.

Isolation Joint. When new slab concrete is placed against existing concrete, such as walls, columns, footings, steps, and abutting driveways and sidewalks, there must be a separation to allow for vertical as well as horizontal movement of the new concrete relative to the old. The same holds true when concrete is placed against any other area of restraint caused by structures of any kind. Horizontal movement is caused mainly by drying shrinkage, or occasionally by thermal contraction and expansion. Vertical movement is usually the result of settlement. The new concrete must be separated or isolated from the old, and the joint is called an isolation joint (sometimes called an expansion joint). The joint is made by inserting premolded joint material between the old and the new construction. (See Figure 16-24.) The joint material must be as wide as the slab is thick, and it must not extend above the slab elevation. It can be attached to the old construction with asphalt or a similar adhesive before the slab concrete is placed. (See Figure 16-25.)

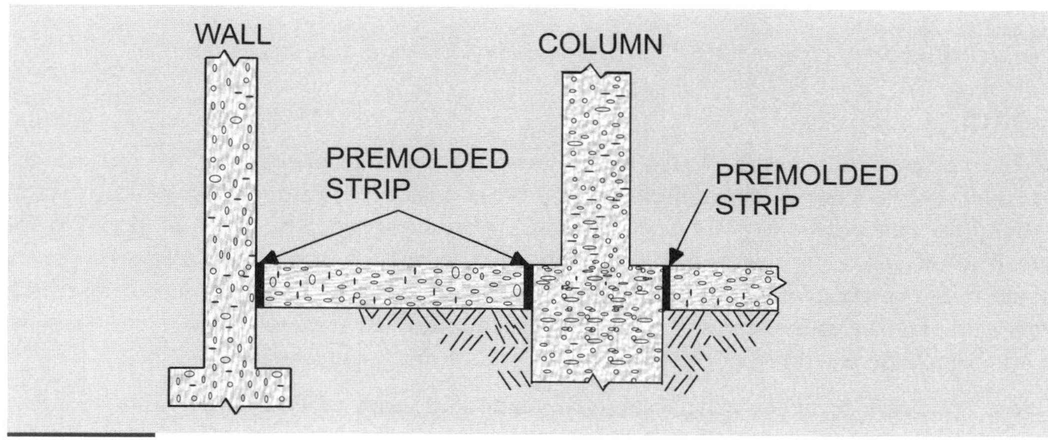

Figure 16-24
Filler for isolation joints is made of bitumen-saturated material that is installed before the concrete is placed.

Figure 16-25
Premolded joint material around column and slab. (Courtesy of PCA)

Joint Layout. A typical joint layout for flatwork around residences is shown in Figure 16-26. Transverse contraction joints in sidewalks should be located every 4 to 5 feet, and 10 to 12 feet in driveways.[16.3]

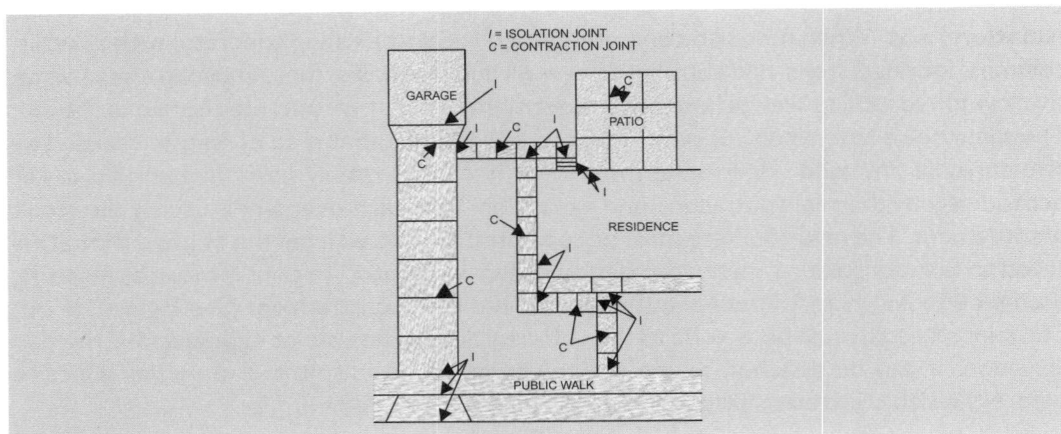

Figure 16-26
Typical locations of isolation and contraction joints in flatwork around residences.

16.3. Floors

Because of the importance of a concrete floor as part of a building, we will give special consideration to floors. The same as for any other structural element, a good floor starts with good specifications. The specifications must designate the type of floor to be constructed, including the materials to be used in the concrete, the mixture, slump, strength of the concrete, subgrade preparation, joint details, grade and thickness of the slab, kind of finish, tolerances and curing. Specifications must state whether reinforcement, vapor barrier or any other special material or treatment is required.

Floors are classified in accordance with the expected usage of that floor. Table 16.1 lists several general categories of floors and their basic requirements. One requirement common to all floors is the requirement for smoothness, that is, how much the surface of the concrete deviates from a true plane. A very close tolerance for a plane surface is a departure of $^1/_8$ inch from the edge of a 10-foot straightedge.[16.4] (See Figure 16-27.)

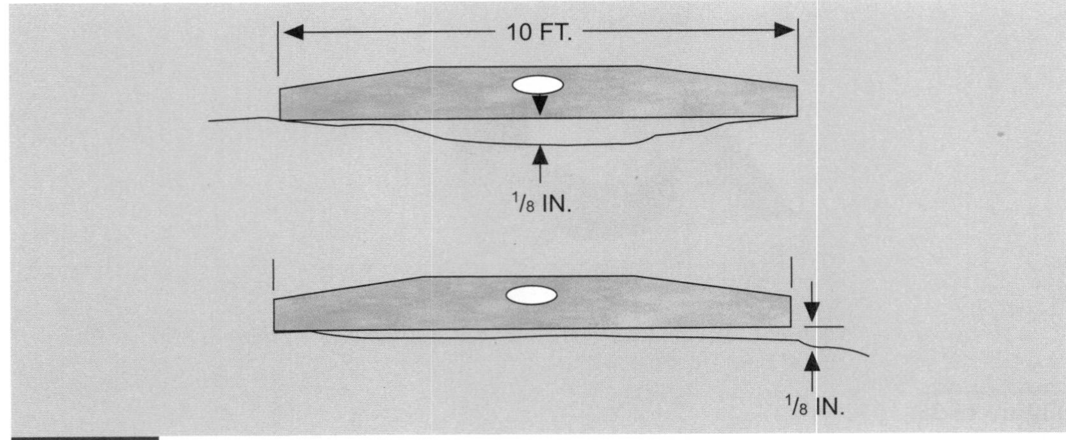

Figure 16-27
A very close tolerance for a floor is a departure of $^1/_8$ inch in 10 feet.

Durability and Corrosion of Floors. A lack of durability can be attributed to a low cement content, which results in a high water-cement ratio and low-strength, wet or high-slump mixes, overvibration of wet mixes, working of the surface when bleed water is present and inadequate curing.

The effect of adequate cement, shown in Figure 16-28, can be nullified by a lack of curing. (See Figure 16-29.) High-slump mixes, overvibration and working of the surface when bleed water is present result in a weak layer of very high water-cement ratio, a soupy material on the surface. (See Figure 16-30.) The thin layer on the surface, consisting of hydrated cement particles, fine sand, silt and voids caused by the high water content, will have strength of but a fraction of the potential strength of the concrete. A good 4000 psi concrete in the slab will be reduced to a 400 or 500 psi layer on the surface. Prolonged curing will not make a good surface out of such concrete, and surface treatments or coatings are of very limited value. A dusty, crumbly floor is inevitable.

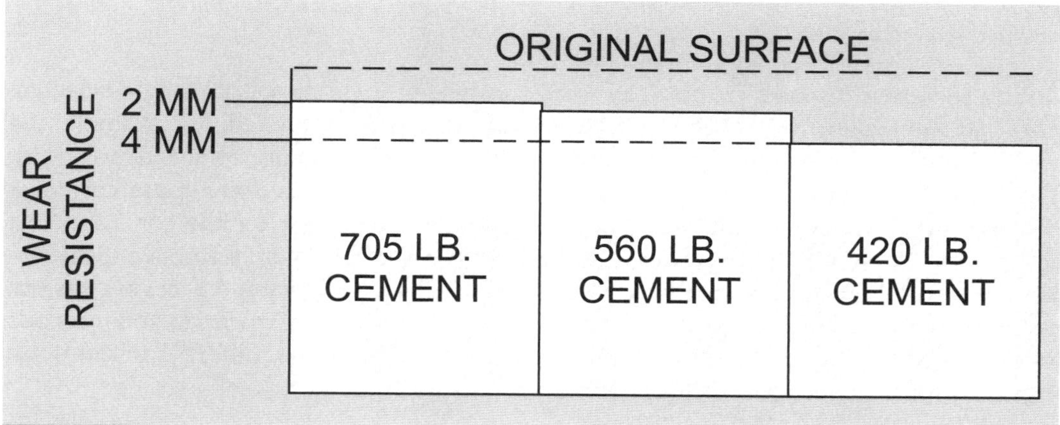

Figure 16-28
The "wear factor," a measure of the hardness or abrasion resistance of the concrete, shows that low cement content is conducive to poor wear resistance.

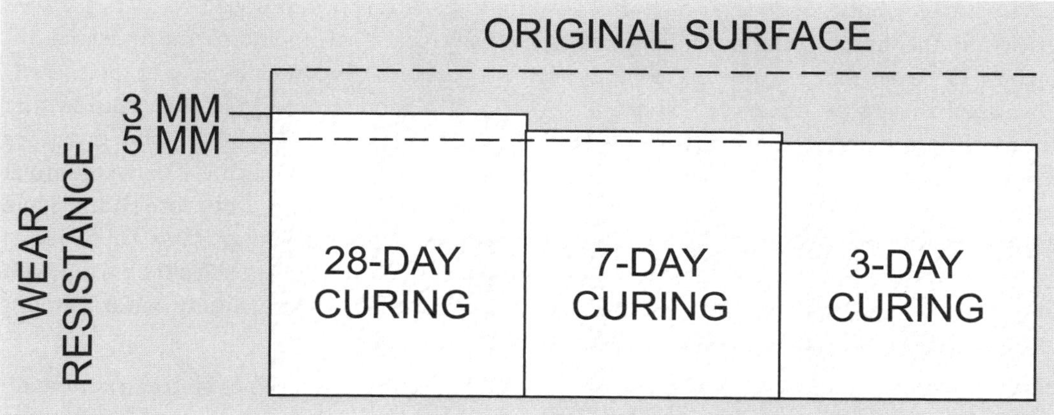

Figure 16-29
Even though the potential hardness or wear resistance of the floor is good, the potential will not be realized if the concrete is not properly cured.

After a good floor has been properly cured, its durability can be improved by permitting the floor to dry thoroughly, at least a day or two, before exposing it to traffic.

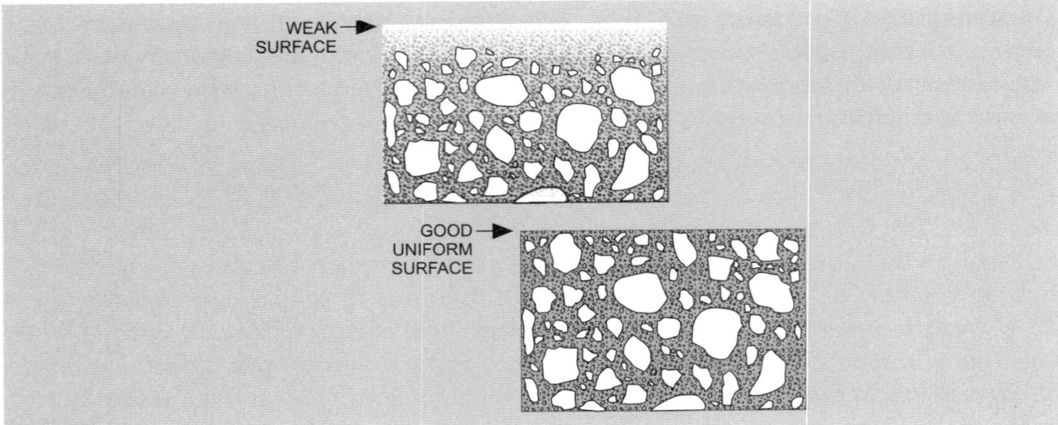

Figure 16-30
Working or finishing the concrete surface at the wrong time results in a weak and soft surface that will soon wear away.

Good concrete is resistant to attack by many substances. By using special care in selecting materials and making the concrete, resistance can be greatly improved. For example, the use of Type V cement makes concrete that withstands an exposure to sulfate solutions. There are, however, some industrial and manufacturing processes that impose corrosive conditions on floors that make some kind of protection necessary for the concrete. The vast number of possibly aggressive materials and the great variety of protective measures are more than can be encompassed in a short chapter. Reference 4.3 covers several hundred common materials that can damage concrete with a list of materials and methods for providing protection. The reader should review Chapter 4 and Table 4.1 of this book regarding durability of concrete in general.

There are a number of manufacturers of protective materials, and the manufacturer's instructions must be followed when using any of these.

Light-Duty Floors are defined as those for residential and light commercial or similar occupancies where loads are not heavy and the concrete is not exposed to an aggressive environment. Local general building codes typically specify a minimum thickness of $3^1/_2$ inches for any slab supported directly on the ground. Normal practices previously discussed should be observed. There is sometimes a tendency to take the attitude that these slabs are relatively unimportant, and therefore observance of the good practices discussed in this book is not necessary. Nothing is further from the truth. All concrete must be carefully and properly made if it is to perform to its best capability. Concrete should have a strength not less than 3000 psi with a slump not greater than 5 inches. (See Table 16.1.) For concrete exposed to freezing and thawing, air-entrained concrete should be used with air content conforming to the limits of Table 16.1C and have a maximum water-cement ratio of 0.45 (roughly equivalent to 4500 psi).

Medium-Duty One-Course Floors. An industrial floor is subject to rough treatment even under the best of circumstances. Steel-wheeled trucks, abrasive action and corrosive liquids are usually present in almost any type of plant or warehouse. Food processing plants such as fruit canneries and meat packing houses subject concrete to especially severe exposure. Floors exposed to the most severe conditions usually consist of a base and special wearing surface. A plain concrete floor, properly placed and finished, can support heavy loads and is entirely satisfactory for many exposure conditions where it is not subjected to severe abrasion or corrosion.

Thickness of the slab is determined on the basis of the strength of the concrete and the expected loading on the floor, with an adequate factor of safety. The slab thickness is based upon the assumption that the slab will be fully supported by the subgrade or base material. A minimum thickness sometimes specified is 5 inches. Theoretically, no reinforcement is required in a fully supported slab. However, there are variations in the quality of the subbase and intensity of loadings applied to the floor that make reinforcement of industrial slabs desirable. Some cracking is apt to occur and the reinforcement serves to prevent the cracks from becoming too wide. The reinforcement that is usually easiest to install consists of welded wire fabric or welded bar mats. (See Figures 18-5 and 18-6.) These should not be placed on the subgrade and then pulled up into the concrete, as this results in improper location of the steel in the concrete. Instead, place concrete up to the level of the steel, then place the reinforcing mats; finally, top off the floor, all in one operation without a cold joint. Even better is to support the steel on mortar blocks or metal chairs at the required elevation and place the concrete in one layer or lift.

Concrete should have a strength of about 4500 psi or more with a maximum slump of about 4 inches. Air-entrained concrete should be used for better workability, durability and watertightness.

Two-Course Heavy-Duty Floors. Heavy-duty, as applied to floors, refers to exposure and wear conditions, and not to load-carrying capacity. Hence, a heavy-duty floor is one exposed to severe abrasion of many passages of steel-wheeled warehouse trucks, abrasion of materials dragged over the surface and similar conditions. Wear resistance is obtained by applying a topping layer $^3/_4$ to 1 inch thick of very stiff, rich concrete to a properly designed and constructed base slab, or by applying a dust coat or dry shake of cement and special hard aggregate to the freshly floated surface of the base slab. These procedures are described in Chapter 17.

Floor on Expansive Soil. In some geographical areas there exist soils that have a large volume change from the dry to the wet condition. These soils are certain clays or soils with a high percentage of clay. When the soil absorbs water, it expands or increases in volume. The expansion exerts considerable force and can damage buildings placed on the expansive soil. Heaving and cracking of the floor can cause cracked walls, doors and windows that will not operate, cracked foundations and other problems.

When expansive soils are present, special provisions may have to be made in design and construction to safeguard against damage. If expansive soils are encountered on a job unexpectedly, the building official should be notified. The checklist in Figure 16-31 will be of assistance to the inspector who finds what appears to be an expansive soil.

16.4. The Suspended Slab

A suspended slab is one that is not supported throughout its area by the subgrade. It is a structural portion of the building designed to transmit its load to walls, beams or columns that compose the structure. Beams, joists or drop panels are frequently cast integrally with the slab, or the slab concrete may be placed on permanent forms that remain in place to become part of the building. Structural slabs of this type are designed as reinforced concrete slabs according to the rules and principles specified in ACI 318 Chapter 13.

Structural requirements of the building, as well as slab finishing requirements, must be considered when proportioning the mix and placing the concrete. Striking off and darbying or bull-floating are performed in the same manner as for a floor or slab on ground. Subsequent finishing operations to prepare the slab in accordance with the specifications are described in Chapter 17.

SWELLING CLAY SOILS CHECKLIST	YES	NO
1. Have similiar soils in this area been known to be expensive?	☐	☐
2. Is there evidence of cracks in footings, walls, curbs, sidewalks or pavements in nearby construction?	☐	☐
3. Are there shrinkage cracks in the soils in dry weather?	☐	☐
4. Does the soil behave as a clay in wet weather? (For example: sticking to shoes and tires, "greasy" in feel between fingers, ponding water to long periods of time.)	☐	☐
QUICK "FIELD TESTS"		
1. Select a small lump of *dry* soil and try to break it between fingers. Is the soil strong and hard to break?	☐	☐
2. Select a large lump of *dry* soil (about2 or 3 pounds), raise it chest high and drop it on a hard surface (pavement), Does the soil stay in one lump, instead of breaking up into several similar pieces? (Disregard the breaking off of thin edges or sharp corners.)	☐	☐
3. Wet some soil in your hand or a dish until it can be easily molded with light finger pressure. Can this soil be rolled out in the palm of your hand into a "thread" about $^1/_8$ inch thick and more than $1^1/_2$ inches long?	☐	☐
4. Mold wet soil (similar to that in No. 3 above) in a ball bigger than 1 inch in diameter. Place it in the cup of your hand and strike the lower part of the hand on your knee with short strokes (2 to 3 inches) 10 to 15 times. Silty material will "bleed" water and show a wet shiny surface. Does the soil you are testing look about the same as when you started, without shiny surface??	☐	☐
5. Take a ball of wet soil (similar to that in No. 4 above), drop it on a piece of smooth, dry glass plate from a height of above 18 inches. Turn glass upside down (soil toward ground), slightly tilted. Tap the top of the glass plate with your fingertips. Does the soil remain stuck in the original position on the glass, instead of falling off or sliding?	☐	☐

A "yes" answer to any of the above questions may indicate a need for further testing. Report the findings to your supervisor.

PROJECT ____________________

LOCATION ____________________ DATE __________ CHECKED BY ______________

REMARKS ____________________

____

____

____

____

Figure 16-31
A checklist will be a help to the inspector if expansive soil is suspected within the building area.

Chapter 17

Finishing and Curing the Concrete

FINISHING

Finishing is defined as the process of "leveling, smoothing, compacting and otherwise treating surfaces of fresh or recently placed concrete to produce desired appearance and service." All unformed surfaces require some degree of finishing to make them acceptable, and formed surfaces are frequently treated to improve appearance, serviceability or watertightness. In addition, there are certain special techniques that are applied to the forms to produce special effects. Treatment of formed surfaces is discussed under "Architectural Concrete" in Chapter 22.

Floors, sidewalks and driveways are typical of the kinds of surfaces that fit into the category of unformed surfaces. In this section we will consider the various methods of achieving desired effects, especially as they apply to floors, in which the finish is obtained by working the surface of the concrete with hand or power tools before final set. A basic requirement is that the amount of surface manipulation that will give the desired finish should be kept to a minimum. A screeded finish may be all that is necessary on a surface to be covered with backfill. Smoothing with a darby or bullfloat should be adequate for a construction joint. More elaborate finishing operations come into the picture when floors and pavements are involved, or when it is necessary to close or seal the small voids in the surface of the concrete.

In Chapter 16 we discussed methods of constructing floors and slabs of various kinds and preparation of the surface for finishing. We will now examine the tools and processes employed to bring the surface to the desired degree of smoothness, hardness and denseness.

17.1. The Tools and How to Use Them

The tools we are concerned with in this section are those that are used by the finisher on horizontal, or nearly horizontal, surfaces. (See Figure 17-1.) These are the surfaces described above as unformed surfaces. Irrespective of whether the slab is a sidewalk, floor or other structure, the finishing operations start following a delay after the surface has been darbied or bullfloated. The delay is necessary to permit the bleed water to leave the surface of the concrete. A basic law of finishing concrete is this: Never use any tools on the fresh concrete while bleed water is present on the surface. To do so is to make a surface that will be weak and dusty. (See Section 2.6.) Timing of finishing is critical; execution of the work must be commenced immediately following the disappearance of the bleed water when the concrete has started to stiffen.

Figure 17-1
At the top is a straight edge, and just below, a darby. For the finisher, from left to right, a finishing trowel, a groover, an edger, and a float. (Courtesy of PCA)

Edging. Not all slabs require edging. If edging is required it is the first operation applied to the concrete in the series of final finishing operations. Edging produces a radius or rounded edge to the concrete (see Figure 17-2) that protects the concrete from chipping or other damage. The tool serving this purpose (see Figure 17-1) is called, appropriately, an edger. Edgers are usually made of steel, although occasionally one made of bronze or malleable iron is found. The radius of the edge formed on the concrete varies from 1/8 inch for floors to 1/2 inch for sidewalks and similar structures. (See Figure 17-2.)

Figure 17-2
The edger makes a rounded edge on the concrete. The edging tool is advanced with the leading edge inclined slightly upward. (Courtesy of PCA)

After all bleed water is off the surface of the slab, the edger is used by running it back and forth along the edge of the slab. Coarse aggregate particles must be covered, and the edge must be smooth. It may be necessary to add small amounts of concrete or mortar to small areas if the edge is too ragged, and care must be taken to avoid making a deep impression in the top of the slab that will be difficult to remove.

Not all slab edges are edged, and the plans should be reviewed. Frequently floors, especially floors to receive a covering of tile or similar material, are not edged. Edging is, however, required along all isolation and construction joints.

Grooving. Following immediately after edging, or even at the same time, the concrete is grooved, or jointed, using a hand tool designed for this purpose. (See Figure 17-1.) The groover, made of steel, bronze or malleable iron, is usually about 6 inches long with the ends upturned slightly to facilitate its use. Widths up to 6 inches are available. The groove-forming bit may be from about $^1/_4$ inch to 1 inch deep. In most slab construction a $^3/_4$-inch or 1-inch bit is employed.

It is difficult to make a straight groove without a guide of some kind. A length of 1-inch board laid on the concrete next to the groove makes a good guide. (See Figure 17-3.) The board must be straight; if it is not, the edge should be planed to a true edge. Grooves, which must be at right angles to the edge of the slab, should be made with the same care as is necessary for edging. If large pieces of aggregate interfere with the tool, they should be removed and the space filled with mortar. The finished groove should be straight, of uniform depth, free of projections of mortar or pebbles, with a symmetrical and smooth radius.

Figure 17-3
Groover in use. The plank is used as a guide. (Courtesy of PCA)

If the grooves are for ornamental purposes only, a shallow bit is adequate, about $^3/_{16}$ or $^1/_4$ inch deep. A shallow bit of this type should never be used to make a control joint, as the depth is not enough to make a weakened plane, and random cracks will occur.

Floating. After edging and grooving, the slab is floated. (See Figure 17-4.) Hand floats come in a variety of sizes and materials, the all-wood float being a common variety. Others are made of magnesium or aluminum. Size varies from 12 to 18 inches long by $3^1/_2$ or $4^1/_2$ inches wide. Special floats for use in making rendered surfaces are faced with carpet, sponge rubber, cork or other materials. Plastic and stainless steel floats are available for use on white concrete to prevent staining of the concrete.

Figure 17-4
After edging and grooving, the surface is hand floated by holding the float flat and moving it in a sweeping arc with a slight sawing motion. (Courtesy of PCA)

The time to start floating depends on many factors— temperature, humidity, wind, slump of the concrete and type of subgrade being the most important. The time might be as short as two hours or as long as eight hours. Judgment based on experience tells the worker when to proceed. There are two important guides: first, the water sheen or shininess must have disappeared from the surface; second, the mix must have stiffened to the point that a person standing on the surface will leave only a slight footprint not over $^1/_4$-inch deep. The time lapse will usually be shorter for air-entrained concrete than for nonair-entrained concrete.

The purpose of floating is to smooth the surface by removing slight rough spots left by the previous operations, to embed large aggregate particles just below the surface, and to slightly consolidate or tighten the surface. When properly done, floating leaves the concrete surface dense and smooth, ready for steel troweling. The marks left by the edger and groover should be removed by floating, unless these marks are desired for decorative purposes, in which case they should be rerun after floating.

It is generally agreed that magnesium or aluminum floats are easier to use than wood floats, especially on air-entrained concrete. Metal floats drag less and make a smoother texture.

Power-driven machines are available for floating, with the finishing floats rotating on a vertical shaft driven by a small gasoline engine. A troweling machine can be fitted with float shoes, or a special machine with a rotating disc can be used. (See Figure 17-5.)

Figure 17-5
Power floating using walk-behind and ride-on equipment. Footprints indicate proper timing. When the bleedwater sheen has evaporated and the concrete will sustain foot pressure with only slight indentation, the surface is ready for floating and final finishing operations. (Courtesy of PCA)

Troweling. After floating, the next finishing procedure is troweling the concrete to produce a hard and dense surface. Trowels, which are made of heat-treated high-carbon steel or stainless steel, are between 3 and 5 inches wide and 10 to 20 inches long. (See Figure 17-6.) There are several special trowels for certain purposes.

Figure 17-6
Steel hand trowel. (Courtesy of PCA)

One kind is made of flexible steel and has rounded ends so that it can be used on curved surfaces. Trowels must be kept clean and in good condition. Although the steel has been hardened, the trowel blade can be damaged by nicks along the edge, or dents. These defects make finishing more difficult and are apt to be reflected as flaws in the troweled surface. A new trowel is easier to use and gives better results after it has been broken in by a few days' use. Because of the flat angle at which the trowel is used for the first troweling, most finishers recommend that a new trowel not be used for the first troweling.

The first troweling is done as soon as the surface has been floated, keeping the blade of the trowel as flat against the surface as possible. (See Figure 17-7.) If the blade is not held flat, it will chatter and produce a rough, washboard effect. A troweled surface should be smooth and free of defects. The smoothness can be improved by additional trowelings. Each troweling is done after a short lapse of time to permit the concrete to take up or increase its set. A smaller trowel should be used for successive trowelings so that the finisher can exert greater pressure on the concrete.

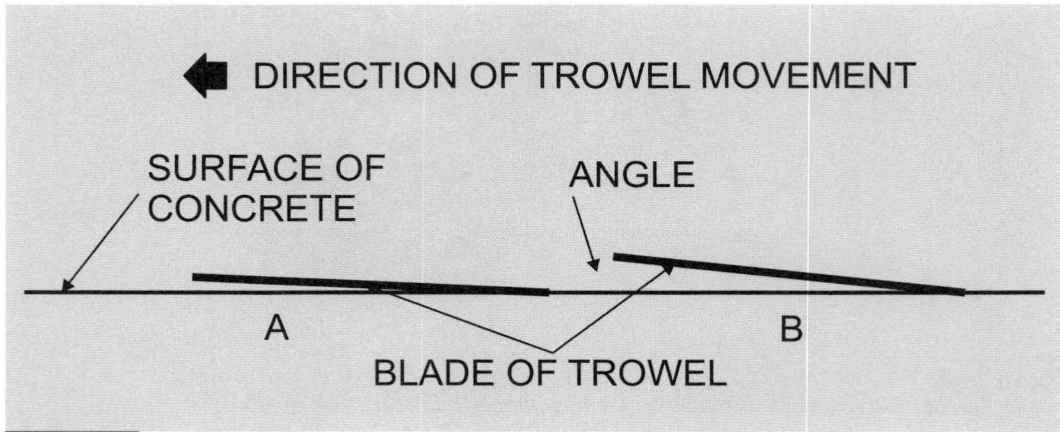

Figure 17-7
For the first troweling, the trowel blade angle should be about as shown at A. The angle shown at B can be used for the final troweling but is too large for the first troweling.

Successive trowelings are done with a blade at a slight angle. (See Figure 17-8.) A final burnishing, using a hard trowel, is done with the blade at an angle and, when properly done, is accompanied by a characteristic ringing sound as the trowel passes over the surface. See Figure 17-9.

Figure 17-8
Trowel must be held firmly and slightly raised in the direction of movement. (Courtesy of PCA)

Figure 17-9
Hand finishing on the final pass is key to good results. (Courtesy of PCA)

Finishing can be done with power tools (See Figures 17-10 and 17-11.) that give the same effect as hand tools but can cover much larger areas in any given time. Some of these machines can be used for floating and troweling all in one operation by the use of adjustable blades and a variable speed motor, making several passes over the concrete. Figure 17-9 shows the use of kneeboards by the finisher to prevent marring the finished concrete.

Figure 17-10
Power trowelling using walk-behind equipment. Note that the workers shoes do not sink into the concrete. (Courtesy of PCA)

Figure 17-11
Rotary troweling machines can efficiently finish very stiff concrete. (Courtesy of PCA)

Sometimes bubbles and blisters will appear in fresh concrete during troweling. This has happened when the partially hardened slab was sprayed with water, then dusted with dry cement. Apparently, water was trapped in the aggregate voids by the cement topping and was released when the troweling was done. This made small bubbles appear in the surface immediately after the troweling. Sprinkling water or cement on the surface to be finished is no way to prepare it and should never be permitted.

Bubbles may appear as a result of too-early power troweling or by operating the trowel with the front edge tilted too high, in effect troweling with the back side of the blades. This causes a too early sealing of the surface with air trapped underneath, hence the bubbles. Prevention is accomplished by troweling at the right time, not too soon, and operating the power trowel with the blades at the proper, flat angle.

Nonslip Finish. The surface resulting from troweling is hard, dense and very smooth. The smoothness may be desirable in some situations, but it will be slippery, especially when the concrete becomes wet. For this reason, a nonslip finish is usually applied. This is done by brooming or brushing the surface. Any degree of roughness can be made, depending on the kind of brush or broom used and the time when the brushing is done.

By drawing a long-handled, soft-bristled push broom over the surface after troweling, a slightly roughened surface is achieved. Stiffer bristles will give a coarser texture. (See Figure 17-12.)

Figure 17-12
A rough texture can be made by drawing a brush or broom over the soft concrete surface. (Courtesy of PCA)

The dry shake method mentioned earlier can be employed to provide a nonslip finish with special hard aggregates. Sawing many parallel shallow grooves has been found effective for pavements that are too smooth. The hardened concrete can be etched with acid, which exposes the aggregate particles and makes a slightly roughened surface.

17.2. Wear Resistance of a Floor

A properly constructed slab, without any special treatment, can be considered to have a wear and hardness factor of 1. This wearing value extends clear through the slab and is the result of good materials and construction, including proper curing. Now assume various treatments to this floor, and note the effect on wearing capacity.

1. No curing at all. Surface is soft and dusty. Wear and hardness factor is 0.3. That is, the floor will give about one-third the service that the properly cured floor will give.
2. Chemical hardening of No. 1. This gives a skin effect that will improve the factor to about 0.5. This nearly doubles the life of No. 1 but is still only about half of the value of a properly cured floor without extra treatment.
3. Application or silica or emery either as a dust coat or $^{3}/_{4}$-inch topping applied to the floor, after which the concrete is cured properly. Factor is 2, which doubles the life of plain concrete, properly cured. The topping can be expected to have a longer life than the dust coat because of its greater thickness—$^{3}/_{4}$ inch compared with about $^{1}/_{8}$ to $^{1}/_{4}$ inch for the dust coat.
4. Application of iron, silicon carbide or aluminum oxide either as a dust coat or $^{3}/_{4}$-inch topping. Hardness factor is about 8 for either treatment, but the topping will wear about three times as long because of its greater thickness.

Dusting. One of the most common faults of concrete floors is dusting, or the wearing away of concrete under traffic. (See Figure 6-26.) Dusting is caused by weak and soft concrete that results from overfinishing, the use of overly fluid or wet mixes (see Figure 2-5), or working of the surface while bleed water is present, causing laitance. Usually it is a combination of these, bringing an excess of water and fines to the surface, producing a weak sand-cement mixture on the wearing surface. The problem finishing may either be in the screeding, floating or troweling operation, as any of these is capable of bringing fines to the surface.

Another cause of dusting is failure to cure adequately, which stops the chemical process of hydration by allowing the water to escape through evaporation, thereby preventing the concrete from attaining its intended strength. Rapid evaporation also causes surface shrinkage, resulting in crazing and surface cracking. During cold weather, carbon dioxide from poorly vented heaters, if it comes in contact with the fresh concrete during the first few hours, will carbonate the concrete, resulting in a soft, chalk-like surface.

Any concrete surface will dust to some extent, especially in heavily traveled areas, although a well-constructed floor made of first-class materials will give satisfactory service under most conditions without special treatment. For exposure to especially severe abrasive conditions it is common practice to apply a dry shake of abrasive material as described below.

Treatments to improve the hardness of a dusting floor include various chemical solutions that are applied to the finished floor. These treatments will not make a good floor out of a poor one and are not a cure-all for poor quality of construction or materials. They have only a limited value in improving high-quality concrete. The solution, applied to the concrete surface, penetrates into the pores of the concrete, forming crystalline or gummy deposits that make the concrete less pervious. Some solutions react with the free lime or calcium carbonate in the concrete, forming calcium silicate, which is much harder and stronger.

There is little that can be done for a floor that is dusting badly. If there is a thin layer of soft, chalky material on the surface, it can be removed with pads of steel wool attached to a scrubbing machine or, in extreme cases, by grinding. The surface is then washed thoroughly, allowed to dry, and a hardening treatment applied.

One of the best of the chemical treatments is made with a mixture of 20 percent zinc fluosilicate and 80 percent magnesium fluosilicate mopped onto the surface. The first coat consists of $^{1}/_{2}$ pound of fluosilicates in one gallon of water, and subsequent coats have 2 pounds per gallon. Apply at least two coats, permitting the concrete to dry between applications (at least four hours). After the final coat, mop the floor to remove encrusted salts.

Commercial sodium silicate (40 percent) is also used by making a solution of 1 gallon of silicate to 3 gallons of water, applied in two or three coats. Be sure each coat is completely dry before applying the next one. Penetration of succeeding applications is aided by scrubbing the surface with stiff fiber brushes between applications. Aluminum sulfate or zinc sulfate can also be used, although the zinc sulfate darkens the floor. The water should be acidulated with a teaspoonful of commercial sulphuric acid per gallon. Aluminum sulfate is difficult to dissolve and should be stirred occasionally for several days. In a wooden or crockery vessel, mix $2^1/_2$ pounds of sulfate per gallon of water. Dilute this solution with an equal volume of water for the first coat. Use the full-strength solution for subsequent coats, allowing 24 hours between coats.

Zinc sulfate solution consists of $1^1/_2$ pounds of sulfate to 1 gallon of acidulated water. It is applied in two coats, with a delay of four hours between applications. Scrub the floor with hot water and dry just before the second application.

17.3. Special Treatments for Floors

The floor in a building is subject to more abuse than any other part of the building. This, however, does not mean that wear-resistant, long-lasting floors cannot be constructed. On the contrary, obtaining a good floor is no secret, although it involves more than usual care in construction. All else being equal, the quality of a floor is directly proportional to the care exercised in designing the floor, selecting the materials and constructing the floor. (See Figure 17-13.)

Heavy-Duty Topping. For a floor that is exposed to especially severe conditions of traffic and abrasion, it is not unusual to construct the floor in two layers.

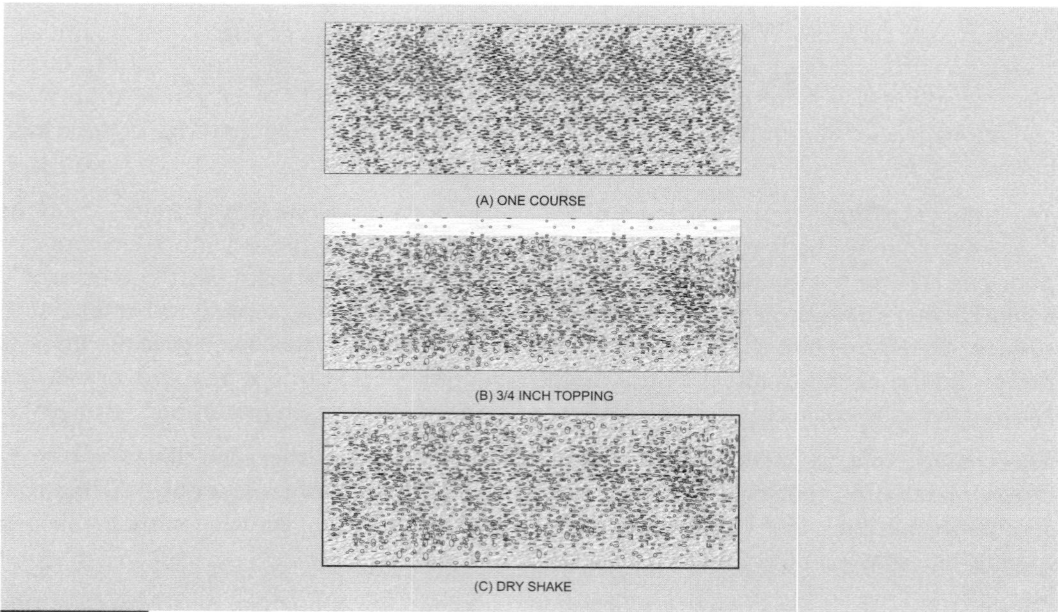

Figure 17-13

Figure 17-13. A heavy-duty floor may consist of a single course of high-quality concrete (A), a base slab with a heavy-duty topping (B) or a slab with a dust-on dry shake of special abrasion-resistant aggregate.

The base slab, prepared as described in Section 16.3, must be designed and constructed to support the expected loading. It should be struck off about $^3/_4$ to 1 inch below the finished grade elevation and should be free of laitance, oil, free water or other contaminants when the topping is applied. The topping can be applied the same day that the base concrete is placed, or it may be delayed until a more convenient time. If placement of the topping is delayed, the base slab, just prior to placing the topping, should be thoroughly cleaned and moistened.

Aggregate must be hard and tough, consisting of quartz, granite or similar natural rock particles, or a manufactured product such as silicon carbide or heat-treated aluminum oxide, depending on the anticipated wear resistance required. Maximum size of aggregate should be $^3/_8$ inch.

Mix proportions are one part cement, one part fine aggregate and one and one-half to two parts of coarse aggregate by volume. After the mix proportions have been determined, batching should be by weight. Sufficient water should be used to provide a very stiff consistency that can just be worked with the straightedge. In any event, no more than 4$^1/_2$ gallons of water per bag of cement should be used (a water-cement ratio of 0.40). The topping should form a ball when squeezed in the hand.

After a thin neat grout consisting of one part cement and one part minus No. 8 sand mixed to a creamlike consistency has been broomed into a small area of the base slab, the topping is placed, straightedged, and tamped or rolled. The area should be as large as can be covered at one time before the grout starts to dry out. The process of spreading grout and applying topping over the small areas is continued until the entire slab is covered, taking special care that edges to be lapped are well bonded and cold joints are avoided. Meanwhile, floating with a power float is commenced as soon as possible. A straightedge is used to detect high and low spots, which are then eliminated. Temperature difference between the base slab and topping should not exceed 10°F when the topping is applied.

When the wearing course is placed on the same day as the base slab, all bleeding water and laitance must be removed from the base slab before the wearing course is applied. Usually, no more than two hours should elapse between finishing the base and applying the wearing course. Spreading, striking-off, compaction and floating are performed in the usual manner, as described above.

After the concrete has hardened sufficiently to prevent working of excess fines to the surface, the final steel troweling is commenced. Dry cement, or cement and dry sand, mixtures must not be spread on the surface to dry up wet spots. A second troweling is sometimes necessary to produce an especially smooth surface if such a surface is desired.

Joints in the base slab must be continued through the wearing course; otherwise the topping will crack. Each joint must be vertical and exactly over the base-slab joint.

Dry Shake. A dry shake or dust coat can be applied to a one- course slab to give it a high resistance to abrasion and impact. (See Figure 17-14.) Application of a dry shake should not be confused with the use of dry material applied to dry a surface when bleed water is present. The latter procedure should not be permitted. A dry shake is spread on the floated slab after the bleed water has disappeared. It is never used when bleed water is present.

Abrasive material for this usage consists of aluminum oxide, silicon carbide or malleable iron particles—the size of particles depending upon traffic conditions and type of finish desired. Particles are seldom larger than those passing a No. 8 screen or smaller than those retained on a 50- to 60-mesh screen. Because of rusting of the iron exposed on the surface,

Figure 17-14
Hand applying and floating dry shake topping. (Courtesy of PCA)

iron aggregate should not be used outdoors or where the floor is exposed to moist conditions.

In using a dry shake, recommendations of the manufacturer should be followed. Lacking such recommendations, a mixture is made consisting of two parts dry abrasive and one part cement. The amount of shake to be used depends upon anticipated traffic conditions, but ranges between $^1/_2$ pound and $1^1/_2$ pounds of aggregate per square foot of floor. About two-thirds of the required quantity of shake should be broadcast over the surface immediately before power floating by sifting it through the fingers or by use of a hand-operated shaker or spreading device. Considerable care is necessary to spread the material evenly over the slab. After the material has been blended into the surface with the power float, the remaining material is spread, taking care to make the complete coating as uniform as possible. The surface is then power floated again, followed by machine troweling, then hand troweling. No water should be used during any of these finishing operations.

The cement and metallic aggregate mixture must not be placed on the concrete base too soon, as this results in some of the metal being worked beneath the surface of the concrete with a thin layer of cement and water on top. This surface cement-water paste will scale or peel later. Overworking of the surface material will also contribute to scaling.

Liquid Hardener. The hardness and wear resistance of some concrete floors can be improved by the application of a liquid hardener. Liquid surface treatments penetrate into the concrete surface and form crystalline or gummy deposits that tend to make the floor less pervious by making the surface harder.

Liquid surface treatments should be considered emergency measures for treatment of deficiencies. Some floor slabs have relatively pervious and soft surfaces that tend to wear and dust rapidly; the hardness and wear resistance of such surfaces can be improved somewhat by the application of a liquid hardener. Liquid surface treatments are, however, not intended to provide additional wear resistance in new, well-designed, well-constructed and cured floors, nor to permit the use of concrete of lower quality. The most effective use of liquid hardeners is on older floors or floors of poor quality that have already started to dust. New floors should be of sufficient quality that such treatments should not be required.

If, for any reason, liquid hardeners are to be applied to new concrete floors, the floor should be moist cured; curing compounds should not be used, because they prevent penetration of the liquid hardener. Liquid surface treatments of magnesium or zinc fluosilicate or sodium silicate should be applied only to concrete floors that are at least 28 days old and that have been thoroughly cured and allowed to air dry.[17.1]

17.4. Decorative Finishes

Concrete lends itself particularly well to the application of decorative finishes. This is especially true of slabs because the concrete is in a plastic and receptive condition during the finishing operations, which is when most of the decorative finishes are applied.

Textured Surface. We have previously discussed the nonslip surface that is obtained by brushing the slab after it has been troweled. Visual interest can be imparted to a brushed surface by brushing in alternate directions or brushing in wavy lines instead of straight.

Light texture can be obtained by holding the trowel flat against the surface and finishing with swirling motions or by moving in a series of short arcs.

Travertine finishes are those that lend texture and color character to a flat slab. These finishes can be used only on concrete that is not exposed to freezing and thawing, because water trapped in the voids formed in these processes is likely to damage the concrete when it freezes. There are two methods used.

In the rock salt method, rock salt is scattered over the troweled surface and pressed into the concrete with a trowel. The salt is coarse sodium chloride water softener salt graded so that 100 percent passes the $^3/_8$-inch sieve and 85 percent is retained on the No. 8. If a slightly coarser texture is desired, the extra-coarse grade of salt can be used. After the slab has been finished in the usual manner, salt is spread on the surface of the concrete at a rate of between 3 and 12 pounds per 100 square feet of area, depending on the texture desired. Uniformity of application is not necessary. The salt crystals are pressed into the concrete with a trowel, using just enough pressure to embed the salt particles so that their tops are barely exposed. The slab is then cured by covering with waterproof curing paper or by one of the water curing methods. Sheet plastic used for a covering might cause discoloration. Clear curing compound can be used, but it makes removal of the salt more difficult. After the slab has been cured for five days, the salt is dissolved by flooding the surface with water, aided by a little gentle brushing. The dissolved salt leaves small pits in the surface.

To make the other type of travertine finish, the slab is edged and darbied in the usual manner. The slab is then broomed with a stiff-bristled broom to assure bond with the finish coat, which is made by mixing one bag white cement and 2 cubic feet of sand with color pigment to give the proper color, usually about $^1/_4$ pound of pigment. The dry materials are mixed thoroughly, then water is added to make a rather soupy mix about the consistency of thick paint. This mortar is vigorously dashed on the surface with ridges about $^1/_4$ to $^1/_2$ inch high. After the surface hardens so that a finisher can get on it with kneeboards, it is troweled, leaving the surface texture partly smooth and partly rough. Variations in texture are made by varying the amount of mortar thrown on the surface and the extent of troweling. The mortar should be of a slightly different color from the base slab. Yellow and brown oxide pigments are commonly used. The finished surface must be properly cured. (See Figures 17-15 and 17-16.)

Flagstone. A simulated flagstone can be made by tooling artificial joints in the concrete in random geometric designs. A regular grooving tool can be used; even better is a piece of $^3/_4$- or $^3/_8$-inch pipe or tubing about 18 inches long bent as shown in Figure 17-17.

Figure 17-15
Example of a travertine texture. (Courtesy of PCA)

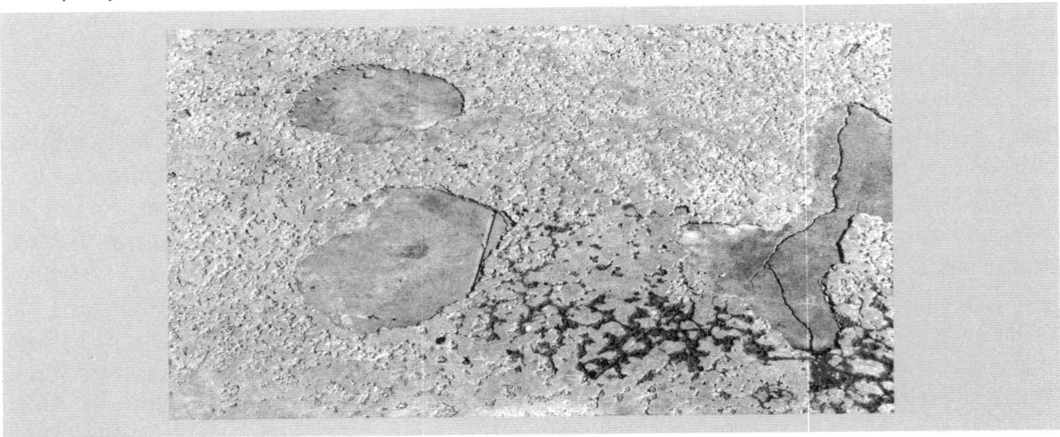

Figure 17-16
This is what happens when an attempt is made to apply a thin travertine surface to an old concrete sidewalk. The old concrete was not properly prepared, and the surfacing is too thin.

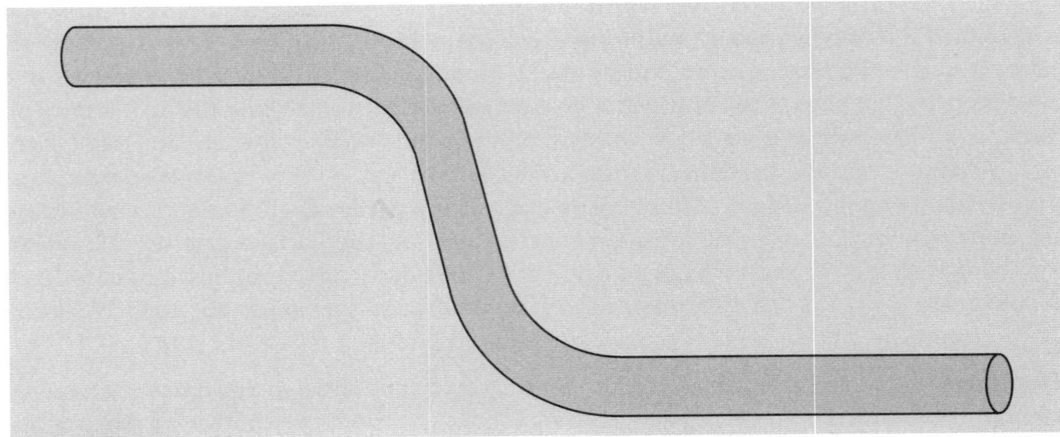

Figure 17-17
A grooving tool for flagstone finish can be made of 3/4-inch tubing.

Just before floating the slab, the pipe tool is worked into the concrete to make grooves about $^3/_4$ inch wide by $^3/_8$ inch deep, using the tool in much the same way that a mason finishes the joints in a masonry wall. After the slab has been floated, the joints should be rerun with the tool to smooth the rough edges. After the slab has been troweled, the joints should then be brushed with a soft bristle brush to clean them.

Interest can be heightened by coloring the joints. Following the first tooling, which should be exaggerated slightly, a small amount of colored mortar is carefully fed into the roughed-out joint, the joint immediately rerun with the tool prior to floating, then finished as described above. The mortar consists of one part cement, two parts fine sand and pigment, with sufficient water to make a soft, plastic mix. Care is necessary to avoid getting the mortar outside of the joint.

Colored Concrete. Color can be imparted to concrete by the surface application of paint or penetrating stains, or by using pigments incorporated in the concrete when it is mixed. Paints provide an infinite range of colors but are subject to weathering and wear if exposed to traffic. Stains penetrate slightly into the concrete and are usually employed to color existing slabs. It is difficult to obtain uniform color with a stain; for this reason the manufacturer's instructions should be followed.

The use of paint and coloring pigments is covered in Chapter 22. With respect to floors, integral color by use of pigments can be accomplished by coloring the entire slab, by placing a thin topping of colored concrete or by a dry-shake method. The color topping for the two-course method is applied to the base slab in the same manner as the heavy-duty topping previously described. This method is used because of its economy as compared with coloring the entire slab.

Material for the dry shake, consisting of white cement, pigment and specially graded silica sand, is best purchased ready to use. Job-mixed material is usually less uniform. The dry mixture is shaken evenly by hand over the surface immediately after the slab is floated. Application must be uniform; otherwise a blotchy surface will result. The first application should use about two thirds of the required total amount of shake. After a few minutes the shake will absorb some water from the concrete, at which time it should be floated. Following this floating, the balance of the shake is evenly spread and floated. Troweling follows the floating. Following a delay to permit the concrete to take up or increase its set, a second troweling is given to the slab. This is usually adequate for exterior work and can be followed by a light brushing to roughen the texture slightly.

Floating and troweling are best done with power tools. For interior slabs, a third troweling by hand is required to give the slab a smooth, dense surface. After the floor has cured and dried it can be given an application of at least two coats of concrete floor wax containing the same color pigment used in the dry shake. Special care is necessary to protect the floor from staining, dirt and traffic until the final waxing has been completed.

Exposed Aggregate. A very attractive surface can be achieved by exposing the coarse aggregate particles. This treatment is especially effective for exterior walks, patios and other surfaces where a somewhat rustic or natural effect is desired. (See Figures 17-18 and 17-19.)

Two methods are employed. In one, the concrete mix contains a higher than normal proportion of coarse aggregate. The second method requires the "seeding" of the darbied surface with coarse aggregate. Exposure in each case is accomplished by washing and brushing the surface of the slab to remove the cement paste coating from the coarse aggregate particles. (See Figure 17-20.) A surface retarder may or may not be used.

Figure 17-18
An exposed aggregate walk, durable and slightly rough, makes for a good appearance. (Courtesy of PCA)

Figure 17-19
There are two ways of obtaining exposed-aggregate finishes: 1 .Seeding a select aggregate into the concrete surface, and 2. the monolithic technique, where a selected aggregate, usually gap-graded, is introduced in the full-depth concrete mix. After the concrete is hard enough to walk on and retain the aggregate, finishers wash and brush the surface to expose the aggregate. (Courtesy of PCA)

Figure 17-20
Finishing exposed aggregate close-up. Water jets remove paste surrounding embedded aggregates. (Courtesy of PCA)

The concrete for the integral method must have a slump not exceeding 3 inches so that the coarse aggregate will remain close to the surface of the slab. Aggregate size should be 1 inch or smaller. On one job the mix contained 3.2 parts of $^{3}/_{8}$-inch pea gravel and 2.3 parts of sand. Working with the darby or bullfloat should be the minimum that will give an acceptably smooth surface. Too much working at this time will force the coarse aggregate too deeply into the concrete. The idea is to have as much aggregate as possible barely below the surface.

After initial set of the concrete when the water sheen disappears from the surface (about an hour, depending on weather conditions), the slab is washed with water while it is brushed with a push broom. The washing removes the cement paste from the coarse particles. Care must be taken to get the particles clean without undercutting or loosening them. An exposure of $^{1}/_{16}$ to $^{1}/_{8}$ inch is a maximum; usually less is adequate. On larger aggregate the depth of exposure can be more.

Instead of scrubbing when the concrete reaches initial set, it may be desirable to delay the washing and brushing until a more convenient time, especially on a large job. This can be accomplished by the use of a surface retarder. The slab is troweled at the proper time after the surface has hardened somewhat. A surface retarding solution is then sprayed on. After a time lapse of 8 to 16 hours, depending on weather conditions, the surface is brushed and washed.

The seeding method starts with placing the concrete in the base slab using a $^{3}/_{4}$-inch MSA with a slump between 2 and 4 inches. Some specifications require that the No. 4 coarse aggregate in the mix be reduced. Because of the volume that will be occupied by the seeded aggregate, the base slab should be finished slightly below finish grade. The amounts shown in Table 17.1 are suggested. After the slab has been darbied or bullfloated, the aggregate is spread by hand or shovel, care being taken to cover the area completely with one layer of stone. The closer the particles are to each other, the better the appearance of the finished job. Slump of the concrete should be about 2 inches when the stone is seeded. Aggregate should be completely embedded by tapping with a float, taking care that none of the aggregate in the base concrete is mixed with the seeded aggregate. Aggregate should be covered with about $^{1}/_{16}$ inch of mortar.

TABLE 17.1
GRADE CORRECTION FOR SEEDED AGGREGATE

SIZE OF AGGREGATE, IN.	DEPTH BELOW FINISH GRADE TO STRIKE OFF, IN.
$^{3}/_{8}$ to $^{5}/_{8}$	$^{1}/_{8}$
$^{1}/_{2}$ to $^{3}/_{4}$	$^{3}/_{16}$
$^{3}/_{4}$ to 1	$^{1}/_{4}$
1 to $1^{1}/_{2}$	$^{5}/_{16}$
$1^{1}/_{4}$ to 2	$^{7}/_{16}$

Example:
If 3/4-inch aggregate is to be seeded, strike off about 3/16 to 1/4 inch below finish grade elevation.

After a suitable delay, the surface is brushed and washed as previously described, either with or without the surface retarder, depending on job conditions. After the exposure has been completed, the slab should be cured, preferably by covering with impervious sheet material, wet burlap or ponding. Curing compound should not be used.

If it is necessary to brighten the surface, it can be washed with weak muriatic acid (10 percent) not less than two weeks after completion of curing. After thoroughly saturating the slab with water, the acid is applied and left on for only a few seconds. The slab is then thoroughly washed with clean water.

The following suggestions apply to any of the methods of exposing aggregates:

1. Uniformity of materials, methods and timing is essential.
2. Timing is dependent on weather conditions. If aggregate is dislodged during washing, the operation should be delayed or a softer brush used.
3. A sample panel should be prepared under job conditions.
4. Do not use calcium chloride in the concrete.
5. If a surface retarder is used, order it from a reliable distributor and follow the manufacturer's instructions.
6. After a reasonable lapse of time, start checking the surface to determine the best time to start scrubbing. Experience is the best guide in this respect.

Terrazzo. The type of floor finish known as terrazzo is a Venetian marble mosaic in a matrix of portland cement mortar. (See Figure 17-21.) Some use has been made recently of an epoxy resin binder in a thin-set terrazzo, which appears to be suitable for toppings with a net thickness as small as $^1/_4$ inch.

Figure 17-21
Installing a terrazzo floor. (Courtesy of PCA)

Terrazzo floors can be made in almost any color and design to suit the artistic requirements of the structure. The use of dividing strips enhances the color pattern and permits intricate detailed designs to be worked into the floor. Size and color of the marble chips used in the floor and the color of the cement matrix can be varied to provide interest and variation. Divider strips are available in several different materials and can be shaped into intricate configurations. (See Figure 17-22.)

Sand-cushion terrazzo is used where structural movement is anticipated. In this method, the structural slab is struck off 3 inches below finish grade for the floor. A layer of dry sand about $^1/_2$ inch thick is spread over the concrete slab, over which an impermeable membrane is placed. A mortar underbed about 2 inches thick is next placed to within $^1/_8$ inch below the finish grade. Next, the divider strips are inserted, and the topping mix is placed.

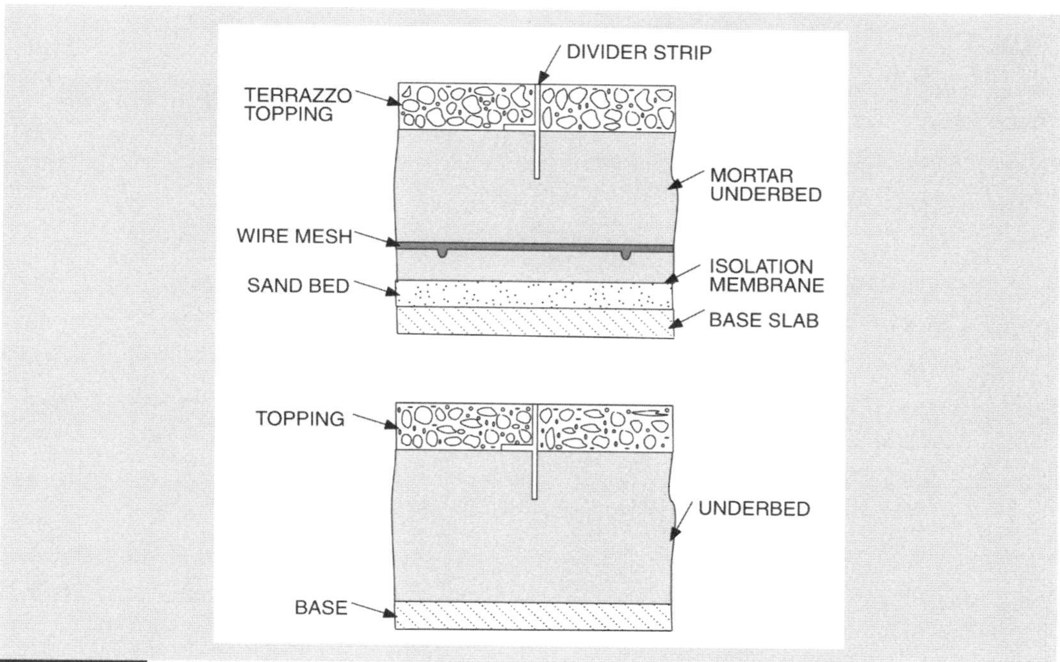

Figure 17-22
The top figure is a cross section of a typical sand-cushion terrazzo. The section in the bottom photo is an example of bonded terrazzo.

Bonded terrazzo has no sand cushion. A mortar underbed 1 inch thick is bonded to the structural slab by the use of a cement slurry broomed onto the slab. Divider strips are inserted, and the topping mix is placed. Overall thickness is usually 1$^{3}/_{4}$ inch, although a thickness up to 2$^{1}/_{2}$ inches might be specified if additional reinforcing is required.

Monolithic terrazzo, the most economical, consists of a $^{5}/_{8}$- inch topping applied directly to the structural slab following the cement slush coat. It provides practically no control over random cracking.

The topping is the mix and materials composing the wearing surface. Standard topping has a minimum thickness, after grinding, of $^{5}/_{8}$ inch and contains No. 1, No. 2 or No. 3 marble chips. (See Table 17.2.) Venetian topping employs chips as large as No. 8 and has a greater thickness than standard. Palladiana or Berliner has a minimum thickness of 1 inch and includes large pieces of marble, up to 140 square inches in area with standard terrazzo dividers. Conductive terrazzo is similar to standard except that acetylene carbon black and isopropyl alcohol are added to the underbed and topping mixes. The use of conductive terrazzo eliminates the danger of sparks in explosive atmospheres.

Mix proportions for the underbed mortar and topping mix, kind and location of divider strips, size and color of marble chips, and finishing details will be found in the job specifications. Dividers control drying shrinkage and other cracks, if they occur, by permitting the crack to form inconspicuously adjacent to the divider. Dividers also permit the laying of different color mixtures accurately in intricate patterns. Because of the extremely dry mix and method of construction, shrinkage of terrazzo is rarely of any consequence. Cracking may occur as a result of structural movement, expansion and contraction, or vibration, and it is this cracking that the dividers control. Divider strips are made of zinc, half-hard brass or colored plastic.

TABLE 17.2
TERRAZZO AGGREGATE SIZES

CHE NO.	PASS SCREEN	RETAINED ON SCREEN
0	1/8 IN.	1/16 IN.
1	1/4	1/8
2	3/8	1/4
3	1/2	3/8
4	5/8	1/2
5	3/4	5/8
6	7/8	3/4
7	1	7/8
8	1 1/2	1
For Venetian terrazzo, the larger chips should be used in the following sizes:		
No. 4 and No. 5	3/4	1/2
Nos. 6, 7 and 8	1 1/8	3/4

The oldest, best and most expensive terrazzo is the sand cushion terrazzo. In all of the methods, after preparation of the base and insertion of dividers, the topping mix is spread. This mixture consists of one bag (94 pounds) of cement, usually white, color pigment as required, and 200 pounds of marble chips. Cement and pigment should be mixed dry, chips added and, finally, not more than 35 pounds of water added. The mix should contain only enough water so that it can be struck off with a straightedge. Immediately following the strike-off, the topping is rolled with heavy rollers, with additional chips sprinkled on the surface so that at least 70 percent of the surface is chips. Floating and troweling with power tools follow rolling, after which the floor is cured.

After curing, the floor is ground or machine rubbed (Figure 17-23), using No. 24-grit abrasive stone for the rough grinding followed by rubbing with No. 80 or finer grit. The floor should be wet, and the ground-off material flushed off. Next, the floor is grouted with a grout made of cement, pigment to match the original mix, and water, by spreading and squeegeeing so that all voids are filled. Surplus grout on the surface should be cleaned off and the floor cured for at least 72 hours. Fine stoning with a No. 80 grit is done while the

Figure 17-23
Buffer-size planetary grinder for polishing concrete (terrazzo) floors. (Courtesy of PCA)

floor is flooded with water. After the floor has been thoroughly cleaned and dried, the penetrating sealer is applied.

A sealer especially formulated for terrazzo should be applied to brighten and protect the floor after cleaning with a mild, neutral cleaner. Varnish, lacquer, shellac and surface waxes should not be used on terrazzo. Floor oils and oily or colored sweeping compounds should not be used for maintenance cleaning.

CURING

17.5. The Need for Curing

Curing has been previously mentioned as a requirement for good, durable concrete. If concrete is not properly cured it has low strength, poor durability and other deficiencies that detract from the usefulness and appearance of the structure. Curing promotes the hydration of the cement, which has been explained as the chemical process of the reaction of cement and water—a reaction that starts as soon as the two materials are brought together and continues at a steadily reducing rate for months or even years under favorable circumstances. As far as the worker on the job is concerned, curing is any method or material that prevents loss of water from the concrete for a reasonable period of time, that maintains the concrete at a reasonable temperature, and that protects it from damage. All concrete must be cured. Figure 17-25 shows what happens to concrete strength if the concrete is not cured. After a year the concrete is only about half as strong as it could be if it had been cured. At 90 days old, concrete not cured at all will have less than half of the strength of concrete cured for 28 days then permitted to dry out; if it is cured for seven days before drying, it will have about 80 percent of the strength of concrete cured for 28 days. Other desirable properties of the concrete suffer in the same proportion. Another important point is that concrete, wet cured after drying, will resume strength gain, although at a reduced rate, and it will not get as strong as it would have if it had been fully cured. The

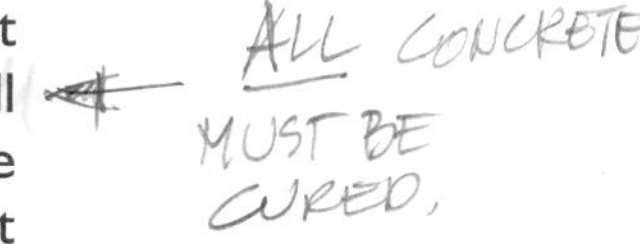

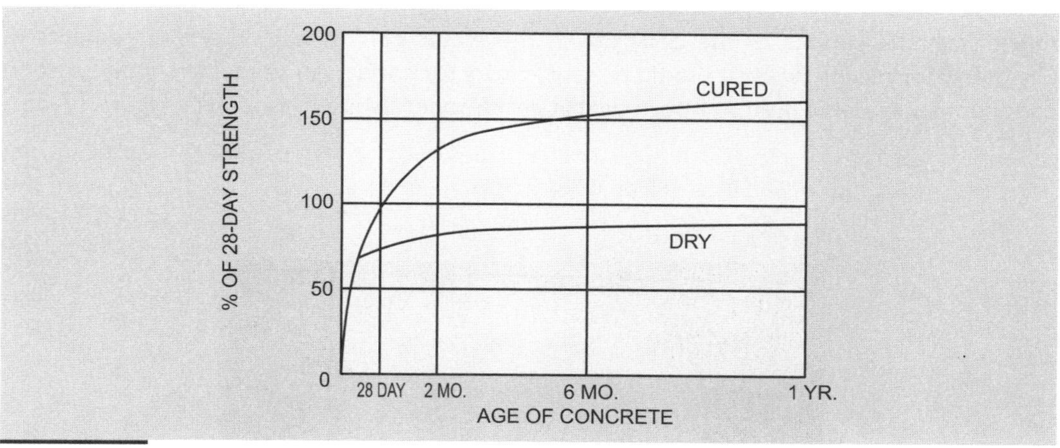

Figure 17-24
Concrete increases in strength almost endlessly if it is kept moist at a reasonable temperature, but dry concrete soon reaches a limit.

initial drying is apt to result in shrinkage cracking and other problems—another reason why concrete should be cured.

Fresh concrete, when placed in the forms, contains more than enough water for combining with the cement during hydration. One function of curing, then, is to preserve or replenish this water for a suitable period of time until the concrete has developed the desired properties. The longer the cure, the better the concrete. From a practical view, the benefits of curing must be balanced against such factors as the cost of curing, the need to put the structure in service or make it available for other crafts, and the availability of curing materials. In other words, there is a practical limit to the length of time concrete can be cured.

17.6. Curing Materials

Materials for curing include water, liquid membrane-forming compounds and various sheet materials and blankets.

Water. The water used for curing concrete should be clean and free of substances that

Figure 17-25
Water curing of sidewalk. Once the concrete has set sufficiently to prevent water erosion, ordinary lawn sprinklers are effective if good coverage is provided. (Courtesy of PCA)

might stain the surface of the concrete. If the water conforms to the requirements of Chapter 9 for mixing water it should be satisfactory for curing. An adequate supply must be continuously available during all periods that green concrete is exposed. (See Figure 17-25.)

Figure 17-26
Lawn sprinklers saturating burlap with water keep the concrete continuously moist. (Courtesy of PCA)

Wet Coverings. Burlap, cotton mats and other moisture-retaining fabrics are commonly used for curing. Burlap reclaimed from other uses or sacks should be avoided unless it has been thoroughly cleaned of any substance that might injure the fresh concrete. Treated burlaps are available that are resistant to rot and fire. New burlap should be washed to remove sizing and make it more absorbent. (See Figure 17-26.)

Curing mats may be made of cotton, jute, sisal or similar material. They usually have a watertight covering that aids in retaining the moisture and are quilted to give them greater strength and durability. They become quite heavy when saturated with water.

Impermeable Sheet Materials. Floors and slabs are frequently cured by covering with an impermeable sheet of some kind. ASTM C 171 covers three types of materials: waterproof paper either regular or white; polyethylene film, clear or white; and white burlap-polyethylene sheet. All materials should be tough, strong and resilient. They should resist normal use on the job without puncturing or tearing.

Figure 17-27
Use of water-proof curing paper is an effective means of curing horizontal surfaces. (Courtesy of PCA)

Waterproof paper consists of two layers of kraft paper reinforced with jute, cotton or plastic yarn embedded in a bituminous cement between the two layers of paper. (See Figure 17-27.) The paper should be light in color. One side is usually white. The paper

Figure 17-28
Polyethylene film is an effective moisture barrier for curing concrete. To minimize discoloration, the film should be keep as flat as possible on the concrete surface. (Courtesy of PCA)

should be factory treated to provide high wet strength. It is available in rolls up to 8 feet wide.

Figure 17-29
Membrane-forming curing compound being applied to driveway with a spray can.

Polyethylene film should consist of a single sheet at least 4 mils (0.004 inch) thick. The clear sheet should be essentially transparent; the white should be opaque. It is available in rolls up to 32 feet wide. (See Figure 17-28.)

Sealing Compounds. The purpose of a concrete sealing compound is to seal the surface of new concrete against the loss of moisture, thus retaining the water in the concrete for hydration of the cement, or curing. (See Figure 17-29.)

This material, variously known as sealing, curing or membrane-forming compound, is a paint-like liquid that, when sprayed on the concrete, forms an impervious membrane over the surface. It is covered by ASTM C 309.

The following types of compounds are covered by ASTM C 309:

Type 1. Clear or translucent without dye.

Type 1-D. Clear or translucent with fugitive dye.

Type 2. White pigmented.

The vehicle should be either:

Class A. No restrictions on vehicle solids material.

Class B. Vehicle solids limited to all resin material.

The dye should render the film distinctly visible for at least four hours after application and should fade completely in not more than one week. All compounds should be of a consistency suitable for spraying, should be relatively nontoxic, should adhere to a vertical or horizontal damp concrete surface when applied at the specified time and coverage, and should not react harmfully with the concrete. The clear compound should not darken the natural color of the concrete.

Curing compounds are ready-mixed when they arrive on the job but need vigorous stirring or agitating before they can be used. Under no circumstances should they be thinned or otherwise altered on the job. Usually, the drums in which sealing compound is shipped are equipped with agitators, which should be operated before and during the time the compound is being sprayed.

Pigmented compound consists of a finely ground pigment in a vehicle consisting of a solvent and waxes, oils or resins. The compound made with an all-wax base is sometimes used, but it is of a heavy consistency that makes it difficult to spray, especially in cold weather. The most commonly used type is made with either a wax and resin base, or a straight resin base. In any pigmented compound, the pigment has a tendency to settle; hence, the need for constant agitation of the material in the drum during application.

The clear compound is formulated with either a wax and resin base or all-resin base. One advantage claimed for the resin type is that it does not tend to segregate as the wax and resin type does; hence, there is less need for agitation.

Sealing compounds should conform to the requirements of ASTM C 309. Sealing compounds are used, without testing or certification, on many jobs. It is good practice, however, to obtain a certification from the manufacturer if tests are not made.

Formerly available only in solvent-release composition, curing compounds are now available from some producers in a water emulsion resin base, either clear or white pigmented. These products, produced to permit their use in areas in which environmental restrictions prohibit the use of solvent-release materials, meet the requirements of ASTM C 309, Type 1, Class B, and Type 2, Class B. The clear compound can be used on surfaces that will receive tile or paint. Both materials should be protected from freezing; however, if freezing occurs, the compound can be used after thorough agitation and remixing. They should not be diluted.

Another curing liquid is designed to serve as a curing compound and parting compound. When applied to a troweled floor, the concrete is effectively sealed to prevent loss of water, and curing is accomplished. The coating also serves as a parting compound to prevent bond of concrete subsequently placed for tilt-up panels. This dual purpose material is available either as a solvent-release or an emulsified solution.

Yet another curing compound is a two-component, water- dispersible epoxy system that forms an emulsion when mixed with water. It can be sprayed on the fresh concrete or exposed aggregate, giving a cure and seal in a single treatment. This coating tends to darken concrete surfaces upon aging.

A recent revision to ASTM C 309 excludes the use of sodium silicate solutions, calling them ineffective for curing concrete: "Sodium silicate solutions are chemically reactive rather than membrane framing"; therefore, they do not meet the intent of the ASTM C 309 specification.

17.7. Time-Duration of Curing

Concrete, on even a single job, is proportioned to have vastly different properties, depending on the requirements of each part of the structure; not only that, but the concrete is placed during widely different weather conditions. For these reasons, it is difficult to arbitrarily set a time limit for the duration of curing. Besides strength, watertightness, surface appearance, resistance to weathering and aggressive substances; surface hardness and other properties of the concrete have to be considered. Weather conditions vary from the hot dry summers of the desert to the wet and freezing weather in northern climates.

The first few days are the most critical in curing concrete. In fact, the first few hours or even minutes after placing can spell the difference between mediocre concrete and good concrete, depending on how well the concrete is taken care of. A dry, hot wind, for example, can cause surface shrinkage and plastic cracking of the fresh concrete unless measures are taken to protect the concrete. Such measures can be thought of as curing. Once the cement sets, development of strength and other properties of the concrete commences, evolving rapidly at first, then increasing at a lessening rate as time continues. Recognizing the differences in required properties of the concrete and differences in the weather will lead to the establishment of several time limitations. As an illustration, for Type III cement, the time period can be shorter than for Type I cement; it can be shorter for rich mixes than for lean, mass mixes. For structural concrete an absolute minimum should be three days; emergency repairs using Type III cement in rich mixes should be cured for 48 hours minimum. Normal concrete in structures and pavement is best cured for seven days. Lean concrete in massive structures requires about two weeks for concrete without pozzolan, or three weeks for pozzolan concrete. During cold weather the concrete requires protection from low temperatures; during hot weather it requires protection from high temperatures and abnormal drying. Special conditions require special treatments. Specifying agencies will consider all sides of the situation and might specify time periods considerably longer or shorter than those given here.

17.8. Methods of Curing

Curing methods can be grouped in two general categories—those that prevent loss of moisture from the concrete by sealing the surface, and those that supply additional moisture to the concrete. Both systems are commonly used, and both are satisfactory when properly applied. This is the problem. Curing of concrete is probably the most abused part of the whole concrete construction endeavor. Any one of the common methods will produce the required results, provided it is properly employed.

Methods that Supply Additional Moisture. In this group we find such methods as sprinkling, ponding and wet coverings of various kinds. In theory, at least, water curing is the ideal way to cure concrete, if a satisfactory temperature can be maintained. In actual practice this ideal is seldom realized because of the difficulty in keeping the concrete continuously wet. Far too often water is sprinkled intermittently by a worker who has too many places to cover or too many other jobs to do or because someone removes the hose to use someplace else. As a result, water curing does not work out as well as intended. Water-saturated mats in close contact with the concrete are probably the best way to keep the entire surface continuously wet. Water should be continuously available by means of sprinklers, soaker hoses or other means. Pieces of burlap loosely draped over the concrete, whether wet or dry, are of little value in curing. Permitting the concrete to be alternately wet and dry can do more harm than good.

Unformed surfaces on the top of walls, girders and similar units should be moistened by wet burlap or other effective means as soon as the concrete has hardened sufficiently to withstand the treatment without damage. This is especially important for construction joints, which cannot be cured with sealing compound.

On large concrete structures, it is sometimes the practice to attach spray pipes to the lower edge of the forms. As the forms are raised for successive lifts, water curing is applied to the freshly exposed concrete. However, water running over the lower portions of the structure for a long period of time is apt to stain the concrete.

Flat slabs can be cured by constructing small earth dikes around the perimeter and flooding the area with water. This method is especially suitable for small slabs on grade. If flooding is done on a suspended slab, the formwork and shoring must be constructed to support the additional weight of the water. Placing of moist backfill on footings or against below-grade walls also provides good curing—as soon as the structure can withstand the earth pressures developed.

Wet coverings of earth and sand have been used and are satisfactory on the small job. Their disadvantages are the danger of staining the concrete, and their relatively high cost. If used, the covering should be at least 2 inches thick, and it must be kept continuously wet. The covering does provide some degree of protection of the concrete to light foot traffic. Cleaning up and disposing of the material can be a problem.

Burlap, cotton mats and other moisture-retaining fabric have proved to be quite satisfactory. They can be applied as soon as the concrete has hardened enough to prevent damage to the surface, must cover all the concrete, including edges, and must be kept wet. Plastic sheets are often placed over wet burlap to retain moisture. The fabric coverings can be used on sloping and vertical surfaces as well as horizontal ones.

Burlap blankets should not be made out of second-hand gunny sacks or similar material. They should be made of at least two layers of new 9-ounce or heavier burlap. Burlap that has been used for curing concrete may be reused, provided it is clean and in good condition.

Mats may be of several materials, such as cotton, sisal or jute. Their efficiency should be at least equal to that of the burlap blankets, and, like the blankets, they should be kept saturated by frequent sprinkling with water. Some mats have a waterproof covering, which is supposed to retain the water underneath. These should be checked periodically to see that the concrete is wet at all times. If the concrete appears to be drying out, additional water should be added immediately.

Care should be taken in placing blankets and mats to ensure that they completely cover the slab, including the edges, and that they are adequately lapped. In placing new mats, allowance should be made for shrinkage. They should be laid on the slab as soon as they can without marring the surface. A fine fog spray may be necessary to keep the concrete moist until it has hardened sufficiently to support the mats. Mats should be kept in place for at least seven days.

Methods that Prevent Loss of Moisture. These can be classified as sealing materials, as their function is to seal the surface of the concrete so that the mix water already present in the concrete will not be lost. They do not provide any additional water. The liquid membrane-forming compounds, commonly called curing or sealing compounds, as described in Chapter 10, are included in this category.

Either the pigmented or clear compound can be used. Usually, because of its heat-reflecting property, the white pigmented material is used on pavements, canal linings and similar slabs with large areas exposed to the sun. The clear compound is used on structures and walks, one reason being that it weathers more uniformly than the white and does not present a mottled appearance after a few weeks.

The compound should be applied to unformed surfaces as soon as the water sheen disappears but while the surface is still moist. If application of the compound is delayed, the surface should be kept wet with water until the membrane can be applied. The compound should not be applied to areas on which bleeding water is standing.

The compound should be applied to formed surfaces immediately after removal of the forms. If the concrete surface is dry, it must be moistened and kept wet until no more water will be absorbed. As soon as the surface moisture film disappears, but while the concrete is still damp, the compound is applied, special care being taken to cover edges and corners. Patching the concrete is done after the compound has been applied. The compound should be sprayed, not brushed, on patched areas.

Brushed application should never be permitted on unformed surfaces, as the surface is still soft enough to be injured by the brush, and the compound will penetrate excessively into the concrete. Brush application may be permitted on formed surfaces in limited instances.

If curing compound is applied to a dry concrete surface, it can strike into or penetrate the surface, resulting in a soft, dusty concrete surface.

Pressure tank equipment should be used for spraying the compound, and the compound should be agitated continually during application. The coverage of about 135 to 150 square feet per gallon should be made in one coat, consisting of two passes of the spray nozzle at right angles to each other. Rough concrete requires slightly more compound than smooth concrete. On rough surfaces, particular care is necessary to ensure complete coverage.

The compound should never be thinned. However, during cold weather, it may be necessary to heat the compound if it becomes too thick for application. Heating should be done in a hot water bath, never over an open flame, and the temperature should never exceed 100°F. The container should be vented and there should be space in the container for the compound to expand. The compound should be agitated during heating.

A specialized compound is available that not only cures but hardens, dustproofs and seals the concrete all in one operation. For use in tilt-up construction, a compound can be sprayed on the floor slab for curing that also acts as a parting compound to prevent bond of the tilt-up concrete to the hardened concrete of the floor.

Sheets of plastic are frequently used for curing and can usually be applied to complex concrete shapes. Polyethylene sheets, either clear or white, come in long rolls in widths up to about 20 feet. Plastic should be applied as soon as the concrete has hardened sufficiently so that it will not be marred. Sometimes, especially on hard-troweled surfaces, the plastic may cause discoloration of the concrete surface.

Many kinds of waterproof papers are used for curing. These come in various widths, some of them reinforced with cotton, jute or glass yarn embedded in a bituminous cement between two layers of paper.

In using any of the impermeable sheets, it is essential that the sheets be well lapped at the joints and that edges of the concrete are well covered. Ends and edges of the sheets must be weighted down with lumber or sand to prevent the sheet from blowing away.

Temporary Curing Film. Rapid evaporation of water from plastic or fresh concrete, especially in flat slabs, caused by low humidity, wind or high temperature, results in excessive shrinkage and plastic cracking. To combat this, a spray-on monomolecular film that reduces the evaporation is available. The material, called Confilm, is sprayed lightly over the surface of the concrete immediately after striking-off or screeding. It is effective only while the concrete is in a plastic state. It is not a substitute for early curing of the hardened concrete and has no effect on subsequent curing methods. Its benefit is in its lessening of evaporation while the slab is being finished.

Research has indicated that low-volume contents of polypropylene fibers (added to ready-mixed concrete at 0.1 percent by volume) reduce cracking potential within the first few hours after placing when exposed to rapid drying conditions (plastic shrinkage, cracking).

High-Temperature Curing. In Chapter 3 we discussed high-temperature curing for accelerating the strength development and hardening of concrete, a method that is used in plants that manufacture precast units of all kinds, such as masonry block, pipe, prestressed beams and girders and, in fact, any items that can be precast. Curing temperature can run as high as 170°F, but most normal cycles include a temperature between 120°F and 160°F.

Under high-temperature conditions, both heat and moisture, which are essential to proper curing, may be lost to the outside through the material used to cover the concrete. Hence, it is important to provide tight covers, insulated if possible. It should also be kept in mind that more moisture is required to maintain 100 percent relative humidity at high temperatures than at low temperatures. At 70°F, 0.016 pound of water is required per pound of air for 100 percent humidity; at 140°F, 0.15 pound is required; and at 180°F, 0.66 pound is required.

Under steam-curing conditions, the heat of hydration serves to raise the concrete temperature in the same way that it does under normal curing. Therefore, there comes a time when the concrete temperature will reach the ambient, or surrounding, temperature within the enclosure, then rise above if control measures are not taken. After the concrete reaches the ambient temperature, the amount of heat should be reduced gradually to minimize the temperature differential between the concrete and the atmosphere within the enclosure and to permit a gradual cooling of the concrete.

When the concrete temperature exceeds the enclosure temperature, moisture may be lost from the concrete. If wet steam is used as a source of heat, moisture loss from the concrete presents no problem, but if heat is provided by other means, there must be a source of additional moisture, such as hot-water sprays.

Concrete should undergo a presetting period of approximately two to three hours at normal temperature after casting before being subjected to high temperature. During cold weather, a small amount of heat may be provided within the enclosure to maintain the temperature near 70°F during preset. Live steam should not play directly on the concrete or forms. Steam is best admitted through many small jets, as would be the case with perforated pipes, rather than through the open end of a hose or pipe.

Optimum conditions for steaming are:

1. A delay of about three hours after placing concrete before steaming
2. A concrete temperature between 120° and 160°F
3. A slow temperature rise, not over 40°F per hour
4. Wet steam
5. Avoidance of thermal shock upon termination of steaming

Moist curing after steaming improves strength and other properties and should be utilized if possible. The greatest advantage of steaming occurs during the first hours, after which it reaches a point of diminishing returns. Temperatures that are too high damage the concrete, and strengths of concrete normally cured for the entire period are usually higher at 28 days and later than those of concrete steam cured during its early age.

The Steel Reinforcement

It was pointed out in Chapter 3 that concrete has excellent strength in compression—that is, it can support heavy loads placed on it—but it is comparatively weak in its ability to resist loads that tend to bend or pull it apart. Loads of this nature are called flexural and tensile. (See Figures 3-1 and 3-3.) Reinforcement is the term used to describe the reinforcing steel inserted in the concrete to give it tensile strength. Other terms are reinforcing bars, resteel or sometimes rebars. Welded wire is sometimes referred to as fabric or mesh, but wire sizes larger than that typically used for wire mesh (slabs-on-ground) are available either to replace reinforcing bars in structural concrete or to be used in combination with bars.

18.1. The Need for Reinforcement

In a concrete beam, the member is loaded in flexure or bending, which induces a tensile stress in one side of the beam. In a horizontal beam this would be at the bottom of the beam at midspan. However, a beam can be vertical; for example, a basement wall supporting an earth load can be a vertical beam. (See Figure 18-1.) Figure 3-3 shows how steel bars embedded in the lower part of a beam enable the beam to carry a heavy flexural load by providing tensile strength in the lower part of the beam.

Reinforcing steel serves purposes other than giving beams flexural strength. In a circular water tank, pressure from the water inside the tank tends to cause the tank to push outward. Steel reinforcement embedded in the concrete gives the structure tensile strength and holds the tank together. In the same way, concrete pipe can resist internal water pressure.

Reinforcement is used to control cracking in slabs caused by expansion and contraction of the concrete that result from temperature and moisture changes. The reinforcement will not prevent cracks, but it will distribute and make them narrower. Reinforcement of this type is called shrinkage and temperature reinforcement, or more commonly crack control reinforcement. The slab is sometimes divided into rectangular panels by contraction joints, without reinforcement, to control cracking (see Section 16.2). A combination of reinforcement and contraction joints is also used for crack control; this combination will allow a wider contraction joint spacing.

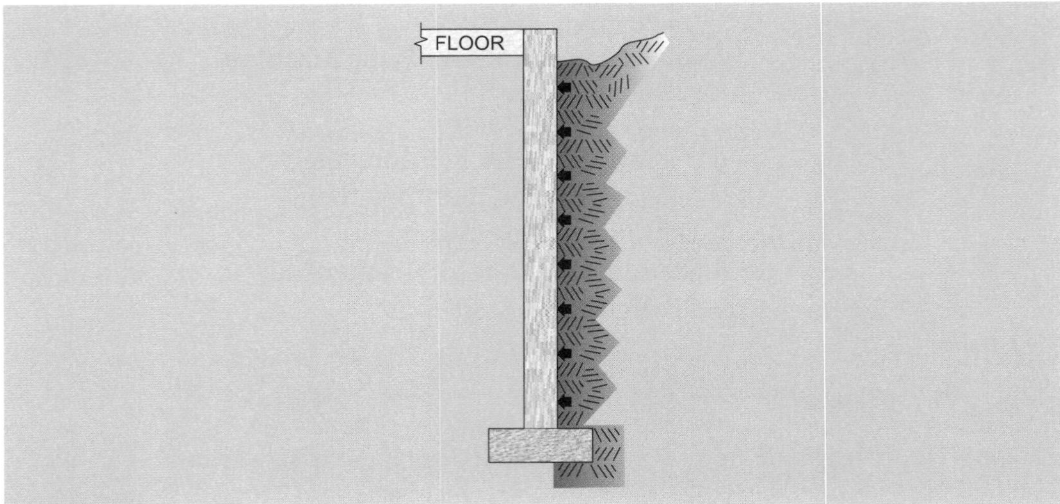

Figure 18-1
A basement wall that supports the earth backfill can be designed as a vertical beam.

Stirrups are U- or W-shaped configurations of small bars placed around the main steel of a beam to resist diagonal tension that develops in the beam as a result of shear. (See Figure 3-1.) Stirrups are usually closer together near the supports. (See Figure 18-2.) Columns are reinforced with ties or spirals at right angles to the longitudinal steel to provide resistance to buckling. As stated earlier, reinforcement is used in concrete to resist tension or pulling. In certain special high-load structures, reinforcing steel is occasionally used for carrying part of the compression load.

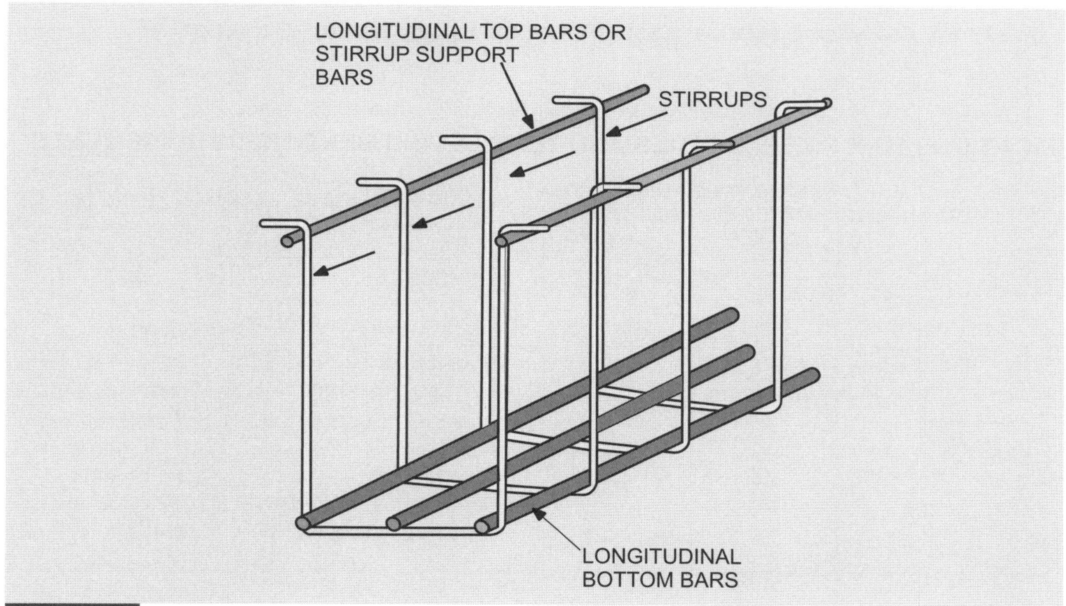

Figure 18-2
A simplified beam steel assembly showing bottom bars, top bars and stirrups. Stirrups come in other shapes and configurations besides the one shown here.

18.2. Reinforcing Steel

Reinforcing Bars. The two types of reinforcing bars are plain bars and deformed bars. Plain bars are smooth, round bars used for certain purposes such as smooth dowels at contraction joints where the bar must be free to slide in the concrete but must provide shear strength across the joint. Column spirals can also be plain wires.

Deformed bars, the standard type of reinforcement used in reinforced concrete, are rolled with deformations consisting of small bumps or ridges on the bar surface to provide good mechanical bond with the concrete. Different manufacturers roll different patterns of deformations on their bars, but they are all required to conform to the ASTM standard specifications for deformed bars.

Table 18.1 lists the bar sizes, grades and tensile and bending requirements for the two most widely used reinforcing bars (carbon-steel and low-alloy steel). The data enclosed in brackets is in metric units; see discussion "Metric Bars". The grade of steel indicates the minimum yield strength of the steel in thousands of pounds per square inch. For example, a Grade 60 carbon-steel bar is specified to have a minimum yield strength of 60,000 psi. Grades of reinforcing steel are specified by the designer and must be indicated on the plans and bar lists. The carbon-steel and low-alloy steel bars are produced by melting steel scrap in an electric-furnace process. The molten steel is continuously cast into reinforcing bars or into billets for later processing into specific bar sizes. Reinforcing bars are hot rolled to form their size and deformations. It should be noted that Grade 60 bars can be manufactured to comply with both ASTM A 615 and ASTM A 706. Such bars are permitted to have dual marks, S and W, to indicate dual compliance. In limited areas of the country, reinforcing bars are rolled from railroad rails (A 996-Rail) or freight car axles (A 996-Axle). Of interest: The older specifications for rail steel (A 616) and axle steel (A 617) were combined into ASTM A 996.

Historically, the size of reinforcing bars, given by the bar number, is the nominal diameter of the bar in eighths of an inch. For example, a #7 bar has a nominal diameter of $^7/_8$ of an inch (0.875 in.). The metric bar numbers are approximations of the bar diameter in millimeters (mm). For example, a #22 metric bar has an approximate diameter of 22 mm. As indicated above, bar numbers are usually shown by the sign "#."

TABLE 18.1
ASTM SPECIFICATIONS-BAR SIZES, GRADES AND TENSILE AND BENDING REQUIREMENTS

TYPE OF STEEL AND ASTM SPECIFICAION	BAR SIZES	GRADES	MINIMUM YIELD STRENGTH, psi (MPa)	MINIMUM TEN-SILE STRENGTH, psi (MPa)	MINIMUM PERCENTAGE ELONGA-TION IN 8 IN. (203.2 mm)	BEND TEST PIN DIAMETER (d = nominal diameter of specimen)
Carbon-Steel A615/A615M	#3 to #6 (#10 to #19)	40 (280)	40,000 (280)	60,000 (420)	#3 (#10) . . . . 11 #4, #5, #6 (#13, #16, #19) 12	#3, #4, #5 (#10, #13, #16) . . . . $3^1/_2d$ #6 (#19) 5*d*
	#3 to #18 (#10 to #57)	60 (420)	60,000 (420)	90,000 (620)	#3, #4, #5, #6 (#10, #13, #16, #19) . . . . 9 #7, #8 (#22, #25) . . . . 8 #9, #10, #11, #14, #18 (#29, #32, #36, #43, #57) . . . 7	#3, #4, #5 (#10, #13, #16) . . . . $3^1/_2d$ #6, #7, #8 (#19, #22, #25) . . . . 5*d* #9, #10, #11 (#29, #32, #36) . . . . 7*d* #14, #18 (#43, #57) 9*d*
	#3 to #18 (#19 to #57)	75 (520)	75,000 (520)	100,000 (690)	3, #4, #5, #6, #7, #8 (#10, #13, #16, #19,#22,#25) #9,#10,#11,#14,#18 (#29, #32, #36, #43, #57) . . . 6	#6,#7, #8 (#19,#22,#25) . . . . 5*d* #9,#10,#11 (#29,#32,#36) . . . . 7*d* #14,#18 (#43,#57) 9*d*
Low-Alloy Steel A706/A706M	#3 to #18 (#10 to #57)	60 (420)	60,000 (420)	80,000 (550)	#3, #4, #5, #6 (#10, #13, #16, #19) . . . . 14 #7, #8,#9,#10,#11 (#22,#25,#29,#32,#36) . . . . . 12 #14,#18 (43,#57) 10	#3, #4, #5 (#10, #13, #16) . . . . 3*d* #6, #7, #8 (#19, #22, #25) . . . . 4*d* #9, #10, #11 (#29, #32, #36) . . . . 6*d* #14, #18 (#43, #57) 8*d*

For low-alloy steel reinforcing bars, the ASTM A 706/A 706M specification prescribes a maximum yield strength of 78,000 psi (540 MPa), and tensile strength must be 1.25 times the actual yield strength.

Bend tests are 180° except ASTM A 615/A 615M permits 90° for bar sizes #14 and #18 (#43 and #57)

When a steel bar is pulled in tension in a testing machine, it stretches a small amount with each increment of load, the stretch being directly proportional to the amount of load up to a certain load. At that certain load, the bar starts to neck down, or get smaller in cross-sectional area. The yield stress is defined as the stress corresponding to the first point on the stress-strain curve at which there is a large increase in strain with no increase in stress. (See Figure 18-3.) The bar continues to yield or stretch until it starts to recover strength, and additional pull is required to produce more stretching or elongation and finally failure of the bar. The load at failure is known as the ultimate tensile strength.

TABLE 18.2
SOFT METRIC BAR SIZES VS INCH-POUND BAR SIZES

METRIC	INCH-POUND
#10	#3
#13	#4
#16	#5
#19	#6
#22	#7
#25	#8
#29	#9
#32	#10
#36	#11
#43	#14

Metric Designations.[18.1] It is important for inspection personnel to be aware of current reinforcing steel industry practice regarding conversion of the inch-pound reinforcing bars to soft metric reinforcing bars. The term soft metric is used in the context of bar sizes and bar size designations. Soft metric conversion means describing the dimensions of inch-pound reinforcing bars in metric units, but not physically changing the bar sizes. In 1997, producers of reinforcing bars began to phase in the production of soft metric bars. Within a few years, the shift to exclusive production of soft metric reinforcing bars was essentially achieved. Virtually all reinforcing bars currently produced and used in the United States are soft metric. Thus, USA-produced reinforcing bars furnished to any construction project most likely will be soft metric. Because soft metric bars are physically the same size as corresponding inch-pound bars, the use of soft metric bars will not require design changes or changes in fabrication and construction practices. It is, however, imperative that field personnel become familiar with the specifications of the new metric bar sizes. The ASTM specifications for reinforcing bars are applicable for either inch-pound units (for example, Specification A 615) or metric units (Specification A 615M), with the ASTM designation: A615/A615M. The "M" is used to denote metric. Within the specification text, the metric units are enclosed within brackets. A one-on-one comparison of soft metric and inch-pound bar sizes is shown in Table 18.2. In Table 18.3, metric and inch-pound data for the Standard ASTM reinforcing bars is listed. As shown, the diameters, cross-sectional areas and weights for the metric bars (except for being expressed in metric units) are exactly the same as those for the inch-pound counterparts. Thus, it is permissible to substitute a metric size bar of Grade 420 for the corresponding inch-pound size bar of Grade 60, and a metric size bar of Grade 520 for the corresponding inch-pound size bar of Grade 75.

Minimum yield strengths in metric units are 280, 350, 420 and 520 MPa (megapascals), which are equivalent to 40,000, 50,000, 60,000 and 75,000 psi, respectively. Metric Grade 420 is the counterpart of standard Grade 60. (See Table 18.1.)

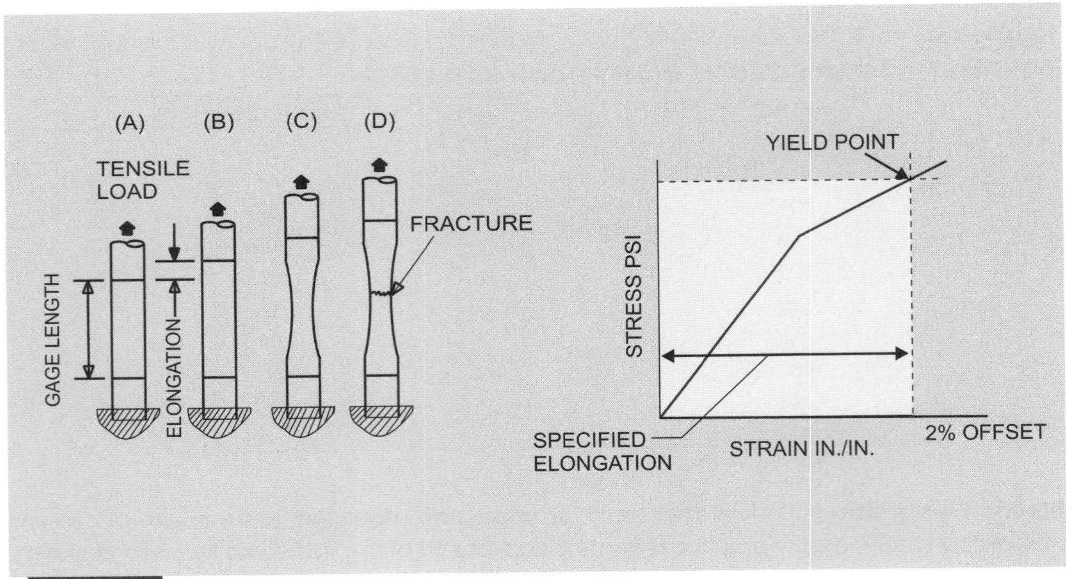

Figure 18-3
When a steel bar is loaded in tension it stretches (or elongates) in an amount directly proportional to the load (B). With continued loading it necks down (C) and finally fails (D). Yield strength is the stress corresponding to the intersection of the stress-strain curve and a line parallel to the elastic part of the curve offset by a strain of 0.2%. Elongation is determined with sensitive gauges by measuring the distance between the gauge marks indicated on the sample (A).

TABLE 18.3
ASTM STANDARD REINFORCING BARS

BAR SIZE	DIAMETER in. [mm]	AREA in.^{2} [mm^2]	WEIGHT lb1ft [kg1m]
#3 [#10]	0.375 [9.5]	0.11 [71]	0.376 [0.560]
#4 [#13]	0.500 [12.7]	0.20 [129]	0.668 [0.944]
#5 [#16]	0.625 [15.9]	0.31 [199]	1.043 [1.552]
#6 [#19]	0.750 [19.1]	0.44 [284]	1.502 [2.235]
#7 [#22]	0.875 [22.2]	0.60 [387]	2.044 [3.042]
#8 [#25]	1.000 [25.4]	0.79 [510]	2.670 [3.973]
#9 [#29]	1.128 [28.7]	1.00 [645]	3.400 [5.060]
#10 [#32]	1.270 [32.3]	1.27 [819]	4.303 [6.404]
#11 [#36]	1.410 [35.8]	1.56 [1006]	5.313 [7.907]
#14 [#43]	1.693 [43.0]	2.25 [1452]	7.650 [11.38]
#18 [#57]	2.257 [57.3]	4.00 [2581]	13.60 [20.24]

Identification Marks.[18.1] The ASTM specifications require identification marks to be rolled into the surface of one side of the bar to denote the producer's mill designation, bar size, type of steel and minimum yield designation. Grade 60 [420] bars show these marks in the following order:

1st—Producing Mill (usually a letter)

2nd—Bar size number (#3 through #11, #14, #18 [#10 through #57])

3rd—Type of steel

- **S** for Billet-steel (A 615/A 615M)
- **W** for Low-alloy steel (A 706/A 706M)
- **I** for Rail-steel (A 996/A 996M)
- **R** for Rail-steel (A 996/A 996M)
- **A** for Axle-steel (A 996/A 996M)

4th—Minimum yield strength designation

The ASTM identification markings for reinforcing bars are illustrated in Figure 18-4.

A mark for minimum yield designation or grade is required for Grade 60 [420] and Grade 75 [520] bars only. Grade 60 [420] bars can either have one single longitudinal line (grade line) or the number 60 [4] (grade mark). Grade 75 [520] bars can either have two grade lines or the grade mark 75 [5].

A grade line is smaller and is located between the two main longitudinal ribs that are on opposite sides of the bar. A grade line must be continued through at least 5 deformation spaces, and it may be placed on the same side of the bar as the other markings or on the opposite side.

Grade 40 [280] and 50 [350] bars are required to have only the first three identification marks; no grade mark or grade line for minimum yield strength is required.

Bar identification marks may also be oriented to read horizontally (at 90° to those illustrated). Grade mark numbers may be placed within separate consecutive deformation spaces to read vertically or horizontally.

Soft metric reinforcing bars are required to be identified in a similar manner as are inch-pound bars. Marking requirements for metric bars are also illustrated in Figure 18-4. For example, consider the marking requirements for a #36, Grade 420 metric bar, which is the counterpart of an inch-pound #11, Grade 60 bar. Regarding the bar size and grade, the ASTM specifications require the number "36" to be rolled onto the surface of the metric bar to indicate its size. For identifying or designating the yield strength or grade, the ASTM specifications provide an option: A mill can choose to roll a "4" (the first digit in the grade number) onto the bar or roll an additional longitudinal rib or grade line to indicate Grade 420.

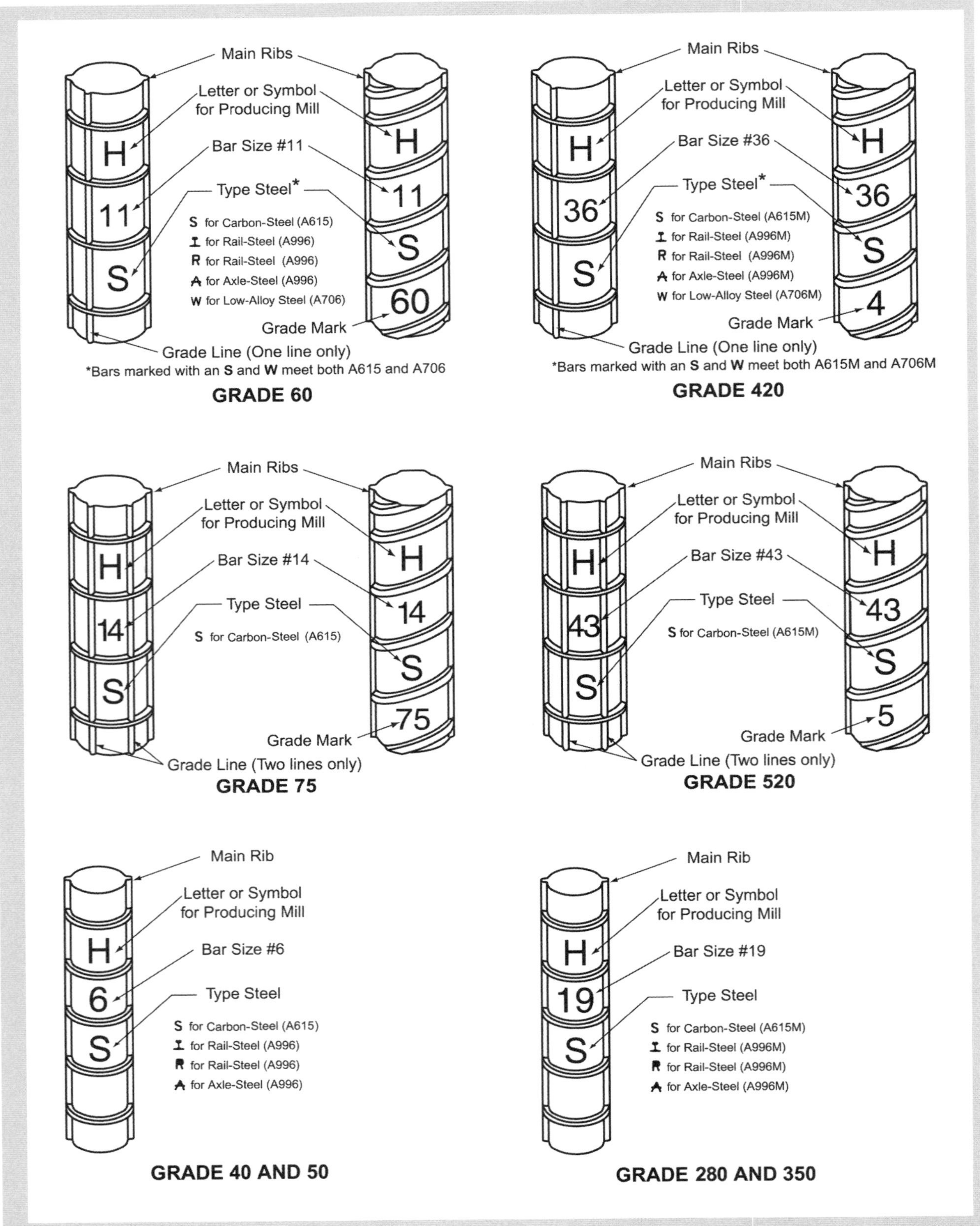

Figure 18-4
The ASTM identification markings for reinforcing bars.[18.1]

Welded Wire Reinforcement (WWR) consists of cold-drawn wire, fabricated in a square or rectangular pattern, resistance welded at all intersections, (see Figure 18-5). The wire used may be either plain (smooth) or deformed. Unlike the protruding deformations on reinforcing bars, the deformations on deformed wire are usually indented. Welded wire reinforcement is designated by the initials WWR. Style identification of WWR is done by denoting plain wire by the letter "W" followed by a number indicating the cross-sectional area in hundredths of a square inch. Deformed wire is denoted by the letter "D" followed by the area number. Industry notation is to use the letters WWR followed by spacing of longitudinal wires, then spacing of transverse wires, and finally by the sizes of longitudinal and transverse wires, respectively.

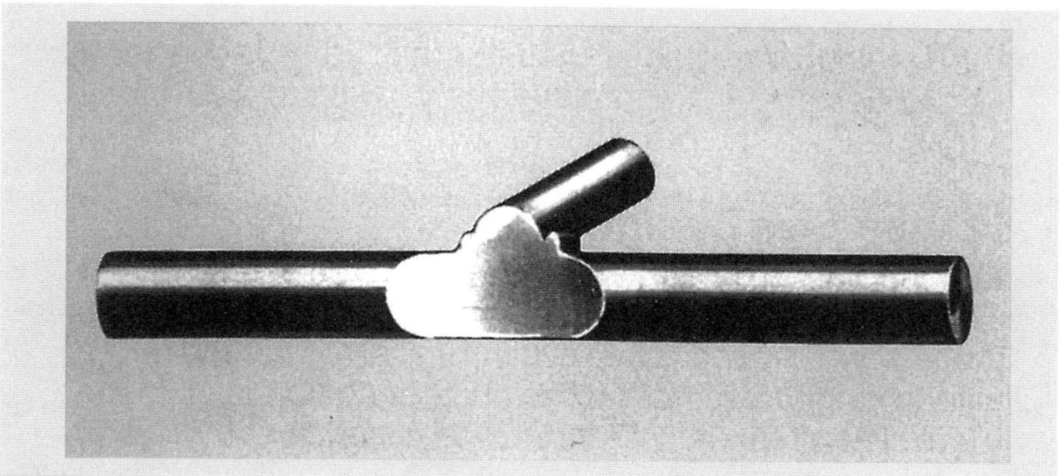

Figure 18-5
Section at weld intersection showing complete fusion of intersecting wires. (Courtesy of WRI)

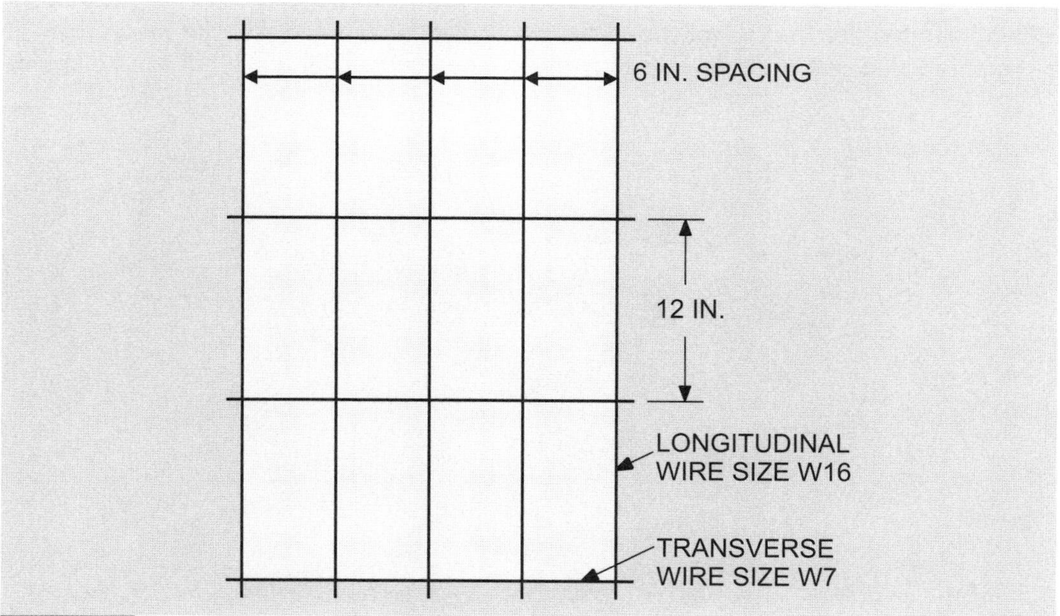

Figure 18-6
Typical rectangular welded wire reinforcement.

For example WWR 6 x 12-W16 x W7 indicates the following described WWR (See Figure 18-6):

Spacing of longitudinal (lengthwise) wires	= 6 inches
Spacing of transverse (cross) wires	= 12 inches
Longitudinal wire size	= W16 (0.16 sq in.)
Transverse wire size	= W7 (0.07 sq in.)

Note that the terms longitudinal and transverse refer to the method of manufacturing; the longitudinal wires are continuously fed through automatic welders with precut transverse wires resistance welded to the longitudinal wires. The terms longitudinal and transverse have no relationship to the position of the sheet in the structure.

If the welded wire sheet had been made with deformed wire instead of plain wire, it would be designated in the same manner except for substituting the appropriate letter D in front of the wire size: WWR 6 x 12-D16 x D7.

Welded wire reinforcement is frequently used in slabs-on-ground (nonstructural) to control shrinkage and temperature cracking. Used in this manner, it is commonly referred to as welded wire fabric, building fabric or just wire mesh, as the wires are usually small in size and the fabric is purchased in rolls rather than in sheets (see Figure 18-7). Unfortunately, the small-size rolled fabric is difficult to properly place and chair-up, and far too frequently ends up at the bottom of the slab, totally out of position to effectively act as crack control reinforcement. The current attitude of the wire reinforcement industry is to encourage the use of larger wire sizes (stiffer) at a larger spacing (to allow workers to step between the wires) and properly chaired so that the crack control reinforcement is properly located in the upper half of the slab (see Figure 18-8). Use of sheets rather than rolls is also advocated, as the rolls are difficult to straighten or flatten out. If rolls are used, it is recommended that the rolls be straightened (see Figure 18-40) and cut to size before

Figure 18-7
Welded-wire "fabric" typically used in light construction as crack control reinforcement. (Courtesy of WRI)

Figure 18-8.
Easier placing can be achieved with step-through styles of WWR (12-inch by 12-inch and larger). Note: The larger wire size at wider spacing will require the flat sheets rather than the customary rolls. (Courtesy of WRI)

placement. See Reference 18.2 and 18.4 for proper placement of welded wire for slab-on-ground in light construction.

Heavier welded wire reinforcement is being specified by engineers (in place of conventional reinforcing bars) as structural reinforcement in structural slabs (see Figure 18-9), shear walls and retaining walls (see Figure 18-10). The heavier WWR is also becoming more common as shear reinforcement in beams (see Figure 18-11), and as tie (confinement) reinforcement in columns (see Figure 18-12). With the varied wire sizes available, from W1.4($^1/_8$-inch diameter) to W or D45 ($^3/_4$-inch diameter), styles of WWR can be specified (and manufactured) in any wire size and spacing required by a structural design. The structural welded wire is available in flat sheets only and is covered by ASTM A 185 for plain welded wire reinforcement and ASTM A 497 for deformed welded wire for concrete reinforcement (see Figures 18-13 and 18-14). Both plain and deformed welded wire are included under the code definition of deformed reinforcement (ACI 318 Section 2.1), and either type may be specified (or substituted for reinforcing bars) as structural reinforcement. Substitution of WWR for reinforcing bars is sometimes done during the bid process of a particular project where the engineer indicated reinforcing bars on the original structural drawings, and a change to WWR is made. The field inspector should be fully aware that a change order from bars to welded wire is acceptable according to the code; however, it does require some redesign considerations by the engineer and engineer approval.

An advantage claimed for WWR is the speed and ease of installation, making a significant savings in time and labor. Field inspection is also simplified; the spacing of the individual wires is exact on account of the precision welding, and it is virtually impossible to leave out some of the reinforcement, as a missing sheet is easily detected. The style identification for WWR, however, favors the engineer more than the field inspector, who must verify the proper wire size as indicated on the structural drawings or placing drawings. Unfortunately, welded wire fabric does not have identification marks rolled into the wire surface as do reinforcing bars.

Figure 18-9
Two-way slab construction using WWR for the top and bottom reinforcement. (Courtesy of WRI)

Figure 18-10.
Vertical and horizontal WWR wall reinforcement. (Courtesy of WRI)

Figure 18-11.
Bent cages of WWR for shear reinforcement in cast-in-place concrete beams are easily handled by two workers.

Figure 18-12
Fabricated column cages using bent sheets of WWR ties. (Note: Construction located in high seismic area!) (Courtesy of WRI)

Figure 18-13.
Heavy "structural" welded wire reinforcement cannot be rolled, so it is manufactured in flat sheets.(Courtesy of WRI)

Metric WWR.[18.5] As in the reinforcing bar industry, the wire reinforcement industry has also converted to a soft metric welded wire production. The same inch-pound die sizes will be used to produce metric WWR. When styles of welded wire reinforcement are called out, both spacings and wire areas will be soft converted and rounded to whole metric numbers:

A typical inch-pound structural WWR style is 12 x 12 – D11 x D11.

The equivalent metric style is 305 x 305 – MD71 x MD71.

A typical inch-pound WWR style for lightly reinforced slab-on-ground is 6 x 6 – W2.9 x W2.9.

The equivalent metric style is 152 x 152 – MW19 x MW19.

The wire spacings are in millimeters (mm) and the wire areas are in square millimeters (mm^2), with the "M" used to identify metric. Metric wire areas and diameters with equivalent inch-pound units are listed in Table 18.4. The common styles of welded wire reinforcement with equivalent metric units are listed in Table 18.5.

WWR Dimensions. Sheet dimensions for WWR are defined as:

Width = Center to center distance between outside longitudinal wires. This dimension does not include overhangs.

Side Overhang = Extension of transverse wires beyond centerline of outside longitudinal wires. If no side overhang is specified, WWR is furnished with overhangs on each side, of no greater than 1 inch (25 mm). Wires cut flush (no overhangs) are specified as (+0", +0"). When specific overhangs are required, they are noted as (+1", +3") or (+6", +6").

Overall Width = Width including side overhangs, in. (or mm). In other words, the tip-to-tip dimension of transverse wires.

Length = Tip-to-tip dimension of longitudinal wires. Whenever possible, this dimension is an even multiple of the transverse wire spacing. The length dimension always includes end overhangs.

End Overhangs = Extension of longitudinal wires beyond centerline of outside transverse wires. Unless otherwise noted, standard end overhangs are assumed to be required, and end overhangs are not specified. Nonstandard end overhangs may be specified for special conditions: preferably, the sum of the two end overhangs should equal the transverse wire spacing.

A typical order for welded wire reinforcement using the above definitions might appear as:

Item	Quantity	Style	Width	Side Overhangs	Lengths
1	1000 Sheets	112 x 12-W11 x W11	90"	(+6", +6")	15'-0"
2	150 Sheets	6 x 6-W4 x W4	60"	(+0", +0")	20'-0"
3	500 Sheets	6 x 12-D10 x D6	96"	(+3", +3")	17'-0"
Metric order might appear as:					
Item	**Quantity**	**Style**	**Width**	**Side Overhangs**	**Lengths**
1	1000 Sheets	305 x305-MW71 x MW71	2286 mm	(+152, +152)	4.6 m
2	150 Sheets	152 x 152-MW26 x MW26	1524 mm	(+0, +0)	6.1 m
3	500 Sheets	152 x 305-MD65 x MD39	2438 mm	(+76, +76)	5.2 m

Bar Mats are similar to WWR mats, except that they are fabricated of reinforcing bars welded at the intersections to form a square or rectangular grid. Deformed bars conforming to ASTM A 615 or A 706, Grades 40 and 60, are used in the manufacture of welded mats. Bar mats are used for structural reinforcement and are covered under ASTM A 184.

TABLE 18.4
INCH-POUND WIRE AREAS AND DIAMETERS (WITH EQUIVALENT METRIC UNITS)

INCH-POUND UNITS			METRIC UNITS		
SIZE	DIAMETER (IN.)	AREA (IN.^{2})	SIZE	DIAMETER (mm)	AREA (mm^2)
W45	0.757	0.450	MW290	19.22	290
W31	0.628	0.310	MW200	16.01	200
W20.2	0.507	0.202	MW130	12.90	130
W18.6	0.487	0.186	MW120	12.40	120
W15.5	0.444	0.155	MW100	11.30	100
W14.0	0.422	0.140	MW90	10.70	90
W12.4	0.397	0.124	MW80	10.10	80
W.10.9	0.373	0.109	MW70	9.40	70
W10.1	0.359	0.101	MW65	9.10	65
W9.3	0.344	0.093	MW60	8.70	60
W8.5	0.329	0.085	MW55	8.40	55
W7.8	0.314	0.078	MW50	8.00	50
W7.0	0.298	0.070	MW45	7.60	45
W6.2	0.283	0.062	MW40	7.10	40
W5.4	0.262	0.054	MW35	6.70	35
W4.7	0.245	0.047	MW30	6.20	30
W4.0	0.226	0.040	MW26	5.70	26
W3.9	0.223	0.039	MW25	5.60	25
W3.1	0.199	0.031	MW20	5.00	20
W2.9	0.192	0.029	MW19	4.90	19
W2.3	0.171	0.023	MW15	4.40	15

TABLE 18.5
COMMON STYLES OF WELDED WIRE REINFORCEMENT (WWR) (U.S. CUSTOMARY STYLES VS METRIC STYLES)

U.S. CUSTOMARY STYLES	As (IN^2/FT)	METRIC STYLE	As (mm^2/m)
4 x 4 - W1.4 x W1.4	0.042	102 x 102 - MW9 x MW9	88.9
4 x 4 - W2.0 x W2.0	0.060	102 x 102 - MW9 x MW13	127.0
4 x 4 - W2.9 x W2.9	0.087	102 x 102 - MW19 x MW19	184.2
4 x 4 - W4.0 x W4.0	0.120	102 x 102 - MW26 x MW26	254.0
6 x 6 - W1.4 x W1.4	0.028	152 x 152 - MW9 x MW9	59.3
6 x 6 - W2.0 x W2.0	0.040	152 x 152 - MW13 x MW13	84.7
6 x 6 - W2.9 x W2.9	0.058	152 x 152 - MW19 x MW19	122.8
6 x 6 - W4.0 x W4.0	0.080	152 x 152 - MW26 x MW26	169.4
4 x 4 - W3.1 x W3.1	0.093	102 x 102 - MW20 x MW20	196.9
6 x 6 - W4.7 x W4.7	0.094	152 x 152 - MW30 x MW30	199.0
12 x 12 - W9.4 x W9.4	0.094	305 x 305 - MW61 x MW61	199.0
12 x 12 -W17.1 x W17.1	0.171	305 x 305 -MW110 xMW110	362.0
6 x 6 - W8.1 x W8.1	0.162	152 x 152 - MW52 x MW52	342.9
6 x 6 - W8.3 x W8.3	0.166	152 x 152 - MW54 x MW54	351.4
12 x 12 - W9.1 x W9.1	0.091	305 x 305 - MW59 x MW59	192.6
12 x 12 -W16.6 x W16.6	0.166	305 x 305 - MW107 x MW107	351.4
6 x 6 - W4.4 x W4.4	0.088	152 x 152 - MW28 x MW28	186.3
6 x 6 - W8 x W8	0.160	152 x 152 - MW52 x MW52	338.7
12 x 12 - W8.8 x W8.8	0.088	305 x 305 - MW57 x MW57	186.3
12 x 12 - W16 x W16	0.160	305 x 305 - MW103 x MW103	338.7
6 x 6 - W4.2 x W4.2	0.084	152 x 152 - MW27 x MW27	177.8
6 x 6 - W7.5 x W7.5	0.150	152 x 152 - MW48 x MW48	317.5

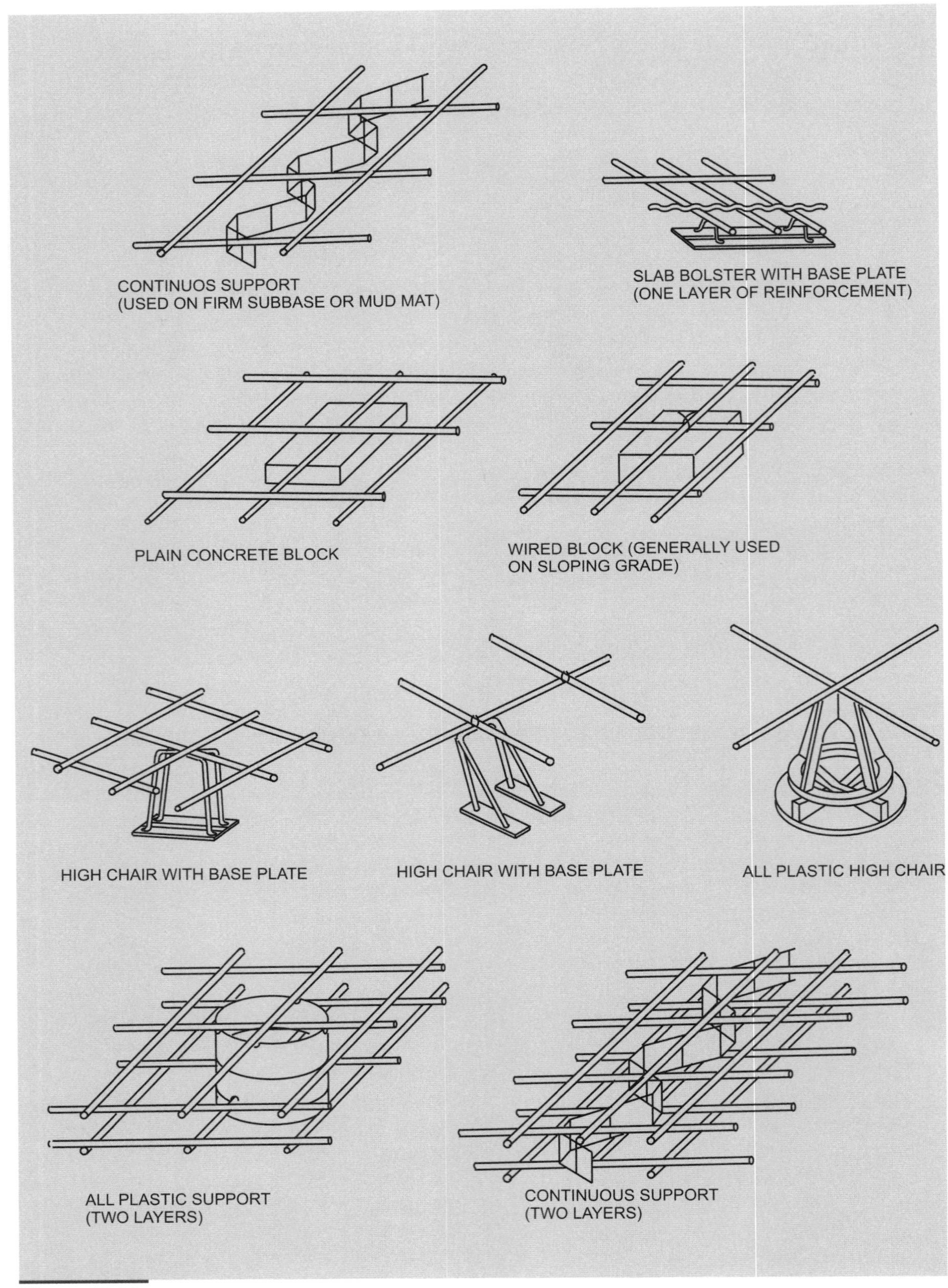

Figure 18-14.
Typical types of support for WWR.[18.6]

18.3. Fabrication

When steel reinforcing bars are manufactured in the steel rolling mill, they are normally cut into lengths of about 60 feet. Some mills will furnish longer bars on special order, but 60-foot bars are the longest that can be handled by truck. Longer bar lengths can be transported by rail or barge. From the manufacturing mill, the bars are shipped to a fabricator who prepares the bars for use in the structure.

Before the reinforcing bars can be used in the structure, individual bars have to be cut to length, many of them must be bent to conform to the design, and all of them must be identified and assembled into bundles that can be handled by a crane. This processing is done by a fabricator.

Placing Drawings and Bar Lists. Placing drawings (sometimes incorrectly called shop drawings), bar lists and schedules are prepared by the detailer in the fabricator's shop. Working from the structural and architectural drawings for the project, the fabricator prepares placing drawings that show how the bars are to be cut and bent and where the bars are located in the structure, and that indicate the necessary accessory materials and items. These drawings classify the reinforcing bars as to number of each kind of bar identified by a certain mark, grade of steel, size and length, and either a bending diagram or reference to standard bends for bent bars. Figure 18-14A illustrates a typical foundation plan such as would be found in a set of structural drawings. The corresponding placing drawing prepared from the structural drawing is shown in Figure 18-14B. The detailer has prepared footing, grade beam and column schedules, which are similar to the design schedules, except that they include the number of like members having identical reinforcement and a description of the bars including the number of pieces, size, length, mark number (if bent bars) and sometimes bending details. Wall footings and walls are usually detailed on the plan or on elevations or sections.

A bar list is a bill of materials or a list of the reinforcement covering a portion of the structure and may cover one or more truckloads. The items listed are taken from the placing drawings. Bars are classified as to size, length and whether they are straight or bent. Identifying marks are shown (when necessary) and dimensions of the bends are indicated. Type of steel is shown. Other information that should be included are name of customer, name and location of job, purchaser's order number, location of the listed material in the job (such as basement columns, first floor slabs and beams, etc.) and a reference to the number of the placing drawing.

For the more complicated jobs, a schedule is sometimes prepared. A schedule, frequently a part of the placing drawing, covers a group of similar items in the structure, such as certain columns or girders that are more or less identical as far as reinforcement is concerned. It is similar to a bar list, but it describes the steel in greater detail and provides information on placing the bars in the forms.

Bending. Upon approval of the placing drawings by the engineer, the fabricator proceeds with the fabrication. Straight bars are cut to length and bundled. Bars to be bent are bent in accordance with the common shapes that have been standardized.[18.5] Special bends are made as required by the detail drawings. All bends are made with the steel at normal room temperature (cold bending). (See Figure 18-15.) Hot bends are made only as specifically designated or approved by the engineer (ACI 318 Section 7.3.1).

The *ACI Detailing Manual*[18.7] illustrates a number of bars bent to different configurations with accompanying identification of the dimensions. (See Figure 18-16.) Any size and length of bar can be bent to any of the standard bends shown by merely stating the identifying number with appropriate values given to the letters designating the several dimensions. This information is given in the bar lists.

When limitations of the structure do not provide enough space to permit sufficient embedment of a straight bar for adequate anchorage, the end of the bar can be hooked. Code requirements for standard end hooks and minimum bend diameters are set forth in ACI 318 Sections 7.1 and 7.2. The term standard hook as used in the code means one of the following (see also Figure 18-17):

1. 180-degree bend plus $4d_b$ extension, but not less than $2^1/_2$ inches at free end of bar
2. 90-degree bend plus $12d_b$ extension at free end of bar
3. For stirrup and tie end hooks in regions of low to moderate seismic risk:

   #5 bar and smaller 90-degree bend plus $6d_b$ extension at free end of bar, or

   #6, #7 and #8 bars, 90-degree bend plus $12d_b$ extension at free end of bar, or

   #8 bar and smaller, 135-degree bend plus $6d_b$ extension at free end of bar
4. For stirrup and tie end hooks in regions of high seismic risk, the reader is referred to the seismic hook definition in ACI 318 Section 2.2.

Minimum diameter of bend measured on the inside of the bar, other than for stirrups and ties, must not be less than the values in Table 18.6. For stirrups and ties, inside diameter of bend must not be less than $4d_b$ for #5 bar and smaller (ACI 318 Section 7.2.2).

Figure 18-15.
A power-operated bar bender, shown bending a 90° bend in the end of three #5 bars.(Courtesy of CRSI)

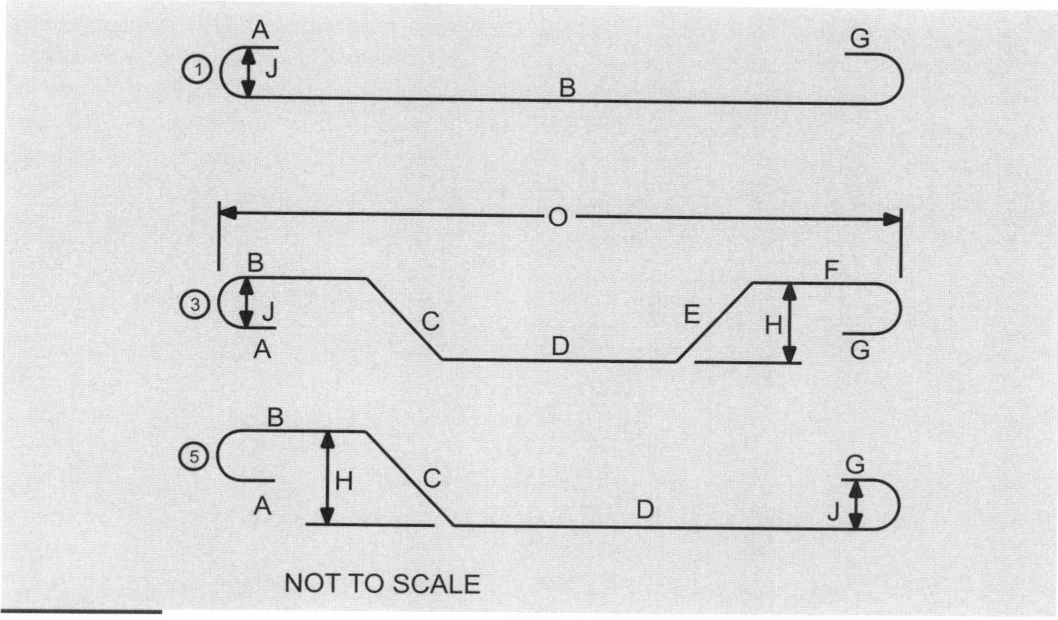

Figure 18-16.
Three examples of standard bent bars, showing how bends and dimensions are indicated

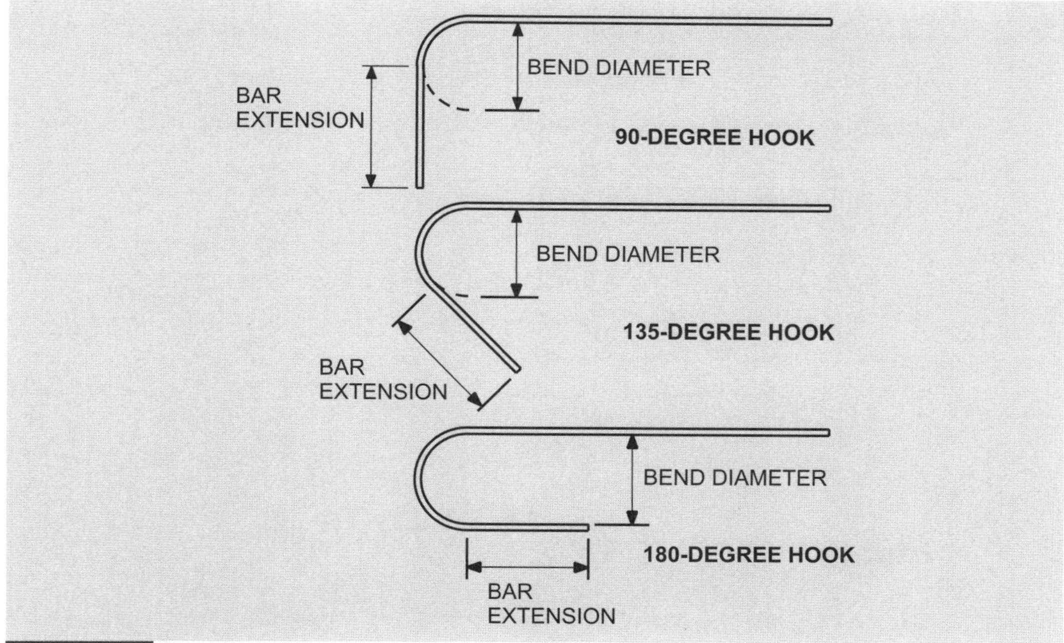

Figure 18-17
Standard end hooks on bars.

TABLE 18.6
MINIMUM DIAMETER OF BEND*

*BAR SIZE	MINIMUM DIAMETER
#3 through #8	$6d_b$
#9, #10 and #11	$8d_b$
#14 and #18	$10d_b$

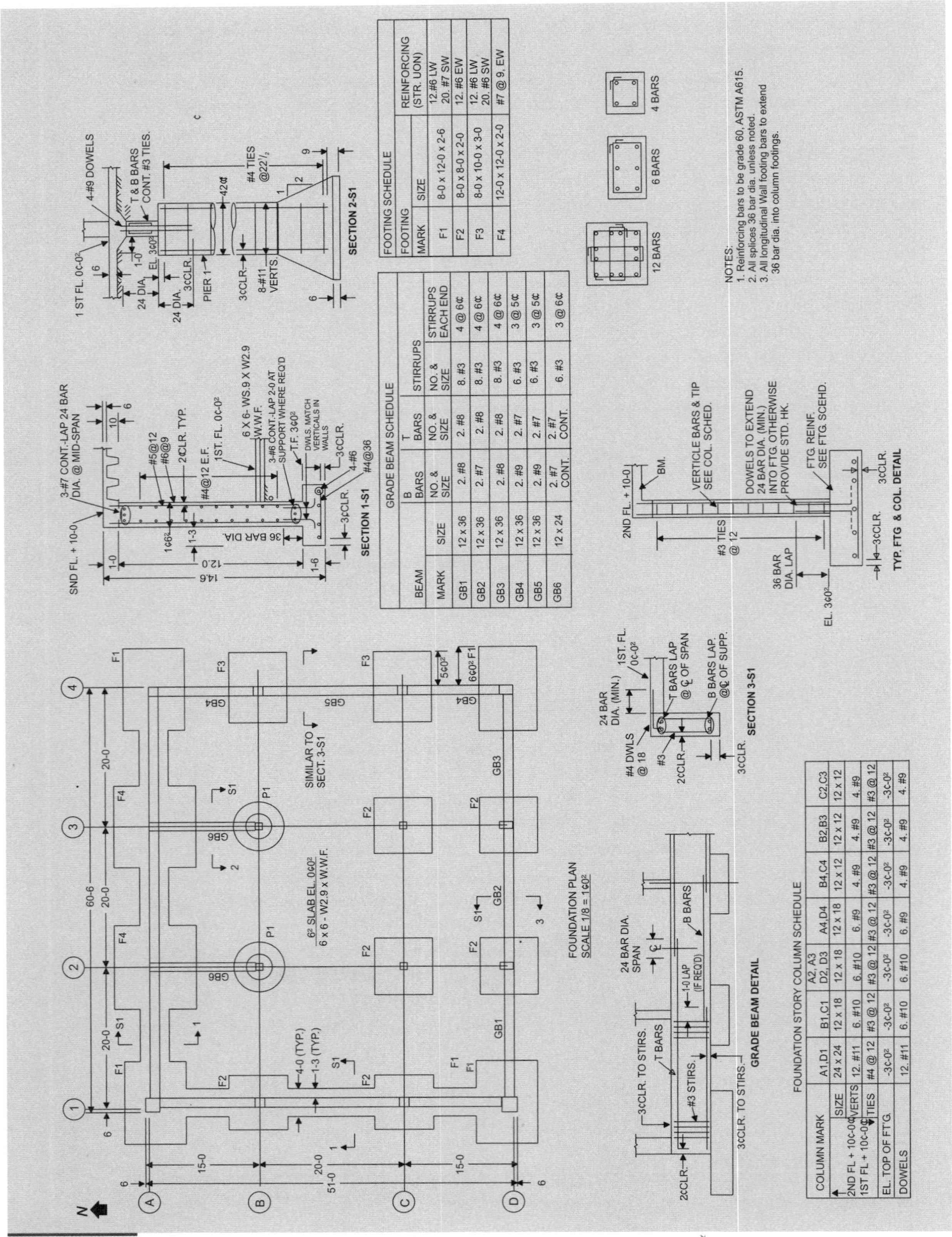
FOUNDATION PLAN
SCALE 1/8 = 1¢0²
6² SLAB EL. 0¢0²
6 x 6 - W2.9 x W2.9 W.W.F.
SIMILAR TO SECT. 3-S1
SECTION 1-S1
SECTION 2-S1
SECTION 3-S1
GRADE BEAM DETAIL
TYP. FTG & COL. DETAIL
GRADE BEAM SCHEDULE
BEAM MARK | SIZE | B BARS NO. & SIZE | T BARS NO. & SIZE | STIRRUPS NO. & SIZE | STIRRUPS EACH END
GB1 | 12 x 36 | 2. #8 | 2. #8 | 8. #3 | 4 @ 6¢
GB2 | 12 x 36 | 2. #7 | 2. #8 | 8. #3 | 4 @ 6¢
GB3 | 12 x 36 | 2. #8 | 2. #8 | 8. #3 | 4 @ 6¢
GB4 | 12 x 36 | 2. #9 | 2. #7 | 6. #3 | 3 @ 5¢
GB5 | 12 x 36 | 2. #9 | 2. #7 | 6. #3 | 3 @ 5¢
GB6 | 12 x 24 | 2. #7 CONT. | 2. #7 CONT. | 6. #3 | 3 @ 6¢
FOOTING SCHEDULE
FOOTING MARK | SIZE | REINFORCING (STR. UON)
F1 | 8-0 x 12-0 x 2-6 | 12. #6 LW 20. #7 SW
F2 | 8-0 x 8-0 x 2-0 | 12. #6 EW
F3 | 8-0 x 10-0 x 3-0 | 12. #6 LW 20. #6 SW
F4 | 12-0 x 12-0 x 2-0 | #7 @ 9. EW
12 BARS
6 BARS
4 BARS
NOTES:
1. Reinforcing bars to be grade 60, ASTM A615.
2. All splices 36 bar dia. unless noted.
3. All longitudinal Wall footing bars to extend 36 bar dia. into column footings.
FOUNDATION STORY COLUMN SCHEDULE
COLUMN MARK | A1,D1 | B1,C1 | A2, A3 D2, D3 | A4,D4 | B4,C4 | B2,B3 | C2,C3
2ND FL + 10¢-0² SIZE | 24 x 24 | 12 x 18 | 12 x 18 | 12 x 18 | 12 x 12 | 12 x 12 | 12 x 12
VERTS | 12. #11 | 6. #10 | 6. #10 | 6. #9 | 4. #9 | 4. #9 | 4. #9
1ST FL + 10¢-0² TIES | #4 @ 12 | #3 @ 12 | #3 @ 12 | #3 @ 12 | #3 @ 12 | #3 @ 12 | #3 @ 12
EL. TOP OF FT'G. | -3¢-0² | -3¢-0² | -3¢-0² | -3¢-0² | -3¢-0² | -3¢-0² | -3¢-0²
DOWELS | 12. #11 | 6. #10 | 6. #10 | 6. #9 | 4. #9 | 4. #9 | 4. #9

Figure 18-17A
Structural drawing for building foundation.

LAP

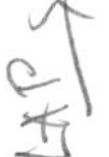

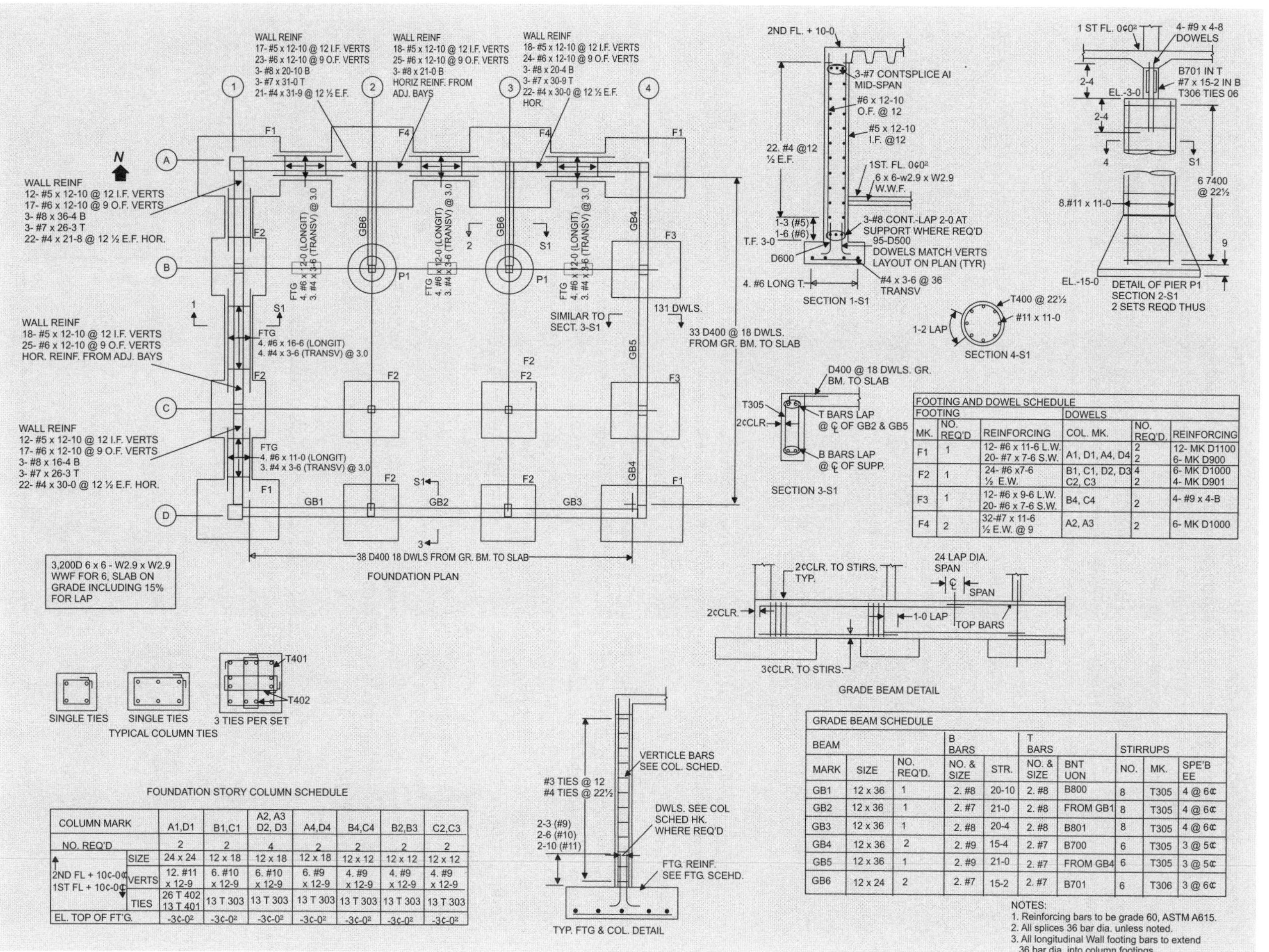

FOUNDATION STORY COLUMN SCHEDULE

COLUMN MARK		A1,D1	B1,C1	A2, A3 D2, D3	A4,D4	B4,C4	B2,B3	C2,C3
NO. REQ'D		2	2	4	2	2	2	2
2ND FL + 10¢-0¢ 1ST FL + 10¢-0¢	SIZE	24 x 24	12 x 18	12 x 18	12 x 18	12 x 12	12 x 12	12 x 12
	VERTS	12. #11 x 12-9	6. #10 x 12-9	6. #10 x 12-9	6. #9 x 12-9	4. #9 x 12-9	4. #9 x 12-9	4. #9 x 12-9
	TIES	26 T 402 13 T 401	13 T 303	13 T 303	13 T 303	13 T 303	13 T 303	13 T 303
EL. TOP OF FT'G.		-3¢-0²	-3¢-0²	-3¢-0²	-3¢-0²	-3¢-0²	-3¢-0²	-3¢-0²

FOOTING AND DOWEL SCHEDULE

FOOTING			DOWELS		
MK.	NO. REQ'D	REINFORCING	COL. MK.	NO. REQ'D.	REINFORCING
F1	1	12- #6 x 11-6 L.W. 20- #7 x 7-6 S.W.	A1, D1, A4, D4	2 2	12- MK D1100 6- MK D900
F2	1	24- #6 x7-6 ½ E.W.	B1, C1, D2, D3 C2, C3	4 2	6- MK D1000 4- MK D901
F3	1	12- #6 x 9-6 L.W. 20- #6 x 7-6 S.W.	B4, C4	2	4- #9 x 4-B
F4	2	32-#7 x 11-6 ½ E.W. @ 9	A2, A3	2	6- MK D1000

GRADE BEAM SCHEDULE

BEAM			B BARS		T BARS		STIRRUPS		
MARK	SIZE	NO. REQ'D.	NO. & SIZE	STR.	NO. & SIZE	BNT UON	NO.	MK.	SPE'B EE
GB1	12 x 36	1	2. #8	20-10	2. #8	B800	8	T305	4 @ 6¢
GB2	12 x 36	1	2. #7	21-0	2. #8	FROM GB1	8	T305	4 @ 6¢
GB3	12 x 36	1	2. #8	20-4	2. #8	B801	8	T305	4 @ 6¢
GB4	12 x 36	2	2. #9	15-4	2. #7	B700	6	T305	3 @ 5¢
GB5	12 x 36	1	2. #9	21-0	2. #7	FROM GB4	6	T305	3 @ 5¢
GB6	12 x 24	2	2. #7	15-2	2. #7	B701	6	T306	3 @ 6¢

NOTES:
1. Reinforcing bars to be grade 60, ASTM A615.
2. All splices 36 bar dia. unless noted.
3. All longitudinal Wall footing bars to extend 36 bar dia. into column footings.

Figure 18-17B
Placing drawing for building foundation.

Tolerances. Cutting and bending must be done accurately. If bars are cut to the wrong length, or bent to the wrong angle, they cannot be placed in the correct position in the forms, and the concrete member may lack strength in a critical area that could lead to failure of the structure. For this reason, accuracy must be emphasized.

Although accuracy is essential, we must nevertheless remember that in any procedure involving measurements, it is necessary to allow for slight inaccuracies in the measurements. These allowances are called tolerances. With respect to fabrication of reinforcing bars, the following tolerances are typical:[18.1]

Overall length of straight bar	± 1 in.
Overall length of bent bar (#3 - #11) (Dimension O in Figure 18-16)	±1 in.
Dimension H, Figure 18-16	+ 0, −$^1/_2$ in.
Outside diameter of a circular tie	±$^1/_2$ in. for diameter ≤ 30 in.
	± 1 in. for diameter > 30 in.
Outside length of a side of a square or rectangular tie or stirrup	±$^1/_2$ in.

In checking reinforcing bars on the job, these tolerances should be allowed, unless the design drawings or project specifications indicate other tolerances.

Tolerances that are allowed for placing steel in the forms are discussed later in this chapter.

Bundling and Tagging. Reinforcing bars are assembled by the fabricator into bundles containing bars usually of one size, type of steel, length and configuration or bend. When only a few bars of any certain length or mark are required, several different groups can be bundled together. Each bundle, which usually weighs in the neighborhood of 3000 pounds, is tied or wired together with No. 12 or No. 9 wire or steel strapping. A durable tag should be attached to each bundle showing complete identification of the bars in the bundle. Tags are frequently made of high-density polyethylene sheet (Tyvek®), which is waterproof, or thin sheet metal with the information embossed thereon. One side of the tag shows shipper's name and address, name of the contractor and job address. The other side (or sometimes a separate tag) describes the reinforcing bars in the bundle, giving grade and bar size, number of pieces, length, identifying mark and buyer's order number. The mark number is the number of the particular group of identical bars on the bar list and placing drawing. (See Figure 18-18.)

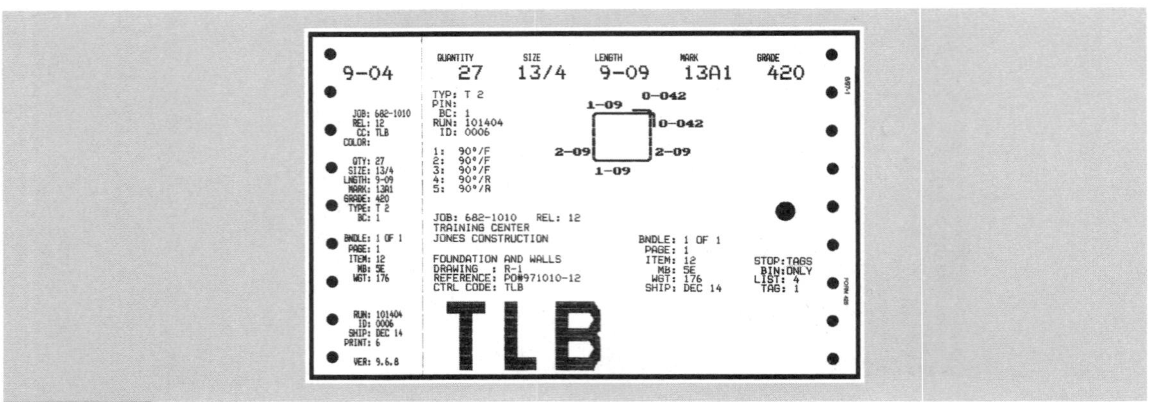

Figure 18-18
A tag attached to a bundle of column ties.(Courtesy of CRSI)

The mark, or mark number, is based on a system that is best suited to the job under construction. The structural drawings will indicate a system of identifying footings, columns, girders, beams, etc., by a series of numbers and letters. The floor or story on which any certain group of bars is to be used will be shown in the mark number. The fabricator uses these numbers as the basis for the marks.

18.4. Handling and Inspection

One of the first things to be done by the steel supervisor and the inspector is to check the structural and architectural drawings with the placing drawings and bar lists so that they can understand the marks and other requirements.

Some jobs are arranged so that the bundles of steel can be hoisted directly from the truck to the area of the structure where they are to be used. This is convenient but requires close scheduling of fabrication, delivery, hoisting equipment and placing to keep the job moving smoothly without delays. If space at the jobsite is available, it is not uncommon to arrange for storage of the reinforcing steel, thus providing a certain degree of flexibility in the operation.

Each shipment of reinforcing steel from the fabricator will be accompanied by a list showing the exact material in the shipment. The list (sometimes called manifest or loading ticket) should immediately be checked with the material received to assure that no errors exist. If the reinforcing steel is to go into storage, it should be piled well off the ground where mud will not splash on it from construction activities. (See Figure 18-19.) If the reinforcing steel is to be stored for a long period of time, it may be necessary to provide some protective covering to keep rain off. Usual practice is to stack the bundles on heavy timbers. Bundles should be arranged so that the tags are accessible without the bundles needing to be moved. The piles should be such that the bundles can be removed in sequence as they are required and placed in the structure without excessive rehandling. Before bundles are hoisted onto the structure, a check should be made to assure that the structure and shoring have sufficient strength to support the weight of the reinforcing steel. Bundles are usually hoisted by means of a bridle and two choker hitches. Cradle hitches are sometimes used, but the choker hitches provide greater safety. Bundles of long bars (about 30 feet long or longer) should be lifted by means of two chokers and a spreader bar.

Figure 18-19
A neat pile of straight bars. (Courtesy of CRSI)

When delivered to the jobsite, reinforcing steel should be free of oil, grease, loose mill scale, paint or other coatings that might interfere with bond (ACI 318 Section 7.4.1). The light coating of rust that is usually present is not detrimental; in fact, it may actually improve bond. (See Figure 18-20.) Strength of the concrete probably has more control over the bond than the surface condition of the bar. Heavy rust, pitting the steel to such an extent as to reduce its cross-sectional area, could be cause for rejection. If a bar appears to have rusted excessively, a sample should be cleaned and weighed to determine compliance with the ASTM specifications.

Figure 18-20
The light coating of rust on these rebars will actually improve the bond between the reinforcement and concrete.

Epoxy coating of steel reinforcement is an acceptable surface condition of reinforcement (see ACI 318 Section 7.4.1). The reduced bond that is due to the epoxy coating is considered in design by requiring longer anchorage and development lengths for the epoxy-coated reinforcement.

Reinforcing steel should be stored on the jobsite on platforms or other supports off the ground to protect it from damage and dirt, in locations where trucks will not splash mud on it and workers will not walk over it. Storage on the site longer than necessary should be avoided because it can result in excessive rusting or contamination.

Reinforcing steel should be cut and bent (fabricated) in the shop rather than on the job. If bending in the field is necessary, it should be done cold (ACI 318 Section 7.3.1).

Heating of the reinforcing steel to facilitate bending may be permitted only when expressly approved by the engineer. If heating is approved, the bar should be heated slowly and should not reach a red heat. After bending, the bar should be slowly air cooled. Rapid cooling in a blast of cold air or in water is detrimental and should not be permitted.

Because of the many steps involved in warehousing, fabrication and other handling, it is sometimes difficult to identify any one shipment of fabricated reinforcing steel with any certain heats of steel from which the material was rolled—although most manufacturers are able to provide this information, especially on large shipments. However, typical mill test reports are available, and check tests for yield, strength, elongation, ultimate tensile strength and bending can be made on samples from the site. Bar samples should be cut from normal bar shipments and should be about 30 inches long.

On some large jobs, the reinforcing steel is inspected at the mill or shop, and bundles shipped to the job are identified with tags from the fabricator and the inspection agency.

The inspector at the jobsite should examine the bars for excessive rust, oil or dirt, and shipping damage such as bending or breakage.

The inspector should see that the reinforcing steel is of the correct grade, size, type and quantity, and that fabricated bars are correctly cut and bent. All bundles of bars should be properly identified with durable tags.

18.5. Placing the Reinforcing Steel

Reinforcement is subject to rough handling and displacement after it has been fixed in place in the forms, hence the need for accurate and secure placement. Workers walk or climb on the steel; loads and bundles may be temporarily stored, or walkways may be supported, on the steel. It is obvious that the reinforcing steel must be well tied and supported.

Layers of bars should be separated by spacers of such shape that they will be easily encased by the concrete. Bars should be separated from horizontal surfaces by spacers. Vertical stirrups should always pass around the main tension members and should be securely attached. The use of pebbles, pieces of broken stone or brick, metal pipe, wooden blocks and similar material for holding reinforcing steel in position should not be permitted. Welding of crossing bars (tack welding) for assembly of reinforcement is prohibited except as specifically authorized by the engineer. (See ACI 318 Section 7.5.4.)

The inspector should inspect the reinforcing steel as early as possible, checking for sizes and bends before the steel is tied rigidly in place, thus helping to avoid expensive corrections. This inspection can frequently be facilitated in deep, thin forms by assembling the steel curtains in place after one side form has been erected, but before the opposite form is set in place.

A practice that should be avoided in placing reinforcing steel for slabs on ground is that of laying the steel mat or mesh on the subgrade ahead of concrete placement, then attempting to lift it through the concrete. (See Figure 18-21.) Equally undesirable is to place the steel on top of the slab and try to push it down. Correct practice is to support the reinforcing steel on chairs or other approved supports at the correct elevation as shown in Figure 18-8 or to place concrete to the level proposed for the steel, lay the steel in place, then complete the slab. A word of caution: the second concrete placement must be placed before the first concrete placement sets; otherwise a cold joint will result.

Figure 18-21
Pulling up mesh while placing concrete. This long established practice does not work! See Figure 18-23. The WWR must be located in the upper half of the slab thickness (ideally $^1/_3$ the depth from top of slab) to be effective as a "crack control" reinforcement.

Figure 18-22
Concrete for this paving slab was placed to the elevation of the WWR, the WWR was laid in place, then the balance of the concrete was placed. Delay between lifts should not be so long that the concrete takes an initial set. (Courtesy of PCA)

Figure 18-23
This 8-inch-diameter core, taken from a slab on grade, shows what happens when the steel is laid on the subgrade ahead of concrete placing. The steel is very rarely pulled up to its correct elevation.

Figure 18-22 shows the second lift of a paving slab being placed on top of welded wire reinforcement. Figure 18-23 shows a core from a slab in which the steel was misplaced because the steel was laid on the subgrade. Figure 18-24 is an interesting placement of WWR; it is possibly to salvage and reuse the wire reinforcement after the badly cracked slab is to be removed and replaced! Proper placement of welded wire reinforcement for slabs on ground is addressed in References 18.2 and 18.4.

Field bending of bars partially embedded in concrete is prohibited by the code, except as shown on the design drawings or permitted by the engineer (ACI 318 Section 7.3.2). Such bending is apt to result in loss of ductility or brittle fracture of the reinforcing steel. If bending is permitted, the steel should be at a temperature of at least 60°F.

Figure 18-24.
How not to locate the slab reinforcement!

Condition of the Steel. The first step in placing reinforcement is to make sure that the correct steel is being used. Having previously checked the drawings and bar lists for the identifying marks, the bundles are now checked to assure that the right bars (or WWR) are being placed. The steel should be free of mud, oil, paint or other coatings that might interfere with bond of the steel to the concrete. Scale forms when bars are rolled in the steel mill as a result of the heating and cooling of the steel. Much of this mill scale, as it is called, comes off the bars during fabrication and handling. The remaining tight mill scale is not detrimental. Steel that has rusted can be used (see Figure 18-21) as long as the rusting has not progressed to the point that the minimum dimensions, including height of deformations, and weight of a hand-wire-brushed specimen are less than the applicable specification requirements. The way this can be checked is to cut a measured length of steel, brush it vigorously with a hand-held wire brush, weigh the specimen, and compare the weight per foot of the brushed specimen with the specified weight shown in Table 18.3.

Steel reinforcement may become contaminated with form oil or other material after it has been placed in the form. These materials must be removed by wiping with a rag soaked in solvent.

Placing in the Forms. The code has very definite requirements regarding the installation or placing of reinforcement in the forms. These are stated in ACI 318 Section 7.5.

In slabs, shallow beams and similar members the steel can be placed after the formwork has been completed. (See Figure 18-25.) It is sometimes advantageous to assemble the steel into cages in which the bars, stirrups and other elements can be tied together at a convenient assembly location. A supporting fixture or jig can be made of lumber on which it is possible to mark off the position of the steel as an aid in assembly. A simple arrangement is shown in Figure 18-26 in which stirrups, support bars and longitudinal steel for a narrow, deep beam is being assembled.

For such structural elements as walls and other deep members, one side of the formwork can be erected, a curtain of steel assembled in place, then the opposite form set in position. Some walls require two curtains—that is, a layer of reinforcement adjacent to each face of the concrete. Steel for a column—or a section of a column in which concrete is to be placed in one operation, such as from one floor to the next—is usually assembled entirely before the formwork is erected.

Figure 18-25.
Reinforcement in place for a suspended slab.

Figure 18-26.
Arranging the beam reinforcement for a one-way joist floor system. After assembly, the entire cage will be set into the beam form and secured in place.

Bars for footings can be assembled into mats in which the bars in each direction are laid out and tied together. The entire mat is then set on supports in the foundation excavation. When a wall (or column) rests on a footing, dowels are usually inserted in the footing, extending up into the wall and lap spliced with the wall's vertical bars. (See Figure 18-27.) Dowels are also used when the wall rests on a grade beam or base of any kind. Splices with the wall vertical bars are usually made by lapping the dowel along the bar. Splices can also be made with mechanical devices or by welding. Dowels are positioned by means of holes drilled through boards attached to the form (see Figure 22-15) or by the use of a single wrap of the column tie. Dowels are used for splicing vertical steel in a column resting on a footing, the same as for walls as shown in Figure 18-27.

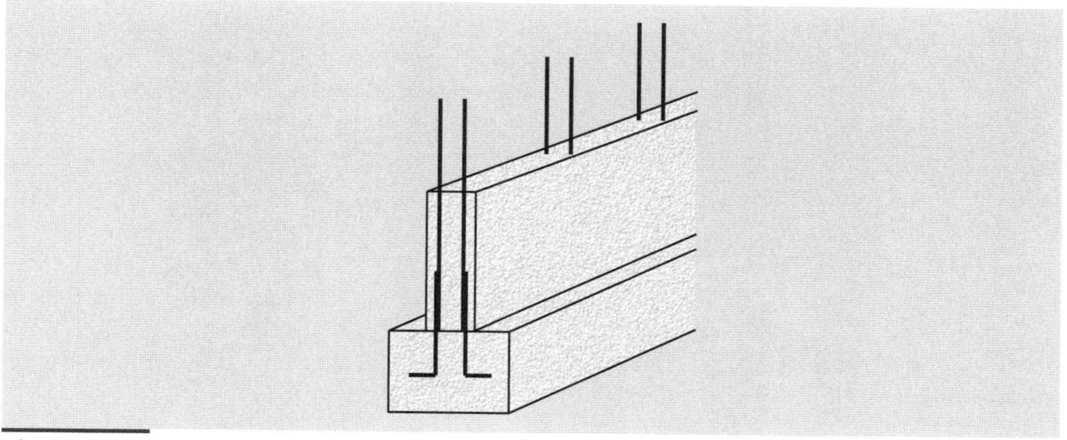

Figure 18-27
Continuity of reinforcement where a column or wall joins a footing is accomplished by placing dowels in the footing concrete at locations where they can be spliced to the vertical reinforcement extending above. Dowels are usually hooked on the lower end.

Splices. Normal handling and placing restraints limit the length of bars that can be used. Vertical bars in columns are usually one story in height, occasionally two stories. Vertical bars in walls are similarly limited. Slabs usually require joining bars in some manner at construction joints. When, because of space or other limitations, it is not possible to use a single bar that extends the full required length, two or more shorter bars joined by splicing must be used.

A splice can be defined as the means by which reinforcing steel is joined to accomplish continuity in the steel. There are three types of splices:

1. Lap splice. In a lap splice, the two bar ends extend past each other so that the bars actually overlap a specified distance.
2. Mechanical splice. Mechanical splices utilize some sort of a threaded or friction mechanism. Several proprietary mechanical connections are available.
3. Welded splice. There are many types of welded splices, including patented self-heating processes.

Splices are permitted by the code, provided they are made only as required or permitted by the design drawings and project specifications. Splices should not be made at any other location without specific approval of the engineer. In a location where several bars are to be spliced, the splices should be staggered; that is, they are not all made at exactly the same cross section of the beam or other structural member, but each is made at a different point in the member.

The inspector should be familiar with ACI 318 Sections 12.14 through 12.19 for requirements regarding splicing. Most of this covers design of splices but is useful for field control also.

Lap splices are the most common and easiest to make. The length of lap varies with the bar size, yield strength of the steel and compressive strength of the concrete, as well as other factors. (See Figure 18-28.) LAP

Several proprietary mechanical splices are commercially available. When splicing bars with mechanical splices, the instructions of the manufacturer should be carefully followed. The work should be done by skilled workers familiar with the proper connection process.

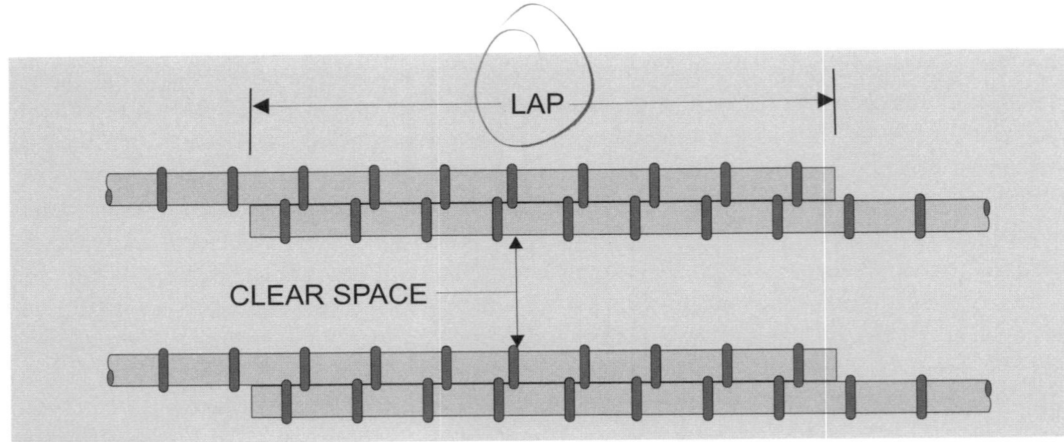

Figure 18-28
A common lap splice. Continuity is achieved by bond of the concrete to the overlapping bars.

Welded splices can be either lap welds or butt welds. There are several techniques and designs of welded splices, the proper ones for any job being described in the design drawings and project specifications. (See Figure 18-29.)

When welding of reinforcing bars is required, the weldability of the steel and the compatibility of the welding procedures need to be considered. The inspector should be ever alert to improper welding techniques and procedures. All reinforcing bar welding must be performed in strict conformance with the American Welding Society "Structural Welding Code—Reinforcing Steel" (AWS D1.4).[18.8] Proper preheat of the bars at the weld location is usually required, dependent on type of steel, chemical composition and bar size. The amount of preheat is indicated in Table 5.2 of the D1.4 code and requires a calculation of a "carbon equivalent" (CE) number based on the chemical composition of the bar as shown in a mill test report. As a general rule, the higher the CE and/or the larger the bar size, the less easily weldable the bar is, and the more preheat is required. This is due to the potential for the formation of a brittle zone next to the weld which is, by definition, nonductile and sensitive to fracture under shock loading (earthquakes). In metallurgical terms, a stress riser (or notch effect) is created. To prevent formation of this brittle zone, preheat the steel around the joint to be welded. This allows the base metal next to the weld, called the heat affected zone (HAZ), to cool more gradually, preventing brittleness. As indicated above, a mill test report stating the chemical composition of the steel is required to calculate a carbon equivalent number. Unfortunately, field realities dictate that once reinforcing steel leaves the tagged mill bundles there is normally no way to identify its heat number and to associate it positively with a mill certificate. The steel loses its traceability and the CE of any given bar cannot be determined, except by independent laboratory testing. If a mill test report is not available to calculate a CE number, the AWS D1.4 permits welding if the following minimum preheat is used:

Carbon Equivalent Unknown
AWS D1.4—Sec. 1.3.4

ASTM	Bar Size	Preheat
A615	#3 - #6	300°F
	#7 - #18	500°F
A706	#3 - #6	None
	#7 - #11	50°F
	#14 & #18	200°F

The field inspector should be aware that the ASTM A 706 bar is intended for welding. By use of chemical composition control in the manufacture of the A 706 bar, the carbon equivalent is limited to 0.55 percent. As such, under most conditions, A 706 bars #11 and smaller do not require any preheating. The reinforcing bar producer is required by the ASTM A 706 specification to report the chemical composition and carbon equivalent in the mill certificate. For the carbon-steel reinforcing bar, the ASTM A 615 specification requires the producer to determine the percentages of carbon and manganese (the two chemical elements necessary to calculate the CE for A 615 bars). However, reporting the chemical composition in the mill certificate is not mandatory. Reporting these material properties should be required in the project specifications so that the contractor will have the carbon equivalent available for the bars if welding is required. Chemical analyses are not ordinarily meaningful for rail-steel and axle-steel reinforcing bars, as welding of these bars is not recommended.

Figure 18-29
Note how the welded splices are staggered at different heights in this column steel. (Courtesy of CRSI)

From a contractor's point of view, preheating takes time (labor), and the materials are costly. The contractor may not be aware of the AWS requirements and/or the consequences of not adhering to the requirements, especially insofar as the consequences are not visually obvious and will probably not be obvious until an earthquake or similar catastrophic type loading occurs. It is noteworthy that the AWS D1.4 welding code requires the contractor to prepare written welding procedure specifications conforming to the requirements of the welding code. Appendix A of the welding code contains a suggested form, which shows the information required for such a specification for each joint welding procedure.

Preheating is usually done by an acetylene torch or other suitable means just prior to welding, with the preheat temperature verified by heat temperature crayon sticks. The bars to be welded must be preheated to the required temperature a minimum of 6 inches each side of the weld location (see Figure 18-29). The inspector should also verify that the correct electrodes (oven dry) are being used:

Electrode Size
AWS D1.4 - Table 5.1

Reinforcing Bar	Bar Tensile Strength	Electrode SMAW
A615 - Grade 40 -Grade 60 - Grade 75	70 ksi 90 ksi 100 ksi	E70 E90 E100
A706 - Grade 60	80 ksi	E80

Electrode size is directly related to the tensile strength of the reinforcing steel. The tensile strength of the electrode (filler material) must be at least equal to the tensile strength of the reinforcing steel being welded (base metal). If, for example, carbon-steel bars (A 615) Grade 60 are being welded, an E90 electrode is required. The letter "E" designates an electrode and the following two digits, 90, designate the tensile strength of the filler material (90,000 psi), which matches the tensile strength of 90,000 psi for the Grade 60 rebar. When joining different grades of steels, the electrode is selected for the lower tensile strength reinforcing bar.

Tack welding or spot welding of crossing bars or other items in contact with the reinforcement should be avoided. Striking an arc by bringing the welding electrode in contact with the reinforcement at random should be avoided. Either of these procedures burns the steel and causes a serious loss of strength.

Bar Supports. Reinforcement must be supported and rigidly fastened in the forms before concrete is placed. Besides holding the steel in place, the supports must be strong enough to support the weight of the steel and of the workers walking on the steel, must be close enough together so that the steel will not sag too much between them, must not cause staining or rusting of the exposed concrete and must not have any other adverse effect in the concrete. Bar supports may consist of steel wire, precast concrete block or plastic. (See Reference 18.1 for specific details of reinforcing bar supports.)

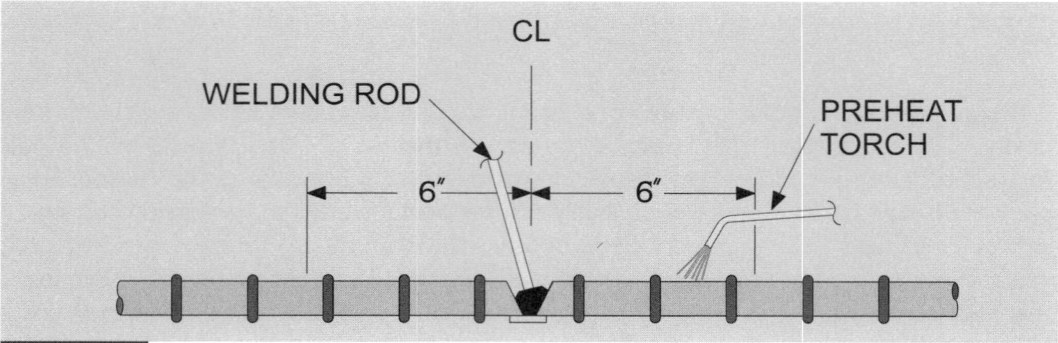

Figure 18-30
Preheat requirement for welding reinforcing bars.

Small precast concrete bar supports commonly known as dobies, usually with protruding wires embedded in them for fastening to the steel, are frequently used, especially in footings and slabs on the ground. (See Figure 18-30.) Factory-made wire bar supports are widely used. These may be made of plain steel wire, galvanized steel wire or stainless steel wire. The lower portion is frequently coated with plastic to prevent rusting that would damage exposed concrete, or it might be fitted with a piece of thin sheet steel called a sand plate to prevent the legs from penetrating into the subgrade.

Figure 18-31.
Precast concrete bar support. Used primarily when placing bars or wire fabric off grade. Generally, the block is cast with embedded 16-gauge tie wire.

Although there are specific supports made for WWR, many of the same supports used for reinforcing bars can be used for welded wire reinforcement. Figure 18-32 shows types of supports common to WWR.

Figure 18-32.
Typical types of support for WWR.[18.6] *(Courtesy of WRI)*

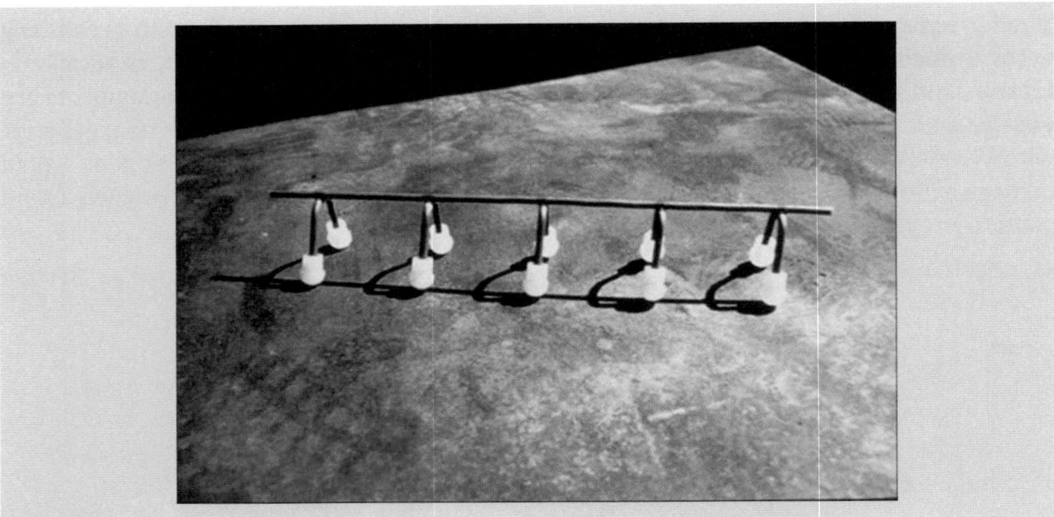

Figure 18-33.
"Slab bolster" made of bright basic wire Note the plastic covering on the feet to prevent rust staining of the surface of the concrete. (Courtesy of CRSI)

Wire bar supports are furnished in four classes depending on their expected exposure and degree of corrosion protection required.[18.1] These classes are:

CLASS 1—MAXIMUM PROTECTION

Plastic-protected wire bar supports—intended for use in situations of moderate to severe exposure and/or situations requiring light grinding or sandblasting of the concrete surface. Legs are protected with a plastic covering. (See Figure 18-33.)

CLASS 1A—MAXIMUM PROTECTION

Epoxy-, vinyl- or plastic-coated bright basic wire bar supports—intended for use in situations of moderate to maximum exposure where no grinding or sandblasting of the concrete surface is required. Class 1A is generally used when epoxy-coated bars are specified.

CLASS 2—MODERATE PROTECTION

Stainless steel-protected wire bar supports—intended for use in situations of moderate exposure and/or situations requiring light grinding or sandblasting of the concrete surface. The bottom of each leg is protected with a stainless steel tip. Class 2 protection is available in either a Type A or B bar support. For a Type A support, a tip of stainless steel is attached to the bottom of each leg such that no portion of the nonstainless steel wire is closer than $^1/_4$ inch from the form surface. Type B stainless steel protected bar supports offer a higher degree of protection with no nonstainless steel wire of the bar support closer than $^3/_4$ inch from the form surface. Stainless steel-protected wire bar supports with protection exceeding $^3/_4$ inch are available by special order.

CLASS 3—NO PROTECTION

Bright basic wire bar supports—with no protection against corrosion. Unprotected wire bar supports are intended for use in situations where surface blemishes can be tolerated or where supports do not come into contact with the exposed concrete surface.

Tie Wire. Reinforcing bars, after their installation in the form, must be tied together to resist movement or displacement of the bars. Tying is accomplished with short pieces of soft annealed wire called tie wire. Size of wire is usually No. $16^1/_2$ or No. 16 or occasionally No. 15 or No. 14. Coated tie wire is used with epoxy-coated reinforcement.

Supporting. The reinforcing steel must be tied and supported so that it is held rigidly in position and will not move under the activities of constructing forms, installing equipment or placing concrete. Distances from the subgrade and forms should be maintained by means of chairs, ties, hangers or other approved supports, some of which are shown in Figure 18-32.

Selection of the appropriate supports and spacers depends upon conditions to which the concrete will be exposed, type of structural element in which the reinforcing steel is used, spacing, clearance and other factors. Supports, shown on the placing drawings, based on the structural drawings for the project, are detailed on the placing drawings and other documents by means of a code that shows the nominal height, length and symbols that give the type of support and class of protection. For example: 2 x 10-6-SB-3 indicates a 2-inch height, 10-foot 6-inch length of slab bolster (symbol SB) made of bright basic wire with no protection from rusting (Class 3).

Wire supports are designed to carry the weight of the reinforcing steel. Workers usually walk on the steel, which imposes an extra load, and occasionally plank runways are supported on horizontal steel in slabs. This practice should be discouraged because, unless special supports have been used to carry the additional weight, it could displace the steel.

For concrete that is to be placed directly on the soil, as in a column or wall footing, the steel can be supported on the concrete bar supports previously described, or special metal supports can be used, consisting of perhaps a normal high chair with a sand plate. (See Figure 18-34.) The sand plate prevents the legs of the support from settling or penetrating into the soil. Another method is to use a steel stake with a cross bar bent or welded at the top to which the reinforcement can be tied.

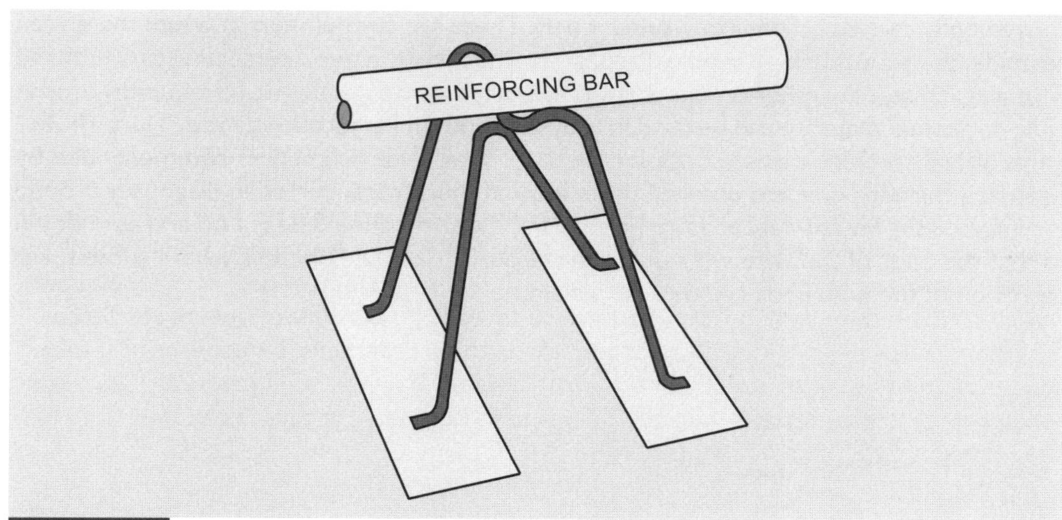

Figure 18-34.
The sand plate, made of light sheet steel, prevents the legs of the chair from settling into the soil.

Precast concrete supports are made either with or without embedded tie wires. Supports without tie wires can be used under horizontal mats; however, for use in providing clear cover from vertical forms, the supports should have wires that can be bent and tied around the steel.

Tying. After the reinforcing steel has been properly spaced and the supports set in place, the steel has to be tied—the bars tied to each other and to the supports so that they will not move. Actually, the job of tying proceeds at the same time as the placing and spacing. (See Figure 18-35.) Tying the steel does not add anything to the strength of the structure; its only object is to hold the steel rigidly in place before the concrete is placed. It is not necessary to tie every intersection of bars; usually tying every fifth or sixth intersection is sufficient.

Figure 18-35.
Epoxy-coated reinforcement with coated tie wire. (Courtesy of PCA)

The 16-gage soft-annealed wire is usually furnished in rolls of 3 or 4 pounds, which can be conveniently carried on the ironworker's belt. There are several ways in which the tie can be made, the easiest being a simple diagonal wrap around the two intersecting bars with the wire ends twisted together on top. Sometimes called a snap tie, this tie is frequently used in tying horizontal mats but can be used in wall steel and similar locations. (See Figure 18-36.) Other more complex ties are occasionally used. Wall reinforcement is commonly tied by wrapping the wire one and one-half times around the vertical bar, then diagonally around the intersecting horizontal bar as in the snap tie. Tie wires should be cut off and bent down so that the ends of the wire will not touch the surface of the concrete where subsequent corrosion of the wire ends could cause a stain.

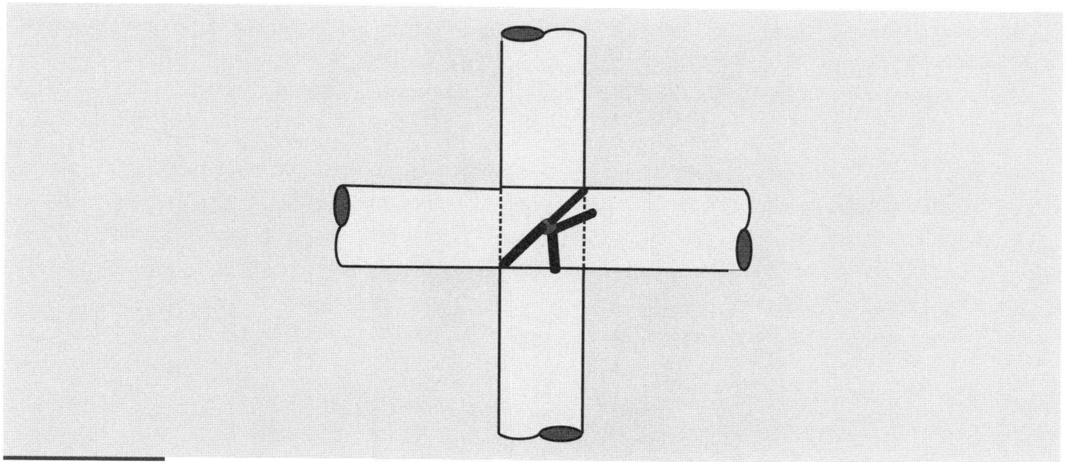

Figure 18-36.
The single tie, sometimes called a snaptie.

Automatic Rebar Tying Machine. Tying steel reinforcing bars has always been tough, manual labor, as reinforcing ironworkers spend much of the day hunched over wrapping and twisting the tie wires by hand. As in other technologies, tying of reinforcing bars has also advanced by new innovated mechanical devices as shown in Figure 18-37. Automatic reinforcing bar tying machines or guns are uniquely designed to replace the manual and backbreaking process of tying reinforcing bars. The machines are battery driven and use a spool of wire fed through the machine to tie the rebar. The ironworker places the machine's jaws on the bar overlap and pushes to close the jaws around the crossed reinforcing bars, then pulls the trigger. The machine feeds the wire around the bar, pulls it, twists it and cuts it. The result is a strong, single strand, double wrap tie. The automatic reinforcing bar tying machine allows ironworkers to work upright and thus avoid the bending, twisting and awkward static trunk posture of manual tying at ground level. (See Figure 18-38.)

Figure 18-37.
Automatic bar-tying machine. (Courtesy of ICM)

Figure 18-38.
Note the difference: working posture while manually tying bar (left) and while working with the bar-tying machine (right). Note the handle extension to allow the ironworker to stand upright for ground-level reinforcing bar installation. (Courtesy of ICM)

Tolerances. We have previously considered tolerances as applied to fabricating the reinforcing bars. Tolerances are also necessary in placing the steel. We must, however, take care that reasonable tolerances are not exceeded. An error of $^1/_2$ inch in the vertical placing of steel in a thin slab could have a serious effect on the load-carrying capacity of that slab. Inspection of structures damaged by earthquakes has revealed serious discrepancies in placing steel. An example is the longitudinal location of hooks in beam steel. A hook on the longitudinal bar should extend into the column a specified distance to provide adequacy of load- carrying capacity where the beam and column join. However, cases have been reported in which the hook scarcely penetrated into the column. (See Figure 18-39.)

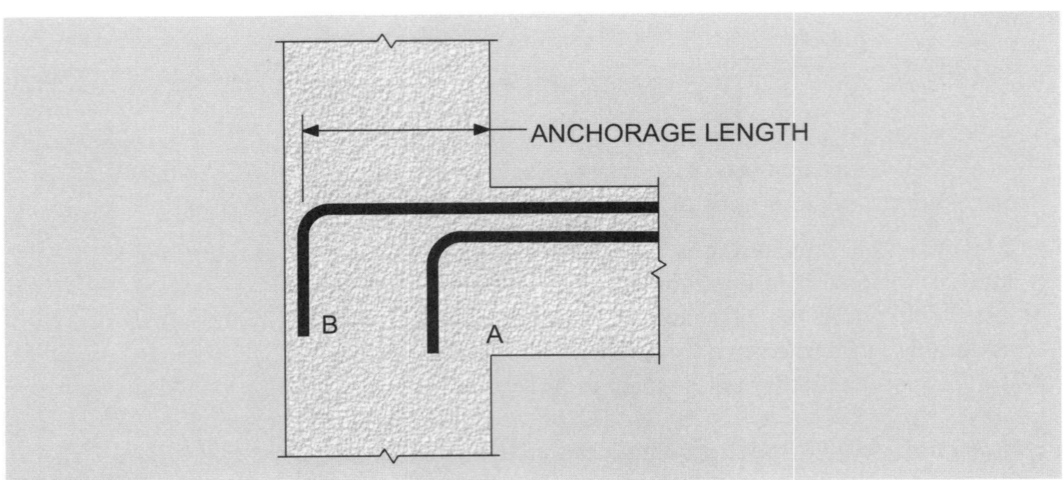

Figure 18-39.
The hook positioned at (A) is of no value in anchoring the beam reinforcement into the column. The hook should be at (B).

The code allows tolerances in placing steel as prescribed in ACI Section 7.5.2. Unless otherwise specified by the engineer, reinforcement must be placed within the tolerances illustrated in Table 18.6.

Tolerance for concrete cover must not exceed minus one-third the minimum concrete cover indicated by the approved plans and specifications. Application of the 1/3 minimum tolerance is illustrated in the following example:

Example—Concrete Cover Tolerance

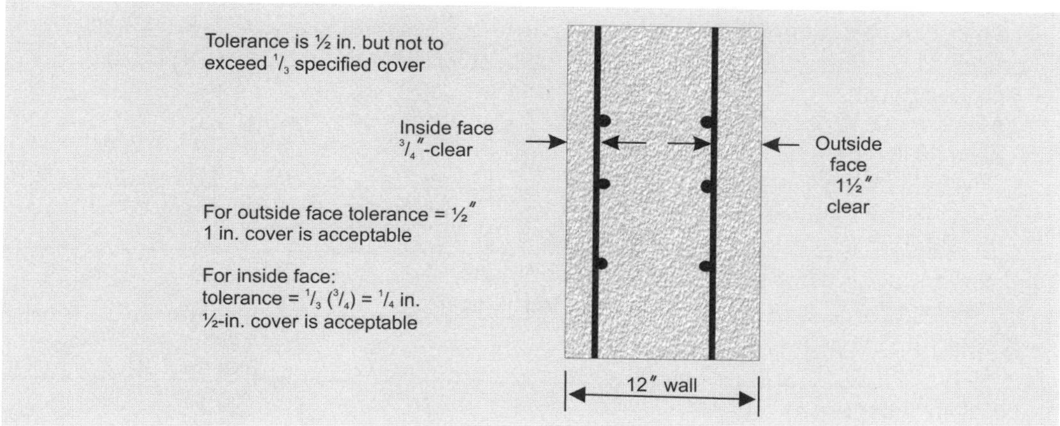

TABLE 18.6
PLACING TOLERANCES FOR REBARS ACI 318—Sec. 7.5.2

Item	Tolerance
• Effective Depth Deviation	
Effective depth 8 in. or less	± 3/8 in.
More than 8 in.	± 1/2 in.
• Concrete Cover	
Effective depth 8 in. or less	- 3/8 in.*
More than 8 in.	- 1/2 in.*
*But not to exceed 1/3 cover	
• Clear Distance to Formed Soffits	- 1/4 in.
• Location of Bar Ends	
General	± 2 in.
Discontinuous ends of brackets and corbels	- ± 1/2 in.

Tolerance for longitudinal location of ends of bars (ACI 318 Section 7.5.2.2) is illustrated in Figure 18-40. Note that tolerance for minimum concrete cover of ACI Section 7.5.2.1 also applies at discontinuous ends of members.

ACI Standard 117 Specifications for Tolerances for Concrete Construction and Materials [18.9] is another valuable reference for the concrete inspector. ACI 117 prescribes acceptable industry tolerances for all phases of concrete construction and materials, and provides considerably more tolerance values than that specified in the ACI 318 Standard. Note that ACI 117 is not mandatory unless specifically referenced in the project specifications. It does, however, provide an excellent guide for acceptable industry tolerances for concrete construction. ACI 117 placing tolerances for reinforcing bars is illustrated in Table 18.7. It is interesting to note that some of the tolerance values are slightly different from that specified in ACI 318. Of course, where a difference exists, ACI 318 governs. For example (not in ACI 318), when placing bars in a slab, the total number of bars in any space or extent of the slab should be the required number, but small differences in the spacing of the individual bars is permissible. In accordance with ACI 117, a spacing tolerance of plus or minus 3 inches from the exact spacing is considered acceptable. (See Figure 18-41.) Bars can be shifted slightly to bypass an insert in the concrete, but extensive relocation of bars should not be made unless approved by the engineer.

TABLE 18.7
PLACING TOLERANCES FOR REINFORCING BARS—ACI 117

• Clear Distance to Forms	
Member size 4 in. or less	+ 1/4 in.
	-3/8 in.
Over 4 in. but not over 12 in.	±3/8 in.
Over 12 in. but not over 24 in.	± 3/8 in.
Over 24 in.	±1
• Concrete Cover	
Member size 12 in. or less	- 3/8 in. *
Over 12 in.	- 1/2 in *
*But not to exceed 1/3 cover	
• Clear Distance Between Bars	clear distance/4 ≤ 1 in.
"Spacing not less than bar diameter of 1 in."	
• Uniform Spacing Between Bars	± 3 in.
"Total number of bars not less than specified number"	
• Spacing of Stirrups	member depth/12
Ties	member size/12
• Location of Bar Ends	
General	± 2 in.
Discontinuous ends of members	± 1 in.
• Embedded and Lap Length of Bars	
#3 – #11	- 1 in
#14 & #18	- 2in.

A final word on tolerances: The inspector needs to recognize that reasonable construction tolerances are necessary and should be permitted. However, a permitted variation from a specified dimension or quantity in one part of construction must not be construed as permitting violation of more stringent requirements of other parts of construction.

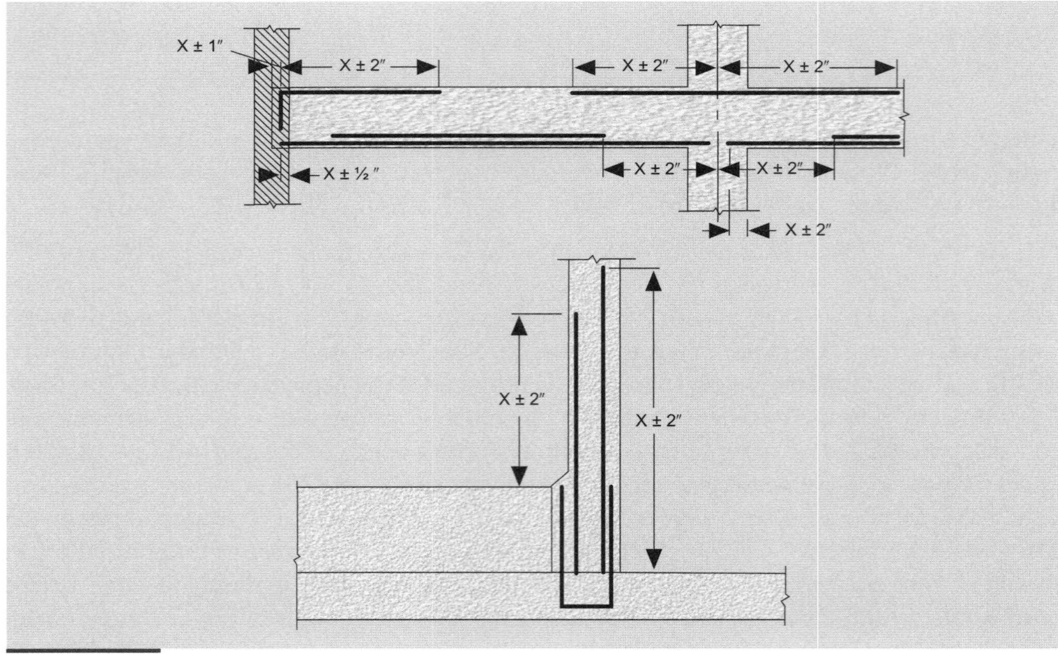

Figure 18-40.
Tolerance for longitudinal location of ends of bars (ACI 318 Section 7.5.2.2).

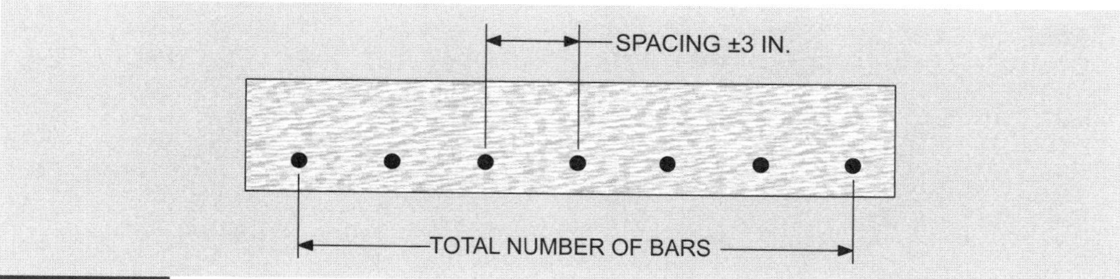

Figure 18-41.
The total number of bars in any width of slab is more important than exact spacing.

Concrete Cover. The clear distance between a reinforcing bar and the surface of the concrete is the concrete cover. (See Figure 18-42.) The primary purpose of adequate cover is primarily to protect the steel from weathering, which causes the steel to rust. Rusting of the steel not only results in decreasing the area of steel, thereby weakening the structure, but also causes an expansion in volume that disrupts the concrete, with consequent cracking and spalling—as we saw in Chapter 6. If the steel is a sufficient distance from the surface of the concrete, and if the concrete is good and sound, rusting will not be a problem. The amount of concrete cover specified for adequate protection is detailed in ACI 318 Section 7.7.

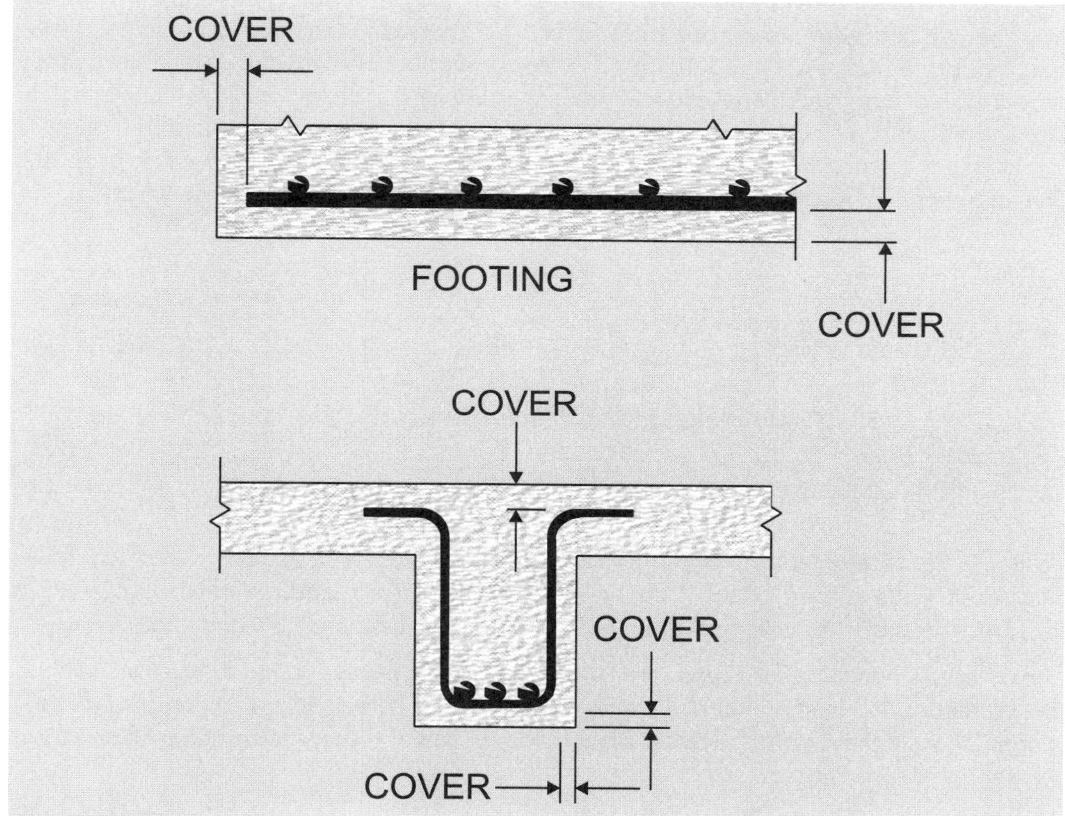

Figure 18-42.
The concrete cover of a reinforcing bar is the space between the bar and the face of the concrete.

For quick reference, the specified cover for cast-in-place concrete with conventional reinforcing bars is summarized in Table 18.8.

TABLE 18.8
CONCRETE COVER FOR REINFORCING BARS IN CAST-IN-PLACE CONCRETE

CONCRETE EXPOSURE		SPECIFIED COVER,* IN.
Cast against ground (without forms)		3
Exposed to weather or ground but cast against forms	#5 bar and smaller	$1\frac{1}{2}$
	Greater than #5 bar	2
Slabs and walls (no exposure)		$\frac{3}{4}$
Beams and columns (no exposure)		$1\frac{1}{2}$

"Measured cover ≥ minimum ± cover tolerance (AC1 318 Section 7.5.2.1).

When the concrete is to be exposed to a corrosive atmosphere or severe conditions, the amount of concrete cover must be suitably increased, and the density and nonporosity of the concrete should be considered. Special protection is sometimes specified. For example, in concrete exposed to seawater a common practice is to specify a concrete cover of 3 inches. Another special exposure condition is fire-resistive construction, as described in the general building code. When the general building code requires a fire-protective cover different from the concrete cover specified in ACI 318 Section 7.7, the greater distance governs. These requirements will be shown on the design drawings or in the project specifications.

Welded Wire Reinforcement. To ensure proper performance of WWR, it is essential that the welded wire be placed on supports to maintain its required position during concrete placement (ACI 318 Section 7.5). Proper support of the welded wire is especially critical in slabs-on-ground applications, where the wire must be located in the upper half of the slab to work effectively as a crack control reinforcement. The supports (concrete blocks, steel or plastic "chair" devices) must be appropriately spaced to work properly. The following can be used as a guide for minimum spacing of supports in each direction: [18.6]

WIRE SIZE	WIRE SPACING	SUPPORT SPACING
W or D9 and larger	12" and greater	4 — 6 ft
W or D5 to W or D8	12" and greater	3 — 4 ft
W or D9 and larger	less than 12"	3 — 4 ft
WorD4 toWorD8	less than 12"	2 — 3 ft
Less than W or D4	less than 12"	2 — 3 ft or less

To avoid displacement by workers walking on the wires prior to and during concrete placement, wide-spaced WWR (12 to 16 inch spacing) can be used. Wide-spaced WWR is recommended by the wire reinforcement industry for all slabs-on-ground applications.

The rolled welded wire has a tendency to curl or coil as it is unwound from the roll owing to the curvature induced when the wire was originally rolled during the manufacturing process. If rolls are supplied, it is essential that a mechanical straightener or other method be used to ensure flatness.

Figure 18-43.
Machines like this have been used to flatten and straighten rolled welded wire fabric to improve placement and proper positioning.

Figure 18-43 illustrates a mechanical straightener manufactured under the name Mesh Runner. No more pulling or tugging to get the mesh flat; the machine unrolls and straightens welded wire fabric at 2 feet per second. The machine also lifts the fabric off the ground by hydraulic arms to eliminate lifting by hand. The Mesh Runner can be adjusted to leave a predetermined curvature in the fabric for spanning over support areas of multispan elevated slabs.

Plain welded wire is spliced by overlapping so that the overlap, measured between the outermost cross wires of each sheet, is not less than the spacing of cross wires plus 2 inches. The amount of lap depends on whether the steel carries more than half of the permissible stress, or less than half (see ACI 318 Section 12.19). Because this is a design consideration, the project drawings should be consulted to determine the correct length of lap. The splice length for deformed wire depends on the strength of the concrete, the yield strength of the steel and other factors as computed by the provisions in ACI 318 Section 12.18.

Regardless of whether reinforcing steel is in the form of welded wire, bar mats or individual bars, it should never be laid on the subgrade and attempts made to raise it into the fresh concrete with hooks or other devices. These procedures never work. The procedures that are satisfactory are (1) supporting the steel on proper supports that hold the steel in position or (2) placing the slab concrete to the level of the steel, laying the steel on the fresh concrete and immediately placing the rest of the concrete in the slab. For proper placement of welded wire for slabs on ground, the inspector should obtain a copy of References 18.4 and 18.6.

18.6. Epoxy-Coated Reinforcement

Epoxy-coated reinforcement (reinforcing bars and welded wire reinforcement) has been in use since 1973 and is now in over 100,000 structures. This system was originally evaluated and proposed by the Federal Highway Administration for use in highway bridge decks, where the concrete is subject to high concentrations of de-icing salts was causing significant deterioration of the nation's bridges. (See Figure 18-44.) Engineers specify the use of epoxy-coated bars in many different reinforced concrete structures that may be subject to corrosive environments. These include marine structures, water-front docks and wharf facilities, cooling towers, pumping stations, parking decks and balconies where the concrete is subjected to chlorides from de-icing chemicals or seawater.

Figure 18-44.
Early use of epoxy-coated reinforcement for the deck top reinforcement. Current practice is to use epoxy-coated reinforcement for all deck reinforcement (top and bottom). Note the use of special coated tie wire and coated bar supports.

Epoxy-coating of reinforcing bars is covered by several ASTM specifications. ASTM A 775 covers epoxy-coating of straight bars that are then fabricated after coating. The straight bars are cleaned, heated and epoxy-coated using a fusion bonded epoxy material prior to shearing and bending to specific job specifications. ASTM A 934 covers epoxy-coating of bars that are coated after shop shearing and bending for a specific construction project. Note that when bars are coated after fabrication according to ASTM A 934 they have limited capability to be successfully field bent or re-bent. If epoxy-coated bars are required to be bent or re-bent in the field, the bars should be epoxy-coated to conform to ASTM A 775.

ASTM A 775 and A 934 provide requirements for coating thickness, adhesion, continuity and flexibility. Prior to shipment, all coating damage that is due to fabrication and handling must be repaired, including all cut ends. Coating damage is to be repaired using a two-part epoxy coating applied in accordance with the patching material manufacturer's recommendations.

Results of many different studies over the past 30 years[18.10] have shown that epoxy-coated rebar provides excellent corrosion protection; however, the amount of damage to the epoxy coating should be minimized. Consequently, just prior to placement of the concrete, the inspector should check for any coating damage to the in-place coated bars. All damage should be repaired with the two-part patching material.

ASTM D 3963 provides a Standard Specification for Fabrication and Jobsite Handling of Epoxy-Coated Steel Reinforcing Bars. This specification covers patching material, handling, fabrication, storage, lifting and handling at the jobsite. Items addressed in this specification include:

1. Use of nylon lifting or padded wire rope slings.
2. Use of spreader bars for lifting bar bundles. Bundles should be lifted at third points with nylon or padded slings, and bundling bands should be made of nylon or be padded.
3. Storage of the bars on padded or wooden cribbing.
4. Moving of the bars. Coated steel should not be dragged over the ground or over other bars.
5. Minimizing of walking on coated steel during or after placing; tools or other construction materials should not be dropped on bars in place.
6. Types of bar supports. Bar supports must be made or coated with a nonconductive material compatible with concrete. This includes wire supports coated with materials such as epoxy or vinyl.
7. Types of tie wire. Epoxy- or plastic-coated tie wire, or nylon-coated tie wire must be used to prevent damage or cutting into the bar coating.
8. Concrete operations. Concrete conveying and placing equipment must be set up, supported and moved carefully to prevent damage to the bar coating.
9. Concrete vibration. Rubber or nonmetallic vibrator heads must be used when consolidating the concrete.

Project specifications should address field repair of epoxy coating after bar placement.

Fading of coating color is not considered damage or cause for rejection; however, bars must be covered if they are to be exposed for more than two months. The ASTM standards place limits on the amount of repair permitted in a section of bar. Note that these limits on repaired damage does not include shaved or cut ends that are coated with patching.

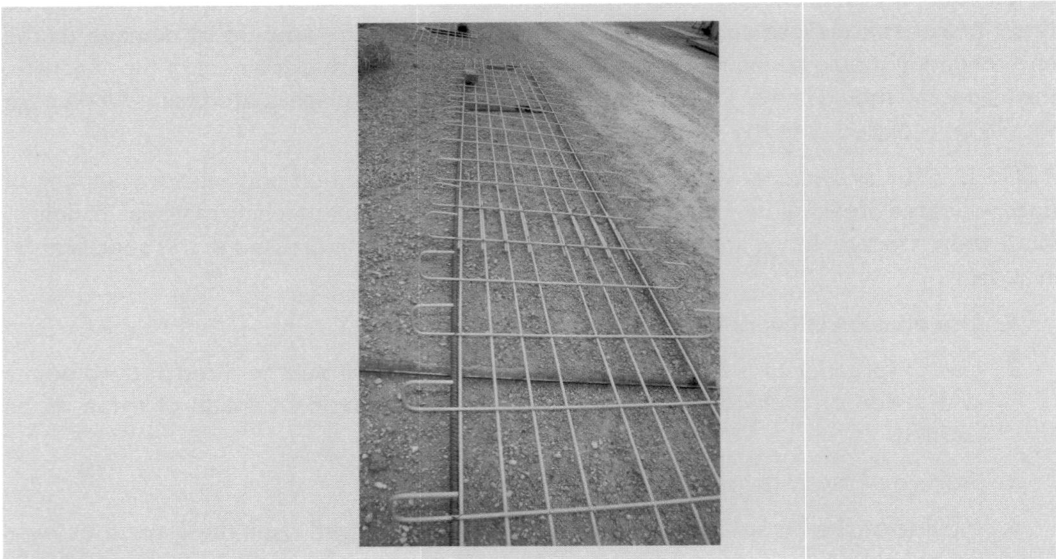

Figure 18-45.
Epoxy-coated WWR, to be used for beam shear reinforcement (Courtesy of WRI)

Epoxy-coated welded wire has also been used. (See Figure 18-45.) ASTM A 884 covers plain and deformed steel welded wire with protective epoxy coating. A Class A coating is intended for use as reinforcement in concrete and a Class B as reinforcement in earth. The same specification limits as discussed for bars also apply for the welded wire. As with bars, coating damage should be repaired prior to concrete placement.

ASTM A 882 prescribes the requirements for epoxy coating of 7-wire prestressing strand. The strand to be coated must conform to ASTM A 416. This material may be used either pretensioned or posttensioned prestressed concrete structures. Use of coated strand in pretensioned applications such as fire-rated construction should be considered with caution on account of the potential of strand slip.

18.7. Galvanized Reinforcement

Galvanized, or zinc-coated, reinforcement has been used in limited amounts for many years, the objective being to minimize corrosion of the steel reinforcement after it has been embedded in the concrete.[18.11] Galvanized reinforcing bars are covered under ASTM A 767. For welded wire reinforcement, the reader should refer to ASTM A 641 for galvanized wire before welding and ASTM A 123 for galvanizing wire after welding.

Construction operations at the jobsite should be conducted so that damage to the galvanized coating is minimized. The project specifications should contain any special handling requirements for galvanized reinforcement, similar to the special handling requirements for epoxy-coated reinforcement. These special provisions should cover compatible bar supports and tie wire, limits on repaired coating damage, and proper repair procedures for repair of damaged coating. Galvanized reinforcement should not be placed in contact with, or in close proximity to, nongalvanized reinforcement or other nongalvanized embedded steel items.

Some architects, because of their concern with the aesthetics of thin, precast wall panels and the greater height of multistory units, have been specifying galvanized reinforcement in order to minimize the danger of rust staining of the panels. Because the steel is close to the surface of these units, and because of the greater cracking potential when handling precast concrete, there is a hazard of internal rusting of the steel.

18.8. Stainless Steel Reinforcement

In 1996, ASTM issued Specification A 955 for stainless steel reinforcing bars (SSRB). Although (SS) have been produced and used for special applications over the years, there had not been a consensus standard until 1996. Stainless steel reinforcing bars are intended for use in highly corrosive environments or in constructed facilities that require nonmagnetic steel reinforcement. Stainless steel reinforcing bars are also being used in applications where corrosion resistance is imperative and concrete repair is costly. Some transportation departments have used stainless steel reinforcing bars on bridge decks of high-volume (see Figure 18-46) urban expressways because the user delay costs of future construction lane closures are not acceptable.

Figure 18-46.
Stainless steel rebars used for renovation of historic building. (Courtesy of PCA)

The requirements of ASTM A 955 parallel those for carbon-steel reinforcing bars in ASTM A 615 with regard to bar sizes, deformations, grades of steel and tensile properties, and bend tests. Use of ASTM A 995 to prescribe stainless steel bars for a construction project will require the engineer to specify the required chemical composition (stainless steel grade), heat treatment conditions, and the application—corrosion resistance or magnetic permeability—and whether either of the two supplementary requirements, S1 and S2, applies. Special corrosion testing is covered by S1. Supplementary requirement S2 prescribes magnetic permeability testing.

In 2002, ASTM issued specification A 1022 for stainless steel plain and deformed welded wire for concrete reinforcement. The same conditions apply to SS welded wire as for SS reinforcing bars.

18.9. Fiberglass Reinforcement

Steel reinforcement has been and continues to be an effective concrete reinforcement but is susceptible to oxidation when exposed to chlorides. Examples of such exposure include marine areas, regions where road salts are used for de-icing and locations where salt contaminated aggregates are used in the concrete mixture. When properly protected from chloride attack, steel reinforcement can last for decades without exhibiting any visible signs of deterioration. However, it is not always possible to provide this kind of corrosion protection. Insufficient concrete cover, poor design or workmanship, poor concrete mix and presence of high amounts of aggressive agents can all lead to corrosion of the steel reinforcing bars and cracking of the concrete. Fiber reinforced plastic (FRP) reinforcement has the potential to address this corrosion deficiency. The FRP composite bar is made from high-strength glass fibers along with a durable vinyl resin. The glass fibers impart strength to the rod, and the vinyl resin imparts high corrosion resistance properties. FRP bars are primarily manufactured using the pultrusion process. Surface deformations that contribute to the bond to concrete are available as ribbed, sand-coated and helically wrapped and sand

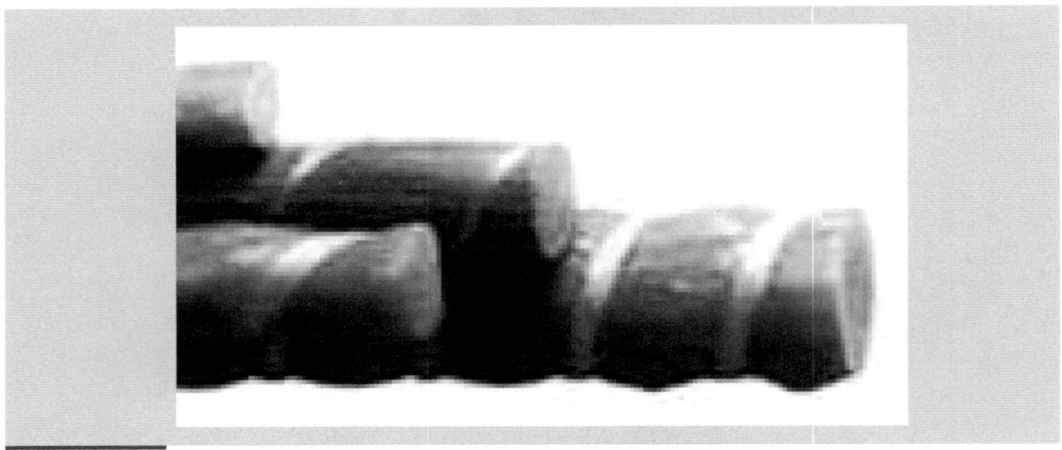

Figure 18-47.
Fiberglass composite rebars. (Courtesy of PCC)

coated. (See Figure 18-47.) The FRP bar is totally resistant to chloride attack, offers a tensile strength of $1^1/_2$ to 2 times that of steel, and weighs only 25 percent of the weight of equivalent size steel reinforcing bars. The FRP bar significantly improves the longevity of concrete structures where corrosion is a major factor. FRP bars are a suitable alternative to steel, epoxy-coated steel and stainless steel bars.

Note that fiberglass reinforcement is not specifically recognized as a concrete reinforcement in the ACI 318 standard for structural concrete; the ACI 318 standard addresses only steel reinforcement. The reader should refer to *Guide for the Design and Construction of Concrete Reinforced with FRP Bars* (ACI 440.1R)[18.12] for additional information on design and construction using FRP bars.

Hot and Cold Weather Concreting

It is not necessary to curtail concrete construction during either hot weather or cold weather if the effects of weather conditions are understood and adequate precautions are taken. Whether concrete is being placed in the summer heat of the desert or during subfreezing temperatures in the north, techniques are available to overcome any adverse conditions that might be encountered. If allowance is made for exposure conditions as they affect water demand, slump loss, rate of strength gain, air content, curing requirements and other values, there is no reason why concrete cannot be placed throughout the year.

Code requirements for hot and cold weather concreting are stated in ACI 318 Sections 5.12 and 5.13.

The inspector should have a copy of Hot Weather Concreting (ACI 305R)[19.1] and Cold Weather Concreting (ACI 306R)[19.2] in his/her reference library. Invariably, if either cold weather or hot weather concreting is a construction consideration, the project specifications will reference ACI 306R or ACI 305R.

HOT WEATHER

The old idea that hot weather concreting requires some sort of magic has long since been abandoned in favor of a realistic approach embodying practical, everyday measures that can easily be taken to assure quality concrete during hot weather as well as at any other time of the year. The question remains: Just when might hot weather concreting possibly be the cause of trouble? Probably when the daytime air temperature approaches 90°F, especially when accompanied by relative humidity values of 25 percent or less (see Figure 19-1), under windy conditions. At that time, some of the precautionary measures that have previously been arranged for on the job should be put into effect. The objective of these precautions is to protect the fresh concrete from damage because of high temperature, to provide adequate curing so that the concrete will develop the strength and durability of which it is capable, and to minimize shrinkage, cracking and dusting.

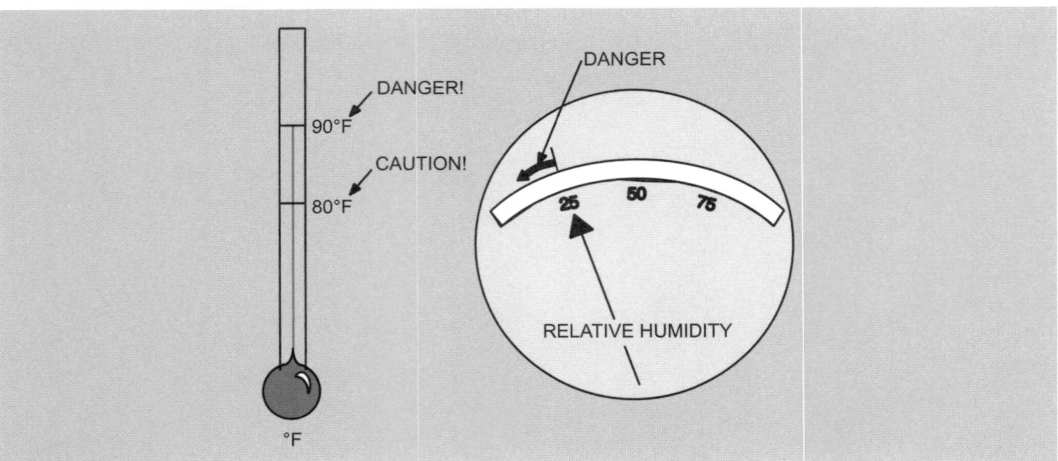

Figure 19-1
When the air temperature passes 80°F, especially when combined with low humidity, it is time to prepare for hot weather protection.

19.1. Hot Weather Effects

In the absence of special precautions, undesirable hot weather effects may include the following:

1. Increased water demand for required consistency
2. Difficulty in control of entrained air
3. Rapid evaporation of mixing water
4. Rapid slump loss
5. Accelerated set
6. Difficulties with normal handling, finishing and curing
7. Greater dimensional change on cooling hardened concrete
8. Increased plastic shrinkage
9. Increased tendency to crack or craze
10. Reduced durability from increased water demand and cracking
11. Reduced strength
12. Variations in the appearance of the concrete surface
13. Reduced bond of concrete to reinforcing steel
14. Increased risk of steel corrosion, from increased permeability and cracking
15. Possible "cold joints"
16. Increased permeability[19.1]

Strength. High temperature can adversely affect the strength, durability, cracking and other properties of the concrete. The ultimate strength may not be as high as that of concrete mixed and cured at moderate temperatures.

Figure 3-11 shows that concrete mixed and-moist cured at high temperatures enjoys a strength advantage for the first few days, but at 28 days or later the concrete placed at normal temperatures has higher compressive strength. Even though the concrete was moist cured at 70°F after two days, the high-temperature specimens soon lost their strength advantage. Figure 19-2, which is based on U.S. Bureau of Reclamation tests, shows this. In all of these tests, the water-cement ratio was held the same for each series of specimens.

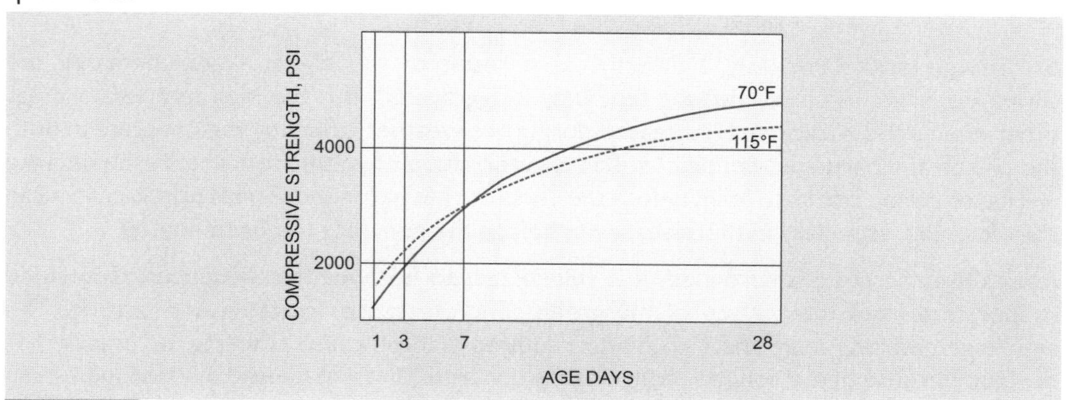

Figure 19-2
Concrete suffers a loss of strength because of high temperature even though the concrete is maintained at 70°F. Specimens were cast, sealed and maintained at temperatures shown for two hours.

If, in addition to the effect of high temperature, the specimens are not cured properly, the effect is even worse because the concrete must be kept moist for proper hydration of the cement.

The net result, as far as job quality of the concrete is concerned, is that the ultimate strength of concrete that is mixed and cured at high atmospheric temperatures is never as good as that of concrete mixed and cured at 70°F or below.

Water Requirement. When the temperature goes up, the mixing water requirement also goes up. For each 10°F rise in temperature the concrete needs about 8 pounds more water (1 gallon) for each cubic yard. (See Figure 19-3.) If the temperature of fresh concrete is increased from 50°F to 100°F, each cubic yard of concrete will require an additional 33 pounds of water to maintain the same slump. This effect is intensified at higher temperatures. Stated another way, on an average, a change of 3 percent in total water will change the slump 1 inch, but as the temperature goes up it takes more water to give the same effect.

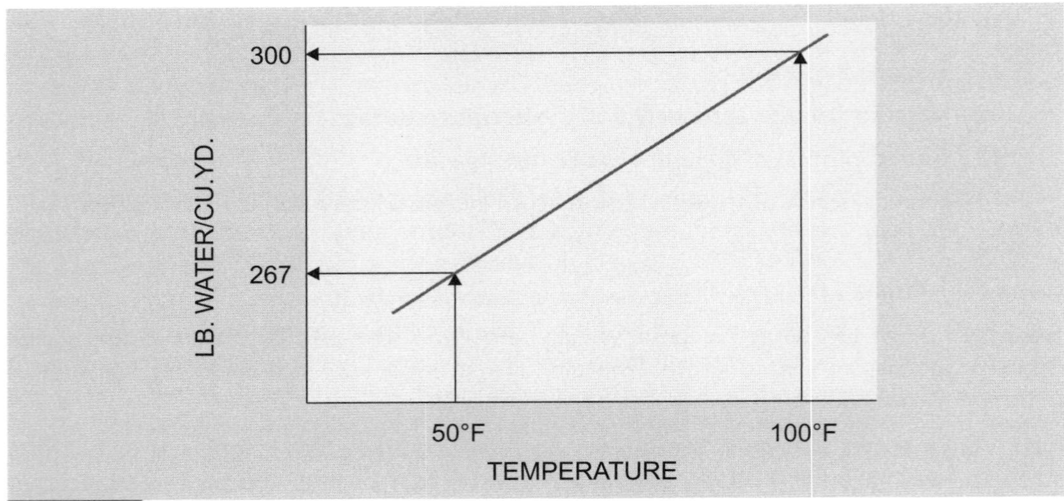

Figure 19-3
Regardless of other influences, the mix water requirement goes up as the temperature rises.

The practical aspect of this is that added water, without the necessary adjustment to the mix proportions, will cause lower strength, impaired durability and all the other problems associated with excess water.

Shrinkage and Cracking. Indirectly, as a result of the higher water demand, the shrinkage, and therefore cracking tendency is aggravated. But the high temperature has other effects too. Cracking is increased during hot weather, affecting the concrete in both the plastic and hardened conditions. Rapid evaporation of water from the fresh concrete will cause plastic cracks to form before the concrete has hardened. These plastic cracks can start forming, especially in thin slabs, even before the concrete has been floated.

After the concrete has hardened, it is still subject to hot weather influences. Continued evaporation from the surface of the green concrete causes cracking and crazing. The shrinkage resulting from the high water requirement will cause cracking, or cracks may develop because of the volume change of the concrete that was caused by cooling from an initial high temperature.

Other Effects. Other hot weather effects deserve special mention. The effect on durability is particularly significant. Resistance to cycles of freezing and thawing, or wetting and drying, and resistance to attack by aggressive solutions, is appreciably lessened. Part of this is due to the higher water-cement ratio (see Figure 4-26), but other hot-weather effects are significant in this respect as well.

The undesirable effects of hot weather can be manifested in the concrete while it is in the fresh, plastic condition, and in the concrete after it has hardened. Difficulties are likely to develop during transporting, placing and finishing the concrete, because of a loss of workability that results from the rapid evaporation of mixing water, loss of slump and accelerated setting of the cement. These hot-weather problems are magnified whenever there are delays in any stage of processing the concrete from the mixer to final finishing.

19.2. Minimizing the Effects of Hot Weather

Trouble-free concrete construction can be achieved by advance planning and preparation so that the job will be ready for hot weather before the hot weather appears. This preparation starts at the batching plant and extends through all steps in handling the concrete until curing has been completed. Basically, what we want to do is keep the concrete as cool as possible through all of these steps, at the same time minimizing the loss of workability and water.

Materials. Control of the temperature of concrete starts with control of the temperature of the ingredients. There is not much that can be done about the cement temperature, except order the cement mill to ship the coolest cement possible. Steps can be taken, however, to improve the temperature of the water and the aggregates, and this is where the greatest benefit can be accomplished. While it takes a change of almost 9°F in the cement to change the concrete temperature 1°F, the same benefit results from a change of about 2°F in the aggregate or 4°F in the water. It is quite obvious that we should keep the aggregates as cool as possible. The sand, being damp, is usually fairly cool, but the coarse aggregate is frequently dry and hot. A common method of cooling coarse aggregate is to sprinkle water on it in the stockpiles. This should be done systematically and uniformly, taking care to avoid fluctuations in moisture content that can give trouble in control of the slump. Cool air jets in the coarse aggregate bins will aid in cooling the damp aggregate. The sand, containing moisture from the processing plant, should not be sprinkled as it is nearly impossible to maintain a uniform water content.

Keeping the mixing water cool is also effective. This can be accomplished by avoiding the use of tanks and pipelines directly exposed to the sun. Insulation or damp coverings are useful. The use of ice is especially advantageous, as the cooling effect is multiplied many-fold by the use of ice. While it takes 1 Btu of heat to raise the temperature of 1 pound of water 1°F, it requires 144 Btu to melt one pound of ice. In some batching plants, blocks of ice will be kept in the water storage tank to lower the water temperature before the water is measured. Other plants have facilities for weighing chipped or flake ice that is admitted to the mixer as part of the water. (See Figure 19-4.) The ice, of course, must all be melted before the concrete is delivered to the job.

The injection of liquid nitrogen into the mixer is another possible alternative. The super cold liquid nitrogen can be added directly into the drum of the truck mixer to lower concrete temperature. (See Figure 19-5.) The addition of liquid nitrogen does not in itself influence the amount of mix water required except that lowering the concrete temperature can reduce water demand.

Figure 19-4
Substituting ice for part of the mixing water will substantially lower concrete temperature. The crusher delivers finely crushed ice to the truck mixer. (Courtesy of PCA)

Figure 19-5
Liquid nitrogen added directly into the drum of the truck mixer is an effective method of reducing concrete temperature during hot-weather concreting. (Courtesy of PCA)

The temperature of concrete can be estimated from the temperature of the ingredients by the use of the nomograph in Figure 13-9.

Careful use of approved admixtures can be of value. Calcium chloride accelerator should not be used. A retarder can be used to advantage to slow or extend the setting time of the cement, this being especially desirable for long hauls or where difficult placing conditions result in slow receiving of the concrete. A water- reducing admixture will be of value in overcoming the abnormally high water demand caused by high temperatures. In many cases it is possible to combine water-reducing and set retarding properties in one admixture. These admixtures should be used under careful engineering control, in accordance with the manufacturer's instructions.

Mixing and Delivery. Virtually all concrete for buildings is ready mixed, and most of the ready-mixed concrete is mixed in transit to the job. The amount of mixing should be the minimum that can achieve the necessary uniformity and quality of the concrete. Overmixing should be avoided. If the truck is on a long haul, or is delayed, the mixer should be stopped before the required number of revolutions, with the remaining mixing done just before discharge. The longer the concrete is in the mixer, the greater the increase in temperature. Heat from the sun passes through the mixer drum, heat is absorbed from the air and heat is produced from the work of mixing heat, all of which raises the concrete temperature. A mixer drum painted white will be as much as 25°F to 30°F cooler than a dark gray one. Aluminum paint is little better than gray. Some operators have provided additional cooling by moistening the exterior of the mixer drum with water dripping from a small tube. It is not necessary to have water running off the drum; just enough should be used to provide a slight dampening. Cooling then results from evaporation of the water.

Mixers should be scheduled so that there is a minimum wait to discharge. Admittedly, it is not possible to anticipate every delay or breakdown, and there will be times, even on a well-organized job, when an unexpected equipment failure can result in a congestion of mixer trucks. However, delays can be minimized by organizing the job properly and scheduling the trucks in accordance with the placing requirements. If feasible, it should be possible for two or more trucks to discharge their loads simultaneously.

The problem of adding tempering water is intensified during hot weather, and the temptation to add extra water must be resisted. Water can be added to the load to adjust to the required slump, provided the specified water-cement ratio is not exceeded, but more than this amount should never be permitted.

Placing and Finishing. All of the requirements for good concrete construction described elsewhere in this book must be followed. Certain of these requirements are more difficult to achieve during hot weather and require special attention. Drying of the concrete and hydration of the cement are more rapid during hot weather, and these facts must be considered when planning operations.

The job must be planned so that the concrete can be received and placed as rapidly as possible. Pumps, vibrators, conveyors and all equipment should be of adequate capacity. Some kind of a standby arrangement should be established so that, in case of the failure of a major item such as a crane or pump, a replacement or alternate equipment will be available with minimum delay. Extra vibrators should be on the job. Sufficient workers of all necessary classes should be on hand.

Subgrades and foundations should have been well soaked beforehand, but should be damp, without puddles of water, when the concrete is placed. All last-minute inspections and adjustments should have been made. If the work is to extend after dark, ample lighting should be available. Arrangements should be made for special protective measures. During very hot weather, pumplines can be kept cool by covering them with damp burlap. Pumps, conveyors, chutes and other handling equipment should be kept shaded from the sun if at all possible. Sprinkling the work area, forms and reinforcing steel lightly just ahead of concrete placing cools the area. Care is necessary to avoid water collecting in puddles on the subgrade or other horizontal areas where concrete is to be placed.

The work area should be protected with sunshades and wind breaks. Sometimes, especially on slab work, it is possible to delay concreting until walls or roof has been erected. Fog nozzles not only raise humidity and lessen evaporation; they also have a cooling effect. When placed on the windward side, fog nozzles are quite effective in cooling the area. Nozzles should make a fine mist or fog. Ordinary spray nozzles are not suitable. The type used in nurseries and poultry houses has been found satisfactory. (See Figure 19-6.) If compressed air is available, atomizers can be used.

Figure 19-6
The fine mist produced by a fog nozzle lowers the temperature and raises the humidity. (Courtesy of PCA)

Figure 19-7
Fogging cools the air and raises the humidity above the concrete surface. (Courtesy of PCA)

Concrete delivery, placing and finishing must all be coordinated to avoid bottlenecks that result in cold joints, rock pockets and other defects. In placing, the amount of concrete exposed should be a minimum to lessen the chance of cold joints. Layers should be shallow and covered with fresh concrete as quickly as possible.

One effect of hot weather is a shortening of the time period between the several finishing operations. This effect becomes more serious if the concrete is hot. Steps to minimize this effect include use of the water fog, which reduces evaporation from the fresh concrete, thereby lessening the tendency toward plastic cracking. Laying polyethylene sheeting over the concrete between each step in finishing can be helpful. Application of a liquid that forms a monomolecular film over the concrete after the strike-off gives a temporary sealing effect that retards evaporation yet has no detrimental effect on finishing.

Curing. Care of the concrete after it has been placed and finished is all too frequently neglected on many jobs. During hot weather this neglect can be especially damaging to the concrete. We know that curing is defined as the protection of the concrete to prevent loss of moisture and to maintain the concrete at a reasonable temperature. When this curing is applied in the right way at the right time, it will minimize many of the deficiencies that develop in the hardened concrete.

The best curing is water curing, if it is done right. Application of water must be prompt and continuous. Permitting the concrete surface to dry between applications of water can lead to cracking and crazing of the concrete. Leaving the forms in place without the application of water will not cure the concrete.

White pigmented curing compounds should be applied to pavements and similar slabs immediately after the water sheen disappears.

Some agencies do not permit the use of curing compounds on important structural concrete during hot weather. When this is the case, water must be used.

Curing of concrete is covered in detail in Chapter 17. A summary of hot weather precautions is given in Table 19.1.

TABLE 19.1
SUMMARY OF HOT WEATHER PRECAUTIONS

1. Plan ahead. Provide the necessary materials and equipment ahead of time, and schedule the job to eliminate delays.
2. Organize the job efficiently, with adequate labor and equipment to handle the concrete, and proper coordination of all steps to prevent delays.
3. Expedite movement of materials and concrete.
4. Sprinkle coarse aggregate stockpiles.
5. Precool aggregates with cold water or cold air jets in the batcher bins.
6. Use chipped ice in the mixing water. All ice should have melted by the time the concrete leaves the mixer.
7. Use a water-reducing retarder in the concrete, under close supervision and control.
8. Do not use an accelerator in the concrete.
9. Paint truck mixers white or light color.
10. Avoid overmixing. If a truck mixer is delayed, stop the mixer and agitate intermittently.
11. Work at night.
12. Shade or insulate water lines and tanks, paint them white, or cover with damp sand or burlap.
13. Investigate the possibility of false set in the cement, and avoid such cement.
14. For a slab on the ground, have the subgrade previously saturated. Subgrade should be damp but not muddy when concrete is placed. Use fog nozzles to lower temperature and raise humidity.
15. Sprinkle the work area to lower the temperature by evaporation. Protect the concrete and work area with sunshades and windbreaks.
16. Start curing as early as possible, applying the selected curing material as soon as the concrete surface will not be marred by such application.
17. If formal curing is delayed, or if concrete is placed during very hot and dry weather, use a fine fog (not spray) over the concrete immediately after final finishing.

COLD WEATHER

When we consider concreting in cold weather, we are not only concerned with temperatures below freezing but also with daytime average temperatures that do not rise above 40°F. The transition periods in the spring and in the fall are sometimes a period of danger. A few relatively warm days in early spring might deceive us into thinking that the cold weather is over, and we get caught with our defenses down when a cold snap comes along. Or, in the fall, a sudden period of cold weather may catch us unprepared.

In the past, it was customary to curtail concrete work during the winter in those areas subject to temperatures below freezing for considerable periods. In recent years, however, with a better understanding of the effects of cold weather, and with improved methods of protection, cold weather is no longer a serious obstacle to construction progress. (See Figures 19-8 and 19-9.) There is no reason why good concrete cannot be made in the winter, provided the effects of cold weather are understood and proper precautions are taken.

Figure 19-8
Snow falling on insulated blankets. (Courtesy of PCA)

Figure 19-9
Heated enclosure for winter concreting work. (Courtesy of PCA)

19.3. Cold Weather Effects

At temperatures above freezing, the rate of hydration of cement decreases as the temperature is lowered. This, of course, affects the strength development. At temperatures below freezing, the fresh concrete is permanently damaged if it becomes frozen. This damage is readily recognizable at any time. It is discussed in Chapters 3 and 4.

In the absence of special precautions, undesirable cold weather effects may include the following:

1. Permanent damage that is due to early freezing
2. Slower setting and slower strength gain
3. Freezing of fresh concrete prior to hardening
4. Reduced durability (same as strength reduction)
5. Freezing of green concrete at edges and corners Dehydrated surface areas that are due to use of space heaters
6. Cracking that is due to sudden temperature change if strength gain is insufficient[19.2]

Strength. The usual effect of low temperatures is shown in Figure 19-10. Specimens made and cured at lower temperatures do not reach the strength of standard cured specimens during their early ages. The trend appears to reverse after an age of three or four weeks. Concrete protected from freezing for two or three days, then exposed to freezing temperatures, undergoes practically no strength gain at such temperatures. (See Figure 19-11.)

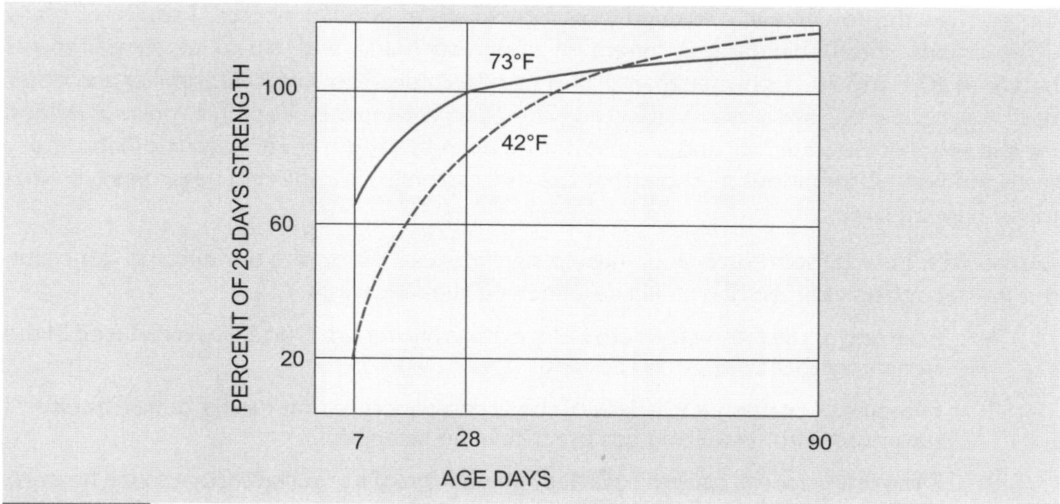

Figure 19-10
Up to an age of three or four weeks, concrete mixed and cured at low temperature is not as strong as concrete made at 73°F. The relationship reverses at three or four weeks and the cold concrete becomes stronger.

The effect of low temperature above freezing varies depending on the method of curing the specimens. As far as job conditions are concerned, we are interested in strength development during the first few days or weeks after the concrete has been placed. When to strip forms, remove shoring or put the structure in service depends on the practical aspects of this early strength development.

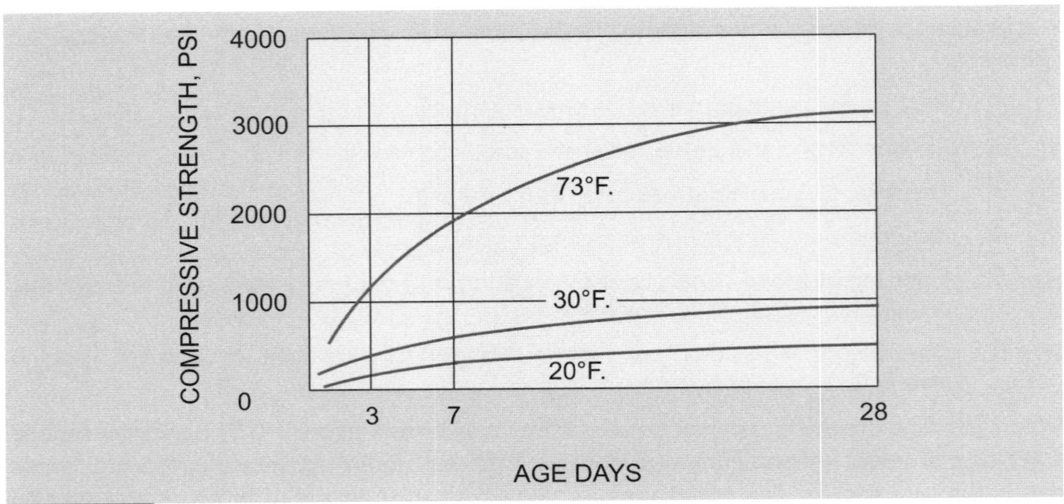

Figure 19-11
Concrete initially protected from freezing for two or three days and then exposed to below-freezing temperatures gains strength slowly.

Other than the lower early strength, cold concrete is good concrete. Results of many observations show that later strength of concrete made and cured at temperatures between 40°F and 70°F is higher than that of standard cured concrete. Durability and other desirable properties are similarly affected. The concrete must be kept in a moist condition for these benefits to accrue, and it must not be permitted to freeze. Results of laboratory tests and field observations all show that the early strength is impaired, regardless of what might happen later.

Durability. Low temperature alone has no significant effect on the durability of concrete, but indirect effects can be harmful. These include the following:

1. Freezing of the fresh concrete before it has hardened. Durability is reduced in the same way that strength is reduced.
2. Freezing of corners and edges of the green concrete that has hardened but is still saturated with water and has practically no strength.
3. Dehydrated areas caused by lack of protection of the surface from space heaters. Cracking is liable to develop in such areas.
4. Cracking and disruption that occur later in the life of the structure as a result of the use of too much calcium chloride, or the use of the chloride in situations where it should not be used. (See Chapter 9.)
5. Cracking as the result of sudden temperature changes (thermal shock) imposed on the concrete before it has sufficient strength.

Other Effects. Adequate protection of the concrete cannot be done without some monetary cost. Heaters, insulation, enclosures and other protective measures cost money. The efficiency of personnel may be reduced. The additional cost must be balanced against the cost of delaying the job to determine whether to proceed with concrete work under the restrictions made necessary by cold weather. Usually, the decision will be to proceed.

When concrete is protected from extreme cold, freezing is prevented; if curing is properly done, good concrete will result.

19.4. Control and Protection

Concrete Production. In climates where freezing rarely occurs but daytime temperatures occasionally get as low as 40°F, it is usual practice to provide some measure of heat to the mixing water. This is usually sufficient. When freezing temperatures are expected at night, then more elaborate means must be provided. Because it is difficult to heat aggregates uniformly or to a predetermined temperature, heating of the aggregates should not be done when heating of the mixing water alone will ensure delivery of concrete of the required temperature. Because the specific heat of water is approximately five times that of aggregate and cement, each pound of water provides five times as much heat as a pound of aggregate or cement. Temperature of the water should not exceed 175°F at the time of mixing with the aggregates.

When the air temperature is below freezing, frozen aggregates should be thawed before entering the mixer. Heating of aggregates is best accomplished by steam or hot water circulating in pipes. Steam jets applied directly to the material should not be used even though they are very effective, because the resulting variable moisture content makes close control difficult. The batching and mixing plant must be heated for proper operation of the equipment and personnel efficiency. Some authorities recommend considering the influence of wind-chill factor on the apparent temperature.

Temperature Limitations. Preparation for cold weather concreting should be made before, not after, the cold weather arrives. Facilities for heating and protection should be on hand, as well as an adequate supply of an accelerating admixture.

Temperature limitations are usually stated in the specifications. Lacking these, the ACI report, "Cold Weather Concreting,"[19.2] contains tables and recommendations for temperatures and protection necessary to assure good concrete. In general, structural concrete for buildings should be placed at a minimum temperature of about 55°F in thin sections, or 50°F in heavier sections such as footings. The temperature of the concrete as mixed should be somewhat higher than the placing temperature because there will be a temperature drop between the batch plant and jobsite. Recommendations are summarized in Table 19.2. Temperature of the concrete as mixed should be maintained at the temperatures shown or not more than 10°F above. Higher temperatures are of no value. In fact, they are detrimental because of higher water demand at the same slump, excessive loss of slump, accelerated set, and possible shrinkage and cracking.

TABLE 19.2
RECOMMENDED CONCRETE TEMPERATURES

AIR TEMPERATURE, °F	SECTION SIZE, MINIMUM DIMENSION, IN.			
	< 12	12-36	36-72	> 72
Minimum concrete temperature as placed, °F				
—	55	50	45	40
Minimum concrete temperature as mixed, °F				
Above 30	60	55	50	45
0 to 30	65	60	55	50
Below 0	70	65	60	55
Maximum temperature drop in first 24 hours after end of protection, °F				
—	50	40	30	20

(Reference ACI 306R-88)

The Concrete Mix. Normal concrete mixtures can be used during the coldest weather without any change in the proportions. Set acceleration is desirable to reduce the length of time the concrete requires protection. Protection of the concrete seeks to prevent freezing damage until the concrete has gained sufficient strength to resist. If the strength development can be speeded up, the period of protection can be shortened.

Type III cement is frequently used for this purpose and can be used in any structural application. Occasionally, use is made of additional regular cement (100 to 200 pounds) to accomplish the same purpose. Set accelerating admixtures are available, the one normally used being calcium chloride. The amount of chloride to use should not exceed 2 percent by weight of the cement; usually $1\frac{1}{2}$ percent is sufficient. Calcium chloride must not be used in prestressed concrete, in concrete for floor or roof decks placed on galvanized sheet steel permanent forms, when aluminum is embedded in the concrete if the aluminum is connected either internally or externally with embedded steel, or in sulfate-resistant concrete. Calcium chloride should be used with caution in reinforced concrete because of potential reinforcement corrosion. Nonchloride accelerators can be used where calcium chloride causes detrimental effects.

Air entrainment is desirable, especially if the concrete will subsequently be exposed to freezing and thawing. A water-reducing admixture may aid in early strength development, especially if it also contains an accelerator.

If calcium chloride is used with an air-entraining agent or other admixture, the admixtures must be admitted to the mixer separately.

All proposed admixtures should be carefully evaluated before they are used, as discussed in Chapter 9. One important fact to remember is that there is no admixture that can lower the freezing temperature of the concrete, at least as far as materials presently known are concerned.

Handling and Placing. Close liaison is necessary between the batch plant and the job to assure handling of the concrete with a minimum of delay. Mixer trucks must be unloaded promptly and the concrete moved into the forms with all possible speed.

All snow, frost and ice must be removed from the forms, reinforcing steel, embedded materials and similar items before concrete placing is begun. Concrete should never be placed on a frozen subgrade, as to do so can result in loss of support when the ground thaws. If concrete is placed against frozen ground without forms, or on a frozen subgrade, the ground may take sufficient heat out of the concrete to cause it to freeze—or the concrete may harden very slowly, even though the air temperature is relatively warm.

Common practice is to enclose the entire structure in tarpaulins, polyethylene, plywood, insulating board, paper or other material that will be tight and strong enough to withstand wind and snow loading. Buildings and other similar types of structures can be enclosed in this manner. (See Figure 19-12.) The enclosure should be at least a foot away from the fresh concrete to provide for air circulation. Heat should be provided before, during and after placing; it is best supplied by releasing moist steam within the enclosure. If dry heat is used, the concrete must be protected from drying by means of curing compounds, blankets or wet burlap.

Figure 19-12
Effective concreting in winter using polyethylene enclosure.

Because of the danger of fire, the use of salamanders and open fires should be discouraged. Hot air blowers, of the type commonly called airplane blowers, located outside the enclosure, can be used to blow hot air into the enclosure. Portable steam generators provide not only heat but also moisture. Pipe coils and radiators are sometimes used. Salamanders, if permitted, must be used with great care because of the fire hazard. Salamanders must be vented to the outside to remove the danger of carbonation of the concrete surface. (See Figure 19-13.) (Carbonation causes a soft, weak concrete surface.) With any of these heaters, protection must be provided to avoid concentrations of heat on concrete surfaces. The exhaust from blowers should not impinge directly on the concrete. Heaters should be elevated or insulated from the slab on which they are standing. See "Soft Surface" in Chapter 6.

Figure 19-13
An indirect-fired heater with vent pipe carrying combustion gases outside the enclosure.

After placement, the concrete should be maintained at a minimum temperature of 55°F, 50°F or 45°F for at least three days for thin, moderate or heavy sections; or two days if an accelerator or Type III cement is used. After this initial protection, it is good practice to maintain a temperature of 40°F for four more days. Discontinuance of protection from freezing must be gradual so that the drop in temperature of any portion of the concrete will not exceed 40°F in 24 hours.

When placing a slab or floor, the tarpaulin or plastic cover should be supported on horses or a framework in such a way that the covering follows closely behind the placing. If it is necessary to open the covering for finishing, only small areas of the slab should be exposed at a time, and then for very short periods.

Insulation. Insulating batts consisting of polyurethane, rock wool, glass wool, fiberglass or balsam wool, covered on each side with plastic or heavy asphalt-impregnated paper, are attached to the outside of the forms, and the normal heat of hydration of the cement will, in many cases, be sufficient to keep the concrete adequately warm, even in zero weather. (See Figure 19-14.) Straw or batts are also used on horizontal surfaces.

Figure 19-14
Concrete footing being covered to retain the heat generated by the hydration of the cement. (Courtesy of PCA)

The insulating material should be attached to the forms with wood cleats or similar means, and should be tight against the forms to prevent circulation of air under the insulation. It should be in place before concrete placing is started and should overlap previously placed concrete by at least 1 foot. Tears and holes in the covering should be patched with waterproof tape. The insulation must be kept dry. Special care is necessary to protect corners and edges of the concrete and thin members.

The ACI report[19.2] contains detailed instructions for insulation of forms and other winter concreting practices. Even in the coldest weather the insulation must be used as directed, as there is actually danger of permitting the concrete to get too hot.

Curing. Maintenance of sufficient moisture for hydration of the cement is a problem because of the use of heat to keep the concrete warm. During cold weather it is usually good practice to leave the forms in place at least until the end of the required curing period. A curing compound can be used on slabs. Water curing is not desirable, because of the problems with ice wherever the water runs out of the enclosure and because of the saturation of the concrete that results from the use of water.

The new concrete should be permitted to dry somewhat before it is exposed to a temperature below 32°F, to minimize the danger of freezing damage.

Measures that can be taken to assure good concrete during cold weather are given in Table 19.3.

TABLE 19.3
SUMMARY OF COLD WEATHER PRECAUTIONS

1. Plan ahead. Provide heaters, enclosures and protective materials before they are needed.
2. Arrange for extra cement, Type III cement or accelerating admixture.
3. Handle concrete with a minimum of delay.
4. Heat the water and aggregates.
5. Use admixtures carefully, as directed.
6. Use air-entrained concrete.
7. Remove snow, ice and frost from subgrade and all surfaces to be in contact with the concrete.
8. Provide insulation or heated enclosures as necessary.
9. Heat enclosures with steam or blowers rather than stoves or salamanders.
10. Protect concrete from drying out until the expiration of curing period.
11. If salamanders are used, vent them outside the enclosure and raise them above the surface of the concrete.
12. Leave forms in place as long as possible.
13. Maintain concrete temperature at 55°F, 50°F or 45°F for thin, moderate or heavy sections for three days; two days for Type III cement or accelerator.
14. Maintain concrete temperature at 40°F for four additional days, permitting the concrete to dry somewhat before exposing it to freezing temperature.
15. Avoid overheating the concrete.

Precast and Prestressed Concrete

Precast concrete is defined in ACI 318 Section 2.2 as "a structural concrete element cast in other than its final position in the structure." This includes those elements (or products/components/members) that are made at a central casting yard some distance from the building site, as illustrated in Figure 20-1, as well as components that are made at the site and subsequently erected into place. A majority of precast concrete products are also prestressed. In this chapter, emphasis is placed on precast, prestressed concrete, with Section 20.6 addressing the special design and construction details of prestressing. The pretensioning method of prestressing, in which the tendons are tensioned before the concrete is placed, is the most common method for precast, prestressed concrete. The post-tensioning method of prestressing, in which the tendons are tensioned after the concrete has hardened and used mostly for cast-in-place concrete, is addressed only briefly in Section 20.10 as it applies to post-tensioned slabs used in buildings. For a more in-depth discussion of detailing and construction procedures for post-tensioned concrete, including inspection guidelines, see References 20.5 and 20.6. Copies of References 20.1 to 20.4 should also be available to anyone employed as a technician or inspector concerned with precast and prestressed concrete construction.

Figure 20-1
Erection of a precast and prestressed concrete building. Note the tagline to control the double-tee unit during erection.(Courtesy of PCI)

Modern high-rise building construction has led to the growth of the precast concrete industry because of the intricacies and economics of the construction industry. Curtain walls, architectural requirements, the need for large clear unobstructed spaces within buildings, and quality control requirements have all enhanced the desirable qualities of precast/prestressed concrete. Small buildings as well benefit from this expansion.

Prefabrication makes good use of concrete because concrete is adaptable to complex forming. Mass production and assembly line techniques lead to economies and quality control measures that are difficult or impossible to achieve on the jobsite. Precast components can be erected rapidly, resulting in further economy.

The use of prestressed concrete for precast construction is also advantageous. To begin with, prestressing increases the span length of members. Prestressed concrete requires less reinforcing steel and concrete for members of equal strength to that of cast-in-place reinforced concrete. Precast/prestressed members are also thinner, have lower depth to span ratios and weigh less—all of which gives more headroom inside the building, a reduction in weight of the building, savings in size of columns and foundations, and less height for the same number of stories.

20.1. Types of Precast Products

There are several ways in which precast products can be classified. They may be made of conventional reinforced concrete, or they may be prestressed; they may be load-bearing or nonload-bearing [a load-bearing member being one that carries part of the dead and live loads of the structure, whereas a nonload-bearing member is one that is attached to and is supported by the building frame (see Figure 20-2) but carries no part of the load]. The following is a general description of some of the available precast components.

Figure 20-2
Nonload-bearing architectural precast cladding (double tee units) being erected on the building frame. Note double tees properly stored on sound dunnage. (Courtesy of PCI)

Structural components consist of a number of girders, beams and joists, stemmed units, and cored or box units. Architectural components consist of various wall and window units, mullions, sills, ornamental bas-relief panels and other specialized components. Miscellaneous items include piles of various cross sections, utility poles, lighting standards and various types of ornamental units.

Framing members are structural load-bearing members such as beams and columns (see Figure 20-3). Beams and girders frequently span 100 feet or more, while hollow-core slabs typically span between 20 and 40 feet. A major advantage of precast concrete is that the product can be molded into almost every shape conceivable. In the early years of the precast industry, single tees, multiple tees, F-slabs, Y, gull-wing and channel sections were used on specific projects. Today, however, these types of sections are used infrequently, mainly because double tees, hollow-core slabs, wall panels, inverted T-beams, L-beams, I-beams, columns and architectural precast shapes can be used very efficiently. Single tees are still used very efficiently. Single tees are still used in storage tanks and specialized structures spanning 100 feet or more. In recent years, sandwich wall panels have become very popular, especially in cold climates for all types of structures. Also, raker beams and seating units are used for stadiums. In the last decade, precast concrete has become increasingly popular in residential buildings and all types of housing. The vast majority of parking structures today use precast prestressed concrete. (See Figure 20-4.) Architectural precast concrete is a common component for cladding low and high-rise buildings. (See Figure 20-7.)

Figure 20-3
Precast beam and column framing component being erected. Note projecting dowels and postgrout tubes for field connections. (Courtesy of PCI)

Figure 20-4
All precast prestressed concrete parking structure. Note the double tee floor units. (Courtesy of PCI)

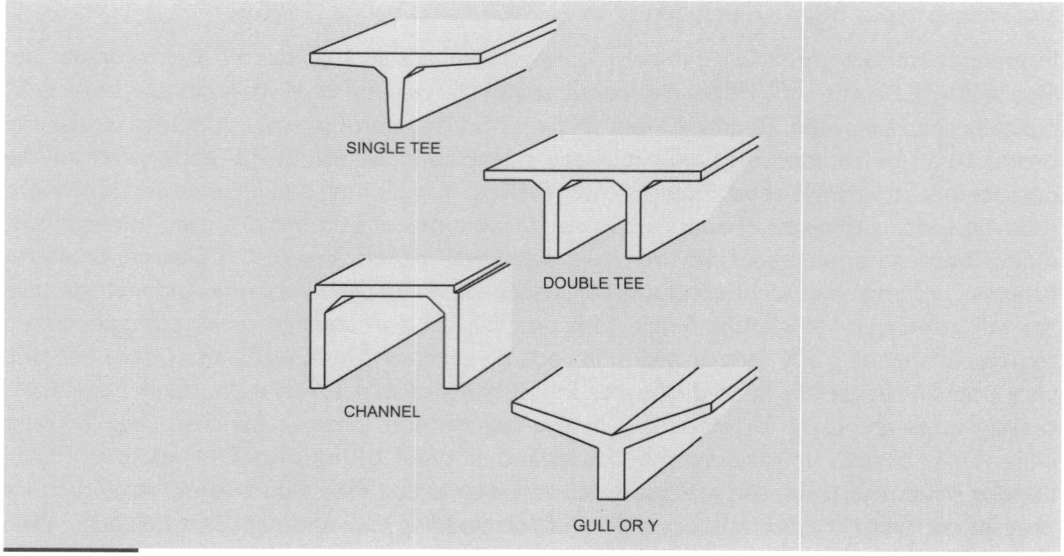

Figure 20-5
Common precast and prestressed concrete products. (Courtesy of PCI)

Figure 20-6
Hollow-core slabs awaiting shipment in storage yard. (Courtesy of PCI)

Architectural precast wall panels include nonload-bearing curtain walls. Such components are normally finished on both sides and usually consist of concrete facings enclosing a layer of insulation, like a sandwich. (See Figure 20-7.) Conduits and other utilities can be accommodated. Window walls comprise a complete wall enclosure ready for glazing. Consisting of mullions, spandrels and related nonstructural elements, window wall units sometimes extend over as many as three stories. Rabbets, formed in the concrete, receive metal window frames or glass set in appropriate gaskets. Facing units are used for modernizing old structurally sound buildings. Attachment to the building is accomplished by means of anchors cast in the concrete of the unit. Grilles or screen walls are highly ornamented units pierced with openings to permit movement of air. They are used for such nonstructural purposes such as solar screens, space dividers and enclosures to hide some aesthetically unpleasant areas.

Figure 20-7
Tilt-frame trailer loaded with nonload-bearing architectural precast window façade units. (Courtesy of PCI)

Precast concrete is sometimes used as a form for cast-in-place concrete, usually where a special architectural effect more easily obtained under plant conditions is desired.

Included in the general category of precast concrete are scores of small off-the-shelf items such as garden furniture, statuary, stepping stones, parking bumpers, and many kinds and sizes of boxes. Pipe is available in a variety of sizes, from 4-inch drain tile to 144-inch water supply and sewer pipe. Many pipes are prestressed. Square, hexagonal and round, solid or hollow piles are usually prestressed. Prestressed concrete utility poles are also available.

These brief descriptions provide a general idea of what the various precast products are. Note, however, that any particular component can often fit into more than one category, functional descriptions overlapping type descriptions. Components may be made of normal-weight concrete or of lightweight concrete; many are prestressed. The description used in the job specifications should govern.

20.2. Shop Drawings

The objective of shop drawings is to clearly show all details and give complete information explaining how the components are made and how they are installed in the building. Prepared by the manufacturer of the precast concrete, they are reviewed by the general contractor, who checks to ensure that the essential information is included and that proper allowance has been made for normal construction and erection tolerances. The drawings should be reviewed to make sure that consideration is given to the requirements of other trades such as air conditioning. The drawings are then passed on to the structural engineer and architect for their review.

The drawings should show size and shape of all components, giving all dimensions and tolerances; they should indicate complete details of the reinforcement, including size, location, grade of steel, type and amount. If special reinforcement, in addition to the specified reinforcement, is required to withstand handling and erection stresses, it should be indicated. Other items include details and location of connections for joining the components to the structure and to each other, covering methods of adjusting the connections, and description of joints and materials to be used in the joints. (See Figure 20-8.) Reference should be made to special erection devices, inserts, reglets and attachments, with necessary details of the items and method of anchoring them in the concrete.

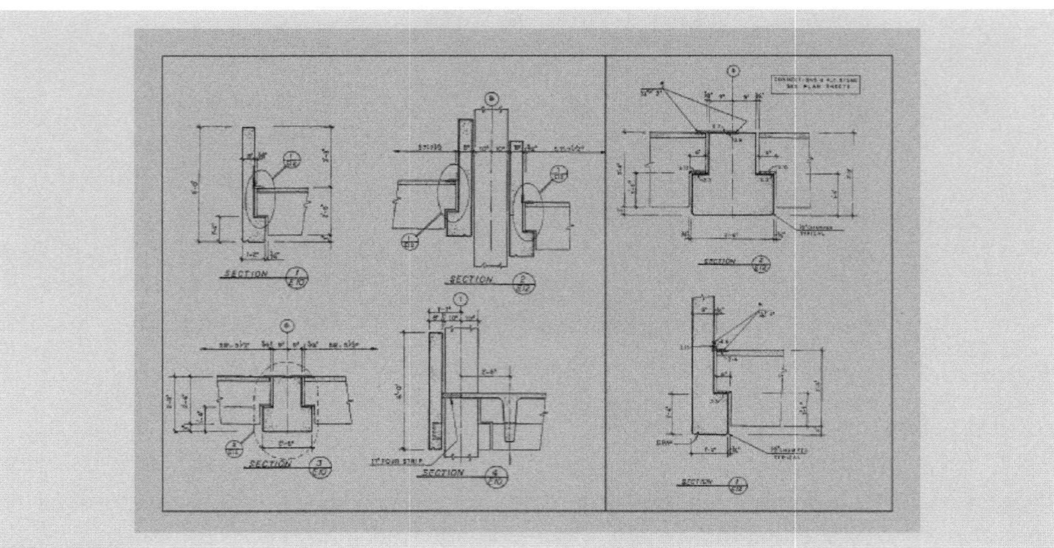

Figure 20-8
Typical shop drawing showing connection details. (Courtesy of PCI)

If components are to be prestressed, the method and materials should be shown, including the allowance for elastic shortening.

Finishes should be indicated and measures outlined for the protection of the components during handling, transportation and erection to avoid damage to the members. If auxiliary supports are required during handling and storage, they should be indicated.

Erection and placing plans should be included, showing the method and sequence of erection, method of plumbing and adjusting the components on the structure. Provision for temporary bracing during erection should be included. Caulking details should be outlined. Identifying marks for the members should be shown on the erection plan.

On some jobs, the erection plans are prepared by the general contractor. Requirements for shop drawings and erection plans are spelled out in the job specifications, and a clear understanding of all responsibilities must be reached before construction commences.

20.3. Forms

The formwork for precast concrete may consist only of a simple pallet or plate on which the concrete is extruded (see Figure 20-9), or long line continuous forms on a long casting bed (see Figure 20-10), or individual forms for single components. The great majority of forms are made of steel. Wood is occasionally used, as well as fiberglass-reinforced plastic such as polyurethane, concrete and plaster. The selection of forming material depends on the configuration of the members, the quantity of identical components to be made, cost and availability of material, and the preference of the precaster.

Figure 20-9
Casting beds for hollow-core slabs. Three extruding beds are shown. Completed slabs are being sawed to length. (Courtesy of PCI)

Wood is easily worked and can be fabricated into intricate details. Wood is not as durable as steel, especially under steam curing conditions. If used, wood should be treated with a sealant for protection of the wood, with special care being given to edges. Even though treated with a sealant, a wood form will perform best if, after each usage, it is cleaned and coated with a parting compound before being used again.

Figure 20-10
A long prestressing bed for double tee units. The prestressing strands are being installed. Note the hold down devices to properly align the strands in the stem of the tee. (Courtesy of PCI)

Fiberglass is an excellent forming material. Strong, versatile and durable, it can be used in elements with considerable intricate detail—detail that would be difficult or impossible to work in metal—and in forms for casting a large number of identical units. Using a single form two hundred times is not unusual.

Fiberglass itself is hard and strong but needs adequate backing to give it rigidity. Concrete cast against new fiberglass may be quite smooth and glossy. Best service will be obtained if a parting compound is used.

By casting against a positive mockup of wood, plaster or concrete, a concrete mold or form is produced. Allowance must be made for shrinkage. After the concrete has cured and dried, it should be given a surface sealing treatment and waxed.

Forms are made of steel when multiple reuses of the form are expected. Steel forms are superior to other materials if they are to be exposed to high-temperature curing. Forms should be constructed so that they will remain rigid after being used many times for containing low-slump concrete subjected to both internal and external vibration. Sheet steel must be sufficiently thick so that it will not deflect or bulge in use. Joints between sheets and rivet heads on the contact areas must be ground smooth and tight so there will be no unsightly offsets or loss of mortar to blemish the appearance of the concrete. Joints at bulkheads and soffit plates on long prestressing beds must be tight and gasketed if necessary. Chamfer strips should be used to form chamfers on the edges of the concrete. Usually, these chamfer strips can serve as gaskets also.

The use of internal form ties should be avoided if possible. Most forms are externally supported, especially those on long pretensioning beds, and internal ties are not needed. If internal ties are used, they should be of the kind that break away or unscrew under the surface of the concrete, leaving no metal on the surface. Forms on casting beds normally include pipes or other provisions for applying heat during the curing cycle. (See Figure 20-11.) Sheet steel contact areas should be checked periodically to detect irregularities that develop in the surface, such as dimpling and buckling.

Figure 20-11
High-temperature steam curing of double-tee units in the casting bed. Note the piping system for applying heat during the curing cycle and the tight covering over the casting bed to retain the heat and moisture. (Courtesy of PCI)

Parting compounds and form oils are available that are especially formulated for precast work, including high-temperature curing. Care is necessary when applying form oils to avoid spilling it on reinforcement, stressing strand or other items to be bonded to the concrete. In long prestressing beds it is sometimes the practice to oil the forms before the steel is placed. When this is done, it is necessary to cover the form, especially the soffit plate, with paper or plastic so as to prevent the oil from getting on the steel. If the oil is applied after the reinforcement and strand have been placed, precautions are necessary to keep the steel clean. If the steel does become contaminated with oil, the oil should be removed with rags soaked in solvent.

20.4. Fabrication

Mixes for precast concrete are usually rich mixes containing as much as 750 pounds of cement per cubic yard. Type III cement is commonly used for high early strength. Natural and artificial lightweight aggregates can be used if weight reduction is important; otherwise, good quality normal aggregates are used. Admixtures should be taken advantage of to provide their special benefits. Water reducers are advantageous in these high-strength concretes, and retarders may occasionally be desirable. Air entrainment is desirable in those units to be exposed to severe conditions, even though there may be a loss of strength. Calcium chloride should never be used in prestressed concrete because of corrosion of the prestressing steel. High-range water reducers (superplasticizers) will be found useful. Because of the relatively short time period between mixing of the concrete and its placement in the form, there is usually no problem with rapid loss of slump. Today, with advancing technology, self-consolidating concrete (SCC) is being used in the industry. (See Figure 20-12.)

Maximum size of aggregate is 1 inch or less, usually $^{3}/_{4}$ inch. Sand content should be the minimum that will provide the necessary workability. Very low-slump concrete is normally used in these plants. Slump rarely exceeds 2 inches and is usually 1 inch or less, with many plants using a "no slump" concrete. Because of this, in many plants pugmill or horizontal pan

Figure 20-12
Self-consolidating concrete (SCC) being used in this precast wall panel. (Courtesy of PCI)

mixers are used instead of the usual rotary mixers. Strength of concrete for transfer of stress (detensioning) will be stated in the specifications. Concrete transfer strength is usually in the range of 3000 to 4000 psi. Through the use of Type III cement, water-reducing admixture, good materials and high-temperature curing, compressive strengths of this magnitude can be reached in less than 15 hours. High early strength is desirable for nonprestressed units as well, so that they can be removed from the forms and put in storage, freeing the forms for the next round of casting. Cylinders cast in metal molds, cured alongside the components, are used for determining these early strengths. Standard cured specimens are used for 28-day strengths, which may be specified as high as 8000 psi. The precast industry almost exclusively uses the 4-inch by 8-inch cylinder for evaluation and acceptance of concrete. (See discussion in Section 13.5.)

Batching of materials is always by weight, except that water and liquid admixtures can be batched by volume. In rare instances concrete might come from a ready-mix plant, but normal practice is for the casting yard to have its own batching and mixing plant. Frequently, both prestressed and nonprestressed members are made at the same facility. Reinforcement can be fabricated and assembled into cages in a part of the yard set aside for this purpose where jigs can be set up on convenient benches. Inserts and other hardware can usually be assembled with the reinforcement in the cages. Most cages can be moved and placed by hand; only the large and heavy ones require a crane.

In-yard transportation of concrete is done by many different methods. A combination of hauling buckets by truck and lifting by crane is common. Forklifts and skip loaders have been used, as well as conveyors and mixer trucks.

Good construction practices must be followed in placing and finishing the concrete. In deep forms such as girders, concrete should be placed in layers not over 16 inches thick, then vibrated, using immersion vibrators, usually assisted by form vibrators.

Operating on long, metal-lined beds, extruding machines can make slabs of various sizes containing continuous cores. Slabs are usually pretensioned. No forms are used; the zero-slump concrete passes into and through the extruder as it moves along the form at a speed of about 3 to 4 feet per minute. After the concrete has been cured, the slab is sawed into the necessary lengths. (See Figures 20-13 and 20-14.)

Figure 20-13
Extruder for placing concrete for cored slabs. Tendons are in place, and the machine is ready to start.

Figure 20-14
After the core slab concrete has been cured, it is cut into appropriate lengths with a diamond saw. (Courtesy of PCI)

In any of these precasting operations, the supervisor and inspector should make a final inspection before concrete is placed in the forms. They should determine that the size, grade, type, quantity and location are correct for reinforcing steel, stressing tendons, inserts and other hardware, and that these items are adequately secured; that the forms are properly aligned, clean and braced; that parting compound is on the form but not on the reinforcement or tendons; and that void formers are accurately positioned and well tied down to prevent them from floating when the concrete is under vibration.

Test cylinders should be made during concreting for field tests and standard 28-day tests. Specimens for field tests should be made in metal molds (heavy steel or cast iron preferred), and those for standard curing can be made in cardboard molds.

20.5. Curing

High-temperature curing is almost universally used in precasting plants to enable a rapid turnover of forms and casting equipment. The minimum curing temperature is 50°F. The maximum curing temperature is limited to 150°F. For this reason, curing compounds are rarely used. Curing is discussed in Chapter 17.

20.6. Prestressed Concrete

In prestressed concrete, tendons consisting of wires, strands or bars are loaded in tension by the application of a pull at the ends of the tendons. Under the tensile load, the tendon elongates or stretches. While the tendon is elongated, the tensile force in the tendon is transferred to the concrete as a compressive force, thereby greatly increasing the load-carrying capacity of the concrete and decreasing the tendency to deflect or sag under load. Transfer of stress is accomplished by one of two methods:

Pretensioning. In this method of prestressing, the tendons are pretensioned; that is, they are tensioned (elongated) prior to placement of the concrete. After the concrete has reached a predetermined strength, the tendons are cut and the prestress force is transferred to the concrete through the bond between the steel and concrete. The prestressing tendons are usually seven-wire strands. The pretensioning method is the most common for precast-prestressed concrete. (See Figure 20-15.)

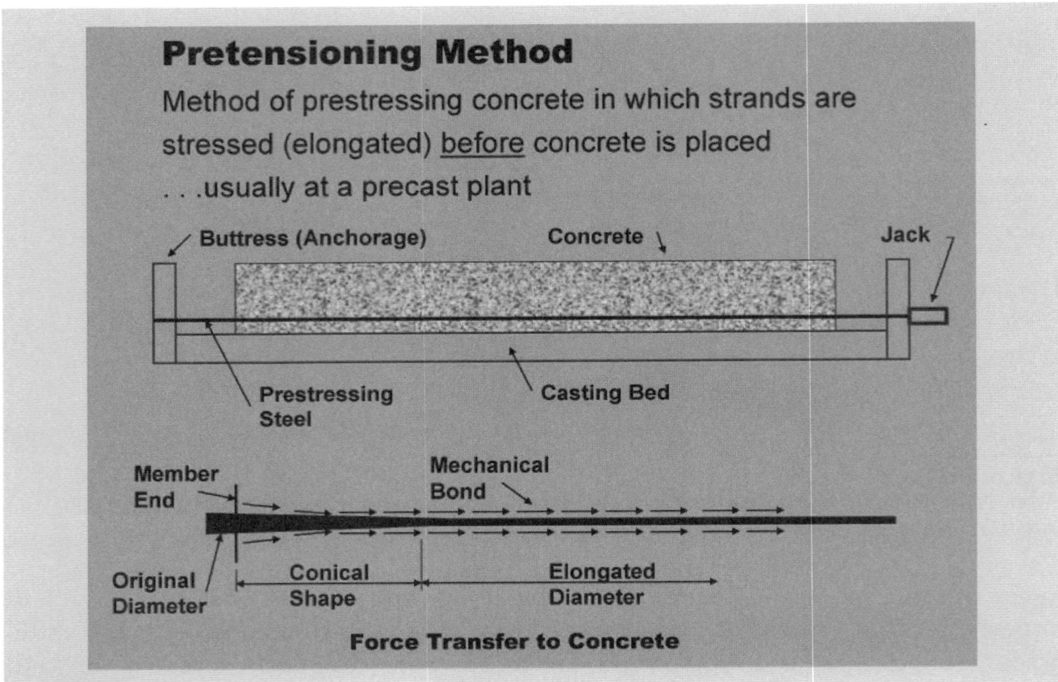

Figure 20-15
Schematic of pretensioning method and force transfer to the concrete.

Post-Tensioning. In this method of prestressing, the tendons are tensioned after the concrete has hardened. Hollow conduit or sleeves containing the unstressed tendons are placed in the forms to the desired profile prior to placing the concrete. The hollow conduit (or ducts) may contain the tendons, or the tendons can be inserted after the concrete is cast. In thin slabs, the tendons are encased in plastic sheathing. The sheathing prevents the concrete from bonding to the steel. The concrete is placed in the usual manner. After the concrete has achieved sufficient strength, the tendons are tensioned by jacking against the concrete member itself and anchored by special fittings at the ends of the member. The prestressing tendons are usually seven-wire strands. Solid steel bars are also used for shorter length applications. (See Figure 20-16.)

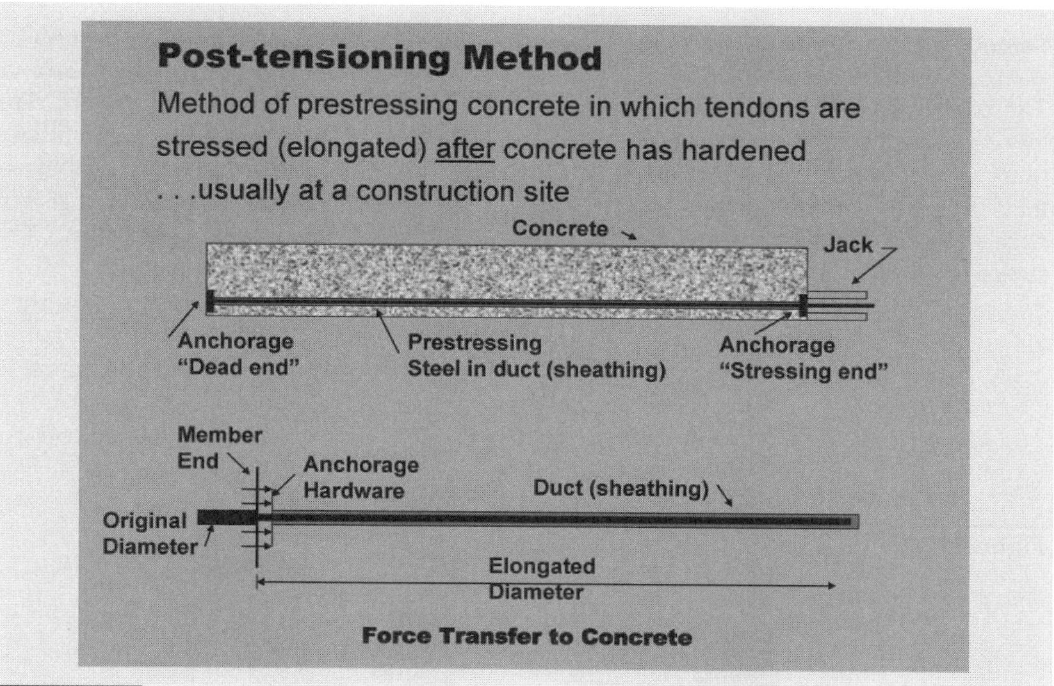

Figure 20-16
Schematic of post-tensioning method and force transfer to the concrete.

Tendons inserted in hollow conduits are normally grouted in their conduits after they are tensioned (bonded prestressing tendons). Tendons encased in plastic sheathing remain unbonded, as bonding of such a tendon by grouting is impossible. Numerous patented post-tensioning systems are available.

Modulus of Elasticity of Concrete. The modulus of elasticity of concrete, Ec, is the ratio of normal stress to corresponding strain in tension or compression. It is the material property that determines the deformability of a concrete member under load. Thus, it is used to calculate deflections, axial shortening and elongation, buckling, and relative distribution of applied forces in composite and nonhomogeneous structural members.

If Ec is not known with certainty, the value obtained from the equation in ACI 318, Section 8.5.1 can be used:

$$E_c = w^{1.5} 33 / f_c'$$

where

E_c = modulus of elasticity of concrete, psi

w = unit weight of concrete, lb per cu ft

f_c' = specified compressive strength of concrete, psi

Types of Prestressing Steel. Steel for prestressed concrete must conform to ASTM A 416, A 421 or A 722. Three types of prestressing steel are covered: seven-wire steel strand (A 416), steel wire (A 421) and high-strength steel bars (A 722). The seven-wire strand consisting of six wires wrapped helically around a smaller center wire (see Figure 20-17) is currently the most widely-used prestressing steel for both pretensioned and post-tensioned concrete in building construction. (See Figure 20-18.) High-strength bars are commonly used for prestressed rock and soil anchor and are sometimes used for precast concrete connections. Prestressing bars are also gaining use in prestressed masonry wall construction.

Sometimes referred to as stranded cable, the seven-wire strand is manufactured to ASTM A 416 specifications. The A 416 specification covers two types and two grades of strand. The two types of strand are low-relaxation and stress-relieved (normal-relaxation). The low-relaxation strand, sometimes known as stabilized wire, is subjected to a continuous thermal-mechanical treatment after stranding to produce improved mechanical properties for design over its counterpart (stress-relieved). The low-relaxation strand is used almost exclusively in today's market because of its improved mechanical properties. The stress-relieved strand will not normally be furnished unless specifically ordered.

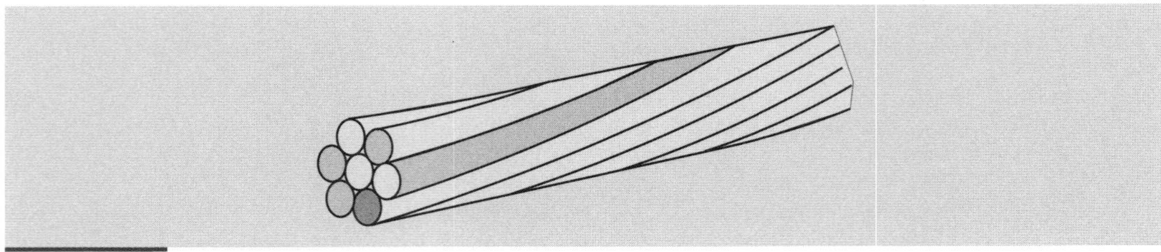

Figure 20-17
Prestressing strand consists of a seven-wire tendon, with six wires wrapped helically around a straight center wire.

Figure 20-18
Reel pack of $^1/_2$-inch seven-wire strand (12,000 ft per reel pack). Note the light rust spots on the strand. This light rusting is not cause for rejection (see ACI 318 Section 7.4.3).

For the most commonly used seven-wire strand, two grades are available: Grade 250 (f_{pu} = 250 ksi) and the more widely used Grade 270 (f_{pu} = 270 ksi). Note that the Grade designations correspond to minimum specified tensile strength (breaking strength) in ksi (kips per square inch) units . . . where one kip = 1000 lb. Thus, 270 ksi = 270,000 psi (lbs per square inch). For the high-strength steel bars, both Grade 145 and 160 are available. Elasticity and other properties of prestressing steel are closely controlled. Of special importance is the elastic modulus, E.

Stress-Strain or Load-Elongation. How steel stretches when a tensile load is applied is discussed in Section 18.2. The ratio of stress to strain, or load to elongation, is the elastic modulus, called E. That is,

$$E = \frac{\text{stress}}{\text{strain}} \text{ or } \frac{\text{load}}{\text{elongation}}$$

Note: Whenever stress-strain data are required or specified, it is to be understood that this implies either stress-strain or load-elongation data. For prestressing steel, the elastic modulus, E_p, averages about 28,500,000 psi. The elastic modulus of prestressing steel must be known within close limits for the purpose of computing the tendon elongation that results from tensioning. Accordingly, ACI 318 Section 8.5.3 states that the "modulus of elasticity E_p for prestressing steel shall be determined by tests or supplied by the manufacturer." Average or typical values of E_p are not acceptable for calculation of tendon elongation. Control of tensioning operations requires accurate data on the physical properties of the actual tendons being tensioned. Each shipment of prestressing tendons should be accompanied by a certified test report for each size of strand, wire or bar, showing results of all tests, including stress-strain or load-elongation data for determining the elastic modulus E_p. As a minimum, the number and frequency of test reports should be:

Strand: one certified test report (one test specimen) should be furnished for each 20-ton production (six reels or coils) of each size of finished strand.

Wire: one certified test report (one test specimen) should be furnished for each 10 coils or less of wire of the same size contained in an individual shipping order.

Bars: Two certified test reports (two test specimens) should be furnished for each bar size rolled from each heat of steel. Whenever one bar size rolled from any one heat exceeds 100 tons, three specimens must be tested.

The above sampling requirements conform to the applicable ASTM specification.

When a sufficient number of stress-strain or load-elongation relationships have been established for a particular type and manufacturer of tendons, it is usually practical to select an average modulus of elasticity that will not be subject to more than 2.5 percent [20.1] variation. Once established, the average modulus of elasticity can be used as long as the resulting computed elongations are within the tolerance limits for measurement of prestressing force.

Protection of Prestressing Steel. Many of the comments relative to regular reinforcing steel also apply to prestressing steel. For example, similar to reinforcing bars, a small amount of rust on the surface of prestressing steel is beneficial to bond. Accordingly, ACI 318 Section 7.4.3 permits a light coating of rust on prestressing steel. However, excessive rust and severe corrosion should not be permitted. Avoidance of corrosion requires somewhat more care for prestressing steel than for reinforcing steel, as higher-strength prestressing steel is more susceptible to corrosion. Severe corrosion may occur if the steel is exposed to galvanic action while in storage.

In pretensioned concrete, the prestress forces are maintained exclusively by bond between the steel and the hardened concrete, hence the importance of maintaining the steel free of deleterious coatings and contamination. Coils and reels of wire or strand should not be stored in the open air but should be kept under cover until time for use.

Figure 20-19
Casting bed for I-girder ready for placement of concrete. Tendons have been tensioned; stirrups and other steel are in place. (Courtesy of PCI)

Casting Beds. Many pretensioned concrete members are made in long casting beds, which may be as long as 650 feet. Long casting beds are practical for producing many units of identical cross-section and strand pattern. Prestressing tendons can be put in place for the entire length of the bed, and spaces for the individual components can be marked off by setting bulkheads at the proper locations along the bed. (See Figure 20-19.) A bed usually consists of a concrete paving slab that serves as a base on which the soffit plates (bottom form) are situated. Soffit plates are frequently supported on steel pilot liners and cross beams to provide access to the underside of the plates for insertion and removal of hardware required in the precast components being made. (See Figure 20-20.)

Figure 20-20
The soffit plate on this bed is elevated on cross beams. The heavy abutment structure is in the foreground. The completed girder is ready for transport to the storage yard. (Courtesy of PCI)

The abutment at each end of the bed must be designed to resist the moment induced when the tendons are elongated. The jacking force imposed on the end abutment for a large girder section can exceed a million pounds. Abutment structures, made of concrete, structural steel or a combination of these materials, are constructed so that the jacking forces can be transferred to the earth or to the concrete of the casting bed. On some beds, especially those for the components of lighter cross section, the bed can be designed so that the forms absorb the reaction to the jacking forces. (See Figure 20-21.)

Figure 20-21
The left photo shows a light abutment for a core-slab extruder bed. The right photo shows a very light bed for thin prestressed planks.

Bulkheads, shaped to fit the cross section of the precast component and define the ends of the components, are usually set with a space of a foot or two between to facilitate subsequent operations. The tendons, of course, must pass through the bulkheads, a situation that is sometimes solved by threading all of the bulkheads for a line of components onto the tendons in a bundle, then distributing them as required along the bed. Segmental bulkheads that can be fitted around the tendons are available. Once located in the bed, the bulkheads are secured to the tendons to prevent movement while concrete is being placed. Bulkheads can be made of steel or plywood, depending on the anticipated number of reuses.

Precasting Operations (Pretensioning). Central casting yard fabrication lends itself particularly well to the pretensioning method of prestressing. Seven-wire strand is universally employed, although occasional use is made of three-wire and four-wire strands, especially in thin or small members. The strands are fixed in special anchors in grillages at each end of the casting bed and are elongated by means of a hydraulic jack equipped with a pressure gauge. (See Figures 20-22, 20-23 and 20-24.)

Figure 20-22
Strand chucks are shown attached to the strands at the stressing end of the abutment. (Courtesy of PCI)

Figure 20-23
Single-strand stressing pump unit and ram with 60-inch stroke. Long rams are required when tensioning long beds to avoid the need to grip strands more than once. (Courtesy of PCI)

Figure 20-24
Disassembled multiple use chuck. Cap and spring keep jaws inline. Strand chucks with cap and spring should always be used on the stressing end of the bed. (Courtesy of PCI)

Strands are either straight or depressed. Straight tendons extend straight through the member from end to end. In a casting bed consisting of several components in line, the strands are straight from live-end anchorage to dead-end anchorage. Design considerations for certain members, especially heavy girders, require that the tendons be modified near the ends to better distribute stresses and minimize cracking in the end areas. One method to accomplish this is to break the bond between the tendon and concrete for several feet at the end (called strand debonding). Another method is to depress the tendon. Called a draped or a depressed tendon, the tendon extends a certain distance near the bottom of the member, equal distance each side of the center of the span, then rises and emerges from the ends of the member near the top. (See Figures 20-25, 20-26 and 20-27.)

Figure 20-25
Heavy structural girders with 30 straight tendons and 10 draped tendons. Note the loop of prestressing strand on top of the girders, which is used for lifting. Clean-up of girder ends in process. (Courtesy of PCI)

Measurement of Prestressing Force. Tension is applied to the tendons by means of hydraulic jacks. Gauge pressure or load indication at the tensioning jack is the principal control used as a tensioning check. Elongation monitoring is a method of checking the tensioning jack during tensioning operations. When the elongation is measured, one is in effect measuring the load being applied to the tendon, as the elongation directly relates to the applied load. The computation of tendon elongation is dependent upon the accurate determination of the prestressing steel properties. For this reason, stress-strain data for prestressing steel are a mandatory portion of a mill certificate received from the tendon manufacturer.

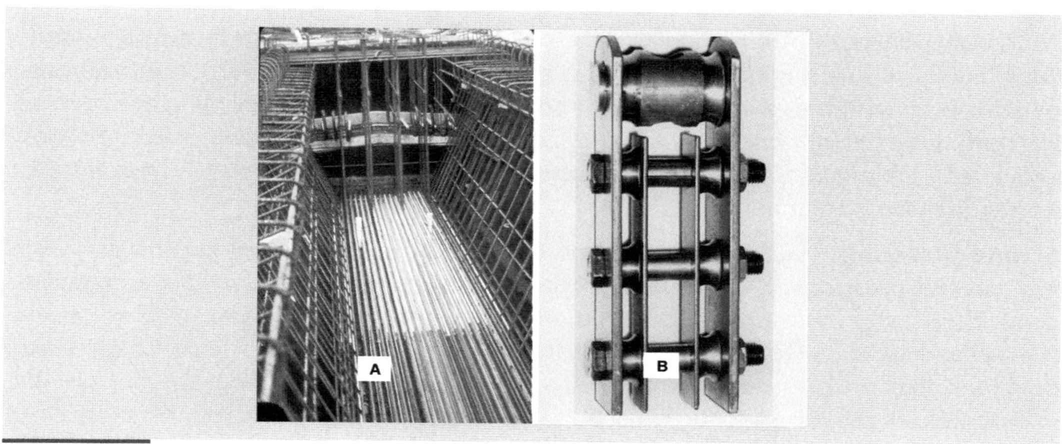

Figure 20-26
Two methods of adjusting for stress distribution are shown. In photo (a) plastic bond-breaking sleeves are installed on several of the strands adjacent to the end bulkhead. In photo (b) some of the strands are draped by passing the strands through a hold-down device, which is embedded in the concrete. (See Figure 20-27.) (Courtesy of PCI)

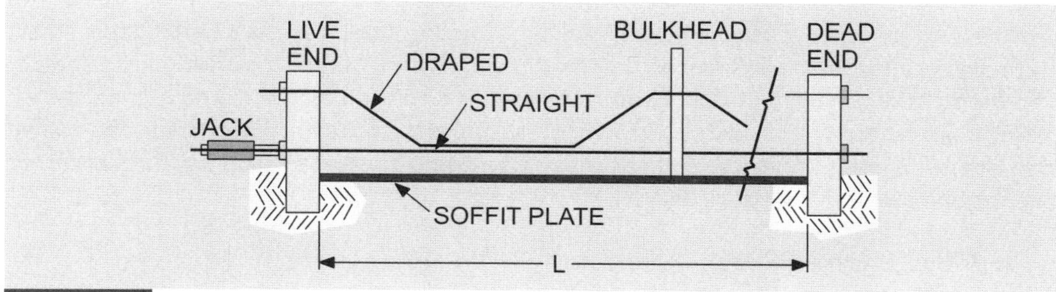

Figure 20-27
Illustration of draped strand layout.

In any prestressing operation there is a small amount of slippage (anchorage seating loss) that develops as the anchors grip the tendons, both at the live end, where the jacks are located, and at the opposite, or dead, end. As the tensioning force is applied to the tendon, there will be a slight movement in the anchors at each end as they grip the strand. The sum of the slippages at each end is the total slippage. Based on ongoing monitoring, anchorage slippage, or seating, is usually an established value for any particular prestressing setup and is in the order of $^1/_8$ inch to $^1/_4$ inch.

The magnitude of the prestressing force to be applied to the strands is a design consideration and will be furnished to the plant personnel along with other tensioning requirements, including physical properties of the prestressing steel. It is a simple matter to calculate the amount the tendons must be elongated when the tensioning requirements, properties of the steel, dimensions of the bed and size of the jacks are known. The jack pressure can be computed to use as a check. For pretensioned members, ACI 318 Section 18.20.1 permits a difference of 5 percent between stress computed from jack pressure and stress computed from measurement of elongation.

Differences in the modulus of elasticity of different production lots of steel, or inaccuracies in pressure gauges on the stressing jacks, are sources of error in measuring jacking forces. These measurements are complicated when depressed or draped strands are used. All equipment should be calibrated at regular intervals, preferably under operating conditions. Measurements should be checked occasionally with a dial gauge extensometer applied to the strand during the prestressing operation.

As noted previously, the modulus of elasticity of prestressing steel averages about 28,500,000 psi. This is apt to vary as much as 8 percent from lot to lot, thus introducing a possible error in stress values computed from strain measurements. By using jack pressure for computing stress, a check is obtained. As long as the two are within reasonably close agreement, the accuracy of the stress measurement is ensured. Each job will have to set up its own standards and tolerances.

Strand Stressing. The following example illustrates stressing and tensioning calculations for precast/prestressed concrete products manufactured under plant-controlled conditions, using the pretensioning method of prestressing. Calculations for the plant-produced products are somewhat more exact than that for the job-site post-tensioned prestressed concrete on account of the higher level of quality control in the manufacture of the plant-produced products. Both calculation procedures give satisfactory results based on the construction methods used and the level of quality control required. Stressing and tensioning calculations for job-site post-tensioned prestressed concrete construction are discussed in Section 20.10.

The example illustrates tensioning calculations for single straight-strand stressing. The strand stress or tensile force is determined on the basis of the physical properties of the prestressing steel being used. For ordinary prestressing steel, the permissible steel stress immediately after prestress transfer is limited to 70 percent of the strand ultimate strength. The modulus of elasticity of the prestressing steel is furnished by the strand manufacturer, having been determined from laboratory testing, and is the basis for determining the change in length (elongation) of the strand as the strand is being tensioned. The basic equation for calculation of strand elongation is

$$\text{Calculated elongation} = \frac{P \times L}{A \times E}$$

where:

A = area of strand

E = modulus of elasticity of strand, supplied by strand manufacturer

L = length of strand being tensioned (chuck-to-chuck)

P = tensioning load (tensile force) applied to the strand

Stressing Calculations. A plant producing precast and prestressed concrete products has a 200-foot-long abutment casting bed (2,400 inches chuck-to-chuck) using hydraulic jacks with a net ram area of 14 square inches. It is planned to stress single $^1/_2$ inch diameter 270k straight strand.

1. Physical properties of strand from mill certificate.

   A = 0.1528 sq in.

   E = 28,500,000 psi

2. Strand to be stressed to 70 percent of ultimate.

   Ultimate = 0.1528 (270,000) = 41,256 lb

   0.70 (41,256) = 28,880 lb

In practice, it is necessary to apply an initial tension in order to take the slack or droop out of the tendon to give a reliable starting point for measuring elongation. Initial tension of 3000 pounds has proven adequate on strand in this casting bed.

3. Basic elongation.

$$\frac{P \times L}{A \times E} = \frac{(28{,}880 - 3000)2400}{0.1528 \times 28{,}500{,}000} = 14.26 \text{ in.}$$

4. Dead end slippage. Based on ongoing monitoring, slippage after initial tension is expected to be $^1/_8$ inch. Add $^1/_8$ inch to elongation. No adjustment in force.
5. Live end seating. Expect $^1/_4$ inch based on past history. Overpull of $^1/_4$ inch required. Adjust force accordingly. Force adjustment, using basic elongation equation:

$$\frac{0.25 \times 0.1528 \times 28{,}500{,}000}{2400} = 454 \text{ lb.}$$

6. Summary.

   Total tensioning load required = 28,880 + 454 = 29,334 lb.

   Jack pressure required = 29,334/14 = 2095 psi

   Total elongation = 14.26 + 0.125 + 0.25 = 14.635 in.

   Tolerance limits for elongation measurement:

   –5% = $13^7/_8$

   +5% = $15^3/_8$

The foregoing computations apply to a bed with straight tendons. Many prestressed components contain depressed, or draped, tendons. Elongation and load computations are made as for a straight tendon. However, it is usually not possible to stress a depressed tendon from one end only, because of friction in the hardware holding the strand down and up at several points through the length of the casting bed. One way to overcome this is to apply the full load to one end of the strand and note the elongation. The jack is then moved to the other end of the bed and the full load applied at that end, the elongation there being noted. The sum of the two elongations should equal the computed total elongation within 5 percent.

Load cells, which employ electric strain gauges (SR-4 gauges), can be used for checking the stressing loads on either straight or depressed strands. When setting up a new tensioning system, especially involving draped strands, strain gauges can be applied directly to the tendons. This is especially desirable when it is necessary to check the uniformity of tension along a depressed strand in a long bed. It is not, however, a method that is practical for routine testing.

Detensioning. The procedure of releasing the tendons from the anchorages is called detensioning. At detensioning, the tension in the tendons is transferred to the concrete, placing the concrete under compression. Strength of the concrete at time of transfer of stress is specified and is determined by field-cured cylinders cured with the components being made.

When multiple strand tensioning has been applied to straight tendons, detensioning can be accomplished by releasing the jacks. Single tendon release is usually accomplished by heating the exposed tendons with a low-oxygen acetylene flame, permitting the tendon to soften and part in a matter of several seconds for each one. (See Figure 20-28.) The presence of depressed tendons imposes conditions that require special procedures that must be developed for each bed.

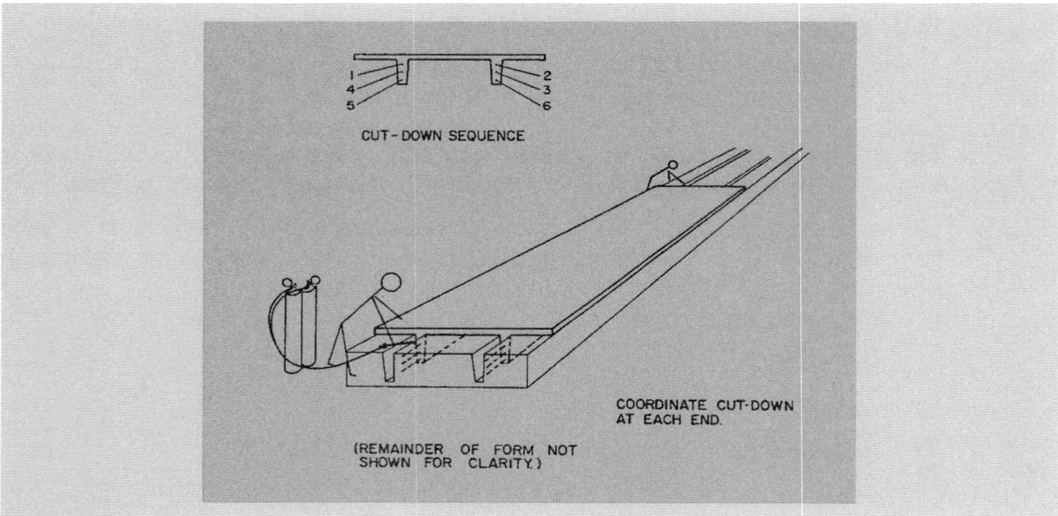

Figure 20-28
Detensioning layout.

The relationship between strength of the concrete, curing cycle, form stripping, release of hold-downs for depressed tendons, and weight of the components is important and must be considered when developing a detensioning pattern to minimize cracking of the concrete. At detension, there is a movement or slippage of the components on the bed, and most components take a camber, or bow. Forms should be released before detensioning. If the components are being cured under high temperature, it is best if the forms can be released early in the curing cycle. Detensioning should be done as rapidly as possible to minimize cracking that might result from unequal stresses in the concrete as it cures.

Discrepancies. Variations in temperature are normally of no importance except as small changes in tension of the steel that result from changes in the temperature surrounding the casting bed. If steel is stressed at a low temperature, there will be a reduction in tension if warm concrete is placed around it. Likewise, a reduction in temperature can cause an increase in tension. Usually these changes are so small that they can be ignored.

As a rule of thumb, for every 10°F of temperature change expected in the strand, a 1-percent difference in stress will occur. That is to say, if the strand temperature rises 10°F from the time of stressing to the time it is surrounded by concrete, then the strand stress will diminish by 1 percent. The opposite would occur if the strand were tensioned at an elevated temperature and the concrete temperature was cooler. The strand would contract and increase in stress by 1 percent for every 10°F temperature drop.[20.2]

The PCI quality control manual[20.1] further states that if a temperature rise of over 25°F is expected, then the strand should be overstressed by an appropriate amount to overcome the expected loss that is due to the expansion of the strand. For example, if a 30°F rise is anticipated between a 30°F ambient temperature at the time the strands are stressed and a 60°F concrete temperature at the time of casting, then the strands should be overstressed by 3 percent to offset this expected loss.

There is a small movement of anchoring abutments and elongation of anchor bolts that should be evaluated at the beginning of a job. Here again, the magnitude is usually so small that it can be ignored, but it should be checked. Strands for pretensioning sometimes become crossed or twisted when they are strung in the form. An occasional twisting can be accepted. One or two wires of a tendon may break during or after stressing. If the number of broken wires is less than 2 percent of the total number of wires, the stressing can be considered acceptable (ACI 318 Section 18.20.4). During single-strand stressing, the strand has a tendency to untwist, which in some cases causes the ram of the jack to revolve. One full revolution can be accepted.

Positioning of the strands is critical. A displacement of only a fraction of an inch can, in some cases, result in deviation of the component from the established line. Tolerances of these measurements are given in the specifications. Position of the tendons should be checked after stressing, before the concrete is placed.

20.7. Handling and Erection [20.4]

Upon completion of the curing period, the precast components can be moved into storage. Cranes, straddle carriers, forklifts and trucks are used for this purpose. In-plant handling is shown in Figures 20-29 and 20-30.

Figure 20-29
Cored slabs and similar units can be handled in many ways. Tongs can be made for gripping the sides of the units, or heavy web straps can be passed around the unit, both permitting the slab to be handled in a horizontal position.

Figure 20-30
The 15-ton bridge crane can handle most units that might be precast. A crane of this kind efficiently covers a large portion of the manufacturing, storage and loading areas. The ingenious apparatus in the middle photo can rotate heavy members through a full circle if necessary to facilitate finishing operations and storage. Forklifts, frequently fitted with special lifting or holding devices, are used in many yards.

When the components are removed from the forms or molds, they should be marked with the identifying numbers shown in the shop drawings or placing plans. Marking should be done with permanent, nonfading paint or ink on surfaces that will not be exposed to view in the finished structure. The job number should be clearly given, as well as the date.

Every plant will have its own particular method of handling products, but there are certain basic requirements that apply to every plant, regardless of the methods and equipment used. Components should be removed from the forms as soon as possible after completion of curing. Units can be moved as soon as detensioning has been completed. Handling should be kept to a minimum because it costs money and exposes the member to possible damage each time it is moved. Pickup points are designated on the components, and it is important to lift the components by attaching lifting equipment at these points only. This is especially important for prestressed members. At no time should a prestressed component be picked up at any point other than at the designated lift points. While in storage, prestressed components must be supported at the bearing points or within a short distance thereof, as approved by the engineer. Prestressed components should not be lifted or stored on their sides.

Most precast concrete members have some type of lifting hardware embedded in the concrete when the component is made. These lifting devices or inserts usually consist of two parts: an anchorage element that is embedded in the concrete and an attachment element that screws into or is joined to the anchorage by some other means. The crane hook fits into the attachment element. Sometimes, a single loop of prestressing strand can be embedded in the concrete, with the loop exposed so that the crane hook can be attached. Choker slings can be used on some units, provided edges of the concrete are protected to prevent chipping and marring. (See Figures 20-16 and 22-5.)

When in storage, precast components should be supported on blocks or runners to raise them above the ground surface to preclude staining or other damage. Components should not be stored on frozen ground without adequate measures to protect them when the ground thaws. Panels and slabs can be stored in a vertical position, tilted back slightly to stabilize them. Location of supports must be carefully designed to avoid localized stressing of the concrete or uneven loading. Stored components in contact with one another must be arranged so that the lifting devices and identifying numbers are accessible, and should be separated by battens or similar material.

Figure 20-31
Double tees loaded on expandable trailer. (Courtesy of PCI)

When loading components on trucks or other conveyances, the same precautions that were used in storage should apply. (See Figure 20-31.) Units hauled flat should be supported so that any warping or distortion of the vehicle bed will not be transmitted to the units; frequently, a three-point support will suffice. Long prestressed girders can be hauled on pole trailers with the prestressed member serving as the pole connecting the tractor and trailing dolly.

Usual practice, especially when handling large and heavy components, is to hoist the components directly from the hauling vehicle and erect them on the structure. For this reason, careful scheduling, from casting to final erection and caulking, is essential. Small components are sometimes stored on the site, but usually the building site is so congested that this is not practical. (See Figures 20-32, 20-33, 20-34 and 20-35.)

Figure 20-32
Modified forklift to erect in area with limited headroom. (Courtesy of PCI)

Figure 20-33
Erection of double tees with straps. (Courtesy of PCI)

Figure 20-34
Chokers to lift hollow-core slabs. (Courtesy of PCI)

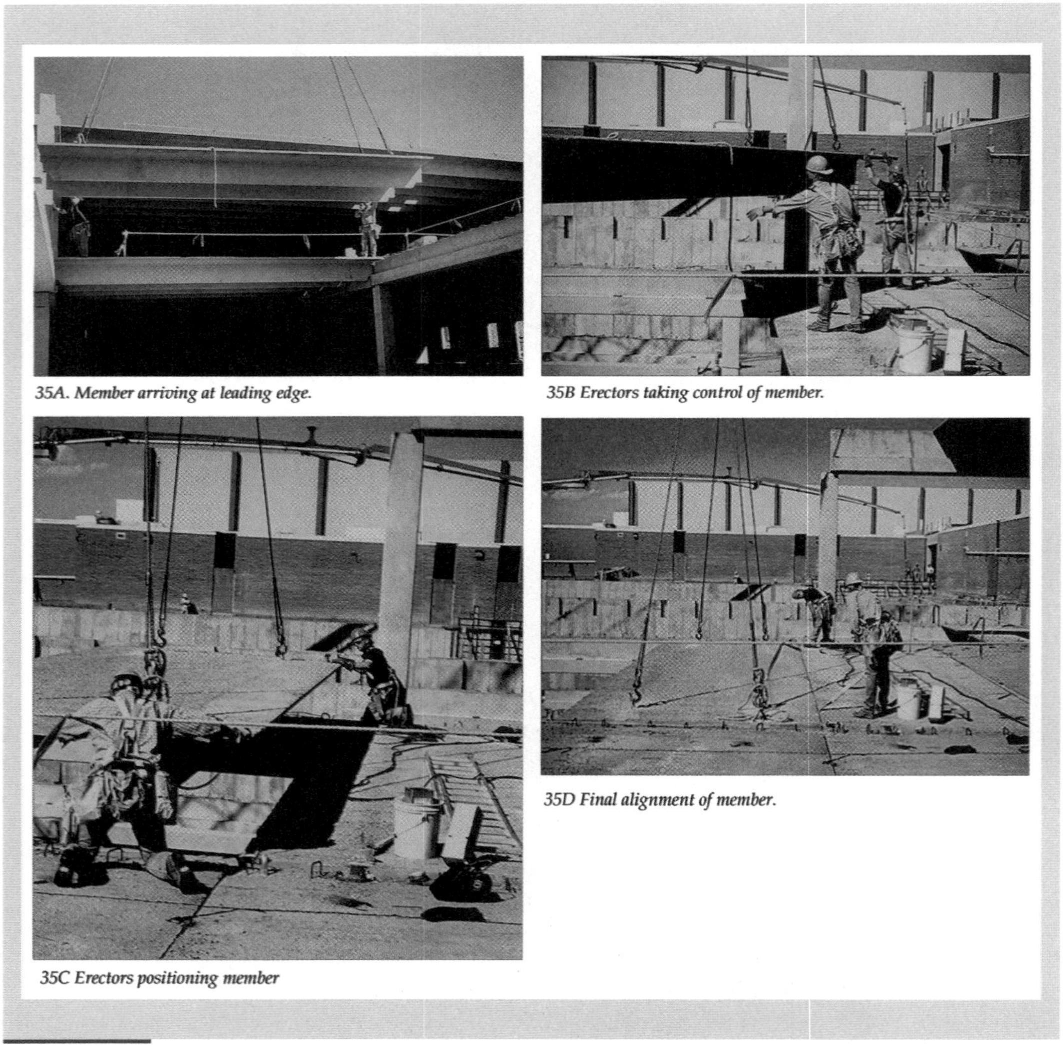

35A. Member arriving at leading edge.

35B Erectors taking control of member.

35C Erectors positioning member

35D Final alignment of member.

Figure 20-35
Erection sequence for a double tee. Erectors working in a controlled access area using a safety monitoring system. (Courtesy of PCI)

After a component has been set in place in the structure, it should be temporarily connected and braced as soon as possible so that the crane can be released. (See Figure 20-36.) It is uneconomical to have a crane standing idle while a connection is welded or otherwise finalized, when a temporary connection can be made. Shims, made of metal, plastic or wood are usually necessary to align or level the components. Wood is apt to be unstable, and steel will rust. Vertical and horizontal alignment is critical. Joints between panels on the exposed face of a building must be of uniform width, and the panels must be carefully aligned in one straight plane or surface.

Figure 20-36
Precast exterior framing with temporary bracing. (Courtesy of PCI)

Connections are defined as the means by which components are attached to the building frame and to each other. No two jobs are ever exactly alike. Regardless of the design of the connection, it must offer safety and stability, both during erection and after the building has been completed. Four qualities are required of any connection:

1. It must be structurally adequate to provide the properties of safety and stability mentioned above.
2. It must possess construction practicability. That is, it must lend itself to fabrication and assembly under conditions and with equipment normally found on a construction site.
3. It must be economical. As much as possible, it should make use of off-the-shelf hardware items, rather than made-to-order components.
4. It must present a pleasing appearance if it is to be exposed. Where possible, the connection hardware should not be exposed to view in the finished building.

After the members have been erected and aligned, most building components such as panels have to be caulked to seal the joints. Sealing compounds are discussed briefly in Chapter 10. Materials and methods for caulking are fully described in the specifications for the job. Caulking should be scheduled for the time of year that the joints are at their normal opening; otherwise they may be underfilled or overfilled. Most manufacturers require that their materials be applied only when the air temperature is above 40°F and that the joint surfaces be clean and dry.

20.8. On-Site Precasting

On those building sites where space is available, some contractors will precast some of the parts for the building. This technique can be adopted for the building in which there are many identical panels, beams or similar components. The benefit is the opportunity to take advantage of assembly-line plant precasting operations to produce many components of uniform quality and appearance.

Casting yard operations are identical with those previously discussed in this chapter. On some projects, the yard layout can be integrated with the building site so that cranes or other hoisting equipment can move components from yard storage directly to their final location in the building.

20.9. Precast Forms and Molds

Forms for cast-in-place concrete can be made of precast concrete. One type is a nonstructural architectural facing, which is attached to the interior of conventional forms. After the cast-in-place concrete has hardened, the formwork is stripped, leaving the precast liner attached to the concrete. More common are forms that become a part of the composite column, wall or beam. A form of this type may be entirely self-supporting, or it can be tied and braced to other parts of the formwork. Anchors are cast in the members to tie them together or attach them to the formwork. One advantage of concrete forms is the uniformity of appearance achieved by plant precasting of the components. Surface defects such as form bolt holes are practically nonexistent. Shoring and external supporting can be reduced, and there are no forms to be removed. Considerable care in erecting the forms is necessary to get them properly aligned. Joints in particular require accuracy to ensure uniformity.

Forms should be handled with the same care as any good precast concrete component. Joints must be caulked or sealed to prevent leakage of mortar out of the concrete placed inside the form. Any mortar or concrete spilled on the exterior of the form should be removed immediately. Usual practice when making a horizontal construction joint is to offset it about 3 inches from any horizontal joint in the form. The only curing necessary is on the concrete exposed at the top.

20.10 Post-Tensioned Slab Construction [20.5]

Post-tensioned slab construction is classified as bonded or unbonded, depending on whether the tendon ducts are filled with grout after stressing (bonded) or whether the tendons are greased and covered with a plastic sheathing (unbonded). Bonded tendons are used principally for beams and primary structural members, utilizing multistrand systems. In a multistrand system, multiple strands are typically installed in a single duct. Strands in a bonded system are typically installed inside the duct without any P/T coating. Once the strands have been stressed, the ducts are grouted to bond the strands to the surrounding concrete and protect the strands from corrosion. (See Figure 20-37.) Unbonded tendons are widely used for one-way and two-way post-tensioned slabs. For slab construction, a large number of single seven-wire strand tendons are normally used.

Figure 20-37
Multistrand post-tensioning systems being installed in a large transfer girder. (Courtesy of PTI)

The emphasis in this section is on detailing and construction procedures, and inspection guidelines, for post-tensioned slabs using unbonded single-strand tendons. A typical post-tensioned slab with tendons in place is shown in Figure 20-38. Other applications of post-tensioning in building construction are discussed in Reference 20.5.

Figure 20-38
A typical one-way beam-and-slab construction with formwork and tendons in place. (Courtesy of PTI)

The field procedures manual for unbonded single-strand tendons, Reference 20.6, was developed by the post-tensioning industry to provide guidance for field personnel involved in the installation, stressing and finishing of unbonded single-strand tendon systems. The manual also provides information for inspection of this type of construction and contains extensive discussion of jobsite troubleshooting. A copy of Reference 20.6 should be readily available to inspection personnel. A word of caution: Reference 20.6 is not a legally adopted reference in the building code or project specifications. It represents the overall industry guidelines. As such, it cannot be quoted or enforced at the jobsite. Most post-tensioning fabricators publish their own set of guidelines, and these are usually referenced in the job specifications for enforcement at the jobsite.

Unbonded Strand Tendons. Unbonded single-strand tendons used in post-tensioned slabs usually consist of $^1/_2$-inch or 0.6 diameter seven-wire strand (see Figure 20-39) with anchorage consisting of a cast wedge plate in which the tendon is gripped by a two-piece wedge. (See Figure 20-40.) It is industry practice to define tendon as the complete assembly as shown in Figure 20-39, consisting of anchorage hardware, strand and sheathing. After the concrete has reached the necessary strength, the strand is stressed and the conical wedge grippers are inserted around the strand in a conical hole in the bearing plate to provide anchorage. The block rubber (or plastic) element to the left of the wedge in Figure 20-39 is used to provide a block-out in the concrete to allow for stressing the tendon. To protect the steel against corrosion, a corrosion-preventive coating (grease) is applied to the strand. The coating also provides lubrication between the strand and the sheathing. The sheathing completely encloses the strand to prevent bond between the prestressing steel and surrounding concrete. The tendon sheathing also provides water-tightness over the entire strand length. Unbonded single-strand tendon systems are required to be protected against corrosion (ACI 318 Section 18.16.4) in accordance with ACI's "Specification for Unbonded Single-Strand Tendons (ACI 423.7)."[20.7]

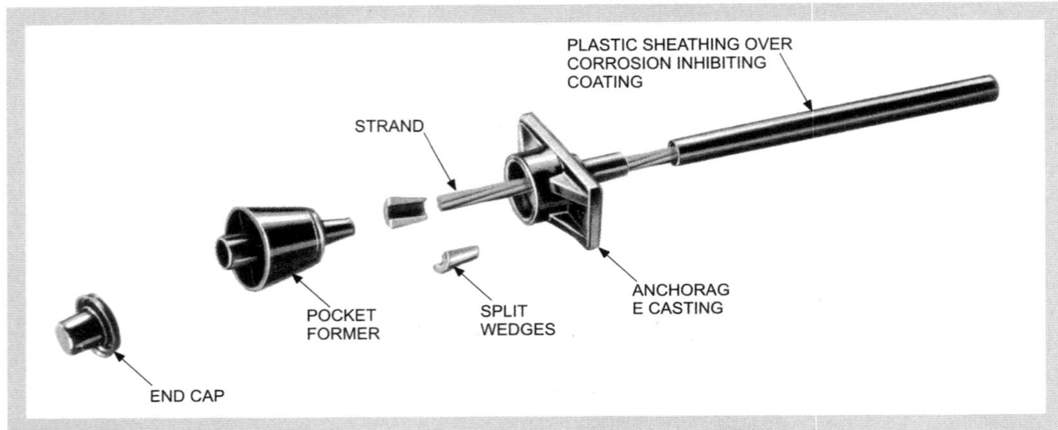

Figure 20-39
Unbonded single-strand tendon system. (Typical of the types of single-strand tendon systems supplied by various fabricators.) (Courtesy of PTI)

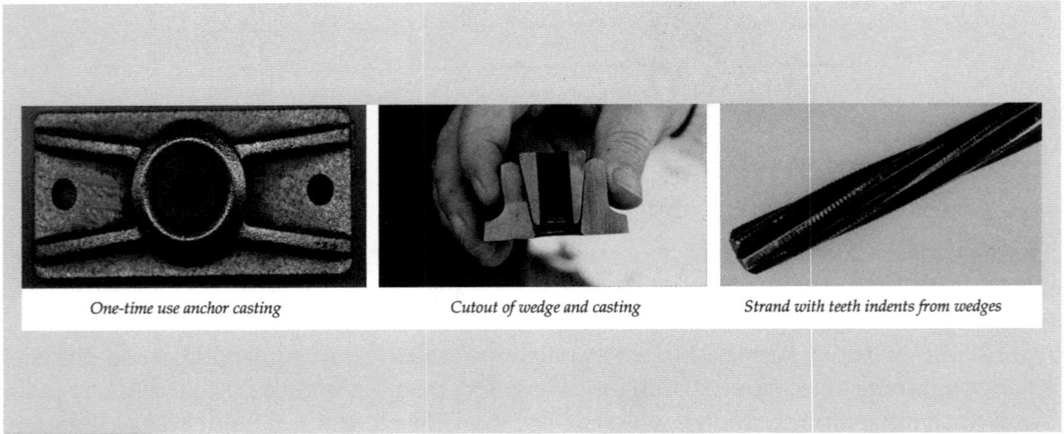

Figure 20-40
Anchorage casting and two-piece wedge details. (Courtesy of PTI)

The Post-Tensioning Institute also offers a training and certification program for field personnel involved in the installation, stressing and inspection of unbonded single strand post-tensioning systems. The reader should contact the Post-Tensioning Institute for further information on this certification and training program. Of special note: certified special inspectors in prestressed concrete (see Section 25.9) may want to further enhance their knowledge and certification credentials by participating in this industry certification program.

Detailing. Construction details for post-tensioned building elements, beams and slabs are first prepared by the design engineer, followed by detailed installation drawings developed by the post-tensioning subcontractor. Installation drawings are normally prepared in greater detail than the design drawings, showing actual tendon layout, tendon placing sequence, stressing and dead ends of the tendons, and estimated elongation of the tendons during stressing. The tendons delivered to the jobsite are color coded as indicated on the installation drawings. A typical flat-plate tendon layout is shown in Figure 20-41. The inspector will need to become very familiar with the specific post-tensioning construction details given on the installation drawings.

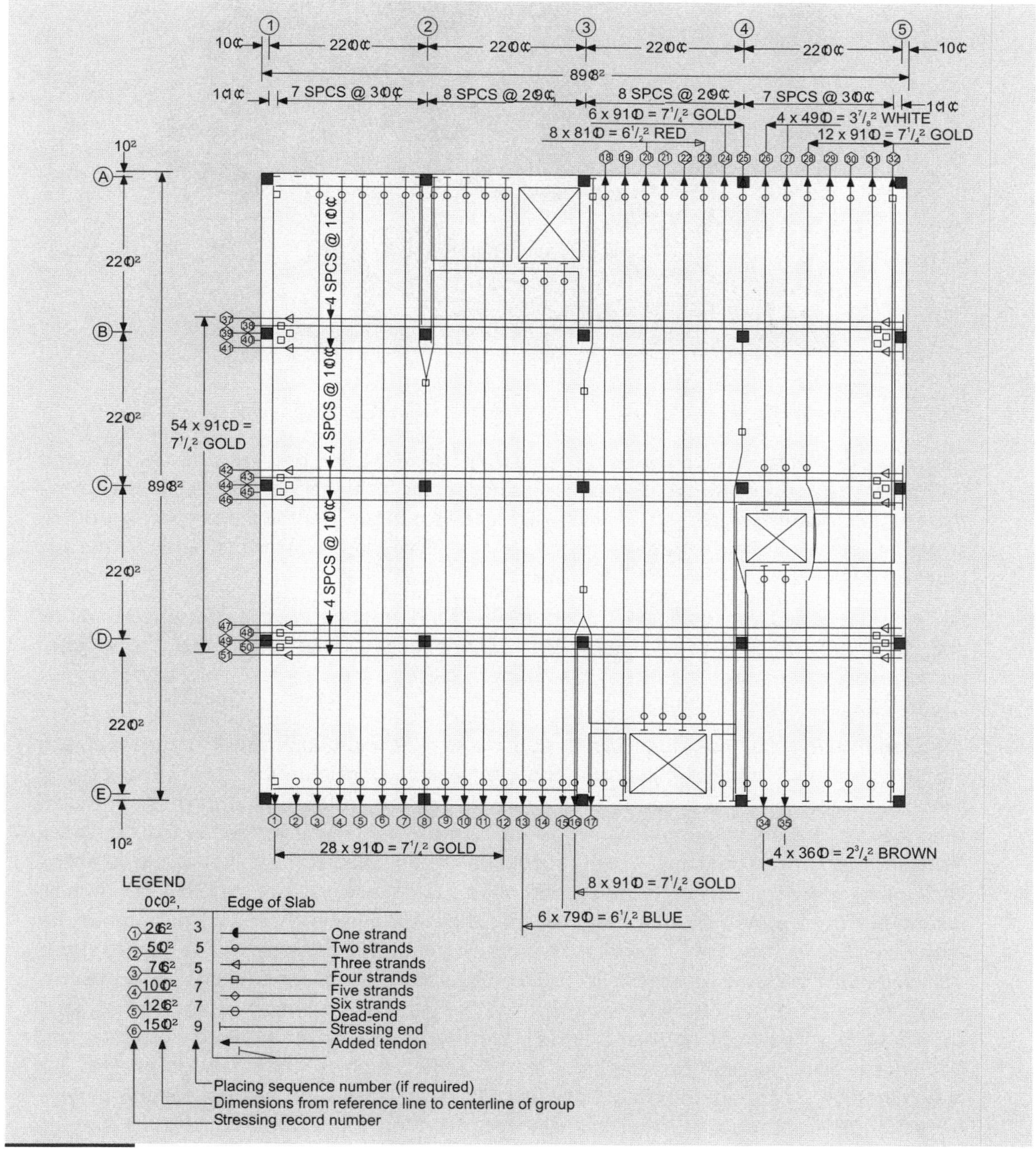

Figure 20-41
Installation drawing for flat-plate tendon layout. (Courtesy of PTI).

Construction. Preassembled tendons are usually shipped to the jobsite in 5-foot-diameter coils, secured by ties at intervals to prevent uncoiling. (See Figure 20-42.) Each tendon is individually marked and identified for proper location in the structure. Care should be exercised in unloading and handling the tendons to prevent damage to the sheathing. If some sheathing damage should occur, it can be repaired in the field with duct tape.

Figure 20-42
Delivery of preassembled tendons with rolls stored vertically. (Courtesy of PTI)

Forms are drilled to receive the tendon-stressing hardware and bearing plates as indicated on the installation drawings. Tendon placing details provided by the post-tensioning materials fabricator will show end anchorage details, tendon identification, spacing profile, stressing data, clearance requirements for the stressing equipment and anchorage block-out dimensions.

For post-tensioned slabs using unbonded single-strand tendons, the preassembled tendons are placed as a unit prior to placing concrete. Supporting ties must be adequate to support tendon weight. Tendons are usually placed before the reinforcing steel, electrical conduit and mechanical work.

Tendon Placing. Each tendon is designed for a specific location in the structure. Tendon placement should normally precede the placement of reinforcing steel. The placing sequence number for slab tendons is indicated on the installation drawings. The slab tendons are placed in numerical sequence based on the placing sequence numbers. Tendons marked for the initial placing sequence are uncoiled starting first at the stressing end, followed by tendons with the second placing sequence number and so forth until all the tendons are unrolled along the path that they will take in their final position in the structure. (See Figure 20-43.) Vertical deviations in tendon location should be kept to about $^1/_4$ inch for slab thickness less than 8 inches, and $^3/_8$ inch to $^1/_2$ inch for thicker slabs or beams. Horizontal plane deviations, which may be necessary to avoid openings, ducts, inserts, etc., should have a radius of not less than 21 feet. Concrete cover between tendons and openings in slabs should normally be at least 6 inches. Single- strand tendon profiles are maintained by tying them to reinforcing steel, chairs or other approved supports with wire ties at about 4-foot centers. A typical single-strand tendon layout is shown in Figure 20-44.

Figure 20-43
Tendons being uncoiled in the approximate location as indicated on the installation drawings. (Courtesy of PTI)

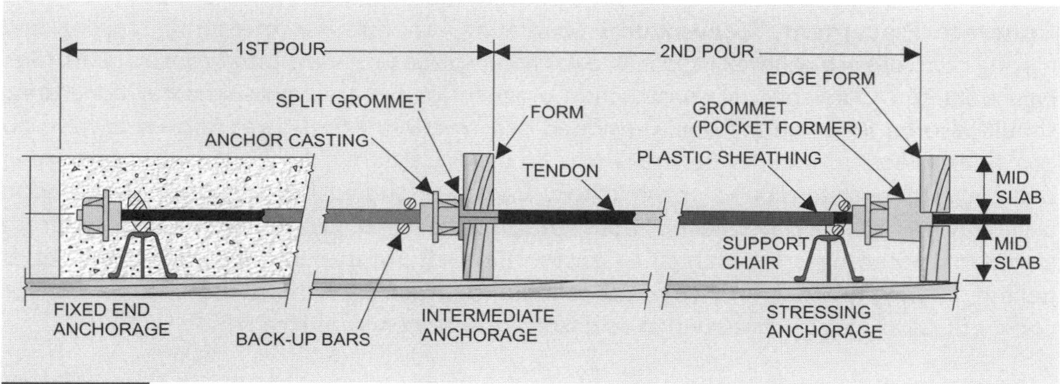

Figure 20-44
Single-strand tendon layout. (Courtesy of PTI)

When welding or burning near tendons, care must be exercised to prevent the tendon from overheating and to keep molten welding slag from coming in contact with the tendons. Grounding of welding equipment to tendons should not be permitted.

Prior to placing concrete, tendon profiles should be checked at critical locations—such as at midspan, inflection points and in negative moment areas over columns or walls—by measuring from the form soffit to the center of the tendon. (See Figure 20-45.) If the tendon sheathing has been damaged, repairs should be made with duct tape to prevent concrete from entering the sheathing and bonding to the tendon. Concrete should not enter pockets or anchorage hardware. Horizontal alignment should be checked to ensure minimum horizontal deviations and proper concrete cover at openings. Workers should be assigned to maintain proper tendon alignment slightly ahead of concrete placement.

Figure 20-45
Measuring tendon profile.

Concrete Placement. Conventional concreting practices for measuring, mixing and placing concrete are generally applicable to cast-in-place post-tensioned construction. (See Figure 20-46.) Conventional construction practice for hot- and cold-weather concreting should also be followed for post-tensioned construction. Admixtures known to have no damaging effects on steel or concrete may be used. Calcium chloride must not be used in concrete for post-tensioned construction. Concrete should be placed so that tendon alignment and reinforcing steel positions remain unchanged. Careful vibration of concrete at tendon anchorages is essential to ensure uniform compaction and to prevent voids behind bearing plates. The concrete should be cured in the same manner as for conventional concrete construction to ensure proper concrete strength.

Figure 20-46
Concrete placement. The same concreting practices are used as in conventional concrete placement, with special care taken to not damage the tendon plastic sheathing.

Tendon Stressing. When tests of field-cured cylinders indicate that the concrete has reached the proper strength (usually 60 to 80 percent of the 28-day strength), the stressing operation may begin. It is essential that the shoring be left in place until the stressing is completed. Tendons should be stressed only when proper stressing data (ram area, forces and gauge readings) for each tendon are determined. The post-tensioning fabricator will establish simple rules and procedures to follow to ensure that stressing is accomplished in a satisfactory and safe manner.

Stressing is monitored in two ways. First, the gauge reading on the pump is transferred into tendon force at the anchorage. Second, the tendon elongation is calculated (see discussion of stressing calculations below). For post-tensioned construction, ACI 318, Section 18.20.1 permits a difference of 7 percent between tendon force measured by gauge pressure and tendon force calculated by elongation measurements.

Stressing equipment used for post-tensioning work has been carefully designed and incorporates reasonable factors of safety. Occasionally, flaws in material are undetected, or the equipment may have been misused. For this reason, extreme caution should be exercised at all times, as stressing is carried out at extremely high pressures. The primary safety rule is to keep workers from directly in back of the stressing equipment or from between the edge of the building and equipment. Failure during the stressing operation may cause serious injury to workers in back of or in the immediate vicinity of the stressing equipment.

Stressing Calculations. The post-tensioning method of prestressing is typically job-site quality control and construction, requiring somewhat less exacting calculation procedures than that used for the plant-produced pretensioned prestressed products. The stressing and tensioning calculations currently used for job-site post-tensioned work have proven over time to give satisfactory results

Typically, the stressing and tensioning data given on installation drawings, prepared by the P/T fabricator, will show stressing and tensioning calculations in terms of either forces or stresses. The inspector needs to realize that either forces or stresses are basically the same data presented in a different format. The following illustrates typical data on installation drawings.

Tendon properties:

270k - $^1/_2$ inch 7-wire strand

Area = 0.153 sq in.

Modulus of Elasticity (E) = 28,500,000 psi (typically assumed value)

Data in terms of tendon forces:

1. Ultimate tendon strength = 0.153(270,000) = 41,300 lb = 41.3 kips*
2. Maximum jacking force = 0.80(41,300) = 33,000 lb = 33 kips
3. Maximum anchoring force = 0.70(41,300) = 28,900 lb = 28.9 kips
4. Tendon elongation (Δ)

$$\Delta = \frac{P \times L}{A \times E} = \frac{28.9\,(Lth \times 12)}{0.153\,(28,500)} = 0.08 \text{ in / ft of tedon length}$$

* 1,000 lbs = 1 kip

Data in terms of tendon stresses:

1. Ultimate tendon stress = 270,000 psi = 270 ksi*

2. Maximum jacking stress = 0.80(270,000) = 216,000 psi = 216 ksi
3. Maximum anchoring force = 0.70(270,000) = 189,000 psi = 189 ksi
4. Tendon elongation (Δ)

$$\Delta = \frac{f \times L}{E} = \frac{189\,(Lth \times 12)}{28{,}500} = 0.08 \text{ in / ft of tendon length}$$

* 1,000 psi = 1 ksi

Note: For 270k – $^1/_2$ inch 7-wire strand, the expected elongation (Δ) = 8 in. per 100 ft of tendon length.

Elongation calculations:

Assume the length of tendon being stressed is 40 ft

Δ = 0.08(40) = 3.20 in. = 3$^1/_4$ in. (measured to nearest $^1/_8$ in.)

Acceptable tolerance on measured elongation = +/– 7%

Maximum measured Δ = 1.07(3.25) = 3.48 = 3$^1/_2$ in.
Minimum measured Δ = 0.93(3.25) = 3.02 = 3 in.

Inspection Guidelines. Three distinct construction phases are involved for all post-tensioning systems:

- Material manufacturing
- Tendon installation
- Tendon stressing

Table 20.1 lists important questions that should be considered by inspectors about each of these phases for unbonded (greased and sheathed) tendons.

Material manufacturing: Most fabrication plants have similar production facilities. Depending on the magnitude of the project, plant inspection may be appropriate. If not, then jobsite material review is in order.

Tendon installation: An experienced inspector should review the process with the placer during installation of the first pour and reach an understanding with the crew regarding critical elements.

Tendon stressing: Jobsite technical instruction on the proper operation of stressing equipment is normally provided in accordance with project specifications by the post-tensioning material fabricator. The inspector must be familiar with the various operations involved in stressing tendons.

TABLE 20.1—GUIDELINES FOR INSPECTION OF UNBONDED TENDONS

MATERIAL MANUFACTURING
Are fixed end wedges evenly and adequately seated in the anchor?
Is excessive sheathing stripped at the fixed end?
Is the plastic sheathing of sufficient and uniform thickness?
Is the grease evenly applied and of consistent texture?
Does the strand appear to be of new quality and free of corrosion when sheathing and grease are removed?
Are the anchors properly cast with smooth wedge holes?
Are the wedges free of rust and steel shavings, and of consistent quality?
Are mill reports and certifications available for the prestressing steel and other components, as required by the specifications?
TENDON INSTALLATION
Are tendon high and low points at the correct elevation?
Are tendon profiles smooth and correctly shaped (parabolic, circular or straight) between reference points?
Do the tendons have excessive horizontal wobble?
Is the sheathing damaged, and if so, has it been repaired?
Does the chair or support-bar system conform to contract documents?
Are the stressing anchors securely fastened to the form with appropriate pocket formers?
Is bursting steel installed behind the anchorages as required by the contract documents?
Has the method of concrete placement been reviewed as to its effect on tendon stability during placement?
Has the conventional steel placement been reviewed?
TENDON STRESSING
Are the stressing anchor wedge holes free of grout, dirt and plastic?
Is a consistent dimension used for the elongation datum mark on the strand?
Is the stressing equipment well maintained, and are calibration charts available?
Is the stressing ram operator careful with the equipment and consistent from tendon to tendon?
Are the tendons stressed slowly enough to allow the strand to overcome as much friction as possible prior to seating?
Are the wedges seated evenly and under pressure?
After elongation approval, are tendon tails cut off well inside the pocket to allow proper grout cover?
Are pocket surfaces sufficiently clean to allow good grout bond during and after patching?

Lightweight and Heavyweight Concrete

Concrete of the type commonly used in engineering structures and buildings, made with gravel and crushed stone aggregate, weighs between 140 and 150 pcf—the exact weight depending on the specific gravity of the ingredients, amount of entrained air and the mix proportions. This concrete is capable of the high strengths presently specified by many engineers.

There are, however, some applications for concrete in which we find special requirements for durability, resistance to attack, high strength and other properties. Usually these special requirements can be met by careful attention to all aspects of good concrete construction with special consideration to strength, water-cement ratio, etc., which are specified. These special concretes are discussed in the parts of this book covering the application or exposure in question.

Among these special applications are structures in which a lower weight of concrete is advantageous for structural reasons or to provide thermal and acoustical insulation. On the other hand, concrete with a high density may be specified for such usage as a shield for an atomic reactor or a counterweight for a movable bridge.

LIGHTWEIGHT CONCRETE

Methods and materials used for making lightweight concrete depend upon the density and end result required. Reduction in weight compared with normal-weight concrete can be obtained by the use of lightweight porous or vesicular aggregate; by omitting the fine aggregate, thus producing a porous, no-fines concrete; by forming bubbles of air or gas in the mix with the addition of a foaming (air-entraining) agent or a preformed foam; or by a chemical action—or a combination of these.

21.1. Kinds of Lightweight Concrete

There are two general types of lightweight concrete: lightweight structural concrete and lightweight insulating concrete. The distinction between the two is made on the basis of weight and strength. Lightweight structural concrete is usually defined as having a compressive strength in excess of 2500 psi at 28 days, and an air-dry density of 85 to 115 pounds per cubic foot. Very lightweight concretes used primarily for insulating purposes have a density of less than 50 pounds per cubic foot, ranging down to about 15 pounds. Compressive strength at 28 days may be as low as 100 psi, ranging up to 1000 psi. Between the limiting values for these two types, there is an overlapping gray area in which a concrete might be designated primarily as either structural or insulating, but with significant value in the other property. This concrete is called a fill concrete.

21.2. Lightweight Structural Concrete

With improvements in materials and techniques, the use of lightweight aggregates for cast-in-place structural concrete is becoming widespread. Strengths of 5000 psi and higher are made without difficulty by concrete weighing 110 pounds per cubic foot. By entraining the proper amount of air, obtaining durable concrete is no problem in those areas subject to freezing and thawing exposures. Air-entrainment also serves to overcome poor workability, segregation and bleeding properties sometimes associated with lean lightweight concrete. (See Figure 21-1.)

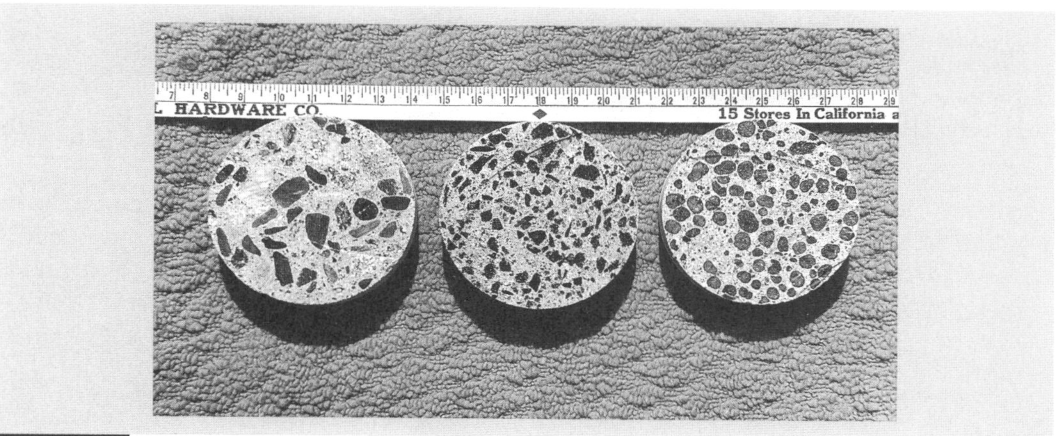

Figure 21-1
Sawed sections of 6-inch concrete cores. On the left is a core of natural gravel concrete; the core in the center was made of volcanic scoria; the core on the right contained manufactured lightweight aggregate.

A primary reason for using lightweight structural concrete is the reduction in weight of the structure, with the resulting savings in cost by permitting smaller footings and a lighter supporting structure. (See Figure 21-2.) Related advantages are its thermal and acoustical insulating values. Values of thermal conductivity (K) of lightweight concretes are between 2 and 4 Btu per hour per square foot (degrees Fahrenheit per inch), compared with about 9 for normal-weight concrete—the lower values being associated with lighter densities.

Figure 21-2
Water Tower Place in Chicago uses structural lightweight concrete in its floors to reduce dead-load weight. (Courtesy of PCA)

Aggregates. Manufactured materials include expanded shale, clay and slate, and expanded blast-furnace slag. (See Figure 21-3.) Natural aggregates locally available in some areas are volcanic scoria, a vesicular lava of rough texture, and pumice, a porous lava similar to scoria. These are sometimes known as volcanic cinders. These materials, though light in weight, are strong and sound, requiring only crushing, washing and screening to prepare them for use in concrete. Lightweight concrete of good strength and other desirable properties can be made from them. Some deposits of diatomite have been found to produce satisfactory aggregate.

Figure 21-3
Lightweight aggregate used to produce structural lightweight concrete. Expanded clay (left) and expanded shale (right). (Courtesy of PCA)

Coal or coke, when burned in high-temperature industrial furnaces and boilers, produce cinders that are suitable for use in concrete. Cinders have been used for many years, especially in concrete block. They should be permitted to age in the stockpile for several weeks before using, to eliminate sulfur compounds. Changes in industrial processes have nearly eliminated cinders as a source of aggregate, and they are rarely used any more.

There are two basic methods of manufacturing lightweight structural aggregates: the rotary kiln process and the sintering process. Good aggregate can be made by either method. In the rotary kiln process, raw material is crushed, screened and fed into the upper end of a rotary kiln, inclined at a slight angle. As the material travels through the kiln toward the burner end, it reaches a temperature of 1800°F to 2000°F and changes into a plastic state. Gases within the material expand to form minute air cells. Material is then discharged, cooled and crushed. Usual sizes are as follows:

Fine:	passing $^3/_{16}$-inch screen (sand)
Intermediate:	$^3/_{16}$ to $^3/_8$ inch
Coarse:	$^3/_8$ to $^3/_4$ inch

The so-called coated aggregates have rather smooth surfaces and lower absorption. Raw material is crushed and screened before being fed into the kiln, or it is formed into pellets on a pelletizing machine or by extruding soft clay or shale through a die. After burning, it is cooled and stored.

In a typical operation, the clay, upon arrival at the processing plant, is pulverized, then screened. In most plants, the next step is to feed the pulverized raw material into a pug mill mixer, where a small amount of moisture is added, then to pass it through an extruder. As the material emerges from the extruder dies, it is cut into small sections that are fed into the kiln, where it is heated to about 1800°F to 2000°F. Passing through the kiln, the feed first gives up its water, then is fused or melted, expanded by evolution of gas, and finally crusted over with a vitreous or ceramic glaze before passing on to the cooler. Subsequent operations consist of screening or classifying, and storage. Variations in the manufacturing process in individual plants may involve only crushing the raw feed or grinding and pelletizing the feed before it goes to the kiln. These manufacturing differences can cause the aggregate from one plant to have an absorption value significantly different from that of an aggregate produced in another plant. Coarse aggregate from any one plant should have a variation of absorption of less than 4 percentage points from the average absorption.

In the sintering process, raw material is crushed and screened, then mixed with a small amount of pulverized coal or coke. This mixture is spread over a traveling grate that passes into a furnace where the fuel is ignited. Continued burning as the grate moves along produces clinkers, which are cooled, crushed and screened. Aggregate sizes are the same as those made in the kiln.

Expanded slag is a lightweight aggregate made by bringing hot, molten blast-furnace slag into contact with controlled amounts of water. The sudden quenching causes rapid cooling of the slag, entrapping large amounts of gas in the slag particles. The expanded slag is subsequently crushed and screened to appropriate sizes.

ASTM C 330 covers lightweight aggregate for structural concrete. Properties of lightweight aggregate for structural concrete are listed in Table 21.1. Gradation requirements for aggregate for structural concrete are shown in Table 21.2.

TABLE 21.1
PROPERTIES OF LIGHTWEIGHT AGGREGATES FOR STRUCTURAL CONCRETE

BULK UNIT WEIGHT, dry loose, should not exceed: for Coarse, 55 pounds per cubic foot; Fine, 70 pounds; Combined, 65 pounds. Unit weight of successive shipments should not differ by more than 10 percent from acceptance sample.

ABSORPTION, will reach 10 percent in one hour, and will exceed 15 percent, for most aggregates, at 24 hours. Absorbed water may exceed 150 pounds per cubic yard of concrete.

PARTICLE SHAPE AND TEXTURE. Natural aggregates, slag and some manufactured aggregates are angular and rough textured. Coated manufactured aggregates are smoother and more nearly rounded in shape.

SOUNDNESS, sodium or magnesium sulfate, ASTM C 88, not a reliable test for lightweight aggregate. Service record of concrete is best indication.

ALKALI REACTIVITY. No problem.

CONTAMINATING SUBSTANCES. Usually result from contamination after manufacture, such as earth scooped up from stockpiles, windblown materials, etc. The staining test detects the presence of iron compounds that may stain the concrete.

ABRASION RESISTANCE. Not significant.

VOLUME STABILITY. No problem.

GRADING. ASTM C 330 sets broad limits. Maximum size of coarse aggregate seldom exceeds $^3/_4$ inch; $^1/_2$ inch maximum is common. Fineness modulus of aggregate in any shipment should not differ by more than 7 percent from that of the acceptance sample

SPECIFIC GRAVITY. Because of the porous nature of these aggregates, leading to difficulties in testing, no attempt is made to control their quality or design concrete mixes on the basis of specific gravity. Specific gravity of coarse aggregate runs between 1.05 and 1.50, and that of fine aggregate, about 1.50 to 1.80.

TABLE 21.2
GRADING REQUIREMENTS FOR LIGHTWEIGHT AGGREGATES FOR STRUCTURAL CONCRETE

SIZE DESIGNATION		PERCENTAGE (BY WEIGHT) PASSING SIEVE HAVING SQUARE OPENINGS									
GRADE OF AGGREGATE	AGGREGATE SIZE	25.0 mm (1 in)	19.0 mm (3/4 in)	12.5 mm (1/2 in)	9.5 mm (3/8 in.)	4.75 mm (No.4)	2.36 mm (No. 8)	1.18 umm (No. 16)	300 umm (No. 50)	150 umm (No. 100)	75 umm (No.200)
Fine	No. 4-0	—	—	—	100	85-100	—	40-80	10-35	5-25	—
Coarse	1"-No. 4	95-100		25-60	—	0-10	—	—	—	—	0-10
	3/4"-No. 4	100	90-100	—	10-50	0-15	—	—	—	—	0-10
	1/2"-No. 4	—	100	90-100	40-80	0-20	0-10	—	—	—	0-10
	3/8"-No. 8	—	—	100	80-100	5-40	0-20	0-10	—	—	0-10
Combined fine and coarse	1/2"-0"	—	100	95-100	—	50-80	—	—	5-20	2-15	0-10
	3/8"-0"	—	—	100	90-100	65-90	35-65	—	10-25	5-15	0-10

Proportioning Mixes. The principles of normal-weight concrete proportioning apply to lightweight concrete and could be followed when use is made of lightweight particles of rounded shape, with sealed or coated surfaces, of low absorption. These principles, however, are difficult to apply to lightweight concrete containing some of the more absorptive aggregates because of the absorption of water, which may be as much as 20 percent by weight of the aggregate; hence, the procedure recommended in ACI 211.2[21.1] should be followed.

In this method, mixes are established by a series of trial mixes proportioned on a cement content basis at the required consistency because the net water-cement ratio cannot be established with sufficient accuracy. Use is made of a specific gravity factor that expresses the relationship between the dry weight of the aggregate and the space it occupies, assuming that no water is absorbed during mixing. All or part of the fine aggregate for concrete of a density greater than 90 pcf will be natural concrete sand. By using the proportioning methods of ACI 211.2, it is not necessary to know the values of absorption and specific gravity of the aggregate, yet this method enables one to proportion and control lightweight concrete with the same degree of precision as normal-weight concrete, provided certain precautions are observed.

Following the suggestions in ACI 211.2, a trial batch is first made, based on dry, loose unit weights of the aggregates and their moisture contents. The first trial batch is made with estimated amounts of the several materials, with sufficient water to produce the required slump. Fine and coarse aggregates should be proportioned in equal volumes, assuming about 28 to 32 cubic feet of dry loose aggregates are required per cubic yard of concrete.

After the first trial batch has been made, corrected quantities per cubic yard are determined, and the "specific gravity factor" for each aggregate is computed. The specific gravity factor expresses the relationship between the dry weight of the aggregate and the space it occupies, assuming that no water is absorbed during mixing. That is, it equals:

$$\frac{\text{weight of aggregate}}{62.4 \times \text{volume occupied in batch}}$$

The volume of aggregate is the difference between 27 cubic feet and the sum of the volumes of cement, water and air in a 1-cubic-yard batch. Subsequent batches are then proportioned at different cement contents.

The producers of lightweight aggregates have developed mix proportioning methods that are uniquely suitable to their particular materials, and the user should consult with the particular lightweight aggregate producer for guidance.

Batching. Production of lightweight concrete presents some difficulties not encountered in normal-weight concrete control. These are not serious and can be kept under control by attention to details. Most of the absorption occurs within the first few minutes of contact with water, but continued absorption causes a significant loss of slump when dry aggregates are batched for concrete. Another disadvantage of using dry aggregates is segregation and loss of fines that occur during handling of the aggregates. If the aggregate contains about three quarters of its potential absorption, its tendency to segregate is greatly reduced, and the slump loss can be kept within reasonable bounds.

Some producers solve this problem by using saturated aggregate, a solution that is satisfactory if the concrete is not to be exposed to freezing and thawing action. Even with air entrainment, concrete made with some lightweight aggregates in a saturated condition does not have good resistance to freezing and thawing, a situation that must be considered when planning to pump lightweight concrete in a cold climate. If, however, the concrete can lose the excess moisture before exposure, it will be satisfactory.

A more recent development is the vacuum treatment of lightweight aggregate, especially desirable for aggregate to be used in concrete to be pumped. (See Chapter 15.) In this process, the dry aggregate is placed in a special chamber and the air exhausted to a 30-inch vacuum. Water is then admitted to the chamber so that the pores in the aggregate particles are filled with water. Water content may be as high as 45 percent by weight. Aggregate can be stored for several weeks after treatment with no serious loss of moisture. Water entrapped in this manner does not affect the water requirement of fresh concrete, nor does it cause bleeding.

Differences in absorbed water result from slight variations in specific gravity of aggregate particles, variations in the time of exposure of the aggregate to moisture, and different mixes. For these reasons, batching lightweight aggregate by weight can cause variations in yield of the concrete. This difficulty can be overcome by batching the coarse aggregate by volume, the principle being that the volume will be constant from batch to batch, and therefore yield will not be affected by variations in absorbed moisture in the aggregate. It is not the absorbed moisture, but the active or free moisture that affects the quality of the concrete.

Equipment for volumetric batching need not be complicated. For example, a batching hopper can be calibrated by painting lines on the interior, indicating the desired volume. If the batch operator cannot see inside the batching hopper, various types of baffles and chutes can be installed to provide adjustable volumes, based on the condition that the hopper is filled to the bottom of the gate or chute each time. A vertical baffle can be installed across the batching bin in such a manner that it can be adjusted vertically. Aggregate flows under the baffle until the level of the aggregate reaches the bottom of the bunker gate and the flow is cut off. Another method is to install a telescoping chute or trunk beneath the bunker gate. By shortening the trunk, more aggregate is permitted to flow into the batching bin. A compartmented wheel is available, consisting of a motor-driven wheel containing compartments of a predetermined volume. Automatic counting of the number of revolutions of the wheel measures the volume of aggregate passed into the batch hopper. (See Figure 21-4.)

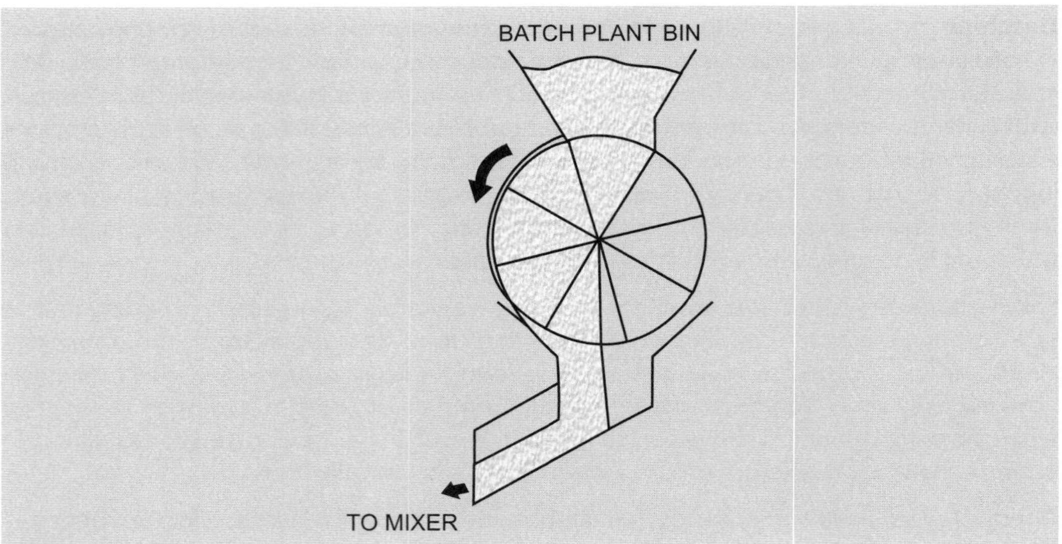

Figure 21-4
A compartmented volumetric batching wheel for measuring volume of aggregates batched. The volume batched is measured by a preset revolution counter.

Mixing and Handling. The appearance of fresh lightweight concrete should be similar to that of normal-weight concrete. Coarse aggregate particles should be evident but not segregated from the mortar. Mortar should be sufficient to coat the coarse aggregate and hold it in suspension, usually requiring that the fine aggregate be between 42 percent and 60 percent of the total aggregate by volume.

The aggregates are lighter in weight than the paste in which they are used, hence the need for careful handling and avoidance of wet consistencies. Because of its lower density, lightweight concrete does not slump as much as normal-weight concrete at the same workability and can therefore be handled with 2 inches less slump than normal-weight concrete. A slump of 4 inches is adequate for any structural work, and 3 inches or less is sufficient for flatwork.

Lightweight concrete may require somewhat more mixing than normal-weight concrete. If truck mixers are being loaded with dry lightweight aggregate, the mixer should first be charged with the aggregate and about three quarters of the water, then mixed for several revolutions until the initial water demand is satisfied. Cement, admixture if any, and the necessary water to give the slump are then introduced into the mixer, and the mixer rotated for 60 additional revolutions at mixing speed. If there is a delay between completion of mixing and discharge, the mixer should be given an additional five revolutions at mixing speed just prior to discharging the concrete.

Checking the unit weight of concrete delivered to the job is an excellent guide to proper batching and yield. As long as the unit weight is uniform, the concrete contains the correct proportions of materials. A change in unit weight indicates a change in aggregate weight or air content. If the air content is unchanged, then the moisture content or density of the aggregates should be checked. Duplicate tests should be run whenever any irregularities are found. Because of the high porosity of lightweight aggregate, air content must be determined by the volumetric method (ASTM C 173). (See Chapter 13.)

Concrete should be consolidated by vibration. In walls, columns and similar structural elements, this is accomplished with internal vibrators. Lightweight concrete moves more readily under vibration than normal-weight concrete; hence, special care is necessary to prevent segregation and the formation of honeycomb. Lightweight concrete can be overvibrated. Vibration should therefore be kept to the minimum that will consolidate the concrete properly. Careful use of a grid tamper (jitterbug) on slabs will help to embed the coarse particles that tend to rise to the surface. This normally follows the strike-off and is followed by darbying or bullfloating. Figure 21-5 is a view of placing lightweight concrete for a floor slab.

Figure 21-5
Lightweight structural concrete is being placed for a floor. (Courtesy of PCA)

Finishing and Curing. The operations of screeding, darbying and bullfloating have a tendency to cause the coarse aggregate particles to float to the surface of the slab and should therefore be kept to a minimum. Most finishers prefer to use a magnesium darby or bullfloat on lightweight concrete. Segregation and a lack of mortar on the surface, sometimes resulting from an undersanded mix, can make finishing difficult.

As in finishing of any concrete, timing is critical. Floating, best done with a magnesium float, should be commenced as soon as the free moisture disappears from the surface, followed immediately by the first flat troweling. If floating is done by machine, normal float shoes will be found suitable. Subsequent finishing is then done in the same manner as for regular-weight concrete.

Good curing is essential and should be continuous for at least seven days. Any of the standard curing methods can be applied, after which the concrete should be permitted to dry before it is put in service.

21.3 Lightweight Insulating Concrete

Concretes possessing densities between 15 and 50 pcf are classified as insulating concrete and are used for thermal insulation in roofs and other areas, as well as fire protection for certain structural portions of buildings. Insulating concrete has but little structural value, as its 28-day compressive strength ranges from less than 100 psi to slightly over 1000 psi, depending on the materials, mix proportions and curing. Thermal conductivity (*k* value) ranges from less than one to about four.

Mineral Aggregate Concrete. Aggregates commonly used are vermiculite and perlite. Vermiculite, a micaceous mineral, expands rapidly to ten or twelve times its original size when heated to about 2000°F, producing a very lightweight product with a density (dry loose) between 6 and 10 pcf. It is extensively used in plaster and as a loose filling in cavities to give improved thermal insulation, as well as in lightweight insulating concrete.

Perlite is a glassy volcanic rock. When heated rapidly to incipient fusion it expands to form cellular material with a density between $7^1/_2$ and 12 pcf. Perlite also is often used in plaster and fills in addition to its use in concrete.

Perlite and vermiculite should meet the requirements of ASTM C 332. They are normally furnished in 4-cubic-foot bags weighing about 32 pounds. Concrete is proportioned as a ratio of loose volume of aggregate per bag of cement, so that a batch containing one bag of each material results in a 1:4 mix. Usual mix proportions are 1:4, 1:6 or 1:8.

When vermiculite is used, a 1:4 mix is recommended for a roof deck placed over form boards of some type, whereas a 1:6 or 1:8 mix would be used for roof insulation placed over structural concrete or steel decking. A 1:6 perlite mix is commonly recommended for general use, and 1:8 for use over structural or precast concrete roof slabs. Lightweight floor fill concrete, at least 2 inches thick, should be a 1:4 mix.

Water requirements vary quite widely because of variations in air content, slump, yield and absorption by the aggregate and are best determined by placing conditions. Both perlite and vermiculite mixes are usually quite plastic, with high slumps, and are usually placed (or "poured," in this case) with a minimum of manipulation. Vermiculite concrete requires about 90 to 100 pounds of water per cubic yard, and perlite requires between 54 and 61 pounds.

Insulating concrete can be mixed either at the site or in transit mixers. For transit-mixed vermiculite concrete, water and cement are placed in the mixer, and the drum rotated slowly while the aggregate is mixed. Mixing should not be done on the way to the job, as overmixing must be avoided, but mixing at maximum speed for about five minutes should take place at the job. Perlite concrete can be mixed at slow speed while in transit for no longer than five minutes, then about five or ten minutes of maximum speed at the job.

These concretes can be handled by conventional concrete conveying equipment, including pumps. Placing and finishing consists of striking off or screeding. (See Figure 21-6.) Tamping, vibrating and troweling are not necessary and should be avoided, as they merely serve to make the concrete denser.

Figure 21-6
Placing and striking off lightweight insulating concrete for a roof.

A common usage of this kind of concrete is for roof and floor decks. The concrete can be placed on a variety of base materials. Figure 21-7 shows a typical roof system. One method is to place the concrete over corrugated or ribbed sheet steel—a method that is especially prevalent for floor construction—or it can be placed on cast-in-place or precast concrete, such as prestressed tees.

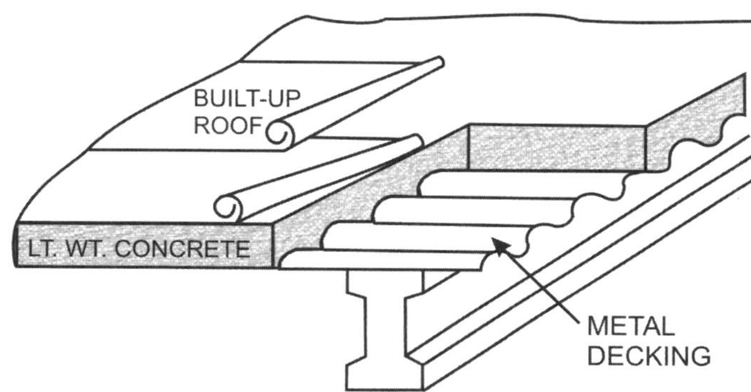

Figure 21-7
A metal-deck insulating concrete roof system consists of lightweight insulating concrete placed on the corrugated or ribbed steel. A built-up roof is applied after the concrete has cured and thoroughly dried.

Another method, when spacing of joists permits, is to lay paperbacked metal lath or ribbed metal lath across the joists and to place the lightweight concrete on either of these materials. Being relatively open on the bottom side, the lath methods are particularly desirable for roof construction, as the concrete dries from both top and bottom, permitting the roofing material to be laid earlier than on the steel or concrete-backed lightweight. The concrete must be completely dry before the roofing can be applied.

Cellular Concrete. Also called aerated or foamed concrete, this concrete contains bubbles of air or gas that are formed in the plastic mortar; the porous structure remains after the material has hardened. Gas is introduced by chemical action or by the addition of a foam or foam-producing substance into the mix. Finely ground aluminum powder, when added in an amount equal to about 0.2 percent by weight of the cement in a slurry, causes the formation of hydrogen gas, which expands the mass and results in a cellular material. Other foaming agents are usually some sort of a hydrolized protein, resin soap, detergent or similar substance.

There are two methods for making mechanically foamed cellular concrete. In the first method the foam is preformed in a special tank by diluting the foaming agent with water. The material passes through a mixing tube from which it is delivered through a flexible tube to the semiplastic cement-water paste in the mixer, preferably a pan or paddle type of mixer. Alternatively, the preformed foam can be placed in the concrete mixer with the water, and the dry cement and fine aggregate, if any, can be added and mixed for a short time. Mixing is continued after the foam has been added until the mass is homogeneous.

In the second method, the cement, aggregate, foaming agent and water are all mixed together in a paddle or pan mixer, sometimes assisted by aeration with compressed air. Both methods require specialized skills and equipment and for this reason are usually done by subcontractors who specialize in this work. Then, too, foamed concrete is apt to suffer high shrinkage, which makes onsite application difficult. It has been used for factory-made precast units with a good degree of success.

HEAVYWEIGHT CONCRETE

Ordinary concrete is a good material for use in the biological shielding of nuclear reactors. It performs quite well for shielding against neutron and gamma rays if a sufficient thickness is used. However, heavy concrete made with special high-density aggregates is more effective and requires less thickness than ordinary concrete. By using special aggregates, concrete weighing as much as 400 pcf can be produced.

21.4. Materials and Properties

The principal aggregates for heavyweight concrete:

1. **Barite.** An ore composed chiefly of barium sulfate, it has a specific gravity of about 4.3. It is relatively soft and, compared with normal aggregate, is of poor physical quality.
2. **Limonite.** Specific gravity ranges from 3.6 to 4.0. Hardness varies from soft, claylike material to hard stone suitable for concrete aggregate. Limonite, which is an iron ore, is somewhat friable and breaks down in handling.
3. **Magnetite.** Another iron ore, it has a specific gravity from 4.9 to 5.2. Magnetite is harder and heavier than limonite.
4. **Iron and Steel.** Scrap steel punchings and graded cast iron are sometimes used for coarse aggregate. These materials have a specific gravity of 7 or 8. It is recommended that they be permitted to rust before using. Slight rusting only is needed.

Most of these heavy aggregates are angular; this, together with their high density, makes a concrete that is less workable than normal concrete of similar characteristics. Mixes should be as stiff as can be placed, using vibration for consolidation.

The amount of mixing should be the minimum that will blend the ingredients; excessive or overmixing breaks down the coarse aggregate particles, especially the soft and friable aggregates. Half-size batches should be used to avoid excessive wear on the equipment. Once mixing of a batch has begun, starting and stopping the mixer frequently should be avoided because of the abnormal loading of the machinery.

Segregation is more serious than for normal-weight concrete because of the differences in specific gravities of the materials. Smaller batches in the mixer are less conducive to segregation.

Good construction practices should be followed in all operations, including design and construction of forms to withstand the greater pressure resulting from heavy concrete. Heavy concrete can be pumped the same as ordinary concrete, but lifts and distances are considerably shorter because of the greater unit weight of the concrete.

Pipes and other embedded items are spaced so close together in some installations that conventional concrete placing techniques cannot be followed, and it is necessary to resort to the intrusion method. In this method, the coarse aggregate is first placed by hand in the spaces to be concreted, and grout pipes are installed at the same time. Subsequently, the space is filled with grout, pumped in under pressure, starting at the lowest portion of the forms. (See Chapter 22.)

Another use for heavy concrete is in such items as counterweights for lift bridges in which it is sometimes necessary to proportion and maintain the concrete at a certain high density within rather close tolerances.

Proportioning of mixes for heavy concrete follows conventional methods, with about the same proportions by volume as for normal concrete. Mixes should be proportioned by volume and the components converted to parts by weight for batching. Parts by weight are, of course, considerably different for the aggregates because of their high density. Cement contents may range from 475 to 650 pounds per cubic yard, with water-cement ratios approximately the same as for normal concrete of similar quality.

Compressive strength and elastic modulus values are similar to those of normal concretes of similar cement contents and water-cement ratios. Durability of heavy concrete exposed to freezing and thawing is rather poor in some cases. For example, barites produce a concrete of low durability. If entrained air can be tolerated, the durability of exposed concrete is greatly improved. Concrete inside buildings, or in a moderate climate where freezing and thawing do not occur, needs no protection.

21.5 Mixes

To determine quantities of materials for a concrete of specified cement content and water-cement ratio required to have a high density, first determine weights and volumes per cubic yard of cement, water, voids, total aggregates and sand. Next determine weight and volume of total coarse aggregate per cubic yard, and finally, determine weight and volume per cubic yard of normal coarse aggregate and heavy aggregate.

Special Concreting Techniques

The great versatility of concrete, together with the unique demands of certain construction problems, have led to the development of a number of special techniques for handling, placing and finishing concrete. The inspector may encounter some of these techniques every day and others never. The knowledgeable inspector should, however, have a general awareness of all special concreting techniques.

22.1. Tilt-Up Construction

This is a type of precast construction in which wall panels are cast in a horizontal position at the site, tilted to a vertical position, and moved into final location as part of the building. (See Figure 22-1.) As a rule, the concrete floor of the building serves as the casting platform. (See Figure 22-2.) Panels may be of solid concrete or of sandwich construction in which relatively thin, high-strength, conventional concrete surfacing layers are separated by a core of low-density insulating material. Tilt-up is especially suitable for commercial and industrial buildings of one or two stories and has been used in residential construction.

Figure 22-1
Tilt-up is the technique of site casting concrete walls, normally on a horizontal surface and then tilting them vertically into place. (Courtesy of TCA)

Figure 22-2
Tilt-up panels being cast on the floor of a new building. (Courtesy of TCA).

Fabrication is accomplished by first placing a bond breaker on the casting floor. Liquids of various types are used. Sheets of plywood, metal or paper have been used but are unsatisfactory because of higher cost, staining or roughness. Liquids consisting of special formulations for this purpose, curing compounds and waxes are applied in two coats, the second coat being applied shortly before the panel concrete is placed. Uniformity of application is important. The inspector must verify that the casting floor is perfectly smooth and uniform, as any imperfection in the floor will be imprinted on the wall. If the panel is to be painted or have other surface treatment later, the bond breaker must be one that will not interfere with the surfacing.

Side forms are usually of wood. Forms for windows and other openings can be metal or wood, with metal preferred because swelling of wood frames makes them difficult to remove and may crack the concrete unless the wood is well saturated before concrete is placed.

Reinforcing steel, inserts, conduits and other services are placed in the form, then the concrete is placed in the same manner as for slabs. Many special finishes can be applied while the concrete is still plastic, including exposed aggregate, embedment of architectural details and ornamentation.

Although various types of cranes and gin holes have been used to lift the panels into place, the mobile truck crane is the choice of most erectors. (See Figure 22-3.) The panels are tilted onto the foundation, plumbed and temporarily braced. Usual practice is to set the panel on two mortar pads previously placed on the foundation and accurately struck-off to the exact elevation for the bottom of the panel. (See Figure 22-4.) Voids between the foundation and the bottom of the panel are later filled with cement mortar or drypack. An alternative method is to set the panel in fresh mortar spread on the top of the foundation.

Tilting induces stresses in the panel that are entirely different from those it will be called upon to resist after it has been erected, and this must be considered in the design. The inspector should make sure that the concrete has reached the required strength with a factor of safety before the panel is tilted. Field-cured cylinders or nondestructive tests will provide this information. Pickup points must be carefully located and lifting equipment designed so as to avoid high localized stresses in the panel that cause cracking, splitting or spalling of the concrete.

Figure 22-3
The mobile truck crane is usually used for lifting tilt-up panels. Note how the lifting gear is rigged to equalize the load on each of the four pickups on the panel. (Courtesy of TCA)

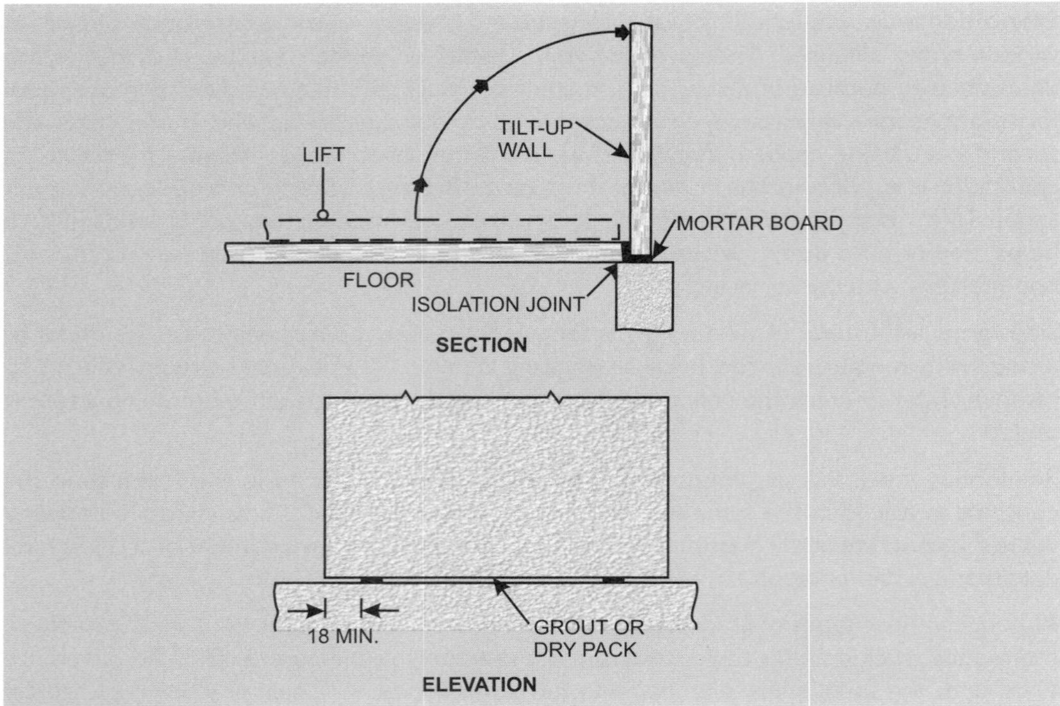

Figure 22-4
The panel is tilted onto two hardened mortar pads on the foundation. Space for an isolation joint is provided between the floor and the panel.

Figure 22-5
Starting a lift using an R-24 (two-high, four-wide) rigging arrangement. (Courtesy of TCA)

The size and location for pickup points on the panel are determined mainly by the size, weight and configuration of the panel. Other factors may enter into the consideration as well. Figure 22-5 shows one typical rigging. Each hoisting line is attached to the panel by means of a lifting device consisting of an anchor that is embedded in the concrete and an attachment element that is bolted into the anchor. Figure 22-6 is one typical insert. Vacuum lifters are sometimes used.

Figure 22-6
The three components of one type of lifting device. A shows the insert that is embedded in the concrete. B shows the swivel lifting plate with the bolt that screws into the insert. C shows the bolt having been tightened and the device ready to be attached to the crane hook.

More lifting force is occasionally required to break the panel loose from the casting floor than is necessary to lift the panel after movement has started. If this happens, it is a good idea to move the panel slightly, if possible, before lifting. This can be accomplished with jacks operating in a horizontal direction, sliding the panel a fraction of an inch, to break the bond.

Temporary bracing is accomplished with braces that are attached to the panel before the panel is tilted. These braces, attached to the panels lying flat as shown in Figure 22-7 can be of wood or metal. The braces shown are made of special heavy steel tubing, with an adjusting jack as shown in Figure 22-8. The lower end of the brace is attached to an insert in the floor. Columns or pilasters between the panels are formed and the concrete placed later in the job. If the column concrete is to be bonded to the panel concrete, tie bars will be cast in the panel to extend into the pilaster. If they are unbonded, contact areas will be coated with a bond breaker, and smooth dowels instead of tie bars will be used. Whether the units are to be bonded is a design consideration, and the construction requirements will be explained in the project plans and specifications.

Figure 22-7
Tilt-up panels ready to be lifted with temporary pipe braces attached. (Courtesy of TCA)

Figure 22-8
By means of this jack the length of the temporary brace can be adjusted to bring the panel into final plumb position.

The role of the special inspector for the tilt-up method of construction is similar to that for any other type of concrete construction, as outlined in Section 25.9. The inspector should, however, be very familiar with the special construction techniques and safety concerns associated with the lifting and erection of tilt-up panels. A comprehensive reference manual published by the Tilt-Up Concrete Association (TCA) is available in Reference 22.1.

22.2. Slipforms

A slipform is a sliding form that moves continuously as concrete is placed. Some slipforms move horizontally; examples are slipforms for canals, pavements, curb and gutter, and highway medians. Other slipforms move vertically, starting at the bottom and rising as the concrete is placed; examples are slipforms that are used for constructing building elements, silos and bins, and tall tubular structures as shown in Figure 22-9.

Vertically Moving Forms. Slipforms are particularly adaptable to constructing the center core of a building, the portion that usually contains stairways, elevators and other service features. They are very useful for constructing storage silos and have been used for constructing the complete wall system for a building. Concrete for the wall under construction is placed continuously in thin layers of 6 to 8 inches, and the form is raised by means of jacks operating on smooth steel rods or pipes that are embedded in the concrete as the structure rises. Jacks are usually hydraulic, although some use has been made of pneumatic jacks. The area at the top of the form is decked over and enclosed to provide a working platform on which the jacking and other equipment is located. Reinforcing steel, inserts and blockout forms are temporarily stored on this platform prior to their installation immediately ahead of the form as it rises. The form consists of an inside and an outside form made of sheet steel with appurtenant bracing, plywood or vertical boards. Forms should be at least 43 inches high and constructed with a slight draft; that is, the space between inner and outer forms is slightly wider at the bottom than at the top, usually about $^1/_8$ inch. The outside form should extend about 6 inches above the inside form. Suspended below the form are platforms on which the finishers work. As the form rises, the finishers are able to repair and finish the concrete surface.

Of great importance is getting the form set up in a perfectly plumb and level condition at the beginning, and locating the jacks so they are all equally loaded. It is extremely difficult to plumb a form that has drifted off the true vertical, although manipulation of the jacks may provide some correction.

Figure 22-9
The slipform method of placing concrete is ideally suited for tall tubular structures. (Courtesy of PCA)

True vertical movement is assured by the use of a center guide that maintains the form in the proper position and by the use of plumb lines or optical plummets. Operation of the jacks can be varied to keep the platform level. A series of interconnected water levels, connected to a central reservoir and situated about the operating deck, enables the operators to regulate the speed of movement of individual jacks, thereby maintaining the deck in a level condition.

It is, of course, impossible to have any projections beyond the face of the wall during slipforming, but there is practically no limit to the variety of inserts, blockouts and openings possible within the limits of the wall. These inserts and openings provide for windows, doorways and attachment of beams and other structural items. Door and window frames can be accurately set in place and anchored by welding or other means ahead of the form, or an oversize blockout can be made and the frame set in place later. The latter method is usually preferred because it permits more accurate adjustment of the frame in its final position.

Slipform construction requires careful inspection and execution of the work by supervisors and workers experienced in this type of work, including farsighted planning of the whole operation. Concrete mixes should be well proportioned, with a slump between 2 and 4 inches, depending on the weather, size of wall, presence of steel or other items in the wall, and similar items. Concrete can be both spaded and vibrated into place, but revibration of previous layers below the one being placed must be avoided. A sufficient and steady supply of concrete must be available and the placement made so there is not much more than an hour's delay between layers, or lifts. Once a placement has been started, it should be continued 24 hours a day until completion. The form should be kept full of concrete the entire time the slip is in progress. The rate of slip varies considerably. Rates as low as 1 inch per hour have been reported, but this is unusually low. Average rates should be about 12 inches per hour. An occasional rate as high as 20 or 24 inches is not unusual.

The normal slipform finish is a rubber float applied to the concrete as it becomes exposed by the vertical movement of the form. Delayed finishes may include a sand finish or cement plaster.

Curing should be applied immediately. Water-curing is satisfactory, but sprayed curing compound is usually employed, provided the compound will be compatible with any coating or treatment to be applied to the concrete later.

Horizontal. Slipforms that move in a horizontal direction are used for placing concrete lining in canals, for curb and gutter, cast-in-place pipe and pavements. Canal liners formerly operated on rails set to line and grade on the berm of the canal, but modern machines operate on crawler tracks, with line and grade controlled by sensors that are actuated by wires set to line and grade.

Concrete is distributed across the forms by means of a bucket and aprons or downspouts. As the form moves ahead, mechanical vibrators consolidate the concrete. Finishers working on outriggers behind the form make the necessary repairs to the surface and make contraction joints. A special carriage, following at the proper time, provides a platform for the application of curing compound. Small self-propelled slipforms are available for continuous placing of concrete in small canals, curb and gutter sections (see Figure 22-10), sidewalks and similar structures. Large slipforms are used for constructing pavements for highways and airports. Pavements as wide as 48 feet and a foot or more in thickness can be formed in one pass.

Figure 22-10
Curb machines continuously extrude low-slump concrete into a shape that immediately stands without support of formwork. (Courtesy of PCA)

22.3. Lift Slabs

In the lift slab technique, the building floor serves as a casting floor, enabling the roof and upper floor slabs to be cast and post-tensioned at ground level. Several slabs may be cast, stacked one on top of the other, each of them raised individually to its required elevation. Raising is accomplished by means of jacks mounted on top of the building columns, openings being left in the slabs to enable them to rise along the columns.

Forming, casting, finishing, curing and tensioning follow closely the procedures for any slab production methods. To raise the slab, suspension rods, reaching down from the jacks on top of the columns, are first attached to collars on the slab, one collar encircling each column. All jacks are synchronized and operate on a short stroke of the magnitude of $^1/_2$ inch to 2 inches. Follower nuts automatically hold the slab when the jacks are retracted. Raising is accomplished by extending the jacks, the slab being held by the nuts while the jacks are retracted, then extending the jacks, repeating the cycle until the slab comes to the required elevation, where the collars are attached to the columns. Usual jacking rate is between 5 and 15 feet per hour. Of special importance to the inspector and construction superintendent is the need to keep the slabs level and to avoid overstressing any portion of the slab.

Lift slab construction is a specialized field, requiring expensive equipment and expert ability on the part of those doing the work. It is nearly always subcontracted to specialists.

22.4. Placing Concrete under Water [22.2]

Concrete should be placed in the dry rather than underwater whenever possible. When it must be placed underwater, certain safeguards and restrictions are necessary to assure the best control possible under the conditions. Regardless of other restrictions, under no circumstances should concrete be placed in running water. In those structures that are located in a current of water, such as a bridge pier foundation in a river, normal practice is to enclose the area within a sheet piling cofferdam or other enclosure that protects the work area from movement of the water.

Most users suggest an increase in cement content over what would be specified for concrete placed in the dry. However, excessive laitance may develop on very rich mixes. Careful handling of the concrete and equipment can keep the loss of cement to a minimum, but an increase of about 10 percent in cement content is good insurance. Entrainment of air and the use of a water-reducing retarder are both beneficial in improving the flow of the concrete and cohesiveness of the mixture. A plastic, workable mix with a slump of 5 to 7 inches is required, as no working or vibration of the concrete can be done. The concrete should not be disturbed after it has been placed. Methods for placing concrete underwater include using the following: tremies, bottom-dump buckets and pumping.

Tremies. A tremie is a smooth pipe, at least 10 inches in diameter, consisting of sections joined together by means of flanged and gasketed couplings, with a hopper section at the top to receive the concrete. (See Figure 22-11.) The tremie should be supported so as to permit free movement of the discharge end over the area in which concrete is being placed, and also to permit rapid vertical movement to regulate or stop the flow of concrete. Usual practice is using one crane to handle the tremie while using another crane to handle concrete buckets and other items. (See Figure 22-12.)

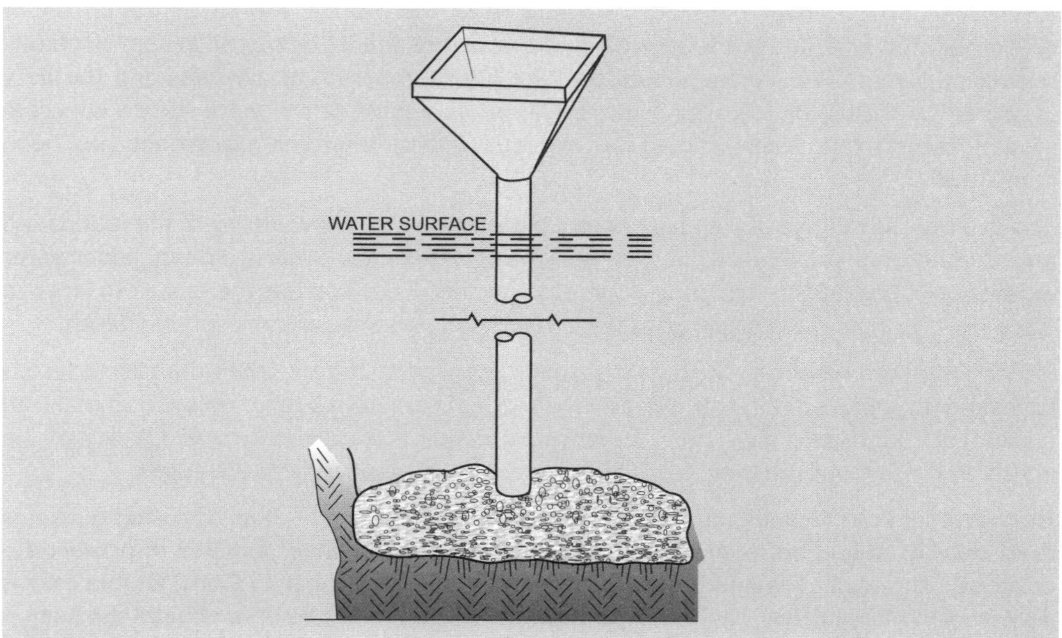

Figure 22-11
The bottom of the tremie pipe must be embedded in the fresh concrete at all times during the placement.

Figure 22-12
The heavy-duty lay-down bucket is feeding concrete to the tremie hopper.

Some kind of end control is necessary at the start of tremie placing of concrete. The best control is achieved with a valve inserted about halfway down the pipe. The pipe is lowered with the valve closed, then air is admitted to the pipe below the valve, forcing the water out, as indicated by bubbles rising to the surface. Concrete is now let into the pipe to slightly exceed the air pressure, and the valve is opened, letting the concrete drop on a cushion of compressed air.

Another method is to use a bottom valve that is shaped like a cone with the point up. To start concrete placing, the valve is closed and several feet of concrete placed in the pipe before the pipe is lowered. After the pipe is in position, the valve is opened and concrete placing continued. The advantage of the cone valve is the control that can be exercised over the flow of the concrete.

To place concrete with a tremie, the bottom of the pipe must be kept continuously submerged in the fresh concrete. Movement of the tremie should be regulated to accomplish this end, and at the same time the concrete should be placed in approximately horizontal layers. The tremie should be kept full of concrete at all times and the flow regulated by raising or lowering the tremie, or operation of the valve. Once concrete placing has started, there should be no interruption until the placement has been completed.

The principal advantage of tremie concrete placement is that dewatering of the foundation area is unnecessary, a feature that permits placing a foundation seal in a deep, underwater excavation. Subsequent structural concrete can then be placed in the dry. Also, tremie placement permits rapid placing of a large volume of concrete at considerable depth.

Inspection of the results is, of course, virtually impossible, except when the top surface is subsequently exposed. On some important work, divers have been employed to make an occasional underwater inspection. Laitance formation is usually excessive. Quality of the concrete throughout the mass is apt to be nonuniform, especially at the edges.

Buckets. Use can be made of a bottom-dump watertight bucket. If an open-top bucket is used, the top should be covered with heavy canvas or other sheet material to protect the concrete. The bucket should be lowered slowly while under water. Considerable care is necessary in manipulating the discharge gate, which should not be opened until the bucket comes into contact with the foundation or previously placed concrete. After the concrete has been discharged, the bucket must be raised slowly until it is well away from the concrete.

Pumping. Development of the mobile concrete pump has made the placing of concrete underwater easier and with better quality control than either the tremie or bucket method. Pumping is now considered to be the best method of placing concrete underwater. For underwater pumping, a temporary plug is placed in the end of the discharge line before the line is lowered into the water. To prevent the concrete from mixing with the water during pumping, the discharge end must be kept continuously submerged in the fresh concrete. The surging action of the pumping will provide some degree of consolidation as the concrete is being placed. The placement must be started slowly to minimize scouring of the bottom. The discharge end should be placed as near as practicable to the surface against which the concrete is to be placed and not raised until a sufficient depth of seal has been established. The discharge line should be lifted slowly to ensure that the seal is not broken.

22.5. Preplaced Aggregate Concrete [22.3]

This is a specialized type of construction in which the forms are filled with coarse aggregate, after which a cement-sand grout, usually with admixtures, is pumped into the form to fill the voids, starting at the bottom. This method has been extensively used for underwater construction of such structures as bridge piers, for repair of deteriorated structures, for especially complicated and congested structures such as nuclear reactor shields and other heavy construction. Another usage is for architectural treatment in which special aggregate is preplaced and held against the form by wire netting while conventional aggregate is preplaced in the center of the beam or other element being made. Special colored grout is then introduced, and after a suitable time the form is stripped and the aggregate exposed by chemical or mechanical means.

Any of the portland cements can be used. The coarse aggregate should conform to ASTM C 33, except for grading. If regular concrete sand is used in the grout, the minimum size of coarse aggregate should be about $1^1/_2$ inches. If plaster sand is used, the coarse aggregate size may as small as $^1/_2$ inch. The aggregate should be graded so it will have the minimum percentage of voids.

A pozzolan is normally used to improve the properties of the cement-sand grout mixture. Air-entraining admixtures and calcium chloride can be used to provide their special benefits.

Preplaced aggregate concrete is a specialized task, and the work should be done by experts in this field.

22.6. Vacuum Concrete

Vacuum concrete is defined as concrete from which water and entrapped air are extracted by a vacuum process before hardening occurs. The vacuum process is accomplished by applying a vacuum to a fresh concrete surface through the medium of special permeable form liners or pads. This patented process removes as much as 40 percent of the water from the surface inch or two of the concrete, and up to 20 percent of the water from a depth of 6 inches below the surface of the concrete, producing, in effect, a case hardened concrete. It also eliminates the large surface voids or bug holes. Removal of air is a surface condition only. The result is higher early strength, improved durability, better wear resistance, less shrinkage and higher density than for comparable untreated concrete.

Another advantage is the saving in time. The mechanical densification of the concrete makes it quite hard, so the forms can be removed in minutes instead of hours, and finishing of slabs can be commenced sooner. It is, of course, an expensive expedient. Vacuum treatment is normally done by subcontractors who specialize in this work and is used only where it is felt that normal good construction practices will not give the quality of concrete surface possible with vacuuming.

22.7. Shotcrete

Shotcrete[22.4] is the term used to designate a mortar or small MSA concrete that is conveyed through a hose in a stream of air and shot (pneumatically projected) onto the surface at high velocity. It is also commonly known as pneumatically placed concrete. Shotcrete requires expert and conscientious construction techniques and should be done under careful supervision and inspection, employing only experienced workers,[22.5] and in accordance with local building code requirements.

Shotcrete is suitable for a variety of new construction and repair work. However, its properties and performance are largely dependent on the conditions under which it is placed, the capability of the particular equipment selected, and especially on the competence of the application crew.

Shotcrete is frequently more economical than conventional concrete because of the reduced forming requirements and because it requires only a small, portable plant for manufacture and placement. Properly applied shotcrete is a structurally adequate and durable material and it is capable of an excellent bond with concrete, masonry, steel, sound rock and some other materials. The uses of shotcrete include the following:

1. New structures—roofs (particularly curved or folded sections), walls, prestressed tanks, reservoir linings, canal linings, swimming pools, tunnels, sewers and shafts. (See Figure 22-13.)
2. Coatings—over brick, masonry, concrete, rock and steel.
3. Encasement—of structural steel for fireproofing and reinforcing.
4. Strengthening—of concrete slabs, and concrete and masonry walls.
5. Repair—of deteriorated concrete in structures such as reservoir linings, dams, tunnels, shafts, elevators, waterfront structures and pipe. Repair of earthquake and fire damage to masonry and concrete structures.
6. Refractory linings—in stacks, furnace walls and boilers.
7. Underground support and slope stabilization.

Figure 22-13
Canal lining being constructed of shotcrete. (Courtesy of PCA)

Shotcreting Processes. There are two basic shotcreting processes: the dry-mix and wet-mix processes.

Dry-mix process—This process consists of the following steps:

1. Cement and damp aggregate are thoroughly mixed.
2. The cement-aggregate mixture is fed into the delivery equipment (a special mechanical feeder or gun).
3. The mixture is metered into the delivery hose by a feed wheel or distributor.
4. This material is carried by compressed air through the delivery hose to a special nozzle. The nozzle is fitted inside with a perforated manifold through which water is introduced under pressure and intimately mixed with the other ingredients.
5. The mortar is jetted from the nozzle at high velocity onto the surface to be shotcreted.

Wet-mix process—This process consists of the following steps:

1. All of the ingredients, including mixing water, are thoroughly mixed.
2. The mortar or concrete is introduced into the chamber of the delivery equipment.
3. The mix is metered into the delivery hose and conveyed by compressed air or other means to a nozzle.
4. Additional air is injected at the nozzle to increase the velocity and improve the gunning pattern.
5. The mortar or concrete is jetted from the nozzle at high velocity onto the surface to be shotcreted.

Shotcrete suitable for normal construction requirements can be produced by either process. However, differences in cost of equipment, maintenance and operational features may make one or the other more attractive for a particular application. Differences in operational features that may merit consideration are given in Table 22.1.

TABLE 22.1
COMPARISON IN OPERATIONAL FEATURES OF DRY- AND WET-MIX PROCESSES

	DRY-MIX PROCESS		WET-MIX PROCESS
1	Mixing water and consistency of mix are controlled at the nozzle	1	Mixing water is controlled at the delivery equipment and can be accurately measured
2	Better suited for placing mixes containing lightweight porous aggregates	2	Better assurance that the mixing water is thoroughly mixed with other ingredients. This may also result in less rebound and waste
3	Capable of longer hose lengths	3	Less dust accompanies the gunning operation

Preparation, Placement and Finishing. Coatings may be applied to deteriorated concrete or masonry, after removal of unsound, deteriorated material; to rock surfaces to prevent scaling or disintegration of newly exposed surfaces; and to steel for fireproofing.

Almost any type of deteriorated concrete surface can be repaired by application of shotcrete.

In new construction, the forms, compacted earth or other surface must be clean and sufficiently rigid to withstand the shooting and stout enough to carry the load. In repair of old concrete or masonry, all the old unsound material must be removed so that the shotcrete is applied to a sound surface. Heavily corroded steel should be sandblasted, using the shotcrete equipment for this purpose. Reinforcing should be securely doweled or bolted in place.

On vertical or overhanging surfaces, the mortar should be applied in layers not exceeding $^{3}/_{4}$-inch thick. On horizontal or nearly horizontal surfaces, when shooting downward, the thickness can be as much as 3 inches. Excessively thick layers will cause the mortar to slough, or sag. Sagging may also occur if too much water is used, or if insufficient time elapses between layers. Time between successive layers should be at least 30 minutes, but should not be so long as to permit the previous layer to set completely.

The nozzle should be held as nearly perpendicular as possible to the surface being treated and should be held uniformly about 3 feet away from the surface at all times.

An adequate supply of air is essential. Pressure should be uniform, at least 45 psi at the gun tank when using up to 100 feet of material hose. The pressure should be increased 5 psi for each additional 50 feet of hose.

A portion of the mortar bounces from the surface where it is being applied, the amount varying with air pressure, quality of sand, placement conditions, and the cement and water contents. The amount of rebound varies from about 30 percent when shooting vertical surfaces to about 20 percent from horizontal or sloping surfaces. Because of its variable quality and low cement content, this rebound material should not be reused. It can, however, be recovered and used as sand in an amount not exceeding 25 percent of total sand requirements if it is screened and free of contaminating material.

Shotcrete can be finished the same as any concrete or mortar, but finishing should be kept to a minimum. After the shotcrete has been applied, it should be cured by any of the standard methods. A light spray of water should be applied as soon as possible without damaging the surface. Water curing should be continued for at least five days. Instead of curing with water, an approved sealing or curing compound can be applied.

Samples for strength tests can be obtained by coring or sawing samples out of the finished work, by shooting a special test panel and coring or sawing samples from the panel, or by shooting directly into molds.

Coring or sawing samples should follow standard procedures for obtaining samples of concrete. When making a test panel, every effort should be made to make a representative panel. All too often the nozzle person will, either consciously or unconsciously, try to make a "good" sample. This must be avoided. Samples can be either small cylinders or cubes.

Test cylinders cannot be made in the molds used for making concrete specimens. Instead, a mold is made of $^{1}/_{2}$-inch mesh hardware cloth shaped into a cylinder. A common size is 6 inches in diameter by 12 inches high, although some users prefer a smaller size. After the mold has been filled, the excess shotcrete is trimmed off the outside. Trimming is done carefully after the shotcrete in the mold has stiffened somewhat but before it has hardened. The mold can be removed after the shotcrete has hardened.

The requirements for shotcrete will be detailed in the project specifications.

22.8. Fastening Base Plates

A base plate is a plate, usually of steel and occasionally of cast iron, that supports a structural column or a piece of machinery. Structural bases are made of flat steel plates, whereas those for machinery may be of flat steel or shaped cast iron. These bases must be set accurately to line and grade, and are usually bolted to the underlying concrete by anchor bolts set in the concrete.

Setting the base to the correct elevation can be accomplished in several ways. One method is to insert wedges of wood or metal at three or more locations under the plate, bearing on the foundation concrete, observing the level of the plate by means of a carpenter's level or surveyor's level. Another method makes use of a washer and nut below the plate and above the plate on each anchor bolt, as shown in Figure 22-14. The nuts can be adjusted to bring the plate to the proper position. On especially heavy bases, adjusting screws may be attached to the plate through threaded holes, or nuts welded to the bottom of the plate. (See Figure 22-15.) These screws bear against a bearing plate resting on the concrete surface. After the base has been adjusted to the correct position, the space underneath is filled with dry-pack mortar or grout. All shims, wedges and adjusting screws should be removed after the grout has hardened.

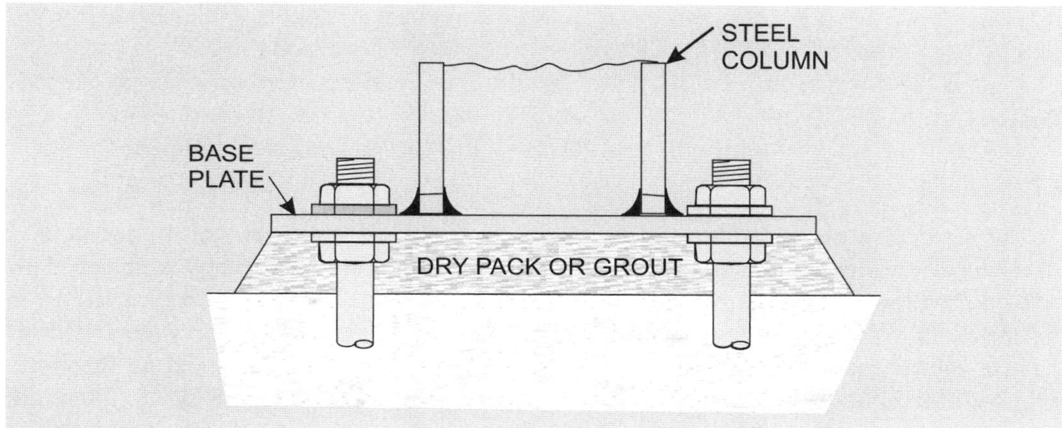

Figure 22-14
A nonshrinking mortar or grout can be used for grouting under a steel column. There must be no free water on the surface or in the bolt holes. After the grout takes its initial set, the edges can be finished as desired. The grout must be cured for a minimum of three days.

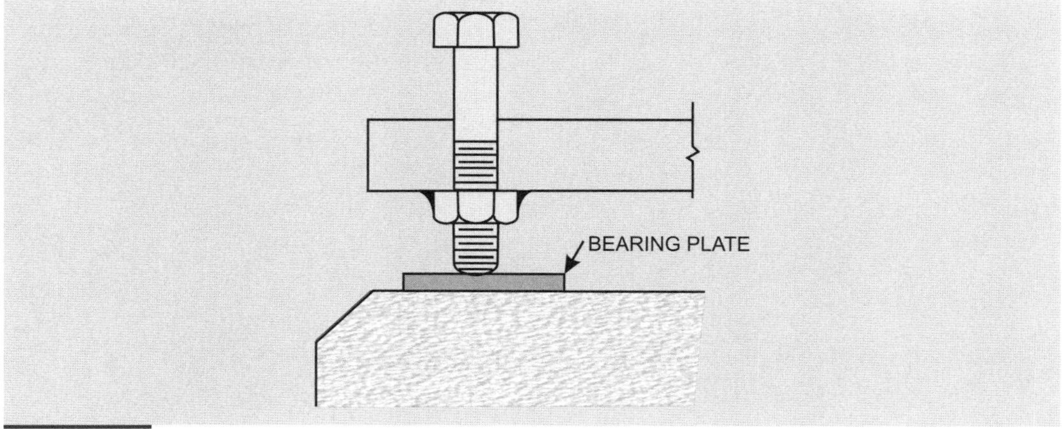

Figure 22-15
Adjustment of heavy base plates can be done with three or four adjusting screws attached to the plate. After the grout under the plate has hardened, the screws can be removed.

Bases are practically always bolted down. Accuracy in setting the anchor bolts can be achieved by supporting them in a template that will hold them while the foundation concrete is being placed. A simple example is shown in Figure 22-16. Templates are usually made of lumber but can be made of plywood or steel. They must be designed and constructed so they cannot be moved by workers stepping on them or by the placing of the concrete. Unless otherwise specified, the bolts should be positioned with an accuracy of $^1/_8$ inch. They should be held rigidly in the template by washers and nuts on both the top and bottom sides of the template. Holes in baseplates are drilled a small amount larger than the bolt diameter to accommodate slight variations in bolt location.

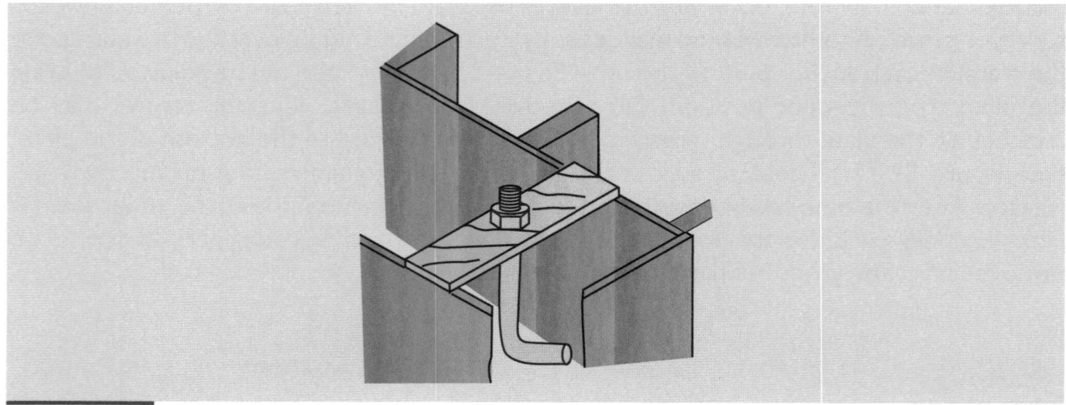

Figure 22-16
A short piece of board fastened to the formwork holds an anchor bolt in position while concrete is being placed.

Straight bars are occasionally used for anchors, provided the length of embedment is determined by the structural engineer. Usually more positive anchorage is achieved by hooking the lower end or by using a headed bolt, sometimes with a washer, as shown in Figure 22-17. To allow for more lateral adjustment, the bolt can be set in a sleeve made of pipe or light sheet metal. The space between the sleeve and the bolt should be filled with grout after the bolt has been lined up with the hole in the plate. See Figure 22-18. There are several patented anchoring devices available that make use of wedges, tapered threads and similar fittings. One of them consists of a sleeve similar to that shown in Figure 22-18 with a stud and coupling flush with the floor. After the machine has been moved into place, the upper fastening stud is inserted into the coupling and secured.

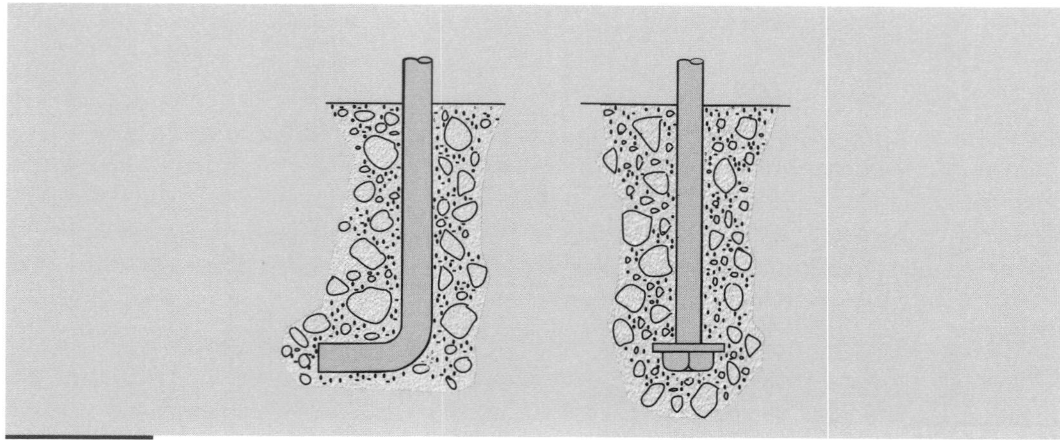

Figure 22-17
Positive anchorage of the anchor bolt can be accomplished by means of a bend or a washer.

Column bearing plates up to 4 inches thick are made of rolled steel, which is either untreated or straightened by pressing. There is usually no difference between top and bottom. Plates 4 inches or more in thickness are planed on the top bearing surface to assure uniform bearing with the structural steel of the column. It is important that the planed surface be on top when the plate is set in position.

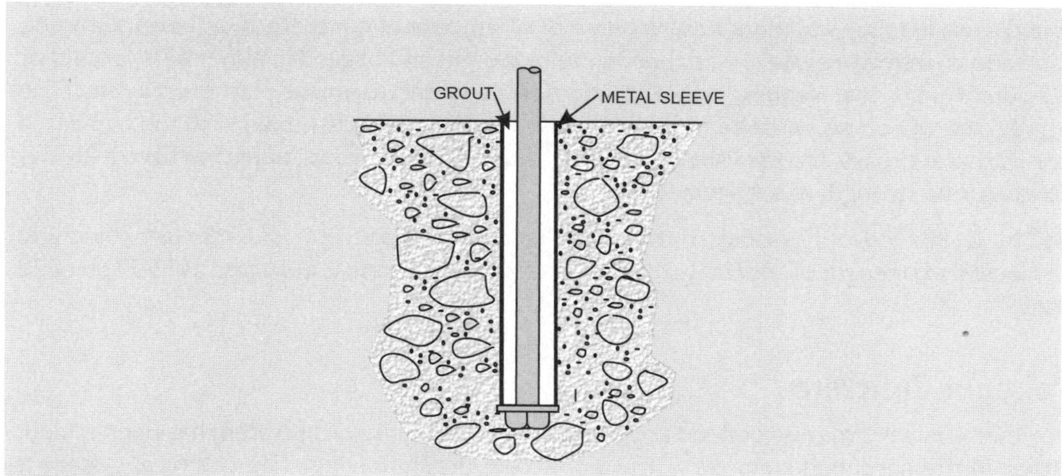

Figure 22-18
By setting the anchor bolt in a sleeve, space is provided for slight movement of the bolt to line up with the hole in the base plate. The sleeve is later filled with grout.

Occasionally, a small column base can be set on the fresh concrete after subsidence occurs but before the concrete has hardened. The base can then be carefully adjusted to final elevation and location. Because of the difficulty of accurate placement, this procedure cannot be used on even a fairly heavy base. Usual practice is to level the foundation concrete from $^3/_4$ inch to 3 inches below the finish elevation of the bottom of the base, then fill the space with mortar or concrete after the foundation has hardened and the plate has been set in place. An example is shown in Figure 22-14. Shrinkage of the mortar or concrete must be avoided or compensated for, which can be done by the use of a proprietary nonshrinking grout (ASTM C 1107), by using a dry-tamped mortar, by keeping the space between the foundation concrete and the base plate as thin as possible, and by using premixed grout.

The proprietary materials are discussed in Chapter 5. Shrinkage-compensating portland cement will be found useful in some cases. These materials should be used as directed by the manufacturer.

When using normal portland cement, a mix of one part cement to two parts sand is frequently specified. At a water-cement ratio of 0.50, grout will be fluid enough to flow into place. If the grout is permitted to stand for about two hours after mixing, its shrinkage can be reduced as much as 90 percent. No additional water should be added, as this will nullify the effect of the delay. A 1:2 dry pack mortar has just enough water so the mass will hold together when squeezed in the hand. When tamped into the space below the plate, the dry-pack will have practically no shrinkage.

Powdered aluminum (not the painting variety) in the amount of 1 teaspoonful to a bag of cement makes an expansive grout. The aluminum should be mixed dry with a small amount of sand before adding to the batch. Tests should be made beforehand, as the kind of cement, kind of aluminum and temperature all affect the timing significantly. The expansion reaches an end point in about 45 minutes. For this reason, small batches should be mixed and the grout placed as rapidly as possible.

22.9. Prebagged Dry Concrete

Available from most building material dealers and lumber yards, dry bagged concrete has found a use in many situations where only a small amount of concrete is required. Cement, sand and coarse aggregate are proportioned by weight and bagged in either $67^1/_2$-pound or 90-pound units that require only the addition of water to make plastic concrete. The aggregates, of course, must be completely dried before they are mixed with the cement, a drying that is usually accomplished in a small kiln. Mixes are proportioned to have a 28-day compressive strength as shown below.

ASTM C 387 covers packaged concrete, lightweight concrete and mortar. Minimum compressive strength of "normal-strength" packaged concrete is indicated as 3500 psi at 28 days.

22.10. Polymer Concrete

Polymer concrete may be defined as concrete to which a chemical system has been added, either in the hardened state or to fresh concrete during mixing. The chemical system is generally a liquid monomer or resin system, which is converted to a solid polymer (plastic) by a polymerization process. Polymerization is accomplished by either radiation or heat. Substantially improved structural and durability properties of the concrete result as compared to conventional concrete. Two types of polymer concrete are available:

Polymer-Impregnated Concrete is hardened concrete impregnated with a liquid monomer or resin system that is subsequently converted to a solid polymer. Results have shown strength values three to four times the strength of comparable untreated concrete, greatly improved resistance to freezing and thawing, greatly increased resistance to sulfate attack, improved abrasion resistance, some improvement in resistance to attack by acids, and decreased water absorption. The concrete may be either completely impregnated or partially impregnated from the surface to a desired depth, depending on the intended use of the concrete and required properties.

Basically, the process consists of first drying the concrete, then soaking it in the monomer solution under slight pressure, and finally polymerization either by radiation or thermal (heat) methods. The radiation process results in a greater improvement in the properties of the concrete, but the thermal drying and curing process is more practical for actual field application.

Surface impregnation and polymerization of concrete in place has been used in a variety of concrete repair and patching applications. It has been used to repair deteriorated concrete railroad structures, for slabs and floors in many kinds of structures, and for sealing and bonding cracks. It can be applied as a thin, self-leveling overlay or as a matrix for bonding coarse aggregate. After a cure of one or two hours, compressive strengths of 5000 to 18,000 psi are common, even when atmospheric temperatures are below freezing.

Polymer-Portland Cement Concrete, also called polymer-modified concrete, is basically normal portland cement concrete to which a polymer or monomer has been added during mixing and subsequently polymerized either in the forms or after form removal. Polymer-portland cement concrete is useful in applications where impermeability and durability are important. Its greatest use is in concrete patching and overlays.

22.11. Fiber-Reinforced Concrete

Fiber-reinforced concrete is normal portland cement concrete reinforced with small fibers of steel, glass, plastic and other natural materials. Fibers are available in a variety of shapes (round, flat, crimped and deformed) and sizes with typical lengths of 0.25 to 3 inches and thickness ranging from 0.0002 to 0.030 inches. A number that is commonly employed in describing a fiber is the aspect ratio, which is the length of the fiber divided by an equivalent fiber diameter, the latter being the diameter of a circle with an area equal to the cross-sectional area of the fiber. For fibers of 0.25 to 3 inches in length, the aspect ratio ranges from about 30 to 150.

Steel-Fiber-Reinforced Concrete is concrete containing dispersed, randomly oriented, steel fibers. The steel fibers have been shown to significantly improve the concrete's flexural strength, impact strength, toughness and resistance to cracking. Efficiency of the wires is improved by crimping the wire or hooking the ends. The fiber aspect ratio should not exceed 100. Steel-fiber-reinforced concrete is used primarily in pavements, industrial floors, overlays and patching work. (See Figure 22-19.) Steel fibers can also be used in fiber-reinforced shotcrete. Steel-fiber-reinforced concrete can be placed by most conventional methods including pumping, as long as the mix is not too wet. Overly wet mixes that are pumped can segregate. Fiber content of 1 percent to 2 percent by volume of concrete is considered the practical upper limit for field placement. The fibers must be dispersed uniformly throughout the concrete mix. This can be done during the mixing phase, preferably before the mix water is added. Also, if the fibers are not added to the mix in proper batching sequence or if the volume percentage of fibers is too high, fibers may clump together or ball up during mixing. Additional information on the proper mixing and placing of steel-fiber-reinforced concrete can be found in Reference 22.6. It is noteworthy that the 2008 edition of ACI Standard 318 (ACI 318-08) recognizes the use of "discontinuous deformed steel fibers" for resisting shear in flexural members. See ACI 318 Section 3.5.1.

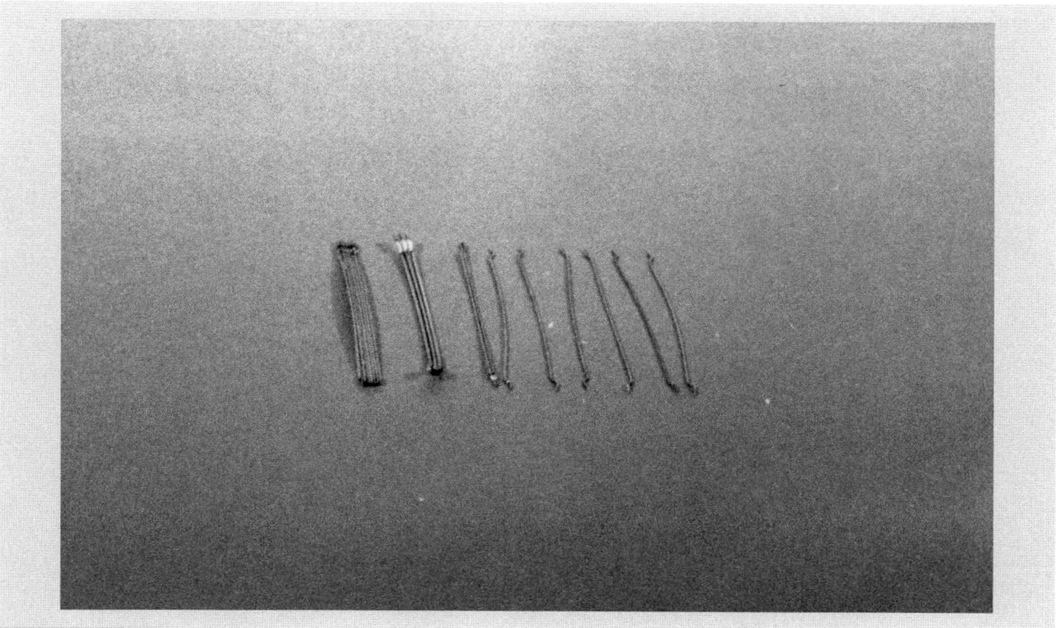

Figure 22-19a
Steel fibers with hooked ends are collated into bundles to facilitate handling and mixing. During mixing, the bundles separate into individual fibers. (Courtesy of PCA)

Figure 22-19b
Concrete with steel fibers for reinforcement. (Courtesy of PCA)

Plastic-Fiber-Reinforced Concrete is used primarily for concrete slabs and other similar applications where plastic shrinkage cracking is a concern. Plastic shrinkage cracking is caused by rapid early drying of the surface of freshly mixed concrete soon after it has been placed and while it is being finished, and it is one of the areas in which research indicates the use of plastic fibers to be an effective deterrent. Proponents also contend that the plastic fibers reinforce concrete against impact forces, abrasion and shattering, and reduce concrete permeability.

The plastic fibers are introduced into the concrete mix in small bundles that unravel and separate into millions of individual fibers. (See Figure 22-20.) The optimum addition of plastic fibers is generally given as 1.5 to 1.6 pounds per cubic yard of concrete. The fibers should be blended into the concrete according to the manufacturer's instructions. Plastic fibers can also be used in fiber-reinforced shotcrete. It is noteworthy that the ACI 318 Standard does not address plastic-fiber-reinforced concrete.

Figure 22-20a
Figure 22-20a. Polypropylene (plastic) fibers. (Courtesy of PCA)

Figure 22-20b
Adding plastic fibers during batching operations. (Courtesy of PCA)

A note of caution: the addition of plastic fibers in a concrete mixture will require more water to maintain a specified slump. The plastic-fiber-reinforced concrete mixture should be designed for the addition of plastic fibers; the additional water required to maintain a specified slump will require more cementitious materials (portland cement) to maintain a design water-cement ratio.

Glass-Fiber-Reinforced Concrete, referred to as GFRC, is the term generally applied to thin-walled architectural panels manufactured by a spray-up process using a cement/aggregate slurry reinforced throughout with glass fibers. In the spray-up process, a continuous strand of glass fiber is fed into a compressed-air-powered gun, where the strand is chopped into predetermined lengths and combined with a sand and cement slurry. Typically, a GFRC panel consists of 5 percent by weight of total mix of alkali-resistant glass fiber combined with a portland cement/sand slurry. Methods of manufacture vary, but spraying either by hand equipment onto a form of the desired shape and size or spraying mechanically on a production line are most common. Glass-fiber-reinforced concrete cladding panels for buildings are manufactured as wall units, window wall units, spandrels, fascia panels and interior feature panels. The panels are custom designed in sizes to suit the modular planning of a particular building. (See Figures 22-21 and 22-22.) Panel manufacture

Figure 22-21
Spray-up process of manufacture of GFRC panels make it easy to create panels for contoured profile buildings. (Courtesy of PCA)

requires expert production techniques and quality control. Reference 22.7 provides recommended practice for glass fiber reinforced concrete, including information on the manufacture and installation of glass-fiber-reinforced concrete panels.

Figure 22-22
Glass-fiber-reinforced concrete panels are light and strong enough to reduce buildings' structural requirements. (Courtesy of PCA)

22.12. Refractory Concrete

Refractory concretes are used in industrial applications where the concrete is exposed to temperatures as high as 3400°F. These refractory concretes are usually nonstructural and are constructed as linings for steel, concrete and masonry structures and chimneys. They are considered to be consumable; that is, they wear out and must be replaced.

Properly designed concrete mixtures with normal portland cement and aggregates are acceptable for temperatures up to about 400°F, portland cement and special aggregates to possibly 2000°F, and calcium aluminate cement and special aggregates for higher temperatures. Aggregates in increasing order of service temperatures include regular structural sand and gravel, limestone, slags, calcined fire clay and bauxite, kaolin and fused alumina. Commercially packaged refractory castables are a frequent and convenient source material, being proportioned for many exposure conditions and requiring only the addition of water to make them usable. Job or field mixes can be used when the size of the project makes them feasible. Shotcrete application is common, as are conventional placing techniques. Adequate curing is necessary, after which the concrete must be permitted to dry completely before firing. The first heatup must be at a reasonably slow rate. The special requirements for refractory concrete must be fully covered in the project specifications.

22.13. Architectural Concrete

Concrete is classified as architectural concrete when the surface of the concrete itself constitutes the final finished surface that is permanently exposed to view in the building. The surface might be formed or unformed concrete, precast or cast in place, plain or ornamental, smooth or of high relief. (See Figure 22-23.) Regardless of the special surface treatments to be applied to the concrete, there are certain fundamentals that apply to all concrete in this category. Of primary importance is uniformity—uniformity of materials, mixtures, mixing, placing, finishing, curing and special treatment. Cement and aggregates should be of the same type from the same source for all of the units of the same kind. Cement content and water-cement ratio should not vary from batch to batch, or during the period that any one type of unit is being made. The addition of admixtures should be uniform.

Figure 22-23
Formed precast architectural concrete cladding panels. (Courtesy of PCA)

After the concrete has been cured, it should be protected with barricades, lagging or similar means to prevent liquids splashed on the surface, damage from impact, or staining—taking care that the protective material itself does not stain. For example, rust may result from iron and steel, or dark stains from wet redwood.

This discussion of architectural concrete will focus on formed concrete. Unformed concrete (floors and pavements) is covered in Chapter 17.

The amount of finishing and dressing to be applied to a concrete surface depends on the kind of surface required and whether it is to be exposed to view. Many times no treatment is required for formed surfaces, because the forming material itself provides the specified finish. Concrete for buildings can conveniently be separated into several categories of surfaces.

Class A surfaces are those that are classified as architectural concrete. These are surfaces that are subject to close public view. (See Figure 22-24.) Forms must be carefully and accurately built by skilled workers, without visible offsets or bulges. Sheathing may be plywood, hardboard, steel, tongue-and-groove boards, or special lining material to provide a special texture to the concrete surface. Smooth surfaces usually require sack rubbing or grout cleaning to provide uniformity of color and texture. Included in this category are the surfaces described in this chapter.

Figure 22-24
Tilt-up concrete building with exposed aggregate finish. (Courtesy of TCA)

A Class B surface is one that is cast against unoiled plywood or form boards. It is a slightly rough surface for bonding to plaster or stucco. Surface treatment includes removal of fins and repair of defective areas.

A Class C surface is one in which roughness is not objectionable, such as surfaces to be backfilled, permanently submerged or otherwise concealed from view. The only surface treatment after removal of forms is filling of tie-rod holes on walls under 12 inches thick, or on all surfaces to be waterproofed (see Chapter 23), and repair of defective concrete. Forms can be quite rough, as long as they do not leak mortar.

Class D is for all permanently exposed surfaces where a higher quality surface is not specified: external portions of bridges (except grade separations); culverts; hydraulic structures such as tunnels, canals, siphons, spillways, dams and retaining walls not subject to close public view; decks and wharves; and certain rough buildings. Forms must be carefully and accurately built, without conspicuous offsets or bulges. Sheathing may be plywood shiplap or steel (not thin sheet steel). Surface treatment includes removal of mortar fins, filling of tie-rod holes, and dressing offsets greater than $^1/_4$ inch and bulges greater than $^1/_2$ inch in 5 feet.

The Sample Panel. The selection of materials and the methods of construction can have a significant effect on the quality of any surface. For this reason, whenever special surfaces are specified it is common practice for the architect to require that the contractor make a sample panel as one of the first things done on the job. The panel should be to full scale, using the materials and equipment that will be used in the job. Sometimes the sample panel can be incorporated in part of the building. The object of the sample is to provide a standard that must be met by all subsequent construction. The sample panel must be formed exactly as the actual structure will be formed, and the concrete mixture must be the one proposed. All steps of placing, vibrating, form stripping, jointing, cleaning, special surface treatment and curing must be those that will be used in the job.

Upon approval by the architect, the panel should remain on the job until construction has been complete.

Integral Color. Pigments incorporated into the concrete impart color to the mortar. By selection of the correct pigment, integral permanent color results. Requirements for pigments are given in Chapter 9. A slight grayish hue is given to the concrete by the cement, which usually is not objectionable. If particularly pure color is desired, especially when using the light pastel shades, white cement and light-colored sand should be used.

Pigmented concrete can be ready-mixed, transported and placed in the same manner as standard concrete. (See Figure 22-25.) There are, however, a few special precautions that must be observed. Only the pure metallic oxide pigments shown in Table 9.4 should be used. The amount should be determined by making test panels (see "The Sample Panel"). The test panels should be viewed after the concrete is thoroughly dry. The use of manufactured, prepackaged pigments will result in good uniformity if the manufacturer's recommendations are followed. Some ready-mix producers are reluctant to make colored concrete because of the small demand and the problems involved. Use of the packaged material usually overcomes this objection. Slump from batch to batch must be consistent, and no water should be added to the concrete after a portion of the batch has been discharged. Concrete should not be retempered. Handling and placing should be done without delay.

Figure 22-25
Integrally colored concrete is made by adding mineral oxide pigments to a mix made with either grey or white cement. (Courtesy of PCA)

Form joints must be tight to prevent loss of water or grout from the concrete. This usually requires taping the joints. Holes left by form ties should be patched as soon as the forms are removed. Because it is nearly impossible to hide them completely, their location should be selected so they will blend into the surface. (See Figure 22-26.) It is sometimes possible to use forms without internal ties. Forming and curing materials that will not stain or discolor the concrete should be used.

Figure 22-26
White cement cast-in-place wall panels with tie holes artistically located. (Courtesy of PCA)

Paint. Almost any type of paint will give satisfactory service when properly applied to concrete. Concrete should be permitted to age several months, if possible, before application of any paint. This results in more uniform suction over the surface and overcomes the effect of residual form oil adhering to the concrete. The paint is less apt to craze, and the danger of efflorescence is lessened.

Concrete should be clean when paint is applied. This requires washing with water and probably some scrubbing. Efflorescence, if any, should be removed with acid. In special cases, sandblasting or rubbing may be necessary. (See Figure 22-27.)

Figure 22-27
Concrete masonry surface is dampened with a water spray just before application of a portland cement paint. (Courtesy of PCA)

Portland cement paint should have a thick, creamy consistency (the first coat may be thinner) and should be applied to the slightly damp concrete with scrub brushes or calcimine brushes. Batches of paint should be used up in three to four hours. However, some commercial paints contain calcium chloride and will have to be used up sooner. Painting is best done during moderate weather.

Curing should be done by fog spraying the surface several times a day for at least two days. Properly applied and cured cement paint should last for several years.

Failures of portland cement paint consist of rapid chalking, caused by low cement content, or poor curing. Other failures are peeling or flaking, resulting from painting a glassy smooth, dirty or dry surface, or unhydrated lime. Pinholes result from insufficient brushing or sprayed coating. Crazing is usually not serious and may result from crazing of the underlying concrete. Efflorescence comes from painting the concrete before it has aged sufficiently.

Other paints include oil-base paints consisting of opaque pigments suspended in a vehicle of drying oils and thinner (linseed oil paints); resin-emulsion paints consisting of water-reducible pigment paste in an emulsified oil-extended resin, usually glycerol phthalate; synthetic rubber paints, either the emulsified synthetic-rubber resin type or the rubber-solution type; and epoxy resin or similar one- or two-component resin paints.

Paint should be applied as directed by the manufacturer, after a proper pretreatment of the surface. The best pretreatment for concrete on which oil paint is to be applied is to permit the concrete to age for about a year. The surface must be clean and dry when the paint is applied. If earlier painting is necessary, a suggested pretreatment consists of a solution of 2 percent zinc chloride and 3 percent phosphoric acid in water applied as a wash. As soon as the surface has dried thoroughly, the paint can be applied.

Clear sealers are available for application to exposed aggregate and other special concrete surfaces. Such coatings should be water clear and should not discolor on exposure to the elements; they are best absorbed into the surface of the concrete so that they do not impart a glossy effect.

Any paint should be applied as directed by the manufacturer. Only paints known to provide a durable coating under the expected exposure conditions should be used.

Exposed Aggregate. Techniques for making exposed aggregate finishes on flat slabs are discussed in Chapter 17. These same methods can be applied to tilt-up panels and, with modification, to vertical surfaces. Other techniques that are applied to horizontal surfaces include sand-bedding, in which the bottom of the form is covered with sand, the depth depending on size of aggregate and amount of exposure. Aggregate particles are hand placed in the sand, the stone and sand given a fine spray of water to settle the sand, after which structural concrete is placed. When the concrete reaches the strength for removal from the form, the panel is raised to a vertical position and the sand washed out with a stream of water. The depth of exposure, deeper than can be achieved by other methods, can be as much as 2 inches.

A similar method is to coat the bottom form with a retarder, hand place the aggregate, then place the structural concrete. As soon as the form can be stripped, the panel is raised to a vertical position and the surface scrubbed with brushes and water to reveal the aggregate. (See Figure 22-28.)

Figure 22-28
Surface of a panel on which the aggregate was exposed by applying the retarder to the form, then washing the concrete when the form was removed. (Courtesy of PCA)

When hand placing aggregate, the aggregate particles should be close together—the closer the better.

Another method is the aggregate transfer method, which is adaptable to formed surfaces. This method consists of attaching the special facing aggregate to plywood liners by means of a special adhesive. The liners are positioned in the forms, after which normal procedures are followed in placing the structural concrete. When the forms are stripped, the liners are removed, leaving the aggregate embedded in the concrete. Depending on the age of the concrete, various methods can be used to reveal the aggregate. Washing and brushing result in a rough surface; sandblasting and bushhammering can be applied at later ages. Grinding produces a smooth surface.

Textures. An almost unlimited variety of surfaces can be made by casting the concrete against special forms or liners made of wood, plaster, plastic or similar materials. One method is to carve the reverse design in polystyrene or similar easily worked material. The possibilities are limited only by the ability of the sculptor to conceive and execute the designs. Surfaces can range from relatively smooth to extreme relief as shown in Figure 22-29. Forms can be lined with such material as rubber float pads and plastics of various textures.

Figure 22-29
White cement wall panels with extreme relief. (Courtesy of PCA)

Another method is to spread rounded, uniform-sized rocks on the bottom of the form, cover the aggregate with polyethylene film, and then place the concrete on the film. This gives the concrete a dimpled effect.

Rubbed Surface. The term rubbing is applied to the process of going over the surface of the hardened concrete to fill surface voids, clean the surface or smooth irregularities in the concrete. Rubbing is applied to vertical, or nearly vertical, formed surfaces. Because of the many materials and processes now available, and because of its high cost, rubbing is not done as much as in the past. However, the processes are still employed, and the workers on the job should be familiar with them.

Rubbing of concrete is looked upon with disfavor in some circles because of the failure of the rubbed surface to produce the desired effect or to maintain its uniform good appearance for any appreciable period of time. These failures can be traced to improper techniques in application or failure to understand the effects of the particular method selected. Rubbing is done to fill the small surface voids, leaving the balance of the surface unchanged; to produce a surface of uniform color and sandy texture; or to remove stains from the hardened concrete.

Rubbing does nothing to improve or change the concrete structurally; its sole purpose is to improve the appearance of the concrete. Several different techniques are available, depending on the method specified or the results desired. None of the techniques described herein should be used for filling or covering rock pockets or honeycomb. These defects, if they are serious enough to weaken the concrete structurally, require careful removal and replacement with sound concrete or mortar.

The time for rubbing and the method used are usually stated in the job specifications. Normally, the probability of better bond of the rubbing grout and the concrete is highest when the forms are stripped and the rubbing performed as early as possible. Grout cleaning, however, should be delayed until the concrete has been cured and construction operations, such as placing concrete in upper stories, have been completed to preclude staining or marking the finished surface.

Forms should be stripped as soon after placing concrete as the specifications or construction operations will permit. The first operation after form removal is pointing up all form bolt holes where such holes are to be filled, removal of mortar fins that projected into cracks in the forms, and repair of rock pockets or honeycomb. The surface, after being thoroughly wetted with water and allowed to drain, should be in a moist condition when further surface treatment as described below is applied. If no surface treatment is to be applied at this time, curing should be commenced.

Sack rubbing is done to fill the minor voids and pits (sometimes called bug holes) in the surface of the concrete. It is best done while the concrete is still green, if possible, although it may be done at any time later. It cannot be applied to a surface that has been coated with curing compound. The concrete surface should be moist but not wet before mortar mixed to a barely damp consistency is spread over the concrete surface with a piece of burlap, rubber float or similar tool. Moistening of the surface is accomplished by thoroughly wetting the concrete, then allowing it to dry to a slightly moist condition. Mortar should consist of one part cement to two or two and one half parts of sand passing the 16-mesh screen (ordinary window screen) with just enough water for the material to stick in a tight ball when squeezed in the hand. Sufficient white cement should be blended with common gray cement so that the patches, when dry, will match the color of the parent concrete. The concrete surface should be rubbed with the burlap or float, making sure that the voids are well filled with a slight excess of mortar. After the mortar takes its initial set, the surface should be rubbed a second time to remove all mortar from the surface, leaving the voids full. The mortar in the void should be neither raised above nor depressed below the concrete surface. Curing should then be applied in the usual manner, either water curing or liquid membrane curing compound. (See Figure 6-17.)

Some agencies recommend the use of a finer sand (passing the 30-mesh screen), making a 1:2 slurry with a consistency resembling thick cream. After rubbing this slurry over the surface with a burlap pad or sponge-rubber float, filling the voids, the surplus material is removed while it is still plastic by rubbing the surface with a dry mix of the same proportions and materials. Curing is then applied.

When properly applied, either of the above treatments will fill the minor surface voids but will have no effect on the general appearance of the concrete, leaving unchanged the color and texture produced by the forms.

A stoned or sand finish produces a surface of uniform color and sandy texture. It requires a considerable amount of rubbing or stoning to blend the grout into the concrete to produce a well-bonded sandy finish. Because of the amount of stoning required, the job is best done with power tools, as workers using hand tools frequently fail to do a thorough job of rubbing, with the result that the grout scales and peels later.

Forms should be stripped, the concrete surface pointed and repaired, and the rubbing commenced as early as possible. If rubbing is delayed, the concrete should be water cured. Immediately prior to starting the rubbing or stoning, the concrete surface should be wetted thoroughly. This wetting is done whether the concrete is stoned while in the green condition or after it has dried out. (If there is curing compound on the surface, see "Grout Cleaning" below.)

A slurry is prepared, consisting of two to two and one-half parts of fine sand passing the 16-mesh screen, to one of cement. The cement should be a mixture of common gray cement and white cement, blended in the proportions to give the correct color to match the existing concrete when dry. Slurry should be of a thick, creamy consistency.

While the concrete surface is still damp, the slurry is spread over the surface with a rubber float or similar tool, care being taken to fill the voids. Vigorous rubbing is then commenced with a No. 16 carborundum stone, bringing up a lather of fines, which mixes with the slurry on the surface. The resulting grout mixture is spread uniformly over the concrete. Rubbing with the No. 16 stone should be continued until the surface skin on the entire area being treated has been removed and replaced by the fines ground out of the concrete, mixed with the slurry and uniformly spread over the surface. The thickness of the grout should be about $^1/_{32}$ inch. The grout is then permitted to harden and hydrate somewhat, after which it is again moistened slightly and ground with a No. 30 carborundum until the surface is uniform and smooth. Normal curing should be commenced as soon as the surface can withstand wetting without streaking or other damage.

Grout Cleaning is a method of removing stains of rust, form oil, curing compound or other material from the walls of a structure or building and providing a uniform color on the surface. The wall should be completed and at least three weeks old before the treatment is attempted.

Slurry or grout for cleaning consists of one part cement with one and one-half to two parts of fine sand passing the 16-mesh screen and sufficient water to produce a consistency of thick paint. Sufficient white cement should be blended with normal gray cement to produce the desired color after the grout has dried. The concrete should be well moistened before the grout is brushed on. The grout is worked into air bubble holes and other voids with the brush, after which the surface of the wall is thoroughly rubbed with a cork float. After the grout has partially set, but while it is still plastic, a sponge rubber float is used to remove the excess grout from the surface without drawing any of the grout out of the voids and holes. Finally, after the surface has completely dried, it is rubbed again, this time with a pack or bundle of dry burlap to remove all dried grout, leaving the wall with no visible skin of grout on the surface.

The work should be so planned as to permit the entire cleaning operation to be completed in one day. Large areas should be broken down into areas that can be handled without interruption in one day so as to avoid leaving the grout on the surface overnight, which would result in an unsatisfactory job.

Minor stains, rough spots and streaks of mortar from higher lifts can sometimes be removed by a light rubbing with a hand stone or with a power sander. Water should be used generously and only a light rubbing done, just enough to remove the roughness without affecting the texture or color of the concrete.

White Concrete. The use of white portland cement (described in Chapter 7) produces a nearly white concrete or mortar. (See Figure 22-30.) With reasonable care, and following good construction practices, a good white concrete job can be achieved. Good quality control is essential. Maximum efficiency and economy result when the producer observes a few simple precautions. Any ready-mix concrete producer or casting yard can make good white concrete.

Subtle differences in texture and color, not objectionable on gray concrete, are apt to stand out on a white concrete surface. This is less likely on plant-manufactured precast concrete than on cast-in-place concrete. The causes of color variation, discoloration and staining include different brands of cement, different aggregates, use of different types of forming materials (for example, new plywood next to old plywood), certain form oils or parting compounds, variations in the mix, different cement contents, different slumps, certain admixtures, variations in curing, sandstreaks and rock pockets, certain sealing and curing compounds, and dirt and staining from construction activities.

Figure 22-30
White portland cement. (Courtesy of PCA)

In many instances, any one of the above-listed factors will have a negligible effect, but all should be considered when setting up a job or when trying to find out what may have happened on a job already completed.

Note that the owner and the architect expect more in quality appearance from white concrete than from ordinary gray concrete. (See Figure 22-31.) The precautions previously mentioned in this chapter, along with the following suggestions, will help satisfy this expectation. Avoid large areas of plain white concrete; break it up with rustication strips, texture or other details. Use the same type of contact forms throughout the job, especially for identical members—for example, all spandrel beams or all column faces. Remember that the second and subsequent usages of plywood and other lumber will probably give a different texture from the first usage. Ensure that reinforcing steel does not touch the form. Do not drop nails, tie wires or debris into the forms. Clean out the mixer drum and all other equipment before handling white cement, aggregates or concrete. Ensure that all tools for placing and finishing are clean and free of gray concrete or rust.

Figure 22-31
White cement, precast walls (Courtesy of PCA)

White sand and coarse aggregate, made by crushing white limestone or white quartz, are available. Because they are 100 percent crushed, these aggregates are apt to produce concrete that is not as workable as that made with the usual river sand and gravel. They are, however, essential if a pure white concrete has been specified. Aggregates should conform to the requirements of ASTM C 33 for concrete aggregates or ASTM C 330 for lightweight aggregate.

Natural sand and coarse aggregate are frequently used for special effects. Sand especially affects the final color, imparting a gray or brown tone to the concrete. Coarse aggregate has less effect, except for surfaces on which the aggregate is to be exposed. On smooth concrete surfaces, dark-colored coarse aggregate will cause a mottled appearance.

A number of special aggregates are available for special effects when the aggregate is to be exposed. Crushed quartz and quartzite come in a variety of colors, especially pink, rose and white. Marble is available in many colors, including red, pink, yellow, green, blue and gray. Granite comes in black, gray and red. There is a great difference in the overall color of river and terrace gravels from different sources, especially in the gray and brown tones. Because of the high cost of special crushed aggregates, it will be well to determine what is available in natural gravels when designing a job.

Lightweight aggregate can be used with white cement successfully. However, most lightweight aggregates are dark in color, except for some natural pumice available in some areas.

White cement responds to admixtures in the same manner as gray cement: retarders retard, accelerators accelerate and water reducers reduce the mix water requirement. As with all such combinations, however, trials should be made before actual use of any admixture to determine just how it reacts under field conditions. Of special importance is the possibility of discoloration resulting from some admixtures.

Pigments, when used with white cement, result in cleaner, brighter-colored concrete than the same pigments with gray cement, especially in the light pastel shades. White cement should always be used for producing light-colored concrete such as pink, rose, buff, ivory, light green and similar pastels. The amount of pigment is usually quite low, sometimes being only 1 percent by weight of cement, or even less. Black is a particularly difficult color to obtain and is not usually uniform in color.

Forming materials must be carefully selected, as some plywoods and hardboards will discolor white concrete when used as forms. Form oils or parting compounds vary considerably in their effect. Tight forms are essential if a smooth white surface is required. Loose or leaky forms will result in sand streaks on the finished surface, of particular importance if off-white sand is used. Do not intermix new and used form liners on the same face of the work. When the preconstruction mock-up is constructed, all forming materials for the job should be used.

Form ties should be of the type that will not leave any metal within 1 inch of the finished surface. They should be located in a pattern approved by the architect. Forms must be well tied and braced, as any bulging or offset is particularly undesirable. Attempts to patch such discrepancies nearly always result in unsightly irregularities of surface and texture.

Clean steel will give no problem when used with a good form oil or parting compounds. Rust on the steel can cause a rough spot or stain. Galvanized steel is liable to stick to concrete. Fiberglass is satisfactory.

Before starting any white job, all equipment for handling white cement and concrete must be thoroughly cleaned. If bulk cement is to be used, the silo, elevator, screw and batcher must be clean. Mixers, buckets, pumps, conveyors, etc., for handling the concrete must be free of accumulations of gray concrete, oil, rust and other possible contaminants. Mixing and curing water must be clean. Even small amounts of foreign material can discolor the finished concrete.

When appearance is of chief importance, the amount of cement per cubic yard is usually more than is necessary for structural purposes. A minimum of 560 pounds (six bags) per cubic yard is suggested. Sand content should be the minimum to provide a workable mix, as excess sand is liable to result in excessive voids or bug holes on vertical surfaces. All materials should be batched by weight, except water and admixtures, which can be batched through a volumetric meter. Variations in the amounts of materials, especially water, can have a pronounced effect on the color of the concrete. Color is affected also by the length of mixing time, and every effort should be made to mix all batches for about the same period of time.

In general, the same finishing techniques that are applied to gray concrete can be used on white concrete. These techniques are well established and need not be described in detail here. There are, however, some refinements that must be observed if the degree of perfection normally expected is to be achieved. These are neither difficult nor expensive, but they are necessary.

The variety of formed surfaces is almost endless, limited only by the ingenuity of the designer and builder. Concrete made with white cement serves as an excellent matrix for exposed aggregate finishes. Pigments are sometimes used. Selection of size, color and hardness of coarse aggregate is important. Carbonate aggregates such as limestone, dolomite and marble should not be used for an acid-etched surfaces. Besides acid etching, other exposure processes are retarded and washed surface, sandblasting, grinding and tooling (usually called bushhammering) with pneumatic chisels, combs and other cutters.

White concrete, like gray concrete, needs to be properly cured if it is to perform satisfactorily, and the same curing procedures required for gray concrete should be observed for white concrete. The need for nonstaining methods and materials must be emphasized, however. Curing can be accomplished with water by sprinkling or ponding, or by covering the concrete with an impervious sheet such as polyethylene or curing paper to retain the moisture and maintain 100 percent humidity around the concrete. If plastic sheets are used for covering a troweled slab, nonuniform color will result if the sheet only partially contacts the slab.

The usual curing compounds are best avoided, as some of them stain the concrete and others take on a blotchy appearance as the material weathers off. Some turn yellow after several months of exposure.

It is sometimes considered desirable to apply a clear sealer to exposed concrete surfaces to lessen contamination of the surface by atmospheric dirt and to brighten exposed aggregate surfaces, as the clear sealers have a tendency to make the concrete appear wet. For this reason, a clear sealer on plain white concrete may cause it to appear dark in color. Unless significant contamination by the atmosphere is expected, it may be advisable to avoid the use of a coating or sealer.

Sandblasting. The surface resulting from sandblasting, as shown in Figure 22-32, is somewhat rough. The degree of roughness depends on the hardness of the aggregate particles, age of the concrete and amount of blasting applied to any unit area. When relatively new concrete is sandblasted, the comparatively soft cement paste is cut away faster than the aggregate, so that a very rough surface results. As the concrete gets older, the paste becomes harder and the difference in rate of cutting decreases.

Figure 22-32
Coarse aggregate particles on this panel have been exposed and highlighted by sandblasting. (Courtesy of PCA)

Wet sandblasting is usually specified because of health and pollution problems with the dry process. The amount of blasting can consist of a light treatment that removes only a thin skin of dry paste, exposing the sand particles, or it may be a deep cut of $^3/_4$ inch. Pleasing results can be achieved by sandblasting standard structural concrete. Some users, however, specify what is called a gap-graded mix, in which some of the intermediate sizes of coarse aggregate are omitted. Special aggregates are frequently used for certain colors or other effects.

Forms for concrete to be sandblasted must be tight to prevent leakage of mortar at the panel joints. Sandblasting will not remove the lines left on the concrete surface by leaking joints. If joints are sealed with a pressure-sensitive tape, there will be no leakage, and the tape marks will be sandblasted off. Uniformity of concrete and of placing are essential for a good surface.

Abrasives consist of blasting grit, silica sand or any hard angular sand. In certain areas, crushed chat, a residue from mining, is used. Faster, deeper cutting is possible with the blasting grit or silica sand.

Sandblasting is usually done while the concrete is quite green, frequently as early as three days, or even one day. Sandblasting at a later age is slower and more expensive but may be necessary because of scheduling problems. All related areas should be blasted at nearly the same age of concrete.

The distance between the nozzle and the surface must be determined by trial. Usually a distance of about 2 feet gives best results.

Bushhammering. Tooling of the surface of the hardened concrete is done with a bushhammer. It gives the concrete an allover rough texture, exposing the aggregate. The degree of roughness depends on whether the surface is lightly or heavily hammered. (See Figure 22-33.)

Figure 22-33a
Bushhammered surface. (Courtesy of PCA)

Figure 22-33b
Close-up of surface texture resulting from bushhammering. (Courtesy of PCA)

The bushhammer consists of a flat-faced tool that fits into the chipping gun. The face of the tool, about $1\frac{1}{2}$ inches in diameter, consists of small pyramidal points over the entire surface. Held in the hands perpendicular to the concrete surface, the gun delivers rapid blows to the surface of the concrete, crushing and spalling a thin skin. Depth of cut depends on the hardness of the concrete and the speed with which the tool is moved over the surface. Hand tools are available and should be used along edges, as the power tool is apt to break or chip the edge.

Most specifications require that the concrete be at least 14 days old when it is hammered. The job should be scheduled so that all areas are worked at very near the same age for best uniformity. The pressure on the gun should be as steady as possible. Care is necessary to avoid cutting too deep, as only a thin skin needs to be removed. It is best to complete a small area at a time. As the cutting teeth on the face of the tool wear down, the texture of the concrete changes slightly. For this reason, bits should be changed frequently.

Another tool used in a similar manner is a single-point chisel or gad, shown in Figure 22-34. This tool makes a somewhat rougher surface than the multiple-point head, forming pits or lines in the surface. (See Figure 22-35.)

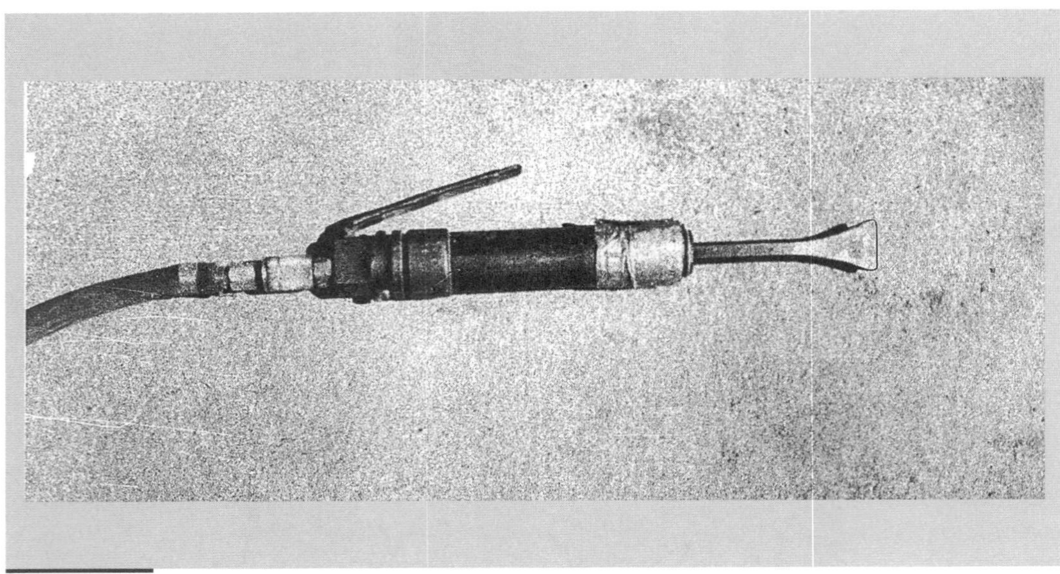

Figure 22-34
A pneumatic hammer with a gad point.

Figure 22-35a
Bushhammered fins. (Courtesy of PCA)

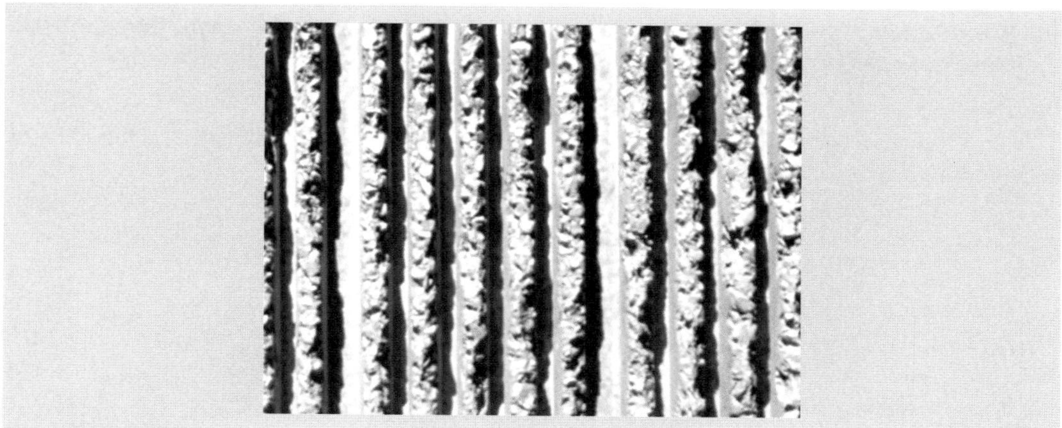

Figure 22-35b
Close-up of bushhammered fins. (Courtesy of PCA)

Acid Etching. When an acid is applied to a concrete surface, the acid reacts with the hydrated cement and etches the surface. The acid will also attack the limestone and marble aggregate. The effect is a slight roughening of the surface and exposure of the aggregate particles. The degree of roughness depends on the age of the concrete, concentration of acid and the length of time the acid remains on the surface. Smooth concrete surfaces are frequently etched. A very slight etching can be applied to other special surfaces, such as exposed aggregate, to remove a thin film of cement and brighten the surface. Etching in precasting plants is frequently accomplished by dipping the precast units in an acid bath.

The concrete surface to be treated must be uniform. Any irregularities of the concrete will still be visible after etching, and some may be made worse. The concrete is first thoroughly wetted with water to saturate it. This is necessary to prevent the acid from penetrating too deeply. The acid, a 10 percent or 15 percent concentration of muriatic (hydrochloric) acid, is mopped or brushed uniformly on the surface and allowed to react for a few seconds, after which the concrete is washed with large quantities of clean water. All of the acid must be completely removed from the concrete.

Trials on sample areas will have to be made to determine the exact procedure to follow. The results differ with different aggregates. Etching can be done as soon as one day after the concrete is placed, or at any time later. All comparable areas should be etched at about the same age.

Acid is a corrosive chemical. Workers should wear rubber protective clothing and should avoid contact with the acid or the treated surface until the area has been thoroughly washed. Breathing the fumes should be avoided. Disposal of the wash water can be a problem in some areas. The acid should not be permitted to come in contact with any surface not to be etched, including metal work, glass, wood, brick or shrubbery.

Grinding. Concrete should be at least 14 days old before it is ground. Grinding should be done with power tools (see Figure 22-36), as hand rubbing or stoning is prohibitively expensive. Pneumatic or electric rotary tools are available. The smoothness of the surface depends on the grit of the stone used in the machine. Normal practice is to start with a coarse stone, about a No. 8, which gives a rather smooth surface. Finer grits can be used if additional polishing is required. Dry grinding is the most rapid and economical method, but safety and health requirements will probably demand the use of wet grinding. Manufacturer's recommendations as to speed of operation and number of grit should be followed.

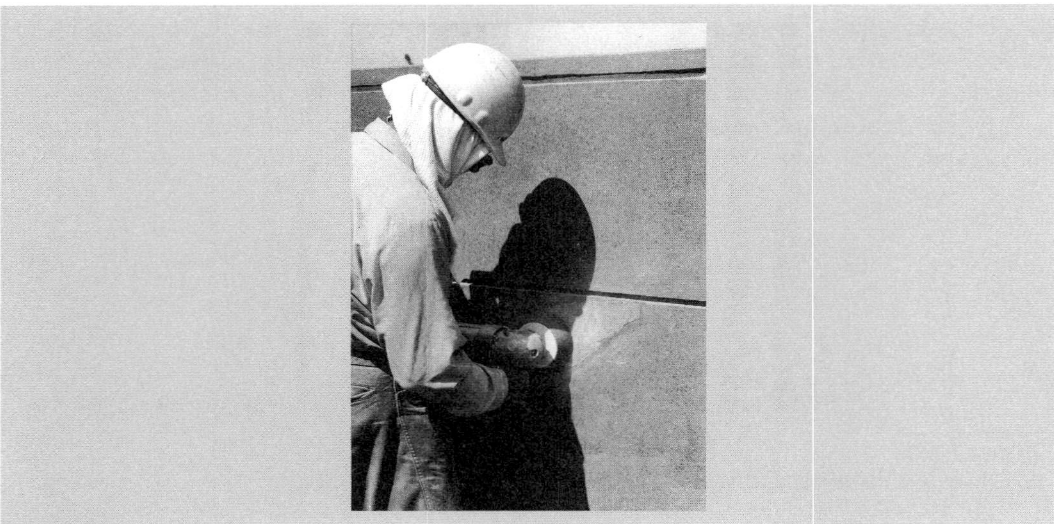

Figure 22-36
Grinding the surface of concrete with a handheld grinder.

22.14. Sulphur Concrete

Sulphur compounds have been used for many years in such applications as grouting anchor bolts. Research has revealed the value of sulphur as a bonding agent for aggregate instead of port- land cement and water. This sulphur concrete is produced by adding hot molten sulphur containing a patented plasticizer to graded aggregate. The mixture is either poured in place or precast into molds. Sulphur content is between 12 and 22 percent. Temperature of sulphur concrete must be between 275°F and 300°F when it is placed. It is remeltable and reusable. It is safe to use when reasonable precautions are taken. Advantages claimed are high-compressive strength in a matter of hours, strength as high as 9000 psi in 24 hours, and excellent corrosion and abrasion resistance. It has been used for floors, walls in a corrosive environment and for precast units. Additional research and field applications are necessary in order to develop standards and specifications to broaden its use.

22.15. Autoclaved Cellular Concrete

Autoclaved cellular concrete, also known as autoclaved aerated concrete, is a special type of lightweight precast concrete building material. It is manufactured from a mortar consisting of portland cement, limestone, powdered silica sand or fly ash, water and aluminum powder. The raw materials are mixed into a slurry and poured into molds. The aluminum powder reacts with calcium hydroxide and water to produce hydrogen gas, which aerates the mixture producing millions of microscopic cells. After the mixture has hardened, it is cut and shaped into building elements (blocks or panels) with precision wires. After cutting, the individual elements are pressure steam cured (autoclaved) at about 375°F and a pressure of 175 psi for a period of 6 to 12 hours. The final ACC product is up to 80 percent air by volume. The finished building elements are commonly shrink-wrapped in plastic and transported directly to the construction site.

This porous building material has densities between 19 and 63 lb/ft^3 and compressive strengths between 300 and 1500 psi. Autoclaved cellular concrete is produced in block or panel form for construction of residential or commercial buildings. (See Figure 22-37.)

Construction is relatively fast and efficient, given the precise, standardized nature of the material. Each of the products— blocks, wall panels, roof/floor slabs and beam elements—are manufactured in a range of sizes depending on the specific application. The ACC building elements can be manufactured with or without reinforcement. Additional information can be found in ACI 523.2R, *Guide for Precast Cellular Concrete.*

Figure 22-37a
Autoclaved cellular concrete block floating in water. (Courtesy of PCA)

Figure 22-37b
Commercial building constructed with autoclaved cellular concrete panels. The ACC wall panels are commonly constructed by joining the panels vertically or horizontally using mortar or glue. (Courtesy of PCA)

22.16. Self-Consolidating Concrete

The construction industry has always longed for a high-performance concrete that can flow into tight and inaccessible spaces without requiring vibration. This need has grown as more designers specify concrete members that are heavily reinforced and require complex formwork. Until recently, the closest the industry came to developing a "self-consolidating" concrete was to add a superplasticizer to a conventionally proportioned concrete mix. Although superplasticizers allow for the use of concrete with a slump of 8 inches or more, such concrete still requires some vibration for adequate consolidation. High doses of superplasticizer create a very fluid concrete, but the mix often segregates because the mortar is too thin to support the coarse aggregate. Today, advances in admixtures and mix proportioning are making self-consolidating concrete a reality.

The key to creating a self-supporting concrete is to produce a very flowable mortar that still has a high enough viscosity to support the coarse aggregate. To produce the desired flowability, newly developed high-range water reducers (polycarboxylate ethers) are typically used to plasticize the mixture. So as to increase the viscosity of the mortar, self-consolidating concrete contains more fine material (fly ash or limestone filler) but essentially the same amount of water as conventional concrete. The low water content ensures high viscosity, so that the coarse aggregate can float in the mortar without segregating.

Self-consolidating concrete (SCC), also called self-compacting concrete, is able to flow and consolidate under its own weight without external vibration. At the same time it is cohesive enough to fill spaces of almost any size and shape without segregation or bleeding. This makes SCC particularly useful wherever placing is difficult, such as in heavily reinforced concrete members or complicated formwork. (See Figure 22-38.)

Figure 22-38
Self-consolidating concrete can flow between and around reinforcement without requiring vibration. (Courtesy of PCA)

Strength and durability of SCC are similar to conventional concrete. However, without proper curing, SCC tends to have higher plastic shrinkage cracking than conventional concrete. The production of SCC is somewhat more expensive than regular concrete; however, construction time is shorter. Furthermore, SCC produces a good surface finish. These advantages make SCC favorable for use in precasting plants.

Because SCC is characterized by special fresh concrete properties, many new tests have been developed to measure flowability, viscosity, blocking tendency, self-leveling and stability of the mixture. One simple test to measure the unblocked flow of the mixture is the J-ring test, which is a modified slump test. The J- Ring (12-inch diameter with circular rods) is added to the conventional slump test (Figure 22-39). The SCC has to pass through the obstacles (circular rods) in the J-ring without separation of paste and coarse aggregates. The slump diameter of a well proportioned SCC is approximately 30 inches, requiring a slump test surface at least 40 inches in diameter. A newly formed ASTM committee is in the process of establishing standard test methods to measure the properties of fresh self-consolidating concrete and to set performance requirements for the material.

Figure 22-39a
J-ring test for SCC. The J-ring simulates reinforcement. (Courtesy of PCA)

Figure 22-39b
Measuring slump flow for SCC. (Courtesy of PCA)

22.17. Controlled Low-Strength Material

Controlled low-strength material (CLSM), also called flowable fill, controlled-density fill (CDF) and various other names, replaces compacted material as structural fill or backfill. It was originally developed as an alternative to backfilling utility trenches with compacted fill. Because CLSM flows and needs no compacting, it is ideal for use in tight or restricted access areas where placing and compacting soil or granular fill is difficult or even impossible. Unlike compacted granular fill material, which must be placed in lifts, spread and compacted after each lift, CLSM is a fluid material that flows easily into place and is self-leveling, and usually requires no onsite equipment other than a truck mixer. (See Figure 22-40.)

Figure 22-40
Flowing down the chute of a ready-mixed truck, CLSM quickly fills the deep utility trench. Because CLSM is self-leveling and self-compacting, the worker simply guides the chute and monitors the flow rate. (Courtesy of PCA)

CLSM is neither a concrete nor a soil-cement but has properties similar to both. CLSM is a fluid cementitious material usually made with portland cement, water and fine aggregate or fly ash, or both. The consistency of CLSM is like that of a slurry or lean grout; it can flow great distances, is self-leveling and doesn't settle after hardening. Typical 28-day compressive strengths range from 50 to 200 psi. Densities range from 115 to 145 lb/ft^3. The material contains many of the same components found in concrete but in different proportions. It is readily available from ready-mixed concrete suppliers, can be batched and mixed using the same equipment that produces concrete, and can be delivered to the jobsite by truck mixers. Ready mixed concrete producers generally have developed standard proportions for CLSM. One precaution: Because flowable CLSM will exert a high fluid pressure against forms or embankments, CLSM should be placed in layers where lateral fluid pressure is of concern, with each layer being allowed to harden before the next layer is placed.

22.18. Pervious Concrete

Although pervious concrete has been in use for more than 50 years in a variety of applications, recent EPA regulations are causing a re-examination of this unique material. Also referred to as no-fines concrete or porous concrete, this material is composed of narrowly graded coarse aggregate, cementitious materials, water and admixtures. Little or no fine aggregate is included in the mixture. Carefully controlled amounts of water and cementitious materials are used to create a paste that forms a thick coating around the coarse aggregate without flowing off during mixing and placing. (See Figure 22-41.) Using just enough paste to coat the aggregate particles maintains a system of interconnected voids (air spaces). The result is a very high-permeability concrete that drains quickly. Because of the high void content, pervious concrete is also lightweight. After placement, pervious concrete resembles popcorn. Its low cement paste content and low fine aggregate content make the mixture harsh, with very low slump. The compressive strength of pervious concrete is limited insofar as the void content is so high; compressive strengths of 500 to 4000 psi are typical and sufficient for many applications. Pervious concrete is not difficult to place but is a bit different from conventional concrete placement. It is a very low workability material, with considerable handwork required for placement. (See Figure 22-42.) The use of a vibrating screed is important for optimum density and strength. After screeding, the material is usually compacted with a hand roller. There are no bull floats, trowels, etc., used in placing pervious concrete. Conventional jointing methods and joint spacing are used. Curing with plastic sheeting must start immediately and continue for at least 7 days. Pervious concrete's main advantage is its ability to pass large amounts of water quickly, and this has dictated traditional applications: drainage media for hydraulic structures, porous base layers under pavements, parking lots, tennis courts and greenhouses. (See Figure 22-44.) The interconnected void structure of pervious concrete allows water to pass through and percolate into the ground. The unique ability of pervious concrete captures rainwater and recharges ground water, reducing storm water runoff and helping owners comply with EPA regulations.

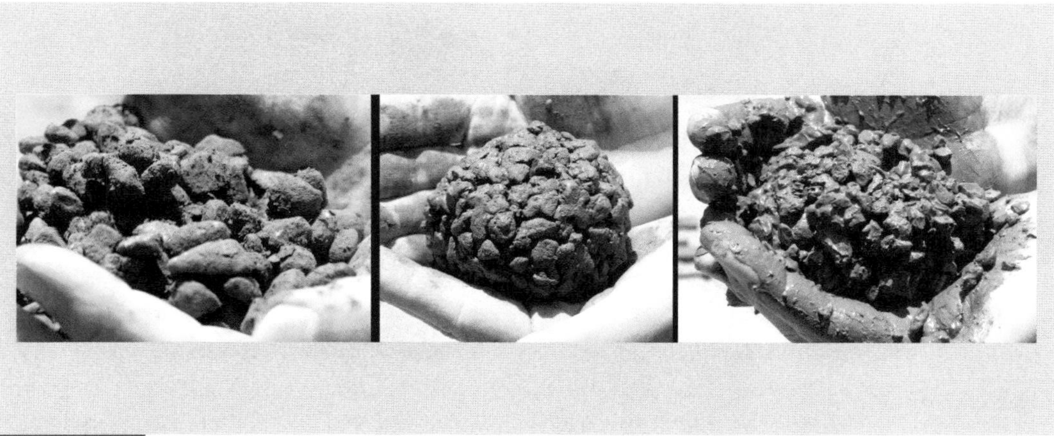

Figure 22-41
Samples of pervious concrete with different water contents formed into a ball: (a), too little water; (b), proper amount of water; (c), too much water. (Courtesy of PCA)

Figure 22-42
Pervious concrete is usually place by hand and then struck off with a vibratory screed, followed by compaction with a hand roller. Prompt curing with a plastic sheeting is required. (See Figure 22-43. (Courtesy of PCA)

Figure 22-43
Plastic sheathing should be used to cover pervious concrete and be installed within minutes of consolidation to prevent moisture loss. (Courtesy of PCA)

Figure 22-44
Pervious concrete allows water to flow through, allowing recharge of groundwater and reducing the need for storm water retention areas. (Courtesy of PCA)

Figure 22-45
Use of UHPC for thin-shelled canopies. (a), Half-canopy in steel form; (b), Canopies ready for transportation. (Courtesy of Lafarge North America)

22.19. Ultra-High Performance Concrete (UHPC)

UHPC, also known as reactive powder concrete (RPC), is a high-strength, ductile material formulated by combining portland cement, silica fume, quartz flour, fine silica sand, high-range water reducer, water, and steel or organic fibers. The material provides compressive strengths up to 29,000 psi. The materials are usually supplied in a three-component premix: powders (portland cement, silica fume, quartz flour and silica sand) preblended in bulk-bags; superplasticizers; and organic fibers. The ductile behavior of this material is a first for concrete, with the capacity to deform and support flexural and tensile loads, even after initial cracking. The use of this material for construction is simplified by the elimination of reinforcing steel and the ability of the material to be virtually self-placing or dry cast.

UHPC technology has a unique combination of superior technical characteristics including ductility, strength and durability, while providing highly moldable (precast) products. (See Figure 22-45.) UHPC offers many advantages, such as speed of construction, improved aesthetics, superior durability and impermeability against corrosion. The material is still in its infancy, and in the next few years much progress is anticipated in construction applications for UHPC.

Waterproofing and Dampproofing

Passage of water through concrete can occur in one of two ways. First, it can occur by the passage of water through channels or openings in the concrete, the water being under a hydrostatic head and in contact with one surface of the concrete. Flow of this type may be of considerable magnitude. The second way is by capillarity. No head is necessary for capillary flow, the flow resulting from a constant supply of moisture on one surface of the concrete, and evaporation from the other side. Flow of the first type, called permeability, is controlled by waterproofing, and flow of the second type is controlled by dampproofing.

Parking structures are especially vulnerable. The structures are open to the weather, and the floor is exposed to de-icing salts and other chemicals brought in on the vehicles. Frequently, the space below the floor is occupied by offices or other facilities, making a watertight floor mandatory. Permeable concrete, cracking, open joints, spalling, scaling and inoperable drains can all contribute to the problem of water passing through the floor.

23.1. Watertight Concrete

The best way to assure watertight impermeable concrete is to incorporate these properties into the structure when it is being built. This is accomplished by building the structure of good, high-quality concrete. Of special importance is the discussion of permeability in Chapter 5 and the coverage of waterproofing admixtures in Chapter 9. For the purpose of emphasis, the following important points should be reviewed.

1. The use of sound, well-graded aggregates of low porosity is necessary. Sand, especially, should consist of rounded particles instead of flat or angular ones in order to produce workable, dense concrete with the minimum water-cement ratio.
2. Concrete should be plastic and workable, thoroughly mixed, with a water-cement ratio not exceeding 0.50. High-slump mixes should be avoided. Entrained air is beneficial by decreasing bleeding and interrupting the water channel structure within the concrete.
3. The concrete should be handled in such a way that segregation and cold joints do not occur. Concrete should be consolidated by means of vibration. Construction joints are a potential source of leakage, especially the horizontal planes between lifts, and require careful placing and vibration of the concrete.
4. Form ties in wall forms should be of the type that can be broken or removed below the concrete surface. Ordinary wire ties should not be used.
5. Finally, the concrete must be thoroughly cured. This should consist of at least seven days of continuous wet curing or the application of a reliable curing compound. Because a curing compound may interfere with subsequent surface treatments, it should be checked out beforehand.

The most important consideration is the application of sound construction practices, including good design, good materials, proper handling and adequate curing. The requirements for workable, durable, strong and crack-free concrete are prerequisites for watertight concrete, and these requirements apply to concrete that is to receive a waterproofing or dampproofing treatment as well as to untreated concrete.

23.2. General

It must be kept in mind that structural movements that produce cracks in the concrete will also crack most surface coatings; hence, the forces that caused the cracks must be neutralized before corrective measures are taken.

Surface treatments consist of relatively thin coatings applied as paints, or heavy membranes of one or more plies, including bituminous coatings, bituminous membranes, bituminous boards, plaster coats, elastomeric two-component compounds, polyurethane, patented panel systems and proprietary compounds of resins and solvents.

There are many varieties of coatings. Bituminous and elastomeric coatings are applied to the surface by painting or mopping, either hot or cold. Linseed oil by itself or in combination with resinous varnishes can be used. Oil paints with a linseed oil or other weather-resistant base are also effective. Portland cement paints, if properly cured, have considerable value in resisting the penetration of water into concrete surfaces under conditions of low pressure. Such paints can be either a mixture of cement and water or specially prepared combinations.

Effective protection of porous concrete against the penetration of water under low pressure can be achieved by means of a plaster coat of portland cement mortar applied to the side of the wall exposed to water pressure. Any cracks in the concrete should be chipped out and filled with dry-pack mortar before applying the plaster coat.

Walls in basements, pits and similar structures should have surface water drained by sloping the ground away from the structure about $^1/_2$ inch in 10 feet and by the use of splash blocks and gutters under down spouts. Subsurface drainage can be provided by means of open-joint tile drains covered with a permeable fill surrounding the structure. There are several ways this can be accomplished, one of which is shown in Figure 23-1. Waterproofing of the concrete is not necessary, but the concrete should be dampproofed. When groundwater is present in the soil, the concrete should be waterproofed, and more elaborate drainage may be necessary.

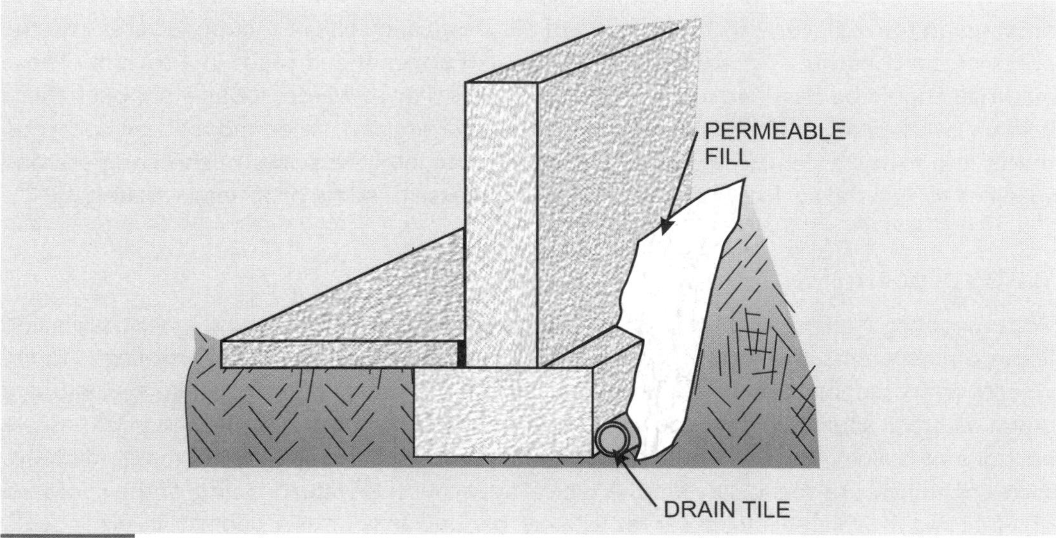

Figure 23-1
One system for drainage of surface water is shown. More elaborate means are usually necessary when groundwater is present.

Any kind of waterproofing or dampproofing should be done by persons expert in such work, using the best materials in accordance with the manufacturer's recommendations. Many of the materials are proprietary systems that include special application techniques. The first requirement is good, dense concrete, followed by adequate knowledge of the proposed protective compounds or processes, gained preferably by a satisfactory service record or by laboratory tests. Field experience should be depended on, however, as laboratory tests can be misleading because of the difficulty in duplicating field conditions.

Good materials and quality of construction are the first consideration in any structure and are essential for watertight concrete. The use of waterproofers or dampproofers, either integral or surface applied, should not be considered compensation for poor construction, lean mixes or deficient materials.

Materials. The membrane type of bitumen consists of layers of saturated felt or fabric, alternated with layers of bituminous material mopped on the surface, usually while hot.

A bituminous coating consists of one or more layers of bitumen, mopped on either hot or cold. Cold-applied material can be reinforced with fibers to improve its strength and elasticity. Fibers include glass, plastic or other inert materials.

The premolded bituminous board consists of an asphalt core-board sealed between two dry liners of asphalt-impregnated paper or felt to which a weather-resistant coating is bonded.

Elastomeric systems consist of a two-component chemically curing compound containing a polysulfide polymer that is cold-applied and forms an impermeable rubbery membrane. Another is a single-component modified polyurethane that forms a similar membrane.

A patented process consists of 4-foot-square panels. Each panel consists of one or more laminations of biodegradable, fluted kraft boards with the flutings filled with bentonite. Soil moisture decomposes the kraft paper, leaving a bentonite gel water barrier. Flexible premolded sheets of neoprene, butyl and other materials are available in sheets as large as 20 by 100 feet.

One type of surface sealant is a cement-base dry material that is mixed with water and brushed on the surface. Another consists of oils, resin and solvent that penetrates into the pores of the concrete. Yet another is a solution of polymerized solids in a solvent. These materials should be classified more as dampproofers than as waterproofers, although there is some overlapping of functions. Sodium silicate (waterglass), when mopped on concrete, reacts and forms a gel-like material that penetrates into the pores of the concrete and solidifies to seal the surface. Sodium silicate is the base of some proprietary materials.

23.3. Waterproofing

Waterproofing is required below grade where groundwater is present against walls and floors of basements and similar spaces of buildings. Above grade, waterproofing is found wherever protection is required against the passage of liquid water from leakage, washing down or other sources. Examples are swimming pools, fountains, decks and plazas above portions of buildings, balconies, air-conditioning ponds, parking garages, malls, kitchens, showers and wet rooms of any kind. A typical application is waterproofing of the roof area of a building in which the roof serves as a car parking area or as a pedestrian plaza deck. Occupied space beneath the deck must be protected from entrance of moisture. Ornamental fountains, planters and swimming pools are sometimes located on these decks, further complicating the problem. Because so many of the waterproofing materials are proprietary systems, we can give only general guidelines here. Manufacturer's instructions must be followed for any specific installation. On large jobs, the manufacturer will provide technical assistance.

When applying any of the following systems, special care is necessary when working around pipes, joints, flashings or any items that extend through the surface of the substrate, to assure that a close, watertight seal is made. Frequently extra plies or layers are used at these areas. Extra thickness should be installed at corners and angles.

Failures of any waterproofing system can usually be traced to improper construction rather than faulty materials. The following points are especially important.

The membrane must be continuous, forming a completely watertight covering. A small leak can give rise to failure by permitting water to penetrate between layers or between the waterproofing and the substrate, causing a head of water to develop, which generates further damage and ultimate failure.

The membrane should be protected as soon as it has been installed. A hammer dropped on the membrane, or a scrape by a piece of lumber being carried, can break the membrane. If the membrane has been punctured, it can be repaired by applying a patch of the membrane material. Membrane applied to a horizontal surface should be protected as directed by the manufacturer. Usually all that is necessary is to barricade it to prevent all foot or wheeled traffic until the topping slab has been placed. Just prior to placing the topping, the membrane should be inspected and any damaged areas repaired.

The membrane may fail by tearing, caused by structural movement of any kind, such as unequal settlement of the structure or movement resulting from thermal expansion or contraction. On an irregular surface the membrane may bridge over low areas if care is not taken to ensure that the membrane fits closely against all surfaces. Bond failure and blisters are the result of placing the membrane on a wet or frozen surface. When working below grade, it is usually necessary to use well points and sumps to take the groundwater away from the work area so that the installation can be made in the dry.

Elastomeric Membrane System. This material consists of two components that are mixed on the job. The resulting mixture has a short pot life, so it must be used within two or three hours. The liquid can be sprayed, brushed, troweled or squeegeed over the surface of the concrete and around pipes, hairline cracks and corners, on either horizontal or vertical surfaces. This polysulfide-based membrane cannot be used where it is exposed to direct sunlight or traffic of any kind.

The surface to receive the waterproofing should be smooth, clean and dry. Water-cured concrete is best, as remnants of curing compound will interfere with adhesion of the membrane to the concrete. Surface voids in the concrete should be filled with mortar, and projections must be removed. A steel trowel finish is desirable. Dust and dirt should be removed from the substrate surface before and during application of the membrane by blowing with air, brushing or vacuum cleaning. Air temperature should be at least 40°F and rising. A coverage at the rate of 20 to 25 square feet per gallon should produce the required thickness of 60 mils. Extra thickness is necessary at flashings, at openings, around pipes and at corners. The membrane must be protected from puncturing or other damage. For vertical walls below grade, placing asphalt-asbestos boards against the membrane will protect it from damage by backfilling.

This material is well suited to waterproofing roof and plaza decks and multideck parking garages in which a two-course slab is constructed. The waterproofing, applied to the structural slab after the concrete has dried, is protected with a covering of saturated felt or asphalt-impregnated board. A topping slab, frequently of lightweight concrete or quarry tile, is then placed. Figure 23-2 shows typical details. More elaborate installations include a insulation layer and permeable percolation layer to collect any water that collects on the surface.

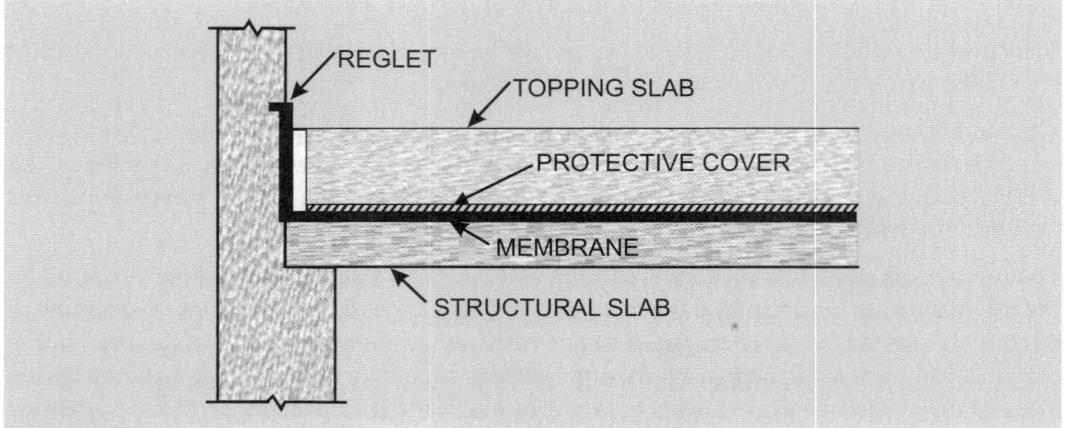

Figure 23-2
Plaza decks must be waterproofed by the installation of an impervious membrane applied to the structural slab. The membrane must be provided with a protective cover before the topping is placed.

Preformed-Sheet Elastomeric Membrane. Preformed, flexible sheets of butyl, neoprene, polyvinyl chloride and similar materials come in thicknesses up to $^1/_8$ inch and in sheets as large as 20 by 100 feet. Substrate must be prepared as for the liquid systems. The membrane is drawn into position without stretching, and is then permitted to lie until relaxed in final position before the splices are made. Bonding of the sheet to the substrate is not normally necessary on horizontal surfaces but may be required in special situations. On vertical surfaces, the adhesive should be applied to both mating surfaces to effect a bond. Splices are made by cleaning the overlapping surfaces, then joining with the cement or tape furnished by the manufacturer.

Single-Component Liquid. Materials in this category include modified polyurethanes and similar materials that can be applied directly to the concrete from the can and spread with a notched squeegee to a thickness of 55 to 60 mils.

Concrete must be smooth, clean and dry. Prior to application of the membrane, all cracks, joints and openings around vents, etc., must be sealed or prestripped by application of the liquid membrane with a notched trowel, making the strip about 55 mils thick, overlapping to a width of 2 inches on each side when applied to nonmoving joints and cracks. Special instructions are given by the manufacturer for treatment of joints that are apt to move. One important point is to prevent adhesion of the membrane to the joint sealant; otherwise the membrane may fail.

Bituminous Membrane. The number of plies varies for different exposures, but the usually recommended minimum consists of two plies of fabric or felt for exterior vertical surfaces; three plies of fabric or felt for interior above-grade surfaces, either horizontal or vertical; and two plies of felt and two plies of fabric for horizontal surfaces below grade or exterior. In many cases, the specifications will require more plies than the minimum if there is standing water on horizontal surfaces, or to withstand high hydrostatic head. Tar and asphalt should not be intermixed; that is, if tar was used as the saturant for the felt or fabric, then tar should be used as the coating material between plies. Likewise, asphalt should be used throughout.

Whether to use felt or fabric is usually a matter of economics. Felt usually costs less than fabric, but fabric is stronger and can be fitted around irregularities more easily. If there is any possibility of unequal settlement or movement of the structure, fabric should be used.

Manufacturers usually specify that the concrete shall be clean, dry and smooth, which means that surface voids must be filled with mortar and all fins and irregularities removed. After preparation of the surface of the concrete, a primer is applied, consisting of a light coating of tar or asphalt, whichever is to be used for the entire system. The primed surface is next coated with the hot bitumen, and the felt or fabric is embedded in the hot bitumen, taking care that the material is smooth without wrinkles, blisters or similar defects. This is followed with alternate moppings of hot bitumen and felt or fabric until the specified number of plies has been installed. Joints between layers of felt or fabric should be staggered so no joints in one layer are directly over joints in an underlying layer. The bitumen should be mopped on carefully so that there will be no pinholes or other breaks in the layer. Corners and junctions should be reinforced with two or more plies of felt or fabric and moppings of bitumen returned 12 or 18 inches on adjacent surfaces. Special care is necessary to assure watertightness around pipes or other items that penetrate through the waterproofing.

The specifications will indicate the number of plies and coverage of bitumen. A minimum commonly specified is 120 pounds of bitumen per 100 square feet for three-ply coverage, going up to 150 pounds for four-ply waterproofing.

Heating of the bitumen should be done in special kettles designed for this purpose. Overheating must be avoided (a yellow smoke usually indicates overheating).

Plaster. The application of a portland cement plaster coat should follow good plastering practices and should comply with local building code requirements. The concrete must be clean and rough. Concrete that has been cast against smooth plywood should be prepared by acid etching or light sandblasting. However, if the concrete is free of traces of curing compound and other coatings, even though smooth, the application of a proprietary bonding agent or cement dash bond coat will suffice. The bonding agent must be one that is certified for use in a moist environment, as some of these materials saponify or otherwise deteriorate in the presence of water. A dash coat can be made of one part portland cement to one and one-half parts plaster sand with enough water to give a plastic consistency. It can be dashed forcibly onto the damp concrete surface with a large brush, such as a whitewash brush, or it can be applied with a plastering machine, and should then be left undisturbed, except for moist curing with a gentle water spray for 24 hours. Plaster is then applied by either hand or machine methods in two coats, each about $^3/_8$-inch thick.

Because of the possible shrinkage and cracking of cement plaster, some use is made of nonshrinking mortar consisting of portland cement, sand, water and a patented shrinkage-control material, the latter usually consisting of powdered iron, a catalyst such as ammonium chloride, and other ingredients. The manufacturer recommends application of a brush bond coat of the material to the roughened concrete, followed immediately by the plaster. Hand application of the plaster is made in two coats, $^1/_4$ inch thick, the second coat troweled on the first as soon as the first has stiffened enough so it will not sag. The finished plaster should be cured for 24 hours.

Any movement of the structure subsequent to application of the plaster resulting from settlement, shrinkage, loading or any other cause that will crack the structure will also crack the plaster and result in failure of the waterproofing.

Some use has been made of cement plaster applied to the interior of a wall. This may be of some value on existing construction where the outside is inaccessible. The wall must be free of water flowing through the concrete or clinging to the surfaces when the plaster is applied. Application of interior plaster under these circumstances should not be done in locations where the exterior surface is exposed to freezing and thawing. Such a plaster coat affords no protection against weathering or attack by substances in the water.

Sheet Lead. Outdoor pools are effectively sealed with a membrane of sheet lead that can be placed before the concrete is placed. Lead is also used in a similar manner for waterproofing under a floor. Sheets can be joined by welding. Because of possible corrosion of the lead when in contact with fresh concrete, the lead should be protected with a coating of bituminous paint.

23.4. Dampproofing

Moisture can pass through most concrete even though there is no hydraulic head or pressure to force it through. The moisture can originate as capillary moisture in the soil, rainwater or other sources. To control the passage of water or vapor under these conditions dampproofing is applied to the concrete. The difference between waterproofing and dampproofing is that dampproofing does not consist of the rather elaborate seal that is accomplished with waterproofing. Dampproofing eliminates the passage of moisture that might create excessive humidity in the interior of the structure or damage interior finishes. Treatments that are acceptable for dampproofing must not be substituted for waterproofing in those situations where hydrostatic pressure, standing surface water or large quantities of free water are present.

The materials that are used for waterproofing can be used for dampproofing as well. Asphalt and tar-base liquids for cold application consist of cutbacks and emulsions. The asphalts usually require a prime coat. Two coats of bitumen are applied by brush, trowel or roller, depending on the consistency; each coat is allowed to dry before the succeeding one is applied. Total thickness should be between $^1/_{16}$ and $^3/_{16}$ inch. Spray application is satisfactory if care is taken to avoid bridging.

Many of the previously mentioned waterproofers, especially the proprietary ones, are also adaptable to the requirements of dampproofers, usually by adjusting the consistency so that the material can be applied as a paint. Some of the clear sealants can be classified as dampproofers.

23.5. Repair of Leaks

Cracks and defective areas on the side of the structure away from the water pressure, such as a basement wall or floor, or the exterior of a tank or pipe, can be effectively sealed against appreciable hydrostatic head. In the hands of a skillful operator, the following method can be used to stop actual flowing leaks.

The leaking area should be prepared by removing all unsound concrete and encrustations. Cracks should be V-grooved. The proper technique is to start at the top and work toward the lowest crack or leak. Small leaks should be plugged before large ones.

There are a number of good proprietary materials available that can be regulated to set in a few minutes, or even in 15 to 30 seconds. Some require only the addition of water to prepare them for use, while others are used with portland cement. Manufacturer's instructions should be followed.

Quick-setting cement can be made by mixing portland cement and aluminous cement or by using Type III cement with calcium chloride. Mortar should consist of one part cement to two parts plaster sand, mixed to a barely moist consistency.

The repair material must be tamped in place because of its barely moist consistency. For large leaks, the material should be shaped and permitted to harden slightly before it is shoved into place, then held in place until it sets. Making repairs of this nature requires working the material with the hands; for this reason, workers should wear rubber gloves. Some of the proprietary materials may be somewhat toxic, requiring that adequate ventilation be provided. Another word of caution: Some materials set rapidly and successfully plug the leak, but their durability is poor. For example, the use of Type III cement with a high percentage of calcium chloride is best considered a temporary expedient.

Introduction to Inspection

When we watch a good baseball game, we see an example of how smooth teamwork can bring success to a group effort. Each person—whether pitcher, shortstop or left fielder—contributes to make the team a winner. The success of the team depends upon each member lending individual expertise and ability to a joint effort.

Similarly, construction requires the efforts of many individuals representing different crafts and professions. These individuals are all part of the construction team. Members are the owner who requires the facility and pays for it; the architect who designs it; the engineers who provide for the foundation and structural adequacy; the contractor and subcontractors who provide the plan and know-how to build the structure; the material suppliers who furnish the materials that are used; the craftsmen and mechanics who intelligently apply their trades in constructing the project; and finally, the inspection and quality control group who assure that the structure is built in accordance with the plans and specifications.

24.1. What Is Inspection?

We can define inspection as the review of the contractor's work to make sure that the specifications and drawings are being followed and that relevant codes and ordinances are being observed. The inspector may represent the owner, a governmental agency, the engineer, the architect, a lending institution or the contractor. The objective of inspection, broadly speaking, is to provide the owner and the building official with a means of assuring quality construction. This is accomplished by (1) seeing that the requirements of the specifications and codes are carried out, (2) seeing that good construction practices are used and (3) recognizing what constitutes good concrete, how to get it and what might happen if you don't get it. Construction that requires a building permit and that is to be performed under the jurisdiction of any city, county or other political subdivision must be inspected in accordance with the requirements of such governmental agency. A typical arrangement is one in which the agency has its own inspection staff directly under the authority of the building official. Inspectors are assigned to certain jobs as necessary for adequate inspection of all parts of the work requiring inspection as specified in the local code.

In some cases, inspection is done on a contract basis by an individual or agency licensed to perform building inspection. Laws of different governmental agencies vary as to the qualifications of the several grades of inspectors regarding experience, education and registration. On projects for which inspection and testing are performed on a contract basis by a commercial testing agency, practice has been to award this contract to the lowest bidder. The fallacy of this practice has long been evident, and it is now more common to negotiate inspection and testing services on a professional basis. After all, it is just as wrong to take bids and award the contract for professional inspection services to the low bidder as it is to take bids and award contracts for any other professional service, such as engineering and medical service, on that basis.

On major projects such as highways and dams, the governmental agency that designs and controls the project takes care of inspections. Inspection thereby becomes an extension of the design responsibility and is directed by the resident engineer of the governmental agency. This traditional method of inspection has passed the test of many years' use on projects large and small.

The more advanced contractors or builders maintain an inspection capability for their own benefit. In so doing, they can control their work, the inspectors serving in the capacity of a quality control staff. Adequate inspection by the contractor can contribute to a smooth-running job with a maximum of harmony with inspectors employed by the owner or governmental jurisdiction. However, the practice of having the contractor employ the testing or inspection service for the owner or building official is neither desirable nor advisable and should be avoided. Regardless of who makes direct payments for inspection, the owner pays for it, and it is far better to have this cost out in the open rather than buried in the contractor's unit costs.

The Inspector. Inspection is not an easy job. The inspector requires certain abilities and training, possibly best described in the following quote:

> A competent inspector is thoroughly conscious of the importance and scope of his work. He is observant, alert and properly trained; he knows both how and why the work is required to be done in a certain way. As designs and specifications change from job to job, it is important that the inspector be thoroughly acquainted with the specifications for the particular work with which he is involved. Armed with this knowledge, and with judgment gained from experience, he will not only detect faulty construction but will also be in a position to forestall it by recognizing the causes in advance and preventing the use of improper procedures.
>
> Although the inspector will require special instruction or advice from his supervisor concerning unusual problems or controversial matters, his initiative is continually brought into play. The inspector should not delay the contractor unnecessarily or interfere with the contractor's methods, unless it is evident that acceptable work will not otherwise be produced. Fairness, courtesy and cooperativeness, coupled with practicality, firmness and a businesslike demeanor, will engender respect and cooperation. Avoidance of needless requirements and restrictions will facilitate the accomplishment of the primary purpose of inspection, which is fulfillment of the specification requirements, and will also enable the contractor to perform his work in the most advantageous and profitable manner.[24.1]

Suggestions and instructions relative to the acceptance or rejection of construction or materials should be given to an authorized representative of the contractor or producer, and not to the workers. The inspector should carefully refrain from giving instructions; otherwise he or she may be considered acting as a supervisor for the contractor's or producer's workers, and may thus be inviting claims by the contractor.

In the course of their duties, inspectors frequently come into possession of confidential information relative to production or composition of materials under their inspection. This information, of course, should not be divulged to competitors or the general public.

Inspectors, in their dealings with contractors and producers, should refrain from criticism of their supervisor or other members of the design and construction team. In the final analysis, the supervisor is responsible for the work, with justifiable reasons for making overall decisions that cannot be revealed to the inspector at the time. In case of a difference of opinion between inspectors, or between an inspector and a supervisor, the inspector should collect the facts and present them to the supervisor and request advice concerning any doubtful procedure or materials.

Adequate supervision of an inspector entails, among other things, proper designation of authority and responsibility. The supervisor should do the following:

1. Outline in advance those matters for which each inspector will be directly responsible.
2. Inform the inspectors as to decisions that they should refer to a higher authority.
3. Ensure that inspectors have access to and read all necessary instructions pertaining to their work, including codes, specifications, plans, letters of instruction, laboratory reports concerning the work, and information concerning special problems and conditions that are apt to arise.
4. Ensure that an inspector is not placed on the job without knowledge of special arrangements or agreements that have been made with the contractor and that affect the work with which the inspector is concerned. Particularly, the inspector should be informed at once of any concessions or special interpretations of the specifications.
5. Ensure that inspectors are made aware that they have full support in all proper execution of the inspection work.
6. Obtain the opinions of inspectors and engineers close to the job before reaching decisions relative to requests and representations made by the contractor.
7. See that every question concerning the job, even though sometimes not pertinent, is promptly and fully answered, and that the inspectors feel free to ask such questions.

Special Inspection. In any work done under the conditions of the code, the building inspector employed by the building official is the primary and official inspector. He or she is required to inspect and approve the job at specific intervals and approve certain work before further construction can be commenced. For example, approval of rebar placement is required before concrete can be placed.

According to the *International Building Code* (IBC),[24.2] special inspection is defined as: "inspection as herein required of the materials, installation, fabrication, erection or placement of components and connections requiring special expertise to ensure compliance with approved construction documents and referenced standards" . . . by a specially qualified special inspector. Knowledge and duties of a special inspector differ from that of a jurisdiction building inspector in that their expertise is very specialized; hence the term special . The special inspector provides continuous and periodic special inspection at all times that the particular construction is in progress. Duties of the special inspector include: review of structural details and verification of compliance with approved construction documents; keeping of accurate records of inspections; furnishing of inspection reports to the building official and registered design professional; notification to contractor of observed discrepancies for correction, notification to the building official and registered design professional of uncorrected discrepancies, and submittal of final inspection report to the building official and registered design professional.

For a more comprehensive discussion of special inspection requirements (job tasks), see Sections 25.9 and 25.10.

24.2. The General Building Code

Most units of government, whether at the state or local level, adopt a general building code to regulate the design and construction of buildings and other structures. Prior to 2000, the majority adopted by reference one of three model building codes, the *National Building Code* (NBC), the *Standard Building Code* (SBC) or the *Uniform Building Code* (UBC). Some jurisdictions wrote their own codes; however, most of these were based on one of the three model codes. A few locally developed building codes are still in use, but the number is dwindling.

In 1994, the three model code organizations that published the NBC, the SBC and the UBC announced the formation of the International Code Council (ICC) for the purpose of publishing the *International Building Code* (IBC)– a single model code to replace the three existing model codes, for adoption nationwide.

Since initial publication in 2000, the IBC has been adopted nationwide by states and local jurisdictions to replace the existing three model codes. To date, the IBC is used almost exclusively to regulate building design and construction nationwide. All governmental agencies at the federal level, and the militaries, are also under the umbrella of the IBC. Like its three predecessors, a new edition of the IBC is published every three years: The 2006 IBC was published in February of 2006 and the current edition was published in February of 2009. Most building code agencies (local, state and federal) that initially adopted the 2000 IBC are now updating to the 2009 IBC.

The *International Building Code* adopts by reference the ACI 318 Standard to regulate concrete design and construction. The 2008 edition of ACI 318 (ACI 318-08) is adopted in Chapter 19 of the 2009 IBC.

Of special note: The code "Special Inspection" requirements for building construction are addressed in Chapter 17 of the IBC. The IBC code requirements for special inspection of concrete work are discussed in Section 25.9.

24.3. The ACI Building Code

The general building code is the primary regulatory document used by governmental agencies (city, county, state, etc.) charged with the enforcement of building regulations within a particular jurisdiction. Many other regulatory documents are referenced in a general building code to regulate proper design and construction of the various materials used in building construction.

Throughout this book we have referred to ACI Standard 318, "Building Code Requirements for Structural Concrete" as our primary code resource document for concrete construction practices. The ACI 318 document, also known as "The ACI Building Code, " is the nationally recognized standard for design and construction of concrete buildings and is written as a legal document so that it can be adopted by reference in a general building code. ACI 318 is adopted as part of most general building codes to regulate proper design and construction of buildings of structural concrete.

The concrete inspector should be thoroughly familiar with the ACI Building Code, principally Chapters 4 through 7 covering concrete construction requirements. As a helpful reference, the ACI 318 includes a commentary on background details and suggestions for carrying out the code requirements or intent. Another good source of information for the inspector is the *Concrete Inspection Handbook*[24.3], published by the Portland Cement Association. The handbook explains, in detail, the purposes of the ACI 318 code requirements for concrete construction (Chapters 4 through 7) and gives additional background information to improve one's understanding of why the code requirements are necessary minimums to ensure quality concrete construction. The inspector also needs to become familiar with the various ASTM specifications for the manufacture, use and methods of tests for the materials used in concrete construction as listed in Chapter 3 of ACI 318. A good source of the pertinent ASTM standards for concrete construction is available in Reference. [24.4.]

24.4. Sources of Authority and Information

Primary Sources. Certain documents pertain specifically to the project under consideration; others are legally a part of the contract documents. In general, there are four primary sources of authority to guide the inspector:

1. The inspector should be governed by what is written in the specifications and shown on the approved plans. These describe what is wanted, and sometimes how to get it—usually in a general way, but at times quite detailed. The object of the specifications and plans is to describe what kind of a structure is required by the owner and what materials are to be used.
2. The inspector should be governed by the general building code, laws and ordinances within the jurisdiction where the project is located.
3. The inspector should be governed by good construction practices. What constitutes good construction practices is something learned by experience, by observing and doing. Methods required by the specifications are based on good construction practices.
4. The inspector should be governed by manuals and instructions furnished by his or her employer. Such instructions serve to guide and assist in obtaining conformance with the first three governing principles. Such manuals and instructions can serve to assist in inspection, but the inspector cannot do an inspection with a guidebook in hand. In the final analysis, the specifications govern.

Other Sources. In addition to primary sources, there are other sources of information available to the inspector. Among these are publications of technical societies, trade associations and manufacturers of construction materials, as well as reference books and texts by various publishers.

Many of the publications of the American Concrete Institute have been referred to in this book. Another useful book is the *ACI Manual of Concrete Inspection*.[24.5]

Manufacturers of admixtures, cement, steel and other materials publish literature describing their products, and many of them furnish good technical reports concerning the use of their materials. Most of this literature is reliable though somewhat promotional, extolling the virtues of the product but soft-pedaling the faults. Often, however, the only source of technical information on a product is that which is available from the manufacturer. Properly evaluated, the manufacturer's literature can contribute significantly to the information needed by the inspector.

Trade associations, such as the Portland Cement Association, National Ready Mixed Concrete Association and Concrete Reinforcing Steel Institute, publish excellent technical reports and standards.

Not to be overlooked are the many governmental agencies that have reports, standards and literature available, including the U.S. Bureau of Reclamation, Corps of Engineers and other agencies whose publications are available from the U.S. Government Printing Office in Washington, D.C.

24.5. Review of the Project

Once an inspector has been assigned to a project, one of the first duties is to become familiar with job requirements that can pertain in any way to the inspection. This may be relatively simple on a small job in which inspection is on a part-time basis, but it is important that the inspector know what the project is all about beforehand. The construction documents should be available for review by the inspector. These include the code, plans, specifications, reference specifications, laboratory reports, concrete mix designs, soils report, purchase orders and any other documents relating to the concrete construction phase of the project.

The inspector should become acquainted with the authorized representatives of the architect, engineer, contractor and building official. The inspector should determine what arrangements have been made for employment of a laboratory, offsite inspection and testing, approval of sources, ready-mixed concrete and subcontractors. Any ambiguities or discrepancies should be reported immediately so that the problem can be clarified before any disagreement arises.

A directory should be compiled showing names, addresses, telephone numbers and names of authorized representatives of all companies and persons involved in concrete work, including building official, owner, architect, engineer, contractor, subcontractors, suppliers, ready-mix supplier and vendors.

24.6. The Laboratory

Whether a laboratory is established on the site depends on the size of the project, the amount of testing to be done and the availability of commercial testing facilities within a reasonable distance from the job. The field laboratory, if one is established, can be simple or elaborate, depending on the above factors. The minimum would be a place where the inspector can make and store test specimens until they can be taken to the testing laboratory. At the other extreme, for a large job in a relatively isolated area, the field laboratory might be a complete unit in itself with curing facilities for a large number of specimens, testing machines, sieve shakers and other testing equipment. For most building jobs, laboratory facilities are usually provided by a commercial testing laboratory, with a minimum of facilities on the site. The central laboratory then provides complete testing and inspection facilities for a number of jobs in the area.

The specifications usually require that the contractor provide space for the field laboratory, but it is up to the testing and inspection agency to furnish the necessary equipment, such as the following:

- Slump cone and rod (ASTM C 143)
- Cylinder molds—either 4- by 8-inch or 6- by 12-inch as indicated by the project specifications (ASTM C 31)
- Apparatus for measuring air content of concrete (ASTM C 173 or C 231)
- Armored thermometer (for concrete temperature) 30°F to 120°F (ASTM C 1064)
- Pocket rule
- Bucket, scoop or shovel
- Waterproof marking crayon or felt-tip pen

The above equipment will enable the inspector to conduct slump tests, make strength test specimens, determine air content and check the temperature of the concrete. These are the minimum tests that should be conducted.

24.7. Approval of Materials[1,2]

The inspector on the job is not usually called upon to approve materials proposed for use in the construction, these approvals being handled beforehand by the building official, architect or engineer. However, the inspector must be satisfied that the material that is used in the structure has been approved and that the particular lot under consideration is of the same composition as the original sample. Any pretested or pre-inspected material is subject to inspection and testing after delivery to the jobsite, regardless of prior inspection or test.

Approval of Sources. It is sometimes desirable to give prior approval to a material source so that tentative approval can be made of material from that source, to enable the contractor to place supply orders with firms that are capable of producing materials conforming to the specifications. Source approval of manufactured materials usually presents no problem. Cement, reinforcing steel, wire and strand for prestressing, and similar items are manufactured to well-established standards and can normally be expected to conform to appropriate ASTM or other standard specifications.

The contractor should designate material sources as early as possible so that preliminary approval of these sources can be made. This preliminary approval may be based on knowledge that the proposed source has a satisfactory record of producing materials to meet project specifications; it may be based on manufacturers' certifications to that effect, or it may be based on results of tests and inspection made under direction of the building official, engineer or architect.

A request for source approval together with pertinent supporting data should be directed by the contractor to the engineer, architect or building official. Supporting data should contain information about the history and service record of the material as well as typical mill, factory or shop tests, including tests by an independent testing laboratory. Any preliminary approval of sources should be considered tentative approval; the contractor is ultimately responsible for furnishing materials that conform to specification requirements.

Preliminary investigation and inspection should not be made on the basis of requests from material suppliers, as this would involve study of numerous materials that would never be used in the job. Only specific sources and materials, as requested by the contractor and intended for use in the project, should be inspected and tested.

Offsite Testing and Inspection. In the case of some projects, especially large ones, or jobs that are in progress for a considerable period of time, it is sometimes desirable to perform inspection of some materials at the point of manufacture. Materials in this category might include cement, structural steel and similar large-volume materials.

Materials that are inspected at the source of supply can be tested, tentatively approved, marked, and either placed in storage or shipped immediately to the project. Such approval is given as a convenience to the contractor to expedite the work, and any materials so approved must comply with the specifications at the time they are incorporated in the work.

Adequate identification of approved materials should be provided in every case. Individual pieces or units can be marked, stamped or tagged. Bulk materials should be placed in bins, tanks, drums or other suitable containers that can be sealed and identified.

Offsite inspection can be handled by a commercial testing agency; this is especially desirable when the engineer or purchaser does not have testing and inspection facilities capable of making inspections in plants located in different parts of the country. To authorize the inspector to enter the plant, the building official should be provided with copies of the contractor's purchase order for the material to be inspected. Upon receipt of a copy of this purchase order, the testing agency contacts the supplier named thereon and arranges for inspection. A statement on the purchase order to the effect that the material is subject to inspection by the laboratory will facilitate this inspection.

Each carload or truckload of approved materials should be accompanied by a tag or card of identification issued by the testing laboratory showing date, source, material, quantity in shipment, identity of transporting unit, destination, job identification, purchase order number and reference to laboratory test report covering the material. These tags should be collected and retained by the inspector receiving the material at the jobsite.

Acceptance on Manufacturer's Certification. In some cases, certain standard materials can be accepted on manufacturer's certification. A request for such approval is made by the prime contractor, the same as for source approval, and should be accompanied by pertinent data to support the acceptance. The supporting data should contain information about the history and service record of the materials, as well as mill or shop tests, and test results by an independent testing agency. The manufacturer of materials accepted on certification will furnish certified tests for each lot or consignment, showing actual test results, a statement that the materials conform to specification requirements (citing the specifications), a statement certifying that the formulation and manufacturing procedures have not changed from those prevailing at the time the original acceptance was made, and an explanation of how the certified materials are identified or marked.

Rejections and Retests. Materials failing to meet the requirements of the specifications should not be used in the work unless specific approval is obtained from the building official for the use of such materials.

Rejected materials should be disposed of in such a manner as to ensure that they will not become intermingled with accepted materials and be incorporated in the structure. Permitting rejected materials to become intermingled with acceptable materials may be cause for rejection of the entire lot, including materials that have been accepted.

There will frequently be requests for retests when materials have been rejected because of not meeting specification requirements. Retests should not be made unless there is some reason to question the accuracy of the original test, and then only with the approval of the building official. If retests are made, at least two samples should be tested for each sample that originally failed. The results of both retests should be within acceptable limits, and the questionable materials should be visually inspected. The report of retests should clearly identify samples to facilitate correlation with the lot of materials that originally failed to meet specification requirements.

The number of tests that constitute a failure is sometimes a bone of contention. The ASTM test methods usually indicate procedures for retests if a test does not meet the standards. For example, one aggregate sieve analysis out of limits in a series of tests, or one low-strength test result, should not be used as a justification for refusal to accept the material or work. A single out-of-limits test result should be viewed with concern but not alarm, and steps should be taken to assure that a trend is not being established. The quality control procedures described in ACI 318 Section 5.6 offer a logical method of evaluation and acceptance of concrete.

24.8. Job Safety

Accident prevention and safety on the job are the responsibility of the contractor. The inspector, however, is involved to the extent that inspection activities on the job must be governed by good, safe practices. Furthermore, when an inspector observes unsafe construction practices, he or she should discuss the situation with the supervisor and, if the condition persists, note it on the daily inspection report. Jobs are subject to regular inspection by federal, state and local safety inspectors. The construction inspector should cooperate with these inspectors in any way that is necessary.

INSPECTION OF MATERIALS

One of the first things for the inspector to do is to determine specification and code requirements regarding certification and testing of cement, aggregates, admixtures and other materials related to the concrete. Questions to be answered include the following:

- Are actual tests required on each lot of material?
- Will a typical analysis be acceptable?
- Is manufacturer's certification necessary or acceptable?
- Should samples be taken from material after delivery?
- Is any offsite inspection being done at the producer's plant?
- If offsite inspection is being done, how will inspected material be identified?

24.9. Portland Cement

Cement is rarely furnished directly to a building job, because practically all concrete comes from a commercial ready-mix manufacturer. However, certain specialty cements (for example, masonry cement for masonry mortar) are usually delivered to the jobsite in bags.

All bags of cement are identified by a rubber-stamp code number imprinted on the bag when it is filled. Codes vary from one manufacturer to another, but they usually denote the day and shift that the bag was filled and occasionally include additional information. It is quite often possible (but not always) for the cement company to determine the lot of cement from which a certain code was shipped, which would enable a laboratory certification to be rendered for that certain lot of cement. If the company is notified beforehand that such lot certification will be required, the certification can then be provided. Another method sometimes practiced is for the producer to furnish a typical certification representing the results of tests of the type of cement in question, such results being typical and not necessarily the results of tests on any particular shipment of bags. Some state and federal agencies require certified test reports of all lots of cement, such reports covering actual tests on the lot in question from which bags were shipped.

When it comes to bulk cement, the procedure is somewhat different. The cement manufacturer always has a record of each lot of cement in each storage silo. This record includes copies of all laboratory tests performed on that cement (see Table 7.2). Because of this, the manufacturer can furnish certification and test reports to accompany any carload or truckload of cement loaded out of a certain silo at any time. If the cement is being delivered direct to the job, the inspector should collect these certifications at the time of delivery.

This practice is sometimes modified by taking samples, either at the source or jobsite, thereby permitting cement to be used on certification pending results of tests of the samples. As long as test results indicate satisfactory compliance with the specifications, this method is quite satisfactory. However, if a test gives results that fail to comply, then a delay until further sampling and testing indicate compliance follows. Some governmental agencies and other users of large quantities of cement sample and test the cement at the time of manufacture. The cement is then stored in reserved silos at the mill, from which shipments are loaded out for the project under inspection by the engineer or owner or a testing agency employed by the engineer or owner.

Cement is manufactured under close quality control and rarely fails to meet specifications. However, wide fluctuations in properties may exist even though the cement meets the specification requirements.

If cement is inspected at the source, each carload or truckload should be accompanied by a tag or card from the inspector, showing the type and brand of cement, source, weight of shipment, date tested, date loaded, name of testing agency, number or name of vehicle and job identification. The jobsite inspector should inspect cars and trucks containing cement for evidence of contamination or damage, supervise the removal of seals and record the seal numbers, and pick up the shipping report that accompanies each shipment.

Cement delivered in bulk should be stored in weathertight bins that protect the cement from dampness. Care should be exercised in transferring cement from carrier to bin to prevent the cement from becoming wet or contaminated with foreign material. Pieces of scrap iron or other material have occasionally been found in bulk cement, originating from the mill or hauling equipment. Such items should be removed, as they can damage concrete-handling equipment. Their presence should be reported to the shipper. Exposure of cement to air should be kept at a minimum, as atmospheric moisture causes partial hydration.

Bagged cement should be stored in a weathertight building so as to permit easy access for inspection, to protect it from dampness and to permit removal in chronological order of receipt. Occasional bags should be weighed as a check on the weight furnished.

Different brands or types should never be mixed. Cement salvaged from spillage around bulk cement bins or from broken bags should not be used. At the time of use, cement should contain no lumps that cannot be broken by light pressure between the fingers.

When ready-mixed concrete is to be furnished to the job, prior arrangements should be made with the producer to assure that the necessary reports and certifications will be available if they are required. A system must be developed so that the portions of the structure can be related to the proper lots of cement.

The minimum specification requirement commonly encountered merely states what type of cement is required. In this case, a statement from the ready-mix producer that the specified cement will be used is all that is necessary.

24.10. Aggregates

The jobsite inspector is rarely called upon to test or inspect sand and gravel, because these materials are delivered to the job in the form of ready-mixed concrete. There are, however, occasional jobs in which concrete is mixed on the site, or times when the inspector finds it necessary to visit the ready-mix batching plant. In these cases, the inspector may become involved in approval of aggregates.

The discussion of aggregates in Chapter 8 covers, besides the aggregates themselves, preliminary approval, stockpiling and sampling of aggregates—items of special importance to an inspector. An inspector assigned to a batching plant, besides having responsibility for control of the batching operations, should check materials delivered to the batch plant and be assured that all materials arriving at the plant have been approved for incorporation into the work. To properly perform inspection duties, the inspector should be informed of all approved sources of materials and expected arrival of shipments.

The inspector should observe unloading of cars, trucks and barges of cement and aggregates, and handling of materials into and out of stockpiles, being watchful to prevent contamination and segregation. The inspector should also collect all shipping reports, test reports, broken car seals and other pertinent information accompanying each shipment. Any handling or storage that results in contamination or deterioration in quality of the materials may be cause for rejection of the materials. Handling and stockpiling aggregates are discussed in Chapter 8.

Aggregates are delivered by truck to most batch plants and dumped into receiving hoppers from which the material is conveyed by belt into the plant bins. There might be only one receiving hopper for all sizes of aggregate, including sand, and one conveyor, or there might be several receiving hoppers in a row side by side, with a belt conveyor in a tunnel underneath. In either case, delivery into the correct plant bin is accomplished through a selective swivel chute at the discharge end of the conveyor. Modern plants are equipped with instruments that show the level of material in each bin, which gate is open, the position of the swivel chute and the operation of the conveyors. These indicators can be interlocked to prevent material from being placed in the wrong bin. Some more elaborate plants have individual receiving hoppers, conveyors and bins for each material.

Sometimes a truck driver or crane operator will place the wrong material in a pile or bin. The only recourse is to remove the improper material, which may require emptying the bin. Such drastic measures can be avoided if the batching plant is equipped with a finish screen that removes undersize material.

Coarse aggregates, in addition to being washed and crushed, are screened into several sizes so that the concrete mixes can be proportioned properly. However, by the time these separated sizes reach the batching plant, they are apt to contain a large amount of undersize caused by numerous handling operations, or they may be seriously segregated. Finish screening of coarse aggregates at the batching plant largely eliminates the accumulations of undersize material and reduces segregation.

The inspector should watch for trash, mud or other contaminants in the aggregates and have such things removed or require that the aggregate be wasted. Materials used in sealing holes in cars, such as paper, rags, boards or straw, are frequently picked up by the unloading equipment. Steps should be taken to prevent such materials from getting into the batch.

When aggregates of different types or sizes are placed in adjoining compartments of the same storage bin, the partition between the two should be built to a height sufficient to prevent material from flowing from one compartment into the other. Partitions should be tight and free from holes through which fine materials might leak from one bin to another.

Part of the inspector's duties includes taking samples of aggregates as described in Chapter 8. Of special importance are aggregate samples for gradation, or sieve analysis. Samples must be representative. In most plants it is possible to sample the entire stream of aggregate as it drops from the bin gate into the weigh batcher. Because of probable segregation in the stream of aggregates, the entire cross section of the stream should be sampled for a short time, rather than taking a portion of the stream for a longer period. Segregation and breakage in the storage bin will be revealed in the results of these tests. Sand, being in a moist condition, is not subject to segregation.

Testing Aggregates. There are standard ASTM test methods for determining all known properties of aggregates, as shown in Table 8.1. Most of these tests are performed in the laboratory, but there are some that might occasionally be made by the job inspector when inspection is required in the batch plant.

A frequently performed test is determination of the moisture content of the aggregate, usually the sand. If the batch plant does not have an electric moisture meter in the sand batcher, moisture content tests are necessary to control the concrete properly and to know what the water-cement ratio is. In a plant equipped with a moisture meter, an occasional check of the moisture in the sand should be made to verify the value shown by the meter.

There are several methods of obtaining the moisture content of aggregates, any of which may be acceptable.

Drying in an oven or over a hot plate is a common method. The sample should be oven dried, and a correction made for absorption. The moisture content is found by the equation

$$\text{Percent Moisture} = \frac{\text{wet weight} - \text{oven dry weight}}{\text{oven dry weight}} \times 100 - \text{Percent Absorption}$$

A slight error is introduced by using oven-dry weight instead of saturated surface-dry weight, but it is not significant and can be ignored. The use of Figure 24-1, based on this formula corrected for surface-dry weight and absorption, will save computation time. All that is necessary is to obtain a sample of 400 grams wet weight, to oven dry and weigh it, and to read the percent moisture from the curve for the correct absorption, which must be known from previous tests.

When no oven or hot plate is available, the aggregate can be dried by the alcohol method. The wet sample is weighed in a large flat pan (cake pan), and 1 or 2 ounces of denatured alcohol mixed with it. The alcohol is ignited and allowed to burn off while the aggregate is being occasionally stirred. If necessary to assure complete drying, the process is repeated. A little practice enables one to estimate the amount of alcohol to use. This should be done in a well-ventilated place, where there is no danger of fire. This method can be used for both fine and coarse aggregate.

Pycnometers and flasks can also be used. An easy-to-use device consists of a pint or quart fruit jar (mason jar) with the top edge ground to a true plane surface, and a small square of glass or clear plexiglass to rest on the top. This is calibrated and used in the same way as the pycnometer (ASTM C 70) except that the jar is filled with water until the water surface, supported by surface tension, is slightly above the edge of the top. The glass or plastic plate is carefully placed over the top by sliding it from one side, taking care to avoid inclusion of air bubbles underneath, and making uniform contact with the top of the jar. Calculations for moisture content are made as described in ASTM C 70.

There are a number of patented devices on the market that give reasonably accurate results. One is a small pressure vessel with a dial pressure gauge on one end. A small weighed sample of sand is placed in the container and the lid fastened. When the instrument is turned upside down, a measured amount of calcium carbide, coming in contact with the sand, reacts with the sand moisture and produces acetylene gas. The pressure created by the gas is a measure of the amount of moisture in the aggregate, which can be read directly on the dial gauge.

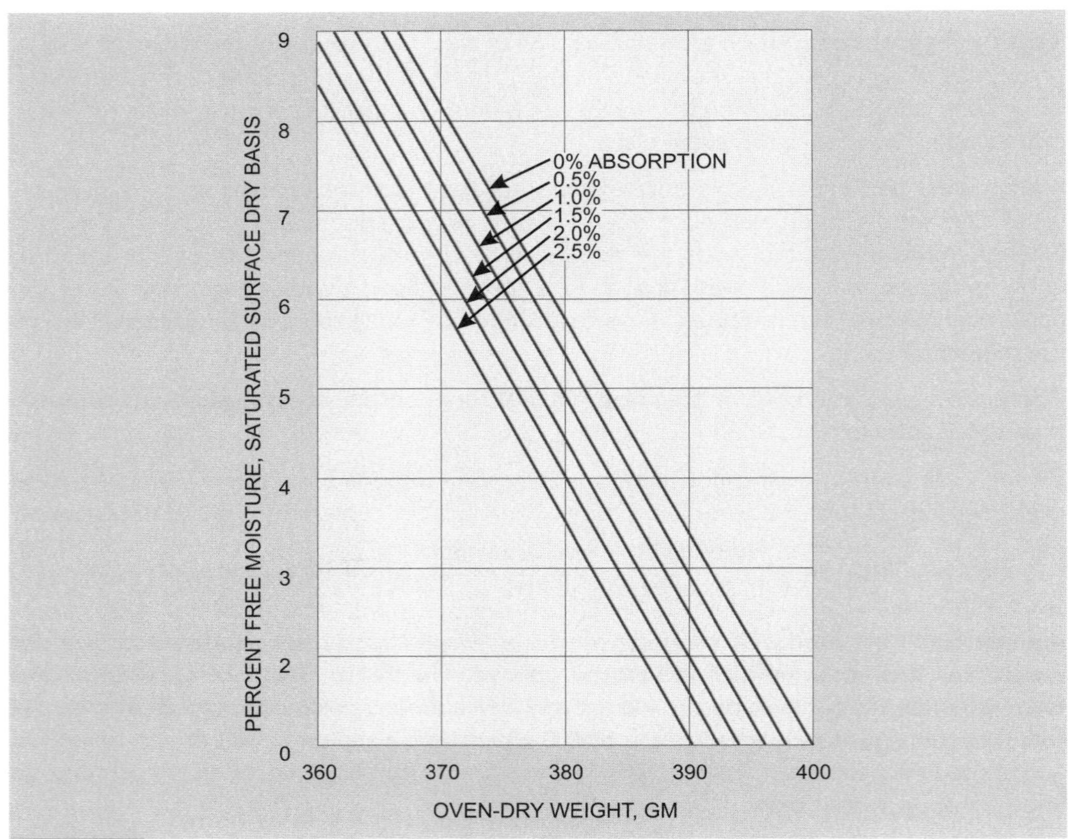

Figure 24-1
Aggregate moisture chart. Weigh out 400 grams of wet aggregate, dry to oven dryness, weigh again. Enter the chart from the dry weight at the bottom, move up to intersect curve for absorption and read percent moisture at the left. If absorption is not known, assume 1.5 percent for sand and 0.5 percent for coarse aggregates.

24.11. Reinforcing Steel

Normally, the jobsite inspector will only need to check that the shipping papers and identification tags properly identify each lot, shipment or truckload of reinforcement. Documents and methods necessary for this identification are discussed in Chapter 18. The specifications may require that samples of steel be obtained for testing in the laboratory. In this case, the sample should be cut from a bar selected at random in the lot being sampled. The sample, which should be at least 30 inches long, can be sawed or burned with an acetylene torch. The inspector should extend his or her observations to include observation of good handling practices as described in Chapter 18.

In case the steel is inspected off-site at the fabricator's shop, the offsite inspector samples and tests each lot of warehouse stock proposed for use. Each consignment of fabricated steel is covered by a certification from the fabricator stating that the steel in the consignment (designated by shipment number on tags, invoice and shipping order) is made from warehouse stock sampled by the inspector. The certification also states that the consignment is made from mill stock covered by the heats referred to in the mill test report forming a part of the certification. This certification forms a part of the report covering the consignment of steel sampled. After fabrication, the inspector attaches the "Approved" tags to the fabricated steel. All tags bear the same number as the test report covering the consignment of steel. Each bundle of bars should have a fabricator's tag giving the fabricator's name, grade and size of bar, length, number of pieces, mark, order number, name of contractor and job address, together with a tag showing report number and indicating approval.

In addition to the cited references for Chapter 18, all of which should be part of an inspector's technical library, the manual *Placing Reinforcing Bars*[24.6] is a valuable field manual on the proper placement of reinforcing steel.

24.12. Other Materials

Admixtures, curing compounds, joint fillers and similar materials are usually accepted on manufacturer's certification. Occasionally the building official or other responsible authority will require copies of test reports by an independent commercial laboratory, especially as supplemental evidence for source approval.

As a matter of convenience in designation on the plans or in the specifications, materials of this nature may be designated by a trade name or name of a manufacturer. This method of designation saves time and confusion. The specifications usually permit the use of alternative material of equal quality, provided the substitution is requested by the contractor, and provided further that the contractor furnishes to the building official the necessary test reports and other information that will enable the building official to judge whether the substitution is in fact equal to the material specified.

An example of this kind of specification might be for a water-reducing retarder to be designated by a certain trade name because the engineer is familiar with that certain brand and knows its basic composition. If a substitute is proposed, the request for substitution must include trade name of proposed substitute, composition, recommendations for use and reports of its effect in or on the concrete.

Inspection of Concrete Construction

The amount and extent of inspection on any job depends upon the specifications; size and character of the job; experience and requirements of the owner, architect and engineer; and especially the inspection requirements of the local governing general building code. In the following discussion, inspection procedures are presented in considerably more detail than is normally required on the average construction job. It is suggested that, for any specific job for which the inspector is retained, a checklist of inspection items be prepared consisting of items requiring inspection as stated in the code and specifications. The details in this chapter will be of considerable help in preparing such a list as well as in making the necessary inspections.

In building construction, the inspector will be called upon to inspect many types of structural and nonstructural concrete elements that compose a building structure and its various access facilities. The more common concrete elements include the following:

I. The foundation
   1. Wall footings
   2. Column footings
   3. Mat footings
   4. Pedestals
   5. Grade beams
   6. Piles
   7. Caissons

II. The floors and roof
   1. Solid slabs
   2. Ribbed slabs (concrete joists)
   3. Waffle slabs
   4. Lift slabs
   5. Slab-beam-and-girder
   6. Arches, shells and domes

III. The walls and columns
   1. Cast-in-place
   2. Precast
   3. Tilt-up walls

IV. The slabs-on-ground
   1. Driveways
   2. Sidewalks
   3. Pavements
   4. Steps

25.1. Preliminary Arrangements

Inspection of concrete begins long before actual placing of the concrete gets underway. There are a number of preliminary questions that must be answered, sources approved, and arrangements made for testing and inspection either at the source, on the jobsite, or both. These arrangements are discussed in Chapter 24. On many jobs the preliminary inspection includes a review of methods and equipment proposed for the job, so as to eliminate last-minute problems concerning the suitability of any items of equipment or their use. Selection of the ready-mixed concrete producer by the contractor determines the source of many of the materials as well as the batching and mixing facilities. The preliminary arrangements include the following:

Approval of Materials—General

Source approval, offsite inspection, sampling, testing, storage, handling methods, rejection

Cement

Source, type, mill tests, temperature, sampling, storage, age, contamination

Aggregates

Source approval, sampling, testing, storage, contamination, segregation

Gradation, specific gravity, absorption, proper size, moisture content, uniformity

Special Aggregates

Lightweight structural

Insulating

Heavy

Unit weight

Admixtures

Air-entraining agents; retarders; accelerators; water reducers; combinations; pigments; workability agents, waterproofers; pozzolans; others

Source, brand name, packaging, recommended dosage, sampling, agitation, protection from freezing

Ready-mix Plant

Calibration of scales, dispensers and batchers; calibration of recorder; moisture meter; mixers clean, with adequate blades; timers or revolution counters; proper water control; rpm of mixers; capacity

Laboratory

Adequately staffed, testing facilities, extent of inspection and testing, proper notification, preparation of mix designs

Concrete Mixture

Strength, maximum size aggregate, appropriate cement, cement content, water-cement ratio, slump, workability, admixtures, unit weight

After the contractor has selected the ready-mix supplier or has made other arrangements for concrete, the laboratory (which should be selected by the owner, architect or engineer) will be in position to make trial mixes and submit mix designs for the concrete to be used.

A large share of the responsibility for quality concrete rests with the ready-mixed concrete producer. An efficient, well-maintained and well-operated plant is essential for production of the kind of concrete required for present-day building construction. As an aid in reaching this goal, the National Ready Mixed Concrete Association has developed a plan for inspection and certification of ready-mixed concrete facilities. By obtaining concrete from a certified plant, the user is assured that the likelihood of deficiencies in quality of concrete is minimized.

Once the preliminaries have been taken care of, the activities associated with inspection of the concrete naturally fall into three stages. The first stage, before concreting, covers the preparation steps, up to the final checkout of the forms. The second stage occurs during the actual batching, mixing and placing of the concrete and extends through the finishing period. Curing and all subsequent steps are included within the third stage.

There is, of course, a certain degree of overlapping of inspection in all of the stages. For example, extending through all stages we find certain items of inspection that are concerned with batching and mixing of the concrete.

25.2. Batching and Mixing

Details of equipment and procedures for batching and mixing concrete, discussed in Chapter 14, should be reviewed when inspection of these facilities is required. Batching (or proportioning as it is sometimes called) will be combined with mixing in a large central plant, but more often it will be found that batches of concrete are proportioned, then dumped into truck mixers. Inspection procedures vary from one plant to another, depending on the type of plant as well as the specification requirements.

The inspector should verify that proportioning and mixing equipment is of sufficient capacity for the size of batches proposed and should make sure that proper materials, as used in the original trial mixes, are being delivered. The mixer or mixers should be checked for conformance with specification requirements and overall condition before concreting operations start. Skips, drums, chutes and other parts should be cleaned of accumulations of cement, aggregates, concrete and other materials. Mixing blades inside the drum should be checked for wear.

At the beginning of each day the inspector should observe and record weather conditions and temperature. The inspector should have all scales balanced, or zeroed, and check operation of batchers, dispensers, meters, valves, gates, etc. If there is a batch counter, the initial reading should be noted. The inspector should make an estimate of the aggregate moisture content and have the scales set accordingly, and calibrate the sand moisture meter if there is one. If there is no moisture meter, a sand sample for moisture test should be taken as soon as the first material has been batched out of the bottom of the storage bin and the moisture content has stabilized. The recorder should be checked and the date, mix number, job identification and inspector's name or initials entered on the chart. The mix should be checked to ensure that the correct concrete is being sent to the job.

In a central mixing plant, the mixer timing meter should be set for the specified time, then checked while the mixer is operating under load. Mixing time is counted from the time all solid materials are in the mixer drum.

During concrete placing, the inspector must do more than merely watch what the contractor is doing. The inspector should confirm that load tickets indicate that the correct mixture is being delivered, and should ensure that the concrete is thoroughly mixed, and that the concrete mixture is used within the specified time limit. The inspector should perform slump and air-content tests, and ensure that the total water and water-cement ratio are not exceeded; make the necessary number of test specimens and see that they are properly protected until the laboratory picks them up; and keep informed of activities in the vicinity that might affect the concrete.

Normal practice is to sample the concrete and perform tests at the point of placement. Insofar as control of the job is concerned, it is mandatory that samples be taken after all water has been added and the concrete is being discharged into the forms. (Discharge into a bucket, pump hopper or buggy may be considered discharge into the forms.)

Now let us return to the three stages of inspection previously mentioned.

25.3. The First Stage—Before Concreting

Before concrete can be placed, the batching and mixing facilities will have been checked out and mixes approved. The excavation, foundation, forms, reinforcement, embedded items, facilities for handling, placing and finishing the concrete, and provision for special conditions must be in order. Included in this stage are the following:

Foundations and Excavation

Line and grade, compaction of soil, clean, no standing water, no frost or ice, drainage

Forms

Line and grade, stability, provision for settlement, ties, braces, proper contact surfaces, inspection and cleanout openings, form oil, blockouts, cleanup

Reinforcement

Grade of steel, size, length, bends, spacing, location, cover, splices, ties, supports, no loose rust, oil, paint or form oil

Embedded Items

Type, size, location, supports and ties, surface condition

Site Facilities

Sufficient workers and equipment, preparation for curing, preparation for hot or cold weather, adequate lighting, notification of ready-mix supplier

25.4. During Concreting

Once the preliminaries have been taken care of, the next step is the actual production and placing of the concrete. Inspection may include occasional observation of operations at the batch plant, although continuous inspection at this location is rarely performed, except in special cases that are clearly defined in the specifications. On most jobs, an inspector is required to be present whenever concrete is actually being placed, and the general building code typically requires continuous inspection on concrete work under certain conditions. Inspection of the concreting phase includes the items in the following list:

Batching Plant

Selection of appropriate mixture, proper scale settings, moisture compensation, accuracy of batching, right size of aggregate, correct type of cement, appropriate admixture, proper batching sequence

Mixing and Delivering

Proper size of mixers, accurate dispatching, load tickets, time of departure, arrival and unloading, elapsed time, avoiding delays, temperature limits, slump control, adding water on the job, disposal of wash water

Handling and Placing

Discharge from mixer, conveying, buckets, pumps, conveyors, chutes, avoiding segregation, avoiding excessive drying or loss of slump, time limit, preparation of construction joints, vertical drop, thickness of layers, minimum horizontal movement, consolidation, under or overvibration, revibration, avoiding rock pockets and honeycomb, removal of internal braces and spacers

Jointing and Finishing

Location of construction joints, preparation, dowels and tie bars, bulkheads, water stops

Contraction and isolation joints, location, type, joint filler, tie bars

Finishing unformed surfaces, minimum use of tools, proper timing, bleeding, plastic cracks, specified finish

Heavy-duty floors, base course, screeding, tamping or rolling, troweling, wearing course, special aggregate

Testing

Where performed, who does sampling and testing, consistency (slump), temperature, strength specimens, cylinders, beams or cores, air content, unit weight, storage of cylinders for standard tests or field tests, moving to laboratory, other tests

25.5. The Final Stage—After Concreting

After the concrete has been placed and finished, we cannot just walk away and forget about it. There is yet much to be done to assure a satisfactory job. It is during the early age of the concrete that much concrete is damaged, especially by failure to cure the concrete properly or failure to protect it from extremes of temperature, either hot or cold. Repairs of rock pockets or other defects should be made as early as possible for the simple reason that it is easier to work on comparatively weak green concrete than on concrete that has matured and become quite hard.

Curing

Wet blankets, hoses and sprinklers, ponding, polyethylene sheets, other coverings, length of curing period, keeping concrete continuously moist

Curing or sealing compound, proper type, agitation in container, spray applicators, time of application, concrete moist at time of application, coverage, weather and temperature

Removal of Forms

Time for removal, protection of concrete, shoring or reshoring, care of forms, cleaning, oiling

Repair of Defects

Rock pockets, sand streaks, leakage through forms, misplaced or offset forms, cracking, filling tie rod holes, bug holes or surface voids

25.6. Special Practices

There are many specialized materials, mixtures and practices for concrete. The foregoing checklist applies to all of these specialties, and portions should be applied to the extent that they are relevant to a particular situation. There are, however, many situations that are unique and have their own individual methods and problems in addition to those common to all.

Prestressed Concrete

Forms: panel joints, no offsets or bulges, chamfer strips, braces, form oil not on rebar or stressing steel, hardware (including inserts and lifting eyes) in place, bulkheads inserted

Steel: rebars correct size, grade, mark, quantity; clean, properly anchored; hold-downs or bond breaker for draped tendons; correct size and grade of strand

Pretensioning: measure tensioning by jack pressure and tendon elongation, broken wires or strands, slippage, twisting, application of prestress, placing concrete, curing, detensioning, releasing hold-downs and forms, form removal, lifting and storage, control of cracking, repair of defects

Post-tensioning: ducts properly placed, correct end anchorages, placing concrete, curing, tensioning, grouting ducts, release of anchors

Concrete: proper mixture, no calcium chloride, slump, adequate vibration, finishing tops of units, preset before application of high-temperature curing, strength specimens cured with units, standard specimens, strength for detensioning or post-tensioning, camber

Hot-Weather Concreting

Temperature limitations; cooling materials, wetting aggregates; ice in mix water; use of retarders; wetting forms; fogging placing area; windbreaks and shades; slump loss; plastic shrinkage cracks; protection from rapid drying; curing (See Table 19.1.)

Cold-Weather Concreting

Temperature limitations, heating materials, hot water, use of accelerator, insulation of forms, enclosures and space heaters, form removal, protection, curing (See Table 19.3.)

Lightweight Concrete

Aggregates: dry, prewetted or vacuum treated, gradation, unit weight, manufactured or natural, absorption

Concrete: structural or insulating, trial mixes, density, strength, proportioning, volumetric batching, special precautions for truck mixers, cellular or foamed concrete, finishing

Heavy Concrete

Aggregates, mixes, batching and mixing problems, density, segregation

Tilt-up

Casting surface; parting compound, forms; special reinforcement; inserts; lifting devices; connections; strength of concrete for tilting; lifting forces; avoidance of sudden jerk or strain; temporary braces; finishes; preparing the foundation; keying to columns

Slipform

Vertical: leveling and plumbing, mixes, uniformity required, slump, thickness of layers, blockouts and inserts, rate of slipping, avoidance of overvibration and revibration, continuity of placing, finishing

Horizontal: propulsion, concrete requirements, finishing

Placing Under Water

Avoid running water, extra cement in mix, high slump required, separate crane for tremie, keep lower end of tremie immersed in concrete, make horizontal layers

Preplaced Aggregate Concrete

Special aggregate grading; minimum voids; placing aggregate; grout pipes, vents, and sounding wells; grout mixture; admixtures; consistency of grout; pumping grout; brooming and finishing top surface

Vacuum Concrete

Forms and pads clean of hardened mortar; lean, low sand mixture; time and duration of vacuuming; inches of vacuum; vibration during vacuuming; form removal

Shotcrete

Dry mix or wet mix, materials, special sand grading, mixes, pressure vessels, water control, forms, ground wires, reinforcement, thickness of layers, rebound, finishing, strength specimens

Fastening Base Plates

Adjusting elevation; setting anchor bolts, tolerances; templates; mortar or grout filler; mixes; compensating for shrinkage; powdered aluminum

Architectural Concrete

Need for uniformity of materials, mixes and treatments; classes of surfaces; sample panel; method of achieving desired surface (See Section 22.13.)

25.7. The Small Job

A large amount of concrete goes into small structures and slabs with very little inspection or no inspection at all. A small amount of concrete, however, should not be looked upon as something to be hurriedly and carelessly produced. In the first place, construction of any kind must be in accordance with the local ordinances and codes. Although inspection may not be required, a permit is usually necessary for practically any type of construction. By observance of the code requirements, and following the basic rules outlined in this book, a good concrete job can be obtained. Further discussion of the small job is given in Section 12.7.

25.8. Records and Reports

Part of the inspector's duty consists of keeping records of activities with which he or she is concerned and reporting these activities. The importance of accurate and complete reports cannot be stressed too strongly. When a situation arises weeks or months, even years, later, requiring information regarding the concrete, the quality and completeness of the reports can make the difference between factual information and estimates sometimes based on little more than guesswork.

Most of the reporting of concrete inspection can be accomplished by the use of form reports. The layout and size of form reports vary from agency to agency, and the inspector should be familiar with the requirements of each particular agency. The general building code specifies only very general requirements, leaving the details up to the building official. There may, of course, be special requirements stated in the specifications. Daily reports are always required. Violations of the code and specifications should be reported in considerable detail, usually requiring a narrative report of some kind. Laboratories routinely furnish reports of their tests and inspections to the building official and others concerned with the project.

In addition to regular reports, the inspector should keep a diary. An inspector simultaneously assigned to several jobs should keep a separate record for each job. Items that should be reported include weather conditions, visitors to the job, materials received, condition and progress of the work, and instructions given to the contractor. The concrete inspection log includes location of concrete placing (portion of structure), time of placing, unusual delays, mixes used, tests performed by the inspector or others, samples of materials and concrete taken, and strength specimens made.

Photographs constitute an important part of the construction record. Photographs taken at regular intervals, and at other times as necessary, can be a valuable asset should disputes arise. The inspector should be authorized to take construction photos as part of his or her inspection duties.

25.9. Special Inspections

Special inspection, as required by the *International Building Code* (IBC) is best defined as the monitoring of the materials and quality of construction that are critical to the integrity of the building structure and therefore warrant special attention. This requires continuous inspection by persons with highly developed skills to verify that the materials and construction comply with the approved plans, specifications and contract documents. Some aspects of the construction may require periodic special inspection only.

Special inspection is one of the significant contributions to the quality of construction in the building field. Special inspection is a joint program of the private construction sector and municipal building departments. Public building inspection personnel perform periodic called inspections. Because of practical budget limitations imposed on these agencies, however, few are expected to provide comprehensive inspections of all critical construction details. Additionally, with the increasing complexity of modern construction, few public inspection agencies are able to staff specialists in the complex and ever-changing specialty practices. The special inspector supplements normal called inspections provided by the building official with special inspection to help ensure that the structure complies with the drawings and construction provisions of the code. The special inspector is a specialist in a more narrow field and may be able to additionally augment the called inspections through a more extensive knowledge in the specialization being inspected.

Special inspectors are approved by the building official and employed by the owner either directly or through the architect or engineer in charge of the design of the structure, or through an independent, approved inspection agency.

In accordance with the IBC, special inspection is required on the following types of concrete work:

Continuous special inspection

- welding of reinforcing steel
- bolts installed in concrete
- sampling/testing fresh concrete
- concrete/shotcrete placement
- prestressing tendon stressing/grouting

Periodic special inspection

- placing reinforcing bars/prestressing tendons
- concrete mix proportions
- concrete curing
- precast concrete erection
- in-situ concrete testing

Required verification and special inspections for concrete construction are detailed in IBC Table 1704.4, including the ACI 318 code sections that address the particular type of construction or operation requiring special inspection.

The special inspector is responsible for the following:

- Observing the work assigned for conformance with approved design drawings and specifications;
- Furnishing inspection reports to the building official, the engineer or architect of record, and any other designated persons;
- Informing the contractor of any discrepancies (if then uncorrected by the contractor, the special inspector also notifies the designer and building official); and
- Furnishing a final signed report stating whether the work requiring special inspection was, to the best of the inspector's knowledge, in conformance with the approved plans and specifications and the applicable workmanship provisions of the code.

The code specifies that "the special inspector shall be a qualified person who shall demonstrate competence, to the satisfaction of the building official, for inspection of the particular type of construction or operation requiring special inspection." In the approval process, the building official should check that the special inspector is employed by the owner, or by the engineer or architect of record acting as the owner's agent. To avoid conflict of interest, the special inspector should not be employed by the contractor. In addition to verifying that applicants have technical competency, the building official should also verify that applicants have related work experience and that they are aware of local code amendments, procedures and requirements.

Building departments and code enforcement agencies are finding it increasingly difficult, because of the complexity in some areas of inspection, to maintain expert staff that can adequately qualify special inspectors. Whereas there are personnel certification programs operated by nationally recognized organizations such as ICC and ACI, the overall system of special inspection rarely involves a single individual. Consequently, in many jurisdictions special inspection agencies are involved in providing special inspection services. Building jurisdictions rely on evidence that these agencies maintain accreditation by nationally recognized accreditation bodies such as the International Accreditation Services (IAS) (a subsidiary of the International Code Council). (See Section 25.11.)

National Certification for Reinforced Concrete Special Inspector. The International Code Council (ICC), in cooperation with the American Concrete Institute (ACI), has developed an exam program for reinforced concrete special inspector certification. By combining the efforts and reputations of these two organizations, this national certification program will become the standard in the coming years of the comprehensive demonstration of the knowledge of codes, standards and practices necessary for the professional competence of special inspectors.

Initially, special inspection certifications were voluntary. However, in recent times, participation in the program has grown dramatically—a trend that can be largely attributed to the use of the programs for employment purposes

Certifications are now used quite commonly for both screening of prospective employees and for merit raises for current employees. Use of certification programs by the private sector has also grown dramatically with expanded use by special inspectors. Additionally, a number of states are considering statewide mandatory certification of construction inspectors. Certification is designed to determine if an individual's knowledge of codes, standards and practices meets or exceeds a prescribed level of competence. Passing the certification exam and subsequent ACI field-testing procedures and experience requirements provides evidence that an individual possesses the critical knowledge of relevant codes, standards and skills necessary for competent practice of the profession.

The requirements for Reinforced Concrete Special Inspector Certification are satisfied by successfully completing the following three criteria:

1. Passing the Reinforced Concrete Special Inspector Exam as provided by the International Code Council (ICC). Successful completion of the exam will provide the candidate with a "Certificate of Completion of the Reinforced Concrete Special Inspector Certification Exam."
2. Combining of (1) the exam and (2) the current certification for the ACI Field Technician Grade I will give the candidate certification as a "Reinforced Concrete Special Inspector Associate."
3. Combining of criteria (1) and (2) above with (3) the ACI verification of minimum education and experience will give the candidate certification by the International Code Council as a "Reinforced Concrete Special Inspector."

A sample of the "Reinforced Concrete Special Inspector Certification" certificate is illustrated in Figure 25-1.

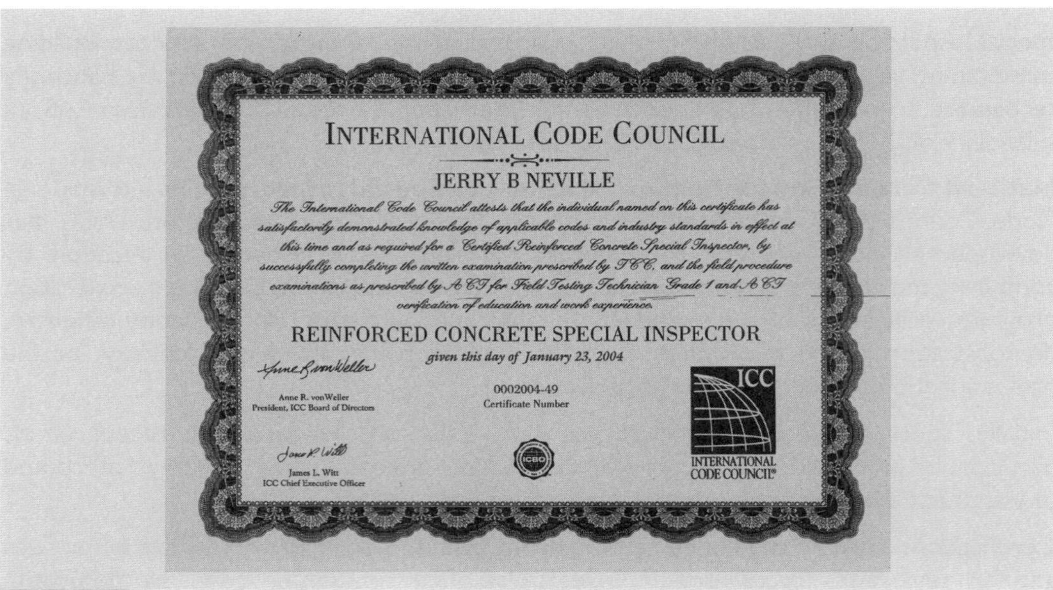

INTERNATIONAL CODE COUNCIL

JERRY B NEVILLE

The International Code Council attests that the individual named on this certificate has satisfactorily demonstrated knowledge of applicable codes and industry standards in effect at this time and as required for a Certified Reinforced Concrete Special Inspector, by successfully completing the written examination prescribed by ICC, and the field procedure examinations as prescribed by ACI for Field Testing Technician Grade 1 and ACI verification of education and work experience.

REINFORCED CONCRETE SPECIAL INSPECTOR

given this day of January 23, 2004

0002004-49
Certificate Number

Anne R. vonWeller
President, ICC Board of Directors

James L. Witt
ICC Chief Executive Officer

ICC
INTERNATIONAL CODE COUNCIL®

Figure 25-1
ICC certification certificate for reinforced concrete special inspector.

The ACI education and experience requirements consist of documentation of two years of work experience in concrete testing and inspection, and completion of at least one of the following education qualifications:

1. At least two years of college or technical school;
2. A high school diploma or its equivalent, plus an additional year of construction testing and/or inspection (not necessarily concrete); and
3. An additional three years of construction testing and/or inspection (not necessarily concrete).

The work experience must be verified by the applicant's employer(s) and will include all of the following:

1. Decision-making responsibility and authority;
2. Verification of compliance with plans, specifications and codes;
3. Evaluation of concrete construction inspection in the field;
4. Documentation and reporting of inspection results; and
5. Proficiency in all appropriate areas of concrete construction inspection.

Information, required references and a candidate bulletin describing the special inspector certification program is available through the ICC offices at any one of the following addresses:

Headquarters (DC) 500 New Jersey Avenue, NW, 6th Floor Washington, DC 20001 [P] 1-888-ICC-SAFE (422-7233) [F] (202) 783-2348	**Birmingham District Office (BIR)** 900 Montclair Road Birmingham, AL 35213 [P] 1-888-ICC-SAFE (422-7233) [F] (205) 599-9871
Chicago District Office (CH) 4051 W Flossmoor Road Country Club Hills, IL 60478 [P] 1-888-ICC-SAFE (422-7233) [F] 1-800-214-7167	**Los Angeles District Office (LA)** 5360 Workman Mill Road Whittier, CA 9060 [P] 1-888-ICC-SAFE (422-7233) [F] (562) 908-5524

or visit the ICC web site at www.iccsafe.org.

Certification for Prestressed Concrete Special Inspector. In addition to the reinforced concrete special inspector certification, the International Code Council (ICC), in cooperation with the Precast Prestressed Concrete Institute (PCI) and the Post-tensioning Institute (PTI), offer a "Prestressed Concrete Special Inspector Certification." The reinforced concrete certification is a prerequisite for obtaining a prestressed concrete certification.

Special Inspector Job Tasks. Table 25.1 describes the typical job tasks involved in performing inspection duties of reinforced and prestressed concrete construction. Actual tasks performed on specific jobs will typically be fewer, limited by contract and/or jurisdictional requirements.

25.10. Special Inspection of "Approved Fabricators"

"Special inspection" is usually not required where the work is done on the premises of a fabricator registered and approved by the building official to perform such work without special inspection. Four general qualification standards for "approved" fabricators are typically required to exempt the need for special inspection:

1. The fabricator must submit a procedural manual that shows quality control procedures;
2. The fabricator's facility and personnel must be verified by an approved inspection or quality control agency;
3. The fabricator's plant must be periodically inspected for conformance to the approved quality control manual by the approved inspection or quality control agency; and
4. The approved inspection or quality control agency is required to notify the approving authority of violations to the procedural manual by the fabricator.

The building official has the choice of requiring special inspection of fabricating plants or of approving the fabricating plant if it complies with the above four quality control guidelines.

A key element in the recognition of a fabricating plant is ongoing inspections by an approved quality control agency approved by the building official.

The *International Building Code* requirements for "approved fabricators" are addressed in Chapter 17 of the IBC.

25.11. International Accreditation Services

Recently, agencies involved in quality control and inspection (inspection agencies) are seeking accreditation by the International Accreditation Services (IAS). This accreditation demonstrates to ICC governmental members and other building officials that the agency has been independently assessed and found to be competent for the inspection services that it provides. Accredited inspection agencies are required to be independent, comply with ISO/IEC Standard 17020 *General Criteria for the Operation of Various Bodies Performing Inspection*, and be subject to periodic on-site assessment. ISO/IEC Standard 17020 has become widely accepted as the preeminent document for accreditation and requires inspection agencies to operate under a quality management system, maintain impartiality and have a sufficient number of permanent personnel with a range of expertise covering the inspection services they provide.

Structural steel and concrete fabricators that have had their internal inspection programs accredited under the IAS Accreditation Criteria for Fabricator Inspection Programs AC172 (steel) or AC157 (concrete), respectively, are also able to demonstrate to building jurisdictions through their accreditation that they meet the four critical criteria outlined in Section 25.10. Fabricators use the services of accredited inspection agencies for periodic inspection of their facilities and to prove that they comply with their own approved quality control manual as required by the IBC.

TABLE 25.1
SPECIAL INSPECTOR JOB TASKS FOR REINFORCED AND PRESTRESSED CONCRETE

I. GENERAL REQUIREMENTS

1. Duties and Responsibilities
Review approved plans and specifications for special inspection requirements. Comply with special inspection requirements of the enforcing jurisdiction.
2. Notification of Discrepancies
Notify the contractor of deviations from approved plans and specifications. If the deviations are uncorrected, notify the architect or engineer of record and the building official of deviations.
3. Inspection Reports
Submit progress reports to the architect or engineer of record and the building official, describing tests which were performed and compliance of work. Submit final summary report stating whether work requiring special inspection was in conformance with the approved plans and applicable provisions of the building code.

II. CONCRETE QUALITY

1. Mix Verification
Verify that individual batch tickets indicate delivery of the approved mix as specified. Verify that concrete ingredients conform to acceptable quality standards. Verify time limits of mixing, total water added, and proper consistency and workability for placement.
2. Testing
Determine the required type, quantity and frequency of tests to be performed on fresh and hardened concrete. Observe sampling of concrete, field testing of fresh concrete, and making of test specimens. Verify that the required concrete strength has been attained prior to tendon stressing.
3. Specimen Handling and Protection
Provide or arrange for proper specimen identification, site storage and protection, and transportation to the testing laboratory.
4. Test Reports
Provide or arrange for communication of field testing results to the architect or engineer of record and to the building official.

III. REINFORCEMENT

1. Quality
Verify that reinforcing and prestressing steel are of the type, grade and size specified and are in conformance with acceptable quality standards. Verify that the prestressing tendon system is fabricated in conformance with acceptable quality standards. Ensure that reinforcing and prestressing steel are free of oil, dirt, and flaking rust and that steel is properly coated and/or sheathed as specified.
2. Tolerances
Verify that reinforcing and prestressing steel are located within acceptable tolerances and are adequately supported and secured to prevent displacement during concrete placement.
3. Cover
Verify that minimum concrete cover is provided.
4. Placement
Verify that placement of reinforcing and prestressing steel (or ducts) complies with required spacing, profile and quantity requirements, as indicated by both the approved plans and placing drawings, or installation drawings.
5. Details
Verify that hooks, bends, ties, stirrups and supplemental reinforcement are fabricated and placed as specified.
6. Splices
Verify that required lap lengths, stagger and offsets are provided. Verify proper installation of approved mechanical connections per the manufacturer's instructions and/or evaluation reports. Ensure that all weldments are as specified, and have been inspected and approved by an approved welding inspector.
7. Prestressed Rock and Soil Anchors
Verify that prestressed rock and soil anchors are fabricated and installed in accordance with national standards or with project specifications.

IV. PRESTRESSING AND GROUTING

1. Stressing
Inspect for proper equipment calibration. Verify that proper stressing (or tensioning) sequences are used, proper jacking forces are applied, and acceptable elongations are attained and recorded.
2. Anchorage Protection
Verify that tendons and anchorages are properly sealed or otherwise protected from corrosive environments.
3. Posttensioning Ducts/Grouting
Verify that ducts are of the required size, are mortar-tight and are nonreactive with concrete, tendons and grout materials. Verify that proper grout materials, strength and grouting pressures are used as specified by the manufacturer.

V. FORMWORK, JOINTS AND EMBEDS

1. Formwork Construction
Verify that formwork will provide concrete elements of the specified size and shape.
2. Construction Joints
Verify that the location and preparation of construction joints are in accordance with the approved plans, specifications and building code requirements.
3. Embeds
Verify that the type, quantity, spacing and location of embedded items are as specified.

VI. CONCRETE PLACEMENT, PROTECTION AND CURING

1. Preplacement
Verify acceptable condition of the place of deposit before the concrete is placed.
2. Placement
Verify that methods of conveying and depositing concrete avoid contamination and segregation of the mix.
3. Consolidation
Verify that concrete is being properly consolidated during placement.
4. Protection
Verify that concrete is protected from temperature extremes and that proper curing is initiated.

Quality Control

When we speak of quality control, we are speaking of a concept or method that has been around for a long time and that is finding ever more application in the concrete industry. Whether we call it quality control, product control, quality assurance or something else, the inspector should become familiar with quality control and its applications. The principles of quality control can be applied to any process in which repetitive testing of one property is done on a series of samples. There are numerous examples in industry and manufacturing. In construction we find application to the production of aggregates, soil density tests and several tests of concrete, of which the strength test is the most common. Although slump tests are made more frequently and could be analyzed statistically, the slump test does not lend itself to the precision of measurement that a strength test does, and the analysis ordinarily would not be as meaningful.

26.1. Quality Control Defined

In any manufacturing process, the producer aims a little high; that is, he or she controls the product by making its critical properties slightly better than the minimum acceptable value to minimize the chance of producing a reject. How high the producer must aim depends on how well the features that affect the critical property can be controlled and to what extent the producer wants to risk a failure or reject. The better the control, the closer the average value of the critical variable can approach the required value, and the more economical the whole process becomes. To find out how good the control is, and how high to aim, the producer applies simple quality control measures. (See Figure 26-1.)

Figure 26-1
Quality control assures maximum efficiency.

Quality control of materials is a relatively old concept as applied to products manufactured at a permanently located factory or mill. In the construction industry, typical examples are control of the manufacture of cement and steel. For these and other such manufactured products, quality control and proper use of materials are essential to ensure that the finished product will conform to established standards.

As applied to construction, quality control is a system through which construction is controlled by scientific methods rather than by chance. Scientific methods of investigating, testing and analyzing provide criteria by which materials are evaluated and used in construction. Specifications for materials, methods of tests and standards of acceptance are established from these criteria. Best use can be made of available materials through realistic specifications designed for each specific job, adequate and impartial inspection of materials and methods, and statistical methods of analysis.

Recent methods effected by several governmental agencies place greater responsibility for inspection and quality control on the contractor and producer. This is good, as it assures that the owner is getting quality materials and construction, and it gives the contractor the satisfaction of using quality control capabilities to control the job. Under these conditions, the owner's activity becomes one of quality assurance, which is to say that the producer (contractor) controls the quality of the product and the purchaser (owner) checks to assure that the product is of the required quality when it is used in the work. Generally, control sampling and acceptance sampling will not be the same. Control samples are taken earlier in the process than assurance samples, and the limits are usually tighter.

To illustrate, assume that the amount of coarse aggregate passing a $^3/_8$-inch sieve is specified not to exceed 30 percent at the batcher. This means that the producer, in order to meet the 30 percent requirement at the batcher, must produce a product with appreciably less material passing the $^3/_8$-inch sieve to allow for breakage and loss during handling between the aggregate plant and the batcher.

For quality control to succeed, there must be a rational system of analyzing the results of tests. This is accomplished by means of statistical quality control, commonly abbreviated as SQC. Statistical computations lend themselves well to computer programming. A continuing analysis procedure can be programmed that provides up-to-the-minute information on concrete strength, aggregate sieve analysis, sand equivalents, cement test and, in fact, any test or analysis that is done on a continuing basis.

Basis of Statistical Quality Control (SQC). Statistical quality control is based on the mathematics of probability. When we consider a large number of test results of a single property, such as strength of a certain concrete mix, or sand equivalent values of a certain sand, we find that all test results are grouped about a central value, which is the average or arithmetic mean. Draw a picture of these numbers and you get a bell-shaped curve called a frequency distribution curve. (See Figure 26-2.)

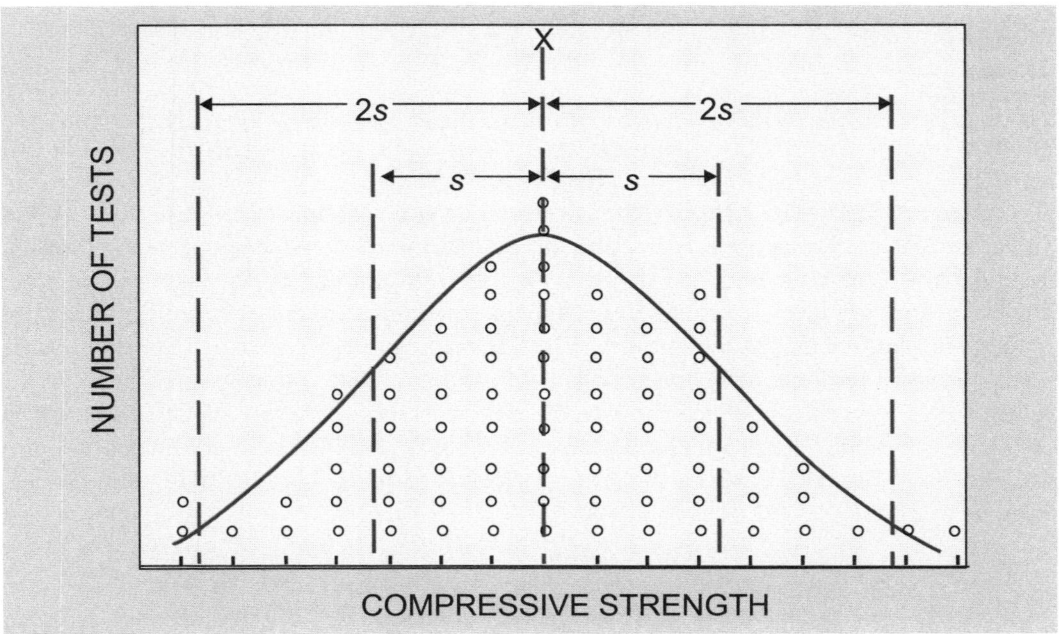

Figure 26-2
A frequency distribution curve showing the strength test results grouped and plotted as dots above the group strength, with a smooth standard curve that shows normal distribution.

This is good as far as it goes, but we do not always have the time to compute and draw a curve, nor does the curve tell us all we want to know. We therefore resort to some simple arithmetic by which we can compute another valuable measure called the standard deviation, which is a measure of central tendency—how our test values vary from the average, or how they spread from low to high. Standard deviation is usually known by the Greek letter sigma, Σ; however, for consistency with its usage in ACI 318 Section 5.3, we will use the letter s in the following discussion. This standard deviation can also be expressed as a percentage of the average, in which case it is called the coefficient of variation, designated by the capital letter V. A spread of one standard deviation on each side of the mean always includes 68 percent of the cases under consideration, and a spread of two standard deviations each way will include approximately 95 percent. Practically 100 percent of results will be included among three standard deviations. (See Figure 26-2.)

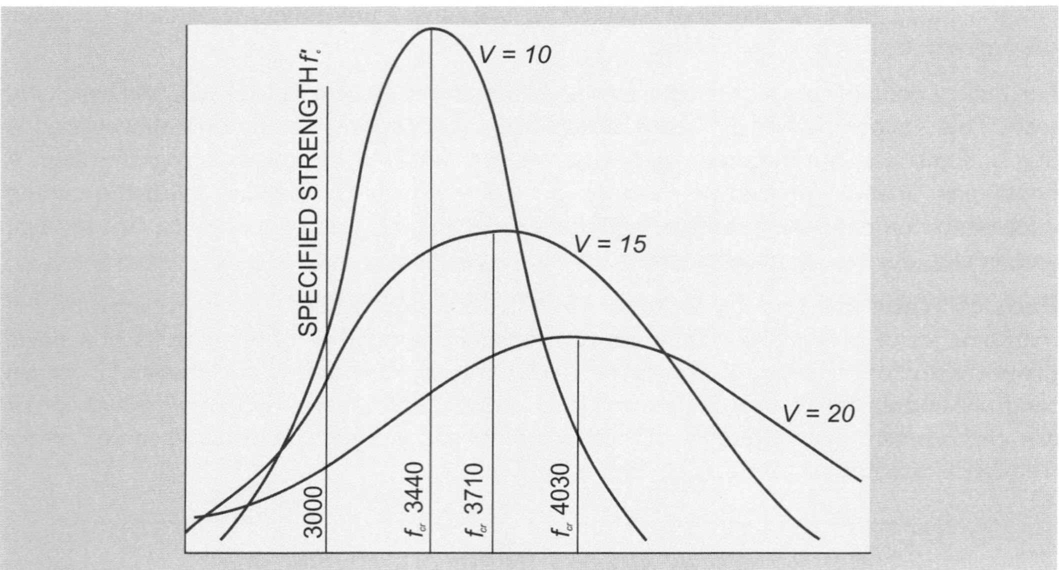

Figure 26-3
The quantity fcr is the average of all tests. If the job is under good control, fcr is about 3440 psi, but if the coefficient of variation goes as high as 20 percent, then the average strength must be

The curves in Figure 26-3 show that, as the standard deviation or coefficient of variation increases, the curve becomes wider and flatter, and the required average strength moves to the right; that is, it becomes higher. We have to aim higher to maintain the same minimum strength level. It is quite obvious that a small standard deviation indicates that the values are grouped closer to the average (the spread or range is smaller); in other words, the process is under closer control than when the standard deviation is high. This is illustrated in Figure 26-3, which shows that to produce concrete to a specified strength of 3000 psi, the required average compressive strength, f_{cr}', is 3440 psi when the coefficient of variation is 10 percent; if the coefficient is permitted to run as high as 20 percent, f_{cr}' must now be 4030 psi.

Another measure is the range that is the spread between the lowest value and the highest value in a group of test values under consideration. The total number of test values under consideration is called the population. Definitions are listed in Table 26.1.

TABLE 26.1
DEFINITIONS

n	Population	All of the test results under consideration. The total number of values.
$\overline{X}$	Average	The arithmetic mean. The sum of all values divided by the number of values. (Called bar *X*.)
s	Standard Deviation	A measure of variation. Derived mathematically from the test results.
V	Coefficient of Variation	The standard deviation expressed as a percentage of the average.
R	Range	The numerical difference between the highest test value and lowest test value in the population.

26.2. Application to Concrete Construction

Experience and tests involving millions of cubic yards of concrete on thousands of projects, large and small, clearly confirm the need for recognizing the variability of materials and for relieving contractors and inspectors of pretending to work to unrealistically tight tolerances. Instead of setting rigid numerical limits beyond which no test result may fall, specifications based on SQC take into account the normal variations in materials, sampling and testing, and base the probabilities on statistical methods. The producer and the engineer are thereby given a tool that permits them to hold to a standard that can be attained; at the same time, the job is protected from materials that are seriously outside design requirements.

Analysis of Concrete Strength Tests. In general, strength is a good index of concrete quality, as most desirable properties are dependent on or related to strength. Strength is usually determined by means of 6- by 12-inch cylinders, made at the time the concrete is placed, and tested in compression at various ages. One might say that the primary function of these compression tests is to serve as a measure of the uniformity and quality of concrete. The magnitude of variations in strength of concrete test specimens depends upon how well the materials, concrete manufacture and tests are controlled.

One of the first steps is a realistic approach to the specified strength. Knowing that concrete is a variable material, the only realistic approach is with the knowledge that some cylinders are going to show unusually high strengths and some will show unusually low strengths.

In writing specifications then, it is more realistic to take a calculated risk, basing the probabilities on statistical methods and permitting a certain percentage of strength tests lower than specified design strength. These variations exist, and a certain percentage of low strengths exists, depending on the average strength level and the quality of control. Recognizing this basic fact does not jeopardize construction quality.

For example, standard specifications of certain governmental agencies for strength of concrete are based on statistical methods of analysis and provide that the minimum average strength shall not be less than the specified 28-day strength. If the minimum strength of an individual test is less than 95 percent of the specified 28-day strength, the contractor must take steps to bring the strength up to the required minimum. This action can be taken if seven-day strengths indicate that the 28-day tests will be low. In addition, any concrete represented by a single test that indicates a strength of less than 85 percent of the specified 28-day strength is subject to rejection. A strength test consists of the average strength of two cylinders taken from a single load of concrete, except that an obviously defective cylinder can be discarded.

To obtain maximum information, a sufficient number of tests should be made to be representative of the concrete produced and appropriate statistical methods should be used to interpret the results. Statistical methods provide the best basis for analyzing these results, for determining potential quality and strength of concrete from any certain plant, and for expressing results in the most useful form. Merely making test specimens and reporting strengths is not enough.

Assume that a structural concrete is required to develop a strength of 4000 psi at age of 28 days. Table 26.2 shows average strength required that will probably enable the operation to meet the quality level shown above each column, for the variation coefficients shown.

TABLE 26.2
REQUIRED AVERAGE COMPRESSIVE STRENGTH AT 28 DAYS FOR 4000 PSI CONCRETE

COEFFICIENT OF VARIATION (PERCENT)	PROPORTION BELOW SPECIFIED STRENGTH		
	1 IN 5	1 IN 10	1 IN 200
5	4200	4300	4600
10	4400	4600	5400
15	4600	5000	6500
20	4800	5400	8300
25	5100	5900	11,200

The coefficient of variation expresses the degree of dispersion on a percentage basis, thus providing a numerical measure of variability of control. A smaller coefficient indicates better uniformity. In field work of this type, the coefficient usually ranges between 10 and 15 percent; occasionally it will be as high as 20 percent and rarely as low as 5 percent. These generalizations assume a sufficiently large number of tests to constitute a representative sample of total concrete placed.

In Table 26.2, strengths in the column headed “1 in 200” show how unrealistic it is to specify an absolute minimum strength. Even though there is an allowance of one-half of 1 percent below 4000 psi, a job under fair control with a coefficient of 15 percent would require an average strength of 6500 psi. With a variation coefficient of 25 percent, the strength requirement is virtually impossible to attain.

A strength of 4000 psi, as used in this example, is commonly specified for certain structural concrete. Strength failures in this type of construction are serious, but an allowance of one in ten below design strength can be permitted. Again using a coefficient of 15 percent, it is seen that required average strength is 5000 psi, which is not difficult to attain. Even with poor control, the required average is within reach.

Control Charts. Analysis of strength tests can be accomplished by means of simplified control charts. Figure 26-4 is a portion of an actual control chart for 3000 psi structural concrete, showing both seven-day and 28-day strength curves, permitting an evaluation of probable 28-day results from seven-day results. This early evaluation is necessary for making decisions before sufficient 28-day strengths become available.

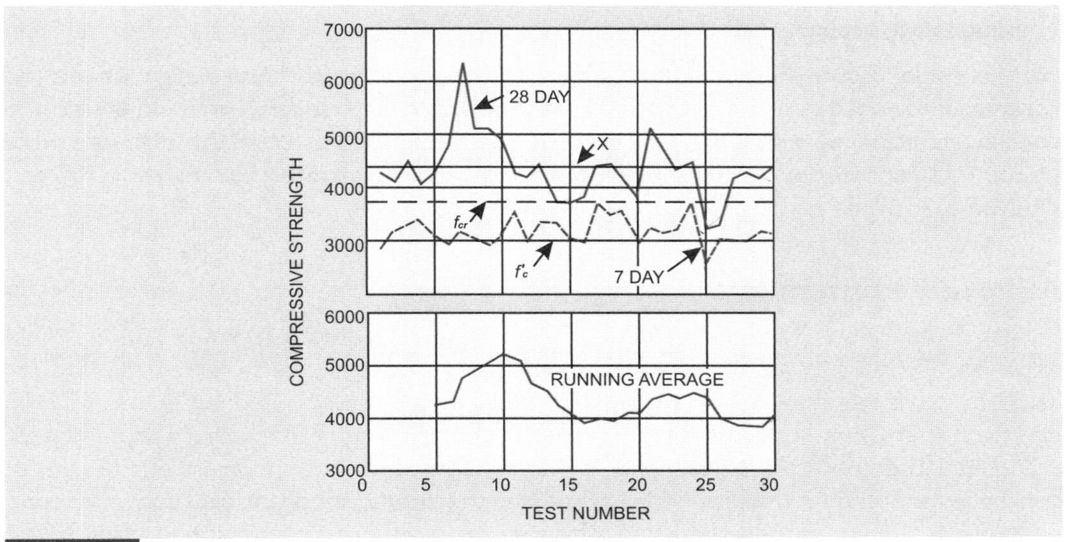

Figure 26-4
Control chart of concrete strength from a job that was poorly controlled. Wide variations in strength, with an average strength far in excess of the required average fcr indicate poor economy.

In the top portion of the chart compressive strengths of pairs of specimens for seven days (shown by dotted line) and 28 days (shown by solid line) have been plotted. These curves show how many 28-day tests are below specified strength and therefore whether specification requirements are being met. From the seven-day strength it is possible, after experience with a given set of materials, to determine whether 28-day strengths are likely to meet specifications.

The bottom of the chart shows moving averages of the 28-day strengths, especially useful for detecting trends and for forecasting from seven-day strengths. The first point for the moving average is computed by averaging the first five tests. The second point is computed by dropping Test I and averaging Tests 2 through 6, then continuing by dropping Test 2 and picking up Test 7, each time dropping one test and picking up a later one. Daily plotting of strength values and averages enables the engineer to analyze trends and develop a course of action, if necessary. Such action might be a change in cement content of the concrete, or a study of job conditions to locate irregularities of operations or materials. This action is sometimes made on the basis of seven-day tests, if its need is clearly demonstrated at that time. The coefficient of variation (14.4 percent) and the standard deviation (620 psi) in this example were about average, bordering on poor, for general construction. In addition, the chart shows that all of the tests were far above the 3000 psi specified, and the actual average (4310 psi) exceeded the required average (3700 psi) by more than 600 psi, both of which indicate inattention to control measures that should have been applied. The lack of control could have been anticipated after the first few seven-day tests were reported, and the mix adjusted to a more economical one. Also, tightening control to diminish the irregularity of job operations (reduce the coefficient of variation) would make further economies possible.

To be statistically reliable, it is necessary to have at least 30 sets of information (in this case, strength tests) for analysis. (Note: ACI 318 Section 5.3.1.2, allows a statistical evaluation of strength tests where a concrete production facility has a test record based on 15 to 29 consecutive tests.) After results of 30 sets have been plotted, the average standard deviation and coefficient of variation are computed. Next, the required average strength is computed for the same 30 sets. Based on this analysis, further adjustments to field operations will probably be necessary.

Now consider Figure 26-5, which is an example of good control. The design for this job specified a required strength, f_c', of 3500 psi. In this case the standard deviation equals 356 psi, the coefficient of variation (V) is 9 percent, and the actual average of the 54 tests is 4070 psi, only 110 percent over the required average of 3960 psi. There was one test (3110 psi) below f_c'.

26.3. ACI 318 Requirements

Standard Deviation. The code (ACI 318 Section 5.3.1) emphasizes the use of "field experience" (statistical evaluation of strength test results) as the preferred method for selecting concrete mixture proportions. The following discussion addresses the ACI 318 statistical approach to establishing the target strength (f_{cr}') required to ensure attainment of the specified compressive strength (f_c'). If a standard deviation s from strength tests of concrete "similar" to that specified is known, the target strength level for which the concrete for the proposed work must be proportioned can be established. This target strength level is then related to a required water-cement ratio to attain the target strength. Concrete used in background tests to determine standard deviation is considered to have been similar to that specified if it was made with the same general types of ingredients, under no more restrictive conditions of control over material quality and production methods than are specified to exist on the proposed work, and if its specified strength did not deviate by more than 1000 psi from the f_c' specified. A change in the type of concrete or a significant increase in the strength level may increase the standard deviation. Such a situation might occur with a change in type of aggregate; i.e., from natural aggregate to lightweight aggregate or vice versa, or with a change from nonair-entrained concrete to air-entrained concrete.

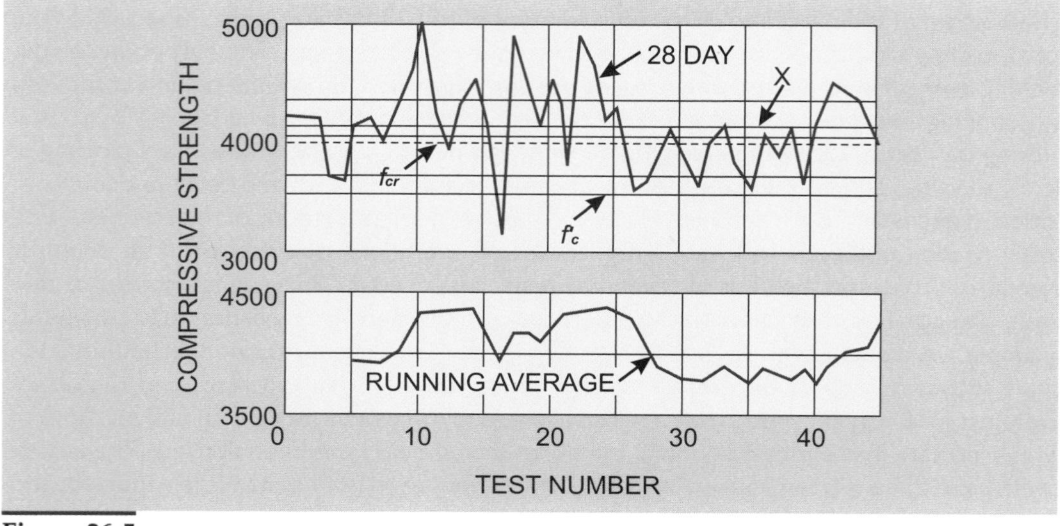

Figure 26-5
A control chart showing good control of concrete strength. Average is only slightly in excess of required average. Note how the strengths leveled out after about 25 tests had been made.

Statistical methods provide valuable tools for assessing the results of strength tests. It is important that concrete technicians understand the basic language of statistics and be capable of effectively utilizing the tool to evaluate strength test results. Figure 26-6[26.1] illustrates several fundamental statistical concepts in the form of a control chart representing six data points. The data points represent six strength test results from a given class of concrete.

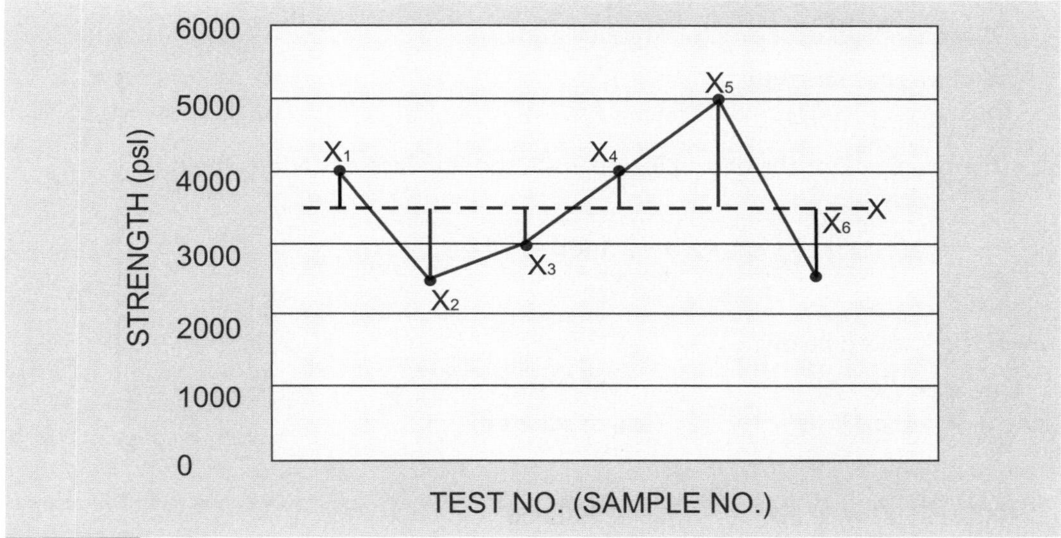

Figure 26-6
Illustration of statistical terms.

The horizontal line represents the average of tests, which is designated . The average is computed by adding all test values and dividing by the number of values summed:

X = (4000 + 2500 + 3000 + 4000 + 5000 + 2500)16 = 3500psi

The average X gives an indication of the overall strength level of the concrete tested.

It would also be informative to have a single number, which would represent the variability of the data about the average. The up and down deviations from the average (3500 psi) are given as vertical lines in Figure 26-6. If one were to accumulate the total length of the vertical lines without regard to whether they are up or down, and divide that total length by the number of tests, the result would be the average length, or the average distance from the average strength:

(500 + 1000 + 500 + 500 + 1500 + 1000)16 = 833 psi

This is one measure of variability. If concrete test results were quite variable, the vertical lines would be long. On the other hand, if the test results were close, the lines would be short.

To emphasize the impact of a few very high or very low test values, statisticians recommend the use of the square of the vertical line lengths. The square root of the sum of the squared lengths divided by one less than the number of tests (some texts use the number of tests) is known as the standard deviation. The measure of variability is designated by the letter s. Mathematically, s is expressed as:

$$s = \sqrt{\frac{\sum (X - \overline{X})^2}{n - 1}}$$

Where

s = standard deviation, psi

Σ = indicates summation

X = an individual strength test results, psi

$\overline{X}$ = average strength, psi

n = number of tests

For example, for the data in Figure 26-6, the standard deviation would be

$$s = \sqrt{\frac{(X_1 - \overline{X})^2 + (X_2 - \overline{X})^2 + \ldots (Xn - \overline{X})^2}{n - 1}}$$

which is calculated below.

Deviation $(X - \overline{X})$ (length of vertical lines)	$(X - X)^2$ (length squared)
4000 – 3500 = + 500	+250,000
2500 – 3500 = -1000	+1,000,000
3000 – 3500 = - 500	+250,000
4000 – 3500 = + 500	+250,000
5000 – 3500 = +1500	+2,250,000
2500 – 3500 = -1000	+1,000,000
Total	+5,000,000

$$s = \sqrt{\frac{5{,}000{,}000}{5}} = 1000 \text{ psi (a very large value)}$$

For concrete strengths in the range of 3000 to 4000 psi, the expected standard deviation, representing different levels of quality control, will range as follows:

STANDARD DEVIATION	REPRESENTING
300 to 400 psi	Excellent Quality Control
400 to 500 psi	Good
500 to 600 psi	Fair
> 600 psi	Poor Quality Control

For the very-high-strength, so-called high-performance concrete (HPC), with strengths in excess of 10,000 psi, the expected standard deviation will range as follows:

300 to 500 psi	Excellent Quality Control
500 to 700 psi	Good
> 700 psi	Poor Quality Control

Obviously, it would be time-consuming to actually calculate s in the manner described above. Most handheld scientific calculators are programmed to calculate standard deviation directly. The appropriate mathematical equations are programmed into the calculator, with the user simply entering the statistical data (test values), then pressing the appropriate function key to obtain standard deviation directly.

The coefficient of variation is simply the standard deviation expressed as a percent of the average value. The mathematical formula is:

$$V = \frac{s}{\overline{X}} \times 100\%$$

For the test results of Figure 26-6:

$$V = \frac{1000}{3500} \times 100 = 29\%$$

Note that the code uses the standard deviation in pounds per square inch instead of the coefficient of variation in percent.

The above example illustrates, for a very limited number of test data, the mathematics involved to compute a standard deviation. A more realistic example would include data from at least 30 strength test results as required by code to establish a representative (valid) standard deviation. Table 26.3 illustrates 30 strength test results from a project calling for column concrete to be normal weight, air-entrained, with a specified strength of 4000 psi. Computation of the mean strength and standard deviation for the 30 strength test results is shown in Table 26.4. The standard deviation of 353 psi represents excellent quality control for the specified 4000 psi concrete. Note that the concrete supplied for this concrete work satisfies the acceptance criteria of ACI 318 Section 5.6.3.3; no single strength test (28-day average) falls below the specified strength (4000 psi) by more than 500 psi (3500 psi), and the average of each set of three consecutive strength tests exceeds the specified strength (4000 psi).

The single low strength test (3950 psi) results from the very low break for cylinder No. 2 (3620 psi). The large disparity between cylinder No. 2 and cylinder No. 1 (4280 psi), both from the same batch, would seem to indicate a possible problem with the handling and testing procedures for cylinder No. 2.

The statistical data from the 30 strength test results can be filed for use on subsequent projects to establish a mix design where the concrete work calls for normal-weight, air-entrained concrete with a specified strength within 1000 psi of the specified 4000 psi value (3000 to 5000 psi). The target strength for mix proportioning would be calculated using the 353 psi standard deviation in ACI 318 Equations (5-1) and (5-2). The low standard deviation should enable the ready-mix company to produce an economical mix for similar concrete work.

TABLE 26.3
STRENGTH TEST DATA

TEST #	DATE OF TEST	28-DAY #1	28-DAY #2	28-DAY AVERAGE	28-DAY AVERAGE (3 CONSECUTIVE)
1	05-March-93	4640	4770	4705	
2	06-March-93	4910	5100	5005	
3	10-March-93	4570	4760	4665	4792
4	12-March-93	4800	5000	4900	4857
5	13-March-93	5000	4900	4950	4838
6	17-March-93	4380	4570	4475	4775
7	19-March-93	4630	4820	4725	4717
8	21-March-93	4800	4670	4735	4645
9	25-March-93	5020	4940	4980	4813
10	28-March-93	4740	4900	4820	4845
11	30-March-93	4300	4110	4205	4668
12	02-April-93	4280	3620	3950	4325
13	05-April-93	4740	4880	4810	4322
14	08-April-93	4870	5040	4955	4592
15	09-April-93	4590	4670	4630	4798
16	15-April-93	4420	4690	4555	4713
17	16-April-93	4980	5070	5025	4737
18	19-April-93	4900	4860	4880	4820
19	20-April-93	5690	5570	5630	5178
20	22-April-93	5310	5310	5310	5273
21	24-April-93	5080	4970	5025	5322
22	28-April-93	4640	4440	4540	4958
23	01-May-93	5090	5080	5085	4883
24	03-May-93	5430	5510	5470	5032
25	07-May-93	5290	5360	5325	5293
26	10-May-93	4700	4770	4735	5177
27	11-May-93	4880	5040	4960	5007
28	15-May-93	5000	4890	4945	4880
29	16-May-93	4810	4670	4740	4882
30	18-May-93	4250	4400	4325	4670

TABLE 26.4
COMPUTATION OF MEAN STRENGTH AND STANDARD DEVIATION

TEST #	28-DAY STRENGTH, X, psi	X - , psi	(X -)^2
1	4705	-130	16,900
2	5005	170	28,900
3	4665	-170	28,900
4	4900	65	4,225
5	4950	115	13,225
6	4475	-360	129,600
7	4725	-110	12,100
8	4735	-100	10,000
9	4980	145	21,025
10	4820	-15	225
11	4205	-630	396,900
12	3950	-885	783,225
13	4810	- 25	625
14	4955	100	10,000
15	4630	-205	42,025
16	4555	-280	78,400
17	5025	190	36,100
18	4880	45	2,025
19	5630	795	632,025
20	5310	475	225,625
21	5025	190	36,100
22	4540	-295	87,025
23	5085	250	62,500
24	5470	635	403,225
25	5325	490	240,100
26	4735	-100	10,000
27	4960	125	15,625
28	4945	110	12,100
29	4740	- 95	9,025
30	4325	-510	260,100
Σ =	145,060		3,607,850

Number of Tests = 30

Maximum Strength = 5620 psi

Minimum Strength = 3950 psi

Mean Strength $= \frac{145{,}060}{30} = 4835 \text{ psi}$

Standard Deviation $= \frac{\sqrt{3{,}607{,}850}}{29} = 353 \text{ psi}$

Required Average Strength. Once the standard deviation has been determined, the required average strength (ACI 318 Section 5.3.2) used as the basis for selecting the concrete mixture proportions (required water-cement ratio) is obtained from the larger of:

For f'_c of 5000 psi or less . . .

$$f'_{cr} = f'_c + 1.34s \quad (5\text{-}1)$$

$$f'_{cr} = f'_c + 2.33s - 500 \quad (5\text{-}2)$$

For f'_c over 5000 psi . . .

$$f'_{cr} = f'_c + 1.34s \quad (5\text{-}1)$$

$$f'_{cr} = 0.90\, f'_c + 2.33s \quad (5\text{-}3)$$

If the standard deviation is unknown, the required average strength f'_{cr}, used as the basis for selecting concrete proportions must be determined from ACI 318 Table 5.3.2.2:

For f'_c		
	less than 3000 psi	$f'_{cr} = f'_c + 1000$ psi
	between 3000 and 5000 psi	$f'_{cr} = f'_c + 1200$ psi
	greater than 5000 psi	$1.10\, f'_c + 700$ psi

Formulas for calculating the required target strengths are based on the following criteria:

1. Eq. (5-1) is based on a probability of 1-in-100 that an average of three consecutive strength tests may be below the specified strength f'_c,

2. Eq. (5-2) is based on a probability of 1-in-100 that an individual strength test may be more than 500 psi below the specified strength f'_c, and

3. Eq. (5-3) is based on the same 1-in-100 probability that an individual test may be less than 0.90 f'_c .

Criterion 1 will produce a higher required target strength than Criterion 2 for low to moderate standard deviations, up to 500 psi. For higher standard deviations, however, Criterion 2 will govern.

The indicated average strength levels are intended to reduce the probability of concrete strength being questioned on the following usual bases: (1) strength averaging below specified f'_c for an appreciable period (three consecutive tests); or (2) an individual test being disturbingly low (more than 500 psi below specified f'_c).

Assuming, for illustration, that the strength test data of Table 26.3, with a standard deviation of 353 psi, represents materials and conditions similar to those expected for proposed concrete work with a specified strength of f'_c = 4500 psi:

1. Normal weight, air-entrained concrete.

2. Specified strength (4000 psi) within 1000 psi of that required for the proposed work (4500 psi).

3. Thirty strength test results.

For a standard deviation of 353 psi, the required average compressive strength (f'_{cr}) to be used as the basis for selection of concrete mixture proportions must be the larger of:

$$f'_{cr} = f'_c + 1.34s = 4500 + 1.34(353) = 5000 \text{ psi, or}$$

$$f'_{cr} = f'_c + 2.33s - 500 \text{ psi} = 4500 + 2.33(353) - 500 = 4800 \text{ psi}$$

Therefore, $f'_{cr} = 5000$ psi.

For this example, with a standard deviation of 353 psi, an allowance of 500 psi is made to account for variations in materials, variations in methods of mixing, transporting and placing the concrete; and variations in making, curing and testing the concrete cylinder specimens. Note that the specified strength, $f'_c = 4500$ psi, is the strength that is expected to be equaled or exceeded by the average of any set of three consecutive strength tests, with no individual test more than 500 psi below the specified 4500 psi strength. (ACI 318 Section 5.6.3.3.)

Once the required average strength f'_{cr} is known, the next step is to select mixture proportions that will produce an average strength at least as great as the required average strength. The concrete producer would select a required water-cement ratio (W/C) reflective of local materials and conditions that would produce a concrete with an average strength of 5000 psi.

Documentation of Average Strength. The documentation of required average strength (ACI 318 Section 5.3.3) may consist of a single strength test record, several strength test records, or suitable laboratory trial mixtures. Generally, if a test record is used for documentation, it will be the same one that was used for computation of the standard deviation. For our example above, the test records used to generate the standard deviation (353 psi) would probably be used by the concrete producer to demonstrate that the concrete mix for which the records were generated will produce the required average strength (5000 psi) for the proposed concrete work. For the purpose of documenting the average strength potential of the concrete mix, the concrete producer need only select 10 consecutive tests from the total of 30 tests that represent a higher average than the required average of 5000 psi. Realistically, the average of the total 30 test results (4835 psi) is close enough to qualify the same concrete mix for the proposed work.

References

CHAPTER 1

1.1 Smiles, Samuel, *Lives of the Engineers*, published by John Murray, Albermarle Street, London, 1861.

1.2 Sections 1.8 and 24.7 adapted from Waddell, Joseph J., *Practical Quality Control for Concrete*, McGraw-Hill Book Co., New York, 1962.

CHAPTER 3

3.1 Adapted from Mercer, L. Boyd, *Ready-Mixed Concrete: Quality Control Requirements*, Cement and Concrete Association, London, 1954.

3.2 "Cold Weather Concreting (ACI 306R-88)," (Reapproved 2002) American Concrete Institute, 1988.

CHAPTER 4

4.1 "Guide to Durable Concrete (ACI 201.2R-08)," American Concrete Institute, 2008.

4.2 Valenta, Oldrich, "General Analysis of the Methods of Testing the Durability of Concrete," Building Research Institute, Prague, 1967.

4.3 "Effects of Substances on Concrete and Guide to Protective Treatments," Portland Cement Association, IS001, 2007.

4.4 Woods, Hubert, "Corrosion of Embedded Material Other Than Reinforcing Steel," RX198, Portland Cement Association, 1966.

4.5 "Diagnosis and Control of Alkali-Aggregate Reactions in Concrete," IS413, Portland Cement Association, 2007.

4.6 "Guide Specification for Concrete Subject to Alkali-Silica Reactions," IS415, Portland Cement Association, 2007

CHAPTER 5

5.1 Lerch, William, "Plastic Shrinkage," Research Department Bulletin, RX081, Portland Cement Association, 1957.

5.2 "Bleeding," RP328, Portland Cement Association, 1994.

CHAPTER 6

6.1 "Concrete Slab Surface Defects: Causes, Prevention, Repair," IS177, Portland Cement Association, 2001.

CHAPTER 8

8.1 "Guide for Use of Normal Weight and Heavyweight Aggregates in Concrete (ACI 221R-96),"(reapproved 2001) American Concrete Institute, 1996.

CHAPTER 9

9.1 Malhotra, V. M., "Superplasticizers: Their Effect on Fresh and Hardened Concrete," *Concrete International*, American Concrete Institute, May 1981.

CHAPTER 11

11.1 "Guide to Formwork for Concrete (ACI 347-04)," American Concrete Institute, 2004.

CHAPTER 12

12.1 "Standard Practice for Selecting Proportions for Normal, Heavyweight, and Mass Concrete (ACI 211.1-91)," (reapproved 2002) American Concrete Institute, 1991.

12.2 "Design and Control of Concrete Mixtures," 14th Edition, EB001, Portland Cement Association, 2002, (rev.) 2008.

12.3 "Guide for Selecting Proportions for No-Slump Concrete (ACI 211.3R-02)," American Concrete Institute, 2002.

CHAPTER 14

14.1 Adapted from "Recommended Guide Specifications for Batching Equipment and Control Systems in Concrete Batch Plants," CPMB Publication No. 102-00, Concrete Plant Manufacturers Bureau, Silver Springs, Maryland, 2000.

14.2 "Concrete Plant Standards of the Concrete Plant Manufacturers Bureau (CPMB 100-07)," Fifteenth Revision, Effective March 20, 2007, Silver Springs, Maryland, 2007.

14.3 "Central Mixing Plants for the Manufacture of Premixed Concrete," *Journal American Concrete Institute*, Volume 21, 1925.

14.4 "The Design and Operation of Central Mixing Plants, A Symposium," *Journal American Concrete Institute*, March 1930.

14.5 Adapted from "Statement of Responsibilities," National Ready Mixed Concrete Association and Associated General Contractors of America, 1980.

CHAPTER 16

16.1 "Guide for Concrete Floor and Slab Construction (ACI 302.1R-04)," American Concrete Institute, 2004.

16.2 J. J. Brooks, P. J. Wainwright, and A. M. Neville, "Time-Dependent Behavior of High-Early-Strength Concrete Containing a Superplasticizer," SP68-5, American Concrete Institute, 1981.

16.3 "Requirements for Residential Concrete Construction and Commentary (ACI 332-04)," American Concrete Institute, 2004.

16.4 "Specifications for Tolerances for Concrete Construction and Materials and Commentary (ACI 117-06)," American Concrete Institute, 2006.

CHAPTER 17

17.1 "Guide for Concrete Floor and Slab Construction (ACI 302.1R-04)," American Concrete Institute, 2004.

CHAPTER 18

18.1 "Manual of Standard Practice," 28th edition, Concrete Reinforcing Steel Institute, 2009.

18.2 "Manual of Standard Practice-Structural Welded Wire Reinforcement," 8th edition, WWR-500, Wire Reinforcement Institute, 2009.

18.3 "Historical Data on Wire, Triangular Wire/Mesh and Welded Wire Concrete Reinforcement Solutions," TF-101-07, Wire Reinforcement Institute, 2007.

18.4 "How to Specify, Order and Use Welded Wire Reinforcement in Light Construction," TF-202-R-03, Wire Reinforcement Institute, 2003.

18.5 "Metric Welded Wire Reinforcement," TF-206-R-03, Wire Reinforcement Institute, 2003.

18.6 "Supports are Needed for Long-Term Performance of Welded Wire Reinforcement in Slabs-on-Grade," TF-702-R2-08, Wire Reinforcement Institute, 2008.

18.7 "ACI Detailing Manual-2004," SP-66(04), American Concrete Institute, 2004.

18.8 "Structural Welding Code-Reinforcing Steel (AWS D1.4/D1.4M: 2005)," American Welding Society, Miami, Florida, 2005.

18.9 "Specifications for Tolerances for Concrete Construction and Materials and Commentary (ACI 117-06)," American Concrete Institute, 2006.

18.10 "Performance of Epoxy-Coated Rebar, Galvanized Rebar, and Plain Rebar with Calcium Nitrite in a Marine Environment," Naval Facilities Engineering Service Center, Port Hueneme, Calif., Douglas F. Burke, July 1994, reprint Concrete Reinforcing Steel Institute, Research Series 2.

18.11 Yeomans, S.R., "Comparative Studies of Galvanized and Epoxy-Coated Steel Reinforcement in Concrete," Durability of Concrete-Second International Conference, Montreal, Canada, Report SP126-19, Vol. 1, 1991, pp. 355–370.

18.12 "Guide for the Design and Construction of Concrete Reinforced with FRP Bars (ACI 440.1R-06)," American Concrete Institute, 2006.

CHAPTER 19

19.1 "Hot Weather Concreting (ACI 305R-99)," American Concrete Institute, 1999.

19.2 "Cold Weather Concreting (ACI 306R-88)," (reapproved 2002) American Concrete Institute, 1988.

CHAPTER 20

20.1 *Manual for Quality Control for Plants and Production of Structural Precast and Prestressed Concrete Products*, 4th edition, MNL-116-99, Precast/Prestressed Concrete Institute, 1999.

20.2 *Quality Control Technician Inspector Level 1 & 2 Training Manual*, 2nd edition, TM-101, Precast/Prestressed Concrete Institute, 2009.

20.3 *Quality Control Personal Certification Level 3 Training Manual*, 1st edition, TM-103, Precast/Prestressed Concrete Institute, 2000.

20.4 Adapted from "*Erector's Manual-Standards and Guidelines for the Erection of Precast Concrete Products*," MNL-127-99, Precast/Prestressed Concrete Institute, 1999.

20.5 *Post-Tensioning Manual*, 6th edition, Post-Tensioning Institute, 2006.

20.6 "Field Procedures Manual for Unbonded Single Strand Tendons," 3rd edition, Post-Tensioning Institute, 2000.

20.7 "Specification for Unbonded Single-Strand Tendon Materials and Commentary (ACI 423.7-07) and Commentary (ACI 423.6R-01," American Concrete Institute, 2007.

CHAPTER 21

21.1 "Standard Practice for Selecting Proportions for Structural Lightweight Concrete (ACI 211.2-98)," (reapproved 2004) American Concrete Institute, 1998.

CHAPTER 22

22.1 "The Tilt-Up Construction and Engineering Manual," 6th Edition, Tilt-Up Concrete Association, 2004-2007.

22.2 Adapted from "Guide for Measuring, Mixing, Transporting, and Placing Concrete (ACI 304R-00)," Chapter 8, Concrete Placed Under Water, American Concrete Institute, 2000.

22.3 Adapted from "Guide for Measuring, Mixing, Transporting, and Placing Concrete (ACI 304R-00)," Chapter 7, Preplaced Aggregate Structural and Mass Concrete, American Concrete Institute, 2000.

22.4 "Guide to Shotcrete (ACI 506R-05)," American Concrete Institute, 2005. Also, "Specification for Shotcrete (ACI 506.2-95)," American Concrete Institute, 1995.

22.5 "Guide to Certification of Shotcrete Nozzlemen (ACI 506.3R-91)," (inactive-historical), American Concrete Institute, 1991.

22.6 "State-of-the-Art Report on Fiber Reinforced Concrete (ACI 544.1R-96)," (reapproved 2002) American Concrete Institute, 1996.

22.7 "GFRC-Recommended Practice," 5th edition, MNL 128-01, Precast/Prestressed Concrete Institute, 2001.

CHAPTER 24

24.1 Adapted from the *Concrete Manual*, 8th edition, U.S. Bureau of Reclamation, Denver, revised 1981.

24.2 "*International Building Code*," Chapter 17, Structural Tests and Special Inspections, International Code Council, Falls Church, VA, 2009

24.3 "Concrete Inspection Handbook," 3rd edition, (revised) EB115.03, Portland Cement Association, 2005.

24.4 "ASTM Standards in ACI 301 and 318," Publication No. SP-71 (08), American Concrete Institute, 2008.

24.5 *ACI Manual of Concrete Inspection*, 9th edition, Publication SP-2 (99), (reapproved 2005) American Concrete Institute, 2005.

24.6 *Placing Reinforcing Bars*, 8th edition, Concrete Reinforcing Steel Institute, 2005.

CHAPTER 26

26.1 "Notes on ACI 318-08 Building Code Requirements for Structural Concrete," Chapter 2, Materials, Concrete Quality, EB702, Portland Cement Association, 2008.

Resource References

For additional information/publications on concrete technology and construction practices, the reader is referred to the following list of concrete industry and technical organizations:

American Concrete Institute (ACI)

38800 Country Club Drive
Farmington Hills, MI 48331
Phone: (248) 848-3800 Fax: (248) 848-3801
Web Site: www.concrete.org

American Society for Testing and Materials (ASTM)

ASTM International
100 Barr Harbor Drive
West Conshohocken, PA 19428-2959
Phone: (610) 832-9585 Fax: (610) 832-9555
Web Site: www.astm.org

Concrete Reinforcing Steel Institute (CRSI)

933 North Plum Grove Road
Schaumburg, IL 60173-4758
Phone: (847) 517-1200 Fax: (847) 517-1206
Web Site: www.crsi.org

National Ready Mixed Concrete Association (NRMCA)

900 Spring Street
Silver Spring, MD 20910
Phone: (301) 587-1400 Fax: (301) 585-4219
Web Site: www.nrmca.com

Portland Cement Association (PCA)

5420 Old Orchard Road
Skokie, IL 60077-1083
Phone: (847) 966-6200 Fax: (847) 966-8389
Web Site: www.cement.org

Precast/Prestressed Concrete Institute (PCI)

209 West Jackson Boulevard, Suite 500
Chicago, IL 60606
Phone: (312) 786-0300 Fax: (312) 786-0353
Web Site: www.pci.org

Post-Tensioning Institute (PTI)

38800 Country Club Drive
Farmington Hills, MI 48331
Phone: (248) 848-3180 Fax: (248) 848-3181
Web Site: www.post-tensioning.org

Tilt-Up Concrete Association (TCA)

P.O. Box 204, 113 First Street West
Mount Vernon, IA 52314
Phone: (319) 895-6911 Fax: (319) 895-8830
Web Site: www.tilt-up.org

Wire Reinforcement Institute (WRI)

942 Main Street, Suite 300
Hartford, CT 06103
Phone: (800) 522-4974 Fax: (860) 808-3009
Web Site: www.wirereinforcementinstitute.org

INDEX

Don't Miss Out On Valuable ICC Membership Benefits. Join ICC Today!

Join the largest and most respected building code and safety organization. As an official member of the International Code Council®, these great ICC® benefits are at your fingertips.

EXCLUSIVE MEMBER DISCOUNTS

ICC members enjoy exclusive discounts on codes, technical publications, seminars, plan reviews, educational materials, videos, and other products and services.

TECHNICAL SUPPORT

ICC members get expert code support services, opinions, and technical assistance from experienced engineers and architects, backed by the world's leading repository of code publications.

FREE CODE—LATEST EDITION

Most new individual members receive a free code from the latest edition of the International Codes®. New corporate and governmental members receive one set of major International Codes (Building, Residential, Fire, Fuel Gas, Mechanical, Plumbing, Private Sewage Disposal).

FREE CODE MONOGRAPHS

Code monographs and other materials on proposed International Code revisions are provided free to ICC members upon request.

PROFESSIONAL DEVELOPMENT

Receive Member Discounts for on-site training, institutes, symposiums, audio virtual seminars, and on-line training! ICC delivers educational programs that enable members to transition to the I-Codes®, interpret and enforce codes, perform plan reviews, design and build safe structures, and perform administrative functions more effectively and with greater efficiency. Members also enjoy special educational offerings that provide a forum to learn about and discuss current and emerging issues that affect the building industry.

ENHANCE YOUR CAREER

ICC keeps you current on the latest building codes, methods, and materials. Our conferences, job postings, and educational programs can also help you advance your career.

CODE NEWS

ICC members have the inside track for code news and industry updates via e-mails, newsletters, conferences, chapter meetings, networking, and the ICC website (www.iccsafe.org). Obtain code opinions, reports, adoption updates, and more. Without exception, ICC is your number one source for the very latest code and safety standards information.

MEMBER RECOGNITION

Improve your standing and prestige among your peers. ICC member cards, wall certificates, and logo decals identify your commitment to the community and to the safety of people worldwide.

ICC NETWORKING

Take advantage of exciting new opportunities to network with colleagues, future employers, potential business partners, industry experts, and more than 50,000 ICC members. ICC also has over 300 chapters across North America and around the globe to help you stay informed on local events, to consult with other professionals, and to enhance your reputation in the local community.

JOIN NOW! 1-888-422-7233, x33804 | www.iccsafe.org/membership

People Helping People Build a Safer World™

Most Widely Accepted and Trusted

Innovative Building Products

Make sure they are up to code with ICC-ES Evaluation Reports

The ICC-ES Solution

ICC Evaluation Service® (ICC-ES®), a subsidiary of ICC®, was created to assist code officials and industry professionals in verifying that new and innovative building products meet code requirements. This is done through a comprehensive evaluation process that results in the publication of ICC-ES Evaluation Reports for those products that comply with requirements in the code or acceptance critera. Today, more code officials prefer using ICC-ES Evaluation Reports over any other resource to verify products comply with codes.

Most Widely Accepted and Trusted

ICC-ES Evaluation Report **ESR-4802**

Issued March 1, 2008

This report is subject to re-examination in one year.

www.icc-es.org | 1-800-423-6587 | (562) 699-0543 *A Subsidiary of the International Code Council®*

DIVISION: 07—THERMAL AND MOISTURE PROTECTION
Section: 07410—Metal Roof and Wall Panels

REPORT HOLDER:

ACME CUSTOM-BILT PANELS
52380 FLOWER STREET
CHICO, MONTANA 43820
(808) 664-1512
www.custombiltpanels.com

EVALUATION SUBJECT:

CUSTOM-BILT STANDING SEAM METAL ROOF PANELS: CB-150

1.0 EVALUATION SCOPE

Compliance with the following codes:

- 2006 *International Building Code®* (IBC)
- 2006 *International Residential Code®* (IRC)

Properties evaluated:

- Weather resistance
- Fire classification
- Wind uplift resistance

2.0 USES

Custom-Bilt Standing Seam Metal Roof Panels are steel panels complying with IBC Section 1507.4 and IRC Section R905.10. The panels are recognized for use as Class A roof coverings when installed in accordance with this report.

3.0 DESCRIPTION

3.1 Roofing Panels:

Custom-Bilt standing seam roof panels are fabricated in steel and are available in the CB-150 and SL-1750 profiles. The panels are roll-formed at the jobsite to provide the standing seams between panels. See Figures 1 and 3 for panel profiles. The standing seam roof panels are roll-formed from minimum No. 24 gage [0.024 inch thick (0.61 mm)] cold-formed sheet steel. The steel conforms to ASTM A 792, with an aluminum-zinc alloy coating designation of AZ50.

3.2 Decking:

Solid or closely fitted decking must be minimum $^{15}/_{32}$-inch-thick (11.9 mm) wood structural panel or lumber sheathing, complying with IBC Section 2304.7.2 or IRC Section R803, as applicable.

4.0 INSTALLATION

4.1 General:

Installation of the Custom-Bilt Standing Seam Roof Panels must be in accordance with this report, Section 1507.4 of the IBC or Section R905.10 of the IRC, and the manufacturer's published installation instructions. The manufacturer's installation instructions must be available at the jobsite at all times during installation. The roof panels must be installed on solid or closely fitted decking, as specified in Section 3.2. Accessories such as gutters, drip angles, fascias, ridge caps, window or gable trim, valley and hip flashings, etc., are fabricated to suit each job condition. Details must be submitted to the code official for each installation.

4.2 Roof Panel Installation:

4.2.1 CB-150: The CB-150 roof panels are installed on roof shaving a minimum slope of 2:12 (17 percent). The roof panels are installed over the optional underlayment and secured to the sheathing with the panel clip. The clips are located at each panel rib side lap spaced 6 inches (152 mm) from all ends and at a maximum of 4 feet (1.22 m) on center along the length of the rib, and fastened with a minimum of two No. 10 by 1-inch pan head corrosion-resistant screws. The panel ribs are mechanically seamed twice, each pass at 90 degrees, resulting in a double-locking fold.

4.3 Fire Classification:

The steel panels are considered Class A roof coverings in accordance with the exception to IBC Section 1505.2 and IRC Section R902.1.

4.4 Wind Uplift Resistance:

The systems described in Section 3.0 and installed in accordance with Sections 4.1 and 4.2 have an allowable wind uplift resistance of 45 pounds per square foot (2.15 kPa).

5.0 CONDITIONS OF USE

The standing seam metal roof panels described in this report comply with, or are suitable alternatives to what is specified in, those codes listed in Section 1.0 of this report, subject to the following conditions:

5.1 Installation must comply with this report, the applicable code, and the manufacturer's published installation instructions. If there is a conflict between this report and the manufacturer's published installation instructions, this report governs.

5.2 The required design wind loads must be determined for each project. Wind uplift pressure on any roof area must not exceed 45 pounds per square foot (2.15 kPa).

6.0 EVIDENCE SUBMITTED

Data in accordance with the ICC-ES Acceptance Criteria for Metal Roof Coverings (AC166), dated October 2007.

7.0 IDENTIFICTION

Each standing seam metal roof panel is identified with a label bearing the product name, the material type and gage, the Acme Custom-Bilt Panels name and address, and the evaluation report number (ESR-4802).

ICC-ES Evaluation Reports are not to be construed as representing aesthetics or any other attributes not specifically addressed, nor are they to be construed as an endorsement of the subject of the report or a recommendation for its use. There is no warranty by ICC Evaluation Service, Inc., express or implied, as to any finding or other matter in this report, or as to any product covered by the report.

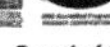

Page 1 of 1

FREE Access to ICC-ES Evaluation Reports!

William Gregory
Building and Plumbing Inspector
Town of Yorktown, New York

"We've been using ICC-ES Evaluation Reports as a basis of product approval since 2002. I would recommend them to any jurisdiction building department, particularly in light of the many new products that regularly move into the market. It's good to have a group like ICC-ES evaluating these products with a consistent and reliable methodology that we can trust."

Becky Baker, CBO
Director/Building Official
Jefferson County, Colorado

"The ICC-ES Evaluation Reports are designed with the end user in mind to help determine if building products comply with code. The reports are easily accessible, and the information is in a format that is useable by plans examiners and inspectors as well as design professionals and contractors."

VIEW ICC-ES EVALUATION REPORTS ONLINE!

www.icc-es.org

09-02246